Lecture Notes in Computer Science 16588

The series Lecture Notes in Computer Science (LNCS), including its subseries Lecture Notes in Artificial Intelligence (LNAI) and Lecture Notes in Bioinformatics (LNBI), has established itself as a medium for the publication of new developments in computer science and information technology research, teaching, and education.

LNCS enjoys close cooperation with the computer science R & D community, the series counts many renowned academics among its volume editors and paper authors, and collaborates with prestigious societies. Its mission is to serve this international community by providing an invaluable service, mainly focused on the publication of conference and workshop proceedings and postproceedings. LNCS commenced publication in 1973.

Santanu S. Dey · Marco Di Summa ·
Domenico Salvagnin
Editors

Integer Programming and Combinatorial Optimization

27th International Conference, IPCO 2026
Padua, Italy, June 17–19, 2026
Proceedings

Springer

Editors
Santanu S. Dey
Georgia Institute of Technology
Atlanta, GA, USA

Marco Di Summa
University of Padova
Padova, Italy

Domenico Salvagnin
University of Padova
Padova, Italy

ISSN 0302-9743 ISSN 1611-3349 (electronic)
Lecture Notes in Computer Science
ISBN 978-3-032-28690-1 ISBN 978-3-032-28691-8 (eBook)
https://doi.org/10.1007/978-3-032-28691-8

This Springer imprint is published by the registered company Springer Nature Switzerland AG
The registered company address is: Gewerbestrasse 11, 6330 Cham, Switzerland

Preface

This volume contains the 33 extended abstracts presented at IPCO 2026, the 27th Conference on Integer Programming and Combinatorial Optimization, held on June 17–19, 2026 at the University of Padova.

The IPCO conference is held under the auspices of the Mathematical Optimization Society. It brings together researchers and practitioners working on a broad range of topics in integer programming and combinatorial optimization, and has become a premier forum for presenting recent advances in theory, computation, and applications. Traditionally, IPCO spans three days of single-track sessions, with no invited talks.

More information on the IPCO conference and its history can be found at https://mathopt.zib.de/?nav=ipco_history. This year, we received 113 submissions. In a double-blind review process, each submission was reviewed by at least three Program Committee members, often with the help of external reviewers. The PC members met in Aussois in January 2026 and, after thorough discussions, selected 33 papers, an acceptance rate of 29%. We regret that many excellent submissions could not be accepted.

As is customary, this volume is jointly edited by the Program Committee Chair, Santanu S. Dey, and the local organizers, Marco Di Summa and Domenico Salvagnin. We expect the full versions of the accepted extended abstracts that appear in this volume to be published in refereed journals. A special issue of Mathematical Programming B featuring these papers is currently in preparation.

The Program Committee selected Stefan Kuhlmann and Robert Weismantel as recipients of the IPCO 2026 Best Paper Award for their paper, "A Threshold Phenomenon for the Shortest Lattice Vector Problem in the Infinity Norm". We warmly congratulate them on this achievement.

We would like to express our sincere gratitude to:

- All authors who submitted extended abstracts to IPCO; we are happy how active all areas of integer programming and combinatorial optimization are;
- The members of the Program Committee, who graciously gave their time and energy;
- The external reviewers, whose expertise was instrumental in guiding our decisions;
- The organizers of the Aussois Workshop: Angelika Wiegele, Britta Peis, and Volker Kaibel, for inviting us to hold the PC meeting in Aussois;
- EasyChair for making the entire reviewing process so pleasant and easy;
- Springer for their efficient cooperation in producing this volume and for sponsoring the Best Paper Award;
- The sponsors:

  • Hexaly
  • Mosek
  • Gurobi
  • AIRO (Italian Operations Research Association)
  • Department of Mathematics "Tullio Levi-Civita" (University of Padova)

- Department of Information Engineering (University of Padova)

– Nicole Megow (Chair, IPCO 2025) for always being ready to respond to our questions;
– The members of the Local Organizing Committee at the University of Padova who made this conference possible;
– The speakers of the summer school preceding IPCO: Margarida Carvalho, Alberto Del Pia, and Oktay Günlük;
– The Mathematical Optimization Society and in particular the members of its IPCO Steering Committee: Karen Aardal, Alberto Del Pia, and Giacomo Zambelli, for their help and advice.

April 2026

Santanu S. Dey
Marco Di Summa
Domenico Salvagnin

Organization

Program Committee

Ahmad Abdi	London School of Economics and Political Science, UK
Merve Bodur	University of Edinburgh, UK
Jose Correa	Universidad de Chile, Chile
Santanu S. Dey (Chair)	Georgia Institute of Technology, USA
Ricardo Fukusawa	University of Waterloo, Canada
Andres Gomez	University of Southern California, USA
Swati Gupta	Massachusetts Institute of Technology, USA
Nathan Klein	Boston University, USA
Alantha Newman	Université Grenoble Alpes, France
Joseph Paat	University of British Columbia, Canada
Thomas Rothvoss	University of Washington, USA
Mohit Singh	Georgia Institute of Technology, USA
László A. Végh	University of Bonn, Germany
Domenico Salvagnin	University of Padova, Italy
Roy Schwartz	Technion - Israel Institute of Technology, Israel
Martin Skutella	Technische Universität Berlin, Germany
Stefan Weltge	Technical University of Munich, Germany
Yu Yokoi	Institute of Science Tokyo, Japan

Local Organizing Committee

Manuel Aprile	University of Padova, Italy
Carla De Francesco	University of Padova, Iyaly
Luigi De Giovanni	University of Padova, Italy
Marco Di Summa (Co-chair)	University of Padova, Italy
Domenico Salvagnin (Co-chair)	University of Padova, Italy

Additional Reviewers

Jason Altschuler
Susanne Armbruster
Georg Anegg
Eleon Bach
Manuel Aprile
Mourad Baiou

János Balogh
Federico Battista
Nathan Benedetto Proença
Amine Bennouna
Kristóf Bérczi
Timo Berthold
Sujoy Bhore
Alexander Black
Jannis Blauth
Joakim Blikstad
Steffen Borgwardt
Miguel Boschcalvo
Niv Buchbinder
Christoph Buchheim
Nairen Cao
Margarida Carvalho
Javier Cembrano
Ben Champion
Jason Chatzitheodorou
Rui Chen
Gerard Cornuéjols
Ryan Cory-Wright
Francisco Criado
Andrés Cristi
Daniel Dadush
Mathieu Dahan
Morteza Davari
Sami Davies
Jesús De Loera
Alberto Del Pia
Anna Deza
Yann Disser
Madison Van Dyk
Franziska Eberle
Friedrich Eisenbrand
Ekin Ergen
Samuel Fiorini
Ayoub Foussoul
Elisabeth Gaar
Ameet Gadekar
Rui Gao
Rohan Ghuge
Ambros Gleixner
Fabrizio Grandoni
Andreas Grothey
Arash Haddadan
Tobias Harks
Guillaume Helbecque
Dorothee Henke
Christoph Hertrich
Ruben Hoeksma
Christopher Hojny
Ymro Hoogendoorn
Florian Hörsch
Chien-Chung Huang
Sophie Huiberts
Ayumi Igarashi
Stavros Ioannidis
Yuni Iwamasa
Afrouz Jabal Ameli
Hongyi Jiang
Ruiwei Jiang
Billy Jin
Sean Kafer
Naonori Kakimura
Sina Kalantarzadeh
Aleksandr M. Kazachkov
Bart de Keijzer
Yulia Kempner
Csaba Király
Tamás Király
Yusuke Kobayashi
Stefan Kober
Burak Kocuk
Zhuan Khye Koh
Martin Koutecký
Jan Kronqvist
Stefan Kuhlmann
Ariel Kulik
Nikhil Kumar
Tim Kunisky
Jannis Kurtz
Monique Laurent
Dabeen Lee
Lisun Lee
Antoine Legrain
Wenzheng Li
Yongchun Li
Tianyi Lin
Alexander Lindermayr

Jeff Linderoth
Siyue Liu
Vasilis Livanos
Andrea Lodi
Georg Loho
Anand Louis
Leonardo Lozano
Ryoga Mahara
Konstantin Makarychev
Enrico Malaguti
Michael Markl
Jannik Matuschke
Carla Michini
Matthias Mnich
Marco Molinaro
Tobias Mömke
Michele Monaci
Jai Moondra
Diego Moran
Viswanath Nagarajan
Martin Nägele
Giacomo Nannicini
Meike Neuwohner
Takashi Noguchi
Timm Oertel
Taihei Oki
Thiago Oliveira
Jim Ostrowski
Matheus Ota
Kanstantsin Pashkovich
Britta Peis
Sebastian Perez-Salazar
Ulrich Pferschy
Marc Pfetsch
Michael Poss
Manish Purohit
Aaron Putterman
Kent Quanrud
Alice Raffaele
Roshan Raj
Victor Reis
Lars Rohwedder
Shinsaku Sakaue
Sriram Sankaranarayanan
Alinson Santos Xavier
Kevin Schewior
Jens Schlöter
Daniel Schmand
Tamas Schwarcz
Matthias Schymura
Sergei Sergeev
Soroosh Shafiee
Neta Singer
Sahil Singla
Tasuku Soma
José A. Soto
Fabian Spaeh
Raphael Steiner
Sebastian Stiller
Lea Strubberg
Nagisa Sugishita
Noriyoshi Sukegawa
Shengding Sun
Kenjiro Takazawa
Erasmo Tani
Tatsuya Terao
Huo Tianming
Alejandro Toriello
Alfredo Torrico
Eduardo Uchoa
Laura Vargas Koch
Antonios Varvitsiotis
Daniel Vaz
Santosh Vempala
Luitgard Veraart
Victor Verdugo
José Verschae
Matias Villagra
Lukas Vogl
Tjark Vredeveld
Lucas Waddell
Matthias Walter
Jiaqi Wang
Justin Ward
Karol Węgrzycki
Noah Weninger
David Williamson
Weijun Xie
Zikai Xiong
Jingye Xu

Luze Xu
Yutaro Yamaguchi
Chiwei Yan
E. Alper Yildirim
Yuichi Yoshida
Kim Yu
Rui Zhang
Yufan Zheng
Zhengjia Zhuo

Contents

On the Complexity of the Odd-Red Bipartite Perfect Matching Polytope

Martin Nägele(✉), Christian Nöbel, and Rico Zenklusen

Department of Mathematics, ETH Zurich, Zurich, Switzerland
{martinn,cnoebel,ricoz}@ethz.ch

Abstract. The odd-red bipartite perfect matching problem asks to find a perfect matching containing an odd number of red edges in a given red-blue edge-colored bipartite graph. While this problem lies in P, its polyhedral structure remains elusive—despite renewed attention to achieving better polyhedral understanding, nurtured by recent advances from two complementary angles. Apart from being a special case of bimodular integer programs, whose polyhedral structure is also badly understood, it is related to one of the most notorious open derandomization questions in theoretical computer science: whether there is a deterministic efficient algorithm for the exact bipartite perfect matching problem, which asks to find a perfect matching with exactly k red edges.

Recent progress towards deterministic algorithms for this problem crucially relies on a good polyhedral understanding. Motivated by this, Jia, Svensson, and Yuan show that the extension complexity of the exact bipartite perfect matching polytope is exponential in general. Interestingly, their result is true even for the easier odd-red bipartite perfect matching problem. For this problem, they introduce an exponential-size relaxation and leave open whether it is an exact description.

Apart from showing that this description is not exact and even hard to separate over, we show, more importantly, that the red-odd bipartite perfect matching polytope exhibits complex facet structure: any exact description needs constraints with large and diverse coefficients. This rules out classical relaxations based on constraints with all coefficients in $\{0, \pm 1\}$, such as the above-mentioned one, and suggests that significant deviations from prior approaches may be needed to obtain an exact description. More generally, we obtain that also polytopes corresponding to bimodular integer programs have complex facet structure.

1 Introduction

Parity-constrained variations of classical combinatorial optimization problems have been studied in the literature for a long time. The arguably most prominent

Parts of this work were done while M. Nägele was employed at University of Bonn. This work was partially supported by the Swiss National Science Foundation (grant no. P500PT_206742) and the Deutsche Forschungsgemeinschaft (DFG, German Research Foundation) under Germany's Excellence Strategy – EXZ-2047/1 – 390685813.
A version of this extended abstract with full proofs is available at [13].

S. S. Dey et al. (Eds.): IPCO 2026, LNCS 16588, pp. 1–16, 2026.
https://doi.org/10.1007/978-3-032-28691-8_1

efficiently solvable example is the minimum odd cut problem, which was first discussed by Padberg and Rao [17]: Here, the goal is to find a cut $C \subseteq V$ minimizing $|\delta(C)|$, i.e., the number of edges crossing the cut, under the extra constraint that $|C|$ is odd.[1] We consider another efficiently solvable[2] parity-constrained problem, namely odd-red perfect matching in bipartite graphs, defined as follows.

Bipartite Odd-Red Perfect Matching: Given a bipartite graph $G = (V, E)$, a subset $R \subseteq E$ of whose edges are red, the task is to decide whether there exists a perfect matching $M \subseteq E$ in G with $|M \cap R|$ odd.

This problem has gained increased attention recently due to its connections to both Integer Programming with bounded subdeterminants and derandomization questions in Theoretical Computer Science, on which we elaborate next.

Recently, parity-constrained problems have come up in an interesting connection to a well-known conjecture in Integer Programming, claiming efficient solvability of integer programs when the constraint matrix fulfills certain subdeterminant bounds. More precisely, an IP $\min\{c^\top x \colon Ax \leq b, x \in \mathbb{Z}^n_{\geq 0}\}$ is called *Δ-modular* (for some $\Delta \in \mathbb{Z}_{\geq 1}$) if the constraint matrix A has full column rank and all its $n \times n$ subdeterminants are bounded in absolute value by Δ. It is conjectured that Δ-modular IPs can be solved in polynomial time for every fixed Δ.[3] One can see that this conjecture implies another well-known variant of it, where instead of requiring A to be Δ-modular, one requires that all subdeterminants of A (not only the $n \times n$ ones) are bounded in absolute value by Δ (these are so-called *totally Δ-modular matrices*).[4] [1] proved the general version of the conjecture to be true for $\Delta = 2$, the so-called *bimodular* case. Their approach prominently features parity constraints: They show how to reduce bimodular IPs to IPs with totally unimodular constraint matrices and a single parity constraint, and then show that such problems can be solved efficiently. Efforts towards generalizing the result of Artmann, Weismantel, and Zenklusen have triggered interest in extending simple parity constraints to more general congruency constraints.

---

1 Given a graph $G = (V, E)$, we denote for $C \subseteq V$ by $\delta(C)$ the set of edges $e \in E$ with exactly one endpoint in C. For $v \in V$, we use the shorthand notation $\delta(v) := \delta(\{v\})$.

2 The bipartite odd-red perfect matching problem can be solved in polynomial time by starting with an arbitraty perfect matching and, if necessary, swapping the role of matching and non-matching edges on an alternating cycle with an odd number of red edges. This has, e.g., been observerd in [1].

3 When talking about polynomial time procedures, we use the standard convention that these procedures are deterministic. When dealing with randomized procedures, we explicitly mention this.

4 Hence, this version does not need the constraint matrix A to have full column rank. Concretely, a totally Δ-modular problem can be interpreded as a Δ-modular problem with full-rank constraint matrix by enlarging the inequality system $Ax \leq b$ through the addition of non-negativity constraints $x \geq 0$. If A is totally Δ-modular, this modification leads to another totally Δ-modular constraint matrix $\binom{A}{-I}$ with then full column rank, which is therefore also a Δ-modular constraint matrix..

Some interesting questions in this context are, for example, asking for a minimum cut of size congruent to k modulo m, for some given $k, m \in \mathbb{Z}$ [15,16], or attempts towards solving totally unimodular IPs with general congruency constraints with modulus $m > 2$, which can be seen as special cases of m-modular IPs [12,14]. Additionally, congruency constraints have recently been studied in matroid base problems, addressing the question of whether, in a matroid whose elements are labelled by integers, there is a basis whose labels sum up to r modulo m, for some given integers r and m [8,10]. Intriguingly, a common theme in all these extensions is that they face barriers when it comes to general moduli m; many approaches are limited to constant prime power moduli, or need to allow for other compromises such as obtaining approximate solutions only or having to involve randomization in the algorithms. The unsatisfactory situation might be explained by a general lack of techniques to approach congruency-constrained problems. Present work typically follows either a partial enumeration approach, or relies on strong proximity results between solutions of the original and the congruency-constrained problem, but these techniques seem to come with inherent limitations. An alternative approach would be to develop a better polyhedral understanding of congruency-constrained problems, which is currently lacking.

The limited polyhedral understanding of Odd-Red Bipartite Perfect Matchings has recently been highlighted by [9] in a completely different context. They were interested in one of the most notorious open derandomization questions in Theoretical Computer Science, namely whether there is a deterministic efficient algorithm for the Bipartite Exact Perfect Matching problem, often also referred to as the Red-Blue Matching Problem. The setting is identical to Bipartite Odd-Red Perfect Matching, except that the goal is to find a perfect matching with exactly some given number k of red edges. This problem can be solved by a celebrated randomized algorithm of [11] in polynomial time, but it is a major open question whether there is a deterministic polynomial-time algorithm for it. There has been no notable progress towards derandomizing the above-mentioned approach until a relatively recent breakthrough of [5] almost completely derandomized the approach of Mulmuley, Vazirani, and Vazirani for the bipartite perfect matching problem (without colors), which had interesting implications in the field of parallel algorithms.[5] Building up on this, derandomizations of the technique of Mulmuley, Vazirani, and Vazirani have been developed for the perfect matching polytope for general graphs [20], matroid intersection [6], and totally unimodular polytopes [7]. What is common in all of these advances is that they rely on a good polyhedral understanding of the underlying problem. However, this is completely lacking for the Bipartite Exact Perfect Matching problem. As a natural first step in this direction, [9] studied the polyhedral structure of the Bipartite Odd-Red Perfect Matching problem. One reason why this is a natural first step is that a description of the Bipartite Exact Perfect Matching polytope could be used to obtain a description (in extended space) of the Bipartite Odd-Red Perfect Matching polytope. This can be achieved by

[5] We refer to an almost complete derandomization because the resulting deterministic parallel algorithm needs quasi-polynomially many processors.

observing that the latter polytope is the convex hull of the union of slices of the former polytope, namely one slice per each odd number of red edges, and by using Balas' disjunctive programming approach (see [2]).

Jia, Svensson, and Yuan showed that the Bipartite Exact Perfect Matching polytope has exponential extension complexity. However, there remains hope that a good (exponential) description of it may still be helpful for derandomization purposes, the same way as it was helpful for derandomizing the approach of Mulmuley, Vazirani, and Vazirani for perfect matchings in general graphs [20], which is also a problem in P with exponential extension complexity [19]. To this end, Jia, Svensson, and Yuan introduced a candidate description for the Bipartite Odd-Red Perfect Matching polytope, and showed that it has the desirable property that it is empty if and only if there is no odd-red perfect matching in the underlying graph. Before we state our results, we briefly review this candidate description.

1.1 The Candidate Description of Jia, Svensson, and Yuan

Let $G = (V, E)$ be a bipartite graph with $|V| = 2n$ vertices for some $n \in \mathbb{Z}_{\geq 1}$, with n vertices on each side of the bipartition, and let $R \subseteq E$ be the set of red edges. We denote by $P_{(G,R)}$ the *bipartite odd-red perfect matching polytope*, i.e.,

$$P_{(G,R)} := \operatorname{conv}(\{\chi^M : M \text{ is a perfect matching in } G \text{ with } |M \cap R| \text{ odd}\}) \ ,$$

where $\chi^M \in \{0,1\}^E$ is the incidence (or characteristic) vector of M.

[9] defined a family of valid constraints for $P_{(G,R)}$ based on the following. Consider a set of $\{0,1\}$ labelings of V given by

$$\mathcal{L}_{\text{all}}(G) := \{L : V \to \{0,1\} \text{ with } |L^{-1}(1)| \equiv n \pmod{*}2\} \ ,$$

and for every $L \in \mathcal{L}_{\text{all}}(G)$, define

$$E_L = \{e \in E \setminus R : L(u) = L(v)\} \cup \{e \in R : L(u) \neq L(v)\} \ .$$

One can observe that for each $L \in \mathcal{L}_{\text{all}}(G)$, every odd-red perfect matching must use at least one edge in E_L. In other words, constraints of the form $x(E_L) \geq 1$ are valid, i.e., they will be satisfied by all incidence vectors of odd-red perfect matchings. We refer to these constraints as *label constraints*. They lead to the following relaxation of $P_{(G,R)}$:

$$Q_{(G,R)} := \left\{ x \in \mathbb{R}^E_{\geq 0} : \begin{array}{r} x(\delta(u)) = 1 \ \ \forall u \in V \\ x(E_L) \geq 1 \ \ \forall L \in \mathcal{L}_{\text{all}}(G) \end{array} \right\} \ .$$

(label constraint relaxation)

As mentioned, it was left open by Jia, Svensson, and Yuan whether the label constraint relaxation $Q_{(G,R)}$ is a correct description of $P_{(G,R)}$, i.e., whether $P_{(G,R)} = Q_{(G,R)}$. Intriguingly, they showed that their relaxation has the remarkable property that $Q_{(G,R)} = \emptyset$ if and only if $P_{(G,R)} = \emptyset$.

1.2 Our Results

In this work, we refute in a strong way the possibility that the label constraint relaxation could capture $P_{(G,R)}$ by showing that any inequality description of $P_{(G,R)}$ must be surprisingly complex. More precisely, any description must deviate significantly not only from the proposed label constraint relaxation $Q_{(G,R)}$, but also from many other classical constraint families for combinatorial optimization problems. Concretely, our main result exhibits an exponentially sized family of facets of $P_{(G,R)}$ that require large and diverse coefficients in any description through a linear constraint. Note that because $P_{(G,R)}$ is a subset of the bipartite perfect matching polytope, it is not full-dimensional. Thus, a facet of $P_{(G,R)}$ may be described by linear inequalities with different normal vectors that are not scaled versions of each other. Theorem 1 shows that any such inequality has complicated structure.

Theorem 1. *For every even $n \in \mathbb{Z}_{>0}$, there exists a bipartite graph $G = (V, E)$ on n vertices, red edges $R \subseteq E$, and an exponentially (in n) sized family $\mathcal{F}$ of facets of the polytope $P_{(G,R)}$ with the following property: For every $F \in \mathcal{F}$ and all $a \in \mathbb{Z}^E$, $b \in \mathbb{Z}$ such that $F = \{x \in P_{(G,R)} : a^\top x = b\}$, we have*

$$\max_{e \in E} |a_e| \geq \frac{n-4}{2} \qquad \text{and} \qquad |\{a_e : e \in E\}| \geq \sqrt{\frac{n-1}{2}}\ .$$

In particular, Theorem 1 excludes any inequality description of $P_{(G,R)}$ consisting of constraints with all coefficients in $\{0, \pm 1\}$, as for example the candidate description by Jia, Svensson, and Yuan. More importantly, though, it underlines that any exact description must involve constraints distinctly different from those typically appearing in most other polytopes associated with combinatorial optimization problems. This is despite the fact that $P_{(G,R)}$ is well-behaved from other points of view, in particular that we can optimize over it in polynomial time [1]. We obtain Theorem 1 by identifying and exploiting a connection to the dominant of the odd cycle polytope, for which we show a similar (actually even slightly stronger) facet-complexity result (see Theorem 4). This puts both the odd-red perfect matching problem and the minimum odd cycle problem in a rare class of natural and efficiently solvable combinatorial problems (not involving any numbers as part of their input) whose associated polytopes have highly complex inequality descriptions. There are other notable examples of problems with that property. This includes, in paticular, the stable set problem in claw-free graphs [18]. Another example, this time of an unbounded polyhedron instead of a polytope, is the dominant of the cut polytope, even for restricted graph classes.[6]

[6] One way this is implied by prior results is as follows: The blocker of the dominant $P^{\uparrow}_{\text{cut}}$ of the cut polytope is the polyhedron $P = \{x \in \mathbb{R}^E_{\geq 0} : x(\delta(C)) \geq 2\ \forall C \subsetneq V, C \neq \emptyset\}$, which is strongly related to the subtour elimination polytope. In particular, any vertex of P is also a vertex of the subtour elimination polytope. The subtour elimination polytope, and therefore also P, can have vertices with many different coefficients (this follows, for example, from [3]). Finally, it remains to observe that the blocker of P is again $P^{\uparrow}_{\text{cut}}$, and each vertex of P appears as the coefficient vector of a facet-defining inequality of $P^{\uparrow}_{\text{cut}}$.

In order to identify complex facets as in Theorem 1, it is enough to consider the bipartite graph G_0 with two parallel edges connecting every two vertices on different sides of the bipartition, and red edges R_0 containing precisely one of the two parallel edges between each such pair of vertices. Indeed, every other bipartite graph G on the same number of vertices with red edges R is a subgraph of (G_0, R_0), and therefore $P_{(G,R)}$ is a face of $P_{(G_0,R_0)}$ that is obtained by setting to zero all variables corresponding to edges in $E(G_0) \setminus E(G)$. Because a description of $P_{(G,R)}$ can be obtained from that of $P_{(G_0,R_0)}$ by simply removing all variables corresponding to edges in $E(G_0) \setminus E(G)$, the complexity of a description of $P_{(G,R)}$ in the sense of Theorem 1 carries over to $P_{(G_0,R_0)}$. Nonetheless, we stick to considering concrete graphs (G, R) for the simplicity of the presentation.

We show that we can carry over our facet complexity result to bimodular integer programs by exploiting a natural bimodular integer programming formulation for bipartite red-odd perfect matchings. In this transfer, one has to take care of the fact that the natural bimodular formulation lives in extended space, whereas the statement we want to obtain is about the original space.

Theorem 2. *For every $m \in \mathbb{Z}_{>0}$, there is an m-variable bimodular integer program whose associated polytope has facets that, in any inequality description, require $\Omega(m^{1/4})$ different coefficients.*

Previously, it was known that bimodular programs capture problems with exponential extension complexity (see, for example, [4,9]). However, prior to our work, the potential existence of clean inequality descriptions with small coefficients was not ruled out.

As a side result, we show that the candidate description $Q_{(G,R)}$ of Jia, Svensson, and Yuan does not admit efficient separation (unless $\mathsf{P} = \mathsf{NP}$).

Theorem 3. *Let $G = (V, E)$ be a bipartite graph with red edges $R \subseteq E$. Given a point $x \in \mathbb{R}^E_{\geq 0}$, it is* NP*-complete to decide whether $x \notin Q_{(G,R)}$. The problem remains* NP*-complete if x is known to be in the perfect matching polytope of G.*

This is in interesting contrast to the fact that the equivalence of optimization and separation implies that polynomial-time separation over $P_{(G,R)}$ is possible, and may arguably be a reason that would make it surprising (though not impossible) to see this constraint family play a major role in an exact description of $P_{(G,R)}$.

1.3 Organization of This Paper

A large part of the remainder of this extended abstract, i.e., Sect. 2, is dedicated to giving an overview of our approach and techniques to prove Theorem 1. Some of the technical proofs are deferred to a long version of this paper. Section 3 shows how to deduce Theorem 2, while detailed proofs of Theorem 3 are deferred to the long version. Finally, a concrete counterexample directly showing $Q_{(G,R)} \neq P_{(G,R)}$ is given in Appendix A.

2 An Overview of Our Approach Towards Complex Facets

We now give an overview of our approach for proving Theorem 1. Concretely, we will explicitly construct a family of facets of $P_{(G,R)}$ with the properties claimed in the theorem. To this end, we exploit a link between the odd-red bipartite perfect matching polytope and the odd cycle polytope. For a (not necessarily bipartite) graph $G = (V, E)$, the *odd cycle polytope* is defined as

$$P_{\text{odd}}(G) = \text{conv}\{\chi^C : C \subseteq E \text{ is an odd cycle in G}\} .$$

To obtain a correspondence between odd cycles and odd-red matchings, we use the following construction: Starting from a graph $G = (V, E)$, we consider a bipartite graph $\widehat{G} = (\widehat{V}, \widehat{E})$, where $\widehat{V}$ consists of two copies $V^+ := \{v^+ : v \in V\}$ and $V^- := \{v^- : v \in V\}$ of V, and $\widehat{E}$ contains all edges $\{v^+, v^-\}$ for $v \in V$, as well as all edges $\{u^+, v^-\}$ and $\{u^-, v^+\}$ for every edge $\{u, v\} \in E$. Moreover, we define $\widehat{R} := \widehat{E} \setminus \{\{v^+, v^-\} : v \in V\}$ to be the subset of red edges in $\widehat{G}$. See Fig. 1 for an illustration of this construction.

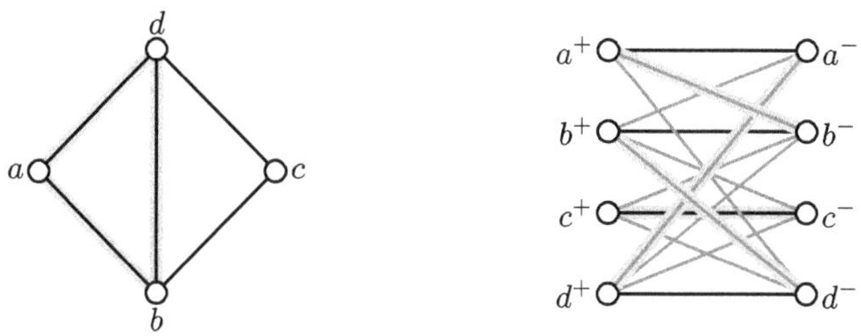

Fig. 1. An initial graph G (left), and the corresponding bipartite graph $\widehat{G}$ with edges of $\widehat{R}$ indicated in red (right). It is highlighted (in gray) how an odd cycle in G gets mapped to an odd-red perfect matching in $\widehat{G}$. (Color figure online)

Now an odd cycle C on vertices $v_1, \ldots, v_k, v_{k+1} = v_1$ in G can be translated, for example, to the odd-red perfect matching M in $\widehat{G}$ with edges $\{v_i^+, v_{i+1}^-\}$ for $i \in [k]$, and $\{v^+, v^-\}$ for all $v \notin C$. Note that we are not claiming a one-to-one correspondence between odd cycles and odd-red perfect matchings here: one could also construct another odd-red perfect matching from the same odd cycle by traversing it in the opposite direction, thus using the edges $\{v_i^-, v_{i+1}^+\}$. Conversely, by reversing the construction above, one can see that in general, an odd-red perfect matching M in $\widehat{G}$ can be mapped to a disjoint union of cycles in G for which the total number of edges is odd (some of the cycles may be degenerate cycles that consist of one edge used twice). Crucially, such a union of cycles will always contain at least one odd cycle, and thus the corresponding incidence vector is contained in the *dominant* of the odd cycle polytope, which is defined as the Minkowski sum

$$P_{\text{odd}}^{\uparrow}(G) = P_{\text{odd}}(G) + \mathbb{R}_{\geq 0}^{E} \ .$$

We will in the following identify a family of complex facets of $P_{\text{odd}}^{\uparrow}(G)$ for complete graphs G, and then transfer these to facets of $P_{(\widehat{G},\widehat{R})}$. The main ingredients for this transformation are laid out next.

2.1 Complex Facets of $P_{\text{odd}}^{\uparrow}(G)$

We begin by showing that $P_{\text{odd}}^{\uparrow}(G)$ belongs to the interesting family of polyhedra arising from combinatorial optimization problems that allow for efficient optimization algorithms, but have complex facets. The result we obtain in terms of facet complexity is even stronger than in Theorem 1, as we can show the need for a linear number of different coefficients.

Theorem 4. *For every odd $n \in \mathbb{Z}_{>0}$, there exists a graph $G = (V, E)$ on n vertices and an exponentially (in n) sized family $\mathcal{F}$ of facets of the polytope $P_{\text{odd}}^{\uparrow}(G)$ with the following property: For every $F \in \mathcal{F}$ and all $a \in \mathbb{Z}^E$, $b \in \mathbb{Z}$ such that $F = \{x \in P_{\text{odd}}^{\uparrow}(G) \colon a^\top x = b\}$, we have*

$$\max_{e \in E} |a_e| \geq n - 4 \qquad \textit{and} \qquad |\{a_e \colon e \in E\}| \geq \frac{n-3}{2} \ .$$

To prove Theorem 4, it is convenient to parameterize the odd size n of the graph by $n = 2k + 3$. Let $G := K_n$ be the complete graph on n vertices with vertex set V and edge set E. We explicitly construct a family of facets $\mathcal{F}$ induced by $(n-2)$-cycles in G, i.e., cycles of length $n - 2 = 2k + 1$ in G. Concretely, consider a $(2k+1)$-cycle $C = (v_1, \dots, v_{2k+1})$ in G, and let s and t be the two vertices of G not contained in C. For an edge $\{v_i, v_j\}$ between two vertices of C, set

$$\ell(\{v_i, v_j\}) = \begin{cases} |j - i| & \text{if } |j-i| \text{ is odd} \\ 2k + 1 - |j - i| & \text{otherwise} \end{cases} \ .$$

Note that equivalently, $\ell(\{v_i, v_j\})$ can be defined as the length of the odd-length path on C from v_i to v_j (also see Fig. 2).

With this notation at hand, we can define the following family of constraints, where $E[S]$, for $S \subseteq V$, denotes the set of all edges with both endpoints in S.

Definition 5 (C-induced constraint). *Let $G = K_n$ for some odd $n \in \mathbb{Z}_{\geq 3}$, and $k = \frac{n-3}{2}$. Consider an $(n-2)$-cycle C in G, and let s and t be the two vertices of G not contained in C. We define the* C-induced constraint *for $P_{\text{odd}}^{\uparrow}(G)$ by*

$$x(\{s,t\}) + k \cdot x(\delta(C)) + \sum_{e \in E[V \setminus \{s,t\}]} \ell(e) \cdot x(e) \geq 2k + 1 \ ,$$

Moreover, denoting by $\mathcal{C}$ the set of all $(n-2)$-cycles in G, we refer to the set of all C-induced constraints for $C \in \mathcal{C}$ as the $\mathcal{C}$-induced constraints *for $P_{\text{odd}}^{\uparrow}(G)$.*

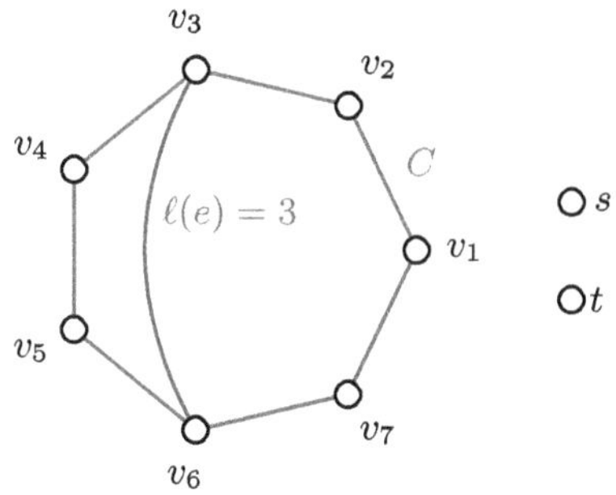

Fig. 2. Partial visualization of the graph G for $k = 3$ ($n = 9$). The edge $e = \{v_3, v_6\}$ has ℓ-value 3 because the odd-length path in the cycle C connecting its endpoints is (v_3, v_4, v_5, v_6) of length 3.

Figure 3 shows the coefficients of a C-induced constraint of $P^{\uparrow}_{\text{odd}}(K_9)$. Note that a C-induced constraint indeed only depends on the choice of a cycle C of length $n-2$ in G, and two different such cycles lead to different constraints. In particular, the C-induced constraints form an exponential family of constraints. Moreover, for $n = 2k+3 \geq 5$, the coefficients of a C-induced constraint are $\{1, 3, 5, \ldots, 2k-1\} \cup \{k\}$, which are at least $k = \frac{n-3}{2}$ distinct values.

The validity of a C-induced constraint can be verified as follows. Let $x = \chi^{\overline{C}}$ be the characteristic vector of an odd cycle $\overline{C}$ in G. If $\overline{C}$ goes through s or t, then the left-hand side term $k \cdot x(\delta(C))$ already contributes $2k$ units, and one more unit is obtained because there must be a third edge in $\overline{C}$, and all edges have strictly positive coefficients. Otherwise, if $\overline{C}$ does not go through s or t, then we can map each edge $e \in \overline{C}$ to the odd-length path P_e in C connecting its endpoints. If the paths P_e for $e \in \overline{C}$ cover all of C, the constraint is fulfilled because the sum of the ℓ-values of the edges in $\overline{C}$, which is part of the left-hand side of the C-induced constraint, is equal to $\sum_{e \in \overline{E}} |P_e|$, which is at least $2k+1$ because the full cycle is covered. A key observation, which finishes the reasoning, is that it is impossible that the paths P_e do not cover all of C. If this happened, then the paths P_e for $e \in \overline{C}$ would correspond to an odd number of odd length intervals on a line, as we may break the cycle open at the uncovered edge without breaking up any of the paths P_e. However, it is impossible that such intervals form a closed walk, because for that, their total length would have to be even.

The intuition behind the choice of the coefficients for a C-induced constraint is—besides the goal of constructing a valid constraint—to ensure that many odd-length cycles are tight for the constraint. The definition of ℓ ensures that the odd-length cycle defined by any chord $\{v_i, v_j\}$ of C together with the even part of C connecting v_i to v_j is tight for the constraint. Additionally, all triangles that contain an edge of C or the edge $\{s, t\}$, together with two appropriate edges crossing the cut $\delta(C)$, are tight for the constraint. Showing that enough of these tight cycles (respectively, their incidence vectors) are linearly independent will allow for showing that a C-induced constraint is indeed facet-defining.

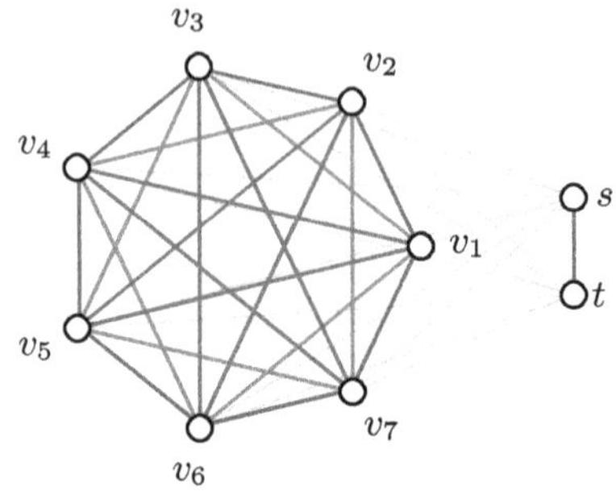

Fig. 3. Visualization of a C-induced-constraint of $P^{\uparrow}_{\text{odd}}(K_9)$. The coefficient of green edges is 1, of blue edges 3, and of red edges 5. Furthermore, the edges between the 7-gon and the vertices s and t, marked in light gray, also have a coefficient of 3. The right-hand side of the constraint is 7. (Color figure online)

Lemma 6. *$\mathcal{C}$-induced constraints are facet-defining for $P^{\uparrow}_{\text{odd}}(G)$.*

A proof of Lemma 6 is deferred to the long version of this paper. Note that, together with the explicit description of $\mathcal{C}$-induced constraints and full-dimensionality of $P^{\uparrow}_{\text{odd}}(G)$, Lemma 6 immediately implies Theorem 4. Indeed, $P^{\uparrow}_{\text{odd}}(G)$ being full dimensional implies that, given a facet, any inequality describing it uses the same normal vector up to scaling. Our goal is to transfer these complex facets of $P^{\uparrow}_{\text{odd}}(G)$ to facets of $P_{(\widehat{G},\widehat{R})}$, while showing that much of their complexity is preserved. We detail how this can be achieved next.

2.2 Transferring Facets of $P^{\uparrow}_{\text{odd}}(G)$ to $P_{(\widehat{G},\widehat{R})}$

Because $P^{\uparrow}_{\text{odd}}(G)$ is a dominant and therefore up-closed, it can be described by so-called *covering inequalities*, i.e., inequalities of the form

$$\sum_{e \in E} a_e x_e \geq b \tag{1}$$

with $a \in \mathbb{R}^E_{\geq 0}$ and $b \in \mathbb{R}$. For any such constraint, we define its *canonical transformation* as

$$\sum_{\{u,v\} \in E} a_{\{u,v\}} \left(y_{\{u^+,v^-\}} + y_{\{u^-,v^+\}} \right) \geq b \ , \tag{2}$$

where y is defined on the edges of $\widehat{G}$. The following is an immediate observation.

Observation 7. *The canonical transformation of any constraint that is valid for $P^{\uparrow}_{\text{odd}}(G)$ is valid for $P_{(\widehat{G},\widehat{R})}$.*

Proof. Consider a valid constraint κ for $P^{\uparrow}_{\text{odd}}(G)$ of the form given in (1), and its canonical transformation as given in (2). Let $y \in P_{(\widehat{G},\widehat{R})}$ be the incidence vector of an odd-red perfect matching M in $\widehat{G}$. As discussed earlier, M can be mapped to a union of cycles in G that contains at least one odd cycle C, by mapping edges $\{v^{\pm}, w^{\mp}\}$ in M back to the edge $\{v, w\}$ in G. Let $x \in P^{\uparrow}_{\text{odd}}(G)$ be the incidence vector of C. Then, the left-hand side of the canonical transformation of κ evaluated at y is at least the left-hand side of κ evaluated at x. Consequently, validity of κ for $P^{\uparrow}_{\text{odd}}(G)$ implies validity of its canonical transformation for $P_{(\widehat{G},\widehat{R})}$. □

For $\mathcal{C}$-induced constraints, which are facets of $P^{\uparrow}_{\text{odd}}(G)$ by Lemma 6, we claim that the canonical transformation not only gives feasible constraints as shown in Observation 7 above, but facet-defining constraints. In order to show that a constraint is facet-defining, we need to identify enough independent points on the facet. Naturally, for a facet κ of $P^{\uparrow}_{\text{odd}}(G)$, the hope is to be able to take independent (characteristic vectors of) odd cycles for which κ is tight, transfer these to odd-red matchings for which the canonical transformation of κ is tight, and argue that these matchings are independent enough to conclude that the canonical transformation of κ is facet-defining for $P_{(\widehat{G},\widehat{R})}$. For carrying out this final argument of independence on the odd-red matching side in an almost black-box way, we exploit the following notion of *expressibility* on the side of odd cycles.

Definition 8 (Expressibility). *Let $G = (V, E)$ be a graph and consider a constraint κ of the form given in* (1) *that is valid for $P^{\uparrow}_{\text{odd}}(G)$. Let*

$$\mathcal{T} := \{C \subseteq E \colon C \text{ is an odd cycle in } G \text{ and } \kappa \text{ is tight for } \chi^C\}$$

be the set of odd cycles in G whose incidence vectors satisfy κ with equality.

(i) For $Q \subseteq E$, we define the Q-expressible edges $E_Q \subseteq E$ through the following iterative procedure: We start with $E_Q = Q$, and as long as there is an edge $e \in E \setminus E_Q$ and a cycle $C \in \mathcal{T}$ with $e \in C$ and $C \setminus \{e\} \subseteq E_Q$, we add e to E_Q.

*(ii) We say that κ is k-*expressing *if there is a set $Q \subseteq E$ of size at most k such $E_Q = E$.*

In other words, a constraint is k-expressing if it suffices to provide k edges through which all others can be expressed using odd cycles as defined above. As one might imagine, this number k is closely related to an upper bound on the co-dimension of the resulting face when a k-expressing constraint of $P^{\uparrow}_{\text{odd}}(G)$ is transformed to a constraint of $P_{(\widehat{G},\widehat{R})}$. For small enough k, exploiting that we know the dimension of $P_{(\widehat{G},\widehat{R})}$, we will hence be able to certify that the resulting constraint is facet-defining. Concretely, our result is the following. (See the full version of this paper for proofs of the following statements.)

Lemma 9. *Let G be an n-vertex graph, and let κ be an $(n-1)$-expressing facet-defining constraint of $P^{\uparrow}_{\text{odd}}(G)$ that is not tight for all of $P_{\text{odd}}(G)$. Then, the canonical transformation of κ is facet-defining for $P_{(\widehat{G},\widehat{R})}$.*

With Lemma 9 at hand, using that for K_n no facet is tight for all odd cycles, all that remains to show is that $\mathcal{C}$-induced constraints of $P^{\uparrow}_{\text{odd}}(G)$ are $(n-1)$-expressing.

Lemma 10. *$\mathcal{C}$-induced constraints are $(n-1)$-expressing.*

Together, Lemmas 9 and 10 immediately yield the following corollary that provides the desired facets of $P_{(\widehat{G},\widehat{R})}$.

Corollary 11. *The canonical transformation of any $\mathcal{C}$-induced constraint is facet-defining for $P_{(\widehat{G},\widehat{R})}$.*

We refer to the facets of $P_{(\widehat{G},\widehat{R})}$ obtained in this way as *transformed $\mathcal{C}$-induced facets* of $P_{(\widehat{G},\widehat{R})}$.

2.3 Asserting Complexity of Transformed $\mathcal{C}$-Induced Facets

Note that transformed $\mathcal{C}$-induced facets of $P^{\uparrow}_{\text{odd}}(G)$ inherit the complexity of the original facets in terms of coefficients in the linear inequality representations, as these remain unchanged through the transformation. However, while facets of $P^{\uparrow}_{\text{odd}}(G)$ have uniquely defined normal vectors (up to scaling) because $P^{\uparrow}_{\text{odd}}(G)$ is full-dimensional, this is not the case for $P_{(\widehat{G},\widehat{R})}$ because

$$\dim(P_{(\widehat{G},\widehat{R})}) = |\widehat{E}| - |\widehat{V}| + 1 \ .$$

In fact, the minimal subspace containing $P_{(\widehat{G},\widehat{R})}$ is determined by the degree constraints $y(\delta(v)) = 1$ for all $v \in \widehat{V}$. In other words, any facet-defining constraint $a^\top y \geq b$ of $P_{(\widehat{G},\widehat{R})}$ can be equivalently expressed in the form

$$\mu \cdot a^\top y + \sum_{v \in \widehat{V}} \lambda_v \cdot y(\delta(v)) \geq \mu \cdot b + \sum_{v \in \widehat{V}} \lambda_v \tag{3}$$

for arbitrary $\mu > 0$ and $\lambda \in \mathbb{R}^{\widehat{V}}$.

Hence, to make sure that a facet of $P_{(\widehat{G},\widehat{R})}$ admits no low complexity representation, we need to show than every such equivalent representation has high complexity. (Note that we typically assume that facets are scaled so that they have integral coefficients and integral right-hand side.)

To this end, we change our viewpoint and treat coefficient vectors $a \in \mathbb{Z}^{\widehat{E}}$ as matrices in $\mathbb{Z}^{V^+ \times V^-}$, with rows indexed by V^+ and columns indexed by V^-. More precisely, for $u \in V^+$ and $v \in V^-$, the matrix entry $a(u,v)$ shall be the coefficient $a_{\{u,v\}}$ of the variable associated with the edge $\{u,v\} \in \widehat{E}$. In Fig. 4, we exemplify this matrix viewpoint by presenting the matrix corresponding to the transformation of the C-induced facet of $P^{\uparrow}_{\text{odd}}(K_9)$ presented earlier in Fig. 3. To see how the matrix representation changes under moving to an equivalent form as in (3), which we write as $(a^{\mu,\lambda})^\top y \geq b^{\mu,\lambda}$ for $\mu > 0$ and $\lambda \in \mathbb{R}^{\widehat{V}}$, observe that for $u \in V^+$ and $v \in V^-$, we have

$$a^{\mu,\lambda}(u,v) = \mu \cdot a(u,v) + \lambda_u + \lambda_v \ . \tag{4}$$

In other words, the matrix $a^{\mu,\lambda}$ can be obtained from the matrix a by scaling with μ, and then adding a value λ_u to all entries in row u and a value λ_v to all entries in column v, for all $u \in V^+$ and $v \in V^-$. Thus, the analysis of equivalent facet representations boils down to analyzing the impact of these simple matrix operations on the matrix obtained from the coefficient vector a.

	v_1^-	v_3^-	v_5^-	v_7^-	v_2^-	v_4^-	v_6^-	s^-	t^-
v_1^+	0	5	3	1	1	3	5	3	3
v_3^+	5	0	5	3	1	1	3	3	3
v_5^+	3	5	0	5	3	1	1	3	3
v_7^+	1	3	5	0	5	3	1	3	3
v_2^+	1	1	3	5	0	5	3	3	3
v_4^+	3	1	1	3	5	0	5	3	3
v_6^+	5	3	1	1	3	5	0	3	3
s^+	3	3	3	3	3	3	3	0	1
t^+	3	3	3	3	3	3	3	1	0

Fig. 4. A matrix representation of the coefficient vector of the translated C-induced facet of $P^{\uparrow}_{\text{odd}}(K_9)$ presented earlier in Fig. 3.

Our complexity results focus on lower bounds on the largest entry in absolute value and the number of different entries that can be achieved through the matrix operations described above. Using specific invariant alternating sums of matrix entries, we can provide a linear lower bound on the first quantity. Interestingly, for the number of different entries, an analysis based on the pigeonhole principle that takes into account only two rows of the matrix already results in a lower bound of order $\Omega(\sqrt{n})$, completing the proof of Theorem 1 (see the full version of this paper for details).

While the precise asymptotic behavior of the smallest achievable number of different coefficients in inequality representations of transformed $\mathcal{C}$-induced facets is left open, we present a way to reduce to $O(n^{2/3})$ many different coefficients (again, we refer to the full version of this paper for details).

3 Bimodular Polytopes With Complex Facets

In this section, we prove Theorem 2 by showing that a natural bimodular formulation of the red-odd perfect matching problem admits the postulated facet complexity. To this end, let $G = (V, E)$ be a bipartite graph with red edges $R \subseteq E$, and observe that the integral solutions of the system

$$\begin{aligned} x(\delta(v)) &= 1 \quad \forall v \in V \\ x(R) - 2y &= 1 \\ x &\in \mathbb{R}^E_{\geq 0} \\ y &\in \mathbb{R}_{\geq 0} \end{aligned} \tag{5}$$

are precisely the pairs (χ^M, r), where M is a red-odd perfect matching in G and $r = \frac{|M \cap R| - 1}{2}$. Also, the system (5) is indeed bimodular: one can observe that all full-rank subdeterminants of the constraint matrix are in $\{-2, 0, 2\}$.

Let P_{bimod} be the convex hull of all integral solutions to (5). By the above,

$$P_{\text{bimod}} = \{(x, (x(R) - 1)/2) : x \in P_{(G,R)}\} ,$$

hence P_{bimod} is obtained from $P_{(G,R)}$ by lifting along the new variable y to the hyperplane defined by $y = \frac{x(R)-1}{2}$. Thus, we get the following.

Observation 12. *For every facet F of P_{bimod}, there is a facet F' of $P_{(G,R)}$ such that*

$$F = \{(x, (x(R) - 1)/2) : x \in F'\} ,$$

and vice versa. In particular, for all $a \in \mathbb{R}^E$ and $b, c \in \mathbb{R}$, we have $F = \{x \in P_{\text{bimod}} : a^\top x + cy = b\}$ if and only if

$$F' = \left\{x \in P_{(G,R)} : \left(a + \frac{c}{2}\chi^R\right)^\top x = b + \frac{c}{2}\right\} .$$

This observation has the following two consequences. First, it allows to directly lift any complex facet F' of $P_{(G,R)}$ to a facet F of P_{bimod}. Second, it implies that any representation of such a lifted facet F through a linear equality can be translated to a representation of the original facet F'. Because the transformation detailed in Observation 12 can at most double the number of different coefficients, and increase the largest absolute value by a factor of at most $\frac{3}{2}$, Theorem 1 directly implies the following strengthening of Theorem 2.

Corollary 13. *For every even $n \in \mathbb{Z}_{>0}$, there exists a bipartite graph $G = (V, E)$ on n vertices, red edges $R \subseteq E$, and an exponentially (in n) sized family $\mathcal{F}$ of facets of the polytope P_{bimod} with the following property: For every $F \in \mathcal{F}$ and all $a \in \mathbb{Z}^E$, $b \in \mathbb{Z}$ such that $F = \{x \in P_{\text{bimod}} : a^\top x = b\}$, we have*

$$\max_{e \in E} |a_e| \geq \frac{n-4}{3} \qquad \text{and} \qquad |\{a_e : e \in E\}| \geq \sqrt{\frac{n-1}{8}} .$$

Note that the number of variables in the bimodular formulation of a red-odd perfect matching problem needs one variables per edge of the bipartite graph plus one additional variable (for r). Hence, a dense n-vertex bipartite graph has $m = \Theta(n^2)$ edges, and the number of different coefficients in Theorem 13 is in $\Omega(\sqrt{n}) = \Omega(m^{1/4})$, as stated in Theorem 2.

Disclosure of Interests. The authors have no competing interests to declare that are relevant to the content of this article.

A Explicit Example For Label Constraint Insufficiency

We give an explicit example of a bipartite graph $G = (V, E)$ and a set of red edges $R \subseteq E$ for which the label constraint relaxation $Q_{(G,R)}$ of [9] is not equal to the odd-red perfect bipartite matching polytope $P_{(G,R)}$. Concretely, we consider the pair (G, R) presented in Fig. 5.

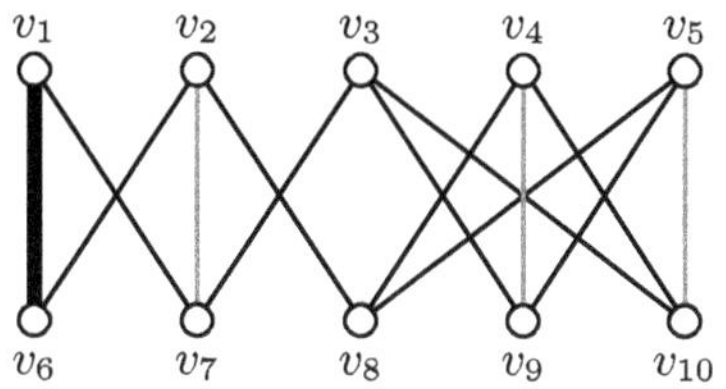

Fig. 5. Example of a graph $G = (V, E)$ with the indicated three red edges $R \subseteq E$ for which the polytope $Q_{(G,R)}$ is not integral. The point y we consider to prove this has a weight of $2/3$ on the edge (v_1, v_6) marked in bold, and $1/3$ on every other edge. (Color figure online)

To be precise, we use $R := \{\{v_2, v_7\}, \{v_4, v_9\}, \{v_5, v_{10}\}\}$. We claim that $P_{(G,R)} \neq Q_{(G,R)}$ for this graph G. To prove this, we define a point $y \in \mathbb{R}^E$ with value $1/3$ on every edge but the edge $\{v_1, v_6\}$, where it has value $2/3$ instead, i.e.,

$$y(e) = \begin{cases} \frac{1}{3} & \text{if } e \neq \{v_1, v_6\} \\ \frac{2}{3} & \text{if } e = \{v_1, v_6\} \end{cases} \quad \text{for all } e \in E \ .$$

To prove the claim, we show that $y \notin P_{(G,R)}$, but $y \in Q_{(G,R)}$. The former is easy to see, as G does not admit an odd-red perfect matching containing $\{v_3, v_7\}$, but y has support on this edge. For the latter, to derive that label constraints $y(E_L) \geq 1$ with $L \in \mathcal{L}_{\text{all}}(G)$ are satisfied, one may exploit the following.

Lemma 14 ([9, **Claim 5.2**]). *Let C be a cycle in G with $|C \cap R|$ odd and let $L \in \mathcal{L}_{\text{all}}(G)$. Then $|E_L \cap C|$ is odd. In particular, the intersection is non-empty.*

Concretely, for any $L \in \mathcal{L}_{\text{all}}(G)$, one can see that at least three edges of the cycles $C = (v_1, v_6, v_2, v_7)$, $C_1 = (v_5, v_{10}, v_3, v_9)$, $C_2 = (v_4, v_{10}, v_5, v_8)$, and $C_3 = (v_3, v_{10}, v_4, v_9)$ contribute to E_L, thus giving the desired $y(E_L) \geq 1$, hence finishing the proof. We refer to the full version of this paper for details.

References

1. Artmann, S., Weismantel, R., Zenklusen, R.: A strongly polynomial algorithm for bimodular integer linear programming. In: Proceedings of the 49th Annual ACM SIGACT Symposium on Theory of Computing (STOC), pp. 1206–1219 (2017)
2. Balas, E.: Disjunctive Programming. Springer, Cham (2018)

3. Boyd, S.: The Subtour Polytope of the Travelling Salesman Problem. PhD thesis, University of Waterloo (1986)
4. Cevallos, A., Weltge, S., Zenklusen, R.: Lifting linear extension complexity bounds to the mixed-integer setting. In: Proceedings of the 29th ACM-SIAM Symposium on Discrete Algorithms (SODA), pp. 788–807 (2018)
5. Fenner, S., Gurjar, R., Thierauf, T.: Bipartite perfect matching is in quasi-NC. In: Proceedings of the 48th ACM Symposium on Theory of Computing (STOC) (2016)
6. Gurjar, R., Thierauf, T.: Linear matroid intersection is in quasi-NC. Comput. Complex. (2020)
7. Gurjar, R., Thierauf, T., Vishnoi, N.K.: Isolating a vertex via lattices: polytopes with totally unimodular faces. SIAM J. Comput. **50**(2), 636–661 (2021)
8. Hörsch, F., Imolay, A., Mizutani, R., Oki, T., Schwarcz, T.: Problems on group-labeled matroid bases. In: Proceedings of the 51st International Colloquium on Automata, Languages, and Programming (ICALP), pp. 86:1–86:20 (2024)
9. Jia, X., Svensson, O., Yuan, W.: The exact bipartite matching polytope has exponential extension complexity. In: Proceedings of the 2023 Annual ACMSIAM Symposium on Discrete Algorithms (SODA), pp. 1635–1654 (2023)
10. Liu, S., Xu, C.: On the congruency-constrained matroid base. In: Proceedings of the 2024 Conference on Integer Programming and Combinatorial Optimization (IPCO), pp. 280–293 (2024)
11. Mulmuley, K., Vazirani, U.V., Vazirani, V.V.: Matching is as easy as matrix inversion. In: Proceedings of the 19th Annual ACM Symposium on Theory of Computing (STOC), pp. 345–354 (1987)
12. Nägele, M., Nöbel, C., Santiago, R., Zenklusen, R.: Advances on strictly Δ-modular IPs. In: Proceedings of the 24th Conference on Integer Programming and Combinatorial Optimization (IPCO), pp. 393–407 (2023)
13. Nägele, M., Nöbel, C., Zenklusen, R.: On the Complexity of the Odd-Red Bipartite Perfect Matching Polytope (2026). arXiv: 2603.18232 [cs.DS]
14. Nägele, M., Santiago, R., Zenklusen, R.: Congruency-constrained TU problems beyond the bimodular case. Math. Oper. Res. **49**(3), 1303–1348 (2023)
15. Nägele, M., Sudakov, B., Zenklusen, R.: Submodular minimization under congruency constraints. Combinatorica **39**(6), 1351–1386 (2019). https://doi.org/10.1007/s00493-019-3900-1
16. Nägele, M., Zenklusen, R.: A new contraction technique with applications to congruency-constrained cuts. Math. Program. **183**(1), 455–481 (2020)
17. Padberg, M.W., Rao, M.R.: Odd minimum cut-sets and b-matchings. Math. Oper. Res. **1**(7), 67–80 (1982)
18. Pêcher, A., Pesneau, P., Wagler, A.K.: On facets of stable set polytopes of claw-free graphs with stability number three. Electron. Notes Disc. Math. **28**, 185–190 (2007)
19. Rothvoß, T.: The matching polytope has exponential extension complexity. J. ACM **64**(6), 1–19 (2017)
20. Svensson, O., Tarnawski, J.: The matching problem in general graphs is in quasi-NC. In: Proceedings of the 58th Annual IEEE Symposium on Foundations of Computer Science (FOCS), pp. 696–707 (2017)

A PTAS for Weighted Triangle-Free 2-Matching

Miguel Bosch-Calvo[1], Fabrizio Grandoni[1], Yusuke Kobayashi[2], and Takashi Noguchi[2](✉)

[1] IDSIA, USI-SUPSI, Lugano, Switzerland
{miguel.boschcalvo,fabrizio}@idsia.ch
[2] RIMS, Kyoto University, Kyoto, Japan
{yusuke,tnoguchi}@kurims.kyoto-u.ac.jp

Abstract. In the *Weighted Triangle-Free 2-Matching* problem (WTF2M), we are given an undirected edge-weighted graph. Our goal is to compute a maximum-weight subgraph that is a 2-matching (i.e., no node has degree more than 2) and triangle-free (i.e., it does not contain any cycle with 3 edges). One of the main motivations for this and related problems is their practical and theoretical connection with the Traveling Salesperson Problem and with some 2-connectivity network design problems. WTF2M is not known to be NP-hard and at the same time no polynomial-time algorithm to solve it is known in the general case (polynomial-time algorithms are known only for some special cases). The best-known (folklore) approximation algorithm for this problem simply computes a maximum-weight 2-matching, and then drops the cheapest edge of each triangle: this gives a 2/3 approximation. In this paper we present a PTAS for WTF2M, i.e., a polynomial-time $(1-\varepsilon)$-approximation algorithm for any given constant $\varepsilon > 0$. Our result is based on a simple local-search algorithm and a non-trivial analysis.

Keywords: Restricted matching · Local search · PTAS

1 Introduction

Let $G = (V, E, w)$, $w\colon E \to \mathbb{Q}_{\geq 0}$, be a simple undirected edge-weighted graph. A 2-matching M of G is a subset of edges $M \subseteq E$ such that at most two edges of M are incident on each node in V. Observe that a 2-matching induces a collection of disjoint paths and cycles. The 2-Matching problem requires to compute a 2-matching of maximum weight $w(M) := \sum_{e \in M} w(e)$: this problem can be solved in polynomial time via a reduction to matching (see, e.g. [35, Chapter 30.6.]).

In this paper we focus on the *Weighted Triangle-Free 2-Matching* problem (WTF2M), a natural generalization of 2-Matching where we further require that the computed matching M does not induce any *triangle* (i.e., cycle with 3 edges). Hartvigsen proposed an (extremely complex) exact algorithm for the unweighted version TF2M in his Ph.D. thesis in 1984 [15], and a (similarly complex) journal

S. S. Dey et al. (Eds.): IPCO 2026, LNCS 16588, pp. 17–31, 2026.
https://doi.org/10.1007/978-3-032-28691-8_2

version of that result was recently published [17]. Recently, Paluch [31] reported another polynomial-time algorithm for the problem (still unpublished). For the sake of simplification in verification and analysis, Bosch-Calvo, Grandoni, and Ameli [8] proposed a simple local search-based Polynomial-Time Approximation Scheme (PTAS) for this problem, i.e., an algorithm that, for any given input constant parameter $\varepsilon > 0$, computes a $(1-\varepsilon)$-approximate solution for the problem in polynomial time. Kobayashi and Noguchi [28] further and substantially simplified the correctness proof of the mentioned PTAS. Exact algorithms and PTASs for TF2M are used as subroutines to design improved approximation algorithms for some network design problems (more details in Sect. 1.2).

Much less is known for the weighted version of the problem. WTF2M is not-known to be NP-hard. Polynomial-time algorithms are known for some special cases of the problem. More specifically, Hartvigsen and Li [18] showed that WTF2M can be solved in polynomial-time when the input graph is subcubic, i.e., the maximum degree is at most 3 (see also [3] for generalizations and [25, 33] for simplifications). Furthermore, Kobayashi [26] proposed a polynomial-time algorithm for the problem when all triangles in the input graph are edge-disjoint. The best-known (folklore) approximation algorithm for WTF2M simply computes a maximum-weight 2-matching and then drops the cheapest edge of each triangle: this gives a 2/3 approximation.

1.1 Our Results and Techniques

In this paper we present a PTAS for WTF2M, the first improvement over the mentioned trivial 2/3 approximation.

Theorem 1. *There is a PTAS for the Weighted Triangle-Free 2-Matching problem.*

In more detail, let $\mathsf{OPT}(G)$ denote some optimal solution to the considered WTF2M instance G, and $\mathsf{opt}(G) = w(\mathsf{OPT}(G))$ denote its weight. We simply use OPT and opt when G is clear from the context.

Our PTAS is based on a simple local-search algorithm, that works as follows. Given a triangle-free 2-matching M, *an augmenting trail P for M* is a trail such that $M \Delta P$ is a triangle-free 2-matching of weight strictly larger than M. Here, Δ denotes the symmetric difference. Given a trail P, by $|P|$ we denote its number of edges. At each step our algorithm finds an augmenting trail P with bounded $|P|$, if any, and uses it to increase the weight of the current solution M (initially empty); see Algorithm 1 for the details.

The key to prove that the above algorithm works is the following theorem.

Theorem 2. *Let $G = (V, E, w)$ be a simple weighted graph, APX be a triangle-free 2-matching of G, and $\varepsilon \in (0, 1]$ be a constant. If $(1-\varepsilon)\mathsf{opt}(G) > w(\mathsf{APX})$, then there exists an augmenting trail P for APX with $|P| \leq 7/\varepsilon$.*

Theorem 2 implies that Algorithm 1 returns a $(1-\varepsilon)$-approximate solution. However, it is not obvious that it runs in polynomial time. To overcome this

Algorithm 1 Local Search for WTF2M.
Input: A constant $\varepsilon \in (0, 1]$ and an undirected edge-weighted graph $G = (V, E, w)$.
Output: A triangle-free 2-matching APX of G satisfying $w(\mathsf{APX}) \geq (1 - \varepsilon)\mathsf{opt}(G)$.

```
APX ← ∅
while there exists an augmenting trail P for APX with |P| ≤ 7/ε do
    APX ← APX Δ P
return APX
```

difficulty, we reduce the problem to the case with polynomially bounded integer weights; see Sect. 2 for details.

Our proof for Theorem 2 follows a line similar to that of [28], which is based on graph reductions. The key ingredient is a generalization of triangle-freeness, which is called $\mathcal{T}$-freeness. In more detail, we generalize the triangle-free constraint by introducing a list $\mathcal{T}$ of triangles that must not be included in the desired 2-matching. By transforming the graph, we reduce the input graph to one with a smaller size of $\mathcal{T}$, and by transforming the trail obtained from the reduced graph back onto the original graph, we construct the required trail (see Sect. 4). The desired result is then obtained by setting $\mathcal{T}$ to the set of all the triangles in G.

As we will see in the technical part, our approach also works for non-simple graphs that contain self-loops but not parallel edges. On such graphs, we obtain a PTAS for the problem of computing a $\mathcal{T}$-free 2-matching for any input set $\mathcal{T}$ of triangles.

It remains an important open problem in the area to find an exact polynomial-time algorithm for WTF2M, if one exists.

1.2 Related Work

We already mentioned the exact algorithms and PTASs for TF2M. An exact algorithm for TF2M was used as a subroutine to design improved approximation algorithms for the $\{1, 2\}$-TSP problem, a well-studied special case of TSP where all edge weights are 1 or 2. A key observation here is that, given a maximum-cardinality 2-matching M on the edges of cost 1, $2n - |M|$ is a lower bound on the minimum cost of a TSP tour. The same lower bound holds if one uses instead a maximum-cardinality triangle-free 2-matching. In [5] a maximum-cardinality triangle-free 2-matching is used as a starting point to build a feasible TSP tour. The triangle-freeness turns out to be a useful feature for technical reasons. A similar approach is also announced in [1].

An exact algorithm for TF2M is also used as a subroutine to design improved approximation algorithms for the 2-Edge-Connected Spanning Subgraph problem (2ECSS). In this problem we are given an undirected graph, and the goal is to compute a subgraph with the minimum number of edges that contains all the nodes and that is 2-edge-connected, i.e., it remains connected after removing one arbitrary edge. This problem is very well studied in terms of approximation algorithms [11,13,21,23,36,38]. The latest improved approximation algorithms [6,20,27] work as follows. They first build a maximum-cardinality triangle-free 2-matching M. Then they convert M into a minimum-cardinality triangle-free 2-edge-cover[1] C. It can be shown that $|C|$ is a valid lower bound on the size of the optimal solution to the input 2ECSS instance. Finally, C is used as a starting point to build a feasible 2ECSS solution. Also in this case, the lack of triangles in C turns out to be a helpful feature in the construction for technical reasons. Indeed, it would be very helpful to avoid even longer cycles, say, with up to 4 or 5 edges. However no exact or sufficiently good approximation algorithms are known for that goal. For similar reasons, TF2M could also be used to simplify (and possibly improve) the approximation factor for the related 2-Vertex-Connected Spanning Subgraph problem [7,14,19,23].

We are not aware of any similar use of the weighted version of the problem, namely WTF2M, for the design of approximation algorithm, though this seems a natural direction for future work.

The Weighted $C_{\leq k}$-Free 2-Matching problem (WCkF2M) is a natural generalization of WTF2M where we wish to compute a maximum-weight 2-matching not inducing any cycle with up to k edges. This problem is NP-hard for $k \geq 4$ even in bipartite graphs [24]. As we already mentioned, WTF2M, i.e., the case $k = 3$, is open. Under the assumption that the weight function satisfies a certain property, called *vertex induced on every square*, Weighted $C_{\leq 4}$-Free 2-Matching problem in bipartite graphs can be solved in polynomial-time (see [29,32,37]).

The situation is slightly different for the unweighted version CkF2M of WCkF2M. Papadimitriou showed that the case $k \geq 5$ is NP-hard (his proof is described by Cornuéjols and Pulleyblank in [12]). The complexity status of the problem is open for $k = 4$. Hartvigsen [16] showed that the case where $k = 4$ is polynomial-time solvable in bipartite graphs (see also [2,34]). Bérczi and Végh [4] showed that the same problem can be solved in polynomial time if the input graph is sub-cubic.

The C_k-Free 2-Matching problem is the problem of finding a maximum-cardinality 2-matching not inducing any cycle with exactly k edges (while shorter cycles are allowed). In particular, TF2M is the special case of the above problem with $k = 3$. Bérczi and Kobayashi [10] showed that the square-free case, i.e., the case $k = 4$, is polynomial-time solvable in sub-cubic graphs. Nam [30] showed that the same problem can be solved in polynomial time if the squares are vertex-disjoint, and this result was generalized by Iwamasa et al. [22].

The problems get easier if we are allowed to take multiple copies of the same edge (i.e., the 2-matching is not simple). In particular, the Weighted Triangle-Free Non-Simple 2-Matching problem can be solved exactly in polynomial time [12].

[1] We recall that a 2-edge-cover is a subset of edges C such that each node has at least 2 edges of C incident to it.

1.3 Preliminaries

We use standard graph notation. Let $G = (V, E, w)$ be an edge-weighted graph with $w \colon E \to \mathbb{Q}_{\geq 0}$. Throughout this paper, we suppose that a graph may have self-loops but has no parallel edges. Given a node $u \in V$ and a subset $F \subseteq E$ of the edges, the relative degree $d_F(u)$ of u w.r.t. F is defined as the number of edges in F incident to v, where self-loops are counted twice, that is, $d_F(u) = |\{uv \mid uv \in F\}| + 1$ if $uu \in F$ and $d_F(u) = |\{uv \mid uv \in F\}|$ otherwise. An edge set $M \subseteq E$ is called a *2-matching* if $d_M(v) \leq 2$ for any $v \in V$. For a subset $F \subseteq E$ of the edges we denote $w(F) = \sum_{e \in F} w(e)$.

A *trail* P (of length $|P| = k - 1$) is a sequence of distinct consecutive edges $u_1u_2, u_2u_3, \ldots, u_{k-1}u_k$. The nodes $u_1, \ldots, u_k$ are not necessarily distinct. We say that u_1 and u_k are the *first* and the *last* node of the trail, resp., and they are both *endpoints* of the trail. Notice that we allow $u_1 = u_k$, in which case we say that the trail is *closed*. For a trail P of G, we denote by $\mathsf{sl}(P)$ the number of self-loops in P. Given a trail P, we will also use P to denote the corresponding set of edges $\{u_1u_2, \ldots, u_{k-1}u_k\}$ (neglecting their order); the meaning will be clear from the context. For two edge subsets $A_1, A_2 \subseteq E$, an *alternating trail w.r.t.* (A_1, A_2) is a (possibly closed) trail in $A_1 \Delta A_2$ that alternates between the edges of $A_1 \setminus A_2$ and $A_2 \setminus A_1$, i.e., $u_iu_{i+1} \in A_1 \setminus A_2$ if and only if $u_{i+1}u_{i+2} \in A_2 \setminus A_1$ for $i = 1, \ldots, k-2$.

A *triangle* is an edge set that forms a cycle of length 3. An edge subset $F \subseteq E$ is called *triangle-free* if it contains no triangle. Let $\mathcal{T}$ be a subset of triangles in $G = (V, E, w)$. An edge set $F \subseteq E$ is called $\mathcal{T}$*-free* if F does not contain any triangle $T \in \mathcal{T}$ as a subset. Note that when $\mathcal{T}$ is the set of all triangles in G, $\mathcal{T}$-freeness is equivalent to triangle-freeness. For a subset $\mathcal{T}$ of the triangles of G and a subset $F \subseteq E$ of the edges, we define $\mathcal{T}(F)$ as the set of all triangles in $\mathcal{T}$ containing at least one edge in F. The following observation is used multiple times throughout our argument.

Observation 1. *Let G be a graph, let $\mathcal{T}$ be a subset of the triangles of G, M be a 2-matching, and T be a triangle of G (possibly, $T \notin \mathcal{T}$). If $|M \cap T| = 2$ then M is $\mathcal{T}(T)$-free.*

Proof. Let $T = \{u_1u_2, u_2u_3, u_1u_3\}$, and assume w.l.o.g. that $T \cap M = \{u_1u_2, u_1u_3\}$. Assume to get a contradiction that M is not $\mathcal{T}(T)$-free, so that M contains a triangle $T' \neq T$ that contains an edge of T, say w.l.o.g. u_1u_2. By the degree constraint on u_1, the only other edge of M incident to u_1 is u_1u_3. Then, since $T' \subseteq M$, T' must also contain u_1u_3 and thus it is equal to T, a contradiction. □

2 Weight Reduction and Proof of Theorem 1

As described in Sect. 1, it is not obvious that Algorithm 1 runs in polynomial time. To resolve this issue, we reduce the problem to the case where the input instance has integer weights in $\{0, \ldots, \lfloor n/\varepsilon \rfloor\}$, where $\varepsilon > 0$ is some given constant. Here, n denotes the number of vertices in the input graph.

Lemma 1. *For any constants $\varepsilon, \varepsilon' \in (0,1]$, given a polynomial-time $(1-\varepsilon')$-approximation algorithm for WTF2M under the assumption that edge weights are integers in $\{0,\ldots,\lfloor n/\varepsilon \rfloor\}$, there is a polynomial-time $(1-\varepsilon)(1-\varepsilon')$-approximation algorithm for WTF2M (with no restrictions).*

Proof. Let $G = (V, E, w)$ be the input instance of WTF2M, and let OPT be some optimal solution for it. Let $e_{\max}$ be an edge with the maximum weight in G, and $W := w(e_{\max})$. We may assume that $W > 0$, since otherwise all edges have weight zero and $w(\mathsf{OPT}) = 0$. Note that $\{e_{\max}\}$ is a feasible solution of WTF2M, so $W \leq w(\mathsf{OPT})$. Next we define for each $e \in E$ the weight $w'(e) = \lfloor \frac{w(e)}{W} \frac{n}{\varepsilon} \rfloor$. To the resulting instance $G' = (V, E, w')$ of WTF2M we apply the $(1-\varepsilon')$-approximation algorithm with bounded integer weights that is given by the assumption of the lemma, and return the resulting solution APX. Let OPT' be an optimal solution to G'. One has that

$$w'(\mathsf{OPT}') \geq w'(\mathsf{OPT}) \geq \sum_{e \in \mathsf{OPT}} \left(\frac{w(e)}{W} \frac{n}{\varepsilon} - 1 \right)$$
$$\geq w(\mathsf{OPT}) \frac{n}{W\varepsilon} - n \geq (1-\varepsilon) w(\mathsf{OPT}) \frac{n}{W\varepsilon},$$

where we have used $|\mathsf{OPT}| \leq n$ and $W \leq w(\mathsf{OPT})$ in the third and the last inequalities, respectively. Hence

$$w(\mathsf{APX}) \geq \frac{W\varepsilon}{n} w'(\mathsf{APX}) \geq \frac{W\varepsilon(1-\varepsilon')}{n} w'(\mathsf{OPT}') \geq (1-\varepsilon)(1-\varepsilon') w(\mathsf{OPT}),$$

which completes the proof. □

By combining Theorem 2 and Lemma 1, we obtain a PTAS for WTF2M and prove Theorem 1 as follows.

Proof (Proof of Theorem1). By Lemma 1 it is sufficient to show that Algorithm 1 is a PTAS for the problem with bounded integer weights. The feasibility of the solution is obvious by construction. Clearly, every iteration of the while loop can be executed in $n^{O(1/\varepsilon)}$ time. Each time this loop is executed, the weight of the solution grows by at least 1 (since all weights are integers). Since the initial solution has weight 0 and no solution can have weight more than n^2/ε (corresponding to n edges of maximum possible weight), the number of iterations of the while loop is bounded by n^2/ε. Then, since in each iteration a trail can be found in $n^{O(1/\varepsilon)}$ time, the overall running time is also bounded by $n^{O(1/\varepsilon)}$. By Theorem 2, as long as $(1-\varepsilon)\mathsf{opt}(G) > w(\mathsf{APX})$, the value of APX is updated. This implies that, when the algorithm terminates, $(1-\varepsilon)\mathsf{opt}(G) \leq w(\mathsf{APX})$. □

What remains is the proof of Theorem 2. In what follows in this paper, we prove the following stronger theorem.

Theorem 3. *Let $G = (V, E, w)$ be a weighted graph. Let $\mathcal{T}$ be a subset of the triangles of G, let A_1, A_2 be two $\mathcal{T}$-free 2-matchings of G, and let $\varepsilon \in (0,1]$ be* *a constant. If* $(1-\varepsilon)w(A_1) > w(A_2)$*, then there exists an alternating trail* P *w.r.t.* (A_1, A_2) *such that: (1)* $A_2 \Delta P$ *is a* $\mathcal{T}$*-free 2-matching, (2)* $w(A_2 \Delta P) > w(A_2)$*, and (3)* $|P| + 2\mathsf{sl}(P) \leq 7/\varepsilon$*.*

Recall that G may have self-loops but has no parallel edges. Notice that we can derive Theorem 2 from Theorem 3 by setting $A_1 = \mathsf{OPT}$ and $A_2 = \mathsf{APX}$. To prove Theorem 3, we first consider the case of $\mathcal{T} = \emptyset$ in Sect. 3, and then consider the general case in Sect. 4.

3 Case with No Forbidden Triangles

In this section, we prove a special case of Theorem 3 where $\mathcal{T} = \emptyset$.

Lemma 2. *Let* $G = (V, E, w)$ *be an edge-weighted graph, let* A_1, A_2 *be two 2-matchings of* G*, and let* $\varepsilon \in (0, 1]$ *be a constant. If* $(1-\varepsilon)w(A_1) > w(A_2)$*, then there exists an alternating trail* P *w.r.t.* (A_1, A_2) *such that: (1)* $A_2 \Delta P$ *is a 2-matching, (2)* $w(A_2 \Delta P) > w(A_2)$*, and (3)* $|P| \leq 3/\varepsilon$*.*

Proof. Let $\mathcal{P}$ be a partition of $A_1 \Delta A_2$ into alternating trails w.r.t. (A_1, A_2) such that $A_2 \Delta P$ is a 2-matching for every $P \in \mathcal{P}$. The existence of such a partition is a well-known fact, see e.g. [28, Lemma 2.1]. It holds that

$$\begin{aligned}\sum_{P \in \mathcal{P}} w(A_2 \cap P) &= w(A_2) - w(A_1 \cap A_2) \\ &< (1-\varepsilon)w(A_1) - w(A_1 \cap A_2) \\ &\leq (1-\varepsilon)(w(A_1) - w(A_1 \cap A_2)) \\ &= \sum_{P \in \mathcal{P}} (1-\varepsilon)w(A_1 \cap P).\end{aligned}$$

Therefore there exists $P^* \in \mathcal{P}$ satisfying that $w(A_2 \cap P^*) < (1-\varepsilon)w(A_1 \cap P^*)$.

Let $m := \lceil \frac{1}{\varepsilon} \rceil$. Since $w(A_1 \cap P^*) > w(A_2 \cap P^*)$, one has that $w(A_2 \Delta P^*) - w(A_2) > 0$, and thus P^* satisfies the claim of the lemma if $|P^*| \leq 2m - 1 \leq 3/\varepsilon$. Now we assume that $P^* = u_1u_2, u_2u_3, \ldots, u_ku_{k+1}$ satisfies $|P^*| > 2m - 1$. We consider the following two cases separately.

Suppose that P^* is closed and even, i.e., $u_1 = u_{k+1}$, and one of u_1u_2 and u_ku_{k+1} is in A_1 while the other is in A_2. Let $\mathcal{S}$ be the set of alternating subtrails P of P^* with $|P| = 2m - 1$ whose first and last edges are in A_2 (including subtrails passing through $u_1 = u_{k+1}$). That is, each trail in $\mathcal{S}$ is of the form $u_iu_{i+1}, u_{i+1}u_{i+2}, \ldots, u_{i+2m-2}u_{i+2m-1}$, where the indices are taken modulo k. It is easy to see that $A_2 \Delta P$ is a 2-matching for any $P \in \mathcal{S}$, because $d_{A_2 \Delta P}(v) \leq d_{A_2}(v) \leq 2$ for any $v \in V$. Every edge in $A_1 \cap P^*$ is contained in exactly $m - 1$ trails in $\mathcal{S}$, while every edge in $A_2 \cap P^*$ is contained in exactly m trails in $\mathcal{S}$.

Therefore,

$$\begin{aligned}\sum_{P\in\mathcal{S}}(w(A_2\Delta P)-w(A_2)) &= \sum_{P\in\mathcal{S}}(w(A_1\cap P)-w(A_2\cap P))\\ &=(m-1)w(A_1\cap P^*)-m\cdot w(A_2\cap P^*)\\ &> m\Big(\Big(1-\frac{1}{m}\Big)-(1-\varepsilon)\Big)w(A_1\cap P^*)\geq 0.\end{aligned}$$

In the last inequality we have used that $m=\lceil\frac{1}{\varepsilon}\rceil$ and in the second-last one the fact that $w(A_2\cap P^*)<(1-\varepsilon)w(A_1\cap P^*)$. This implies the existence of $P\in\mathcal{S}$ such that $w(A_2\Delta P)-w(A_2)>0$, which satisfies the claim of the lemma.

Suppose now that P^* is not closed or not even, i.e., $u_1\neq u_{k+1}$, or $u_1=u_{k+1}$ and both u_1u_2 and u_ku_{k+1} are in A_i for some $i\in\{1,2\}$. Let $\mathcal{S}$ be the set of alternating subtrails P of P^* satisfying one of the following conditions:

- $|P|=2m-1$, and both the first and last edges of P are contained in A_2; or
- $|P|<2m-1$, where either the first or last edge of P coincides the first or last edge of P^*, and the other extremal edge of P is contained in A_2 (possibly, $|P|=1$).

Note that each trail in $\mathcal{S}$ is of the form $u_iu_{i+1}, u_{i+1}u_{i+2},\ldots,u_{i+\ell-1}u_{i+\ell}$ for some $\ell\leq 2m-1$, where the indices are *not* taken modulo k. One sees that $A_2\Delta P$ is a 2-matching for every $P\in\mathcal{S}$, because $d_{A_2\Delta P}(v)\leq\max\{d_{A_2}(v),d_{A_2\Delta P^*}(v)\}\leq 2$ for every $v\in V$. Every edge in $A_1\cap P^*$ is contained in exactly $m-1$ trails in $\mathcal{S}$, while every edge in $A_2\cap P^*$ is contained in exactly m trails in $\mathcal{S}$. Then, by the same argument as in the previous case, we see that there exists $P\in\mathcal{S}$ such that $w(A_2\Delta P)-w(A_2)>0$, which satisfies the claim of the lemma. □

This lemma together with the following observation, proves the special case of Theorem 3 in which $\mathcal{T}=\emptyset$.

Observation 2. *For a trail P satisfying the condition in Lemma 2, it holds that $|P|+2\mathsf{sl}(P)\leq\frac{7}{\varepsilon}$.*

Proof. Since G has no parallel edges, $\mathsf{sl}(P)$ is at most the number of nodes of P, which is equal to $|P|-\mathsf{sl}(P)+1$. Therefore, $\mathsf{sl}(P)\leq\frac{|P|+1}{2}$, so $|P|+2\mathsf{sl}(P)\leq 2|P|+1\leq\frac{7}{\varepsilon}$. □

4 Simple Cases for Theorem 3

We prove Theorem 3 by induction on $|\mathcal{T}|$. The base case, i.e., when $\mathcal{T}=\emptyset$, is given by Lemma 2 and Observation 2. Suppose that $|\mathcal{T}|>0$ holds. We consider the following cases, assuming at each case that the previous ones do not hold:

(1) There exists a triangle $T\in\mathcal{T}$ not contained in $A_1\cup A_2$.
(2) There exists a triangle $T\in\mathcal{T}$ such that $|A_1\cap T|=|A_2\cap T|=2$.
(3) There exists a triangle $T\in\mathcal{T}$ such that $|A_1\cap T|=2$ and $A_1\Delta T$ is $\mathcal{T}$-free.

(4) There exists a triangle $T \in \mathcal{T}$ such that $|A_2 \cap T| = 2$ and $A_2 \Delta T$ is $\mathcal{T}$-free.
(5) There exists a triangle $T \in \mathcal{T}$ such that $|A_1 \cap T| = 2$.
(6) There exists a triangle $T \in \mathcal{T}$ such that $|A_2 \cap T| = 2$.

For each $T \in \mathcal{T}$ with $T \subseteq A_1 \cup A_2$, it holds that $|A_1 \cap T| = 2$ or $|A_2 \cap T| = 2$, because A_1 and A_2 are $\mathcal{T}$-free. Thus, Cases **(1)**, **(5)**, and **(6)** cover all possible cases, and thus Theorem 3 follows from these six cases. Notice that the fact that Cases **(2)**, **(3)**, and **(4)** do not hold is necessary for the proof of Cases **(5)** and **(6)**.

In this section, we discuss Cases **(1)**, **(2)**, and **(3)**, while the other cases are dealt with in the full version [9] due to the space limitation. In the figures that we will use to illustrate the different cases, solid lines represent edges in A_1, while dashed lines represent those in A_2.

(1) There Exists a Triangle $T \in \mathcal{T}$ not Contained in $A_1 \cup A_2$. Define $\mathcal{T}' := \mathcal{T} \setminus \{T\}$. Since A_1 and A_2 are both $\mathcal{T}'$-free 2-matchings and $|\mathcal{T}'| < |\mathcal{T}|$, by the induction hypothesis, there exists an alternating trail P w.r.t. (A_1, A_2) such that $|P| + 2\mathsf{sl}(P) \leq 7/\varepsilon$, $w(A_2 \Delta P) > w(A_2)$, and $A_2 \Delta P$ is a $\mathcal{T}'$-free 2-matching of G. Since T is not contained in $A_1 \cup A_2$, $A_2 \Delta P$ does not contain T, hence $A_2 \Delta P$ is $\mathcal{T}$-free. Altogether P satisfies the claim of the theorem.

(2) There Exists a Triangle $T \in \mathcal{T}$ such that $|A_1 \cap T| = |A_2 \cap T| = 2$. Let $T = \{u_1u_2, u_2u_3, u_1u_3\}$. Since both A_1 and A_2 are $\mathcal{T}$-free and Case **(1)** does not hold, we can assume w.l.o.g. that $u_2u_3 \in A_1 \cap A_2$, $u_1u_2 \in A_1 \setminus A_2$, and $u_1u_3 \in A_2 \setminus A_1$ (see Fig. 1 left). We define an auxiliary weighted graph $G' = (V', E', w')$ as follows:

- $V' := V \cup \{u_1'\}$, $E' := (E \setminus \{u_1u_2, u_1u_3\}) \cup \{u_1'u_2, u_1'u_3, u_1u_1'\}$,
- $w'(e) = w(e)$ for all $e \in E \setminus \{u_1u_2, u_1u_3\}$,
- $w'(u_1'u_2) = w(u_1u_2)$, $w'(u_1'u_3) = w(u_1u_3)$, $w'(u_1u_1') = 0$.

We also define:

- $A_1' := (A_1 \setminus \{u_1u_2\}) \cup \{u_1'u_2, u_1u_1'\}$, $A_2' := (A_2 \setminus \{u_1u_3\}) \cup \{u_1'u_3, u_1u_1'\}$,
- $\mathcal{T}' = \mathcal{T} \setminus \mathcal{T}(\{u_1u_2, u_1u_3\})$.

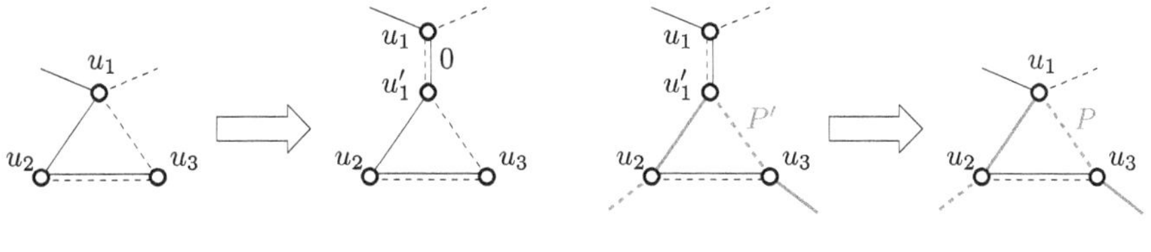

Fig. 1. Reduction for Case (2).

Observe that $(1-\varepsilon)w'(A_1') = (1-\varepsilon)w(A_1) > w(A_2) = w'(A_2')$. We claim that A_i' is a $\mathcal{T}'$-free 2-matching of G', for $i \in \{1, 2\}$. Since $d_{A_i'}(u) = d_{A_i}(u) \leq 2$

for every node $u \in V$, and $d_{A'_i}(u'_1) = 2$, A'_i is a 2-matching of G'. We also observe that the only edges in A'_i that are not present in A_i are incident to $u'_1 \notin V$, so they are not edges of triangles in $\mathcal{T}$, and thus they are not edges of triangles in $\mathcal{T}'$. This, together with the fact that A_i is $\mathcal{T}$-free implies that A'_i is $\mathcal{T}'$-free, for $i \in \{1,2\}$. Thus, A'_i is a $\mathcal{T}'$-free 2-matching of G', for $i \in \{1,2\}$. Also, since T contains both u_1u_2 and u_1u_3, one has $|\mathcal{T}'| < |\mathcal{T}|$. One can apply the induction hypothesis to find an alternating trail P' w.r.t. (A'_1, A'_2) such that $|P'| + 2\mathsf{sl}(P') \leq 7/\varepsilon$, $w'(A'_2 \Delta P') > w'(A'_2)$, and $A'_2 \Delta P'$ is a $\mathcal{T}'$-free 2-matching of G'.

Let P be the trail in G resulting from replacing u'_1u_2 by u_1u_2 and u'_1u_3 by u_1u_3 in P', if any of them is present in P' (see Fig. 1 right). Notice that P is an alternating trail w.r.t. (A_1, A_2). We claim that P satisfies the conditions of the theorem.

Notice that $|P| + 2\mathsf{sl}(P) = |P'| + 2\mathsf{sl}(P') \leq 7/\varepsilon$. We now prove that $w(A_2 \Delta P) > w(A_2)$. Observe that A_2 (resp., $A_2 \Delta P$) in G is obtained from A'_2 (resp., $A'_2 \Delta P'$) in G' by contracting $u_1u'_1$ (i.e., by removing $u_1u'_1$ and by identifying u'_1 with u_1). Since $w'(u'_1u_2) = w(u_1u_2)$, $w'(u'_1u_3) = w(u_1u_3)$, and $w'(u_1u'_1) = 0$, this observation shows that $w(A_2 \Delta P) = w'(A'_2 \Delta P') > w'(A'_2) = w(A_2)$.

Let us now show that $A_2 \Delta P$ is a 2-matching. Since $A_2 \Delta P$ is obtained from $A'_2 \Delta P'$ by contracting $u_1u'_1$, for every node $u \in V \setminus \{u_1\}$, $d_{A_2 \Delta P}(u) = d_{A'_2 \Delta P'}(u) \leq 2$. Since the edge $u_1u'_1$ counts toward both $d_{A'_2 \Delta P'}(u_1)$ and $d_{A'_2 \Delta P'}(u'_1)$, but is not present in $A_2 \Delta P$, one has that $d_{A_2 \Delta P}(u_1) = d_{A'_2 \Delta P'}(u_1) - 1 + d_{A'_2 \Delta P'}(u'_1) - 1 \leq 2$.

Finally, it is left to show that $A_2 \Delta P$ is $\mathcal{T}$-free. Since $A'_2 \Delta P'$ is $\mathcal{T}'$-free, $A_2 \Delta P$ is $\mathcal{T}'$-free. Now, notice that the edge u'_1u_2 belongs to P' only if u'_1u_3 also belongs to P' because $d_{A'_2 \Delta P'}(u'_1) \leq 2$. Thus, if $u'_1u_2 \in P'$, both $u_1u_2, u_1u_3 \in P$ and $(A_2 \Delta P) \cap T = \{u_1u_2, u_2u_3\}$, so $A_2 \Delta P$ is $\mathcal{T}(T)$-free by Observation 1 (implying that it is $\mathcal{T}(\{u_1u_2, u_1u_3\})$-free). Assume now $u'_1u_2 \notin P'$, so $u_1u_2 \notin P$ and thus $u_1u_2 \notin A_2 \Delta P$. If $u_1u_3 \in A_2 \Delta P$ then $(A_2 \Delta P) \cap T = \{u_1u_3, u_2u_3\}$, so $A_2 \Delta P$ is $\mathcal{T}(T)$-free by Observation 1 (implying that it is $\mathcal{T}(\{u_1u_2, u_1u_3\})$-free). On the other hand, if $u_1u_3 \notin A_2 \Delta P$, then one has that $u_1u_2, u_1u_3 \notin A_2 \Delta P$, and thus $A_2 \Delta P$ is $\mathcal{T}(\{u_1u_2, u_1u_3\})$-free. The claim follows.

(3) There Exists a Triangle $T \in \mathcal{T}$ such that $|A_1 \cap T| = 2$ and $A_1 \Delta T$ is $\mathcal{T}$-free.

Let $T = \{u_1u_2, u_2u_3, u_1u_3\} \in \mathcal{T}$. Since Cases **(1)** and **(2)** do not hold, we can assume w.l.o.g. that $\{u_1u_2, u_1u_3\} \subseteq A_1 \setminus A_2$ and $u_2u_3 \in A_2 \setminus A_1$. We consider the following subcases.

(3.1)$u_1u_1 \in A_2$ If $w(u_1u_2) + w(u_1u_3) > w(u_2u_3) + w(u_1u_1)$ then let $P := \{u_1u_2, u_2u_3, u_1u_3, u_1u_1\}$. Let us show that P satisfies the claim of the theorem. Indeed, since $w(u_1u_2) + w(u_1u_3) > w(u_2u_3) + w(u_1u_1)$, $w(A_2 \Delta P) > w(A_2)$. One has $|P| + 2\mathsf{sl}(P) = 6 < 7/\varepsilon$, and since P is an even closed alternating trail, $A_2 \Delta P$ is a 2-matching. Moreover, since $(A_2 \Delta P) \cap T = \{u_1u_2, u_1u_3\}$, $A_2 \Delta P$ is $\mathcal{T}(T)$-free by Observation 1. Since the only edge in P not in T is u_1u_1, this implies that $A_2 \Delta P$ is $\mathcal{T}$-free.

From now on we assume that $w(u_1u_2) + w(u_1u_3) \leq w(u_2u_3) + w(u_1u_1)$. We set (see Fig. 2):

- $A_1' := (A_1 \setminus \{u_1u_2, u_1u_3\}) \cup \{u_2u_3, u_1u_1\}$, $A_2' := A_2$,
- $w'(e) := w(e)$ for all $e \in E$,
- $\mathcal{T}' := \mathcal{T} \setminus \mathcal{T}(\{u_1u_2, u_1u_3\})$.

Notice that A_1' and A_2' are both 2-matchings of G. Clearly, A_2' is $\mathcal{T}'$-free, and $A_1 \Delta T$ is $\mathcal{T}$-free by this case assumption, so $A_1' = (A_1 \Delta T) \cup \{u_1u_1\}$ is also $\mathcal{T}'$-free. Thus, A_i' is a $\mathcal{T}'$-free 2-matching of G, for $i \in \{1, 2\}$. Also, $w'(A_2') = w(A_2)$ and $w'(A_1') = w(A_1) + w(u_2u_3) + w(u_1u_1) - w(u_1u_2) - w(u_1u_3) \geq w(A_1)$, because $w(u_1u_2) + w(u_1u_3) \leq w(u_2u_3) + w(u_1u_1)$. By the assumption of the theorem, one has $(1 - \varepsilon)w'(A_1') \geq (1 - \varepsilon)w(A_1) > w(A_2) = w'(A_2')$.

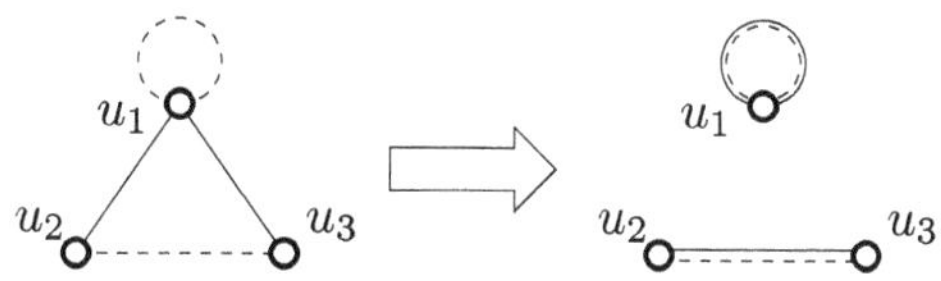

Fig. 2. Reduction for Case (3.1).

Since A_1' and A_2' are both $\mathcal{T}'$-free 2-matchings and $(1-\varepsilon)w'(A_1') > w(A_2')$, one can apply the induction hypothesis to find an alternating trail P' w.r.t. (A_1', A_2') such that $|P'| + 2\mathsf{sl}(P') \leq 7/\varepsilon$, $w'(A_2' \Delta P') > w'(A_2')$, and $A_2' \Delta P'$ is a $\mathcal{T}'$-free 2-matching of G'. Let $P := P'$, and notice that P is an alternating trail w.r.t. (A_1, A_2). We claim that P satisfies the conditions of the theorem. Observe that $|P| + 2\mathsf{sl}(P) = |P'| + 2\mathsf{sl}(P') \leq 7/\varepsilon$ and $w(A_2 \Delta P) = w'(A_2' \Delta P') > w'(A_2') = w(A_2)$.

Since $A_2' = A_2$ and $P' = P$, it holds that $d_{A_2 \Delta P}(u) = d_{A_2' \Delta P'}(u) \leq 2$ for every $u \in V$, so $A_2 \Delta P$ is a 2-matching. Moreover, since $A_2' \Delta P'$ is $\mathcal{T}'$-free, so is $A_2 \Delta P$. Also, since $u_1u_2, u_1u_3 \notin A_1' \cup A_2'$, one has $u_1u_2, u_1u_3 \notin P$, and thus $u_1u_2, u_1u_3 \notin A_2 \Delta P$. Therefore, $A_2 \Delta P$ is $\mathcal{T}(\{u_1u_2, u_1u_3\})$-free. The claim follows.

(3.2)$\boldsymbol{u_1u_1 \notin A_2}$. Notice that $u_1u_1 \notin A_1$, because of the degree constraints on u_1. We define

- $A_2' := A_2$,
- $w'(e) := w(e)$ for all $e \in E \setminus \{u_1u_1\}$,
- $\mathcal{T}' := \mathcal{T} \setminus \mathcal{T}(\{u_1u_2, u_1u_3\})$.

If $w(u_1u_2) + w(u_1u_3) - w(u_2u_3) \geq 0$, in case $u_1u_1 \notin E$, we add u_1u_1 to E with some arbitrary non-negative weight. Furthermore, we set

- $A_1' := (A_1 \setminus \{u_1u_2, u_1u_3\}) \cup \{u_2u_3, u_1u_1\}$,
- $w'(u_1u_1) := w(u_1u_2) + w(u_1u_3) - w(u_2u_3)$.

Otherwise, if $u_1u_1 \in E$, we remove it from E. Furthermore, we set

– $A'_1 := (A_1 \setminus \{u_1u_2, u_1u_3\}) \cup \{u_2u_3\}$.

Notice that A'_1 and A'_2 are both 2-matchings of G. Clearly, A'_2 is $\mathcal{T}'$-free, and $A_1 \Delta T$ is $\mathcal{T}$-free by this case assumption, so $A'_1 \subseteq (A_1 \Delta T) \cup \{u_1u_1\}$ is also $\mathcal{T}'$-free. Note also that $w'(A'_2) = w(A_2)$ and $w'(A'_1) \geq w(A_1)$.

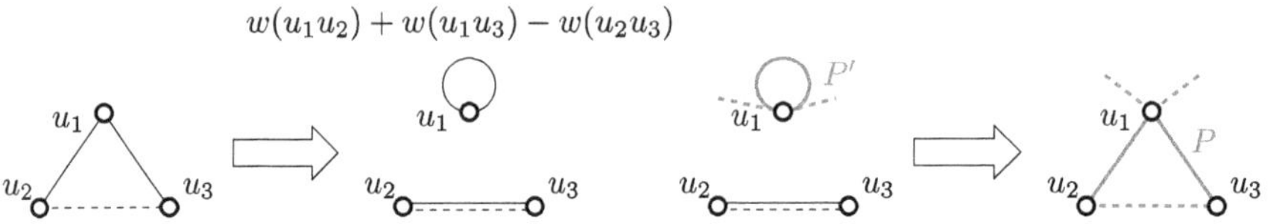

Fig. 3. Reduction for Case (3.2) (Color figure online).

Since A'_1 and A'_2 are both $\mathcal{T}'$-free 2-matchings and $w'(A'_2) = w(A_2) < (1 - \varepsilon)w(A_1) \leq (1-\varepsilon)w'(A'_1)$, one can apply the induction hypothesis to find an alternating trail P' w.r.t. (A'_1, A'_2) such that $|P'| + 2\mathsf{sl}(P') \leq 7/\varepsilon$, $w'(A'_2 \Delta P') > w'(A'_2)$, and $A'_2 \Delta P'$ is a $\mathcal{T}'$-free 2-matching of G'. If $u_1u_1 \in P'$, then let P be the alternating trail resulting from inserting the trail u_1u_2, u_2u_3, u_1u_3 in place of u_1u_1 (see Fig. 3 right). If $u_1u_1 \notin P'$, then let $P := P'$. Notice that P is an alternating trail w.r.t. (A_1, A_2). We claim that P satisfies the conditions of the theorem, i.e., $A_2 \Delta P$ is a $\mathcal{T}$-free 2-matching, $|P| + 2\mathsf{sl}(P) \leq 7/\varepsilon$, and $w(A_2 \Delta P) > w(A_2)$.

Notice that $|P| + 2\mathsf{sl}(P) = |P'| + 2\mathsf{sl}(P') \leq 7/\varepsilon$ holds regardless of whether $u_1u_1 \in P'$ or not. If $u_1u_1 \notin P'$, since $A_2 = A'_2$ and $P = P'$, we have $w(A_2 \Delta P) = w'(A'_2 \Delta P') > w'(A'_2) = w(A_2)$. On the other hand, if $u_1u_1 \in P'$, then $A_2 \Delta P = ((A'_2 \Delta P') \setminus \{u_1u_1, u_2u_3\}) \cup \{u_1u_2, u_1u_3\}$. Therefore,

$$\begin{aligned} w(A_2 \Delta P) &= w'(A'_2 \Delta P') - w'(u_1u_1) - w'(u_2u_3) + w(u_1u_2) + w(u_1u_3) \\ &= w'(A'_2 \Delta P') > w'(A'_2) = w(A_2), \end{aligned}$$

where in the second equality we have used that $w'(u_1u_1) = w(u_1u_2) + w(u_1u_3) - w(u_2u_3)$ and $w'(u_2u_3) = w(u_2u_3)$. In both cases, it holds that $w(A_2 \Delta P) > w(A_2)$.

It is left to show that $A_2 \Delta P$ is a $\mathcal{T}$-free 2-matching. Recall that $A'_2 = A_2$. If $u_1u_1 \in P'$, then $P' = P$ and thus $d_{A_2 \Delta P}(u) = d_{A'_2 \Delta P'}(u) \leq 2$ for all $u \in V$. On the other hand, if $u_1u_1 \notin P'$, then $A_2 \Delta P = ((A'_2 \Delta P') \setminus \{u_1u_1, u_2u_3\}) \cup \{u_1u_2, u_1u_3\}$. Therefore, $d_{A_2 \Delta P}(u) = d_{A'_2 \Delta P'}(u) \leq 2$ for all $u \in V$. Thus, $A_2 \Delta P$ is a 2-matching.

Since $A'_2 \Delta P'$ is $\mathcal{T}'$-free, $A_2 \Delta P \subseteq (A'_2 \Delta P') \cup \{u_1u_2, u_1u_3\}$, and neither u_1u_2 nor u_1u_3 is contained in a triangle in $\mathcal{T}'$, one has that $A_2 \Delta P$ is $\mathcal{T}'$-free. Now, if $u_1u_1 \in P'$, then $u_1u_2, u_1u_3, u_2u_3 \in P$ and thus $(A_2 \Delta P) \cap T = \{u_1u_2, u_1u_3\}$, so $A_2 \Delta P$ is $\mathcal{T}(\{u_1u_2, u_1u_3\})$-free by Observation 1. If $u_1u_1 \notin P'$, then since $P = P'$ contains neither u_1u_2 nor u_1u_3, $A_2 \Delta P$ is also $\mathcal{T}(\{u_1u_2, u_1u_3\})$-free. Therefore, $A_2 \Delta P$ is $\mathcal{T}$-free. The claim follows.

Acknowledgments. This work was supported by the joint project of Kyoto University and Toyota Motor Corporation, titled "Advanced Mathematical Science for Mobility Society", by JSPS KAKENHI Grant Numbers JP22H05001 and JP24K02901, and by JST SPRING, Grant Number JPMJSP2110. The first 2 authors are partially supported by the SNF Grant 200021_200731/1.

Disclosure of Interests. The authors have no competing interests to declare that are relevant to the content of this article.

References

1. Adamaszek, A., Mnich, M., Paluch, K.: New approximation algorithms for (1,2)-TSP. In 45th International Colloquium on Automata, Languages, and Programming (ICALP 2018), volume 107 of Leibniz International Proceedings in Informatics (LIPIcs), pp. 9:1–9:14 (2018)
2. Babenko, M.A.: Improved algorithms for even factors and square-free simple b-matchings. Algorithmica **64**(3), 362–383 (2012)
3. Bérczi, K.: The triangle-free 2-matching polytope of subcubic graphs. Technical Report TR-2012-02, Egerváry Research Group, Budapest (2012). http://egres.elte.hu
4. Bérczi, K., Végh, L.A.: Restricted b-matchings in degree-bounded graphs. In: Proceedings of the 14th Conference of Integer Programming and Combinatorial Optimization (IPCO 2010), volume 6080 of LNCS, pp. 43–56 (2010)
5. Bläser, M., Ram, L.S.: An improved approximation algorithm for TSP with distances one and two. In: Liśkiewicz, M., Reischuk, R., (eds.), Fundamentals of Computation Theory, pp. 504–515. Springer, Berlin, Heidelberg (2005)
6. Bosch-Calvo, M., Garg, M., Grandoni, F., Hommelsheim, F., Ameli, A.J., Lindermayr, A.: A 5/4-approximation for two-edge connectivity. In: Koucký, M., Bansal, N., (eds.), Proceedings of the 57th Annual ACM Symposium on Theory of Computing, STOC 2025, Prague, Czechia, 23-27 June 2025, pp. 653–664. ACM (2025)
7. Bosch-Calvo, M., Grandoni, F., Ameli, A.J.: A 4/3 approximation for 2-vertex-connectivity. In: Etessami, K., Feige, U., Puppis, G., (eds.), 50th International Colloquium on Automata, Languages, and Programming, ICALP 2023, 10-14 July 2023, Paderborn, Germany, volume 261 of LIPIcs, pp. 29:1–29:13. Schloss Dagstuhl - Leibniz-Zentrum für Informatik (2023)
8. Bosch-Calvo, M., Grandoni, F., Ameli, A.J.: A PTAS for triangle-free 2- matching. arXiv preprint arXiv:2311.11869 (2023)
9. Bosch-Calvo, M., Grandoni, F., Kobayashi, Y., Noguchi, T.: A PTAS for weighted triangle-free 2-matching. arXiv preprint arXiv:2603.09144 (2026)
10. Bérczi, K., Kobayashi, Y.: An algorithm for (n-3)-connectivity augmentation problem: jump system approach. J. Comb. Theory Seri. B **102**(3), 565–587 (2012)
11. Cheriyan, J., Sebő, A., Szigeti, Z.: Improving on the 1.5-approximation of a smallest 2-edge connected spanning subgraph. SIAM J. Discret. Math. **14**(2), 170–180 (2001)
12. Cornuéjols, G., Pulleyblank, W.: A matching problem with side conditions. Discret. Math. **29**(2), 135–159 (1980)
13. Garg, M., Grandoni, F., Ameli, A.J.: Improved approximation for two-edge-connectivity. In: Proceedings of the 34th Annual ACM-SIAM Symposium on Discrete Algorithms (SODA 2023), pp. 2368–2410 (2023)

14. Garg, N., Vempala, S.S., Singla, A.: Improved approximation algorithms for biconnected subgraphs via better lower bounding techniques. In: Ramachandran, V., (eds.), Proceedings of the Fourth Annual ACM/SIGACT-SIAM Symposium on Discrete Algorithms, 25-27 January 1993, Austin, Texas, USA, pp. 103–111. ACM/SIAM (1993)
15. Hartvigsen, D.: Extensions of Matching Theory. PhD thesis, Carnegie Mellon University (1984). https://david-hartvigsen.net
16. Hartvigsen, D.: Finding maximum square-free 2-matchings in bipartite graphs. J. Comb. Theory Seri. B **96**(5), 693–705 (2006)
17. Hartvigsen, D.: Finding triangle-free 2-factors in general graphs. J. Graph Theory **106**(3), 581–662 (2024)
18. Hartvigsen, D., Li, Y.: Triangle-free simple 2-matchings in subcubic graphs (extended abstract). In: Proceedings of the 12th International Conference on Integer Programming and Combinatorial Optimization, IPCO '07, pp. 43–52. Springer-Verlag, Berlin, Heidelberg (2007)
19. Heeger, K., Vygen, J.: Two-connected spanning subgraphs with at most $\frac{10}{7}OPT$ edges. SIAM J. Discret. Math. **31**(3), 1820–1835 (2017)
20. Hommelsheim, F., Lindermayr, A., Liu, Z.: A better-than-5/4-approximation for two-edge connectivity. In: Proceedings of the 36th Annual ACM-SIAM Symposium on Discrete Algorithms (SODA 2026), pp. 1468–1520 (2026) arXiv:2509.19655
21. Hunkenschröder, C., Vempala, S., Vetta, A.: A 4/3-approximation algorithm for the minimum 2-edge connected subgraph problem. ACM Trans. Algorithms **15**(4), 1–28 (2019)
22. Iwamasa, Y., Kobayashi, Y., Takazawa, K.: Finding a maximum restricted t-matching via Boolean edge-CSP. In Proceedings of the 32nd Annual European Symposium on Algorithms (ESA 2024), volume 308 of LIPIcs, pp. 75:1–75:15 (2024)
23. Khuller, S., Vishkin, U.: Biconnectivity approximations and graph carvings. J. ACM **41**(2), 214–235 (1994)
24. Király, Z.: Restricted t-matchings in bipartite graphs. Technical Report QP-2009-04, Egerváry Research Group, Budapest (2009). http://egres.elte.hu
25. Kobayashi, Y.: A simple algorithm for finding a maximum triangle-free 2-matching in subcubic graphs. Discret. Optim. **7**(4), 197–202 (2010)
26. Kobayashi, Y.: Weighted triangle-free 2-matching problem with edge-disjoint forbidden triangles. Math. Program. **192**(1–2), 675–702 (2022)
27. Kobayashi, Y., Noguchi, T.: An approximation algorithm for two-edge-connected subgraph problem via triangle-free two-edge-cover. In 34th International Symposium on Algorithms and Computation (ISAAC 2023), volume 283 of Leibniz International Proceedings in Informatics (LIPIcs), pp. 49:1–49:10 (2023)
28. Kobayashi, Y., Noguchi, T.: Validating a PTAS for triangle-free 2-matching via a simple decomposition theorem. In: 2025 Symposium on Simplicity in Algorithms (SOSA), pp. 281–289 (2025)
29. Makai, M.: On maximum cost $K_{t,t}$-free t-matchings of bipartite graphs. SIAM J. Discret. Math. **21**, 349–360 (2007)
30. Nam, Y.: Matching Theory: Subgraphs with Degree Constraints and Other Properties. PhD thesis, University of British Columbia (1994)
31. Paluch, K.: Triangle-free 2-matchings. arXiv preprint arXiv:2311.13590 (2023)
32. Paluch, K., Wasylkiewicz, M.: Restricted t-matchings via half-edges. In: Proceedings of the 29th Annual European Symposium on Algorithms (ESA 2021), volume 204 of LIPIcs, pp. 73:1–73:17 (2021)

33. Paluch, K., Wasylkiewicz, M.: A simple combinatorial algorithm for restricted 2-matchings in subcubic graphs - via half-edges. Inf. Process. Lett. **171**, 106146 (2021)
34. Pap, G.: Combinatorial algorithms for matchings, even factors and square-free 2-factors. Math. Program. **110**(1), 57–69 (2007)
35. Schrijver, A.: Combinatorial optimization: polyhedra and efficiency. Springer Science & Business Media (2003)
36. Sebő, A., Vygen, J.: Shorter tours by nicer ears: 7/5-approximation for the graph-tsp, 3/2 for the path version, and 4/3 for two-edge-connected subgraphs. Combinatorica **34**(5), 597–629 (2014)
37. Takazawa, K.: A weighted $K_{t,t}$-free t-factor algorithm for bipartite graphs. Math. Oper. Res. **34**, 351–362 (2009)
38. Vempala, S., Vetta, A.: Factor 4/3 approximations for minimum 2-connected subgraphs. In: Jansen, K., Khuller, S. (eds.) APPROX 2000. LNCS, vol. 1913, pp. 262–273. Springer, Heidelberg (2000). https://doi.org/10.1007/3-540-44436-X_26

Stronger Hardness for Maximum Robust Flow and Randomized Network Interdiction

Jannik Matuschke(✉)

Research Center for Operations Management, KU Leuven, Leuven, Belgium
jannik.matuschke@kuleuven.be

Abstract. We study the following fundamental network optimization problem known as MAXIMUM ROBUST FLOW (MRF): A planner determines a flow on s-t-paths in a given capacitated network. Then, an adversary removes k arcs from the network, interrupting all flow on paths containing a removed arc. The planner's goal is to maximize the value of the surviving flow, anticipating the adversary's response (i.e., a worst-case failure of k arcs). It has long been known that MRF can be solved in polynomial time when $k = 1$ [2], whereas it is NP-hard when k is part of the input [10]. However, the complexity of the problem for constant values of $k > 1$ has remained elusive, in part due to the structure of the natural LP description preventing the use of the equivalence of optimization and separation. This paper introduces a reduction showing that the basic version of MRF described above encapsulates the seemingly much more general variant where the adversary's choices are constrained to k-cliques in a compatibility graph on the arcs of the network. As a consequence of this reduction, we are able to prove the following results: (1) MRF is NP-hard for any constant number $k > 1$ of failing arcs, answering an open question raised in [10]. (2) When k is part of the input, MRF is $P^{NP[\log]}$-hard. (3) The integer version of MRF is Σ_2^P-hard.

1 Introduction

Network flows are a fundamental concept in combinatorial optimization and operations research. They form the central ingredient of numerous optimization models with applications concerning communication, energy transmission, logistics, road traffic, and many more. In many of these applications, the underlying network infrastructure is susceptible to failures, e.g., due to technical malfunctions, unforeseen load peaks, or intentional sabotage, causing flow using links affected by such failures to be interrupted. This motivates the study of robust network optimization models [1,2,4–6,10,12,20,21], which aim to produce resilient solutions by anticipating worst-case failure scenarios.

A basic model for robustness of network flows is the MAXIMUM ROBUST FLOW (MRF) problem proposed by Aneja, Chandrasekaran, and Nair [2]. This problem can be seen as a game played between a *planner* and an *interdictor*,

S. S. Dey et al. (Eds.): IPCO 2026, LNCS 16588, pp. 32–46, 2026.
https://doi.org/10.1007/978-3-032-28691-8_3

an adversary emulating worst-case failure scenarios in the network. The input to MRF consists of a digraph $D = (V, A)$, arc capacities $u \in \mathbb{Q}_+^A$, a source $s \in V$, a sink $t \in V$, and an interdiction budget $k \in \mathbb{N}$. The planner wants to send flow from s to t with the goal of maximizing the amount surviving a worst-case failure of k arcs selected by the interdictor, where any flow on paths containing a failed arc is lost. Formally, let $\mathcal{P} \subseteq 2^A$ denote the set of s-t-paths in D, let $X := \{x \in \mathbb{Q}_+^{\mathcal{P}} : \sum_{P:a\in P} x_P \leq u_a \ \forall\, a \in A\}$ denote the set of s-t-flows in D respecting the capacities u, and let $\mathcal{S}_k := \{S \subseteq A : |S| = k\}$ denote the set of possible failure scenarios. The planner then wants to solve

$$\max_{x\in X} \min_{S\in\mathcal{S}_k} \textstyle\sum_{P\in\mathcal{P}:P\cap S=\emptyset} x_P.$$

Note that the use of path flows in the problem definition is crucial, as the amount of flow interrupted by the failure of multiple arcs does not depend on the arc-flow values alone (see, e.g., [5, Fig. 6]). Note further that $|\mathcal{P}|$ might be exponential in the size of the digraph D given as input.

MRF is closely related to the extensively studied NETWORK INTERDICTION problem [9,16,25,26], in which the players act in the reverse order, i.e., the interdictor removes a set of k arcs so as to minimize the max-flow value in the remaining network. Bertsimas et al. [4] observed that the optimal value of MRF equals that of RANDOMIZED NETWORK INTERDICTION (RNI), in which the interdictor selects a probability distribution over possible failure sets and the planner chooses a flow knowing the distribution, i.e.,

$$\min_{z\in\Delta(\mathcal{S}_k)} \max_{x\in X} \textstyle\sum_{P\in\mathcal{P}} (1 - \sum_{S\in\mathcal{S}_k:P\cap S\neq\emptyset} z_S) \cdot x_P$$

where $\Delta(\mathcal{S}_k) := \{z \in [0,1]^{\mathcal{S}_k} : \sum_{S\in\mathcal{S}_k} z_S = 1\}$ denotes the probability distributions over $\mathcal{S}_k$.

Indeed, MRF and RNI can be equivalently described by the following pair of primal and dual linear programs (see [19] for a detailed discussion):

$$\begin{array}{llrcll}
(P) & \max & \sum_{P\in\mathcal{P}} x_P & - & \lambda & \\
 & \text{s.t.} & \sum_{P\in\mathcal{P}:a\in P} x_P & \leq & u_a & \forall\, a \in A \\
 & & \sum_{P\in\mathcal{P}:P\cap S\neq\emptyset} x_P & \leq & \lambda & \forall\, S \in \mathcal{S}_k \\
 & & x & \geq & 0 &
\end{array}$$

$$\begin{array}{llrcll}
(D) & \min & \sum_{a\in A} u_a y_a & & & \\
 & \text{s.t.} & \sum_{a\in P} y_a + \sum_{S\in\mathcal{S}_k:P\cap S\neq\emptyset} z_S & \geq & 1 & \forall\, P \in \mathcal{P} \\
 & & \sum_{S\in\mathcal{S}_k} z_S & = & 1 & \\
 & & y, z & \geq & 0 &
\end{array}$$

Note that, at optimum, variable λ equals the amount of flow lost in a worst-case scenario. However, both (P) and (D) feature an exponential number of variables.

In this paper, we study the complexity of MRF and RNI. For this purpose, we will refer to the corresponding decision versions of the problems:

MRF*: Given an instance $(D = (V, A), s, t, u, k)$ of MRF and $L \in \mathbb{Q}_+$, decide whether $\max_{x \in X} \min_{S \in \mathcal{S}_k} \sum_{P \in \mathcal{P}: P \cap S = \emptyset} x_P \geq L$.

RNI*: Given an instance $(D = (V, A), s, t, u, k)$ of MRF and $L \in \mathbb{Q}_+$, decide whether $\min_{z \in \Delta(\mathcal{S}_k)} \max_{x \in X} \sum_{P \in \mathcal{P}} (1 - \sum_{S \in \mathcal{S}_k: P \cap S \neq \emptyset} z_S) \cdot x_P < L$.

Note that MRF* and RNI* are complements of one another. Moreover, for constant values of k, the number of constraints in (P) is polynomial, implying that it has an optimal solution with polynomial support (though the number of variables may still be exponential). Hence, when restricting to instances where k is bounded by a constant, MRF* is in *NP* and RNI* is in *coNP*. In general, however, MRF* and RNI* are not known to be contained in *NP* or *coNP*.

1.1 Previous results on the complexity of MRF/RNI

Aneja et al. [2] observed that for $k = 1$, the worst-case amount of flow lost in any solution to MRF equals the maximum flow through any arc. Thus, MRF and RNI can be solved in polynomial time by projecting (P) to arc-flow variables. Interestingly, by a simple rounding argument, this result extends to the case where the flow is required to be integral.

An alternative way to obtain the result from [2] lies in solving the separation problem of (D), which for $k = 1$ reduces to a standard shortest path problem. However, Du and Chandrasekaran [11] showed that the dual separation problem becomes *NP*-hard already for $k = 2$. They suggested that this implies, via the equivalence of optimization and separation [13], that MRF/RNI for $k = 2$ are likewise *NP*-hard. Unfortunately, Disser and Matuschke [10] later observed that this implication does not hold: The equivalence of optimization and separation implies that it is *NP*-hard to optimize *arbitrary* linear objectives over the polyhedron defined by (D). However, the linear objectives that arise from MRF/RNI are not arbitrary, as the z-variables do not appear in the objective. Indeed, the instances constructed in [11] to establish *NP*-hardness of dual separation contain s-t-cuts of cardinality 2. For such instances, both (P) and (D) have trivial solutions of value 0. Therefore, the authors of [10] conclude: "If Maximum Robust Flow for $k = 2$ is indeed *NP*-hard, a reduction that shows this would need to construct considerably more involved instances of the problem."

For the case that k is part of the input, Disser and Matuschke [10] show that MRF is strongly *NP*-hard. Their result relies heavily on the hardness of the interdictor's best-response problem, i.e., solving $\max_{S \in \mathcal{S}_k} \sum_{P: P \cap S \neq \emptyset} x_P$ for given $x \in X$, which however can be solved by enumeration when k is constant.

See [19] for a discussion of the approximability of MRF/RNI and work on related problems, such as Network Interdiction and variants of MRF/RNI.

1.2 Our contribution

We introduce a generalization of MRF that we show to be equivalent to the original problem via a series of careful reductions. This facilitates the analysis of MRF/RNI, leading to a series of new strong hardness results for both problems.

Hardness Results for MRF/RNI.

As mentioned above, the complexity of MRF/RNI for constant values of $k > 1$ is an open question. We answer this question by showing that the problems are in fact hard already when $k = 2$.

Theorem 1. *Let $k \in \mathbb{N}$ with $k > 1$. Then* MRF* *restricted to instances with interdiction budget k is NP-complete, and* RNI* *restricted to instances with interdiction budget k is coNP-complete.*

For the general case that k is not restricted to a constant, we show that both problems are at least as hard as any problem in $P^{NP[\log]}$, the class of all decision problems that can be solved in polynomial time by an algorithm that makes a logarithmic number of calls to an *NP*-oracle. [1]

Theorem 2. MRF* *and* RNI* *are $P^{NP[\log]}$-hard.*

Theorem 2 gives a strong indication that the two problems are neither contained in *NP* nor *coNP*. In fact, the result implies that it is not even possible to describe general instances of MRF or RNI by boolean formulas over constantly many instances of problems from *NP*, unless the polynomial hierarchy collapses to its third level [8]. From a practical point of view, this rules out, e.g., compact mixed integer programming formulations for the problems. Theorem 1 and 2 also reveal a surprising contrast between RNI and its integer version, Network Interdiction (NI). Note that a given solution $S \in \mathcal{S}_k$ to NI can be evaluated simply by computing the maximum flow value in the digraph $(V, A \setminus S)$. Hence, NI is in *NP*, and for constant values of k, it can be solved in polynomial time by enumeration. Thus, RNI is an interesting case of a problem where the convexity introduced by allowing randomized strategies makes the problem harder rather than easier.

We also consider the integral version of MRF, which requires the flow value on every path to be an integer. For this problem, we obtain hardness at the second level of the polynomial hierarchy.

Theorem 3. *Integral* MRF* *is Σ_2^P-hard.*

We remark that Theorem 3 does not follow from recent meta theorems on the hardness of robust/interdiction versions of *NP*-hard problems [14,15]. Indeed, as maximum integral s-t-flows can be computed in polynomial time, this problem marks a curious example of a problem whose complexity jumps at least by two levels in the polynomial hierarchy when going to the robust version.

Technical Overview: MRF **with Restricted Interdiction.** At the core of all three results mentioned above is a generalized version of MRF, which we call Maximum Robust Flow with Restricted Interdiction (MRF-R), in

[1] An *NP*-oracle is a subroutine that, given an instance of a problem from *NP*, returns the correct answer in polynomial time. The class $P^{NP[\log]}$ (also known as θ_2^P [7,17]) is situated between *NP* and Σ_2^P and contains the so-called boolean hierarchy [22].

which we are given an additional *compatibility graph* $H = (A, E_H)$ on the arcs of D and where the interdictor's choices are restricted to

$$\mathcal{S}_{H,k} := \{S \subseteq A : |S| \leq k, S \text{ is a clique in } H\}.$$

For our purposes, it will suffice to consider the following decision variant of the problem on directed acyclic graphs (DAGs) with unit capacities:

MRF-R*: Given an instance $(D = (V, A), s, t, u, k)$ of MRF, where D is a DAG and $u \equiv 1$, a compatibility graph $H = (A, E_H)$, and a demand $\theta \in \mathbb{Q}_+$, decide whether there exists $x \in X$ such that $\sum_{P \in \mathcal{P}} x_P = \theta$ and $\max_{S \in \mathcal{S}_{H,k}} \sum_{P \in \mathcal{P}: S \cap P \neq \emptyset} x_P \leq k - 1$.

Our main technical contribution is the following reduction, showing that any MRF-R* instance can in fact be modeled by an MRF* instance.

Theorem 4. *There is a polynomial transformation from* MRF-R* *to* MRF* *that preserves the value of k. The same transformation also applies to the integral versions of the problems. The capacities in the resulting* MRF* *instance are polynomial in the size of the network.*

Our proof of Theorem 4 consists of two major steps. First, we show that we can model the compatibility graph in MRF-R* by introducing a multicommodity version of MRF (called MRF-M*), where the commodities are used to ensure that incompatible arcs from the MRF-R* instance necessarily share a significant portion of flow on a common path, making it unattractive for the interdictor to simultaneously interdict any pair of such incompatible arcs. We discuss this reduction from MRF-R* to MRF-M* in Section 3.

Second, we show that the multicommodity problem MRF-M* can be reduced to the single-commodity problem MRF* by carefully building a "wrapper" around the multicommodity network that ensures that any optimal single-commodity flow has to respect the source-sink pairings of the commodities in order to keep the loss at worst-case interdiction within a certain bound. This reduction is discussed in Sect. 4.

Note that the reduction in Theorem 4 is robust in the sense that it preserves the value of k as well as integrality of flows. In order to establish the hardness results for MRF discussed earlier, it thus suffices to show the same hardness results for MRF-R*. Accordingly, we provide the following three reductions.

1. To prove Theorem 1, we provide a reduction from FRACTIONAL GRAPH COLORING, an *NP*-complete LP relaxation of the classic GRAPH COLORING problem, to MRF-R* with $k = 2$. Here, we crucially use the flexibility won by the introduction of the compatibility graph to force the planner to send flow only along paths corresponding to feasible color classes. By the reduction from FRACTIONAL GRAPH COLORING, whose hardness stems from the PCP theorem [18], our proof avoids the issue of the inapplicability of the equivalence of optimization and separation that has been a major obstacle in previous attempts to resolving the complexity of MRF for constant k. We discuss this reduction in Sect. 2.

2. To prove Theorem 2, we make use of Wagner's Theorem [24], a meta theorem that establishes $P^{NP[\log]}$-hardness by reducing from the question whether certain monotone sequences of instances of an *NP*-complete problem contain an odd or even number of YES instances. To obtain our result, we reduce the even instances in the sequence to MRF-R* (using the preceding result), while we encode the odd instances as CLIQUE problems in the compatibility graph. In the full version [19] we discuss how this can be done in such a way that a single MRF-R* instance suffices to decide Wagner's parity problem.
3. To prove Theorem 3, we reduce the Σ_2^P-complete CLIQUE INTERDICTION problem, which asks for a small subset of vertices to be removed from a graph G so as to reduce its clique number, to INTEGRAL MRF-R*. The flexibility of MRF-R* allows us to encode the cliques of G directly into the compatibility graph, and a careful construction forces the planner to send flow along a path that intersects with every sufficiently large clique in the compatibility graph. See the full version [19] for details.

A schematic overview of all reductions in this paper is given in Fig. 1.

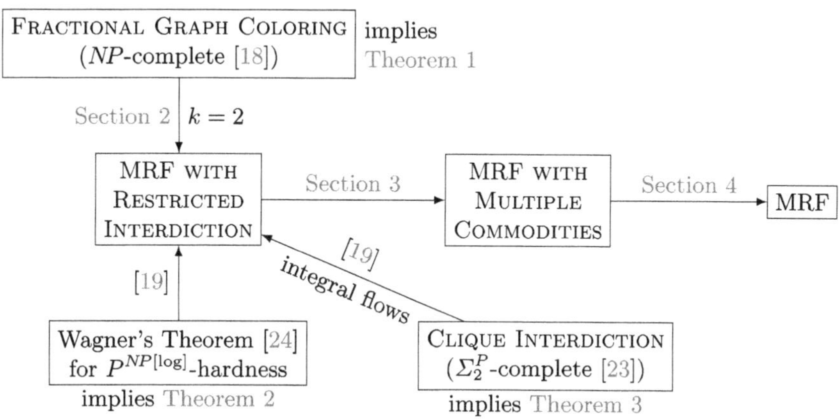

Fig. 1. Overview of the reductions presented in this paper.

1.3 Notation

We introduce some notation that we will use throughout the paper. For $n \in \mathbb{N}$, we let $[n] := \{1, \dots, n\}$. For a digraph $D = (V, A)$, we denote the set of arcs leaving/entering $U \subseteq V$ by $\delta_D^+(U) := \{(v, w) \in A : v \in U, w \in V \setminus U\}$ and $\delta_D^-(U) := \{(v, w) \in A : v \in V \setminus U, w \in U\}$. Finally, for any flow $x \in X$, we use

$$x[a] := \sum_{P \in \mathcal{P}: a \in P} x_P \quad \text{and} \quad \lambda(x, S) := \sum_{P \in \mathcal{P}: P \cap S \neq \emptyset} x_P$$

to denote the flow on arc $a \in A$ and the flow lost at failure of $S \subseteq A$.

2 *NP*-completeness of MRF-R* for $k = 2$

In this section, we establish the following theorem, which together with Theorem 4 implies Theorem 1. Its proof is based on a reduction from Fractional Graph Coloring, which introduce before describing the reduction.

Theorem 5. *Let $k \in \mathbb{N}$ with $k > 1$. Then* MRF-R* *restricted to instances with interdiction budget k is NP-complete.*

Fractional Graph Coloring. The *fractional chromatic number* of an undirected graph $G = (V, E)$ is given by

$$\begin{aligned} \chi_{\mathrm{f}}(G) := \min \quad & \textstyle\sum_{I \in \mathcal{I}(G)} y_I \\ \text{s.t.} \quad & \textstyle\sum_{I \in \mathcal{I}(G): v \in I} y_I = 1 \quad \forall\, v \in V \\ & y \geq 0 \end{aligned}$$

where $\mathcal{I}(G) \subseteq 2^V \setminus \{\emptyset\}$ is the set of independent sets in G. It is well-known that $\chi_{\mathrm{f}}(G)$ logarithmically approximates the chromatic number of G, from which it inherits $|V|^{-\delta}$-hardness of approximation for some constant $\delta > 0$ [18]. For our purposes the following consequence of this hardness suffices:

Theorem 6 ([18, **Theorem 2.9**]). *The following problem is NP-complete: Given a graph G and an integer $\ell \in \mathbb{N}$, decide whether $\chi_f(G) \leq \ell$.*

Constructing an MRF-R* **Instance.** Let $G = (V, E)$ be an undirected graph and let $\ell \in \mathbb{N}$ (w.l.o.g. $\ell \geq 2$). We construct a digraph $D = (V', A)$ with source s and sink t and a compatibility graph H. Let $e_1, \ldots, e_m$, with $m := |E|$, be an arbitrary ordering of the edges of G. We define $V' := \{z_1, \ldots, z_{m+1}\}$ and let $s := z_1$ and $t := z_{m+1}$. The arc set A of D is comprised by bundles of parallel arcs A_i for each $i \in [m]$, where bundle A_i contains ℓ arcs from z_i to z_{i+1}. For each bundle A_i, representing edge $e_i = \{v, w\}$ in G, we fix two arbitrary but distinct arcs a_{iv} and a_{iw}. We further let $E_H := \big\{\{a_{iv}, a_{jv}\} : i, j \in [m], i \neq j, v \in e_i \cap e_j\big\}$ be the edge set of the compatibility graph $H = (A, E_H)$, i.e., H contains an edge between any two arcs representing the same vertex of G. See Fig. 2 for a depiction of the construction.

Theorem 5 follows from the following lemma, which shows that the MRF-R* instance with $k = 2$ and $\theta = \ell$ is a YES instance if and only if $\chi_{\mathrm{f}}(G) \leq \ell$.

Lemma 7. *There is an s-t-flow x in D for capacities $u \equiv 1$ with $\sum_{P \in \mathcal{P}} x_P = \ell$ and $\max_{S \in \mathcal{S}_{H,2}} \lambda(x, S) \leq 1$ if and only if $\chi_f(G) \leq \ell$.*

Proof. We show how to construct a fractional coloring from a given flow x with the properties stated in the lemma. For the reverse direction see [19]. We start by defining $A_v := \{a \in A : a = a_{iv}$ for some $i \in [m]\}$ for $v \in V$ and $I_P := \{v \in$

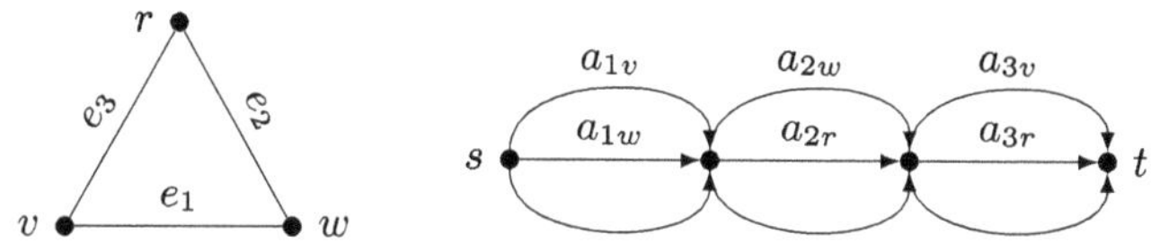

Fig. 2. Illustration of the reduction from FRACTIONAL GRAPH COLORING (graph on the left) to MRF-R* with $k = 2$ (digraph on the right). The compatibility graph of the constructed MRF-R* instance has the edges $\{a_{1v}, a_{3v}\}$, $\{a_{1w}, a_{2w}\}$, and $\{a_{2r}, a_{3r}\}$.

$V : A_v \subseteq P\}$ for $P \in \mathcal{P}$. Note that $I_P \in \mathcal{I}(G)$ for every $P \in \mathcal{P}$, as P cannot contain both parallel arcs a_{iv} and a_{iw} for any edge $e_i = \{v, w\} \in E$.

Define $y_I := \sum_{P \in \mathcal{P}: I_P = I} x_P$. Note that $\sum_{I \in \mathcal{I}(G): v \in I} y_I = \sum_{P \in \mathcal{P}: v \in I_P} x_P$ for all $v \in V$. We show that the latter equals 1, which implies that y is a fractional coloring of G and hence $\chi_f(G) \leq \sum_{I \in \mathcal{I}(G)} y_I = \sum_{P \in \mathcal{P}} x_P = \ell$.

Let $v \in V$ and fix any $a \in A_v$. Note that every arc in D is on an s-t-cut of capacity ℓ and hence $\sum_{P \in \mathcal{P}} x_P = \ell$ implies $x[a'] = u_{a'} = 1$ for all $a' \in A$. We obtain $1 \geq \lambda(x, \{a, a'\}) = \sum_{P \in \mathcal{P}: P \cap \{a, a'\} \neq \emptyset} x_P = 2 - \sum_{P \in \mathcal{P}: a, a' \in P} x_P$ for all $a' \in A_v \setminus \{a\}$, where the first inequality follows from $\{a, a'\} \in E_H$ (as A_v is a clique in H) and the final inequality follows from $x[a] + x[a'] = 2$. Rearranging yields $\sum_{P \in \mathcal{P}: a, a' \in P} x_P \geq 1$ for all $a' \in A_v \setminus \{a\}$, which must hold with equality because $x[a] = x[a'] = 1$. Thus, if $a \in P$ for some $P \in \mathcal{P}$ with $x_P > 0$, then $a' \in P$ for all $a' \in A_v$, or in other words: $v \in I_P$. Hence $\sum_{P \in \mathcal{P}: v \in I_P} x_P = x[a] = 1$. $\square$

3 Reducing MRF-R* to MRF-M*

For our reduction from MRF-R* to MRF*, we introduce a multi-commodity variant of MRF (MRF-M) as an intermediate problem. The input to this problem is a digraph $D = (V, A)$ with capacities u, interdiction budget $k \in \mathbb{N}$, and a set of commodities K with source s_i, sink t_i, and demand d_i for each $i \in K$. For $i \in K$, let $\mathcal{P}_i$ denote the set of s_i-t_i-paths in D and let $\mathcal{P} = \bigcup_{i \in K} \mathcal{P}_i$. We let

$$X_K := \{x \in \mathbb{Q}_+^{\mathcal{P}} : \textstyle\sum_{P \in \mathcal{P}: a \in P} x_P \leq u_a \ \forall a \in A, \ \sum_{P \in \mathcal{P}_i} x_P = d_i \ \forall i \in K\}$$

denote the set of multi-commodity flows in D for commodities K that send d_i units of flow from each s_i to t_i while respecting arc capacities u. For our purposes it will again be useful to consider a particular decision variant of the problem:

MRF-M*: Given an input as described above, decide whether there is $x \in X_K$ with $\max_{S \in \mathcal{S}_k} \sum_{P \in \mathcal{P}: P \cap S \neq \emptyset} x_P \leq kM - 1$, for $M := \max_{a \in A} u_a$.

We show how to transform instances of MRF-R* to instances of MRF-M* with some additional properties that will be useful later when we continue the reduction in Section 4 to finally obtain an MRF* instance.

Theorem 8. *There is a polynomial transformation from* MRF-R* *to* MRF-M* *that preserves the value of* k. *The same transformation also applies to the integral versions of the problems. Moreover, the resulting* MRF-M* *instance has the following additional properties:*

1. *Capacities are integral and* $M := \max_{a \in A} u_a$ *is polynomial in* $|A|$.
2. *All* $i \in K$ *fulfill* $\sum_{a \in \delta^+(s_i)} u_a = \sum_{a \in \delta^-(t_i)} u_a = d_i$ *and* $\delta^-(s_i) = \delta^+(t_i) = \emptyset$.
3. *There is a commodity* $i_0 \in K$ *such that* $\sum_{i \in K \setminus \{i_0\}} d_i \leq M - 3$.
4. *For any* $x \in X_K$ *it holds that* $\sum_{P \in \mathcal{P}_i : a \in P} x_P \geq 2$ *for all* $i \in K \setminus \{i_0\}$ *and all* $a \in A$ *with* $u_a \geq M - 2$.
5. *There is* $U \subseteq V$ *such that* $\{s_i : i \in K\} \subseteq U$, $\{t_i : i \in K\} \subseteq V \setminus U$, $\sum_{a \in \delta^+(U)} u_a = \sum_{i \in K} d_i$, *and* $|\{a \in \delta^+(U) : u_a = M\}| = k - 1$.

Constructing an MRF-M* **Instance.** To illustrate the main principle behind the reduction, we describe a simplified construction which ignores properties 1. to 5.. The complete proof of Theorem 8 can be found in [19]. Given an instance of MRF-R* with DAG $D = (V, A)$, $s, t \in V$, interdiction budget k, demand θ, and compatibility graph $H = (A, E_H)$, we construct an instance of MRF-M* with the same interdiction budget k and a digraph $D' = (V', A')$ with capacities u' and commodities K as follows.

We first fix a linear order $\succ$ of the arcs of D that is consistent with the DAG structure in the sense that for $a, a' \in A$ with $a \succ a'$ there is no path in D from the tail of a to the head of a'. We define $F := \{(a, a') : \{a, a'\} \notin E_H, a \succ a'\}$, i.e., F contains all pairs of incompatible arcs, internally ordered by $\succ$. To construct D', we subdivide every arc $a = (v, w)$ of D by introducing the nodes v_a^+ and v_a^- and replacing a with the path v-v_a^+-v_a^--w. We further introduce the arc $(v_a^-, v_{a'}^+)$ for every $(a, a') \in F$. We set the capacities u' in D' to $M := |A|$ for the arcs of the form (v_a^+, v_a^-) for $a \in A$, and to 1 for all other arcs. The set of commodities is $K := \{0\} \cup F \cup A$, where $s_0 = s$, $t_0 = t$, $d_0 = \theta$, $s_{a,a'} = v_a^+$, $t_{a,a'} = v_{a'}^-$, $d_{a,a'} = 1$, and $s_a = v_a^+$, $t_a = v_a^-$, $d_a = |\{a' \in A : \{a, a'\} \in E_H\}|$. See Fig. 3 for a depiction of the construction.

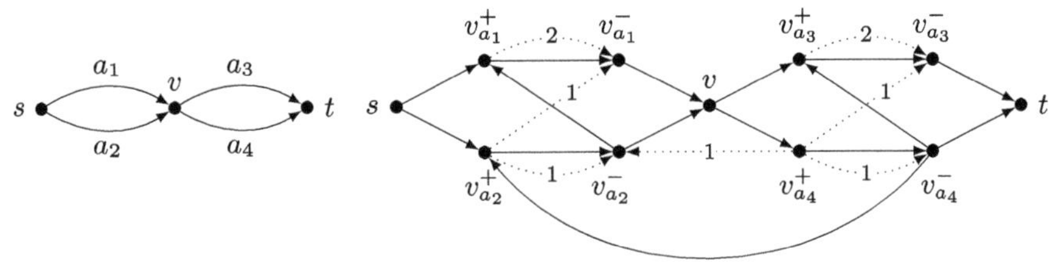

Fig. 3. From the MRF-R* instance with the digraph on the left and compatibility graph with edges $\{a_1, a_3\}, \{a_1, a_4\}, \{a_2, a_3\}$, we obtain the MRF-M* instance depicted on the right. The solid arcs represent the network D', the dotted arcs indicate demands, i.e., a dotted arc (s_i, t_i) represents a commodity i with source s_i, sink t_i, and demand d_i as indicated on the label (we omit commodity 0). Capacities are 1 except for the four arcs $(v_{a_1}^+, v_{a_1}^-), \ldots, (v_{a_4}^+, v_{a_4}^-)$, each of which has capacity $M = 4$.

Analysis. Let X_K denote the set of multicommodity flows in D' for capacities u' fulfilling demands d for commodities K and let $\mathcal{S}'_k := \{S' \subseteq A' : |S'| \leq k\}$. The following lemma completes the proof of Theorem 8 (ignoring properties 1. to 5.).

Lemma 9. *There is* $x' \in X_K$ *with* $\max_{'S \in \mathcal{S}'_k} \lambda(x','S) \leq kM - 1$ *if and only if there is* $x \in X$ *with* $\sum_{P \in \mathcal{P}} x_P = \theta$ *and* $\max_{S \in \mathcal{S}_{H,k}} \lambda(x, S) \leq k - 1$.

Proof. Let $x \in X$ with $\sum_{P \in \mathcal{P}} x_P = \theta$ and $\max_{S \in \mathcal{S}_{H,k}} \lambda(x, S) \leq k - 1$. Let $\phi(x)$ be the s-t-flow in D' obtained from x by replacing each arc $a = (v, w) \in A$ with its corresponding subpath v-v_a^+-v_a^--w. Moreover, let $\hat{x}$ be the flow on commodities in $K \setminus \{0\}$ that routes the entire demand of commodity $a \in A$ along the one-arc path (v_a^+, v_a^-) and routes the entire demand of commodity $(a, a') \in F$ via the path v_a^+-v_a^--$v_{a'}^+$-$v_{a'}^-$. Note that $\hat{x}[(v_a^+, v_a^-)] = M - 1$ for all $a \in A$ and that $\hat{x}[(v_a^-, v_{a'}^+)] = 1$ for all $(a, a') \in F$. In particular, $\hat{x}$ leaves spare capacity exactly 1 on each arc corresponding to the subdivision of D, while saturating all other arcs. Thus, $x' := \phi(x) + \hat{x} \in X_K$ is a feasible multicommodity flow (note that $\phi(x)[a] \leq 1$ for all $a \in A$ and $\phi(x)$ sends θ units from s_0 to t_0).

Now assume by contradiction that there is $S' \in \mathcal{S}'_k$ with $\lambda(x', S') > kM - 1$. Note that S' must consist of k arcs of capacity M. Hence $S' = \{(v_a^+, v_a^-) : a \in S\}$ for some $S \subseteq A$ with $|S| = k$. Moreover, if there is $(a, a') \in F$ with $a, a' \in S$, then S' intersects the path v_a^+-v_a^--$v_{a'}^+$-$v_{a'}^-$ of commodity (a, a') twice, which would imply $\lambda(x', S') \leq kM - 1$. Thus, $\{a, a'\} \in E_H$ for all $a, a' \in S$ with $a \neq a'$, which implies $S \in \mathcal{S}_{H,k}$ and therefore $\lambda(x, S) \leq k - 1$. We obtain the contradiction $\lambda(x', S') = \lambda(\hat{x}, S') + \lambda(\phi(x), S') = k(M - 1) + \lambda(x, S) \leq kM - 1$.

Conversely, one can show that in fact any $x' \in X_K$ is of the form $\phi(x) + \hat{x}$ for some $x \in X$ with $\sum_{P \in \mathcal{P}} x_P = \theta$ and that any $S \in \mathcal{S}_{H,k}$ with $\lambda(x, S) > k - 1$ induces a set $S' \in \mathcal{S}'_k$ with $\lambda(S', x') > kM - 1$. □

4 Reducing MRF-M* to MRF

We prove the following theorem, which with Theorem 8 implies Theorem 4.

Theorem 10. *There is a polynomial transformation from* MRF-M*, *restricted to instances fulfilling properties 1. to 5. from Theorem 8, to* MRF*. *The transformation preserves the value of k. The same transformation also applies to the integral versions of the problems.*

Constructing an MRF***Instance.**

Consider an instance of MRF-M* given by a digraph $D = (V, A)$, capacities u and commodities K and interdiction budget k, and that fulfills properties 1. to 5. stated in Theorem 8. For simplicity of notation, assume that $K = \{0, 1, \ldots, m\}$ for $m = |K| - 1$, with $i_0 = 0$ for the commodity i_0 given by property 3.

We construct the following MRF instance on digraph $D' = (V', A')$ with source s', sink t', capacities u', and the same interdiction budget k. To simplify the description, we make use of *immune arcs* that cannot be interdicted (this is without loss of generality in the context of this reduction; see [19] for details). Define $K' := K \setminus \{0\}$ and let $M' := M + 2k - 3$. The set of nodes is $V' := V \cup \{s', t', \bar{s}\} \cup \{s'_i : i \in K'\}$. The arc set A' contains the following arcs:

- all arcs of A, with their original capacities u,
- immune arcs (s', s_0) and (t_0, t'), each of capacity d_0,
- an immune arc (s', s_1) of capacity M',
- for each $i \in \{2, \dots, m\}$, an immune arc (s', s_i), each of capacity $M' - 2k$,
- an immune arc $(s', \bar{s})$ of capacity $2k \cdot (M-2)$,
- a bundle $\bar{A}$ of $2k$ arcs from $\bar{s}$ to t', each with capacity $M-1$;
- moreover, for every $i \in K'$:
  - an arc (s'_i, s_i) with capacity M',
  - an immune arc (s_i, s'_{i+1}) of capacity $2k$ (where $s'_{m+1} := \bar{s}$),
  - for each $j \in K' \setminus \{i\}$, an immune arc (s_i, t_j) of capacity d_j,
  - a bundle A_i of d_i arcs from t_i to t', each with capacity $M-1$,
  - an immune arc (s_i, t) of capacity $\xi_i^s := M' - 2k - \sum_{j \in K'} d_j$,
  - an immune arc (s', t_i) of capacity $\xi_i^t := d_i \cdot (M - 1 - m)$.

The construction is depicted in Fig. 4. Note that $\xi_i^s \geq 0$ due to property 3. and that with the exception of s', t', and the internal nodes of D, the in-capacity of a node equals its out-capacity.

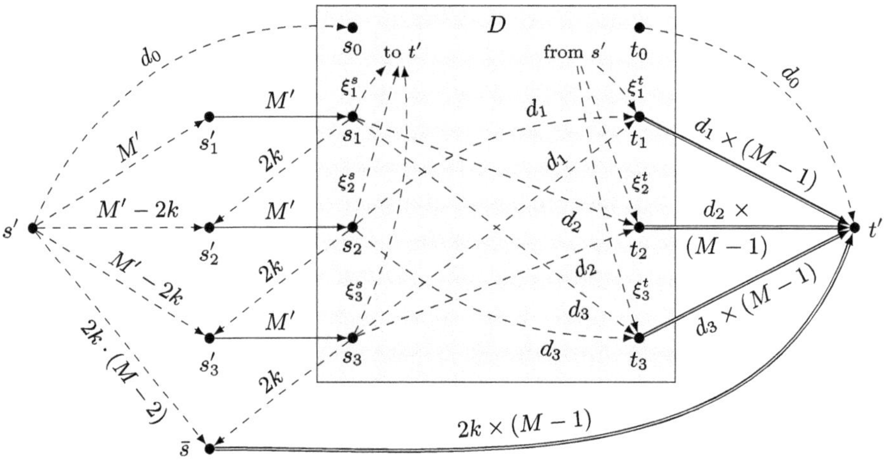

Fig. 4. Illustration of the reduction from MRF-M* with digraph D and four commodities to MRF. Dashed arcs represent immune arcs, with the label indicating the capacity. Single-lined solid arcs represent regular arcs, with the label indicating the capacity. Double-lined arcs labeled $\ell \times c$ represent bundles of ℓ arcs with capacity c each. The figure omits the internal nodes and arcs of the digraph D.

Analysis. Let $\mathcal{S}_k := \{S \subseteq A : |S| \leq k\}$ and $\mathcal{S}'_k := \{S \subseteq A' \setminus A_{\mathrm{I}} : |S| \leq k\}$, where A_{I} is the set of immune arcs. Let X_K be the set of multicommodity flows in D for capacities u, sending d_i units from s_i to t_i for all $i \in K$. Let $\mathcal{P}'$ be the set of s'-t'-paths in D' and let X' be the set of s'-t'-flows in D' for capacities u'. To complete the reduction, we prove the following lemma, where $\Delta := \sum_{a \in \delta^+_{D'}(s')} u'_a$.

Lemma 11. *There is* $x \in X_K$ *with* $\max_{S \in \mathcal{S}_k} \lambda(x, S) \leq kM - 1$ *if and only if there is* $x' \in X'$ *with* $\min_{S \in \mathcal{S}'_k} \sum_{P \in \mathcal{P}' : P \cap S = \emptyset} x'_P \geq \Delta - (kM - 1)$.

We sketch the main ideas for proving the lemma; see [19] for details. We construct the following "base flow" $\bar{x}$ that sends

- $2k$ units of flow along paths s'-s'_1-s_1-s'_2-s_2-...-s'_m-s_m-$\bar{s}$-t',
- $2k \cdot (M - 2)$ units of flow along paths s'-$\bar{s}$-t',
- ξ^s_i units of flow along the path s'-s'_i-s_i-t' for each $i \in K'$,
- ξ^t_i units of flow along paths s'-t_i-t' for each $i \in K'$,
- d_j units of flow along paths s'-s'_i-s_i-t_j-t' for each $i, j \in K'$ with $i \neq j$,

where we split flow equally among the arcs of the bundle if the path described by the node sequence contains such a bundle.

Note that $\bar{x}$ sends $\Delta - \sum_{i \in K} d_i$ units of flow from s' to t' and that it does not use any arc in $A \cup \{(s', s_0), (t_0, t')\}$. Moreover, for $i \in K'$, it leaves a spare capacity of d_i along the path s'-s'_i-s_i and a spare capacity of 1 on each of the d_i arcs in A_i. We can thus combine $\bar{x}$ with a multicommodity flow $x \in X_K$, whose paths we appropriately extend to obtain an s'-t'-flow $\psi(x)$: For every $P \in \mathcal{P}_i$ and $a \in A_i$ for $i \in K'$, let $Q_{P,a}$ be the path derived from P by attaching the path s'-s'_i-s_i as a prefix and $a \in A_i$ as a suffix, and set $\psi(x)_{Q_{P,a}} := \frac{x_P}{d_i}$; for every $P \in \mathcal{P}_0$, set $\psi(x)_{R_P} := x_P$, where R_P is the concatenation $(s', s_0) \circ P \circ (t_0, t')$.

Lemma 12. *Let* $x \in X_K$ *with* $\max_{S \in \mathcal{S}_k} \lambda(x, S) \leq kM - 1$ *and let* $x' := \psi(x) + \bar{x}$. *Then* $x' \in X'$ *and* $\min_{S \in \mathcal{S}'_k} \sum_{P \in \mathcal{P}' : P \cap S = \emptyset} x'_P \geq \Delta - (kM - 1)$.

Proof (Sketch). As argued above, $\psi(x)$ fits into the spare capacity left by $\bar{x}$ and thus $x' \in X'$ with $\sum_{P \in \mathcal{P}'} x_P = \Delta$. Moreover, we establish that $\lambda(x', S) \leq kM - 1$ for any $S \in \mathcal{S}'_k$. If $S \subseteq A$, this follows from $\lambda(x', S) = \lambda(x, S) \leq kM - 1$. If $S \not\subseteq A$, then $\lambda(x', S) > kM - 1$ is only possible if S contains at least one arc of capacity M', which must be of the form (s'_i, s_i). For this case, one verifies that our construction ensures that (s'_i, s_i) shares a sufficiently high amount of flow on common paths with any other arc a so that the marginal contribution of a to $\lambda(x', S)$ is bounded by $M - 2$ and hence $\lambda(x', S) \leq M' + (k-1)(M-2) = kM - 1$. For $a \in A$, e.g., this follows from property 4. of Theorem 8, while for $a \in A_i$ this follows from the fact that there is one unit of flow on paths containing both (s'_i, s_i) and a as extensions of s_i-t_i-paths from x. □

Conversely, one can show that every $x \in X'$ with the desired properties must actually be a combination of $\bar{x}$ with such an extended multicommodity flow.

Lemma 13. *Let* $x' \in X'$ *with* $\min_{S \in \mathcal{S}'_k} \sum_{P \in \mathcal{P}' : P \cap S = \emptyset} x'_P \geq \Delta - (kM - 1)$. *Define* x *by* $x_P := \sum_{Q \in \mathcal{P}' : P = Q \cap A} x'_Q$ *for each* $P \in \mathcal{P}_i$ *and* $i \in K$. *Then* $x \in X_K$ *and* $\max_{S \in \mathcal{S}_k} \lambda(x, S) \leq kM - 1$.

Proof (Sketch). We can show that x' must saturate all arcs in $A' \setminus A$, as otherwise we can construct sets $S \in \mathcal{S}'_k$ whose failure leaves less than $\Delta - (kM - 1)$ units of flow surviving (for constructing these sets we make use of a cut derived from property 5. of Theorem 8 and the bundle $\bar{A}$). In fact, x' must use the same paths as $\bar{x}$ on $A' \setminus A$. In addition, for each $a \in A_i$, the flow x' must send one unit on paths containing both (s_i', s_i) and a, as otherwise $\lambda(x', S) > kM - 1$ for any set S containing (s_i', s_i), a, and $k-2$ arcs from $\bar{A}$. Because $\bar{x}$ does not send any flow on paths containing both (s_i', s_i) and a, these $|A_i| = d_i$ units of flow must traverse s_i-t_i-paths in A for $i \in K'$. By a capacity argument, one obtains the same for $i = 0$. Thus $x \in X_K$, and $\lambda(x, S) \leq \lambda(x', S) \leq kM - 1$ for all $S \in \mathcal{S}_k$. □

5 Conclusion

Several interesting questions concerning MRF/RNI remain. Notably, the approximability of the problems is wide open: All known approximation algorithms [3–5] achieve approximation ratios linear in k, with no hardness of approximation results known (except for strong *NP*-hardness ruling out fully polynomial-time approximation schemes). Our transformation from MRF-R* to MRF* might serve as a promising starting point for obtaining stronger inapproximability results. However, some significant extensions would be necessary so that the constructions can work for near-optimal instead of optimal solutions.

While our results indicate that MRF* and RNI* are unlikey to be contained in *NP* or *coNP*, pinpointing their exact complexity remains an intriguing question. At the moment, it is not even known whether the problems are contained in PSPACE. This question seems closely related to the size of the support of optimal solutions to the LPs (P) and (D). Are there instances where optima necessarily have exponential support?

Finally, MRF/RNI are an example of a combinatorial zero-sum game (where strategy sets are of exponential size and given only implicitly), and exponentially sized LPs arise frequently in this context. At the moment, showing hardness for such problems is challenging, and identifying more "LP-type" problems to reduce from (such as FRACTIONAL COLORING, MRF, and RNI) could facilitate the analysis of such games.

Acknowledgements. The author thanks Martijn van Ee for pointing him to the complexity class Δ_2^P, which led to the proof of Theorem 2. This work was supported by the special research fund of KU Leuven (project C14/22/026).

Declaration of Competing Interests. The author has no competing interests to declare that are relevant to the content of this article.

References

1. Aggarwal, C.C., Orlin, J.B.: On multiroute maximum flows in networks. Networks **39**, 43–52 (2002)
2. Aneja, Y.P., Chandrasekaran, R., Nair, K.P.K.: Maximizing residual flow under an arc destruction. Networks **38**, 194–198 (2001)

3. Baffier, J.F., Suppakitpaisarn, V., Hiraishi, H., Imai, H.: Parametric multiroute flow and its application to multilink-attack network. Discret. Optim. **22**, 20–36 (2016)
4. Bertsimas, D., Nasrabadi, E., Orlin, J.B.: On the power of randomization in network interdiction. Oper. Res. Lett. **44**, 114–120 (2016)
5. Bertsimas, D., Nasrabadi, E., Stiller, S.: Robust and adaptive network flows. Oper. Res. **61**, 1218–1242 (2013)
6. Bertsimas, D., Sim, M.: Robust discrete optimization and network flows. Math. Program. **98**, 49–71 (2003)
7. Buss, S.R., Hay, L.: On truth-table reducibility to sat. Inf. Comput. **91**, 86–102 (1991)
8. Chang, R., Kadin, J.: The boolean hierarchy and the polynomial hierarchy: a closer connection. SIAM J. Comput. **25**, 340–354 (1996)
9. Chestnut, S.R., Zenklusen, R.: Hardness and approximation for network flow interdiction. Networks **69**, 378–387 (2017)
10. Disser, Y., Matuschke, J.: The complexity of computing a robust flow. Oper. Res. Lett. **48**, 18–23 (2019)
11. Du, D., Chandrasekaran, R.: The maximum residual flow problem: NP-hardness with two-arc destruction. Networks **50**, 181–182 (2007)
12. Gottschalk, C., Koster, A.M., Liers, F., Peis, B., Schmand, D., Wierz, A.: Robust flows over time: models and complexity results. Math. Program. **171**, 55–85 (2018)
13. Grötschel, M., Lovász, L., Schrijver, A.: Geometric algorithms and combinatorial optimization, vol. 2. Springer Science & Business Media (2012)
14. Grüne, C., Wulf, L.: On the complexity of recoverable robust optimization in the polynomial hierarch (2025). https://doi.org/10.4230/LIPIcs.MFCS.2025.52
15. Grüne, C., Wulf, L.: Completeness in the polynomial hierarchy for many natural problems in bilevel and robust optimization. In: Integer Programming and Combinatorial Optimization, Lecture Notes in Computer Science, vol. 15620, pp. 256–269 (2025)
16. Guruganesh, G., Sanita, L., Swamy, C.: Improved region-growing and combinatorial algorithms for k-route cut problems. In: Proceedings of the Twenty-Sixth Annual ACM-SIAM Symposium on Discrete Algorithms, pp. 676–695. SIAM (2014)
17. Hemachandra, L.A.: The strong exponential hierarchy collapses. In: Proceedings of the Nineteenth Annual ACM Symposium on Theory of Computing, pp. 110–122 (1987)
18. Lund, C., Yannakakis, M.: On the hardness of approximating minimization problems. J. ACM **41**, 960–981 (1994)
19. Matuschke, J.: Stronger hardness for maximum robust flow and randomized network interdiction (2025). https://arxiv.org/abs/2511.06505
20. Matuschke, J., McCormick, S.T., Oriolo, G.: Rerouting flows when links fail. SIAM J. Discret. Math. **34**, 2082–2107 (2020)
21. Matuschke, J., McCormick, S.T., Oriolo, G., Peis, B., Skutella, M.: Protection of flows under targeted attacks. Oper. Res. Lett. **45**, 53–59 (2017)
22. Riege, T., Rothe, J.: Completeness in the boolean hierarchy: exact-four-colorability, minimal graph uncolorability, and exact domatic number problems. Electron. Colloquium Comput. Complex (2006)
23. Rutenburg, V.: Propositional truth maintenance systems: classification and complexity analysis. Ann. Math. Artif. Intell. **10**, 207–231 (1994)
24. Wagner, K.W.: More complicated questions about maxima and minima, and some closures of NP. Theoret. Comput. Sci. **51**, 53–80 (1987)

25. Wood, R.K.: Deterministic network interdiction. Math. Comput. Model. **17**, 1–18 (1993)
26. Zenklusen, R.: Network flow interdiction on planar graphs. Discret. Appl. Math. **158**, 1441–1455 (2010)

Approximation Schemes for Planar Graph Connectivity Problems

Meike Neuwohner[1], Vera Traub[2], and Rico Zenklusen[3](✉)

[1] CNRS & DIENS, ENS Paris - PSL Research University, Paris, France
meike.neuwohner@ens.fr
[2] Department of Computer Science, ETH Zurich, Zurich, Switzerland
vtraub@ethz.ch
[3] Department of Mathematics, ETH Zurich, Zurich, Switzerland
ricoz@ethz.ch

Abstract. Finding a smallest subgraph that is k-edge-connected, or augmenting a k-edge-connected graph with a smallest subset of given candidate edges to become $(k + 1)$-edge-connected, are among the most fundamental Network Design problems. They are both APX-hard in general graphs. However, this hardness does not carry over to the planar setting, which is not well understood, except for very small values of k. One main obstacle in using standard decomposition techniques for planar graphs, like Baker's technique and extensions thereof, is that connectivity requirements are global (rather than local) properties that are not captured by existing frameworks.

We present a novel, and arguably clean, decomposition technique for such classical connectivity problems on planar graphs. This technique immediately implies PTASs for the problems of finding a smallest k-edge-connected or k-vertex-connected spanning subgraph of a planar graph for arbitrary k. By leveraging structural results for minimally k-edge-connected graphs, we further obtain a PTAS for planar k-connectivity augmentation for any constant k. We complement this with an NP-hardness result, showing that our results are essentially optimal.

Keywords: Approximation Algorithms · Planar Graphs · Network Design · Connectivity Augmentation · Baker's Technique

1 Introduction

The field of Network Design in the context of approximation algorithms has been largely shaped by basic questions on how to create graphs that fulfill certain connectivity requirements in the most economical way. Some of the most

Part of the research was done while Meike Neuwohner was affiliated with the London School of Economics and Political Science. During this time, Meike Neuwohner was supported by EPSRC grant EP/X030989/1.

S. S. Dey et al. (Eds.): IPCO 2026, LNCS 16588, pp. 47–63, 2026.
https://doi.org/10.1007/978-3-032-28691-8_4

fundamental such questions are the k-Edge/Vertex-Connected Spanning Subgraph problems and augmentation problems. They aim at creating a graph with a prescribed connectivity. The k-Edge/Vertex-Connected Spanning Subgraph problems ask about creating such a graph from scratch, whereas augmentation problems start with an existing graph and enlarge it by adding edges.

More formally, let $k \in \mathbb{Z}_{\geq 1}$. In the k-Edge-Connected Spanning Subgraph problem (k-ECSS), we are given a graph $G = (V, E)$, and the goal is to find a spanning subgraph of G that is k-edge-connected with as few edges as possible. The k-Vertex-Connected Spanning Subgraph problem (k-VCSS) requires k-vertex-connectivity instead.[1] In the k-Connectivity Augmentation problem (k-CAP), we are given a k-edge-connected graph $G = (V, E)$ together with a set $L \subseteq \binom{V}{2}$ of candidate edges to add, also called *links*, and the task is to find a subset $F \subseteq L$ of minimum size so that $(V, E \dot{\cup} F)$ is $(k+1)$-edge-connected.[2] In the weighted counterparts of these problems, k-WECSS, k-WVCSS, and k-WCAP, respectively, edges/links have non-negative costs and the goal is to minimize the total cost of the selected edges/links.

The study of these and related problems, on some of which we expand later, has a long history and led to a rich and steadily growing body of results and new techniques. We refer the interested reader to [38,45,64–66] and references therein for further information.

The problems k-WECSS, k-WVCSS, and k-WCAP are well-known to be hard to approximate, in particular APX-hard, in essentially all natural variations (see [23,28,52,59]). Therefore, the gold standard for approximation algorithms in general graphs is achieving strong constant-factor guarantees.

Another approach to these basic problems is to understand whether better approximation guarantees can be obtained in certain relevant graph classes. One particularly interesting, and arguably among the most studied, classes in this context, are planar graphs, which is also the focus of this paper. Contrary to general graphs, k-ECSS and k-VCSS, as well as further variations, are only known to be NP-hard (instead of APX-hard) in planar graphs [36], and no hardness result for k-CAP in planar graphs was known prior to this work. An additional reason to hope for stronger results in planar graphs is that many graph optimization problems that are NP-hard in general graphs, admit polynomial time approximation schemes (PTASs) in planar graphs.

Indeed, important progress has been made on PTASs for these problems in planar graphs. Unfortunately, this progress has largely been limited to small connectivity requirements. The most relevant prior results in this context for our work are the following. Provan and Burk [60] showed that planar 1-WCAP can be solved exactly in polynomial time. Moreover, Czumaj et al. [22] pre-

[1] Observe that, for $k = 1$, the problems k-ECSS and k-VCSS coincide and correspond to the Minimum Spanning Tree problem.

[2] Note that k-ECSS seeks a k-edge-connected graph whereas k-CAP aims for $(k+1)$-edge-connectivity. Despite this k vs. $k+1$ discrepancy, we stick to these definitions of k-ECSS and k-CAP as they are common in the literature.

sented PTASs for planar 2-ECSS and planar 2-VCSS.[3] The first result was later strengthened by Berger and Grigni [5], who presented a linear-time PTAS for planar 2-ECSS and a PTAS for planar 2-WECSS. Both Czumaj et al. [22] and Berger and Grigni [5] proceed by repeatedly buying cheap cycles, which allows for splitting the connectivity requirement into a requirement inside the cycle and one outside the cycle. This technique is specific to 2-edge/vertex-connectivity, and it is unknown how to generalize it to higher connectivities. More recently, Zheng [67] showed that also both planar 3-ECSS and planar 3-VCSS admit a PTAS. While some of the used techniques have parallels to ours, namely the creation of overlapping layers, the analysis remains tailored to connectivity 3, and is based on properties that fail for higher connectivities.

One of the arguably most generic frameworks to obtain PTASs for planar graph problems, which also inspired our work, grew out of Baker's technique [3] and extensions thereof, for example by Klein [50]. These approaches have been hugely successful, leading to PTASs for a large class of problems on planar graphs, including maximum independent set, minimum vertex cover, minimum dominating set, a linear time PTAS for the Traveling Salesman Problem (TSP), and many others. For optimization problems where edges have to be picked, the approaches can be divided into four main steps (where not all steps are always needed):

(i) *Sparsifier step:* Delete some edges of the input graph while approximately preserving the optimal value.
(ii) *Decomposition step:* Decompose the resulting sparsified graph into independent subproblems of bounded treewidth, by fixing/buying a cheap substructure of the solution.
(iii) *Dynamic programming step:* Solve each subproblem optimally using dynamic programming.
(iv) *Combining step:* Combine the solutions of the subproblems to obtain a solution for the whole graph.

The decomposition step typically considers the planar dual to partition the graph into layers of constant width, defined by their distance from the outer face. Such layers are known to have constant treewidth. To achieve this decomposition, one buys a small constant fraction of all edges. To make sure that this constant fraction is small in cost, the sparsifier step is often necessary, whose goal is to make sure that the cost of all edges is at most a constant factor higher than the cost of an optimal solution.

Unfortunately, this approach is known to require a certain *locality* of the considered problem, which is not satisfied by many problems with more global requirements. This need for locality has been repeatedly highlighted as a major obstacle of Baker's technique (see, for example, [25,31,50,67]), in particular when more global connectivity requirements are needed, like in k-ECSS. There are some notable exceptions, which, however, do not apply to our setting of

[3] They also presented PTASs for the weighted versions when the ratio of the total edge cost to the optimum solution cost is bounded.

higher edge-connectivities. One is the framework of bidimensionality introduced by Demaine and Hajiaghayi [24], which leads to a PTAS for connected dominating set, through a technique that can be interpreted as a non-trivial extension of Baker's approach. Another one is the ubiquity framework of Cohen-Addad et al. [21] and an extension thereof by Fomin et al. [30], which, if certain conditions are met by the problem under consideration, leads to a PTAS for planar graphs. As mentioned, a Baker-type approach has been used to obtain a linear time PTAS for planar TSP [50], improving on prior PTASs [2,40] not relying on Baker's technique. Klein avoids the use of higher connectivities by phrasing TSP as the problem of finding a (singly) connected graph with only even degrees. In summary, even 2-edge-connectivity falls outside the classical realm of Baker's framework (and is also neither captured by the bidimensionality [24] nor the ubiquity framework [21,30]).

The goal of this paper is to give a novel and arguably clean Baker-type technique that, for the first time, handles larger connectivity requirements, thus overcoming the locality issue of Baker's approach for these settings.

1.1 Our Results

We start with our main result for planar connectivity augmentation, which works also for weighted instances as long as the *cost ratio*, the ratio between the maximum and minimum cost of any link, is bounded by a constant.

Theorem 1. *For any fixed $k \in \mathbb{Z}_{\geq 1}$, there is a PTAS for planar k-WCAP for bounded cost ratio instances.*

Prior to our work, an exact algorithm was known for planar 1-WCAP [60]. For larger k, we are unaware of any planar k-WCAP algorithms that improve over the best-known algorithms for general graphs [63,64], which only give constant-factor guarantees.

We complement the above with the following hardness result for planar k-CAP, which we derive through a reduction from planar 3-SAT.

Theorem 2. *For any $k \geq 2$, planar k-CAP is NP-hard.*

Prior to our work, it was unknown whether planar k-CAP is NP-hard for any $k \geq 2$. The presented PTAS together with the above hardness result therefore essentially settles the approximability of planar k-CAP.

Interestingly, classical Baker-type approaches for (even unweighted) planar k-CAP, as outlined on page 3, face two important issues. First, it is highly unclear how to sparsify a planar k-CAP instance. This is a major obstacle, because an optimal k-CAP solution may cost a negligible amount compared to the cost of all links. (See discussion and example in Sect. 3.) Second, as previously mentioned, prior decomposition techniques do not apply to higher connectivities. This is why also for the (unweighted) k-ECSS problem, where sparsification turns out to be easy, a PTAS was only known for $k \leq 3$. We resolve these issues through a combination of structural insights and a new decomposition technique.

Our new decomposition also readily applies to planar k-ECSS and its vertex-connectivity counterpart, planar k-VCSS, leading to the following result.

Theorem 3. *For any fixed $k \in \mathbb{Z}_{\geq 1}$, planar k-WECSS and planar k-WVCSS, both with bounded cost ratios, admit linear-time PTASs.*

Note that a planar graph can be at most 5-vertex-connected because it contains a vertex with at most 5 neighbors.[4] Hence, a PTAS for planar k-VCSS is only relevant for $k \leq 5$.

Combining the above result with known k-ECSS techniques that are strong for large k, we can lift the restriction of k being a fixed constant at the expense of losing linear-time efficiency.

Corollary 4. *Planar k-WECSS admits a PTAS when the cost ratio is bounded and k is part of the input.*

Prior to our work, PTASs for planar k-ECSS and planar k-VCSS were only known for $k \leq 3$ [5,22,67]. (Where [5] presented a PTAS for planar 2-WECSS, without restrictions on the cost ratio.)

1.2 Further Related Results

Augmentation problems have been heavily studied in general graphs and many special graph classes (see [1,12–14,16,17,20,27,29,32,39,47,53–55,57,58,63,64] and references therein), including planar graphs [41,61]. The (unweighted) augmentation variant for planar graphs studied in [41] is a variation of k-CAP. In particular, they fix a subset of the vertices $W \subseteq V$ and allow edges to be added between *any* pair of vertices in W as long as planarity of the whole graph is preserved. Their goal is to make the subgraph induced by W (singly) connected. Similarly, Rutter and Wolff [61] show NP-hardness for making a planar graph 2-edge-connected if one is allowed to add edges between *any* pair of vertices, as long as the resulting graph remains planar.

In general (not necessarily planar) graphs, k-ECSS, for any $k \in \mathbb{Z}_{\geq 2}$, is known not to admit a PTAS unless P = NP [23,28,59]. For the weighted version, k-WECSS, the best constant-factor approximation algorithm is an over four decades old 2-approximation [32,33], which can also be matched with more recent techniques, like primal-dual [38] or iterative rounding [45]. Progress beyond the factor 2 has been achieved for (the unweighted) 2-ECSS, where successive improvements [11,18,37,44,48,51,62] led to the currently best approximation factor of slightly below $5/4$ [43].

For arbitrary $k \geq 2$, it is known how to get $(1 + \mathcal{O}(1/k))$-approximation algorithms [19,34,35].

We note that the difficulties of obtaining results for larger connectivities faced here are not shared by some other variants of the problem, for example

[4] And there are planar graphs that are actually 5-vertex-connected, like the icosahedral graph. This is the graph formed by the vertices and edges of an icosahedron, one of the 5 platonic solids.

when multiple copies of the same edge can be bought, which is known as the (weighted) k-edge-connected spanning multi-subgraph problem (k-(W)ECSM). Indeed, one can for example extend the idea to contract cheap edge sets to simplify the instance. More precisely, if one can identify an edge set U of cost at most $\mathcal{O}(\varepsilon/k\, c(\mathrm{OPT}))$, whose contraction simplifies the instance, then one can buy k many times each edge of U and contract U. Exploiting this, Demaine et al. [26] presented a PTAS for planar k-WECSM, for any fixed k. Together with known k-WECSM results for the non-planar case, like the $(1 + \mathcal{O}(1/k))$-approximation algorithm of [42], this implies a PTAS for planar k-WECSM even when k is part of the input.[5]

Moreover, we would like to highlight that there has been a very fruitful line of work on further survivable network design problems and related problems on planar graphs (and graphs of bounded genus), including Steiner Tree and generalizations thereof (we refer the interested reader to [4,6–10,49] and references therein).

Finally, we note that the generality of Baker's technique has led to an interesting attempt to provide a syntactic characterization of problems that admit a PTAS on planar graphs through components based on Baker's technique [46].

1.3 Organization of the Paper

In Sect. 2, we provide an overview of our new decomposition technique, which we call *k-edge-safe cover*, using planar k-ECSS as an example. This allows us to use Baker-type techniques for higher connectivities. Section 3 discusses how further structural insights together with this technique can be used to obtain our augmentation result, Theorem 1. Due to space constraints, missing proofs are deferred to the long version of the paper (see https://arxiv.org/abs/2512.21128).

2 Instance Decomposition via *k*-Edge-Safe Covers

We now exemplify our decomposition technique for connectivity problems on planar graphs, which can be thought of as an extension of Baker's framework. For the sake of simplicity of presentation, we focus on planar k-WECSS in this overview. The same approach can be used for k-WVCSS and we will use it, together with additional techniques, to obtain our main result on planar connectivity augmentation problems.

Let $k \in \mathbb{Z}_{\geq 1}$, and let $G = (V, E)$ be a planar graph with positive edge costs $c \colon E \to \mathbb{R}_{>0}$. Moreover, we denote by $\Delta := \max_{e \in E} c(e)/\min_{e \in E} c(e)$ the max-to-min cost ratio of the instance. As in the classical framework of Baker, we aim at decomposing the problem into independent instances of bounded treewidth, which are solvable in linear time by a result of Chalermsook et al. [15].

[5] Together with the above-mentioned $(1+\mathcal{O}(1/k))$-approximation for k-WECSM, this leads to a PTAS for planar k-WECSM (with k part of the input) as follows. If ε is larger than the $\mathcal{O}(1/k)$ term, then one can simply use the $(1+\mathcal{O}(1/k))$-approximation algorithm for planar k-WECSM; otherwise, k is a constant and one can use the PTAS for planar k-WECSM by Demaine et al. [26].

Theorem 5 (see [15], Theorem 3). *For any fixed $k \in \mathbb{Z}_{\geq 1}$, planar k-WECSS and planar k-WVCSS can be solved in linear time in graphs of bounded treewidth.*

Continuing with classical Baker, we will buy a small set of edges $E_{\mathcal{U}}$ (the choice of notation will become clear soon), so that the problem decomposes into independent instances of small treewidth. Interestingly, despite the fact that they can be solved independently of each other, they are actually not disjoint in terms of edges/vertices; this is in contrast to classical applications of Baker's framework. Another key difference to prior approaches is that the edges we buy do not have the desired property themselves, i.e., the connected components of $E_{\mathcal{U}}$ are not necessarily k-edge-connected. Instead, we take a different perspective. Recall that a graph is k-edge-connected if every cut has at least k crossing edges. We want to buy an edge set $E_{\mathcal{U}}$ and decompose the graph into (overlapping) parts, so that every cut that is not fully contained in one of the parts has at least k edges in $E_{\mathcal{U}}$ that cross it. Hence, it remains to take care of cuts fully contained in one of the parts, which means that it suffices to k-edge-connect the parts, which decomposes the problem. We now formalize this idea.

We start by observing that, to check k-edge-connectivity of a connected graph, it suffices to check that every *connected cut* is crossed by at least k edges. Connected cuts, also sometimes called *simple cuts*, are defined as follows.

Definition 6 (Connected Cut). *Let $G = (V, E)$ be a graph. A nonempty vertex set $S \subsetneq V$ is a connected cut (in G) if both induced subgraphs $G[S]$ and $G[V \setminus S]$ are connected graphs.*

A connected graph G is k-edge-connected if and only if every connected cut has at least k crossing edges.[6] Based on this observation, we introduce the notion of a *k-edge-safe cover*, which allows us to decompose a k-WECSS instance.

Definition 7. *Let $G = (V, E)$ be a graph and let $\mathcal{U} = (U_i)_{i \in I}$ be a collection of nonempty subsets of V that cover V. Let $V_{\mathcal{U}}$ be the set of vertices that appear in at least 2 distinct U_i's and let $E_{\mathcal{U}}$ be the set of edges incident to a vertex in $V_{\mathcal{U}}$.*

We call $\mathcal{U}$ a k-edge-safe cover of G *if for every connected cut S, we have*

(i) $|\delta(S) \cap E_{\mathcal{U}}| \geq k$, *or*
(ii) $\delta(S) \subseteq E[U_i]$ *for some $i \in I$.*

The definition of a k-edge-safe cover aims at providing the conditions we need to decompose the problem after buying the edges $E_{\mathcal{U}}$. In the lemma below, $(V, F_i)/(E[V \setminus U_i])$ denotes the graph obtained from (V, F_i) by contracting all edges in $E[V \setminus U_i]$, i.e., we contract all edges with both endpoints in $V \setminus U_i$.[7]

[6] To see this, note that in a connected graph $G = (V, E)$, an arbitrary cut, which is a nonempty vertex set $S \subsetneq V$, is connected if and only if it is minimal in the following sense: there is no other cut $\overline{S}$ with $\delta(\overline{S}) \subsetneq \delta(S)$, where $\delta(W)$, for $W \subseteq V$, contains all edges with precisely one endpoint in W.

[7] An edge contraction merges the endpoints of the edge into a single vertex and deletes resulting loops.

Lemma 8. *Let $G = (V, E)$ be a connected graph, and let $\mathcal{U} = (U_i)_{i \in I}$ be a k-edge-safe cover of G. Let $(F_i)_{i \in I}$ be edge sets such that $(V, F_i)/(E[V \setminus U_i])$ is k-edge-connected. Let $F := E_{\mathcal{U}} \cup \bigcup_{i \in I}(F_i \cap E[U_i])$. Then (V, F) is k-edge-connected.*

Lemma 8 shows that, after we buy the edges in $E_{\mathcal{U}}$, it indeed suffices to find k-edge-connected subgraphs independently in each graph $(V, E)/(E[V \setminus U_i])$. For such a solution to be competitive with respect to an optimum solution in the original graph (V, E), we need that both (i) $E_{\mathcal{U}}$ is cheap, and (ii) after having bought $E_{\mathcal{U}}$, the total cost of optimum k-ECSS solutions in $(V, E)/(E[V \setminus U_i])$ compares to the cost of an optimum solution in (V, E). We show that the second condition is satisfied by any k-edge-safe cover. (The first one will be fulfilled by choosing the k-edge-safe cover appropriately, which we discuss afterwards.)

Lemma 9. *Let $G = (V, E)$ be a k-edge-connected graph with edge costs $c \colon E \to \mathbb{R}_{\geq 0}$, and let $\mathcal{U} = (U_i)_{i \in I}$ be a k-edge-safe cover of G. For $i \in I$, let $G_i := (V, E)/(E[V \setminus U_i])$ and obtain c_i from c by setting the costs of all edges in $E_{\mathcal{U}}$ and in $E(G_i) \setminus E[U_i]$ to zero. Let OPT_i be an optimum k-WECSS solution for (G_i, c_i). Then, $F := E_{\mathcal{U}} \cup \bigcup_{i \in I}(\mathrm{OPT}_i \cap E[U_i])$ is a k-WECSS solution in G of cost*

$$c(F) \leq c(\mathrm{OPT}) + c(E_{\mathcal{U}}),$$

where OPT *is an optimum k-WECSS solution in (G, c).*

Proof (Sketch of Proof). By Lemma 8, F is a feasible k-WECSS solution in G. Moreover, for each $i \in I$, $T_i := \mathrm{OPT} \setminus E[V \setminus U_i]$ is a feasible k-WECSS solution in (G_i, c_i), implying $c_i(\mathrm{OPT}_i) \leq c_i(T_i) = c((\mathrm{OPT} \cap E[U_i]) \setminus E_{\mathcal{U}})$. Thus, using $E[U_i] \cap E[U_j] = E[U_i \cap U_j] \subseteq E_{\mathcal{U}}$ for every $i, j \in I$ with $i \neq j$, we get

$$c(F) \leq c(E_{\mathcal{U}}) + \sum_{i \in I} c((\mathrm{OPT}_i \cap E[U_i]) \setminus E_{\mathcal{U}}) \leq c(E_{\mathcal{U}}) + c(\mathrm{OPT}).$$

□

Hence, decomposing the problem using k-edge-safe covers allows us to find a solution that is competitive with respect to an optimum solution in the original graph, as long as we ensure that $c(E_{\mathcal{U}})$ is small. Moreover, to be able to solve the resulting instances efficiently, we want the graphs G_i to have small treewidth. The following statement shows that such k-edge-safe covers do exist, even without requiring G to be k-edge-connected.[8] In terms of the cost of $E_{\mathcal{U}}$, the statement only shows that it can be made arbitrarily small compared to $c(E)$, which we compare afterwards against $c(\mathrm{OPT})$.

Theorem 10. *Let $\varepsilon \in (0, 1)$. Given a planar connected graph $G = (V, E)$ with edge costs $c \colon E \to \mathbb{R}_{\geq 0}$, we can, in time $\mathcal{O}(|V| + |E|)$, compute a k-edge-safe cover $\mathcal{U} = (U_i)_{i \in I}$ of G, together with the graphs $G_i := (V, E)/(E[V \setminus U_i])$ for $i \in I$, such that*

[8] However, if the graph G is not k-edge-connected, then we won't be able to find a feasible k-ECSS solution in each of the G_i.

(i) $c(E_{\mathcal{U}}) \leq \varepsilon \cdot c(E)$, *and*
(ii) G_i *has treewidth at most* $\frac{26k}{\varepsilon}$ *for each* $i \in I$.

Our construction of a k-edge-safe cover to obtain Theorem 10 can be interpreted as an extension of Baker's framework. Loosely speaking, each U_i is a ring of vertices whose planar-dual distance to a fixed root lies in a bounded range. In order to be able to extend our results to the vertex-connectivity setting, we use the following slightly modified version of the dual graph: the *vertex-face graph* D_G for a fixed planar embedding of $G = (V, E)$ has vertex set $V \cup F$, where F is the set of faces, and there is an edge between $v \in V$ and $f \in F$ if vertex v is on face f. Then we fix a root $r \in V$ and partition V into rings R_j (for $j = 1, 2, \dots$) of the form $R_j = \{v \in V : \mathrm{dist}_{D_G}(r, v) \in [2\alpha_j, 2\alpha_{j+1})\}$, where the integers α_j are chosen such that the *width* $\alpha_{j+1} - \alpha_j$ of each ring R_j is $\mathcal{O}(k)$. Each U_i is then defined as the union of $\mathcal{O}(\frac{1}{\varepsilon})$ many consecutive rings R_j, where U_i and U_{i+1} are overlapping in one ring R_j.

To put the pieces together, it remains to show $c(E) = \mathcal{O}(c(\mathrm{OPT}))$. In general, for k-WECSS, even with bounded cost ratio, this is not true, as there could be many parallel edges that increase the cost of E without contributing to OPT. However, we can assume without loss of generality that a k-WECSS instance has at most k parallel edges between any pair of vertices. Under this additional assumption, we have $c(E) \leq 6 \cdot \Delta \cdot c(\mathrm{OPT})$ because any simple planar graph on n vertices has fewer than $3n$ edges and OPT must have at least $\frac{n \cdot k}{2}$ edges.

Combining these results yields our PTAS for planar k-WECSS with fixed k and bounded cost ratio.

3 A PTAS for Planar *k*-WCAP

As k-WCAP constitutes a special case of $(k+1)$-WECSS, it appears tempting to follow the same strategy as in the previous section: Given an instance $(G = (V, E), L, c)$ of planar k-WCAP, first apply Theorem 10 to obtain a $(k+1)$-edge-safe cover $\mathcal{U} = (U_i)_{i \in I}$ for $G + L := (V, E \dot{\cup} L)$ such that the total cost of $L_{\mathcal{U}}$, the set of links incident to $V_{\mathcal{U}}$, is bounded by $\varepsilon \cdot c(L)$. Next, for $i \in I$, solve the k-WCAP optimally in the subinstance (G_i, L_i, c_i) of bounded treewidth that arises from (G, L, c) by contracting $(E \dot{\cup} L)[V \setminus U_i]$. Finally, combine the obtained solutions with the set $L_{\mathcal{U}}$ to a solution to the whole instance.

Unfortunately, this approach does not result in a PTAS, even when all links have unit costs. The reason for this is that $c(L)$, the total cost of the link set, can be much larger than the cost of an optimum solution, rendering it too expensive to purchase all of $L_{\mathcal{U}}$ for our solution.[9] To see this, consider the instance of planar 3-CAP shown in Fig. 1. Adding a single link suffices to render the graph 4-edge-connected; however, the total number of links may be linear in the number of vertices.

[9] Note that this does not contradict the results in the previous section: An instance of k-CAP corresponds to an instance of $(k+1)$-WECSS with $\{0, 1\}$-costs; such an instance does not have a bounded cost ratio.

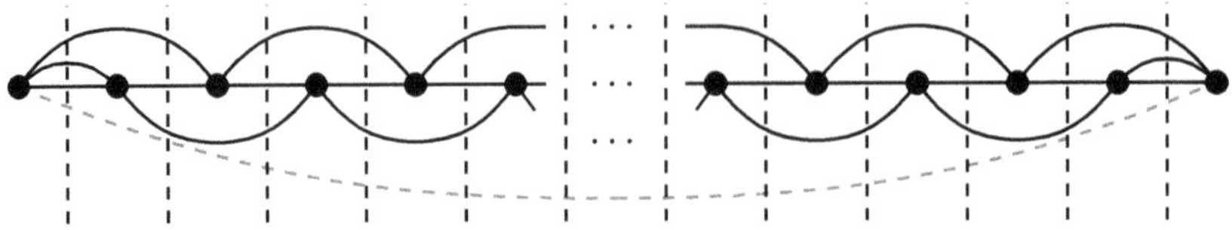

Fig. 1. A 3-edge-connected planar graph G, with 3-cuts indicated by dashed vertical lines. One link (dashed, red) suffices to cover all 3-cuts in G. (Color figure online)

We show that structures similar to the one shown in Fig. 1, which we call *snug paths*, form the only obstacle in applying the above strategy to obtain a PTAS for planar k-WCAP with bounded cost ratio. More precisely, we prove that if an instance of planar k-WCAP with bounded cost ratio does not contain any snug paths, then the total cost of all links can be bounded in terms of the cost of an optimum solution. To obtain this result, we will assume that a given instance (G, L, c) of planar k-WCAP is feasible (i.e., $G + L$ is $(k+1)$-edge-connected) and has the property that G is *minimally k-edge-connected*, meaning that every edge of G appears in a cut of size k. Both assumptions are without loss of generality: Feasibility can be checked in polynomial time and if G is not minimally k-edge-connected, we can contract all edges that do not appear in any k-cut. Note that contractions preserve planarity.

3.1 Identifying the Problem: Snug Vertices and Snug Paths

For the remainder of this section, we fix a planar k-WCAP instance $(G = (V, E), L, c)$, where G is minimally k-edge-connected, and an arbitrary *root* $r \in V$. We call a vertex $v \in V \setminus \{r\}$ *snug* if $\deg(v) \geq k+1$ (where $\deg(v)$ denotes the degree of v) and there exist two distinct k-cuts $S_1, S_2 \subseteq V \setminus \{r\}$ with $S_2 = S_1 \cup \{v\}$. We denote the set of snug vertices by V_{snug}. One can show that for $v \in V_{\text{snug}}$, there exists a *unique* pair of distinct k-cuts (S_1^v, S_2^v) with $S_1^v, S_2^v \subseteq V \setminus \{r\}$ and $S_2^v = S_1^v \cup \{v\}$, which we call the *snug shores* for v. In Fig. 1, every vertex except for the outer two vertices is snug, its snug shores are the k-cuts drawn to its left and right; the snug vertex "snugly fits in-between".

To capture relations between snug vertices and their snug shores, we define the directed *chain graph* $G_{\text{chain}} = (V_{\text{snug}}, A)$, where $(u, v) \in A$ iffx $S_2^u = S_1^v$ (the outer snug shore of u equals the inner snug shore of v). In Fig. 1, G_{chain} is a single directed path from left to right, visiting all but the two outer vertices. Moreover, G_{chain} is computable in polynomial time, e.g., by enumerating all minimum cuts of G [56]. We show the following.

Lemma 11. *G_{chain} is a vertex-disjoint union of directed paths.*

Lemma 12. *The underlying undirected graph of G_{chain} is a subgraph of G.*

Lemma 13. *An arc $(u, v) \in A$ does not cross any k-cut other than $S_2^u = S_1^v$ (and $V \setminus S_2^u$).*

We call the maximal paths in G_{chain} the *snug paths* of G and denote their collection by $\mathcal{P}^G_{\text{chain}}$. Lemma. 14 formalizes our claim that snug paths constitute the only obstacle in bounding $c(L)$.

Lemma 14. *We have* $|V \setminus V_{\text{snug}}| < 4 \cdot n_k(G)$ *and* $|\mathcal{P}^G_{\text{chain}}| < 2 \cdot n_k(G)$, *where* $n_k(G)$ *denotes the number of vertices of* G *of degree* k.

Observe that if G_{chain} does not contain any arcs, i.e., every snug path consists of a single snug vertex, then Lemma 14 and planarity of $G+L$ allow us to linearly bound $|L|$ in terms of $n_k(G)$. On the other hand, in an optimum solution OPT, every vertex of degree k must be incident to at least one link in OPT, implying $2 \cdot |\text{OPT}| \geq n_k(G)$. As the cost ratio of the instance is bounded, $c(L)$ can then be at most a constant factor larger than $c(\text{OPT})$.

3.2 Solving the Problem: Contracting Snug Paths

Lemma 14 tells us that in the graph $G/\mathcal{P}^G_{\text{chain}}$ that arises from G by *contracting every snug path into a single vertex*, the total number of vertices will be in $\mathcal{O}(|\text{OPT}|)$. Note that by Lemma 12, these contractions preserve planarity. By Lemma 13, contracting $\mathcal{P}^G_{\text{chain}}$ preserves all k-cuts except for snug shores.

This suggests the following modified strategy: Apply Theorem 10 to $\widetilde{G}_{\mathcal{P}} := (V_{\mathcal{P}}, E_{\mathcal{P}} \dot{\cup} L_{\mathcal{P}}) := (G+L)/\mathcal{P}^G_{\text{chain}}$, the graph arising from $G+L$ by the contraction of snug paths. This yields a $(k+1)$-edge-safe cover $\mathcal{U}^{\mathcal{P}} = (U_i^{\mathcal{P}})_{i \in I}$ of $\widetilde{G}_{\mathcal{P}}$ such that each subinstance $\widetilde{G}_{\mathcal{P}}/((E_{\mathcal{P}} \dot{\cup} L_{\mathcal{P}})[V_{\mathcal{P}} \setminus U_i^{\mathcal{P}}])$ has bounded treewidth and such that the set of links incident to $\bigcup_{i,j \in I : i \neq j}(U_i^{\mathcal{P}} \cap U_j^{\mathcal{P}})$ is cheap. Lift $\mathcal{U}^{\mathcal{P}}$ to a "cover" $\mathcal{U}$ for G.

We remark that we cannot hope for $\mathcal{U}$ to be a $(k+1)$-edge-safe cover for $G+L$; this is because we cannot control the interaction between snug paths and connected cuts of size more than k. Instead, we introduce the weaker notion of a *k-augmentation-safe cover*, which is tailored to the augmentation setting and only considers k-cuts of G rather than all connected cuts.

To obtain a good k-augmentation-safe cover by lifting a $(k+1)$-edge-safe cover in $\widetilde{G}_{\mathcal{P}}$, there are multiple challenges. First of all, we have to discuss how to solve the k-WCAP in each of the subinstances we obtain in the uncontracted instance, which may no longer have bounded treewidth.

To this end, we introduce the concept of *snug-treewidth* and show that instances with constant snug-treewidth can be solved efficiently.[10]

Definition 15. *The* snug-treewidth $\text{snugtw}(G, L)$ *of an instance* (G, L, c) *of k-WCAP is the treewidth of* $(G+L)/\mathcal{P}^G_{\text{chain}}$.

Theorem 16. *For any constant* $\tau \in \mathbb{Z}_{\geq 0}$, *k-WCAP can be solved in polynomial time on instances of snug-treewidth at most* τ.

---

[10] We remark that $\mathcal{P}^G_{\text{chain}}$ depends on the (fixed) choice of the root that we implicitly make. While this dependence can be avoided, it simplifies the presentation.

While Theorem 10 allows us to compute a $(k+1)$-edge-safe cover $\mathcal{U}^{\mathcal{P}}$ in $\widetilde{G}_{\mathcal{P}}$ such that each piece $\widetilde{G}_{\mathcal{P}}/((E_{\mathcal{P}}\dot{\cup}L_{\mathcal{P}})[V_{\mathcal{P}}\setminus U_i^{\mathcal{P}}])$ has bounded treewidth, we point out that its lift $\mathcal{U}$ to G will not necessarily have bounded snug-treewidth in each piece. The reason for this is that a snug path of G contained in $G/((E\dot{\cup}L)[V\setminus U_i])$ does not necessarily constitute a snug path of this subinstance. The latter can only be guaranteed for snug paths that are "far enough from the boundary of U_i", ensuring that the corresponding snug shores also correspond to k-cuts in $G/((E\dot{\cup}L)[V\setminus U_i])$. Similarly, we cannot even guarantee the properties of a k-augmentation-safe cover for snug shores too close to the "boundaries" of the U_i. To deal with these issues, we have to "buy" a set of additional links covering the snug shores of the collection $\mathcal{Q}$ of snug paths situated close to the "boundaries" of the sets $(U_i)_{i\in I}$, making sure that we can safely keep these paths contracted.

Another challenge we need to address is how to bound the total cost of the links present in the contracted instance by $\mathcal{O}(c(\mathrm{OPT}))$ to make sure that the link set $L_{\mathcal{U}}$ is cheap. We point out that this is not an immediate consequence of Lemma 14. The reason for this is that the contraction produces parallel links, and to ensure feasibility after un-contracting, it is not sufficient to keep only a cheapest copy. Instead, we work with a carefully "thinned" link set $L^{\star}$, ensuring that $L^{\star}_{\mathcal{U}}$ is cheap, and combining it with solutions to the subinstances guarantees feasibility. Our main technical result is Theorem 17.

The link set $L^{\star\star}$ in the theorem below contains both $L^{\star}_{\mathcal{U}}$ and the above-mentioned links we buy to cover the snug shores of the collection $\mathcal{Q}$ of snug paths situated close to the boundaries.

Theorem 17. *Let $\varepsilon\in(0,1)$. Given an instance (G,L,c) of planar k-WCAP with cost ratio Δ, we can, in polynomial time, compute $\mathcal{Q}\subseteq\mathcal{P}^{G}_{\text{chain}}$, link sets $L^{\star},L^{\star\star}\subseteq L$ such that the graph $(G+L^{\star})/\mathcal{Q}$ is $(k+1)$-edge-connected, and a k-augmentation-safe cover $\mathcal{U}=(U_i)_{i\in I}$ for the instance $(G_{\mathcal{Q}}=(V_{\mathcal{Q}},E_{\mathcal{Q}}),L^{\star}_{\mathcal{Q}},c_{\mathcal{Q}}):=(G,L^{\star},c)/\mathcal{Q}$ such that:*

(i) $|I|\in\mathcal{O}(|V(G)|)$;
(ii) *For $i\in I$, the subinstance $(G^i_{\mathcal{Q}},L^i_{\mathcal{Q}},c^i_{\mathcal{Q}}):=(G_{\mathcal{Q}},L^{\star}_{\mathcal{Q}},c_{\mathcal{Q}})/((E_{\mathcal{Q}}\dot{\cup}L^{\star}_{\mathcal{Q}})[V_{\mathcal{Q}}\setminus U_i])$ has snug-treewidth $\mathcal{O}(\frac{k\Delta^2}{\varepsilon^2})$;*
(iii) *For each $i\in I$, let $\mathrm{OPT}_i\subseteq L^{\star}$ correspond to an optimum solution to the subinstance $(G^i_{\mathcal{Q}},L^i_{\mathcal{Q}},c^i_{\mathcal{Q}})$. Then $F:=(\bigcup_{i\in I}\mathrm{OPT}_i)\cup L^{\star\star}$ is a solution to the instance (G,L,c) with $c(F)\le(1+\varepsilon)\cdot c(\mathrm{OPT})$.*

Data Availability Statement. No data are associated with this article. Data sharing is not applicable to this article.

No Competing Interest Statement Required By Publisher. The authors have no competing interests to declare that are relevant to the content of this article.

References

1. Adjiashvili, D.: Beating approximation factor two for weighted tree augmentation with bounded costs. ACM Tran. Algorithms **15**(2), 19:1–19:26 (2018). https://doi.org/10.1145/3182395
2. Arora, S., Grigni, M., Karger, D., Klein, P., Woloszyn, A.: A polynomial-time approximation scheme for weighted planar graph TSP. In: Proceedings of the 9th Annual ACM-SIAM Symposium on Discrete Algorithms (SODA), pp. 33–41 (1998)
3. Baker, B.S.: Approximation algorithms for NP-complete problems on planar graphs. J. ACM **41**(1), 153–180 (1994)
4. Bateni, M., Hajiaghayi, M., Marx, D.: Approximation schemes for steiner forest on planar graphs and graphs of bounded treewidth. J. ACM **58**(5), 21:1–21:37 (2011). https://doi.org/10.1145/2027216.2027219
5. Berger, A., Grigni, M.: Minimum weight 2-edge-connected spanning subgraphs in planar graphs. In: Automata, Languages and Programming (ICALP), pp. 90–101 (2007). https://doi.org/10.1007/978-3-540-73420-8_10
6. Borradaile, G., Kenyon-Mathieu, C., Klein, P.: A polynomial-time approximation scheme for Steiner tree in planar graphs. In: Proceedings of the 18th Annual ACM-SIAM Symposium on Discrete Algorithms (SODA), pp. 1285–1294 (2007)
7. Borradaile, G., Klein, P., Mathieu, C.: An O(n log n) approximation scheme for Steiner tree in planar graphs. ACM Trans. Algorithms **5**(3), 31:1–31:31 (2009). https://doi.org/10.1145/1541885.1541892
8. Borradaile, G., Zheng, B.: A PTAS for three-edge-connected survivable network design in planar graphs. In: Proceedings of Approximation, Randomization, and Combinatorial Optimization. Algorithms and Techniques (APPROX/ RANDOM 2017). vol. 81, pp. 3:1–3:13 (2017). https://doi.org/10.4230/LIPIcs.APPROX-RANDOM.2017.3
9. Borradaile, G., Demaine, E.D., Tazari, S.: Polynomial-time approximation schemes for subset-connectivity problems in bounded-genus graphs. Algorithmica **68**(2), 287–311 (2012). https://doi.org/10.1007/s00453-012-9662-2
10. Borradaile, G., Klein, P.: The two-edge connectivity survivable-network design problem in planar graphs. ACM Trans. Algorithms **12**(3), 30:1–30:29 (2016). https://doi.org/10.1145/2831235
11. Bosch-Calvo, M., Garg, M., Grandoni, F., Hommelsheim, F., Ameli, A.J., Lindermayr, A.: A 5/4-approximation for two-edge connectivity. In: Proceedings of the 57th Annual ACM Symposium on Theory of Computing (STOC), pp. 653–664 (2025). https://doi.org/10.1145/3717823.3718275
12. Byrka, J., Grandoni, F., Jabal Ameli, A.: Breaching the 2-approximation barrier for connectivity augmentation: a reduction to steiner tree. In: Proceedings of 52nd ACM Symposium on Theory of Computing (STOC) (2020). https://doi.org/10.1145/3357713.3384301
13. Cecchetto, F., Traub, V., Zenklusen, R.: Bridging the gap between tree and connectivity augmentation: unified and stronger approaches. In: Proceedings of 53rd Annual ACM Symposium on Theory of Computing (STOC), pp. 370–383 (2021). https://doi.org/10.1145/3406325.3451086
14. Cecchetto, F., Traub, V., Zenklusen, R.: Better-than-$\frac{4}{3}$-approximations for leaf-to-leaf tree and connectivity augmentation. Math. Program. **207**(1), 515–549 (2024). https://doi.org/10.1007/s10107-023-02018-3

15. Chalermsook, P., Das, S., Even, G., Laekhanukit, B., Vaz, D.: Survivable network design for group connectivity in low-treewidth graphs. In: Approximation, Randomization, and Combinatorial Optimization. Algorithms and Techniques (APPROX/RANDOM 2018). vol. 116, pp. 8:1–8:19 (2018). https://doi.org/10.4230/LIPIcs.APPROX-RANDOM.2018.8
16. Cheriyan, J., Gao, Z.: Approximating (unweighted) tree augmentation via lift-and-project, Part I: stemless TAP. Algorithmica **80**(2), 530–559 (2017). https://doi.org/10.1007/s00453-016-0270-4
17. Cheriyan, J., Gao, Z.: Approximating (Unweighted) tree augmentation via lift-and-project, Part II. Algorithmica **80**(2), 608–651 (2017). https://doi.org/10.1007/s00453-017-0275-7
18. Cheriyan, J., Sebő, A., Szigeti, Z.: Improving on the 1.5-approximation of a smallest 2-edge connected spanning subgraph. SIAM J. Discret. Math. **14**(2), 170–180 (2001)
19. Cheriyan, J., Thurimella, R.: Approximating minimum-size k-connected spanning subgraphs via matching. SIAM J. Comput. **30**(2), 528–560 (2000). https://doi.org/10.1137/S009753979833920X
20. Cohen, N., Nutov, Z.: A $(1 + \ln 2)$-approximation algorithm for minimum-cost 2-edge-connectivity augmentation of trees with constant radius. Theoret. Comput. Sci. **489**, 67–74 (2013). https://doi.org/10.1016/j.tcs.2013.04.004
21. Cohen-Addad, V., Colin de Verdière, É., Klein, P.N., Mathieu, C., Meierfrankenfeld, D.: Approximating connectivity domination in weighted bounded-genus graphs. In: Proceedings of the 48th Annual ACM Symposium on Theory of Computing (STOC), pp. 584–597 (2016). https://doi.org/10.1145/2897518.2897635
22. Czumaj, A., Grigni, M., Sissokho, P., Zhao, H.: Approximation schemes for minimum 2-edge-connected and biconnected subgraphs in planar graphs. In: Proceedings of the 15th Annual ACM-SIAM Symposium on Discrete Algorithms (SODA), pp. 496–505 (2004)
23. Czumaj, A., Lingas, A.: On approximability of the minimum-cost k-connected spanning subgraph problem. In: Proceedings of the 10th Annual ACM-SIAM Symposium on Discrete Algorithms (SODA), pp. 281–290 (1999)
24. Demaine, E.D., Hajiaghayi, M.: Bidimensionality: new connections between FPT algorithms and PTASs. In: Proceedings of the Sixteenth Annual ACM-SIAM Symposium on Discrete Algorithms (SODA), pp. 590–601 (2005)
25. Demaine, E.D., Hajiaghayi, M.: Approximation schemes for planar graph problems. In: Encyclopedia of Algorithms, pp. 59–62. Springer, Boston, MA (2008). https://doi.org/10.1007/978-0-387-30162-4_32
26. Demaine, E.D., Hajiaghayi, M., Mohar, B.: Approximation algorithms via contraction decomposition. Combinatorica **30**(5), 533–552 (2010). https://doi.org/10.1007/s00493-010-2341-5
27. Even, G., Feldman, J., Kortsarz, G., Nutov, Z.: A 1.8 approximation algorithm for augmenting edge-connectivity of a graph from 1 to 2. ACM Trans. Algorithms **5**(2), 21:1–21:17 (2009). https://doi.org/10.1145/1497290.1497297
28. Fernandes, C.G.: A better approximation ratio for the minimum size$k$-edge-connected spanning subgraph problem. J. Algorithms **28**(1), 105–124 (1998). https://doi.org/10.1006/jagm.1998.0931
29. Fiorini, S., Groß, M., Könemann, J., Sanità, L.: Approximating weighted tree augmentation via chvátal-gomory cuts. In: Proceedings of the 29th Annual ACM-SIAM Symposium on Discrete Algorithms (SODA), pp. 817–831 (2018). https://doi.org/10.1137/1.9781611975031.53

30. Fomin, F.V., Lokshtanov, D., Saurabh, S., Zehavi, M.: Approximation schemes via width/weight trade-offs on minor-free graphs. In: Proceedings of the 2020 ACM-SIAM Symposium on Discrete Algorithms (SODA), pp. 2299–2318 (2019). https://doi.org/10.1137/1.9781611975994.141
31. Fox-Epstein, E., Klein, P.N., Schild, A.: Embedding planar graphs into low-treewidth graphs with applications to efficient approximation schemes for metric problems. In: Proceedings of the 30th Annual ACM-SIAM Symposium on Discrete Algorithms (SODA), pp. 1069–1088 (2019)
32. Frederickson, G.N., JáJá, J.: Approximation algorithms for several graph augmentation problems. SIAM J. Comput. **10**(2), 270–283 (1981). https://doi.org/10.1137/0210019
33. Frederickson, G.N., Jájá, J.: On the relationship between the biconnectivity augmentation and traveling salesman problems. Theoret. Comput. Sci. **19**, 189–201 (1982)
34. Gabow, H.N., Goemans, M.X., Tardos, É., Williamson, D.P.: Approximating the smallest it k-edge connected spanning subgraph by LP-rounding. Networks **53**(4), 345–357 (2009). https://doi.org/10.1002/net.20289
35. Gabow, H.N., Gallagher, S.R.: Iterated rounding algorithms for the smallest k-edge connected spanning subgraph. SIAM J. Comput. **41**(1), 61–103 (2012). https://doi.org/10.1137/080732572
36. Garey, M.R., Johnson, D.S.: Computers and intractability; a guide to the theory of NP-completeness. W. H. Freeman & Co. (1990)
37. Garg, M., Grandoni, F., Jabal Ameli, A.: Improved approximation for two-edge-connectivity. In: Proceedings of the 2023 Annual ACM-SIAM Symposium on Discrete Algorithms (SODA), pp. 2368–2410 (2023). https://doi.org/10.1137/1.9781611977554.ch92
38. Goemans, M.X., Williamson, D.P.: A general approximation technique for constrained forest problems. SIAM J. Comput. **24**(2), 296–317 (1995)
39. Grandoni, F., Kalaitzis, C., Zenklusen, R.: Improved approximation for tree augmentation: saving by rewiring. In: Proceedings of 50th ACM Symposium on Theory of Computing (STOC), pp. 632–645 (2018). https://doi.org/10.1145/3188745.3188898
40. Grigni, M., Koutsoupias, E., Papadimitriou, C.: An approximation scheme for planar graph TSP. In: Proceedings of IEEE 36th Annual Foundations of Computer Science (FOCS), pp. 640–645 (1995). https://doi.org/10.1109/SFCS.1995.492665
41. Gutwenger, C., Jünger, M., Leipert, S., Mutzel, P., Percan, M., Weiskircher, R.: Subgraph induced planar connectivity augmentation. In: Graph-Theoretic Concepts in Computer Science (WG), pp. 261–272 (2003). https://doi.org/10.1007/978-3-540-39890-5_23
42. Hershkowitz, D.E., Klein, N., Zenklusen, R.: Ghost value augmentation for k-edge-connectivity. In: Proceedings of the 56th Annual ACM Symposium on Theory of Computing (STOC), pp. 1853–1864 (2024). https://doi.org/10.1145/3618260.3649715
43. Hommelsheim, F., Lindermayr, A., Liu, Z.: A Better-Than-5/4-approximation for two-edge connectivity. In: Proceedings of the 2026 Annual ACM-SIAM Symposium on Discrete Algorithms (SODA), pp. 1468–1520 (2026)
44. Hunkenschröder, C., Vempala, S., Vetta, A.: A 4/3-Approximation algorithm for the minimum 2-edge connected subgraph problem. ACM Trans. Algorithms **5**(4), 55:1–55:28 (2019)
45. Jain, K.: A Factor 2 approximation algorithm for the generalized steiner network problem. Combinatorica **21**, 39–60 (2001). https://doi.org/10.1007/s004930170004

46. Khanna, S., Motwani, R.: Towards a syntactic characterization of PTAS. In: Proceedings of the Twenty-Eighth Annual ACM Symposium on Theory of Computing (STOC), pp. 329–337 (1996). https://doi.org/10.1145/237814.237979
47. Khuller, S., Thurimella, R.: Approximation algorithms for graph augmentation. J. Algorithms **14**(2), 214–225 (1993). https://doi.org/10.1006/jagm.1993.1010
48. Khuller, S., Vishkin, U.: Biconnectivity approximations and graph carvings. J. ACM **41**(2), 214–235 (1994). https://doi.org/10.1145/174652.174654
49. Klein, P.N., Mathieu, C., Zhou, H.: Correlation clustering and two-edge-connected augmentation for planar graphs. Algorithmica **85**(10), 3024–3057 (2023). https://doi.org/10.1007/s00453-023-01128-w
50. Klein, P.: A linear-time approximation scheme for planar weighted TSP. In: 46th Annual IEEE Symposium on Foundations of Computer Science (FOCS), pp. 647–656 (2005). https://doi.org/10.1109/SFCS.2005.7
51. Kobayashi, Y., Noguchi, T.: An approximation algorithm for two-edge-connected subgraph problem via triangle-free two-edge-cover. In: 34th International Symposium on Algorithms and Computation (ISAAC), pp. 49:1–49:10 (2023). https://doi.org/10.4230/LIPIcs.ISAAC.2023.49
52. Kortsarz, G., Krauthgamer, R., Lee, J.R.: Hardness of approximation for vertex-connectivity network design problems. SIAM J. Comput. **33**(3), 704–720 (2004)
53. Kortsarz, G., Nutov, Z.: A simplified 1.5-approximation algorithm for augmenting edge-connectivity of a graph from 1 to 2. ACM Trans. Algorithms **12**(2), 23:1–23:20 (2016). https://doi.org/10.1145/2786981
54. Kortsarz, G., Nutov, Z.: LP-relaxations for tree augmentation. Discret. Appl. Math. **239**, 94–105 (2018). https://doi.org/10.1016/j.dam.2017.12.033
55. Nagamochi, H.: An approximation for finding a smallest 2- edge-connected subgraph containing a specified spanning tree. Discret. Appl. Math. **126**(1), 83–113 (2003). https://doi.org/10.1016/S0166-218X(02)00218-4
56. Nagamochi, H., Nishimura, K., Ibaraki, T.: Computing all small cuts in an undirected network. SIAM J. Discrete Math. **10**(3), 469–481 (1997). https://doi.org/10.1137/S0895480194271323
57. Nutov, Z.: On the tree augmentation Problem. Algorithmica **83**(2), 553–575 (2020). https://doi.org/10.1007/s00453-020-00765-9
58. Nutov, Z.: Approximation algorithms for connectivity augmentation problems. In: Proceedings of the 16th International Computer Science Symposium in Russia (CRS), pp. 321–338 (2021). https://doi.org/10.1007/978-3-030-79416-3_19
59. Pritchard, D.: K-edge-connectivity: approximation and LP relaxation. In: Approximation and Online Algorithms (WAOA), pp. 225–236 (2011). https://doi.org/10.1007/978-3-642-18318-8_20
60. Provan, J.S., Burk, R.C.: Two-connected augmentation problems in planar graphs. J. Algorithms **32**(2), 87–107 (1999). https://doi.org/10.1006/jagm.1999.1005
61. Rutter, I., Wolff, A.: Augmenting the connectivity of planar and geometric graphs. Electron. Notes Discrete Math. **31**, 53–56 (2008). https://doi.org/10.1016/j.endm.2008.06.009
62. Sebő, A., Vygen, J.: Shorter tours by nicer ears: 7/5-approximation for the graph-TSP, 3/2 for the path version, and 4/3 for two-edge-connected subgraphs. Combinatorica **34**(5), 597–629 (2014). https://doi.org/10.1007/s00493-014-2960-3
63. Traub, V., Zenklusen, R.: A $(1.5+\varepsilon)$-approximation algorithm for weighted connectivity augmentation. In: Proceedings of the 55th Annual ACM Symposium on Theory of Computing (STOC), pp. 1820–1833 (2023)

64. Traub, V., Zenklusen, R.: Better-Than-2 approximations for weighted tree augmentation and applications to steiner tree. J. ACM **72**(2), 16:1–16:40 (2025). https://doi.org/10.1145/3722101
65. Vazirani, V.V.: Approximation Algorithms. Springer (2001)
66. Williamson, D.P., Shmoys, D.B.: The Design of Approximation Algorithms, Cambridge University Press (2011)
67. Zheng, B.: Linear-time approximation schemes for planar minimum three-edge connected and three-vertex connected spanning subgraphs (2017). https://doi.org/10.48550/arXiv.1701.08315

Degree Realization with Minimum Dominating Set

Amotz Bar-Noy[1], Igor Kalinichev[2(✉)], David Peleg[2], and Dror Rawitz[3]

[1] City University of New York (CUNY), New York, USA
amotz@sci.brooklyn.cuny.edu
[2] Weizmann Institute of Science, Rehovot, Israel
{igor.kalinichev,david.peleg}@weizmann.ac.il
[3] Bar Ilan University, Ramat-Gan, Israel
dror.rawitz@biu.ac.il

Abstract. The Degree Realization problem requires, given a sequence d of n positive integers, to decide whether there exists problems in which the goal is to compute a realization that optimizes some quality measure. Efficient algorithms are known for the problems of finding a realization with the maximum clique, the maximum independent set, or the minimum vertex cover. In this paper, we focus on the Degree Realization with Min Dominating Set problem for which such algorithm was not known. The goal is to find a realization whose minimum dominating set is minimized among all the realizations of the given sequence d. We present polynomial time realization algorithm for this problem.

A related problem of interest and importance is *characterizing* the sequences with a given value of the optimized function. This leads to an efficient computation of the optimized value without providing the realization that achieves that value. This paper provides a succinct characterization of sequences with a dominating set of a given size.

Keywords: Degree sequences · Graph algorithms · Graph realization

1 Introduction

Given a non-increasing sequence $d = (d_1, \ldots, d_n)$ of positive integers, the Degree Realization (DR) problem requires to decide if d is the degree sequence of some n-vertex (simple undirected) graph $G = (V, E)$, that is, $\deg_G(i) = d_i$ for every $i \in [1, n]$, and to construct such a graph if it exists. In such a case, we say that d is *graphic*. For instance, the sequence $d = (4, 3, 2, 1, 1)$ cannot be realized since $\sum_i d_i$ is odd, and $d' = (4, 3, 1, 1, 1)$ cannot be realized despite the fact that $\sum_i d'_i$ is even. In contrast, $d'' = (4, 3, 2, 2, 1)$ is graphic.

The two key questions studied extensively in the past concern identifying characterizations (or, necessary and sufficient conditions) for a sequence to be

Supported by US-Israel BSF grant 2022205.

S. S. Dey et al. (Eds.): IPCO 2026, LNCS 16588, pp. 64–80, 2026.
https://doi.org/10.1007/978-3-032-28691-8_5

graphic, and developing effective and efficient algorithms for finding a realizing graph for a given sequence if it exists. A necessary and sufficient condition for a given sequence of integers to be graphic (also implying an $O(n)$ decision algorithm) was presented by Erdős and Gallai in [8]. (For alternative proofs see [1,5,7,26–28].) Havel [16] and Hakimi [15] described an $O(\sum_i d_i)$-time algorithm that given a sequence d of integers proves that the given sequence is not graphic or computes an m-edge graph realizing it, where $m = \frac{1}{2}\sum_{i=1}^{n} d_i$.

A more challenging variant of the problem arises when the given sequence has many possible realizing graphs, but some realizations are more desirable than others in various ways. In such a case, it is of interest to look for a realizing graph that also optimizes some quality measure. Hereafter, we refer to such problems as *optimized realization* problems. For example, consider the classical MAX CLIQUE (MC) problem. Given a graph $G = (V, E)$, a clique is a set $Q \subseteq V$, such that $(v, w) \in E$, for every $v, w \in Q$. The size of the maximum clique in G is denoted $\mathrm{MC}(G) = \max\{|Q| \mid Q \text{ is a clique in } G\}$. It may be desirable to find, for a given graphic sequence d, a realizing graph G with the largest possible clique. The resulting optimized realization problem is defined as follows. The MAX CLIQUE DEGREE REALIZATION (MC-DR) problem requires to find a realizing graph G attaining $\mathrm{MC}(d)$, where $\mathrm{MC}(d) = \max\{\mathrm{MC}(G) \mid G \text{ is a realization of } d\}$. Similarly, one may define optimized realization problems corresponding to other graph optimization problems, including MAX INDEPENDENT SET, MIN VERTEX COVER, MAX MATCHING, and MIN DOMINATING SET. Their optimal values on a given graph G are defined as $\mathrm{MIS}(G)$, $\mathrm{MVC}(G)$, $\mathrm{MM}(G)$, and $\mathrm{MDS}(G)$, resp. (Recall that $D \subseteq V$ is a *dominating set* if either $v \in D$ or v has a neighbor in D, for every $v \in V$.) The corresponding optimized realization problems on degree sequences are named MIS DEGREE REALIZATION (MIS-DR), MVC DEGREE REALIZATION (MVC-DR), MM DEGREE REALIZATION (MM-DR), and MDS DEGREE REALIZATION (MDS-DR), and their optimal values on a given sequence d are denoted by $\mathrm{MIS}(d)$, $\mathrm{MVC}(d)$, $\mathrm{MM}(d)$, and $\mathrm{MDS}(d)$.

As is well known, the graph versions of most of these problems are NP-hard. In contrast, the optimized realization versions of the MC, MVC, and MIS have been shown to be polynomial-time solvable. Rao [22] gave a characterization of sequences which can be realized by a graph containing K_ℓ (i.e., a clique of size ℓ). This result was based on a phenomenon collectively known as the *prefix lemma*, which for MC says that if d has a realization containing K_ℓ, then it has a realization such that the ℓ vertices of maximum degree induce a clique K_ℓ. This result implies a polynomial-time realization algorithm for MC-DR. Recall that given a graph $G = (V, E)$, MIS is equivalent to MC in the complement graph $\bar{G} = (V, \bar{E})$, where $\bar{E} = \{(u, v) \mid (u, v) \notin E\}$. Consequently, a realization algorithm for MIS-DR can be derived directly from the MC-DR algorithm by applying it to the transformed sequence $(n-1-d_n, \ldots, n-1-d_1)$. Furthermore, since $V \setminus I$ is a vertex cover, if I is an independent set, it follows that the MIS-DR realization algorithm also works for MVC-DR. Note also that a prefix lemma applies to MVC. Moreover, this prefix property is satisfied by MDS [13]. However, so far it was not known how to exploit this property in order to derive a polynomial time algorithm for MDS-DR. The special case of this problem where $d_1 = O(1)$ was handled in [13], but the general problem was left open.

Our Results. This paper provides polynomial time realization algorithm for MDS-DR. It makes use of an existing prefix lemma (Theorem 1 in [13]). In addition, we develop Erdős-Gallai like characterization for MDS-DR. This characterization can be used to efficiently compute the size of the minimum dominating set of a sequence without providing a realization by searching for the size of the minimum dominating set. Interestingly, our characterization for the MDS-DR problem is based on the construction and correctness of our realization algorithm, and in turn, the characterization help us reducing the complexity of the realization algorithm, because the realization algorithm does not need to search for the size of the minimum dominating set.

As mentioned above, there are several known algorithms for finding a graph G realizing d, including the well-known Havel–Hakimi algorithm [15,16]. For our purposes, though, we need to use a somewhat lesser-known but highly versatile algorithm due to Fulkerson, Hoffman and McAndrew [11], hereafter named the FHM algorithm, which is based on (i) realizing (d, d) by a *bipartite* graph $\hat{G}$, (ii) converting $\hat{G}$ to a half-integral general (non-bipartite) realization G^{ω} for d, and (iii) rounding G^{ω} to an integral general realization G. Our approach is based on modifying step (i) so that the bipartite realization $\hat{G}$ has an optimal dominating set, and preserving this property during the transformations of steps (ii) and (iii). The challenging obstacle is that the rounding process of step (iii) is oblivious to the issue of size, so a bipartite $\hat{G}$ with a small dominating set might be transformed into a general G with a large dominating set. Hence, it is necessary to modify the FHM algorithm in non-trivial ways in order to solve MDS-DR[1].

Related Work. One related direction involves network realization with a given subgraph. Kundu [19] showed that the sequence d and the component-wise difference $d - d'$, such that $d'_i \in \{k, k+1\}$ and $d' \leq d$ (component-wise), are graphic only if there exists a graph G realizing d that has a subgraph G' realizing d'. This result can be used to decide whether a given sequences has a perfect matching by assigning $d'_i = 1$ for every i. Kleitman and Wang [18] gave an algorithm for computing a realization of d containing a subgraph which realizes d'. This algorithm can be used to compute a realization of d that contains a perfect matching, if one exists. Extensions of the above result were presented in [18,21].

Rao and Rao [23] and Kundu [19] gave a characterization of sequences that can be realized by a Hamiltonian graph. Chungphaisan [6] gave an algorithm, that given a sequence d, constructs a realization with a Hamiltonian cycle (or path), if one exists. Rao [22] characterized sequences that can be realized by a graph containing K_ℓ (a clique of size ℓ). This result was based on the Prefix Lemma for MC. An alternative proof was given in [17], and a constructive proof was presented in [30]. It follows that MC-DR can be solved in polynomial

[1] The same approach can provide an alternative algorithm for MC-DR and for MVC-DR (or MIS-DR). We omit the details.

time by checking if d has a realization containing K_ℓ, for $\ell \in [2, n]$, where $[i, j] = \{i, \ldots, j\}$, for $i \leq j$. In contrast, it is NP-hard to approximate MC within a ratio of $O(n^{1-\varepsilon})$, for any $\varepsilon > 0$ [31]. Gould, Jackson and Lehel [14] extended Rao's result by showing that if d has a realization containing H as a subgraph (but not necessarily an induced sub-graph), then there exists a realization of d containing H such that the vertices of H have the $|V(H)|$ largest degrees. Yin [29] used this result to further extend Rao's result by giving a characterization of graphic sequences that contains a split graph $S_{r,s}$ composed of a clique is of size r and an independent set of size s. (Observe that $S_{r,1}$ is a clique of size $r+1$.)

Gentner, Henning and Rautenbach [13] proved the prefix lemma for MDS-DR and gave a realization algorithm for MDS-DR on sequences with $d_1 = O(1)$, leaving the general case open. They also provided characterizations for MIS-DR and MDS-DR in forests. Gentner, Henning and Rautenbach [12] gave characterizations to realizations that minimize the maximum independent set and that maximize the minimum dominating set in forests. Note that MDS is not approximable within $\alpha \log n$, for some $\alpha > 0$, unless P = NP [25], and within $(1-\varepsilon)\log n$, for any $\varepsilon > 0$, unless NP $\subseteq$ DTIME$(n^{\log\log n})$ [10].

Bock and Rautenbach [4] studied MM-DR in trees and bipartite graphs, where the partition is given. Recently, Erdös et al. [9] studied MM-DR in general graphs. Given a sequence d and an integer ν, they presented an Erdős-Gallai type characterization, based on a system of $O(n)$ inequalities, which is satisfied if and only if d has a realization with a matching of size ν. Applying these characterizations to a given degree sequence, the maximum size of a matching can be computed efficiently, although the specific realization is not provided.

Fulkerson, Hoffman and McAndrew [11] obtained conditions for the existence of an f-factor that are applicable only to the family of multi-graphs that satisfy the so called *odd cycle condition*. Kundu [20] used this result to provide simplified conditions for the factorization of graphs that satisfy the odd cycle condition. Rao [24] and Yin [29] also used the result of [11]. Anstee used the technique of [11] to provide an algorithmic proof of the f-factor theorem [2] and a simplified characterization for the existence of a (g, f)-factor for special cases [3].

2 Realization with Minimum Dominating Set

In this section, we describe an algorithm for constructing a realization of a given graphic sequence d, which in addition has a dominating set D of the minimum size γ among all the possible realizations. For this, we employ the Prefix Lemma 1 (Theorem 1 in [13]) and a suitable modification of the FHM realization algorithm [11].

Our algorithm proceeds in several steps, presented in the coming subsections. First, the MDS-DR problem over general graph is reduced to the same problem over bipartite graphs. It is done by a modification of the FHM realization algorithm, that ensures preservation of a dominating set. Then, the MDS-DR problem over bipartite graphs is reduced to a maximum flow problem. To be more specific, given a candidate size γ for the minimum dominating set and a degree sequence pair (d, d), we construct a bipartite flow graph $G_{d,\gamma}$, such that if the maximum flow attained in $G_{d,\gamma}$ equals $\sum_{i=1}^{n} d_i$, then it corresponds to a realization $\hat{G}$ of (d, d) with a dominating set of size 2γ. The Prefix Lemma 1 narrows down the search of the dominating set to a polynomial number of candidates, which allows solving MDS-DR in polynomial time.

2.1 Prefix Lemma

Define a *γ-prefix-dominated realization* of the sequence d to be a realization, where the vertices with the γ highest degrees form a dominating set.

Lemma 1 (Prefix Lemma [13]). *If a sequence d has a realization with a minimum dominating set of size γ, then d has a γ-prefix-dominated realization.*

Although the prefix lemma states that if $\gamma = \text{MDS}(d)$ then there is a realization with dominating set on γ vertices of highest degrees, it is not immediately clear how to select edges so as to obtain this realization, hence additional ideas are needed.

2.2 Reduction to a Bipartite Sequence Pair

A bipartite graph $\hat{G} = (V, W, \hat{E})$, where $V = \{v_1, v_2, \ldots, v_n\}$ and $W = \{w_1, w_2, \ldots, w_n\}$, is a *$\gamma$-prefix-dominated* realization for the sequence pair (d, d) if it satisfies the following properties.

(D1) $\hat{G}$ realizes the sequence pair (d, d).
(D2) $(v_i, w_i) \notin \hat{E}$ for every $i \in [1, n]$.
(D3) $\hat{D} = \hat{D}_V \cup \hat{D}_W$ is a dominating set in $\hat{G}$, where $\hat{D}_V = \{v_1, \ldots, v_\gamma\}$ and $\hat{D}_W = \{w_1, \ldots, w_\gamma\}$ are prefixes of V and W, respectively.

We describe a polynomial time algorithm based on the FHM algorithm [11], that given a γ-prefix dominated realization $\hat{G}$ for the sequence pair (d, d) produces a γ-prefix dominated realization G for d.

Step 1: Compute a half-integral solution.

a. For all $i, j \in [1, n]$, let $y_{ij} = \begin{cases} 1, & \{v_i, w_j\} \in \hat{E}, \\ 0, & \text{otherwise}, \end{cases}$ and $\omega(i, j) = \frac{1}{2}(y_{ij} + y_{ji})$.
b. Define a weighted graph $G^\omega = (V^\omega, E^\omega, \omega)$ with vertex set $V^\omega = [1, n]$ and an edge $e = (i, j)$ of weight $\omega(e) = \omega(i, j)$ for every $i, j \in V^\omega$. Clearly, w is half-integral.
c. Define the *weighted degree* of a vertex $i \in V^\omega$ to be $d^\omega(i) = \sum_{j \in V^\omega} \omega(i, j)$. Note that G^ω realizes d in the *weighted* sense, namely,
$d^\omega(i) = \sum_{j \in V^\omega} \omega(i, j) = \frac{1}{2}\left(\sum_{j=1}^{n} y_{ij} + \sum_{j=1}^{n} y_{ji}\right) = d_i$, for any $i \in V^\omega$.

d. Partition the vertex set V^ω of G^ω into $D = [1, \gamma]$ and $S = [\gamma+1, n]$. Note that by construction, D is a dominating set for G^ω, and moreover, the total weight of the edges connecting any nondominating vertex in S with its dominating neighbours in D is at least one. Indeed, for any $s \in S$, by property (D3),

$$\sum_{x \in D} \omega(s, x) = \sum_{j \in [1,\gamma]} \omega(s, j) = \frac{1}{2} \sum_{j \in [1,\gamma]} y_{sj} + \frac{1}{2} \sum_{j \in [1,\gamma]} y_{js} \geq 1 \quad (1)$$

Step 2: Preparing for discarding non-integral weights while keeping the degrees. Construct a graph $G^{1/2} = (V^{1/2}, E^{1/2})$ by removing from G^ω the edges of integral weight and keeping only those of weight $1/2$. Formally, $V^{1/2} = V^\omega$ and $E^{1/2} = \{e \in E^\omega \mid \omega(e) = 1/2\}$.

Observation 1. *The graph $G^{1/2}$ is even (i.e., all its vertex degrees are even).*

Proof. Each vertex $i \in V^\omega$ has an integral weighted degree $d^\omega(i)$, so the number of edges of weight $1/2$ incident to i must be even. □

The first two steps are similar to the FHM algorithm [11], and the main changes w.r.t. that algorithm occur in the subsequent steps, and pertain to the process of modifying G^ω and getting rid of non-integral weights while keeping the degrees unchanged without violating Inequality (1). The modifications happen concurrently on the graphs G^ω and $G^{1/2}$ (i.e., whenever changing the weight of some edge e in G^ω from 0 or 1 to $1/2$ or vice versa, $G^{1/2}$ is modified accordingly, adding or removing the edge e).

Specifying the modifications require the following definition. A 4-vertex path $P[a, s, b, c]$ in $G^{1/2}$ is a *2-dom path* if $s \in S$ and $a, b \in D$, i.e., the nondominating s has two neighboring dominators.

Step 3: Eliminate 2-dom paths.
The next step in the algorithm transforms the weights in G^ω until it is free of 2-dom paths. This is done as follows. While there is a 2-dom path in $G^{1/2}$, apply one of the following three *modification rules* to the edge weights in G^ω, according to a weight of the edge (a, c) in G^ω. The different possible situations are visualized in Fig. 1. All figures in this section maintain the convention that black nodes are dominating, white nodes are nondominating and gray nodes can be either. Furthermore, solid lines represent edges of positive weight and dashed lines represent edges of weight 0.

(MR1) If $\omega(a, c) = 0$: set $\omega(a, s) \leftarrow 0$, $\omega(s, b) \leftarrow 1$, $\omega(b, c) \leftarrow 0$, $\omega(a, c) \leftarrow 1/2$.
(MR2) If $\omega(a, c) = 1/2$: set $\omega(a, s) \leftarrow 1$, $\omega(s, b) \leftarrow 0$, $\omega(b, c) \leftarrow 1$, $\omega(a, c) \leftarrow 0$.
(MR3) If $\omega(a, c) = 1$: set $\omega(a, s) \leftarrow 1$, $\omega(s, b) \leftarrow 0$, $\omega(b, c) \leftarrow 1$, $\omega(a, c) \leftarrow 1/2$.

Note that the modifications preserve the weighted degree of every vertex in G^ω and the total weight of the edges between any nondominating vertex and its dominating neighbors. Moreover, each time a modification is applied, the

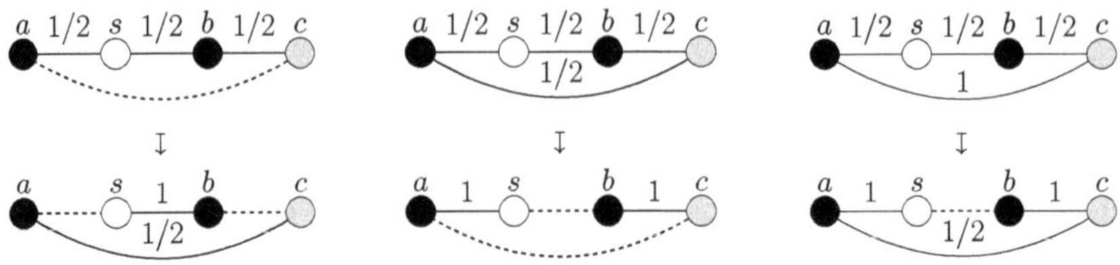

(a) MR1 (b) MR2 (c) MR3

Fig. 1. Illustration of the modification rules

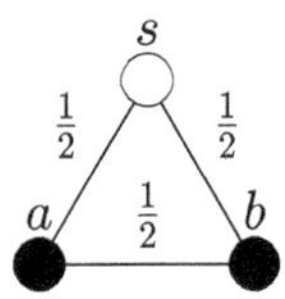

Fig. 2. A cycle in $\mathscr{C}^{\Delta}$

number of edges in $G^{1/2}$ decreases. Thus, the 2-dom paths are eliminated from $G^{1/2}$ (with the corresponding paths eliminated from G^{ω}) within a polynomial number of steps.

Step 4: Separate special cycles.
At the start of this step, $G^{1/2}$ is free of 2-dom paths. Let S' be the set of all the nondominating vertices in G^{ω} that are connected to a dominator vertex with at least one edge $\hat{e}$ with $\omega(\hat{e}) = 1$. Choose such an edge $\hat{e}_s$ arbitrarily for each vertex $s \in S'$ and denote the collection of these edges by $E' = \{\hat{e}_s \mid s \in S'\}$.

Next consider the set $S^{\Delta} = S \setminus S'$ of the remaining nondominating vertices. Since every $s \in S^{\Delta}$ is not connected to any vertex in D with an edge of weight one in G^{ω}, it has at least two dominating neighbours in $G^{1/2}$ by Inequality (1). Choose arbitrarily two such neighbours $a_s, b_s \in D$ for every $s \in S^{\Delta}$.

Observation 2. *For every $s \in S^{\Delta}$, a_s and b_s are not connected to any vertices in $G^{1/2}$ besides possibly each other and s.*

Proof. Having another vertex c neighboring, say, b_s, in $G^{1/2}$ would imply the existence of a 2-dom path $P[a_s, s, b_s, c]$ in $G^{1/2}$, leading to a contradiction. □

In all modifications performed in subsequent steps of the algorithm, edges are only removed from (but never added to) $G^{1/2}$, so Observation 2 continues to hold until the end of the algorithm's execution.

Since a_s and b_s have integral weighted degrees in G^{ω}, Observation 2 implies that the edge (a_s, b_s) must exist in $G^{1/2}$. It follows that there is a cycle $C[a_s, s, b_s]$ in $G^{1/2}$ for each $s \in S^{\Delta}$ (see Fig. 2). Let $\mathscr{C}^{\Delta} = \{C[a_s, s, b_s] \mid s \in S^{\Delta}\}$ be a set of all such cycles.

Observation 3. *Any two different cycles $C, C' \in \mathscr{C}^{\Delta}$ have disjoint vertices.*

Proof. Consider cycles $C = C[s, a_s, b_s]$ and $C' = C[s', a'_s, b'_s]$. By construction, $s \neq s'$, so C and C' can only intersect in a_s or b_s, but this cannot happen by Observation 2. □

Step 5: Partition into cycles.
Denote the set of all the edges of the cycles in $\mathscr{C}^\Delta$ by

$$E(\mathscr{C}^\Delta) = \{(a_s, b_s), (a_s, s), (s, b_s) \mid C[a_s, s, b_s] \in \mathscr{C}^\Delta\}$$

Consider a subgraph H of $G^{1/2}$ with the same vertices, $V(H) = V^{1/2}$, and edges $E(H) = E^{1/2} \setminus E(\mathscr{C}^\Delta)$. Since $G^{1/2}$ is an even graph by Observation 1 and the cycles in $\mathscr{C}^\Delta$ are disjoint by Observation 3, it follows that H is an even graph. So there is a polynomial time algorithm for partitioning edge set of H into disjoint cycles, such that each cycle contains an entire connected component. Denote the set of these cycles by $\mathscr{C}'$ and let $\mathscr{C} = \mathscr{C}' \cup \mathscr{C}^\Delta$. We call a cycle *even* (resp., *odd*) if it has an even (resp., odd) number of edges.

Observation 4. *The number of odd cycles in $\mathscr{C}$ is even.*

Proof. Observe that $\sum_{e \in E^\omega} \omega(e) = \frac{1}{2} \sum_{i=1}^{n} d_i = m$, where m is the number of edges, which is an integer. Thus, the number of edges with weight 1/2 must be even. Since the cycles in $\mathscr{C}$ cover all of $E^{1/2}$ and are disjoint, the observation follows. □

Observation 5. *If cycles $C \in \mathscr{C}^\Delta$ and $C' \in \mathscr{C}'$ have a non-empty intersection, then $C \cap C' = \{s\}$ for $s \in S^\Delta$ corresponding to the cycle C.*

Proof. Let $C = C[a_s, s, b_s]$ with $a_s, b_s \in D$ and $s \in S^\Delta$. By Observation 2, a_s and b_s have degree zero in H, so they do not belong to any cycle in $\mathscr{C}'$. □

Step 6: Eliminate even cycles.
For every even cycle $C \in \mathscr{C}$ do the following.

1. Traverse C starting from an arbitrary vertex $x \in C$ and continuing along the cycle until returning to x. Denote the resulting sequence of edges by $E(C) = (e_1, e_2, \ldots, e_\ell)$.
2. Increase (respectively, decrease) the weights of the edges on even (resp., odd) positions in the sequence $E(C)$ by 1/2. That is, for every $i \in [1, \ell]$ set

$$\omega(e_i) \leftarrow \begin{cases} 1, & i \text{ is even,} \\ 0, & i \text{ is odd.} \end{cases}$$

Note that this procedure does not change the weighted degrees in G^ω or the weights of the edges in E' and does not affect other cycles in $\mathscr{C}$. Hereafter, we refer to such modifications as *neutral.*

Step 7: Eliminate odd cycles.
Finally, arrange the odd cycles in $\mathscr{C}$ (whose number is even by Observation 4) in pairs. For every pair (C, C'), proceed according to Case 1 below if the cycles intersect and according to Case 2 otherwise.

Case 1: There is a vertex $x \in C \cap C'$:

1. As before, traverse both cycles starting at x. Denote the resulting edge sequences in C and C' by $E(C) = (e_1, e_2, \ldots, e_\ell)$ and $E(C') = (e'_1, e'_2, \ldots, e'_k)$.
2. Cycles in $\mathscr{C}'$ are disjoint by definition and cycles in $\mathscr{C}^\Delta$ are disjoint by Observation 3. Thus one of the cycles (C, C') belongs to $\mathscr{C}'$ and the other to $\mathscr{C}^\Delta$. Redefine them so $C \in \mathscr{C}^\Delta$ and $C' \in \mathscr{C}'$.
3. For every $i \in [1, \ell]$, $j \in [1, k]$, modify the edge weights in the cycles as follows:

$$\omega(e_i) \leftarrow \begin{cases} 0, & i \text{ is even}, \\ 1, & i \text{ is odd}. \end{cases} \qquad \omega(e'_j) \leftarrow \begin{cases} 1, & j \text{ is even}, \\ 0, & j \text{ is odd}. \end{cases}$$

Note that this modification is neutral.

By Observation 5, $x \in S^\Delta$ and $C = C[x, a_x, b_x]$ for some $a_x, b_x \in D$. Note that x is connected to a_x and b_x in G^ω with edges of weight 1 after the modification. Neither a_x nor b_x belongs to any other cycle in $\mathscr{C}$ by Observation 2, so the weights $\omega(x, a_x)$ and $\omega(x, b_x)$ are not modified further by the algorithm. Let E_1^Δ contain an edge (x, a_x) for every cycle $C \in \mathscr{C}^\Delta$ that was processed in this case.

Case 2: $C \cap C' = \emptyset$**:**

First, we have the following lemma.

Lemma 2. *For any two disjoint odd cycles $C, C' \in \mathscr{C}$, there exist $x \in C$ and $y \in C'$, such that $(x, y) \notin E'$, $\omega(x, y) \neq 1/2$, and if C or C' belongs to $\mathscr{C}^\Delta$, then the corresponding vertex does not belong to S^Δ.*

Proof. Recall that each cycle in $\mathscr{C}^\Delta$ contains exactly one vertex from S^Δ. Choose vertices $x \in C$ and $y' \in C'$, so if $C \in \mathscr{C}^\Delta$ (respectively, $C' \in \mathscr{C}^\Delta$), then $x \notin S^\Delta$ (respectively, $y' \notin S^\Delta$). If $(x, y') \notin E'$, then the chosen x and $y = y'$ satisfy the Lemma (the condition $w(x, y) \neq 1/2$ is proved later). Otherwise, one of the vertices x, y' must be dominating and the other nondominating. Without loss of generality assume $y' \in D$ and $x \in S$. Replace y' with $y \in C'$, such that $y \neq y'$ and if $C' \in \mathscr{C}^\Delta$, then $y \notin S^\Delta$. Since each cycle has at least three vertices, this is always possible. Each edge in E' corresponds to a unique nondominating vertex and (x, y') corresponds to x, so E' does not contain (x, y).

Next we prove that $\omega(x, y) \neq 1/2$. Assume that $C, C' \in \mathscr{C}'$. Then by construction, they are not connected by any edge in $E^{1/2} \setminus E(\mathscr{C}^\Delta)$. Assume, towards contradiction, that they are connected by an edge $e \in E(\mathscr{C}^\Delta)$, so e belongs to a cycle $C[s, a_s, b_s]$ with $a_s, b_s \in D$ and $s \in S^\Delta$. It follows that either C or C' contains either a_s or b_s, but it is not possible by Observation 5, a contradiction. Now assume that $C = C[s, a_s, b_s] \in \mathscr{C}^\Delta$ with $a_s, b_s \in D$ and $s \in S^\Delta$. Since $x \in C \setminus S^\Delta$, x is either a_s or b_s. However a_s and b_s do not have any edges outside C in $G^{1/2}$ by Observation 2, a contradiction. The lemma follows. □

Next we describe how to modify a pair (C, C').

1. Choose vertices $x \in C$ and $y \in C'$ according to Lemma 2.

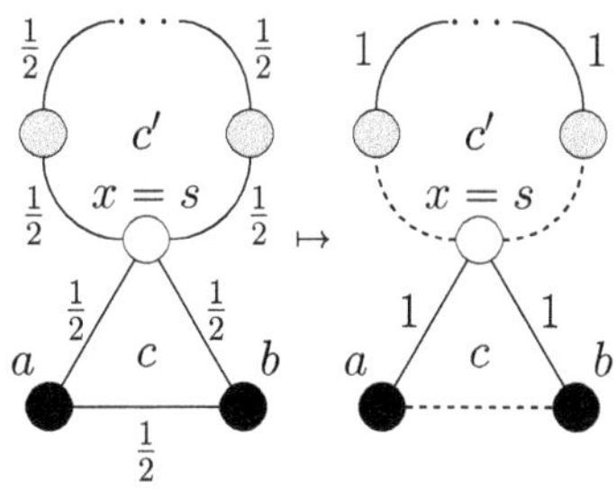

(a) Case 1.

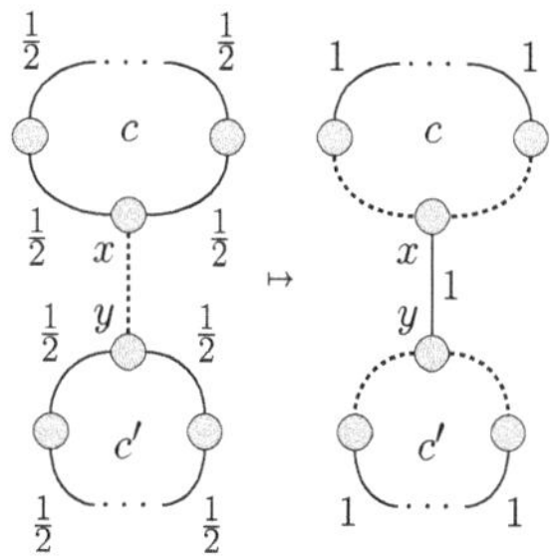

(b) Case 2, $\xi = 0$.

Fig. 3. Illustrations of Step 7

2. Traverse both cycles starting in x and y accordingly and denote the resulting sequences of edges in C and C' by $E(C) = (e_1, e_2, \ldots, e_\ell)$ and $E(C') = (e'_1, e'_2, \ldots, e'_k)$ respectively.
3. It follows that $\omega(x, y) \in \{0, 1\}$. Let $\xi = \omega(x, y)$ and perform the following. Set $\omega(x, y) \leftarrow 1 - \xi$ and for every $i \in [1, \ell]$ and $j \in [1, k]$ modify the edge weights in the cycles as follows

$$\omega(e_i) \leftarrow \begin{cases} 1 - \xi, & i \text{ is even,} \\ \xi, & i \text{ is odd,} \end{cases} \qquad \omega(e'_j) \leftarrow \begin{cases} 1 - \xi, & j \text{ is even,} \\ \xi, & j \text{ is odd.} \end{cases}$$

Note that this modification is neutral.

If $\mathscr{C}^\Delta$ contains C or C', then the corresponding $s \in S^\Delta$ is connected to a dominator vertex with an edge $\hat{e}_s$, such that $\omega(\hat{e}_s) = 1$ after applying the modification. Indeed, s was not chosen as a starting vertex x or y, so it is connected to one of its two neighbors a_s or b_s in the corresponding cycle with an edge of weight one. But both its neighbors are dominating vertices, so $\hat{e}_s$ exists. Note that neither a_s nor b_s belongs to any other cycles in $\mathscr{C}$ by Observation 2, so the weight $\omega(\hat{e}_s)$ is not modified further by the algorithm. Let E_2^Δ contain all edges $\hat{e}_s$ for $s \in S^\Delta$ that were processed in this case (Fig. 3).

Step 8: Generate the output G.
Denote $E^\Delta = E_1^\Delta \cup E_2^\Delta$. For each cycle $C_s \in \mathscr{C}^\Delta$, the corresponding $s \in S^\Delta$ is connected to a dominator vertex with an edge $\hat{e}_s \in E^\Delta$ of weight one. Since there is one-to-one correspondence between $\mathscr{C}^\Delta$ and S^Δ, E^Δ covers all the vertices of S^Δ, in the sense that for every $s \in S^\Delta$ exists the corresponding edge $\hat{e}_s \in E^\Delta$, such that $s \in \hat{e}_s$.

After applying Steps 3–8, each edge in G^ω has weight either one or zero. On the other hand, each step preserves the weighted degrees in G^ω, so G^ω is still a weighted realization of d. It follows that G^ω transforms into a simple graph G with an edge (i, j) whenever $\omega(i, j) = 1$ in G^ω. Clearly, G realizes d.

Finally, since $S = S' \cup S^{\Delta}$, every nondominating vertex $s \in S$ is connecting to the dominating set by some edge in $E' \cup E^{\Delta}$. These edges have weight one in G^{ω}, so G contains them all. Therefore, D is a dominating set in G.

Lemma 3. *There is a γ-prefix-dominated realization $\hat{G}$ of (d,d) if and only if there is a γ-prefix-dominated realization G of d.*

Proof. ($\Rightarrow$) If there γ-prefix-dominated realization of (d,d), then the aforementioned algorithm produces a γ-prefix-dominated realization of d.

($\Leftarrow$) Let $G = (V,E)$ with $V = [1,n]$ be a realization of d, where $\deg(i) = d_i$ for $i \in [1,n]$, with a dominating set $D = [1,\gamma]$. Consider the graph $\hat{G} = (\hat{V},\hat{W},\hat{E})$ with $\hat{V} = \{v_1,\ldots,v_n\}$, $\hat{W} = \{w_1,\ldots,w_n\}$ and $(v_i,w_j) \in \hat{E}$ if and only if $(i,j) \in E$. Clearly, $\hat{G}$ realizes (d,d) (satisfying property (D1)) and does not have edges (v_i,w_i) for every $i \in [1,n]$ (satisfying (D2)). Let $\hat{D} = \hat{D}_V \cup \hat{D}_W$, where $\hat{D}_V = \{v_1,\ldots,v_\gamma\}$ and $\hat{D}_W = \{w_1,\ldots,w_\gamma\}$ are prefixes of $\hat{V}$ and $\hat{W}$, respectively. Since D is a dominating set in G, every vertex in $\hat{W} \setminus \hat{D}_W$ is connected to some dominator vertex in $\hat{D}_V$ and every vertex in $\hat{V} \setminus \hat{D}_V$ is connected to some dominator vertex in $\hat{D}_W$. Thus, $\hat{D}$ is a dominating set in $\hat{G}$ (satisfying (D3)). □

2.3 Reduction to Flow

We have reduced the initial problem of finding a general realization with a dominating set of a certain type to the same problem in a bipartite setting. Hence, we are left with the task of constructing, for a given sequence d with $\mathrm{MDS}(d) = \gamma$, a γ-prefix-dominated realization $\hat{G} = (V,W,\hat{E})$ for the sequence pair (d,d).

To find the desired realization minimizing the dominating set, we go through every possible value of γ and check if a suitable realizing graph $\hat{G}$ exists by reducing the problem to a flow problem. Specifically, given d and a value γ, we construct a directed bipartite flow graph $G_{d,\gamma}$ with edge set $\mathscr{E}$ as follows. Corresponding to V and W, we have the sets $X = \{x_1,\ldots,x_n\}$ and $Y = \{y_1,\ldots,y_n\}$. The node set of the flow graph is $V = X \cup Y \cup X'_S \cup Y'_S \cup \{s,t\}$, where the nodes are organized into the following categories.

- Candidate dominators: $X_D = \{x_i \mid i \in [1,\gamma]$ and $Y_D = \{y_j \mid j \in [1,\gamma]\}$,
- Nondominating nodes: $X_S = \{x_i \mid i \in [\gamma+1,n]$ and $Y_S = \{y_j \mid j \in [\gamma+1,n]\}$,
- Supplementary nodes: $X'_S = \{x'_i \mid i \in [\gamma+1,n]$ and $Y'_S = \{y'_j \mid j \in [\gamma+1,n]\}$.
- Source and sink: {s,t}.

The source s is connected to the nodes of X. Similarly, the nodes of Y are connected to the sink t. In addition, the supplementary nodes of X'_S and Y'_S are used to create the connections and flow capacities and enforce the degree and domination constraints. See Figure 4, in which the flow from s to t goes left-to-right.

The edges of $\mathscr{E}$ are capacitated and directed (from left to right). We use the notation (α,β,δ) for an edge that leads from the node α to the node β and can carry up to δ units of flow. The source s has edges leading to every node $x_i \in X$, with capacity d_i, to enforce the degree constraint for the corresponding vertex

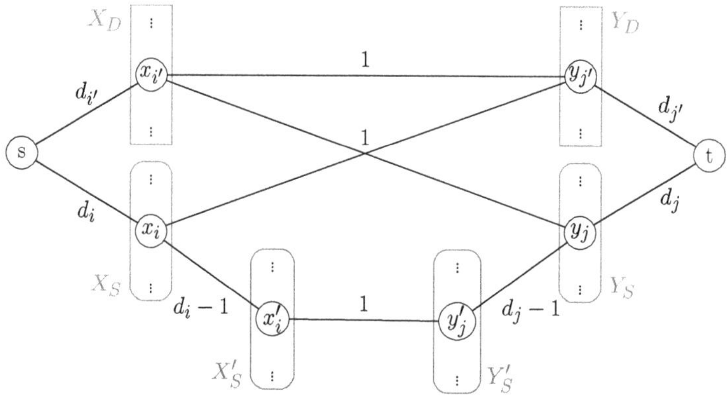

Fig. 4. The flow graph

v_i in the realizing graph G. Similarly, there are edges leading from every node $y_j \in Y$ to the sink t, with capacity d_j, to enforce the degree constraint for the vertices w_j in G.

Additional edges are used to ensure the correctness of the reduction. We write $\tilde{X} \between \tilde{Y}$ to indicate that the vertices $\tilde{X}$ and $\tilde{Y}$ induce an almost complete bipartite graph in $G_{d,\gamma}$, missing only the edges mentioned in property (D2) (i.e., $\{(x_i, y_i) \mid i \in [1,n]$ and $\{(x'_i, y'_i) \mid i \in [\gamma+1, n]$, which are not present in $G_{d,\gamma}$). The various groups of V are connected by $\mathscr{E}$ in the following way: $X_D \between Y_D$, $X_D \between Y_S$, $X_S \between Y_D$ and $X'_S \between Y'_S$ with each edge having capacity one. Finally, every $x_i \in X_S$ is connected to the corresponding $x'_i \in X'_S$ with capacity $d_i - 1$, and similarly, every $y'_j \in Y'_S$ is connected to the corresponding $y_j \in Y_S$ with capacity $d_j - 1$. Denote these connections by $X_S \equiv X'_S$ and $Y'_S \equiv Y_S$. The connections are depicted in Fig. 4 and all the edges are listed below.

$$\begin{aligned}\mathscr{E} \;=\; & \{(s, x_i, d_i) \mid i \in [1,n]\} \;\cup\; \{(y_j, t, d_j) \mid j \in [1,n]\} \\ & \cup \{(x_i, y_j, 1) \mid i, j \in [1,\gamma],\ i \neq j\} \cup \{(x_i, y_j, 1) \mid i \in [1,\gamma],\ j \in [\gamma+1, n]\} \\ & \cup \{(x_i, y_j, 1) \mid i \in [\gamma+1, n],\ j \in [1,\gamma]\} \cup \{(x_i, x'_i, d_i - 1) \mid i \in [\gamma+1, n]\} \\ & \cup \{(y'_j, y_j, d_j - 1) \mid j \in [\gamma+1, n]\} \cup \{(x'_i, y'_j, 1) \mid i, j \in [\gamma+1, n],\ i \neq j\}\,.\end{aligned}$$

Next, we compute a maximum flow from the source s to the sink t in Since all the capacities are integral, there is an integral maximum flow, which can be computed with corresponding algorithms. Hence, we may assume that the computed maximum flow is integral. Let $D = D_A \cup D_B$ with $D_A = \{v_1, \ldots, v_\gamma\}$ and $D_B = \{w_1, \ldots, w_\gamma\}$. Note that by definition, $D_A = \{v_i \mid x_i \in X_D\}$ and $D_B = \{w_j \mid y_j \in Y_D\}$

Lemma 4. *There exists a bipartite graph $\hat{G} = (V, W, \hat{E})$ that realizes the degree sequence pair (d, d) such that D is a dominating set in $\hat{G}$ if and only if the maximum $s - t$ flow in $G_{d,\gamma}$ is $\sum_{i=1}^{n} d_i$.*

Proof. ($\Leftarrow$) Suppose the maximum s-t flow in $G_{d,\gamma}$ is $\sum_{i=1}^{n} d_i$, and let flow be an integral maximum flow. Define the bipartite graph $\hat{G} = (V, W, \hat{E})$, where $V = \{v_1, \ldots, v_n\}$ and $W = \{w_1, \ldots, w_n\}$, by adding an edge (v_i, w_j) to $\hat{E}$, for every $i, j \in [1, n]$ such that there is an edge e in $G_{d,\gamma}$ of the form (x_i, y_j) or (x'_i, y'_j) with $\mathsf{flow}(e) = 1$. Since the flow in integral, its values on the edges (x_i, y_j) and (x'_i, y'_j) are either 0 or 1.

To verify that $\hat{G}$ correctly realizes (d, d), we first observe that the selected edges form a graph (rather than a multigraph). Indeed, for any pair $i, j \in [1, n]$, the flow graph $G_{d,\gamma}$ contains either only the edge (x'_i, y'_j) (in case $x_i \in X_S$ and $y_j \in Y_S$) or only the edge (x_i, y_j) (otherwise), but not both.

Next we show that $\deg_G(v_i) = d_i$ for every $i \in [1, n]$ and $\deg_G(w_j) = d_j$ for every $j \in [1, n]$. Note that the total flow from s to t is $\sum_{i=1}^{n} d_i$, which means that all edges from vertices y_j to t and from s to x_i are saturated. Four cases should be considered.

Case A: If $v_i \in D_A$, then the degree $\deg_{\hat{G}}(v_i)$ of v_i in $\hat{G}$ is set by the number of edges (x_i, y_j) in $\mathscr{E}$ with $\mathsf{flow}(x_i, y_j) = 1$ for $j \in [1, n]$. Since x_i receives a flow of d_i from the source s, by flow conservation it outputs the same amount (in one unit flows) through edges (x_i, y_j), hence its degree is d_i.

Case B: If $w_j \in D_B$, then $\deg_{\hat{G}}(w_j)$ is determined by the number of edges (x_i, y_j) in $\mathscr{E}$ with $\mathsf{flow}(x_i, y_j) = 1$ for $i \in [1, n]$, and its analysis is similar to Case A.

Case C: If $v_i \in V \setminus D_A$, then $\deg_{\hat{G}}(v_i)$ is determined by the number of edges (x_i, y_j) and (x'_i, y'_j) in $\mathscr{E}$ carrying a flow unit for $j \in [1, n]$. We divide the output flow of v_i into two terms: $\mathsf{flow}(s, x_i) = \sum_{j \in D_B} \mathsf{flow}(x_i, y_j) + \mathsf{flow}(x_i, x'_i)$. The first term increases the degree by the corresponding amount of flow, because each edge (x_i, y_j) has capacity one. The rest of the flow goes to x'_i. Since x'_i outputs flow through the edges (x'_i, y'_j) with unit capacity, it increases the degree of v_i by $\mathsf{flow}(x_i, x'_i)$. Hence, the degree of v_i is $\deg_{\hat{G}}(v_i) = \mathsf{flow}(s, x_i) = d_i$.

Case D: If $w_j \in W \setminus D_B$, then $\deg_{\hat{G}}(w_j)$ is determined by the number of edges (x_i, y_j) and (x'_i, y'_j) in $\mathscr{E}$ carrying a flow unit for $i \in [1, n]$, and its analysis is similar to Case C.

Therefore, $\hat{G}$ correctly realizes the degree sequence pair (d, d), so property (D1) holds. Also, since $G_{d,\gamma}$ contains neither $\{(x_i, y_i) \mid i \in [1, n]\}$ nor $\{(x'_i, y'_i) \mid i \in [\gamma + 1, n]\}$, it follows that $\hat{G}$ does not have edges $\{(v_i, w_i) \mid i \in [1, n]\}$, hence it satisfies property (D2).

It remains to show property (D3), namely, that D is a dominating set for $\hat{G}$. Since $y_j \in Y_S$ conserves flow and $\mathsf{flow}(y'_j, y_j) \leq d_j - 1$, there is at least one $x_i \in X_D$ with $\mathsf{flow}(x_i, y_j) = 1$. Similarly, for every $x_i \in X_S$ there exists at least one $y_j \in Y_D$ such that $\mathsf{flow}(x_i, y_j) = 1$. It follows that at least one vertex from D is connected to each $v_i \in V \setminus D$ and $w_j \in W \setminus D$.

($\Rightarrow$) Conversely, suppose that $\hat{G} = (V, W, \hat{E})$, where $V = \{v_1, \ldots, v_n\}$ and $W = \{w_1, \ldots, w_n\}$, is a bipartite graph realizing (d, d), such that D is a dominating set. We need to define a flow function on $G_{d,\gamma}$ and prove that the maximum $s - t$

flow equals to $\sum_{i=1}^{n} d_i$. Obviously, the flow cannot exceed $\sum_{i=1}^{n} d_i$, so it suffices to prove that $\sum_{i=1}^{n} d_i$ is achievable. The flow is defined as follows.

1. **Saturating all edges from the source and to the sink**:
   Set $\mathsf{flow}(s, x_i) \leftarrow d_i$ and $\mathsf{flow}(y_j, t) \leftarrow d_j$ for every $i, j \in [1, n]$.
2. **Flow to supplementary X vertices**:
   For every $i \in [\gamma+1, n]$, set $N_i \leftarrow |\{(v_i, w_j) \in \hat{E} \mid j \in [\gamma+1, n]\}|$.
   Set $\mathsf{flow}(x_i, x'_i) \leftarrow N_i$ for every $x_i \in X_S$
   (Note that since v_i is connected with at least one dominating vertex, $N_i \leq d_i - 1$, so the flows satisfy the capacity constraints.)
3. **Flow from supplementary Y vertices**:
   For every $j \in [\gamma+1, n]$, set $M_j \leftarrow |\{(v_i, w_j) \in \hat{E} \mid i \in [\gamma+1, n]\}|$.
   Set $\mathsf{flow}(y'_j, y_j) \leftarrow M_j$ for every $y_j \in Y_S$.
   (Note that again the flows satisfy the capacity constraints.)
4. **Flow through edges**:
   For every pair $i, j \in [1, n]$ such that $(v_i, w_j) \in \hat{E}$:
   (a) If $x_i \in X_S$ and $y_j \in Y_S$, then set $\mathsf{flow}(x'_i, y'_j) \leftarrow 1$,
   (b) If either $x_i \in X_D$ or $y_j \in Y_D$, then set $\mathsf{flow}(x_i, y_j) \leftarrow 1$.
5. **Completing the flow function**:
   Set the flows of all other edges to 0.

Flow Legality. By construction, the total flow out of the source s and into the sink t is $\sum_{i=1}^{n} d_i$. It remains to verify that all other vertices conserve flow.

Node x_i: By construction, the incoming flow at x_i is $\mathsf{flow}(s, x_i) = d_i$. As for the outgoing flow, there are exactly d_i indices j for which $(v_i, w_j) \in \hat{E}$.
- If $i \in [1, \gamma]$, then the flows emanating from x_i are $\mathsf{flow}(x_i, y_j) = 1$ for precisely those indices. Thus, the outgoing flow is $\sum_{j=1}^{n} \mathsf{flow}(x_i, y_j) = d_i$.
- If $i \in [\gamma+1, n]$, then the outgoing flows from x_i are $\mathsf{flow}(x_i, y_j) = 1$ for $j \in [1, \gamma]$ and $\mathsf{flow}(x_i, x'_i) = N_i = |\{(v_i, w_j) \in \hat{E} \mid j \in [\gamma+1, n]|$. Thus, the outgoing flow is $\sum_{j \in [1,\gamma]} \mathsf{flow}(x_i, y_j) + \mathsf{flow}(x_i, x'_i) = d_i$.

Node y_j: By construction, the outgoing flow at y_j is $\mathsf{flow}(y_j, t) = d_j$. As for the incoming flow, the analysis is similar to outgoing flow at Node x_i.

Node x'_i: The incoming flow at x'_i is N_i by construction. The outgoing flow is $\sum_{j \in [\gamma+1,n]} \mathsf{flow}(x'_i, y'_j) = N_i$ by definition.

Node y'_j: The outgoing flow from y'_j is M_j by construction. The incoming flow is $\sum_{i \in [\gamma+1,n]} \mathsf{flow}(x'_i, y'_j) = M_j$ by definition.

Hence, the flow function defined above is valid, and the maximum $s - t$ flow in $G_{d,\gamma}$ equals to $\sum_{i=1}^{n} d_i$, implying the existence of a bipartite graph G with D as a dominating set.

□

The next theorem states the main result of this section.

Theorem 1. *There exists a polynomial-time algorithm for constructing a realization of a given graphic sequence d that also has a dominating set D of minimum size (among all possible realizations of d).*

3 An Erdős–Gallai Type Characterization

In this section we complement the previous result by providing an Erdős–Gallai Type characterization for the degree sequences d that have a realization with a dominating set of size γ.

It was proven in the previous section that a sequence d has a realization with a dominating set of size γ if and only if the flow graph $G_{d,\gamma}$ admits a flow of size $\sum_{i=1}^{n} d_i$. Thus, by the max–flow min–cut theorem, d has a realization with a dominating set of size γ if and only if the capacity of every cut in $G_{d,\gamma}$ is at least $\sum_{i=1}^{n} d_i$. This can serve as a basis for the sought characterization. Unfortunately, a naive implementation yields a characterization of length exponential in the input size. To overcome that difficulty, we apply the following strategy. As a first step, we show that a minimum cut belongs to one of three families of cuts $\mathscr{F}_1$, $\mathscr{F}_2$, and $\mathscr{F}_3$. Each cut in the above families induces a constraint whose size is at least $\sum_{i=1}^{n} d_i$. Next, we decrease the number of constraints by showing that it is sufficient to consider $O(n)$ constraints per family. Specifically, we obtain the following theorem.

Theorem 2. *A sequence d has a realization with a dominating set of size γ if and only if the following constraints systems are satisfied.*

$$\sum_{i=\gamma+1}^{\gamma+k} (d_i - 1) \le k(k-1) - (n-\gamma) + \sum_{i=1}^{\gamma} d_i + \sum_{i=\gamma+k+1}^{n} \min\{k, d_i - 1\}, \forall k \in [0, n-\gamma]$$

$$\sum_{i=1}^{k} d_i \le k(k-1) + \sum_{i=k+1}^{n} \min\{k, d_i\}, \forall k \in [0, n]$$

$$\sum_{i=\gamma+1}^{\gamma+k} d_i \le k(k-1) + \sum_{i=1}^{\gamma} \min\{k, d_i\} + \sum_{i=\gamma+k+1}^{n} \min\{k, d_i - 1\}, \forall k \in [0, d_1]$$

Notice that the second system coincides with the Erdős-Gallai conditions.

The proof of the theorem and a running time analysis of our realization algorithm are omitted for luck of space.

Disclosure of Interests. The authors have no competing interests to declare that are relevant to the content of this article.

References

1. Aigner, M., Triesch, E.: Realizability and uniqueness in graphs. Discr. Math. **136**, 3–20 (1994)
2. Anstee, R.P.: An algorithmic proof of Tutte's f-factor theorem. J. Alg. **6**(1), 112–131 (1985)
3. Anstee, R.P.: Simplified existence theorems for (g, f)-factors. Discr. Appl. Math. **27**(1–2), 29–38 (1990)

4. Bock, F., Rautenbach, D.: On matching numbers of tree and bipartite degree sequences. Discr. Math. **342**(6), 1687–1695 (2019)
5. Choudum, S.A.: A simple proof of the Erdös-Gallai theorem on graph sequences. Bull. Austral. Math. Soc. **33**(1), 67–70 (1991)
6. Chungphaisan, V.: Construction of hamiltonian graphs and bigraphs with prescribed degrees. J. Comb. Theory B **24**(2), 154–163 (1978)
7. Dahl, G., Flatberg, T.: A remark concerning graphical sequences. Discr. Math. **304**, 62–64 (2005)
8. Erdős, P., Gallai, T.: Graphs with prescribed degrees of vertices [hungarian]. Mat. Lapok (N.S.) **11**, 264–274 (1960)
9. Erdős, P.L., Kharel, S.R., Mezei, T.R., Toroczkai, Z.: New results on graph matching from degree-preserving growth. Mathematics **12**(22) (2024)
10. Feige, U.: A threshold of $\ln n$ for approximating set cover. J. ACM **45**(4), 634–652 (1998)
11. Fulkerson, D.R., Hoffman, A.J., McAndrew, M.H.: Some properties of graphs with multiple edges. Can. J. Math. **17**, 166–177 (1965)
12. Gentner, M., Henning, M.A., Rautenbach, D.: Largest domination number and smallest independence number of forests with given degree sequence. Discr. Appl. Math. **206**, 181–187 (2016)
13. Gentner, M., Henning, M.A., Rautenbach, D.: Smallest domination number and largest independence number of graphs and forests with given degree sequence. J. Graph Theory **88**(1), 131–145 (2018)
14. Gould, R.J., Jackson, M.S., Lehel, J.: Potentially g-graphical degree sequences. In: Combinatorics, Graph Theory, and Algorithms, vol. 1, pp. 451–460 (1999)
15. Hakimi, S.L.: On realizability of a set of integers as degrees of the vertices of a linear graph - I. SIAM J. Appl. Math. **10**(3), 496–506 (1962)
16. Havel, V.: A remark on the existence of finite graphs. Casopis Pest. Mat. **80**, 477–480 (1955)
17. Kézdy, A., Lehel, J.: Degree sequences of graphs with prescribed clique size. In: Graph Theory, Combinatorics, Algorithms, and Applications, pp. 535–544 (1998)
18. Kleitman, D.J., Wang, D.L.: Algorithms for constructing graphs and digraphs with given valences and factors. Discr. Math. **6**, 79–88 (1973)
19. Kundu, S.: The k-factor conjecture is true. Discr. Math. **6**(4), 367–376 (1973)
20. Kundu, S.: A factorization theorem for a certain class of graphs. Discr. Math. **8**(1), 41–47 (1974)
21. Kundu, S.: Generalizations of the k-factor theorem. Discr. Math. **9**(2), 173–179 (1974)
22. Rao, A.R.: The clique number of a graph with a given degree sequence. In: ISI Lecture Notes, vol. 4, pp. 251–267 (1979)
23. Rao, A.R., Rao, S.B.: On factorable degree sequences. J. Comb. Theory B **13**, 185–191 (1972)
24. Rao, S.B.: Towards a theory of forcibly hereditary P-graphic sequences. In: Combinatorics and Graph Theory, pp. 441–458. Springer (1981)
25. Raz, R., Safra, S.: A sub-constant error-probability low-degree test, and a sub-constant error-probability PCP characterization of NP. In: 29th ACM STOC, pp. 475–484 (1997)
26. Tripathi, A., Tyagi, H.: A simple criterion on degree sequences of graphs. Discr. Appl. Math. **156**, 3513–3517 (2008)
27. Tripathi, A., Venugopalan, S., West, D.B.: A short constructive proof of the Erdös-Gallai characterization of graphic lists. Discr. Math. **310**, 843–844 (2010)

28. Tripathi, A., Vijay, S.: A note on a theorem of Erdös & Gallai. Discr. Math. **265**, 417–420 (2003)
29. Yin, J.H.: A rao-type characterization for a sequence to have a realization containing a split graph. Discr. Math. **311**(21), 2485–2489 (2011)
30. Yin, J.H.: A short constructive proof of A. R. Rao's characterization of potentially K_{r+1}-graphic sequences. Discr. Appl. Math. **160**(3), 352–354 (2012)
31. Zuckerman, D.: Linear degree extractors and the inapproximability of max clique and chromatic number. Theory Comput. **3**(1), 103–128 (2007)

1.64-Approximation for Chromatic Correlation Clustering via Chromatic Cluster LP

Chenglin Fan[1(✉)], Dahoon Lee[2], and Euiwoong Lee[3]

[1] Department of Computer Science and Engineering, Seoul National University, Seoul 08826, South Korea
fanchenglin@snu.ac.kr
[2] Department of Mathematical Sciences, Seoul National University, Seoul 08826, South Korea
dahoon46@snu.ac.kr
[3] Computer Science and Engineering Division, University of Michigan, Ann Arbor, MI 48109, USA
euiwoong@umich.edu

Abstract. Chromatic Correlation Clustering (CCC) generalizes Correlation Clustering by assigning multiple categorical relationships (colors) to edges and imposing chromatic constraints on the clusters. Unlike traditional Correlation Clustering, which only deals with binary (+/−) relationships, CCC captures richer relational structures. Despite its importance, improving the approximation for CCC has been difficult due to the limitations of standard LP relaxations. We present a randomized 1.64-approximation algorithm to the CCC problem, significantly improving the previous factor of 2.15. Our approach extends the cluster LP framework to the chromatic setting by introducing a chromatic cluster LP relaxation and an rounding algorithm that utilizes both a cluster-based and a greedy pivot-based strategy. The analysis bypasses the integrality gap of 2 for the CCC version of standard LP and highlights the potential of the cluster LP framework to address other variants of clustering problems.

Keywords: Correlation Clustering · Approximation · LP

1 Introduction

Correlation Clustering (CC) is a foundational unsupervised learning problem that partitions a signed graph while minimizing disagreements with the provided edge labels. Given an undirected graph $G = (V, E = E^+ \uplus E^-)$, where edges are labeled as either similar ('+') or dissimilar ('−'), the goal is to find a node partition (namely, a clustering) $\mathcal{C} = \{C_1, \ldots, C_k\}$ ($\biguplus_{i=1}^k C_i = V$) that minimizes the total number of disagreements:

S. S. Dey et al. (Eds.): IPCO 2026, LNCS 16588, pp. 81–96, 2026.
https://doi.org/10.1007/978-3-032-28691-8_6

$$\text{cost}(\mathcal{C}) = \underbrace{|\{uv \in E^+ : u \in C_i, v \in C_j, i \neq j\}|}_{\text{Positive edges between clusters}} + \underbrace{|\{uv \in E^- : \exists C_i, u, v \in C_i\}|}_{\text{Negative edges within a cluster}}. \tag{1}$$

The first term counts similar pairs incorrectly separated, and the second counts dissimilar pairs incorrectly grouped together [5].

To find approximate solutions, CC is often formulated as a linear program (LP). The standard **edge-based LP** relaxation uses fractional variables $x_{uv} \in [0, 1]$ to represent the probability that nodes u and v are separated. This formulation enforces triangle inequality constraints ($x_{uw} \leq x_{uv} + x_{vw}$) to ensure the separation variables induce a valid pseudometric, while minimizing the objective:

$$\text{minimize} \quad \sum_{(u,v)\in E^+} x_{uv} + \sum_{(u,v)\in E^-} (1 - x_{uv}) \tag{LP}$$

$$\text{subject to} \quad x_{uv} + x_{vw} \geq x_{wu} \qquad \forall u, v, w \in V \tag{2}$$

$$x_{uv} \in [0, 1] \qquad \forall uv \in \binom{V}{2} \tag{3}$$

A standard approximation for CC, the *pivot-based algorithm*, leverages the optimal solution to the edge-based LP. The procedure selects a vertex uniformly at random to serve as a *pivot* and then iteratively and probabilistically clusters other vertices with the pivot based on the LP values and the signs of the underlying edges. Since edge-based LP is known to have an integrality gap of at least 2, algorithms relying solely on this relaxation were fundamentally limited by this factor.

Recently, a stronger formulation known as the **cluster LP** relaxation [14] has been introduced. This relaxation assigns a variable z_S to every nonempty subset $S \subseteq V$, $S \neq \emptyset$. Although the number of variables in this LP is exponential in $|V|$, which prohibits a direct computation of the optimal solution, [14] demonstrated a method to construct a near-optimal solution in polynomial time. Using this stronger fractional solution, they achieved a significantly improved approximation factor of $1.485 + \varepsilon$.

While standard CC excels at capturing binary relationships, many real-world applications involve richer, multi-categorical data. For instance, social networks encode diverse interaction types (e.g., professional, familial), and biological networks annotate proteins with distinct functional labels. In such settings, clusters must satisfy both structural coherence and categorical homogeneity. This motivates **Chromatic Correlation Clustering (CCC)** [10], a powerful generalization of CC.

CCC operates on edge-colored graphs, where each color represents a distinct relationship type. The problem imposes two key requirements on the final clustering:

- **Chromatic Clusters**: Each cluster is now assigned a color (or label) as well.
- **Color-Consistent Edges**: Edges *within* a cluster should ideally match the cluster's assigned color, while edges *between* different clusters should ideally be labeled with the special color γ, indicating dissimilarity.

The objective in CCC is to find a partition that minimizes the number of disagreements—edges that violate these conditions. Formally, given the color assignment $\phi : \binom{V}{2} \to L \cup \{\gamma\}$ ($|L| < \infty$), the goal is to find a node partition $\mathcal{C}$ along with the color assignment $\Phi : \mathcal{C} \to L$ that minimizes the following:

$$\text{cost}((\mathcal{C}, \Phi)) = \underbrace{|\{uv : \phi(uv) \in L, (u \in C_i, v \in C_j, i \neq j \text{ or } \exists C_i, u, v \in C_i, \Phi(C_i) \neq \phi(uv)|}_{\text{Positive edges with violated assignment}} + \underbrace{|\{uv : \phi(uv) = \gamma, \exists C_i, u, v \in C_i\}|}_{\text{Negative edges within any cluster}}.$$

This makes CCC NP-hard yet highly valuable for applications such as detecting communities with uniform relationship types in network analysis [10], merging records in databases under categorical constraints [3], and clustering proteins with shared functional annotations [28].

1.1 Our Contributions

We present a significant improvement over previous theoretical works on CCC by extending the cluster LP framework to the chromatic setting.

Theorem 1 (Main result). *For any $\varepsilon > 0$, there exists a $(\frac{18}{11} + \varepsilon)$-approximation algorithm for CCC with time complexity $n^{\text{poly}(1/\varepsilon)} \cdot \text{poly}(L)$.*

This improves upon the previous best-known approximation factor of 2.15 [24] for this problem.

Our contributions on top of the main result are the following:

- **Flexibility on the Extension of the Cluster LP Framework:** We extend the powerful cluster LP framework in [14] to the chromatic setting, maintaining intuition for the framework. This demonstrates the robustness and generality of the cluster LP approach.
- **A Strengthened Standard LP Relaxation for CCC:** Derived from the extension of cluster LP to the chromatic setting, we also provide that the standard LP for CCC can be further strengthened in a natural way (see Condition (15) compared to Condition (8)).

1.2 Overview of Techniques

Extension of Cluster LP. In Sect. 2.1, we will extend the following (cluster LP) to the chromatic setting.

$$\text{minimize} \quad \sum_{uv \in E^+} x_{uv} + \sum_{uv \in E^-} (1 - x_{uv}) \qquad \text{(cluster LP)}$$

$$\text{subject to} \quad \sum_{S \ni u} z_S = 1 \qquad \forall u \in V \tag{4}$$

$$\sum_{S \ni uv} z_S = 1 - x_{uv} \qquad \forall uv \in \binom{V}{2} \tag{5}$$

$$z_S \geq 0 \qquad \forall S \subseteq V,\, S \neq \emptyset \tag{6}$$

Here, $z_S \in [0, 1]$ indicates the number of cluster S within the clustering. Therefore, constraint (4) ensures that each node is contained in some unique cluster, while constraint (5) ensures that the separation variables x_{uv} are consistent with the probability of nodes u and v belong to the same cluster.

These ideas can be well preserved in the extension to the chromatic setting by presenting the decomposition of z_S into variables $\{z_S^c\}_{c \in L}$ (see Sect. 2.1).

Extension of Near-Optimal Solution Construction. In Sect. 4, we will also make a generalization of the whole procedure suggested by [14] of constructing the near-optimal solution for the cluster LP towards the chromatic setting. The following describes the high-level idea on each of the procedures with our extension conducted:

1. **Preclustering – See the full version for details:** Starting from clustering such that obj $= O(1) \cdot$opt, the preclustering step identifies *locally near-optimal* (i.e., mostly + internally and mostly − externally) disjoint non-singleton subsets (denoted as *atoms*) and which pairs of atom (or singleton) are strongly far apart (denoted as non-*admissible edges*) [12].
In the chromatic setting, each atom and admissible edge now have a color: The color of atom should be unique, while multiple edges with different colors between two vertices are possible. Since every atom is entirely contained within a single cluster of the optimal clustering and also within a single cluster of the target near-optimal clustering, both clusters having the same color, it follows that any admissible edge incident to an atom must have the same color as that atom.
2. **Bounded Sub-Cluster LP Relaxation – See the full version for details:** Based on the information about atoms and admissible edges from the previous step, *bounded sub-cluster LP* is then formulated to find the $(1+\varepsilon)$-approximate clustering. The bounded sub-cluster LP consists of variables y_S^s, indicating the number of clusters with size s containing S. By restricting the upper bound of s as the polynomial of $1/\varepsilon$, the number of variables in the LP is then polynomial by n, thus solving the LP takes $n^{\text{poly}(1/\varepsilon)}$ time.
In the chromatic setting, the natural extension for the variable is then $y_S^{s,c}$, indicating the number of clusters with size s of color c containing S. Since the number of variable is then multiplied by L, solving the LP takes $n^{\text{poly}(1/\varepsilon)} \cdot \text{poly}(L)$ time.
3. **Conversion to the Solution for Cluster LP –See the full version for details:** The optimal solution for the bounded sub-cluster LP is then converted to the solution for the original cluster LP, based on the fact that the

cluster inside the near-optimal solution can be effectively sampled from $\{y_S^s\}$ using the Raghavendra-Tan rounding technique [30]. Following the sampling of (s, u) with a probability proportional to y_u^s, the technique is applied to $\{y_{Su}^s\}$ normalized by y_u^s.
In the chromatic setting, the c from $y_S^{s,c}$ is yet another categorical property just as s; therefore, the process can be generalized by sampling (s, c, u) with a probability proportional to $y_u^{s,c}$, followed by the utilization of the Raghavendra-Tan rounding technique on $\{y_{Su}^{s,c}\}$ normalized by $y_u^{s,c}$.

Extension of Clustering Algorithm with Triple-based Analysis. In Sect. 2.2, we introduce a $\frac{18}{11}$-approximation algorithm for CCC from the solution of the chromatic cluster LP, which would be constructed from the previous step. The algorithm is based on the idea suggested by [14], which randomly selects the execution between the cluster-based algorithm and the pivot-based algorithm.

The original cluster-based algorithm samples the cluster S with a probability proportional to z_S, which can easily be generalized to the chromatic setting by sampling (S, c) with a probability proportional to z_S^c.

The pivot-based algorithm used in the algorithm of [14] not only uses the specific deterministic rounding function but also exploits more information from cluster LP solution by sampling the cluster during the pivoted subroutine. In this work, we will instead use a purely pivot-based algorithm for CCC with the *color pre-categorization* [31] and *rounding function* [16,24] technique. Although the color pre-categorization technique allows for simpler analysis of the algorithm, this has the limitation that the approximation factor cannot be less than 1.5, which would be discussed in Sect. 3.

We choose rounding functions for the pivot-based algorithm that correspond to the greedy selection on the pivot edges, which then yields the promising $\frac{18}{11}$-approximation in total. This can be proven by the usual triple-based analysis for CCC, with the budget non-linearly depending on the fractional edge separation value x of the LP. This is because the performance of the cluster-based algorithm depends on the value of x, which is between $[1, 2]$ [14]. The proof of such an approximation factor is given in Sect. 3.

Based on this result, we would provide an intuition in Sect. 2.2 about when we would prefer to choose the pivot-based algorithm over the cluster-based algorithm, depending on the given input ϕ. Also, since the rounding functions were naively chosen in the pivot-base algorithm, we believe that an even better approximation factor can be achieved by fine-tuning the rounding functions, or even further by exploiting more information from the cluster LP solution as in the pivot-based algorithm by [14].

1.3 Related Work

The *Correlation Clustering* (CC) problem, introduced by Ben-Dor and Yakhini [9], is APX-hard, with subsequent work delivering progressive improvements to its approximation guarantees. Early efforts established constant-factor approximations: Bansal et al. [5] laid the groundwork for such results, which were later refined to a 4-approximation by Charikar et al. [15], a 3-approximation by Ailon et al. [2], and a break through 2.06-approximation by Chawla et al. [16]. Subsequent advances using LP-rounding techniques broke the 2-approximation barrier [22]; more recently, Cohen-Addad et al. leveraged both Sherali-Adams hierarchies [22] and preclustering [21] to achieve a $(1.73+\varepsilon)$ -approximation. The current state-of-the-art, a $(1.485+\varepsilon)$ -approximation, stems from Cao et al.'s *cluster LP* framework [14]—a unified approach that subsumes previous relaxations and enables efficient solution computation. Most recently, Cao et al. [12] presented a breakthrough method for solving the Correlation Clustering LP in sublinear time.

Multi-relational network analysis motivated *Chromatic Correlation Clustering* (CCC), which extends traditional correlation clustering via colored edges. While early work by Bonchi et al. [10] developed heuristics, theoretical foundations emerged when Anava et al. [3] proposed an LP-based method, achieving a 4-approximation guarantee. Klodt et al. [28] later connected CCC to classical results, showing Pivot maintains a 3-approximation. Recent advances in approximation algorithms for CCC have achieved new theoretical milestones. Xiu et al. [31] developed an LP-based method that improved the approximation factor to 2.5. Fan et al. [24] advanced the state-of-the-art by developing a 2.15-approximation algorithm for Chromatic Correlation Clustering through an innovative combination of standard LP relaxation with problem-specific rounding techniques. Their work also established an inherent limitation of this approach, proving that no approximation ratio better than 2.11 can be achieved within the same framework. In contemporary data analysis, the application of correlation clustering frequently encounters practical limitations, such as restricted memory or the need to process data as it arrives in a continuous stream. This growing demand for efficiency has catalyzed extensive research into the development of specialized correlation clustering algorithms designed for dynamic, streaming, online, and distributed environments [4,6–8,11,13,17–20,23,25–27,29].

After the arXiv version of this paper appeared, Abbasi et al. [1] posted a manuscript independently achieving a 2-approximation for CCC. They use a cluster LP-based technique similar to ours.

2 Overview of the Framework

2.1 Chromatic Cluster LP

The standard LP relaxation for CCC suggested by [10] is as follows:

$$\text{minimize} \quad \sum_{\phi(uv)\neq\gamma} x_{uv}^{\phi(uv)} + \sum_{\phi(uv)=\gamma}\sum_{c\in L}(1-x_{uv}^{c}) \tag{CCC-LP}$$

$$\begin{aligned}
\text{subject to} \quad & x_{uv}^c \geq x_u^c,\, x_v^c & \forall uv \in \binom{V}{2},\, c \in L \quad (7)\\
& x_{uv}^c + x_{vw}^c \geq x_{wu}^c & \forall u, v, w \in V,\, c \in L \quad (8)\\
& \sum_{c \in L} x_u^c = |L| - 1 & \forall u \in V \quad (9)\\
& x_u^c,\, x_v^c,\, x_{uv}^c \in [0, 1] & \forall uv \in \binom{V}{2},\, c \in L \quad (10)
\end{aligned}$$

Here, each $1 - x_u^c$ and $1 - x_{uv}^c$ represents the fractional assignment of the vertex u and edge uv to a cluster of color c, respectively.

Applying the idea of cluster LP, we can formulate the chromatic version of the cluster LP as (chromatic cluster LP).

$$\text{minimize} \quad \sum_{\phi(uv) \in L} x_{uv}^{\phi(uv)} + \sum_{\phi(uv)=\gamma} \sum_{c \in L} (1 - x_{uv}^c) \qquad \text{(chromatic cluster LP)}$$

$$\begin{aligned}
\text{subject to} \quad & \sum_{S \ni u} z_S^c = 1 - x_u^c & \forall u \in V,\, c \in L \quad (11)\\
& \sum_{c \in L} (1 - x_u^c) = 1 & \forall u \in V \quad (12)\\
& \sum_{S \ni uv} z_S^c = 1 - x_{uv}^c & \forall uv \in \binom{V}{2},\, c \in L \quad (13)\\
& z_S^c \geq 0 & \forall S \subseteq V,\, S \neq \emptyset,\, c \in L \quad (14)
\end{aligned}$$

Here, z_S^c indicates the fractional count of cluster S of color c within the clustering. Therefore, constraint (12) ensures that each node is contained in some unique cluster, while constraint (11) and (13) ensures that x_u^c and x_{uv}^c are consistent with the probability that node u belongs to the cluster of color c, or nodes u and v belong to the same cluster of color c, respectively.

We can also check that the solution from the cluster LP induces the feasible solution for the standard LP: (7) can be deduced from (11) and (13), (9) can be deduced from (12), and (10) can be deduced from all 4 inequality constraints—(11) to (14)—in (chromatic cluster LP).

When it comes to the condition (8), even stronger constraint can be deduced, which is the following:

$$x_{uv}^c + x_{vw}^c \geq x_{wu}^c + x_v^c \quad \forall uvw \in \binom{V}{3},\, c \in L. \qquad (15)$$

In fact, this constraint is natural regarding the interpretation of variables x_{uv}^c as *bottleneck distances* [24]: Since x_{uv}^c is a bottleneck distance $d(u, v)$ from u to v conditioned by the use of edges uu_c and vv_c,

$$\begin{aligned} x_{uv}^c + x_{vw}^c &= \max\{d(u, u_c), d(u_c, v_c), d(v_c, v)\} + \max\{d(v, v_c), d(v_c, w_c), d(w_c, w)\} \\ &= \max\{\max\{d(u, u_c), d(u_c, v_c)\}, d(v_c, v)\} \\ &\quad + \max\{d(v, v_c), \max\{d(v_c, w_c), d(w_c, w)\}\} \\ &\geq d(v_c, v) + \max\{\max\{d(u, u_c), d(u_c, v_c)\}, \max\{d(v_c, w_c), d(w_c, w)\}\} \\ &\geq d(v_c, v) + \max\{d(u, u_c), d(u_c, w_c), d(w_c, w)\} \\ &= x_v^c + x_{uw}^c \end{aligned}$$

where the first inequality comes from the fact that $\max\{a, c\} + \max\{b, c\} \geq c + \max\{a, b\}$.

Since the number of variables in the chromatic cluster LP is $|L| \cdot 2^{|V|}$, we obtain the near-optimal solution rather than the actual optimal solution for the LP, whose procedure is described in Sect. 4. Thus, we have the following theorem:

Theorem 2. *The feasible solution* $\{z_S^c\}$ *for* (chromatic cluster LP) *such that* $\text{obj}(z) = (1+\varepsilon)\,\text{opt}$ *can be obtained in* $n^{\text{poly}(1/\varepsilon)} \cdot \text{poly}(L)$ *time.*

2.2 Algorithmic Framework

We can naturally extend the original cluster-based rounding algorithm by [14] to the chromatic setting, which is described in Algorithm 1.

Algorithm 1. Cluster-Based Rounding

Input: Near-optimal (chromatic cluster LP) solution $\{z_S^c\}$
Output: Chromatic clustering Φ

```
Φ(V²) ← γ, V' ← V
while V' ≠ ∅ do
    Choose (c, S) with probability proportional to z_S^c over all c ∈ L, ∅ ≠ S ⊆ V
    if V' ∩ S ≠ ∅ then
        Φ((V' ∩ S)²) ← c, V' ← V'\S
    end if
end while
```

Using Algorithm 1 exclusively results in $2+\varepsilon$-approximation algorithm, given that the solution to (chromatic cluster LP) is $(1 + O(\varepsilon))$-optimal.

Theorem 3. *Algorithm 1 yields a* $(2+\varepsilon)$*-approximation algorithm for CCC with time complexity* $n^{\text{poly}(1/\varepsilon)} \cdot \text{poly}(L)$.

Proof. In Algorithm 1, for every $uv \in \binom{V}{2}$, the probability of vertices being separated is $\frac{\sum_{c\in L}(1-x_{uv}^c)}{2-\sum_{c\in L}(1-x_{uv}^c)} \leq \sum_{c\in L}(1 - x_{uv}^c)$, while the probability of vertices not being in the same cluster of color c' is $1 - \frac{1-x_{uv}^{c'}}{2-\sum_{c\in L}(1-x_{uv}^{c})} = \frac{2x_{uv}^{c'}-\sum_{c\neq c'}(1-x_{uv}^{c})}{2-\sum_{c\in L}(1-x_{uv}^{c})} \leq \frac{2x_{uv}^{c'}}{1+x_{uv}^{c'}}$. Therefore, the probability of violation for the edge uv is bounded above by twice the value of the corresponding term in the objective. Since $\mathrm{obj}(\{z_S^c\}) = (1 + O(\varepsilon))\,\mathrm{opt}$, this finalizes the proof.

According to the proof, the bottleneck of the approximation factor occurs on the + edge whose x-value is close to 0. This behavior is different from the usual pivot-based algorithm, whose bottleneck usually occurs when some x-value is large with the tight triangle inequality. Thus, this factor can be moderated by randomizing the execution of algorithm between the cluster-based algorithm (Algorithm 1) and the pivot-based algorithm (Algorithm 2 [24,31].) This is formally described in Algorithm 3 [14].

Applying the derandomization, the algorithm would choose the better option out of two: Intuitively, the algorithm would choose the pivot-based algorithm if the input ϕ is *almost well-clustered* (i.e., clusters of the optimal clustering consist of a very low number of violating internal edges) and choose the cluster-based algorithm otherwise.

Algorithm 2. Pivot-Based Rounding

Input: Near-optimal (chromatic cluster LP) solution $\{z_S^c\}$, rounding functions $f^+, f^-, f^\circ : [0,1] \to [0,1]$
Output: Chromatic clustering Φ

$\Phi(V^2) \leftarrow \gamma$, $V_c \leftarrow \emptyset$ for all $c \in L$
for $u \in V$ **do**
  if $\exists c \in L$ s.t. $x_u^c < \frac{1}{2}$ **then**
    $V_c \leftarrow V_c \cup \{u\}$
  end if
end for
for $c \in L$ **do**
  while $V_c \neq \emptyset$ **do**
    Choose $u \in V_c$ uniformly random
    $C \leftarrow \{u\}$
    for $v \in V_c \backslash \{u\}$ **do**
      Set p_{uv} as following:

$$p_{uv} = \begin{cases} f^+(x_{uv}), & \phi(uv) = c; \\ f^-(x_{uv}), & \phi(uv) = \gamma; \\ f^\circ(x_{uv}), & \text{otherwise.} \end{cases}$$

      $C \leftarrow C \cup \{v\}$ with probability $1 - p_{uv}$
    end for
    $\Phi(C^2) \leftarrow c$, $V_c \leftarrow V_c \backslash C$
  end while
end for

Algorithm 3. Cluster/Pivot-based Rounding

Input: Near-optimal (chromatic cluster LP) solution $\{z_S^c\}$, target approximation factor $\alpha \in [1.5, 2]$
Output: Chromatic clustering Φ

return Algorithm 1 with probability $\alpha/2$, Algorithm 2 otherwise

Indeed, Algorithm 3 achieves the significantly smaller approximation factor of $\frac{18}{11} + \varepsilon \approx 1.64 + \varepsilon$, even with a greedy selection of pivoted edges that only gather edges of the same color with the pivot vertex. This corresponds to

$$f^+ = 0,\ f^- = f^\circ = 1.$$

The proof of the factor is provided in Sect. 3.

2.3 Clustering as a Binary Function

For analysis purposes, we provide an alternative definition for chromatic clustering.

Proposition 1 (Chromatic Clustering as a Binary Function). *Chromatic Clustering $\Phi : \mathcal{C} \to L$ has a one-to-one correspondence with the symmetric function $\Phi : V \times V \to L \cup \{0, \gamma\}$ such that*

1. $\Phi(u, v) = 0 \iff u = v$,
2. $\Phi(u, v) \in L \implies \Phi(u, w) = \Phi(v, w) \in \{\Phi(u, v), \gamma\},\ \forall w \in V \backslash \{u, v\}$.

This interpretation is analogous to the ultrametric from the UMVD problem. When comparing the order structure of the codomain between CCC and UMVD, as shown in Fig. 1, elements of the codomain for the CCC problem are organized horizontally, while those for the UMVD problem are organized vertically.

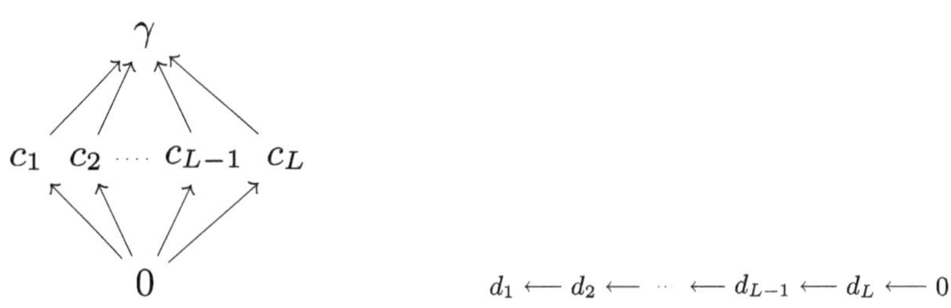

(a) CCC (b) UMVD

Fig. 1. Comparison in order structure of codomain for two problems. The directional edge $a \to b$ indicates $a \leq b$. Self-loop on each vertex is omitted.

Next, we provide some terminology helpful for analyzing local cost or its difference.

Definition 1. *For a vertex v, define $\Phi_v := \Phi(v, \cdot)$.*

Definition 2. *For $f, g : V \to R$ and $K \subseteq V$, define $d_0(f,g) := |\{v \in V \mid f(v) \neq g(v)\}|$ and $d_0^K(f,g) := d_0(f,g) + |\{v \in V \backslash K \mid f(v) \neq g(v)\}|$.*

Note that the weight given for the vertices outside K is twice larger than for those inside K in the definition of d_0^K. This terminology is especially helpful when analyzing the difference in objective value between two clusterings whose difference occurs only when at least one of the vertex in the pair is included in K.

Proposition 2. $\mathrm{obj}(\Phi) = \frac{1}{2}\sum_{u \in V} d_0(\phi_u, \Phi_u)$.

Proposition 3. *If $\{(u,v) \mid \Phi^1(u,v) \neq \Phi^2(u,v)\} \subseteq K \times V \cup V \times K$, then*

$$\mathrm{obj}(\Phi^1) - \mathrm{obj}(\Phi^2) = \frac{1}{2}\sum_{u \in K} d_0^K(\phi_u, \Phi_u^1) - d_0^K(\phi_u, \Phi_u^2).$$

Proposition 4. $d_0(f,g) - d_0(f,h) = |\{v \in V \mid g(v) \neq f(v) = h(v)\}| - |\{v \in V \mid g(v) = f(v) \neq h(v)\}|$.

The following definitions provide alternative definitions of basic terminologies on chromatic clustering from the perspective of a binary function.

Definition 3. *The cluster C of Φ, denoted as $C \in \Phi$, is a maximal subset of V such that $C^2 \subseteq \Phi^{-1}(L \cup \{0\})$.*

Definition 4. *If $S \subseteq \exists C \in \Phi$ for a nonempty subset S of V (that is, a subset of some cluster), we define $\Phi(S)$ as $\max \Phi(C^2)$.*

3 Triple-Based Analysis

We first prove that Algorithm 3 achieves an approximation factor of $(\frac{18}{11} + \varepsilon)$. Recall our selection of rounding functions:

$$f^+(x) = 0, \quad f^-(x) = f^\circ(x) = 1. \tag{16}$$

The goal is to ensure that the Algorithm 3 with the target approximation factor α actually achieves the $\alpha + \varepsilon$-approximation if $\alpha \geq \frac{18}{11}$.

Let any $uv \in \binom{V}{2}$ be given. If $\nexists c \in L$ such that $u, v \in V_c$, then $\Phi(u,v) = \gamma$ if Algorithm 2 is used, due to the first subroutine of the algorithm. On the other hand, any of the values x_{uv}^c is bounded below by $1/2$, since either x_u^c or x_v^c is bounded below by $1/2$. Therefore, if $\phi(uv) \in L$, then the cost incurred on the edge uv is

$$\frac{\alpha}{2} \cdot \frac{2x_{uv}^{\phi(uv)} - \sum_{c \neq \phi(uv)}(1 - x_{uv}^c)}{2 - \sum_{c \in L}(1 - x_{uv}^c)} + \left(1 - \frac{\alpha}{2}\right) \cdot 1 \leq \frac{\alpha}{2} \cdot \frac{2x_{uv}^{\phi(uv)}}{1 + x_{uv}^{\phi(uv)}} + 1 - \frac{\alpha}{2} = 1 - \frac{\alpha}{2} \cdot \frac{1 - x_{uv}^{\phi(uv)}}{1 + x_{uv}^{\phi(uv)}}$$

whose difference with $\alpha x_{uv}^{\phi(uv)}$ is

$$\alpha x_{uv}^{\phi(uv)} - \left(1 - \frac{\alpha}{2} \cdot \frac{1 - x_{uv}^{\phi(uv)}}{1 + x_{uv}^{\phi(uv)}}\right) = \frac{\alpha}{2} \cdot \left(2(1 + x_{uv}^{\phi(uv)}) + \frac{2}{1 + x_{uv}^{\phi(uv)}} - 3\right) - 1 \geq \frac{\alpha}{2} \cdot \frac{4}{3} - 1 \geq 0$$

since $x_{uv}^{\phi(uv)} \geq 1/2 > 0$ and $\alpha \geq 1.5$. Thus, the cost incurred on the edge is bounded above by α times the corresponding LP value. Since the inequality is tight, this also implies that Algorithm 3 cannot achieve the approximation factor below 1.5 due to the strict majority selection in the first subroutine of Algorithm 2.

If $\phi(uv) = \gamma$, then the cost incurred on the edge uv is

$$\frac{\alpha}{2} \cdot \frac{\sum_{c \in L}(1 - x_{uv}^c)}{2 - \sum_{c \in L}(1 - x_{uv}^c)} + \left(1 - \frac{\alpha}{2}\right) \cdot 0 \leq \frac{\alpha}{2} \cdot \sum_{c \in L}(1 - x_{uv}^c) \leq \alpha \cdot \sum_{c \in L}(1 - x_{uv}^c),$$

which also implies that the cost incurred on the edge is bounded above by α times the corresponding LP value.

Therefore, it is sufficient to check on edges uv where both x_u^c and x_v^c are smaller than 1/2 for some $c \in L$, i.e., the vertex u and v have the same strict majority color as c. We can apply the usual triangle analysis to analyze Algorithm 2 since the second subroutine of the algorithm is exactly the color-wise execution of the pivot algorithm [24,31].

Now, fix any $c \in L$.

Definition 5 (Edge sign assignment with respect to color c**).** $E^+ = \phi^{-1}(c) \cap \binom{V_c}{2}$, $E^- = \phi^{-1}(\gamma) \cap \binom{V_c}{2}$, $E^\circ = \phi^{-1}(L \backslash \{c\}) \cap \binom{V_c}{2}$.

Suppose:

$$e.cost_w(uv) + e.cost_u(vw) + e.cost_v(wu) \leq e.lp_w(uv) + e.lp_u(vw) + e.lp_v(wu) \quad (17)$$

for all $u, v, w \in V_c$, where

$$e.cost_w(u,v) = \begin{cases} p_{uw}(1 - p_{vw}) + (1 - p_{uw})p_{vw}, & uv \in E^+; \\ (1 - p_{uw})(1 - p_{vw}), & uv \in E^-; \\ 1 - p_{uw}p_{vw}, & uv \in E^\circ; \end{cases} \quad (18)$$

$$e.lp_w(u,v) = (1 - p_{uw}p_{vw}) \cdot b(uv), \quad (19)$$

$$b(uv) = \begin{cases} \frac{\alpha}{1-\alpha/2} \cdot \frac{{x_{uv}^c}^2}{1+x_{uv}^c}, & uv \in E^+; \\ \frac{\alpha}{1-\alpha/2} \cdot \frac{(1+2x_{uv}^c)(1-x_{uv}^c)}{2(1+x_{uv}^c)}, & uv \in E^-; \\ \frac{\alpha}{1-\alpha/2} \cdot \frac{\max\left\{\frac{1}{2}, 1-x_{uv}^c, 1-x_{vw}^c, 1-x_{wu}^c\right\}^2}{1+\max\left\{\frac{1}{2}, 1-x_{uv}^c, 1-x_{vw}^c, 1-x_{wu}^c\right\}}, & uv \in E^\circ. \end{cases} \quad (20)$$

Here, $e.cost_w(u,v)$ denotes the probability of a violation on the edge uv conditioned on the pivot w, while $e.lp_w(u,v)$ denotes the probability of at least one of u and v contained in the cluster C conditioned on the pivot w multiplied by the charging factor b. Then the following is the sufficient condition to achieve the approximation factor of $\alpha + \varepsilon$ [14]:

$$\phi(uv) \in L \implies \frac{\alpha}{2} \cdot \frac{2x_{uv}^{\phi(uv)} - \sum_{c \neq \phi(uv)}(1 - x_{uv}^c)}{2 - \sum_{c \in L}(1 - x_{uv}^c)} + \left(1 - \frac{\alpha}{2}\right) \cdot b(uv) \leq \alpha x_{uv}^{\phi(uv)},$$

$$\phi(uv) \in \gamma \implies \frac{\alpha}{2} \cdot \frac{\sum_{c \in L}(1 - x_{uv}^c)}{2 - \sum_{c \in L}(1 - x_{uv}^c)} + \left(1 - \frac{\alpha}{2}\right) \cdot b(uv) \leq \alpha \cdot \sum_{c \in L}(1 - x_{uv}^c),$$

which turns out to be true.

Lemma 1. *If inequality* (17) *holds for all possible configurations, then Algorithm 3 yields an $\alpha + \varepsilon$-approximation algorithm for CCC.*

Proof. Consider the following three cases:

1. $\phi(uv) = c$

$$\frac{\alpha}{2} \cdot \frac{2x_{uv}^{\phi(uv)} - \sum_{c' \neq \phi(uv)}(1 - x_{uv}^{c'})}{2 - \sum_{c' \in L}(1 - x_{uv}^{c'})} + \left(1 - \frac{\alpha}{2}\right) \cdot b(uv) \leq \frac{\alpha}{2} \cdot \frac{2x_{uv}^c}{1 + x_{uv}^c} + \left(1 - \frac{\alpha}{2}\right) \cdot b(uv) = \alpha x_{uv}^c.$$

2. $\phi(uv) = \gamma$
   Let $h(x) = \frac{(3-2x)x}{2(2-x)}$ $(x \in [0,1])$. Then h is an increasing function since $h'(x) = \frac{(1-x)(3-x)}{(2-x)^2} \geq 0$ in the domain. Since $0 \leq 1 - x_{uv}^c \leq \sum_{c' \in L}(1 - x_{uv}^{c'}) \leq 1$,

$$\frac{1 - \alpha/2}{\alpha} \cdot b(uv) = h(1 - x_{uv}^c) \leq h\Big(\sum_{c' \in L}(1 - x_{uv}^{c'})\Big) = \frac{(3 - 2 \cdot \sum_{c' \in L}(1 - x_{uv}^{c'})) \cdot \sum_{c' \in L}(1 - x_{uv}^{c'})}{2(2 - \sum_{c' \in L}(1 - x_{uv}^{c'}))}.$$

   Therefore,

$$\frac{\alpha}{2} \cdot \frac{\sum_{c \in L}(1 - x_{uv}^c)}{2 - \sum_{c \in L}(1 - x_{uv}^c)} + \left(1 - \frac{\alpha}{2}\right) \cdot b(uv) \leq \frac{\alpha}{2} \cdot \frac{4 - 2 \cdot \sum_{c' \in L}(1 - x_{uv}^{c'})}{2 - \sum_{c' \in L}(1 - x_{uv}^{c'})} \cdot \sum_{c' \in L}(1 - x_{uv}^{c'})$$
$$= \alpha \cdot \sum_{c' \in L}(1 - x_{uv}^{c'}).$$

3. $\phi(uv) \in L \backslash \{c\}$
   Let $g(x) = \frac{x^2}{1+x}$ $(x \in [0,1])$. Then g is an increasing function since $g'(x) = \frac{x(2+x)}{(1+x)^2} \geq 0$ in the domain. Since $0 \leq \max\left\{\frac{1}{2}, 1 - x_{uv}^c, 1 - x_{vw}^c, 1 - x_{wu}^c\right\} \leq x_{uv}^{\phi(uv)}$ [31],

$$\frac{1 - \alpha/2}{\alpha} \cdot b(uv) = g\left(\max\left\{\frac{1}{2}, 1 - x_{uv}^c, 1 - x_{vw}^c, 1 - x_{wu}^c\right\}\right) \leq g(x_{uv}^{\phi(uv)}) = \frac{{x_{uv}^{\phi(uv)}}^2}{1 + x_{uv}^{\phi(uv)}}.$$

   Therefore,

$$\frac{\alpha}{2} \cdot \frac{2x_{uv}^{\phi(uv)} - \sum_{c' \neq \phi(uv)}(1 - x_{uv}^{c'})}{2 - \sum_{c' \in L}(1 - x_{uv}^{c'})} + \left(1 - \frac{\alpha}{2}\right) \cdot b(uv) \leq \alpha \cdot \frac{x_{uv}^{\phi(uv)} + {x_{uv}^{\phi(uv)}}^2}{1 + x_{uv}^{\phi(uv)}}$$
$$= \alpha \cdot x_{uv}^{\phi(uv)}.$$

Therefore, the sufficient condition for the approximation factor of $\alpha + \varepsilon$ is satisfied.

Now, it is sufficient to show that the inequality (17) is valid.

Lemma 2. *The ineqaulity* (17) *is true with rounding functions given as* (16) *and* $\alpha = \frac{18}{11} \approx 1.64$. *The proof is deferred to appendix.*

This finalizes the proof of Theorem 1.

4 Construction of Near-Optimal Solution to Chromatic Cluster LP

In this section, we describe the whole procedure of constructing the near-optimal solution to (chromatic cluster LP). This procedure comprises four key components: preclustering, bounded sub-cluster LP relaxation, cluster sampling, and the construction of the LP under the chromatic setting. Due to space constraints, detailed elaborations are deferred to the full version of this paper on arXiv.

Acknowledgments. We thank the anonymous reviewers for their valuable feedback. The work of Chenglin Fan and Dahoon Lee on this project was supported by the National Research Foundation of Korea (NRF) grant funded by the Korean government (MSIT) (No. RS-2026-25484051), as well as by the New Faculty Startup Fund from Seoul National University.

Disclosure of Interests. The authors have no competing interests to declare that are relevant to the content of this article.

References

1. Abbasi, F., An, H., Byrka, J., Lee, C., Shin, Y.: Chromatic correlation clustering via cluster LP (2025). https://arxiv.org/abs/2510.13446
2. Ailon, N., Charikar, M., Newman, A.: Aggregating Inconsistent Information: Ranking and Clustering, vol. 55. Association for Computing Machinery, New York, NY, USA (2008). https://doi.org/10.1145/1411509.1411513
3. Anava, Y., Avigdor-Elgrabli, N., Gamzu, I.: Improved theoretical and practical guarantees for chromatic correlation clustering. In: Proceedings of the 24th International Conference on World Wide Web, WWW 2015, pp. 55–65. International World Wide Web Conferences Steering Committee, Republic and Canton of Geneva, CHE (2015). https://doi.org/10.1145/2736277.2741629
4. Assadi, S., Wang, C.: Sublinear time and space algorithms for correlation clustering via sparse-dense decompositions. In: Proceedings of the 13th Innovations in Theoretical Computer Science Conference (ITCS), pp. 10:1–10:20 (2022)
5. Bansal, N., Blum, A., Chawla, S.: Correlation clustering. Mach. Learn. **56**(1–3), 89–113 (2004). https://doi.org/10.1023/B:MACH.0000033116.57574.95
6. Bateni, M., et al.: Optimal fully dynamic k-center clustering for adaptive and oblivious adversaries. In: Proceedings of the 2023 Annual ACM-SIAM Symposium on Discrete Algorithms (SODA), pp. 2677–2727 (2023)

7. Behnezhad, S., Charikar, M., Ma, W., Tan, L.: Almost 3-approximate correlation clustering in constant rounds. In: 63rd IEEE Annual Symposium on Foundations of Computer Science (FOCS), pp. 720–731 (2022)
8. Behnezhad, S., Charikar, M., Ma, W., Tan, L.: Single-pass streaming algorithms for correlation clustering. In: Proceedings of the 2023 ACM-SIAM Symposium on Discrete Algorithms (SODA), pp. 819–849 (2023)
9. Ben-Dor, A., Yakhini, Z.: Clustering gene expression patterns. In: Proceedings of the Third Annual International Conference on Computational Molecular Biology, RECOMB 1999, pp. 33–42. Association for Computing Machinery, New York, NY, USA (1999). https://doi.org/10.1145/299432.299448
10. Bonchi, F., Gionis, A., Gullo, F., Tsourakakis, C.E., Ukkonen, A.: Chromatic correlation clustering. ACM Trans. Knowl. Discov. Data **9**(4) (2015). https://doi.org/10.1145/2728170
11. Braverman, V., Dharangutte, P., Pai, S., Shah, V.: Fully dynamic adversarially robust correlation clustering in polylogarithmic update time. In: Proceedings of the 28th International Conference on Artificial Intelligence and Statistics (AISTATS) (2025)
12. Cao, N., et al.: Solving the correlation cluster LP in sublinear time. In: Proceedings of the 57th Annual ACM Symposium on Theory of Computing, STOC 2025, pp. 1154–1165. Association for Computing Machinery, New York, NY, USA (2025). https://doi.org/10.1145/3717823.3718181
13. Cao, N., et al.: Static to dynamic correlation clustering. CoRR abs/2504.12060 (2025)
14. Cao, N., Cohen-Addad, V., Lee, E., Li, S., Newman, A., Vogl, L.: Understanding the cluster linear program for correlation clustering. In: Proceedings of the 56th Annual ACM Symposium on Theory of Computing, STOC 2024, pp. 1605–1616. Association for Computing Machinery, New York, NY, USA (2024). https://doi.org/10.1145/3618260.3649749
15. Charikar, M., Guruswami, V., Wirth, A.: Clustering with qualitative information. J. Comput. Syst. Sci. **71**(3), 360–383 (2005). https://doi.org/10.1016/j.jcss.2004.10.012, learning Theory 2003
16. Chawla, S., Makarychev, K., Schramm, T., Yaroslavtsev, G.: Near optimal LP rounding algorithm for correlation clustering on complete and complete k-partite graphs. In: Proceedings of the 47th Annual ACM Symposium on Theory of Computing (STOC), pp. 219–228 (2015)
17. Cohen-Addad, V., Hjuler, N., Parotsidis, N., Saulpic, D., Schwiegelshohn, C.: Fully dynamic consistent facility location. In: Proceedings of the 33rd Annual Conference on Neural Information Processing Systems (NeurIPS) (2019)
18. Cohen-Addad, V., Lattanzi, S., Maggiori, A., Parotsidis, N.: Online and consistent correlation clustering. In: Proceedings of the 39th International Conference on Machine Learning (ICML), pp. 4157–4179 (2022)
19. Cohen-Addad, V., Lattanzi, S., Maggiori, A., Parotsidis, N.: Dynamic correlation clustering in sublinear update time. In: Proceedings of the 41st International Conference on Machine Learning (ICML), pp. 9230–9270 (2024)
20. Cohen-Addad, V., Lattanzi, S., Mitrovic, S., Norouzi-Fard, A., Parotsidis, N., Tarnawski, J.: Correlation clustering in constant many parallel rounds. In: Proceedings of the 38th International Conference on Machine Learning (ICML), vol. 139, pp. 2069–2078 (2021)
21. Cohen-Addad, V., Lee, E., Li, S., Newman, A.: Handling correlated rounding error via preclustering: a 1.73-approximation for correlation clustering. In: 2023 IEEE

64th Annual Symposium on Foundations of Computer Science (FOCS), pp. 1082–1104 (2023). https://doi.org/10.1109/FOCS57990.2023.00065
22. Cohen-Addad, V., Lee, E., Newman, A.: Correlation Clustering with Sherali-Adams . In: 2022 IEEE 63rd Annual Symposium on Foundations of Computer Science (FOCS), pp. 651–661. IEEE Computer Society, Los Alamitos, CA, USA (2022). https://doi.org/10.1109/FOCS54457.2022.00068
23. Dalirrooyfard, M., Makarychev, K., Mitrovic, S.: Pruned pivot: correlation clustering algorithm for dynamic, parallel, and local computation models. CoRR abs/2402.15668 (2024). https://doi.org/10.48550/arXiv.2402.15668
24. Fan, C., Lee, D., Lee, E.: Improved approximation algorithms for chromatic and pseudometric-weighted correlation clustering. CoRR abs/2505.21939v2 (2025). https://doi.org/10.48550/ARXIV.2505.21939, to appear at NeurIPS 2025
25. Fichtenberger, H., Mirrokni, V., Zadimoghaddam, M.: Correlation clustering in data streams. In: Proceedings of the 33rd Annual ACM Symposium on Parallelism in Algorithms and Architectures (SPAA) (2021)
26. Guo, A., Mitrovic, S., Vassilvitskii, S.: Distributed correlation clustering. In: Advances in Neural Information Processing Systems (NeurIPS) (2021)
27. Jaghargh, M., Vassilvitskii, S., Lattanzi, S.: Scalable correlation clustering: an empirical study. In: Proceedings of the 25th ACM SIGKDD International Conference on Knowledge Discovery & Data Mining (2019)
28. Klodt, N., Seifert, L., Zahn, A., Casel, K., Issac, D., Friedrich, T.: A color-blind 3-approximation for chromatic correlation clustering and improved heuristics. In: Proceedings of the 27th ACM SIGKDD Conference on Knowledge Discovery & Data Mining, KDD 2021, pp. 882–891. Association for Computing Machinery, New York, NY, USA (2021). https://doi.org/10.1145/3447548.3467446
29. Lattanzi, S., Vassilvitskii, S.: Consistent k-clustering. In: Proceedings of the 34th International Conference on Machine Learning (ICML), pp. 1975–1984 (2017)
30. Raghavendra, P., Tan, N.: Approximating CSPs with Global Cardinality Constraints Using SDP Hierarchies, pp. 373–387 (2012). https://doi.org/10.1137/1.9781611973099.33
31. Xiu, Q., Han, K., Tang, J., Cui, S., Huang, H.: Chromatic correlation clustering, revisited. In: Advances in Neural Information Processing Systems 35: Annual Conference on Neural Information Processing Systems 2022, NeurIPS 2022, New Orleans, LA, USA, November 28 - December 9 (2022)

O(log n)-Approximation Algorithms for Bipartiteness Ratio

Tasuku Soma[1,2(✉)], Mingquan Ye[3,4], and Yuichi Yoshida[3]

[1] The Institute of Statistical Mathematics, Tokyo, Japan
[2] RIKEN AIP, Tokyo, Japan
soma@ism.ac.jp
[3] National Institute of Informatics, Tokyo, Japan
yyoshida@nii.ac.jp
[4] University of California, Santa Cruz, Santa Cruz, USA
mye9@uic.edu

Abstract. We propose an $O(\log n)$-approximation algorithm for the bipartiteness ratio of undirected graphs introduced by Trevisan [54], where n is the number of vertices. Our approach extends the cut-matching game framework for sparsest cut to the bipartiteness ratio, and requires only poly $\log n$ many single-commodity undirected maximum flow computations. Therefore, with the current fastest undirected max-flow algorithms, it runs in almost linear time. Along the way, we introduce the concept of well-linkedness for skew-symmetric graphs and prove a novel characterization of bipartiteness ratio in terms of well-linkedness in an auxiliary skew-symmetric graph, which may be of independent interest.

As an application, we devise an $\widetilde{O}(mn)$-time algorithm for the minimum uncut problem: given a graph whose optimal cut leaves an η fraction of edges uncut, we find a cut that leaves only an $O(\log n \log(1/\eta)) \cdot \eta$ fraction of edges uncut, where m is the number of edges.

Finally, we propose a directed analogue of the bipartiteness ratio, and we give a polynomial-time algorithm that achieves an $O(\log n)$ approximation for this measure via a directed Leighton–Rao-style embedding. We also propose an algorithm for the minimum directed uncut problem with a guarantee similar to that for the minimum uncut problem.

Keywords: Approximation Algorithms · Bipartiteness Ratio · Cut-matching Game

1 Introduction

Let $G = (V, E, w)$ be a connected undirected graph with n vertices, m edges, and a positive edge weight $w : E \to \mathbb{Z}_{++}$, where $\mathbb{Z}_{++}$ is a set of positive integers. The *(normalized) Laplacian matrix* of G is given by $\mathbf{I}_n - \mathbf{D}^{-1/2}\mathbf{A}\mathbf{D}^{-1/2}$, where $\mathbf{I}_n \in \mathbb{R}^{n\times n}$ is the $n \times n$ identity matrix, $\mathbf{A} \in \mathbb{R}^{n\times n}$ is the weighted adjacency

S. S. Dey et al. (Eds.): IPCO 2026, LNCS 16588, pp. 97–111, 2026.
https://doi.org/10.1007/978-3-032-28691-8_7

matrix of G, and $\mathbf{D} \in \mathbb{R}^{n\times n}$ is the diagonal matrix with $\mathbf{D}_{ii}$ being equal to the weighted degree of vertex i. The Laplacian matrix is symmetric and positive semidefinite, and its eigenvalues satisfy $0 = \lambda_1 \leq \cdots \leq \lambda_n \leq 2$. A classical result in spectral graph theory states that G is bipartite if and only if $\lambda_n = 2$. Trevisan [54] proved a quantitative version of this result. For a nonzero vector $\mathbf{x} \in \{0, \pm 1\}^V$, let

$$\beta(\mathbf{x}) := \frac{\sum_{e=(i,j)\in E} w(e) \cdot |x_i + x_j|}{\sum_{i\in V} \deg(i) \cdot |x_i|},$$

where $\deg(i) = \sum_{(i,j)\in E} w(i,j)$ is the weighted degree of vertex i. The *bipartiteness ratio* of G is then defined by

$$\beta(G) := \min_{\mathbf{x}\in\{0,\pm1\}^V \setminus \{\mathbf{0}\}} \beta(\mathbf{x}). \tag{1.1}$$

Since each non-zero $\{0, \pm 1\}$-vector $\mathbf{x}$ corresponds to a tripartition (L, R, Z) of V such that $L = \{i \in V \mid x_i = 1\}$, $R = \{j \in V \mid x_j = -1\}$, and $Z = \{k \in V \mid x_k = 0\}$, we can represent

$$\beta(G) = \min_{(L,R,Z)\text{: tripartition of } V} \frac{2w(E(L)) + 2w(E(R)) + w(E(L \cup R, Z))}{\mathrm{vol}(L \cup R)},$$

where $E(L)$ (resp., $E(R)$) is the set of edges whose endpoints are within L (resp., R), and $E(L \cup R, Z)$ is the set of edges connecting $L \cup R$ and Z, and $\mathrm{vol}(L \cup R) = \sum_{i\in L\cup R} \deg(i)$ is the volume of $L \cup R$. Obviously, $\beta(G) = 0$ if and only if G is bipartite. Trevisan [54] showed[1] that the bipartiteness ratio is closely related to the largest eigenvalue λ_n of the Laplacian matrix, specifically,

$$\frac{2 - \lambda_n}{2} \leq \beta(G) \leq \sqrt{2(2 - \lambda_n)}.$$

Furthermore, Trevisan also present a simple algorithm that finds a nonzero vector $\mathbf{x} \in \{0, \pm 1\}^V$ such that $\beta(\mathbf{x}) \leq \sqrt{2(2 - \lambda_n)}$ given an eigenvector corresponding to λ_n.

Trevisan's inequality can be regarded as an analogue of the *Cheeger's inequality* [3,4], which relates the second smallest eigenvalue with the *conductance* of graphs. For a vertex subset $\emptyset \subsetneq S \subsetneq V$, let

$$\phi(S) := \frac{w(E(S, \overline{S}))}{\min\{\mathrm{vol}(S), \mathrm{vol}(\overline{S})\}},$$

where $E(S, \overline{S})$ is the set of edges connecting S and $\overline{S}$. Then the conductance of G is defined as

$$\phi(G) := \min_{\emptyset\subsetneq S\subsetneq V} \phi(S).$$

[1] Although Trevisan [54] states the Cheeger-type inequality for unweighted graphs, the same argument carries over directly to the weighted case with only minor adjustments.

Cheeger's inequality states that

$$\frac{\lambda_2}{2} \leq \phi(G) \leq \sqrt{2\lambda_2}.$$

Furthermore, there is a simple algorithm that finds S such that $\phi(S) \leq \sqrt{2\lambda_2}$ given an eigenvector corresponding to λ_2.

Just as Cheeger's inequality captures a graph's expansion properties through the second smallest eigenvalue of its Laplacian, the bipartiteness ratio quantifies a graph's deviation from bipartiteness via the largest eigenvalue of the normalized Laplacian. Indeed, Trevisan's inequality (and its variants) is sometimes called the *dual Cheeger inequality* [10,49]. As an application of his inequality, Trevisan [54] showed a purely spectral algorithm for max cut with a nontrivial approximation ratio better than 1/2. Recently, bipartiteness ratio has found applications in network analysis in a similar fashion to graph conductance and sparsest cut [9,46,56].[2] Despite these rich connections and their algorithmic promise, obtaining nontrivial approximation guarantees for the bipartiteness ratio remains an open challenge: no polynomial-time approximation algorithm is currently known. Developing such an algorithm would yield new spectral and combinatorial tools both for classical maximum cut derivatives and for the burgeoning field of network analysis.

1.1 Our Contributions

We present the first $O(\log n)$-approximation algorithm for the bipartiteness ratio of undirected graphs. More precisely, we study the following *b-bipartiteness ratio*, which generalizes the original bipartiteness ratio [54]. Let $b : V \to \mathbb{Z}_{++}$ be a positive vertex weight. For a vector $\mathbf{x} \in \{0, \pm 1\}^V \setminus \{\mathbf{0}\}$, let

$$\beta_b(\mathbf{x}) := \frac{\sum_{(i,j)\in E} w(i,j) \cdot |x_i + x_j|}{\sum_{i\in V} b(i) \cdot |x_i|}.$$

Then, we define the b-bipartiteness ratio of G by $\beta_b(G) := \min_{\mathbf{x}\in\{0,\pm 1\}^V\setminus\{\mathbf{0}\}} \beta_b(\mathbf{x})$. Note that the original bipartiteness ratio $\beta(G)$ (see (1.1)) is the special case of $\beta_b(G)$ where $b(i) = \deg(i)$ $(i \in V)$.

Here is our main result.

Theorem 1. *There is a randomized $O(\log n)$-approximation algorithm for the b-bipartiteness ratio of an undirected graph. That is, the algorithm finds a nonzero vector $\mathbf{x} \in \{0, \pm 1\}^V$ such that $\beta_b(\mathbf{x}) \leq O(\log n) \cdot \beta_b(G)$ with probability at least $1 - 1/\mathrm{poly}(n)$. The time complexity is $O(\log(w(E) \cdot b(V)) \cdot \log^3 n \cdot \max\{\log^2 n, \log b(V)\} \cdot \min\{b(V), n^2\})$ arithmetic operations and $O(\log(w(E) \cdot b(V)) \cdot \log^2 n)$ single-commodity max-flow computations on an auxiliary undirected graph of size $O(m + n)$.*

[2] Indeed [9,46,56] studied even a more general network model called *signed graphs* and analyzed spectral clustering. Since the main motivation of the present paper is in approximation algorithms for the bipartiteness ratio, we focus on ordinary undirected graphs rather than signed graphs.

By employing the almost-linear–time algorithm for exact maximum flow [18], our algorithm runs in $\widetilde{O}(\min\{b(V), n^2\} + m)$ time. In the case of the original bipartiteness ratio, where $b(V) = O(m)$, this simplifies to a total running time of $\widetilde{O}(m)$.

1.2 Application to Min Uncut

The *maximum cut* and *minimum uncut* problems are extensively studied classical combinatorial optimization problems. In fact, they are closely related to the bipartiteness ratio problem. In the both problems, we are given a undirected graph $G = (V, E)$ and the goal is to find a bipartition $(S, \overline{S})$ of V. In the maximum cut problem, the objective is maximizing the weight of edges crossing the cut $E(S, \overline{S})$, whereas in the minimum uncut problem, the objective is minimizing the weight of edges that remain uncut $E(S) \cup E(\overline{S})$. To see that, recall that each non-zero vector $\mathbf{x} \in \{0, \pm 1\}^V$ corresponds to a partition of V such that $L = \{i \mid x_i = 1\}$, $R = \{j \mid x_j = -1\}$, and $Z = \{k \mid x_k = 0\}$. Suppose $Z \subseteq V$ is given, that is, we set the subvector $\mathbf{x}_Z$ of $\mathbf{x}$ corresponding to Z to zero, then we have

$$
\begin{aligned}
&\min_{\mathbf{x} \in \{0,\pm 1\}^V \setminus \{\mathbf{0}\},\ \mathbf{x}_Z = \mathbf{0}} \frac{\sum_{(u,v) \in E} w(u,v) \cdot |x_u + x_v|}{\sum_{u \in V} b(u) \cdot |x_u|} \\
&= \min_{L \cup R = V \setminus Z,\ L \cap R = \emptyset} \frac{2 \cdot w(E(L)) + 2 \cdot w(E(R)) + w(E(L \cup R, Z))}{b(L \cup R)} \\
&= \min_{L \cup R = V \setminus Z,\ L \cap R = \emptyset} 1 - \frac{2 \cdot w(E(L,R))}{b(L \cup R)} \\
&= 1 - 2 \cdot \max_{L \cup R = V \setminus Z,\ L \cap R = \emptyset} \frac{w(E(L,R))}{b(L \cup R)},
\end{aligned}
\tag{1.2}
$$

which is reduced to computing the maximum cut of the subgraph induced by the vertex subset $V \setminus Z$ since Z is given.

Likewise, given an approximate algorithm for bipartiteness ratio, we can develop some approximate algorithm for max cut [39,54]. Given a non-zero vector $\mathbf{x} \in \{0, \pm 1\}^V$ corresponding to the partition $V = L \cup R \cup Z$, we have

$$
\beta(\mathbf{x}) = \frac{2 \cdot w(E(L)) + 2 \cdot w(E(R)) + w(E(L \cup R, Z))}{\mathrm{vol}(L \cup R)} = 1 - 2 \cdot \frac{w(E(L,R))}{\mathrm{vol}(L \cup R)}.
$$

If we have an upper bound for $\beta(\mathbf{x})$, that naturally gives rise to a lower bound for $\frac{w(E(L,R))}{\mathrm{vol}(L \cup R)}$. Consequently, we can lower bound the fraction of cutting edges in the subgraph induced by the vertex subset $L \cup R$, and continue this process on the remaining subgraph of Z. The following is our result for min uncut.

Theorem 2. *Given a graph whose minimum uncut fraction is η, there exists a randomized algorithm that returns a cut leaving at most an $O(\log n \log(1/\eta)) \cdot \eta$ fraction of uncut edges with probability at least $1 - 1/\mathrm{poly}(n)$ in $\widetilde{O}(mn)$ time.*

Table 1 compares our result with the state-of-the-art min uncut algorithms with the same style guarantees [2,26,39,54]. Compared with [30,39], our approximation guarantee is independent of the eigenvalue of the normalized Laplacian matrix. For [54], although our method is similar to it, our algorithm depends roughly linearly on η, while [54]'s work depends on $\sqrt{\eta}$. Moreover, compared with the linear programming (LP) and semidefinite programming (SDP) based approximation [2,26], although our method performs worse on the approximation ratio, the running time is $\widetilde{O}(mn)$, which significantly outperforms their running time even with latest fast solvers for LP and SDP [35,36].

Table 1. Summary of known min-uncut algorithms and our work. Here α_k denotes the k-th smallest eigenvalue of the matrix $2\mathbf{I}_n - \boldsymbol{\mathcal{L}}$ and λ_{n-k} denotes the $(n-k)$-th smallest eigenvalue of $\boldsymbol{\mathcal{L}}$. The parameter $\varepsilon \in (0,1)$ in [30] is an arbitrary fixed constant.

Reference	Uncut fraction	Time complexity
Trevisan [54]	$O(\sqrt{\eta})$	Spectral decomposition
Kwok et al. [39]	$O(\frac{k}{\alpha_k}\log\frac{\alpha_k}{k\eta})\cdot\eta$	Spectral decomposition
Guruswami and Sinop [30]	$\frac{1+\varepsilon}{\lambda_{n-k}}\cdot\eta$	$2^{O(k/\varepsilon^3)}n^{O(1/\varepsilon)}$
Garg, Vazirani, and Yannakakis [26]	$O(\log n)\cdot\eta$	$\widetilde{O}(m^\omega)$ [36]
Agarwal et al. [2]	$O(\sqrt{\log n})\cdot\eta$	$\widetilde{O}(m^\omega)$ [35]
This work	$O(\log n\log(1/\eta))\cdot\eta$	$\widetilde{O}(mn)$

1.3 Directed Bipartiteness Ratio

Let $G=(V,E)$ be a directed graph. We say that G is *(strongly) bipartite* if there exists a bipartition (L,R) of V such that all arcs in E go from L to R. Inspired by the bipartiteness ratio of undirected graphs, we define the *directed b-bipartiteness ratio* of G as follows. First, for $\mathbf{x}\in\mathbb{R}^V$, let

$$\beta_b(\mathbf{x}) = \frac{\sum_{(i,j)\in E}\psi_{ij}(\mathbf{x})}{\sum_{i\in V} b(i)|x_i|}, \quad \text{where } \psi_{ij}(\mathbf{x}) := \begin{cases} |x_i + x_j| & \text{if } x_i \geq x_j, \\ |x_i| + |x_j| & \text{otherwise.} \end{cases} \tag{1.3}$$

Then, the directed b-bipartiteness ratio of G is defined as

$$\beta_b(G) = \min_{\mathbf{x}\in\{0,\pm1\}^V\setminus\{\mathbf{0}\}} \beta_b(\mathbf{x}).$$

Intuitively, we look for a large witness of bipartiteness: a tripartition (L,R,Z) with L and R playing the roles of the two sides, whose violations of the forward orientation are few compared with the total b-weight of vertices that participate. When $\mathbf{x}\in\{0,\pm1\}^V$ encodes such a tripartition (L,R,Z) as above,

$$\beta_b(\mathbf{x}) = \frac{2\cdot(w(E(R,L)) + w(E(L)) + w(E(R)) + w(E(L\cup R,Z)) + w(E(Z,L\cup R))}{b(L\cup R)},$$

where $E(R, L)$ denotes the set of arcs from R to L and $E(L)$ denotes the set of arcs within L. Thus a small ratio certifies the existence of a large almost strongly bipartite subgraph, and the ratio becomes zero exactly when G contains a non-trivial strongly bipartite subgraph. Similar to (1.2), the directed bipartiteness ratio has a close relation with directed min uncut. For a tripartition (L, R, Z) of $\mathbf{x} \in \{0, \pm 1\}^V$ with $Z \subset V$ being fixed, we have

$$\min_{L \cup R = V \setminus Z,\ L \cap R = \emptyset} \beta_b(\mathbf{x}) = 1 - 2 \cdot \max_{L \cup R = V \setminus Z,\ L \cap R = \emptyset} \frac{w(E(L, R))}{b(L \cup R)}, \tag{1.4}$$

which is equivalent to solving the directed min uncut of subgraph $G[L \cup R]$. Moreover, we can verify that the function ψ_{ij} is *bisubmodular* [25] by an exhaustive case analysis.

We show that we can approximate $\beta_b(G)$ within a factor of $O(\log n)$ in polynomial time:

Theorem 3. *There is a randomized polynomial-time $O(\log n)$-approximation algorithm for the directed b-bipartiteness ratio of a directed graph. That is, the algorithm finds a nonzero vector $\mathbf{x} \in \{0, \pm 1\}^V$ such that $\beta_b(\mathbf{x}) \le O(\log n) \cdot \beta_b(G)$ with probability at least $1 - 1/\mathrm{poly}(n)$.*

The *minimum directed uncut* problem is the directed analogue of minimum uncut: given a directed graph $G = (V, E)$ and a bipartition $(S, \bar{S})$ of the vertex set, an edge is uncut if it does not go from S to $\bar{S}$ (that is, it stays inside S or $\bar{S}$, or points from $\bar{S}$ to S). The goal is to choose $(S, \bar{S})$ that minimizes the total weight of such uncut edges, or equivalently maximizes the weight of the directed cut $E(S, \bar{S})$. We have the following theorem using Theorem 3.

Theorem 4. *Given a graph with the fraction of the minimum directed uncut being η, there is a randomized polynomial-time algorithm for finding a cut that leaves the fraction of edges $O(\log n \log(1/\eta)) \cdot \eta$ with probability at least $1 - 1/\mathrm{poly}(n)$.*

The best known approximation ratio of minimum directed uncut is $O(\sqrt{\log n})$ based on SDP [2], so our approximation ratio is weaker. The primary advantage of our approach is its simplicity: we obtain it via a recursive reduction to approximating the directed bipartiteness ratio, mirroring the undirected setting. If we could approximate the directed bipartiteness ratio in $\widetilde{O}(m)$ time, as in the undirected case, then our algorithm would run in $\widetilde{O}(mn)$ time, outperforming [2]. Whether such a fast approximation exists is currently unclear, and we leave this as an open problem.

1.4 Our Technique

We briefly describe our techniques here.

Cut-Matching Game for Bipartiteness Ratio. Our approach for the approximation algorithm for bipartiteness ratio is extending the *cut-matching game* framework [37], which is originally designed for sparsest cut (i.e., $b \equiv 1$), to bipartiteness ratio.

Let us first review the original cut-matching game for sparse cut. Suppose that we want to check whether $\phi_b(G) \geq 1$ or not. The key fact of the cut-matching game is that $\phi_b(G) \geq 1$ if and only if G is *well-linked*, i.e., for any disjoint vertex subsets $A, B \subseteq V$ with $|A| = |B|$, there exists an A–B flow in G such that it satisfies edge capacity w and unit flow goes out from and in to every vertex in A and B, respectively; see, e.g., [15]. Such a flow is said to be *saturating*. The cut-matching game is a repeated game of two players, the cut player and the matching player. Let H be an empty multigraph on V. In each round, the cut player generates a bipartition $(S, \overline{S})$ of the vertex set V. Without loss of generality, we assume that $|S| \leq |\overline{S}|$. If S is not well-linked to some subset in $\overline{S}$—which can be checked in a single-commodity flow computation—then the game ends; by the above characterization, $\phi_b(G) < 1$. Indeed, we can even find a sparse cut S as well by finding a minimum cut. If S is well-linked, there must exist a flow that saturates S by definition. The matching player finds such a flow and adds its demand graph to H (as a multigraph).[3] The game ends once H becomes an $O(1)$-expander, i.e., $\phi_b(H) \geq \Omega(1)$. Suppose that the game ends after T rounds by finding an expander H. Then, since H is embeddable to G with congestion $O(T)$, we have $\phi_b(G) \geq \Omega(1/T)$. Thus, we achieve an $O(T)$-approximation for sparsest cut. Khandekar, Rao, and Vazirani [37] showed that there exists a randomized strategy of the cut player such that the game ends after $T = O(\log^2 n)$ rounds with high probability. This is later improved to $O(\log n)$-approximation by Orecchia et al. [47]. Furthermore, Arora and Kale [7] provides a very systematic interpretation and analysis of the cut-matching game with *matrix multiplicative weight update (MMWU)* and proves $T = O(\sqrt{\log n})$, which is the current best approximation ratio.

To design a cut-matching game for bipartiteness ratio, we first prove an analogous characterization of bipartiteness ratio in terms of flows in an *auxiliary graph* G'. Let V^+ and V^- be the disjoint copies of V. We denote the copies of a vertex i in V^+ and V^- by i^+ and i^-, respectively. Let E' be the set of edges between V^+ and V^- such that for each edge $(i, j) \in E$, there are two corresponding edges $(i^+, j^-) \in E'$ and $(i^-, j^+) \in E'$. Let $G' = (V^+ \cup V^-, E')$ be the resulting undirected graph. Note that G' is bipartite by construction. For $X \subseteq V$, denote by X^+ (resp. X^-) the corresponding set of X in V^+ (resp. V^-). For G and G', see Fig. 1 for an illustration. Given a vertex weight b on V, we can naturally induce a vertex weight b' on V' such that $b'(u) = b(i)$ for $i \in V$

[3] For the unweighted case (i.e., $w \equiv 1$), the demand graph is simply a matching between S and $\overline{S}$ of size $|S|$, hence the name of the matching player.

and $u \in \{i^+, i^-\}$. With a slight abuse of notation, we also denote the induced vertex weights on V' by b, and similarly, the induced edge weights on E' by w.

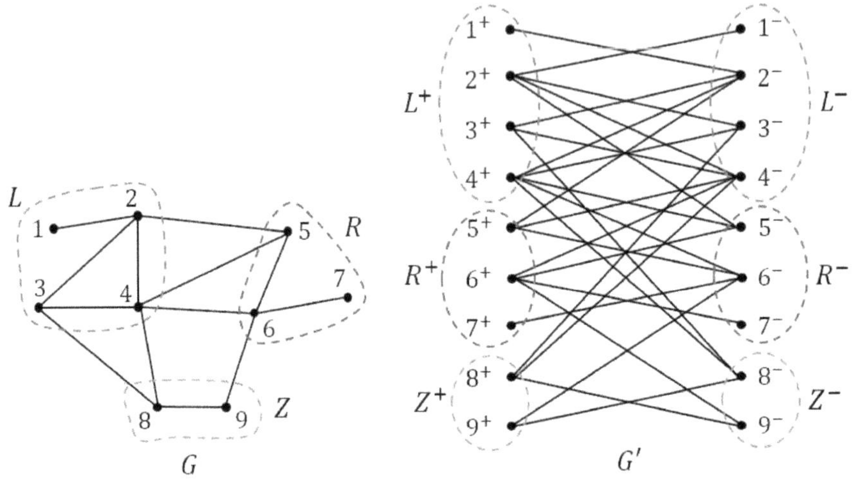

(a) (b)

Fig. 1. (a) the original graph $G = (V, E)$. Given a vector $\mathbf{x} = [1, 1, 1, 1, -1, -1, -1, 0, 0]^\top$, we have a corresponding partition of $V = L \cup R \cup Z$, where $L = \{1, 2, 3, 4\}$, $R = \{5, 6, 7\}$, and $Z = \{8, 9\}$. (b) the corresponding bipartite graph $G' = (V^+ \cup V^-, E')$, where each edge $(u, v) \in E$ corresponds to two edges (u^+, v^-) and (v^+, u^-) in E'. Moreover, the three subsets L, R, and Z of V correspond to the subsets L^+, L^-, R^+, R^-, and Z^+, Z^- of V', respectively.

The auxiliary graph reveals that bipartiteness ratio has a structure quite similar to sparsest cut. For any $\mathbf{x} \in \{0, \pm 1\}^V \setminus \{\mathbf{0}\}$ corresponding to tripartition (L, R, Z) of V, let $S := L^+ \cup R^-$ and $\overline{S} := V' \setminus S$. Then we can show

$$\beta_b(\mathbf{x}) = \frac{w(E'(S, \overline{S}))}{b(S)},$$

and thus we can represent the b-bipartiteness ratio as

$$\beta_b(G) = \min_{S = L^+ \cup R^-, \text{ disjoint } L, R \subseteq V} \frac{w(E'(S, \overline{S}))}{b(S)}.$$

This formula is the same as generalized sparsest cut (1.5), except for the additional constraint on S. Note that if $S = L^+ \cup R^-$, then $\overline{S} \supseteq L^- \cup R^+$ and therefore $\min\{b(S), b(\overline{S})\} = b(S)$.

Furthermore, the auxiliary graph G' has the following symmetry: $(i^+, j^-) \in E'$ if and only if $(j^+, i^-) \in E'$. This is a special case of *skew-symmetric* graphs [28]. Taking this symmetry into account, we define that G' is well-linked if any *symmetric* source-sink pair $A = L^+ \cup R^-$, $B = L^- \cup R^+$ for disjoint $L, R \subseteq V$ admits a saturating flow in G'. Then, we show that one can characterize the bipartiteness ratio in terms of well-linkedness: $\beta_b(G) \geq 1$ if and only if G' is well-linked. Again, this is parallel to the aforementioned characterization of sparsest cut in terms of well-linkedness.

Using our novel characterization of bipartiteness ratio, we propose the following cut-matching game for bipartiteness ratio. Again, let H be an empty multigraph on V. In each round, the cut player generates a tripartition (L, R, Z) of V. Let $A = L^+ \cup R^-$ and $B = L^- \cup R^+$ be the corresponding symmetric source-sink pair in G'. If (A, B) is not well-linked in G', then the game ends; we know $\beta_b(G) < 1$. Again, we can even find $\mathbf{x} \in \{0, \pm 1\}^V$ such that $\beta_b(\mathbf{x}) \leq 1$ by finding a minimum cut. Otherwise, there exists a saturating flow in G' from A to B, which the matching player finds. Then, we add to H a multigraph on V induced from the demand graph of the flow. The game ends if H satisfies $\beta_b(H) \geq \Omega(1)$. Using a MMWU analysis similar to [7], we present a randomized strategy of the cut player such that the game ends after $T = O(\log n)$ rounds with high probability. Since H is embeddable to G with congestion $O(T)$, we achieve the desired $O(\log n)$-approximation.

Each iteration of the cut matching game amounts to finding a Gram decomposition of a certain positive definite matrix provided by MMWU and a single-commodity max-flow on the auxiliary graph G'. Roughly speaking, the former task is to compute a Gram decomposition of the matrix exponential of an $n \times n$ real symmetric matrix. Although the standard method takes $O(n^3)$ time, there is a technique for computing *approximate* Gram decompositions much more efficiently [7,41], which we can adopt to our setting.

Finally, we remark that the same construction of G' was used in the context of approximating min uncut [2], which is to find a partition (L, R) (i.e., $Z = \emptyset$) minimizing uncut $w(E(L) \dot{\cup} E(R))$. Note that since $Z = \emptyset$, minimizing uncut is equivalent to minimizing the ratio objective β_b. We generalize their result to show that the same construction works for bipartiteness ratio, which is to find a tripartition (L, R, Z) minimizing β_b.

Application to Min Uncut. Building on our approximation algorithm for bipartiteness ratio, we propose an approximation algorithm for the minimum uncut problem. Our method is conceptually similar to those in [39,54], which showed a close connection between min uncut and bipartiteness ratio. Briefly, we invoke our approximation algorithm for bipartiteness ratio, which returns a partition of the vertex set $V = L \cup R \cup Z$, and then recursively apply this process to the subgraph induced by the vertex subset Z, where the entries of $\mathbf{x}_Z$ are zero. Since the size of vertex set decreases in every iteration, this method would take $\widetilde{O}(mn)$ time.

Directed Bipartiteness Ratio. To establish Theorem 3, we follow a two-step blueprint in the spirit of the Leighton–Rao algorithm. We first build a skew-symmetric auxiliary graph G' of the input digraph by replacing every arc with a constant-size gadget. This lets us represent the directed b-bipartiteness ratio as the minimum ratio-cut value over consistent subsets of G', generalizing the undirected cut formulation. We then solve a directed semimetric relaxation whose optimal value upper-bounds $\beta_b(G)$. Applying the directed ℓ_1 weak-embedding theorem [14], we round the solution by sampling a random directed cut in the embedded space and projecting it to a consistent cut; this yields the desired $O(\log n)$ approximation.

Applying the same recursion as in our min uncut algorithm, but now equipped with the directed bipartiteness ratio approximation, yields the algorithm for minimum directed uncut.

1.5 Related Work

Trevisan [54] defined the formulation of the bipartiteness ratio for an undirected, unweighted graph G, and proved an analog of Cheeger's inequality involving λ_n, the largest eigenvalue of the normalized Laplacian of G. Additionally, Trevisan proposed an algorithm that given a graph G with the fraction of the maximum cut edges being $1-\eta$, returns a vector $\mathbf{x}$ in polynomial time such that $\beta(\mathbf{x}) \leq 2\sqrt{\eta}$. By recursively applying this algorithm to the subgraph induced by the zero entries of $\mathbf{x}$, one can find a cut that cuts at least a $1-\Theta(\sqrt{\eta})$ fraction of the edges. Recently, Pokharanakar [49] extended the concept of the bipartiteness ratio from finite graphs to the setting of graphons. In this work, the author formulated an appropriate definition of bipartiteness ratio for graphons and proved a Cheeger-type inequality that relates this ratio to the top of the spectrum of the associated graphon Laplacian.

For a positive vertex weight $b : V \to \mathbb{Z}_{++}$, the *generalized sparsest cut* problem[4] is to compute

$$\phi_b(G) := \min_{\emptyset \subsetneq S \subsetneq V} \frac{w(E(S,\overline{S}))}{\min\{b(S), b(\overline{S})\}}, \tag{1.5}$$

where $b(S) := \sum_{i \in S} b(i)$. The *conductance* corresponds to the case where $b(i) = \deg(i)$ for all $i \in V$. For $b(i) = 1$ for all $i \in V$, the problem is known as the *sparsest cut* problem. For $w(e) = 1$ $(e \in E)$ and $b(i) = 1$ $(i \in V)$, $\phi_b(G)$ is called the *edge expansion* of G. Computing the conductance or (generalized) sparsest cut of a graph is NP-hard [52], so research has focused on efficient approximation algorithms. Khandekar, Rao, and Vazirani [37] introduced a purely combinatorial $O(\log^2 n)$-approximation via the cut–matching game. Leighton and Rao [42] achieved an $O(\log n)$-approximation by solving a multicommodity flow relaxation, and Orecchia et al. [47] showed the same guarantee can be obtained with a sequence of single-commodity max-flow calls. SDP relaxations yield better ratios: Arora, Rao, and Vazirani [8] gave an $O(\sqrt{\log n})$-approximation by rounding an SDP. Sherman [53] achieved an $O(\sqrt{\log(n)/\varepsilon})$-approximation with $\tilde{O}(n^{\varepsilon})$

[4] The most general form of generalized sparsest cut allows arbitrary demand graphs; see [38, Section 19.3]. Here, we only consider the case such that the demand graph is a complete graph and b is arbitrary, which is relevant to bipartiteness ratio.

max-flow computations for any $\varepsilon > 0$. Furthermore, Arora et al. [6,7] obtained the $O(\sqrt{\log n})$ ratio in $\widetilde{O}(n^2)$ time. In [6], they did so by efficiently constructing expander flows, whereas in [7], they introduced a general primal–dual framework for solving SDPs via MMWU.

The cut-matching game, originally proposed by Khandekar et al. [37], provides a fast combinatorial framework for approximating the sparsest cut in undirected graphs through flow-based techniques. In their work [37], they designed a cut player strategy capable of constructing a graph with edge expansion $\Omega(1)$ within $O(\log^2 n)$ rounds. This was later improved by Orecchia et al. [47], which developed a cut player strategy that achieves edge expansion $\Omega(\log n)$ in the same number of rounds. Moreover, Louis [44] introduced a cut player strategy for cut-matching game on directed graphs, and leveraged it to develop an $O(\log^2 n)$-approximate algorithm for directed sparsest cut using $O(\log^2 n)$ max-flow computations. Subsequently, Lau et al. [41] improved upon the cut-matching game framework for directed graphs introduced in [44], leading to an $O(\log n)$-approximate algorithm for directed edge expansion that runs in almost linear time. Additionally, it is worth noting that the cut-matching game has evolved into a versatile algorithmic primitive, finding applications in a variety of domains such as edge-disjoint paths [5,19,23], dynamic graph algorithms [11,12,20–22,24,29,45], hypergraph ratio cuts [16,55], expander decompositions [13,17,32,33,43,51], hierarchical decomposition [29,50], network flows [31,48], etc.

For the maximum cut problem, which is equivalent to the minimum uncut problem, a simple deterministic algorithm achieves a 1/2-approximation in polynomial time. Under the *unique games conjecture*, the best approximation ratio is the Goemans–Williamson bound of $\alpha_{\text{GW}} \approx 0.878$ [27], obtained via an SDP relaxation followed by randomized hyperplane rounding. Suppose a maximum cut leaves an η fraction of edges. Trevisan [54] gave a purely spectral algorithm that, in this regime, finds a cut leaving an $O(\sqrt{\eta})$ fraction of edges. Kwok et al. [39] generalized this approach by considering higher-order eigenvalues: if α_k denotes the k-th smallest eigenvalue of the matrix $2\mathbf{I}_n - \boldsymbol{\mathcal{L}}$, then their algorithm produces a cut that leaves $O(\frac{k}{\alpha_k}\log\frac{\alpha_k}{k\eta})\cdot\eta$ fraction of edges. Building on SDP hierarchies, Guruswami and Sinop [30] showed that, for any $\varepsilon \in (0,1)$, one can achieve a fraction of uncut edges $O(\frac{1+\varepsilon}{\lambda_{n-k}})\cdot\eta$ in time $2^{O(k/\varepsilon^3)}n^{O(1/\varepsilon)}$, where λ_{n-k} is the $(n-k)$-th smallest eigenvalue of the normalized Laplacian and the rounding is performed on an SDP from the Lasserre hierarchy [40]. Since the minimum uncut is equivalent to the maximum cut under complementarity, one may also leverage approximation algorithms for the former. Garg et al. [26] reduced min-uncut to the minimum multicut problem, yielding an $O(\log n)$-approximation for min-uncut. Arora et al. [2] later improved this to an $O(\sqrt{\log n})$-approximation by solving SDP relaxations recursively.

It is a natural progression to generalize algorithms developed for undirected graphs to directed settings. Hajiaghayi and Räcke [34] devised an $O(\sqrt{n})$-approximation for directed sparsest cut via a new LP-rounding technique for the fractional relaxation, and Agarwal, Alon, and Charikar [1] later improved this to

$\widetilde{O}(n^{11/23})$ with a refined randomized rounding scheme. Yoshida [57] introduced a nonlinear Laplacian for digraphs and proved a Cheeger-type inequality. For directed edge expansion, Lau, Tung, and Wang [41] combined triangle inequalities with the reweighted eigenvalue formulation to obtain an almost-linear-time $O(\sqrt{\log n})$-approximation via a strengthened SDP relaxation.

Full Version. Due to the space limitations, technical details and full proofs are deferred to the full version: https://doi.org/10.48550/arXiv.2507.12847

Acknowledgments. TS is supported by JSPS KAKENHI Grant Numbers JP24K21315 and JP19K20212, and JST, PRESTO Grant Number JPMJPR24K5, Japan. YY is supported by JSPS KAKENHI Grant Number JP24K02903 and JP22H05001. MY is supported by Japan Science and Technology Agency (JST) as part of Adopting Sustainable Partnerships for Innovative Research Ecosystem (ASPIRE), Grant Number JPMJAP2302. Part of this research was done while MY was supported by a UC Santa Cruz start-up grant under Vaggos Chatziafratis.

Disclosure of Interests. The authors have no competing interests to declare that are relevant to the content of this article.

References

1. Agarwal, A., Alon, N., Charikar, M.S.: Improved approximation for directed cut problems. In: Proceedings of the 39th Annual ACM Symposium on Theory of Computing (STOC), pp. 671–680 (2007)
2. Agarwal, A., Charikar, M., Makarychev, K., Makarychev, Y.: $O(\sqrt{\log n})$ approximation algorithms for Min UnCut, Min 2CNF deletion, and directed cut problems. In: Proceedings of the 37th Annual ACM Symposium on Theory of Computing (STOC), pp. 573–581 (2005)
3. Alon, N., Milman, V.: λ_1, isoperimetric inequalities for graphs, and superconcentrators. J. Comb. Theory Ser. B **38**(1), 73–88 (1985)
4. Alon, N.: Eigenvalues and expanders. Combinatorica **6**(2), 83–96 (1986)
5. Andrews, M.: Approximation algorithms for the edge-disjoint paths problem via raecke decompositions. In: Proceedings of the IEEE 51st Annual Symposium on Foundations of Computer Science (FOCS), pp. 277–286 (2010)
6. Arora, S., Hazan, E., Kale, S.: $O(\sqrt{\log n})$-approximation to sparsest cut in $\widetilde{O}(n^2)$ time. SIAM J. Comput. **39**(5), 1748–1771 (2010)
7. Arora, S., Kale, S.: A combinatorial, primal-dual approach to semidefinite programs. J. ACM **63**(2), 1–35 (2016)
8. Arora, S., Rao, S., Vazirani, U.: Expander flows, geometric embeddings and graph partitioning. J. ACM **56**(2), 1–37 (2009)
9. Atay, F.M., Liu, S.: Cheeger constants, structural balance, and spectral clustering analysis for signed graphs. Discret. Math. **343**(1), 111616 (2020)
10. Bauer, F., Jost, J.: Bipartite and neighborhood graphs and the spectrum of the normalized graph Laplace operator. Comm. Anal. Geom. **21**(4), 787–845 (2013)
11. Bernstein, A., Gutenberg, M.P., Saranurak, T.: Deterministic decremental reachability, SCC, and shortest paths via directed expanders and congestion balancing. In: Proceedings of the IEEE 61st Annual Symposium on Foundations of Computer Science (FOCS), pp. 1123–1134 (2020)

12. Bernstein, A., Gutenberg, M.P., Saranurak, T.: Deterministic decremental SSSP and approximate min-cost flow in almost-linear time. In: 2021 IEEE 62nd Annual Symposium on Foundations of Computer Science (FOCS), pp. 1000–1008 (2021)
13. Chang, Y.J., Saranurak, T.: Deterministic distributed expander decomposition and routing with applications in distributed derandomization. In: Proceedings of the IEEE 61st Annual Symposium on Foundations of Computer Science (FOCS), pp. 377–388 (2020)
14. Charikar, M., Makarychev, K., Makarychev, Y.: Directed metrics and directed graph partitioning problems. In: Proceedings of the 17th Annual ACM-SIAM Symposium on Discrete Algorithms (SODA), pp. 51–60 (2006)
15. Chekuri, C.: Lecture notes on topics in graph algorithms. University of Illinois (2024). https://courses.grainger.illinois.edu/cs598csc/fa2024/Notes/lec-cut-matching.pdf
16. Chen, A., Orecchia, L., Tani, E.: Submodular hypergraph partitioning: metric relaxations and fast algorithms via an improved cut-matching game. In: 52nd International Colloquium on Automata, Languages, and Programming (ICALP), pp. 49–1 (2025)
17. Chen, D., Meierhans, S., Gutenberg, M.P., Saranurak, T.: Parallel and distributed expander decomposition: simple, fast, and near-optimal. In: Proceedings of the 2025 Annual ACM-SIAM Symposium on Discrete Algorithms (SODA), pp. 1705–1719 (2025)
18. Chen, L., Kyng, R., Liu, Y.P., Peng, R., Gutenberg, M.P., Sachdeva, S.: Maximum flow and minimum-cost flow in almost-linear time. In: Proceedings of the 63rd Annual IEEE Symposium on Foundations of Computer Science (FOCS) (2022)
19. Chuzhoy, J.: Routing in undirected graphs with constant congestion. In: Proceedings of the 44th Annual ACM Symposium on Theory of Computing (STOC), pp. 855–874 (2012)
20. Chuzhoy, J.: A distanced matching game, decremental APSP in expanders, and faster deterministic algorithms for graph cut problems. In: Proceedings of the 2023 Annual ACM-SIAM Symposium on Discrete Algorithms (SODA), pp. 2122–2213 (2023)
21. Chuzhoy, J., Gao, Y., Li, J., Nanongkai, D., Peng, R., Saranurak, T.: A deterministic algorithm for balanced cut with applications to dynamic connectivity, flows, and beyond. In: Proceedings of the IEEE 61st Annual Symposium on Foundations of Computer Science (FOCS), pp. 1158–1167 (2020)
22. Chuzhoy, J., Khanna, S.: A new algorithm for decremental single-source shortest paths with applications to vertex-capacitated flow and cut problems. In: Proceedings of the 51st Annual ACM SIGACT Symposium on Theory of Computing (STOC), pp. 389–400 (2019)
23. Chuzhoy, J., Li, S.: A polylogarithmic approximation algorithm for edge-disjoint paths with congestion 2. In: Proceedings of the 53rd Annual IEEE Symposium on Foundations of Computer Science (FOCS), pp. 233–242 (2012)
24. Chuzhoy, J., Zhang, R.: A new deterministic algorithm for fully dynamic all-pairs shortest paths. In: Proceedings of the 55th Annual ACM Symposium on Theory of Computing (STOC), pp. 1159–1172 (2023)
25. Fujishige, S., Iwata, S.: Bisubmodular function minimization. SIAM J. Discret. Math. **19**(4), 1065–1073 (2005)
26. Garg, N., Vazirani, V.V., Yannakakis, M.: Approximate max-flow min-(multi) cut theorems and their applications. In: Proceedings of the 25th annual ACM Symposium on Theory of Computing (STOC), pp. 698–707 (1993)

27. Goemans, M.X., Williamson, D.P.: .879-approximation algorithms for max cut and max 2sat. In: Proceedings of the 26th Annual ACM Symposium on Theory of Computing (STOC), pp. 422–431 (1994)
28. Goldberg, A.V., Karzanov, A.V.: Path problems in skew-symmetric graphs. Combinatorica **16**(3), 353–382 (1996)
29. Goranci, G., Räcke, H., Saranurak, T., Tan, Z.: The expander hierarchy and its applications to dynamic graph algorithms. In: Proceedings of the 2021 ACM-SIAM Symposium on Discrete Algorithms (SODA), pp. 2212–2228 (2021)
30. Guruswami, V., Sinop, A.K.: Lasserre hierarchy, higher eigenvalues, and approximation schemes for quadratic integer programming with PSD objectives. arXiv preprint arXiv:1104.4746 (2011)
31. Haeupler, B., Hershkowitz, D.E., Li, J., Roeyskoe, A., Saranurak, T.: Low-step multi-commodity flow emulators. In: Proceedings of the 56th Annual ACM Symposium on Theory of Computing (STOC), pp. 71–82 (2024)
32. Haeupler, B., Hershkowitz, D.E., Tan, Z.: New structures and algorithms for length-constrained expander decompositions. In: Proceedings of the 65th Annual IEEE Symposium on Foundations of Computer Science (FOCS), pp. 1634–1645 (2024)
33. Haeupler, B., Huebotter, J., Ghaffari, M.: A cut-matching game for constant-hop expanders. In: Proceedings of the 2025 Annual ACM-SIAM Symposium on Discrete Algorithms (SODA), pp. 1651–1678 (2025)
34. Hajiaghayi, M.T., Räcke, H.: An $O(n)$-approximation algorithm for directed sparsest cut. Inf. Process. Lett. **97**(4), 156–160 (2006)
35. Huang, B., Jiang, S., Song, Z., Tao, R., Zhang, R.: Solving SDP faster: a robust IPM framework and efficient implementation. In: Proceedings of the IEEE 63rd Annual Symposium on Foundations of Computer Science (FOCS), pp. 233–244 (2022)
36. Jiang, S., Song, Z., Weinstein, O., Zhang, H.: A faster algorithm for solving general LPs. In: Proceedings of the 53rd Annual ACM SIGACT Symposium on Theory of Computing, pp. 823–832 (2021)
37. Khandekar, R., Rao, S., Vazirani, U.: Graph partitioning using single commodity flows. In: Proceedings of the 38th Annual ACM Symposium on Theory of Computing (STOC), pp. 385–390 (2006)
38. Korte, B., Vygen, J.: Combinatorial Optimization: Theory and Algorithms, vol. 21, 6th edn. Springer, Heidelberg (2018)
39. Kwok, T.C., Lau, L.C., Lee, Y.T., Oveis Gharan, S., Trevisan, L.: Improved Cheeger's inequality: analysis of spectral partitioning algorithms through higher order spectral gap. In: Proceedings of the 45th Annual ACM Symposium on Theory of Computing (STOC), pp. 11–20 (2013)
40. Lasserre, J.B.: An explicit equivalent positive semidefinite program for nonlinear 0–1 programs. SIAM J. Optim. **12**(3), 756–769 (2002)
41. Lau, L.C., Tung, K.C., Wang, R.: Fast algorithms for directed graph partitioning using flows and reweighted eigenvalues. In: Proceedings of the 2024 Annual ACM-SIAM Symposium on Discrete Algorithms (SODA), pp. 591–624 (2024)
42. Leighton, T., Rao, S.: Multicommodity max-flow min-cut theorems and their use in designing approximation algorithms. J. ACM **46**(6), 787–832 (1999)
43. Long, Y., Saranurak, T.: Near-optimal deterministic vertex-failure connectivity oracles. In: Proceedings of the 63rd Annual IEEE Symposium on Foundations of Computer Science (FOCS), pp. 1002–1010 (2022)
44. Louis, A.: Cut-matching games on directed graphs. arXiv preprint arXiv:1010.1047 (2010)

45. Nanongkai, D., Saranurak, T.: Dynamic spanning forest with worst-case update time: adaptive, las vegas, and $O(n^{1/2-\varepsilon})$-time. In: Proceedings of the 49th Annual ACM SIGACT Symposium on Theory of Computing (STOC), pp. 1122–1129 (2017)
46. Neumann, S., Peng, P.: Sublinear-time clustering oracle for signed graphs. In: International Conference on Machine Learning, pp. 16496–16528 (2022)
47. Orecchia, L., Schulman, L.J., Vazirani, U.V., Vishnoi, N.K.: On partitioning graphs via single commodity flows. In: Proceedings of the 40th Annual ACM Symposium on Theory of Computing (STOC), pp. 461–470 (2008)
48. Peng, R.: Approximate undirected maximum flows in $O(m\text{polylog(n)})$ time. In: Proceedings of the 27th Annual ACM-SIAM Symposium on Discrete Algorithms (SODA), pp. 1862–1867 (2016)
49. Pokharanakar, M.M.: The dual Cheeger-Buser inequality for graphons. arXiv preprint arXiv:2502.15213 (2025)
50. Räcke, H., Shah, C., Täubig, H.: Computing cut-based hierarchical decompositions in almost linear time. In: Proceedings of the 25th Annual ACM-SIAM Symposium on Discrete Algorithms (SODA), pp. 227–238 (2014)
51. Saranurak, T., Wang, D.: Expander decomposition and pruning: faster, stronger, and simpler. In: Proceedings of the 13th Annual ACM-SIAM Symposium on Discrete Algorithms (SODA), pp. 2616–2635 (2019)
52. Shahrokhi, F., Matula, D.W.: The maximum concurrent flow problem. J. ACM **37**(2), 318–334 (1990)
53. Sherman, J.: Breaking the multicommodity flow barrier for $O(\sqrt{\log n})$-approximations to sparsest cut. In: Proceedings of the 50th Annual IEEE Symposium on Foundations of Computer Science (FOCS), pp. 363–372 (2009)
54. Trevisan, L.: Max cut and the smallest eigenvalue. SIAM J. Comput. **41**(6), 1769–1786 (2012)
55. Veldt, N.: Cut-matching games for generalized hypergraph ratio cuts. In: Proceedings of the ACM Web Conference (WWW), pp. 694–704 (2023)
56. Xiao, H., Ordozgoiti, B., Gionis, A.: Searching for polarization in signed graphs: a local spectral approach. In: Proceedings of the Web Conference (WWW), pp. 362–372 (2020)
57. Yoshida, Y.: Nonlinear laplacian for digraphs and its applications to network analysis. In: Proceedings of the 9th ACM International Conference on Web Search and Data Mining (WSDM), pp. 483–492 (2016)

Approximating Multiple-Depot Capacitated Vehicle Routing via LP Rounding

Zachary Friggstad[1] and Tobias Mömke[2](✉)

[1] University of Alberta, Alberta, Canada
zacharyf@ualberta.ca
[2] University of Augsburg, Augsburg, Germany
moemke@informatik.uni-augsburg.de

Abstract. In *Capacitated Vehicle Routing with Multiple Depots* (CVRP-MD) we are given a set of client locations C and a set of depots R located in a metric space with costs $c(i, j)$ between $u, v \in C \cup R$. Additionally, we are given a capacity bound k. The goal is to find a collection of tours of minimum total cost such that each tour starts and ends at some depot $r \in R$ and includes at most k clients and such that each client lies on at least one tour. Our main result is a 3.9365-approximation based on rounding a new LP relaxation for CVRP-MD.

1 Introduction

In *Capacitated Vehicle Routing with Multiple Depots* (CVRP-MD) we are given a set of client locations C and a set of depots R located in a metric space with costs $c(i, j)$ between $u, v \in C \cup R$. Additionally, we are given a capacity bound k. The goal is to find a collection of tours of minimum total cost such that each tour starts and ends at some depot $r \in R$ and includes at most k clients and such that each client lies on at least one tour.

The problem is well-studied in the Operations Research literature, as it has relevant applications such as delivering goods from multiple production facilities to consumers. For an overview of applied results we refer to [14]. Notably, in the overview they write that most exact algorithms for solving the classical vehicle routing are difficult to be adapted for solving CVRP-MD.

As CVRP-MD generalizes capacitated vehicle routing (CVRP), it is APX-hard even for $k = 3$ [2]. We therefore focus on approximation algorithms. The first approximation algorithm for CVRP-MD was a $(2\alpha + 1)$-approximation by Li and Simchi-Levi [13] where α is the approximation guarantee for the Traveling Salesman Problem (TSP) in the corresponding metric. Later, a 4-approximation using a completely different approach was given by Harks, König and Matuschke [10] in the course of studying a more general problem. We give a

Z. Friggstad—Research supported by an NSERC Discovery Grant.
T. Mömke—Partially supported by DFG Grant 439522729 (Heisenberg-Grant).

S. S. Dey et al. (Eds.): IPCO 2026, LNCS 16588, pp. 112–127, 2026.
https://doi.org/10.1007/978-3-032-28691-8_8

more detailed overview of these algorithms in Sect. 1.2 as they are both relevant to our results.

Combining modern TSP results [11,12] with [13] gives a $(4 - 10^{-36})$-approximation for CVRP-MD. To date, the best CVRP-MD approximation is by Zhao an Xiao [16]. The authors combine a recent approximation algorithm for CVRP of Blauth, Traub and Vygen [4] with the results above and obtain an approximation ratio of $4 - 2 \cdot \delta \approx 3.9993$ where $\delta \approx 1/3000$ is a constant used in the work of Blauth, Traub, and Vygen. Our main result is a new approximation algorithm with improved approximation ratio.

Theorem 1. *There is a polynomial-time randomized algorithm for* CVRP-MD *that finds a solution whose expected cost is at most* 3.9365 *times the optimum solution's cost.*

Our improved approximation algorithm combines several key ideas from previous work in CVRP-MD plus branching decompositions of preflows [3]. In fact, our new approach is compatible with the approach by Blauth, Traub, and Vygen and can be slightly improved by an amount in the order of 10^{-3} just like the CVRP-MD improvement in [16].

1.1 Notation and Preliminaries

We view the given metric as a complete graph with vertices $V = C \cup R$ and edge distances given by $c(u, v), u, v \in C \cup R$. Throughout, we let OPT denote an optimal solution and opt the optimum solution cost. For a collection of edges S (perhaps with multiple copies of an edge) we let $c(S)$ denote the total cost of all copies of edges in S. We view a tour T as the multiset of edges, if T only visits one client then it would contain two copies of the edge from the depot to the client. For brevity, we let $r_T \in R$ denote the root of a tour T. For any multiset of edges F, we let C_F denote all clients that are the endpoint of at least one edge in F. We will only use this notation when F is a tour or a tree.

Most CVRP approximations in general metrics rely on the following *radial lower bound*:

$$\mathrm{lb} := \frac{2}{k} \sum_{v \in C} c(v, R)$$

where $c(v, R)$ denotes $\min_{r \in R} c(v, r)$. The following was proven by Haimovich and Rinnooy Kan [9] for the single-depot case, the proof extends immediately to the multiple-depot case as noted in [13].

Lemma 1 (Haimovich, Rinnooy Kan [9]). *lb* $\leq$ *opt*

For each client $v \in C$, we say that $\frac{2}{k} \cdot c(v, R)$ is the radial lower bound contribution of v.

1.2 Review of Previous Approximations

Our approach incorporates elements of two previous approximations. We briefly summarize them below plus the way that the CVRP algorithm by Blauth,

Traub, and Vygen [4] was adapted to get the slightly improved approximation for CVRP-MD by Zhao and Xiao [16].

Tree Splitting by Harks, König and Matuschke [10].
The algorithm starts with computing the minimum-cost R-rooted spanning forest, i.e., a collection of trees spanning $C \cup R$ where each tree contains a single $r \in R$ which we view as the root of that tree. Note that the cost of this forest is at most opt since the optimum solution, viewed as a collection of edges, ensures that each client is in the same component as some depot. If some tree T in the forest includes more than k clients we repeat the following until it does not.

First, pick the deepest node v of T whose subtree contains more than k clients. Let $u_1, \ldots, u_a$ be the children of v, so each contains at most k nodes in their subtree by our choice of v. Group these subtrees together arbitrarily so each group has at most k nodes but two groups cannot be merged into a single group (i.e., any two groups together have more than k clients). There must be at least one group with at least $k/2$ clients: otherwise, if there were two groups, they could be merged and if there is one group with less than $k/2$ clients then the total size of the subtree rooted at v (including v itself) is an integer less than $k/2+1$, so it must be at most k which contradicts our choice of v.

For any group with at least $k/2$ clients, we turn it into a tour spanning all clients in the trees by doubling the trees including the parent edges to v. Graft this tour into the depots by picking a client v' on the tour uniformly at random and adding the two copies of the edge $v'r$, where r is the depot nearest to v'. Now shortcut the resulting Eulerian tour past repeated occurrences of nodes and also past *all* occurrences of v (so v is not on the tour). Also remove the subtrees of these groups from the tree T. After repeating this pruning step until T has at most k clients, we get one last tour by doubling all edges on T and letting the root of T serve as the depot for this tour.

For a tour that was split off of the tree, each client v' has a probability of at most $2/k$ of being picked to connect to its nearest depot r. Two copies of the edge $v'r$ were added in this case, so the expected total cost of all edges added in this way is at most $\frac{2}{k} \cdot \sum_{v' \in C} 2 \cdot c(v', R) = 2 \cdot \text{lb}$. Since the forest has cost at most opt and since we only doubled its edges to form tours, the total cost of the final solution is at most $2 \cdot \text{lb} + 2 \cdot \text{opt} \leq 4 \cdot \text{opt}$ in expectation. The algorithm is easy to derandomize by picking the nearest depot to each tour that is split off rather than of picking a client randomly and connecting it to its nearest depot.

Tour Splitting by Li and Simchi-Levi [13]
This algorithm is more along the line of the classic tour-splitting algorithm for single-depot CVRP. It starts by considering the metric obtained by contracting all of R to a single node v_R and recomputing the shortest path distances. It uses an α-approximation for TSP in this contracted metric. Replacing each edge of the TSP tour by its shortest path in the contracted graph yields a collection of v_R-rooted tours. These tours lift to walks in the original graph whose endpoints lie in R (some walks may start and end at different depots). The cost of these walks is at most $\alpha \cdot \text{opt}$ and each client lies on at least one walk. Next, shortcut

each of these and remove the endpoint (a depot) to get a collection of paths $\mathcal{P}$, each starting at some $r \in R$ and each client lying on exactly one path.

Finally, standard tour partition is performed: for each such path P we pick every k-th client along P starting with an initial random offset from the depot on P. For each chosen client v a tour is formed by starting at the depot closest to v and visiting the next $k-1$ clients after v on P (or to the end of P if there are fewer clients). The remaining clients before the first one that is picked are connected to the start node r' of P (a depot) by doubling the subpath from r' to these remaining clients.

The total cost of the doubled subpaths is at most $2\alpha \cdot \mathrm{opt}$ and the total cost of the doubled edges connecting sampled clients to their nearest depot is at most $\frac{2}{k} \cdot \sum_{v \in C} c(v, R) = \mathrm{lb}$ in expectation. The total cost is then at most $2\alpha \cdot \mathrm{opt} + \mathrm{lb}$ which, given the current state of the art for TSP, is at most $(4 - 10^{-36}) \cdot \mathrm{opt}$ [11,12].

Combining Approaches

Zhao and Xiao [16] note that the improved algorithm for CVRP by Blauth, Traub and Vygen [4] yields an improved approximation for CVRP-MD as follows. If $\mathrm{lb} \le (1-\delta) \cdot \mathrm{opt}$ for some constant $\delta > 0$ then the tree splitting approach is already better[1]. Otherwise, in the contracted metric from the tour splitting approach it is possible to use the algorithm by [4] to get a TSP solution whose cost is at most $(1+f(\delta)) \cdot \mathrm{opt}$ where $f(\delta)$ is some function that vanishes as $\delta \to 0$. Following the rest of the tour splitting approach finds a solution with cost at most $2 \cdot (1 + f(\delta)) \cdot \mathrm{opt} + \mathrm{lb} \le (3 + 2 \cdot f(\delta)) \cdot \mathrm{opt}$. By picking $\delta = 1/3000$ as in [4] they get a $(4 - 1/1500)$-approximation.

1.3 Our Approach

Our approach combines elements of tree splitting and tour splitting along with a bidirected cut relaxation for the problem. First, notice that the tour splitting approach would yield a much better approximation if it was possible to get a TSP tour whose cost is considerably better than $\alpha \cdot \mathrm{opt}$. That is essentially what is done in [16] to get their $(4-1/1500)$-approximation but in their approach the tradeoff between the two approaches based on how close lb is to opt requires lb to be very close to opt to see any improvement in the final approximation guarantee.

We do considerably better in this tradeoff, but at the expense of not spanning all nodes. For now, let δ be such that $\mathrm{lb} = (1 - \delta) \cdot \mathrm{opt}$. Our actual definition of δ is slightly different but the above simple definition serves to explain the high level approach we take.

That is, using an LP relaxation and a preflow-splitting result by Bang-Jensen et al. [3], we sample a collection of paths rooted at the depots whose expected cost is $(1+\delta) \cdot \gamma \cdot \mathrm{opt}$ for some constant γ close to $1/2$. These paths may not span all nodes, but we ensure any client $v \in C$ has a probability at most $e^{-\gamma}$ of not

[1] As is tour splitting, but the improvement is greater with tree splitting given the $1.5 - 10^{-36}$ TSP approximation.

being spanned. Using an adaptation of the Bridge Lemma for vehicle routing problems in [5], we can graft in the missing clients into the sampled paths with a forest whose cost is at most $e^{-\gamma} \cdot \text{opt}$ in expectation.[2]

We are therefore left with R-rooted paths spanning some clients and a forest of trees that grafts the missing nodes into these paths. Our high-level goal is to turn these paths and trees into feasible tours. We introduce extensions of both tree splitting and tour splitting to obtain these tours while charging their cost to:

- At most 2 times the cost of all edges on the paths.
- At most 1 times the radial lower bound contribution of clients covered by the paths.
- At most 2.5 times the cost of the forest.
- At most 1.694323 times the radial lower bound contribution for clients covered by the forest.

Some intricate cancellation happens: recall that in expectation, the cost of the paths is at most $\gamma \cdot (1 + \delta) \cdot \text{opt}$ where $\gamma < 1/2$ and the cost of the forest is at most $e^{-\gamma} \cdot \text{opt}$. All clients' radial lower bounds are charged at least once, and an additional at most 0.694323 times if they are not covered by the paths, which happens only with probability at most $e^{-\gamma}$. Using $\text{lb} \leq (1 - \delta) \cdot \text{opt}$, a calculation shows that the positive dependence on δ charged to the edge costs is overshadowed by the negative dependence on δ in the radial lower bound charges. For the optimal choice of γ, this leaves us with an approximation guarantee of $3.9365 - 0.49826 \cdot \delta$.

New Tree Splitting

For our tree pruning procedure, we first consider pruning a group of subtrees if the cost of the tour can be charged to 2.5 times the edge cost in these subtrees plus 1.694323 times the radial lower bound of the clients. If this is not possible, we consider a new step that tries to cover exactly k clients by completely covering all clients in one group of subtrees and an appropriate number of carefully chosen clients in another subtree: if this can be done in a way that admits the same charging scheme to edges in the completely-removed subtrees and radial lower bound contribution of covered clients, we do it. We show that when this is not possible, the tree has fewer than k clients plus additional properties on its costs that enable the next step.

New Tour Splitting

Once all trees are "small" after tree pruning, we consider each path P and the short trees rooted in P. Using a careful DFS ordering of the trees, we obtain a

[2] The approach in [5] is to use the preflow-splitting result of [3] and to apply an argument related to the bridge lemma of [6] in order to collect vertices not covered by branchings. In our new result, we use a similar starting point, but we obtain the preflows from a more involved LP and use the result from [5] as a black box for collecting the missing vertices. A similar approach of sampling a tree and grafting in missing nodes is also taken in [1].

single path P' that is split in a manner similar to [13], except additional care is taken. That is, if a tour split occurs within one of the short trees rooted in P then some of the edges of the tree would be charged more than twice in the final tours. We take a more careful approach to how we deal with a split in the middle of a tree to limit the extent to which its edges are charged more than twice, this is enabled through properties of the short trees that are produced by the tree pruning step and may result in some clients in the short trees being selected more than once to connect to their nearest depot. Ultimately, in this step we still ensure that each edge of a small tree is charged at most 2.5 times and each client covered by these trees has its radial lower bound charged to an extent of 1.69432, just like with the tree pruning phase. Furthermore, the edges on the paths and radial lower bounds of clients on the paths will be charged to an extent of 2 and 1, respectively.

We finish this overview by remarking that our approach is compatible with the Blauth, Traub, and Vygen approach to CVRP [4]. That is, if lb is smaller than opt by some constant factor then our new algorithm is even better than our worst-case analysis and if lb is very close to opt then an extension of [4] to CVRP-MD described in [16] will also give an approximation guarantee better than our algorithm. So our final approximation guarantee can further be improved by a small constant in the order of 10^{-3}.

2 CVRP-MD Algorithm

We begin by giving an LP relaxation and describing how to sample paths covering some clients. After this is presented, in Sect. 2.3 we provide pseudocode for the entire algorithm which calls on two subroutines that are fully described and analyzed in later sections.

We can assume $k \geq 3$ as the case $k = 1$ is trivial and the case $k = 2$ can be solved by a minimum-cost matching algorithm (the cost of a matching between clients being the cheapest tour that serves them both) allowing for a client to be matched with itself. We also assume $c(v, R) > 0$ for each client $v \in C$ as we could trivially cover clients with $c(v, R) = 0$ using 0-cost singleton tours. For a client $v \in C$ we let $\ell_v = \frac{2}{k} \cdot c(v, R)$ denote its radial lower bound contribution and for $C' \subseteq C$ we let $\ell(C') = \sum_{v \in C} \ell_v$ be the contribution of clients in C' to lb. In particular, $\ell(C) = \text{lb}$.

Consider the following LP relaxation in the bidirected graph obtained by considering both directions of any edge to be different directed edges. Here, variable $z^r_{v,v}$ indicates that there is a tour from depot $r \in R$ that has v as its furthest node (breaking ties arbitrarily) and $z^r_{v,u}$ for $u \neq v$ indicates u is covered by the corresponding tour from r that has v as its furthest node. We view all such tours as being two $r - v$ paths and we model them in the LP as 2 units of flow from r to v in the directed graph. For a directed edge e, let $x^r_{v,e}$ indicate that edge e is traversed in the forward direction among the 2 paths from r to v.

$$\textbf{minimize} \quad \sum_{r\in R}\sum_{v\in C}\sum_e c(e)x^r_{v,e} \qquad \textbf{(LP-CVRP-MD)}$$

$$\text{s.t.} \quad x^r_v(\delta^{out}(u)) = \begin{cases} 2\cdot z^r_{v,v} & \text{if } u = r \\ 0 & \text{if } u = v \\ z^r_{v,u} & \text{otherwise} \end{cases} \quad \forall\, r\in R, u,v\in V$$

$$x^r_v(\delta^{in}(u)) = \begin{cases} 0 & \text{if } u = r \\ 2\cdot z^r_{v,v} & \text{if } u = v \\ z^r_{v,u} & \text{otherwise} \end{cases} \quad \forall\, r\in R, u,v\in V$$

$$x^r_v(\delta^{in}(S)) \geq z^r_{v,u} \qquad \forall\, r\in R, u,v\in V, \{u\}\subseteq S\subseteq V-\{r\}$$

$$\sum_{r\in R}\sum_{v\in C} z^r_{v,u} = 1 \qquad \forall\, u\in V$$

$$z^r_{v,u} \leq z^r_{v,v} \qquad \forall\, r\in R, u,v\in C$$

$$z^r_{v,u} = 0 \qquad \forall\, r\in R, u,v\in C \text{ s.t. } c(u,r) > c(v,r)$$

$$\sum_{u\in C} z^r_{v,u} \leq k\cdot z^r_{v,v} \qquad \forall\, r\in R, v\in V$$

$$x, z \geq 0$$

Here and throughout the rest of the description of the algorithm, we define δ to be such that $\sum_{r,v} 2c(v,r)\cdot z^r_{v,v} = (1-\delta)\text{opt}_{LP}$ where opt_{LP} refers to the optimum solution value of the LP relaxation. For each $r\in R, v\in C$, since the vector x^r_v is sending $2z^r_{v,v}$ units of $r-v$ flow then its cost is at least $2c(v,r)$. So $\delta\in[0,1]$.

Lemma 2. *$opt_{LP} \leq opt$ and $lb \leq (1-\delta)\cdot opt_{LP}$.*

Proof. The natural $\{0,1\}$-solution corresponding to the optimum CVRP-MD solution is feasible. That is, each tour can be naturally viewed as two units of flow from the depot of that tour to the farthest node. It is routine to verify that all constraints hold under this assignment and that the cost of this $\{0,1\}$ solution is opt.

For the second part of the lemma, consider any $r\in R, v\in C$ and any u supported by the flow x^r_v, so $c(u,r)\leq c(v,r)$. Then the LP constraints imply

$$\sum_{u\in C} z^r_{v,u}\cdot 2c(r,u) \leq k\cdot 2c(v,r)\cdot z^r_{v,v} = (1-\delta)\text{opt}_{LP}.$$

Summing over all $r\in R, v\in C$ and using $\sum_{r,v} z^r_{v,u} = 1$ for each $u\in C$ completes the proof. □

Finally, we recall the following decomposition by Bang-Jensen, Frank, and Jackson [3] which we paraphrase in a way that is convenient for us. An (out) r-branching in a directed graph is a tree rooted at r that is oriented away from r but does not necessarily span all nodes of the graph. Recall also that an r-preflow in a directed graph is an assignment f of values to edges such that $f(\delta^{in}(v))\geq f(\delta^{out}(v))$ for all nodes $v\neq r$.

Theorem 2. *Let $G = (V, E)$ be a directed graph. Let f be an r-preflow with rational-valued entries for some $r \in V$. Suppose each $v \in V - \{r\}$ has $r \to v$ connectivity λ_v under f. For any rational value $K \geq 0$, there are r-branchings B_i with associated weights $\mu_i \geq 0$ with $\sum_i \mu_i = K$ such that each $e \in E$ lies on at most an f_e-weight of branchings and each $v \in V - \{r\}$ lies on at least a $\min\{K, \lambda_v\}$-weight of branchings.*

Post and Swamy [15] showed how to find the branchings $\{B_i\}$ and corresponding weights $\{\mu_i\}$ in polynomial time. In particular, the decomposition involves only a polynomial number of branchings.

2.1 Step: Preflow Rounding

Consider some constant $0 < \gamma \leq 1/2$. We will eventually fix $\gamma := 0.46821$ to optimize our analysis. We use the following procedure to sample a collection of R-rooted paths from an optimal extreme point (hence rational) solution (x, z) to (**LP-CVRP-MD**).

- For each $r \in R, v \in C$ consider the r-v (pre)flow given by values $\gamma \cdot x^r_{v,e}$, which is sending $2\gamma \cdot z^r_{v,v}$ units of flow from r to v. Using $K_{r,v} := 2\gamma \cdot z^r_{v,v}$ in Theorem 2, we obtain a collection of branchings $\{B_i\}_{i \geq 1}$ with corresponding nonnegative weights μ_i summing to $K_{r,v}$ such that each directed edges lies on at most a $\gamma \cdot x^r_{v,e}$-fraction of branchings. Furthermore, v lies on each of these branchings and every other $u \in C - \{v\}$ lies on at least a $\gamma \cdot z^r_{v,u}$-fraction of these branchings (noting $\gamma \cdot z^r_{v,u} \leq K_{r,v}/2$ by the LP constraints). Let $\mathcal{B}_{r,v}$ be the subset of these branchings $\{B_i\}_{i \geq 1}$ obtained by independently sampling each one to lie in $\mathcal{B}_{r,v}$ with probability μ_i which is at most 1 since $\mu_i \leq K_{r,v}$ and $\gamma \leq 1/2$.
- Next, for each $r \in R$ and $v \in C$ we turn each branching $B \in \mathcal{B}_{r,v}$ into an r-v path by adding the reverse direction of all edges not on the r-v path in B and shortcutting the resulting Eulerian walk. Call the resulting path $P(B)$.
- Finally, we simplify the resulting collection of paths as follows. While there is some $v \in C$ that lies on at least two paths $P(B), P(B')$ remove the occurrence of v from one of these paths, choosing arbitrarily, by either shortcutting the path if v was not the last node on the path or truncating the path to end at the node just before v if v was the last node on the path. Discard any paths that no longer contain any clients.

For $r \in R$ let $\mathcal{P}_r$ be the set of the resulting paths that start at r and let $\mathcal{P} = \cup_{r \in R} \mathcal{P}_r$ be the set of all such paths.

The resulting collection of paths have the following properties: a) each path $P \in \mathcal{P}_r$ begins at r and, otherwise, only contains clients, b) each path in $\mathcal{P}$ visits at least one client, c) no client lies on more than one path in $\mathcal{P}$ (though multiple paths may originate from the same depot), d) a client lies on a path in $\mathcal{P}$ if and only if it lies on at least one sampled branching.

Lemma 3. $\mathbb{E}[\sum_{P \in \mathcal{P}} c(P)] \leq \gamma \cdot (1 + \delta) \cdot opt_{LP}$

Proof. For a branching B we have $c(P(B)) \leq 2 \cdot c(B) - c(v, r)$ because the path was obtained by doubling all edges *not* on the r-v path in B and then shortcutting the resulting Eulerian walk. So the expected cost of $\mathcal{P}$ is at most

$$\sum_{r \in R} \sum_{v \in C} \sum_{B_i \in \mathcal{B}_{r,v}} \mu_i \cdot (2 \cdot c(B) - c(v,r)) \leq 2 \cdot \gamma \cdot \text{opt}_{LP} - \sum_{r \in R} \sum_{v \in C} c(v,r) \sum_{B_i \in \mathcal{B}_{r,v}} \mu_i$$
$$= 2 \cdot \gamma \cdot \text{opt}_{LP} - \sum_{r \in R} \sum_{v \in C} K_{r,v} \cdot c(v,r)$$

where the inequality is because for each $r \in R$ and $v \in C$ we have each edge e lying on at most a $\gamma \cdot x^r_{v,e}$-fraction of branchings in $\mathcal{B}_{r,v}$.

Recalling $K_{r,v} = 2\gamma \cdot z^r_{v,v}$, the last term $\sum_r \sum_v K_{r,v} \cdot c(v,r)$ equals $\gamma \cdot \sum_{r,v} 2c(r,v) \cdot z^r_{v,v} = (1 - \delta) \cdot \text{opt}_{LP}$ so the expected cost of $\mathcal{P}$ is seen to be at most $2 \cdot \gamma \cdot \text{opt}_{LP} - \gamma \cdot (1 - \delta) \cdot \text{opt}_{LP} = \gamma \cdot (1 + \delta) \cdot \text{opt}_{LP}$. □

2.2 Step: Covering the Uncovered Nodes with Trees Rooted in $\mathcal{P}$

Recall that $\mathcal{P}$ is the set of all R-rooted paths sampled in the previous step. Let U be all clients that do not lie on a path, this is the same as the set of clients that did not lie on any branching that was sampled.

The next step is to compute a cheapest spanning forest F rooted at nodes in $R \cup (C - U)$. That is, each component of F includes precisely one node in $R \cup (C - U)$. Intuitively, the trees of F are being used to graft in uncovered clients into the paths. Before converting $\mathcal{P}$ and F into a feasible CVRP-MD solution, we note a few properties of U and F.

Lemma 4. $\mathbb{E}[\ell(U)] \leq e^{-\gamma} \cdot lb$

Proof. For each $r \in R, v \in V$ if we view each branching B_i obtained from Theorem 2 as a set covering its clients, then each client $u \in C$ lies on a total weight of $\gamma \cdot z^r_{v,u}$ of such sets. Summing over all r, v we see that each $u \in C$ lies on a total weight of $\sum_{r,v} \gamma z^r_{v,u} = \gamma$ such sets. Since the branchings were sampled independently, the usual analysis for randomized-rounding set cover algorithms using the arithmetic-geometric mean inequality shows that a node $u \in C$ is not covered by any sampled branching with probability at most $e^{-\gamma}$. □

To bound the expected cost of F, we use the following adaptation of the Bridge Lemma of [7] to vehicle routing problems.

Theorem 3 (Böhm et al. [5]). *Let $G = (V, E)$ be an undirected graph with edge costs c. For any $S \subseteq V$, let c_S be the minimum-cost of a spanning forest whose trees are rooted in S. Let $T \subseteq V$ be nonempty and let $\mu : 2^V \to \mathbb{Q}_{\geq 0}$ be a probability distribution over subsets of $V \setminus T$. Also suppose there is some $\gamma \geq 0$ such that for that any $v \in V \setminus T$ we have $\mathbf{Pr}_{S \sim \mu}[v \notin S] \leq \gamma$. Then $\mathbb{E}[c_{T \cup S}] \leq \gamma \cdot c_T$.*

Using this, we bound the expected cost of F as follows.

Lemma 5. $\mathbb{E}[c(F)] \leq e^{-\gamma} \cdot opt_{LP}$

Proof. In Theorem 3, the graph is the undirected metric over $R \cup V$, $T := R$, and the probability distribution μ is given by the subset of clients that are covered by a path in $\mathcal{P}$. From the analysis of Lemma 4, we know for each client v that $\mathbf{Pr}_{S \sim \mu}[v \notin S] \leq e^{-\gamma}$. So by Theorem 3, $\mathbf{E}[c(F)] \leq e^{-\gamma} \cdot c_R$.

We conclude by showing $c_R \leq \text{opt}_{LP}$. To do that, consider the following natural cut-based LP relaxation for the minimum-cost R-rooted directed spanning forest in the bidirected graph obtained from our given metric over $R \cup V$.

$$\min\{\textstyle\sum_e c(e) \cdot y_e : y(\delta^{in}(S)) \geq 1 \ \forall \ \emptyset \subsetneq S \subseteq V, y \geq 0\}.$$

Here, we are letting y_e be a variable for each directed edge and $c(e)$, for a directed edge is the same as the cost of the underlying undirected edge.

It is well known that that extreme points of this relaxation are integer, e.g., by contracting R to a single node an using integrality of the corresponding directed arborescence LP relaxation [8]. It is also easy to verify that the vector $\sum_{r \in R, v \in C} x_v^r$ is feasible for this LP relaxation. So the cheapest R-rooted spanning forest has cost at most opt_{LP}. That is, $c_R \leq \text{opt}_{LP}$. □

Throughout the rest of the algorithm, we remove clients from the set U as they are covered. In the description of the algorithm, we will say a client $u \in C$ is *uncovered* if $u \in U$ when that step is executed.

2.3 Algorithm Summary

The entire algorithm for vehicle capacity $k \geq 3$ is summarized in Algorithm 1. Here, we use $\gamma = 0.46821$. The two subroutines mentioned in Algorithm 1 will be covered in the sections below.

2.4 Step: Tree Pruning

Recall F is a forest where each tree has a single node that is covered by the paths in $\mathcal{P}$. That is, each tree can be viewed as rooted in $R \cup (C - U)$. All other nodes in the tree are initially uncovered. The goal of this section is to splice off some tours to cover all but a bounded number of uncovered clients in each tree T, i.e., we want each tree rooted in a node rooted in $\mathcal{P}$ to be *small*.

Each time we cover some clients in a tree T, we will cut out some subtrees of T and declare some uncovered nodes of T to now be covered. The cost of the tour will be charged to both the cost of the edges removed from T and also to the radial lower bound for the newly-covered clients.

Throughout the algorithm, for a node v in tree T we let T_v be the subtree rooted under v and $U(T_v)$ be the uncovered nodes in T_v. For a node u of T, let $u_1, u_2, \ldots$ denote its children. If each T_{u_i} has at most k uncovered nodes, we can consider the following **grouping scheme**. Initially let each T_{u_i} lie in a group of its own. Then while it is possible to merge two groups so the resulting group still has at most k uncovered clients, do so. For such a group G, let $c(G)$ denote the

Algorithm 1 Multiple-Depot CVRP Algorithm

Let (x, z) be an extreme point optimal solution to (**LP-CVRP-MD**).
$\mathcal{P} \leftarrow \emptyset$
for each $r \in R, v \in C$ **do**
  Apply Theorem 2 to preflow $\gamma \cdot x_v^r$ using $K := 2\gamma \cdot z_{v,v}^r$ to get branchings $\mathcal{B}_{r,v} = \{B_i\}$ with corresponding weights $\{\mu_i\}$. {Recall v will be on each such B_i.}
  for each $B_i \in \mathcal{B}_{r,v}$ **do**
    Let P_i be an $r - v$ path obtained by doubling edges of B_i not on the $r - v$ path in B_i and shortcutting the resulting Eulerian $r - v$ walk past repeated clients.
    With probability μ_i add P_i to $\mathcal{P}$.
$U \leftarrow$ clients in C not lying on any path in $\mathcal{P}$.
Shortcut paths in $\mathcal{P}$ so each $v \in C - U$ lies on exactly one path.
$F \leftarrow$ min-cost spanning forest of $R \cup C$ rooted in $R \cup (C - U)$.
for each tree T of F **do**
  Apply Algorithm 2 in Section 2.4 to cover some clients of T with tours and prune it to a shorter tree T'.
for each path $P \in \mathcal{P}$ **do**
  Apply the tour partitioning procedure in Section 2.5 to P and the trees T' rooted in P to cover the remaining clients.

total cost of all subtrees in the group plus the edges connecting these subtrees to u and $U(G)$ be the uncovered clients lying on a subtree in G.

For a group G we let $\mathcal{T}(G)$ be a tour obtained by doubling the edges on all subtrees in G (including the edge connecting subtrees T_{u_i} to u) and adding two copies of the cheapest edge connecting some depot $r' \in R$ to some client in $U(G)$, and then shortcutting the resulting Eulerian tour so it only visits $\{r'\} \cup U(G)$.

For two different groups G, G' we also consider the following tour $\mathcal{T}(G, G')$ that covers exactly k uncovered clients.

- Let A be the $k - |U(G)|$ clients in $U(G')$ with largest ℓ_v-values. This is possible since $|U(G)| + |U(G')| > k$ as G and G' cannot be merged.
- Let $\mathcal{T}(G, G')$ be the tour spanning $U(G) \cup A$ and a depot r' obtained by first doubling all edges in the subtrees of both G and G' along with the parent edges of these subtrees conneecting them to u, then adding two copies of the cheapest edge connecting some $r' \in R$ to some client in $U(G) \cup A$, and finally shortcutting the resulting Eulerian tour so it only visits $\{r'\} \cup U(G) \cup A$.

Finally, we introduce two constants: (a) $\beta = 0.5902302342$ governs the behaviour of the algorithm and (b) $\Delta = 1.6353454381$ is used in the analysis. These parameters were chosen in a way to optimize the approximation guarantee obtained from our analysis while satisfying the following bounds that will be used at various points in our arguments.

- $2 - 0.5/\Delta \leq 1/\beta$
- $1.5 + (\beta - 0.5)/\Delta \leq 1/\beta$
- $1/(1 - \beta) < 3$
- $0 \leq \frac{\Delta - (1-\beta) \cdot 1.5}{3 - 2\beta} \leq 1$ and $0 \leq \frac{2 + 3\beta/\Delta}{2 + 1/(1-\beta)} \leq 1$.

Algorithm 2 summarizes the tree pruning procedure. Notice if the second case within the while loop is executed, the subtrees in G_j are not pruned from the tree. Figure 1 illustrates this. Also, this case is only executed if the cost of the resulting tour can be charged against only $c(G_i)$ and $\ell(U(G_i) \cup A)$, i.e. the edges in the subtrees from G_j will not be charged.

Algorithm 2 Pruning a tree T from the forest F rooted at a node $v_T \in R \cup (C - U)$.

---

while the following removes some clients from U_T **do**
  Let u be any deepest node in T such that $|U(T_u)| > k$, using $u = v_T$ if no such node exists
  Group the subtrees $T_{u_1}, \ldots$ into groups $G_1, G_2, \ldots$ using the procedure discussed above
  if $c(\mathcal{T}(G_i)) \leq 2.5 \cdot c(G_i) + \frac{1}{\beta} \cdot \ell(U(G_i))$ for some group G_i **then**
    Include tour $\mathcal{T}(G_i)$ in the final answer
    Remove all subtrees in G_i from T
    $U \leftarrow U - U(G_i)$
  else if $c(\mathcal{T}(G_i, G_j)) \leq 2.5 \cdot c(G_i) + \frac{1}{\beta} \cdot \ell(U(G_i) \cup A)$ for two distinct groups G_i, G_j **then**
    {Here, A is the set of $k - |U(G_i)|$ clients in $U(G_j)$ with largest ℓ_v-values}
    Include tour $\mathcal{T}(G_i, G_j)$ in the final answer
    Remove only subtree G_i from T {Clients in A remain in T, but are still regarded as covered}
    $U \leftarrow U - (U(G_i) \cup A)$

In each iteration of the loop, the node u, while uncovered at that time, will not be considered covered by any tour that is found that iteration as such tours only cover nodes in subtrees rooted under children of u even though the parent edge to u of such a subtree is used in the construction of the tour.

The following is immediate from the algorithm.

Corollary 1. *Let F^{init} be the initial forest rooted and F' the forest obtained after pruning each tree using Algorithm 2. Similarly, let U^{init} be the initial set of clients that were not covered by $\mathcal{P}$ and U' be the set of clients that remain uncovered after applying Algorithm 2 to each tree of F^{init}.*

Let $\mathcal{T}_1, \mathcal{T}_2, \ldots, \mathcal{T}_m$ be the tours that were found over all applications of Algorithm 2 to each tree in F^{init}. Then $\sum_{i=1}^{m} c(\mathcal{T}_i) \leq 2.5 \cdot (c(F^{init}) - c(F')) + \frac{1}{\beta} \cdot (\ell(U^{init}) - \ell(U'))$.

Proof. Each time a tour $\mathcal{T}_i$ is included in the solution, its cost is bounded by 2.5 times the cost of the subtrees that are removed plus $1/\beta$ times the radial lower bound of the uncovered clients that were just now covered by $\mathcal{T}_i$. □

We also show that the remnants of trees of the forest that remain after this step are "small" and satisfy an additional property relating their cost to their uncovered clients' radial lower bound contribution if they are not too small. These properties will be helpful in the final stage of the algorithm.

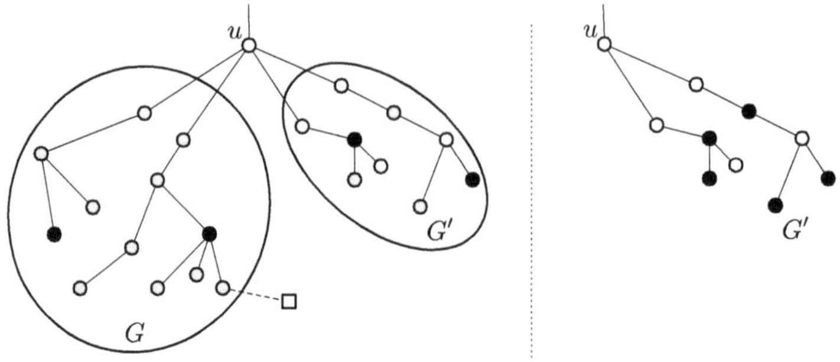

Fig. 1. Left: The black nodes were already covered by a previous pruning step and the grey nodes are the ones to be covered in the tour $\mathcal{T}(G_i, G_j)$ (so $k = 10$). All nodes of $U(G)$ will be covered and the grey nodes in G' are precisely A. The square node is the depot closest to a client in $U(G_i) \cup A$. Tour $\mathcal{T}(G_i, G_j)$ is obtained by doubling all edges shown in the picture and shortcutting to only include $U(G_i) \cup A$ and the depot. Right: All subtrees in G_i are pruned, but all subtrees in G_j remain. The nodes of A are now covered. (Color figure online)

Lemma 6. *Let T be any tree in the forest F and T' be the result of applying Algorithm 2 to T. Then $|U(T')| \leq \beta \cdot k$ and, further, if $|U(T')| > k/2$ then $\ell(U(T')) \geq \Delta \cdot c(T')$.*

2.5 Step: Forming the Final Tours

Let F' be the portion of the initial forest F that remains after applying Algorithm 2 to each tree T of F. Recall that U denotes the clients that have not yet been covered by the algorithm, so at this point U is the set of all clients not lying on a path in $\mathcal{P}$ and not covered by a tour that was produced the various calls to Algorithm 2. Each tree T' in F' remains rooted in a path of $\mathcal{P}$ and all clients currently in U lie on one of these trees. These trees T' satisfy the properties in Lemma 6.

For each path $P \in \mathcal{P}$, let C_P denote the clients on P. If $U = \emptyset$ at this point, we would conclude the algorithm by taking each path in $\mathcal{P}$ and applying the tour splitting procedure from [13]. This charges each edge on each path at most twice and plus an additional cost of $\ell(C_P)$. But there may be "small" trees left in the forest F'. One could try to graft them into the path P they are rooted in naively by doubling their edges and shortcutting the Eulerian tour to get a path that can be grafted into P. But then when applying tour splitting to the path, we might charge an edge of a tree T' up to four times. In particular, if the tour is split at a node v in a tree T' then the root-to-v path in T' would be doubled twice in the final analysis.

The main idea in this section is to handle this problem with care by selecting a careful depth-first search (DFS) ordering of each tree T' and being careful to only charge a cheap portion of each tree more than twice. The key concept here is to look at how deep the tree is. For a tree T' in what is left of the forest, let

$v_{T'}$ be the root of T' and define $\texttt{trunk}(T')$ to be the highest-cost path from $v_{T'}$ to a leaf node in T'.

Consider a DFS of T' starting at $v_{T'}$ which first recurses along the path to the leaf node defining $\texttt{trunk}(T)$ but all other decisions in the DFS can be made arbitrarily. Output the uncovered nodes of T' in a post-order traversal in this DFS. In particular, the DFS will not output any node until after it has traveled down along the entire trunk and it will not output the root $v_{T'}$ itself.

To create the final tours we introduce one final concept. Namely, we classify a tree T' in one of two ways.

- **Short** if $c(\texttt{trunk}(T')) \leq \frac{c(T)}{2} + \frac{\ell(U(T'))}{4} \cdot \frac{|U(T')|}{k}$.
- **Tall** otherwise.

For each path $P \in \mathcal{P}$ with, say, $P = v_1, v_2, \ldots, v_a$ (here $v_1 \in R$) construct a final ordering of P and all trees of F' rooted in P as follows: after v_i but before v_{i+1}, output the nodes of the tree T' of F' rooted at v_i in the post-order DFS traversal mentioned above. See the top part of Fig. 2 to see an example of this ordering. Let P' be the resulting sequence $r = u_1, u_2, u_3, u_3, \ldots, u_{a'}$. Pick an offset $\tau \in \{1, 2, \ldots, a\}$ and let $S_{P'} = \{u_\tau, u_{\tau+k}, u_{\tau+2k}, \ldots\}$ be nodes about which we will *split* the sequence P'. For brevity r_i be the depot nearest to u_i for each i.

We create tours as follows. First, partition P' into subsequences using this random offset, i.e., for each integer $b \geq 0$ consider the sequence $P'_b := u_{b \cdot k+\tau}, u_{b \cdot k+\tau+1}, \ldots, u_{(b+1)\cdot k+\tau-1}$ (omitting indices outside the range $[1, a']$). So all but, perhaps, the first and last subsequence have length exactly k.

For each subsequence P'_b, if it is not the last subsequence and if the next client $u_{(b+1)\cdot k+\tau}$ (which is in $S_{P'}$) lies in $U(T')$ for a tall tree T', we further split P'_b as follows. Say u_j is the base of T', notice u_j lies in P'_b because $|U(T')| \leq \beta \cdot k < k$. Split P'_b into two subsequences: the prefix up to and including u_j and the suffix after u_j. Finally, append $u_{(b+1)\cdot k+\tau-1}$ to the second subsequence. After doing this for all P'_b, note the following properties: (a) each subsequence contains exactly one node in $S_{P'}$ or the depot u_1 at the start of P, (b) each node $S_{P'}$ lies in two subsequences only if it is in $U(T')$ for a tall tree T', otherwise it appears on exactly one subsequence.

Finally, for each of the resulting subsequences we obtain a tour spanning by doubling all edges of $P \cup F'$ lying between clients of the subsequence and adding a doubled edge connecting the unique node of $\{u_1\} \cup S_{P'}$ in the subsequence to its nearest depot. Shortcut the resulting tour past all nodes except the depot and those in the subsequence. If the subsequence was the suffix of a split subsequence of the form P'_b, then also shortcut the tour past $u_{(b+1)\cdot k+\tau}$ as it will be covered in the next subsequence. This way, every tour spans at most k clients and all uncovered clients grafted onto P are now covered. This procedure is depicted in Fig. 2.

To analyze the cost of this tour, for each subsequence we double all edges of P and the various trees T' rooted in P that lie between two nodes of the subsequence. Intuitively, every edge of P is used by at most one subsequence and every edge of each tree T' is as well with two exceptions:

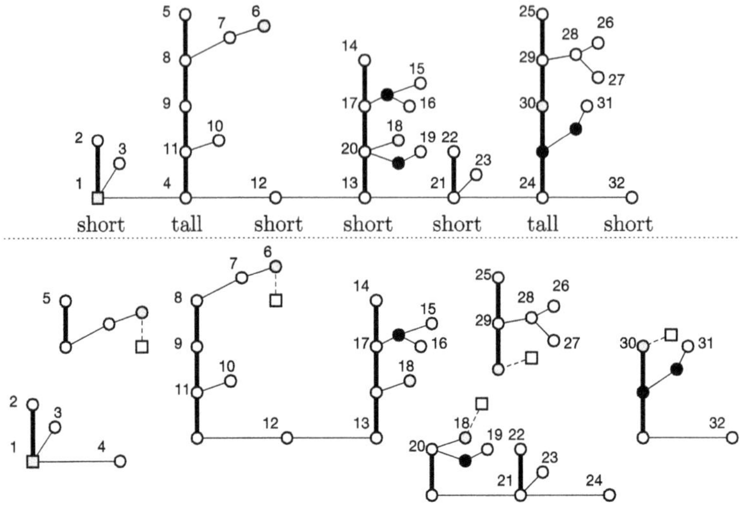

Fig. 2. **Top**: A path $P \in \mathcal{P}$, depicted horizontally, and the trees T' of F' with a root in P. The thick vertical edges are the trunks of the trees. The numbers indicate the final ordering of the nodes. The black nodes were covered earlier by tours in the tree pruning stage. With $k = 12$ and offset $\tau = 6$, the set consisting of the depot and $S_{P'}$, depicted with grey nodes, would be nodes $1, 6, 18$ and 30 in this ordering. **Bottom**: The edges that were doubled to form the various tours. The white square nodes are depots nearest each node of $S_{P'}$. Notice each edge lies on at most one subtree with the following exceptions: for a large tree T' some edges from a single path between $S_{P'}$ and the trunk appear on two subtrees and for a small tree T' some edges on a single path from $S_{P'}$ to the root of the tree appear on two subtrees. A client that is not numbered in means we do not consider it as being covered by the corresponding tour.

- In short trees T', if some node of $U(T')$ lies in $S_{P'}$ then edges lying on some path to the root $v_{T'}$ will be used by two subsequences. This happens only with probability $|U(T')|/k$ and if it does, the cost of the edges that are used by two subsequences is at most $c(\texttt{trunk}(T'))$.
- In tall trees T', if some node of $U(T')$ lies in $S_{P'}$ then edges lying on some path from $S_{P'}$ to a node on $\texttt{trunk}(T')$ will be used by two subsequences. This happens only with probability $|U(T')|/k$ and if it does, the cost of the edges that are used by two subsequences is at most $c(T') - c(\texttt{trunk}(T'))$. But in this case we also have to remember that the node $v \in U(T') \cap S_{P'}$ will be used to connect two subsequences to its nearest depot (i.e. we will charge ℓ_v twice in this case).

Disclosure of Interests. The authors have no competing interests to declare that are relevant to the content of this article.

References

1. Armbruster, S., Mnich, M., Nägele, M.: A (3/2 + 1/e)-approximation algorithm for ordered TSP. In: Kumar, A., Ron-Zewi, N. (eds.) Approximation, Randomization, and Combinatorial Optimization. Algorithms and Techniques, APPROX/RANDOM 2024, London School of Economics, London, UK, 28–30 August 2024. LIPIcs, vol. 317, pp. 1:1–1:18. Schloss Dagstuhl - Leibniz-Zentrum für Informatik (2024)
2. Asano, T., Katoh, N., Tamaki, H., Tokuyama, T.: Covering points in the plane by k-tours: towards a polynomial time approximation scheme for general k. In: ACM Symposium on Theory of Computing (STOC), pp. 275–283 (1997)
3. Bang-Jensen, J., Frank, A., Jackson, B.: Preserving and increasing local edge-connectivity in mixed graphs. SIAM J. Disc. Math. **8**, 155–178 (1995). https://api.semanticscholar.org/CorpusID:17614575
4. Blauth, J., Traub, V., Vygen, J.: Improving the approximation ratio for capacitated vehicle routing. Math. Program. **197**(2), 451–497 (2023)
5. Böhm, M., Friggstad, Z., Mömke, T., Spoerhase, J.: Approximating traveling salesman problems using a bridge lemma. In: Proceedings of the 2025 Annual ACM-SIAM Symposium on Discrete Algorithms (SODA). SIAM (2025)
6. Byrka, J., Grandoni, F., Rothvoß, T., Sanità, L.: Steiner tree approximation via iterative randomized rounding. J. ACM **60**(1), 6:1–6:33 (2013)
7. Byrka, J., Grandoni, F., Rothvoss, T., Sanità, L.: Steiner tree approximation via iterative randomized rounding. J. ACM **60**(1) (2013)
8. Edmonds, J.: Optimum branchings. J. Res. Natl. Bureau Stand. B **71**(4), 233–240 (1967)
9. Haimovich, M., Rinnooy Kan, A.H.G.: Bounds and heuristics for capacitated routing problems. Math. Oper. Res. **10**(4), 527–542 (1985)
10. Harks, T., König, F.G., Matuschke, J.: Approximation algorithms for capacitated location routing. Transp. Sci. **47**(1), 3–22 (2013)
11. Karlin, A.R., Klein, N., Oveis Gharan, S.: A (slightly) improved approximation algorithm for metric tsp. In: Proceedings of the 53rd Annual ACM SIGACT Symposium on Theory of Computing (STOC), pp. 32–45 (2021)
12. Karlin, A.R., Klein, N., Oveis Gharan, S.: A deterministic better-than-3/2 approximation algorithm for metric TSP. In: Pia, A.D., Kaibel, V. (eds.) Proceedings of 24th International Integer Programming and Combinatorial Optimization (IPCO'23), pp. 261–274 (2023)
13. Li, C.L., Simchi-Levi, D.: Worst-case analysis of heuristics for multidepot capacitated vehicle routing problems. ORSA J. Comput. **2**(1), 64–73 (1990)
14. Montoya-Torres, J.R., Franco, J.L., Isaza, S.N., Jiménez, H.F., Herazo-Padilla, N.: A literature review on the vehicle routing problem with multiple depots. Comput. Ind. Eng. **79**, 115–129 (2015)
15. Post, I., Swamy, C.: Linear programming-based approximation algorithms for multi-vehicle minimum latency problems. In: Proceedings of 26th Annual ACM-SIAM Symposium on Discrete Algorithms (SODA'15), pp. 512–531 (2015)
16. Zhao, J., Xiao, M.: Improved approximation algorithms for multidepot capacitated vehicle routing. In: COCOON (2). Lecture Notes in Computer Science, vol. 14423, pp. 378–391. Springer, Heidelberg (2023). https://doi.org/10.1007/978-3-031-49193-1_29

The Integrality Gap of the Traveling Salesman Problem is 4/3 if the LP Solution Has at Most $n + 6$ Non-Zero Components

Tullio Villa[1,3](✉), Eleonora Vercesi[2,3], Janos Barta[1,3], and Monaldo Mastrolilli[1,3]

[1] Scuola Universitaria Professionale della Svizzera Italiana (SUPSI), Lugano, Switzerland
{tullio.villa,janos.barta,monaldo.mastrolilli}@supsi.ch
[2] Università della Svizzera Italiana (USI), Lugano, Switzerland
eleonora.vercesi@usi.ch
[3] Dalle Molle Institute for Artificial Intelligence USI-SUPSI, Lugano, Switzerland

Abstract. We address the classical Dantzig-Fulkerson-Johnson formulation of the symmetric metric Traveling Salesman Problem and study the integrality gap of its linear relaxation, namely the Subtour Elimination Problem (SEP). This integrality gap is conjectured to be 4/3. We prove that, when solving a problem on n nodes, if the optimal SEP solution has at most $n+6$ non-zero components, then the conjecture is true. To establish this result, we devise a new methodology that combines theoretical analysis and computational verification.

Keywords: Symmetric Traveling Salesman Problem · Linear Programming Relaxation · Integrality Gap

1 Introduction

Let $K_n = (V_n, E_n)$ be the complete graph on n nodes, and let $\boldsymbol{c} \in \mathbb{R}^{E_n}$ be a non-negative cost vector. The *Traveling Salesman Problem* (TSP) consists of finding a Hamiltonian tour of minimum total cost. This paper focuses exclusively on the *symmetric* formulation, where the graph K_n is undirected, hence we use ij and ji interchangeably to denote the edge connecting nodes i and j. Additionally, we only consider *metric* cost vectors, satisfying the triangle inequalities. Henceforth, we simply use the term TSP to indicate the symmetric metric TSP.

The TSP is typically formulated as an Integer Linear Program (ILP). Among the most prominent formulations is the one of Dantzig-Fulkerson-Johnson (DFJ) [9], whose associated relaxed Linear Problem (LP) is known in the literature as the Subtour Elimination Problem (SEP), or Held-Karp relaxation:

S. S. Dey et al. (Eds.): IPCO 2026, LNCS 16588, pp. 128–143, 2026.
https://doi.org/10.1007/978-3-032-28691-8_9

$$\text{DFJ}: \min_{\boldsymbol{x}\in P^n_{\text{SEP}}\cap\mathbb{Z}^{E_n}} \sum_{e\in E_n} c_e x_e\ , \qquad \text{SEP}: \min_{\boldsymbol{x}\in P^n_{\text{SEP}}} \sum_{e\in E_n} c_e x_e\ .$$

The polytope on which they are described is the *subtour elimination polytope*:

$$P^n_{\text{SEP}} := \left\{ \boldsymbol{x}\in\mathbb{R}^{E_n} \mid \sum_{e\in\delta(v)} x_e = 2\ \ \forall v\in V_n,\ \sum_{e\in\delta(S)} x_e \geq 2\ \ \forall S\in\mathcal{S},\ 0\leq x_e\leq 1\ \ \forall e\in E_n \right\},$$

where $\mathcal{S} := \{S \subseteq V_n \mid 3 \leq |S| \leq n-3\}$ and, for any subset $S \subseteq V_n$, $\delta(S)$ is the set of edges having exactly one node in S[1]. The term *subtour elimination constraints* refers to the second set of constraints in the definition of P^n_{SEP}.

For a given cost vector $\boldsymbol{c}$, we denote the optimal solutions of the integral and relaxed problem by $\text{TSP}(\boldsymbol{c})$ and $\text{SEP}(\boldsymbol{c})$, respectively. The *integrality gap* is the ratio of these two values; it can be evaluated for a single instance cost $\boldsymbol{c}$, as $\alpha(\boldsymbol{c})$, or considered in a worst case scenario, as α:

$$\alpha(\boldsymbol{c}) := \frac{\text{TSP}(\boldsymbol{c})}{\text{SEP}(\boldsymbol{c})}\ , \qquad \alpha := \sup_{c\ \text{metric}} \frac{\text{TSP}(\boldsymbol{c})}{\text{SEP}(\boldsymbol{c})}\ .$$

An alternative granularity of the integrality gap can also be defined by focusing on a specific vertex $\boldsymbol{x} \in P^n_{\text{SEP}}$. We denote this by $\text{Gap}(\boldsymbol{x})$, and it is given by:

$$\text{Gap}(\boldsymbol{x}) := \sup\left\{ \frac{\text{TSP}(\boldsymbol{c})}{\text{SEP}(\boldsymbol{c})} \mid \boldsymbol{c}\ \text{metric}\ ,\ \boldsymbol{x} \in \arg\min \text{SEP}(\boldsymbol{c}) \right\}\ .$$

The exact value of α is currently unknown; in order to determine it, various approaches have been explored, with increasing interest over the past decades. Wolsey [24] was the first to establish a $3/2$ upper bound for α, with an analysis built on the Christofides' classic algorithm [7]. This bound remained untouched for nearly fifty years. The first breakthrough was achieved only in 2021, when Karlin, Klein, and Oveis Gharan proposed a better-than-$3/2$ approximation algorithm for the general TSP [17]. Its analysis [18] lead to the corresponding improvement on the upper bound for the integrality gap: $\alpha \leq 3/2 - \varepsilon$, where $\varepsilon \sim 10^{-36}$. The best known lower bound for α is $4/3$, achieved by families of instances that where already known in 1995, when Goemans [10] formulated the "$4/3$-conjecture", which asserts that the integrality gap is exactly $4/3$. Despite several attempts, nobody was able to disprove the conjecture, nor to settle it conclusively.

Other than these lower and upper bounds, only results for specific subclasses of instances are available in the literature. Benoit and Boyd [1,3] exploited exhaustive enumeration of the vertices of P^n_{SEP} to compute the exact value of the integrality gap for TSP instances with $n \leq 12$. Schalekamp, Williamson, and van Zuylen [21] conjectured that the maximum integrality gap is attained on *half-integer* instances, that is, cost vectors whose optimal SEP solutions have all the entries in $\{0, 1/2, 1\}$. For this class of instances, the integrality gap was

[1] The brackets denoting singletons are omitted for brevity, e.g., $v = \{v\}$; this abuse of notation is used throughout this work, when the context does not lead to ambiguity.

shown to be strictly less than 3/2 by Karlin et al. in [16]. This achievement was improved in [11] and subsequently in the as-yet unpublished work [19]. In recent years, promising lines of research have examined subclasses of SEP solutions $\boldsymbol{x}$ characterized by specific properties of the so called *support graph*, defined as the undirected weighted graph $G_x = (V_n, E_x)$ such that $ij \in E_x \Leftrightarrow x_{ij} > 0$, and the weight on edge ij is given by x_{ij}. Notice that the number of edges in the support graph G_x is, by definition, the number of non-zero components of $\boldsymbol{x}$. Boyd and Carr [2] proved that the integrality gap is 4/3 when the support graph of the SEP solution contains disjoint 1/2 triangles. For graph-TSP instances[2], Boyd et al. [6] proved that the 4/3-conjecture holds for cubic graphs. Mömke and Svensson [20] improved this result showing that the integrality gap is 4/3 for graph-TSP restricted either to half-integral solutions or to a class of graphs that contains subcubic and claw-free graphs. Boyd and Sebő [5] proved that the integrality gap is at most ∼ 1.4286 for instances having as SEP solution one of the so-called Boyd-Carr points (firstly defined in [2]). Jin et al. [15] proved the 4/3-conjecture for instances having as SEP solution cycle-cut points, that is, points for which every non-singleton tight set can be written as the union of two tight sets.

Different approaches have also been attempted to improve the lower bound. In [1,13,14,25], families of TSP instances with high integrality gap, asymptotically tending to 4/3 were shown. In [22], the authors proposed an approach to heuristically generate instances with a high integrality gap: no instance with an integrality gap greater than 4/3 was found.

Contribution and Outlook. This work presents both a theoretical contribution and a novel methodology that opens new research directions in the study of the integrality gap. We prove the 4/3-conjecture for all the vertices of P^n_{SEP} whose support graph contains at most $n+6$ edges. Remarkably, this class is rich enough to contain vertices that are not covered by any previously known result in the literature. Figure 1 shows an example of such a vertex.

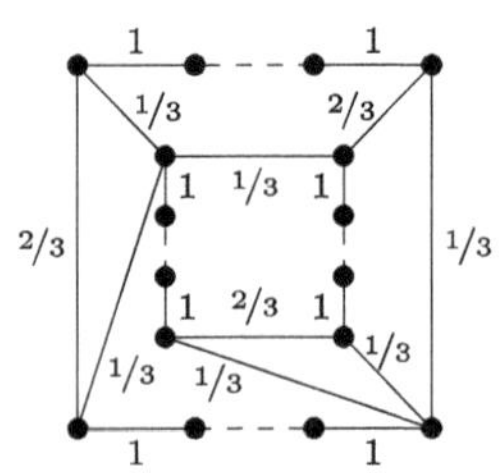

Fig. 1. Example of a vertex considered in this work but not in previous literature. Dashed edges may be replaced by paths of edges of weight 1 of arbitrary length.

[2] Graph-TSP is a subclass of metric TSP where the distance between the nodes is computed as the minimum number of edges separating them.

To establish our result, we define, for each integer k, the family $\mathcal{F}_k$ which comprises, across all polytopes P^n_{SEP} for $n \in \mathbb{N}$, the vertices whose support graphs contain exactly $n+k$ edges. Although the family $\mathcal{F}_k$ is infinite, we identify a finite subset $\mathcal{A}_k$ of its elements, which we call *ancestors*, that allows us to make overall considerations on the entire family. We devise the *Gap-Bounding (GB) algorithm* and prove that its application on these ancestors returns upper bounds on the integrality gap of all the costs whose SEP solution is a vertex of $\mathcal{F}_k$. The application of the GB algorithm on the elements of $\mathcal{A}_k$ with $k \leq 6$ yields a computer-aided proof of the desired result.

The methodology presented constitutes in itself a novel way to approach the integrality gap problem, combining theoretical analysis and computational verification. Unlike earlier computational proofs (see [1,3]) that were limited to enumerating vertices for a fixed dimension n, our framework fixes a small parameter k and derives results valid for *all* n, hereby achieving a form of universality through a computational approach. To the best of our knowledge, this is the first time that a computer-aided proof is implied in bounding the integrality gap for *infinitely many* vertices.

We conclude by noting that the vast majority of instances used in the literature to establish lower bounds on the integrality gap are special cases of our result (see, e.g., [1,13,14,25]: counting nodes and edges in the constructions, one may see that the instances in [1,13,25] belong to $\mathcal{F}_3$ and the instances in [14] belong to $\mathcal{F}_6$). Other families, however, fall outside the scope of our analysis: the donut instances proposed in [5] have a non-constant surplus of edges over nodes, and the lower bounds on their integrality gap provided in the article converge to $4/3$. Interestingly, for each donut instance, the mentioned lower bound is consistently dominated by the integrality gap of an instance of $\mathcal{F}_3$ with the same number of nodes[3]. This observation aligns with the empirical evidence that, for a small number of nodes $n \leq 12$, the value of the integrality gap of a vertex correlates with the number of edges in its support graph (more details in Sect. 7). Taken together, these insights motivate the following hypothesis, which highlights the significance of the class of vertices considered in this work.

Few Edges Hypothesis. *For every n, among the vertices of P^n_{SEP}, the maximum integrality gap is realized on a vertex with the minimum possible number of edges in the support graph, that is $n+3$.*

In summary, our work contributes to three different fronts.

1. It establishes the $4/3$ conjecture for newly identified infinite classes of vertices.
2. It demonstrates the potential of computer-aided proofs to advance our understanding of the integrality gap across infinite vertex classes, proposing a flexible framework that can be adapted to other contexts.
3. It opens the compelling research direction of investigating the correlation between the integrality gap and the number of edges in the support graph.

---

[3] It can be checked comparing the explicit formulas for the lower bounds on the integrality gap in [1] and [5].

A proof of the Few Edges Hypothesis, together with the contribution of this work, would result in definitively solving the 4/3 conjecture.

Outline. Section 2 introduces key definitions from the literature. Section 3 presents the concept of ancestors together with the procedure to retrieve them. Section 4 provides the theoretical basis for the GB algorithm, whose design is detailed in Sect. 5. Section 6 explains how to apply the GB algorithm to bound the integrality gap for vertices with up to $n+6$ non-zero components. Section 7 explores future research directions. Proofs and technical details, omitted in this manuscript for brevity, may be found in the extended version [23].

2 Background Material

2.1 Preliminaries from Graph Theory

Before giving the list of definitions used in this work, we clarify that, to avoid confusion, given a graph $G=(V,E)$, the term *node* refers to the elements of V; the term *vertex* is reserved exclusively for the extreme points of P^n_{SEP}.

Definition 1 (Hamiltonian walk, Hamiltonian tour). *Let G be an undirected graph. A* Hamiltonian walk *is a closed walk in G that visits every node at least once, possibly traversing the same edge multiple times. If every node is visited exactly once, we call the walk* Hamiltonian tour. *For a given walk $\boldsymbol{w}$, $w_{ij} \in \mathbb{N}$ denotes the* multiplicity *of ij in $\boldsymbol{w}$, that is, the number of times the walk $\boldsymbol{w}$ uses the edge ij (the analogous notation $t_{ij} \in \{0,1\}$ is used for a tour $\boldsymbol{t}$).*

For brevity, in what follows the terms *walk* and *tour* always indicate Hamiltonian walks and tours. Moreover, we use the same literal to denote both the walk $\boldsymbol{w}$ (respectively the tour $\boldsymbol{t}$) and its *characteristic vector*, that is, the vector of w_{ij} (respectively t_{ij}) entries. When the underlying graph G is not specified, we assume it is the complete one.

Remark 1 . Since we have interest in the shortest possible walks $\boldsymbol{w}$, we assume that every edge ij is traversed at most twice (i.e. $w_{ij} \in \{0,1,2\}$)[4]. Notice that, under this assumption, the number of walks considered on a graph becomes finite.

2.2 Preliminaries from the Literature on the Integrality Gap

In this section, we collect from the literature (mainly from [1]) the definitions and results are used in this manuscript.

Theorem 1 ([1,4]). *Let $\boldsymbol{x} \in P^n_{SEP}$ be a vertex. Then $|E_{\boldsymbol{x}}| \leq 2n-3$.*

[4] It is well known (see, e.g. [8]) that, when an edge is used strictly more than twice by a walk, it is possible to get rid of two copies of that edge to obtain a shorter walk.

Definition 2 (1-edge, 1-path, from [1]). *Let $\boldsymbol{x} \in P^n_{SEP}$ and let e be an edge of K_n; e is called* 1-edge *of $\boldsymbol{x}$ if $x_e = 1$. When $\boldsymbol{x}$ is fractional (that is, not a tour), we call* 1-path *of $\boldsymbol{x}$ a maximal path of 1-edges in the support graph $G_{\boldsymbol{x}}$; the nodes of degree 2 in the 1-path are called* internal nodes, *the two remaining nodes are called* end nodes.

Theorem 2 ([1]). *Let $\boldsymbol{x} \in P^n_{SEP}$ be a fractional vertex. Then $\boldsymbol{x}$ has at least three distinct 1-paths.*

We now introduce a construction presented in [1] and give it a name using the authors' initials.

Definition 3 (BB-move, from [1]). *Let $\boldsymbol{x} \in P^n_{SEP}$ and let ab be one of its 1-edges. We call* BB-move *the construction of a new point $\boldsymbol{x}' \in P^{n+1}_{SEP}$ defined as follows, where v is the new node added ($V_{n+1} = V_n \cup \{v\}$):*

$$\begin{aligned} x'_{ab} &= 0 \ , & x'_e &= 0 \ \ \forall e \in \delta(v) \setminus \{av, vb\} \ , \\ x'_{av} &= x'_{vb} = 1 \ , & x'_e &= x_e \ \forall e \notin \delta(v) \cup \{ab\} \ . \end{aligned}$$

We denote this construction by $\text{BB}(\boldsymbol{x}, ab) := \boldsymbol{x}'$. *When it is clear from the context or not strictly necessary, we omit the edge ab in the notation:* $\text{BB}(\boldsymbol{x})$.

Theorem 3 ([1]). *Let $\boldsymbol{x} \in P^n_{SEP}$ and let ab be one of its 1-edges. Then $\text{BB}(\boldsymbol{x}, ab)$ is a vertex of P^{n+1}_{SEP} if and only if $\boldsymbol{x}$ is a vertex of P^n_{SEP}.*

3 Families of Vertices with a Given Number of Edges

In this section, we study the vertices of P^n_{SEP} considering, in their support graph, the relation between the number of edges and the number of nodes.

Lemma 1. *Let $\boldsymbol{x}$ be a fractional vertex of P^n_{SEP}. Then $|E_{\boldsymbol{x}}| \geq n + 3$.*

Our aim now is to describe, for a given k, all possible fractional vertices with n nodes and $n + k$ edges. For this purpose, we define the families of vertices $\mathcal{F}_k$:

$$\mathcal{F}_k := \{\boldsymbol{x} \text{ fractional vertex } \mid \boldsymbol{x} \in P^n_{\text{SEP}} \text{ for some } n, \ |E_{\boldsymbol{x}}| = n + k\} \ .$$

Notice that Lemma 1 implies $k \geq 3$, hence $\mathcal{F}_1 = \mathcal{F}_2 = \emptyset$. Observe also that a BB-move increases both the number of nodes and edges by 1, thus preserving their difference k. Therefore, in virtue of Theorem 3, a vertex is in $\mathcal{F}_k$ if and only if its image under a BB-move is in $\mathcal{F}_k$ as well. This brings us to define the *ancestors*, the vertices of $\mathcal{F}_k$ that can not be obtained as results of a BB-move, that is, vertices without internal nodes in the 1-paths.

Definition 4 (Ancestor of order k). *An* ancestor of order k *is a vertex $\boldsymbol{x}$ of $\mathcal{F}_k$ with no node of degree 2. We denote the set of ancestors of order k as*

$$\mathcal{A}_k := \{\boldsymbol{x} \in \mathcal{F}_k \mid \boldsymbol{x} \textit{ has no node of degree } 2\} \ .$$

Definition 5 (Successor). *Given a vertex $\boldsymbol{x}$, we call* successor *of $\boldsymbol{x}$ any vertex $\boldsymbol{x}'$ obtained by sequential applications of the* BB*-move.*

We can thus recover the whole family $\mathcal{F}_k$ by taking the elements of $\mathcal{A}_k$ and replacing any 1-edge with a 1-path of arbitrary length; $\mathcal{F}_k$ may be seen as the set of successors of $\mathcal{A}_k$. At this point, to give a complete description of $\mathcal{F}_k$, it only remains to retrieve all the elements of $\mathcal{A}_k$.

Lemma 2. *Let $\boldsymbol{x}$ be an ancestor in $\mathcal{A}_k$ and let n be its number of nodes. Then $k+3 \leq n \leq 2k$.*

Therefore, the elements of $\mathcal{A}_k$ are finite and can be recovered from the lists of vertices of P^n_{SEP} with n up to $2k$ (provided we have them at our disposal). We simply need to scroll through the lists of vertices of P^n_{SEP} for all the $n = k+3, \ldots, 2k$, and extract the ones with exactly $n+k$ edges and no node of degree 2. Since, at the time being, an exhaustive list of fractional vertices (up to isomorphism[5]) is already available for n up to 12 [1,3], it is possible to completely determine $\mathcal{A}_3, \mathcal{A}_4, \mathcal{A}_5, \mathcal{A}_6$ (up to isomorphism) by Lemma 2. For instance, Fig. 2 shows all the ancestors in $\mathcal{A}_4$: as $k = 4$, they can be extracted from the lists of vertices of P^7_{SEP} and P^8_{SEP}. All the vertices of $\mathcal{F}_4$ are either of these shapes or with the 1-edges replaced with 1-paths of arbitrary length. Similarly, simply examining the structure of the sole ancestor in $\mathcal{A}_3$, one can deduce that the family $\mathcal{F}_3$ consists precisely of those vertices formed by two 1/2-triangles connected by three 1-paths. In virtue of this characterization, one may see that Conjecture 4.1 in [1] supports the Few Edge Hypothesis.

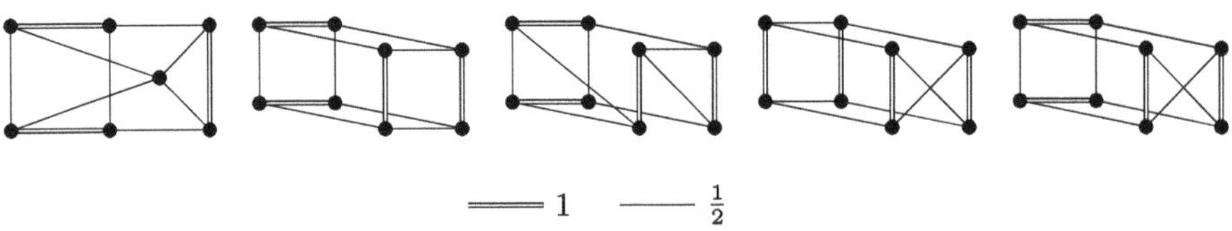

Fig. 2. Vertices of $\mathcal{A}_4$, up to isomorphism.

4 Redefining the Gap problem

In this section, we formulate a relaxed definition of Gap.

Definition 6 (Gap$^+$). *Let $\boldsymbol{x} \in P^n_{SEP}$ be a vertex. The* Gap$^+$ *of $\boldsymbol{x}$ is*

$$Gap^+(\boldsymbol{x}) := \sup \left\{ \frac{TSP(\boldsymbol{c})}{\boldsymbol{c}\boldsymbol{x}} \mid \boldsymbol{c} \text{ metric} \right\}.$$

[5] An isomorphism between vertices is a relabeling of their nodes.

By removing the requirement $\boldsymbol{x} \in \arg\min \mathrm{SEP}(\boldsymbol{c})$, we obtain the inequality:

$$\mathrm{Gap}^+(\boldsymbol{x}) \geq \mathrm{Gap}(\boldsymbol{x}) \; . \tag{1}$$

Our objective is therefore to provide an upper bound on Gap^+, which consequently yields a valid upper bound for Gap. To compute the inverse of $\mathrm{Gap}(\boldsymbol{x})$, Benoit and Boyd, in [1], designed an LP named $\mathrm{OPT}(\boldsymbol{x})$. Adopting the same strategy, for a given a vertex $\boldsymbol{x}$, we consider a corresponding linear problem $\mathrm{OPT}^+(\boldsymbol{x})$.

$$\begin{aligned}
&\mathrm{OPT}^+(\boldsymbol{x}): \\
&\quad \text{minimize} \sum_{ij \in E_n} x_{ij} c_{ij} && && (2)\\
&\text{subject to:} c_{ik} + c_{jk} - c_{ij} \geq 0 && \forall ij \in E_n,\ k \neq i,j, && (3)\\
&\qquad \sum_{ij \in E_n} t_{ij} c_{ij} \geq 1 && \forall \boldsymbol{t} \text{ tour}, && (4)\\
&\qquad c_{ij} \geq 0 && \forall ij \in E_n. && (5)
\end{aligned}$$

Constraints (3) force $\boldsymbol{c}$ to be a metric cost; constraints (4) normalize $\mathrm{TSP}(\boldsymbol{c}) = 1$; minimizing $\boldsymbol{cx}$ is then equivalent to maximizing $1/\boldsymbol{cx} = \mathrm{TSP}(\mathbf{c})/\boldsymbol{cx}$, thus

$$\frac{1}{\mathrm{OPT}^+(\boldsymbol{x})} = \mathrm{Gap}^+(\boldsymbol{x}) \; . \tag{6}$$

The following lemma is the first result that studies the behavior of Gap^+ under the application of the BB-move.

Lemma 3. *Let* $\boldsymbol{x} \in P^n_{SEP}$. *Then* $\mathrm{Gap}^+(\mathrm{BB}(\boldsymbol{x})) \geq \mathrm{Gap}^+(\boldsymbol{x})$. *That is, the* BB-*move is* Gap^+*-increasing.*

Remark 2. The two definitions Gap and Gap^+ are actually different: there exist vertices $\boldsymbol{x}$ realizing the strict inequality $\mathrm{Gap}(\boldsymbol{x}) < \mathrm{Gap}^+(\boldsymbol{x})$, as illustrated in Fig. 3. The relaxed version Gap^+ has been introduced because the previous lemma, which is of crucial relevance throughout our work, fails for Gap: the vertex in Figure 3 serves as a counterexample for this fact as well. The reason behind this discrepancy is the following. An optimal cost $\boldsymbol{c}$ achieving $\mathrm{Gap}^+(\boldsymbol{x})$ need not lie in the optimal normal cone of $\boldsymbol{x}$ (i.e., one may have $\boldsymbol{x} \notin \arg\min \mathrm{SEP}(\boldsymbol{c})$). Accordingly, while it is easy to modify $\boldsymbol{c}$ and obtain a cost for $\mathrm{BB}(\boldsymbol{x})$ attaining at least an equal Gap^+, the same modification may yield an unfeasible cost for $\mathrm{Gap}(\mathrm{BB}(\boldsymbol{x}))$, not belonging to the normal cone of $\mathrm{BB}(\boldsymbol{x})$. Nevertheless, despite the dissimilar pointwise behavior, both definitions lead to the same global limit: α. Since every cost belongs to the normal cone of at least one vertex, the integrality gap may be obtained as:

$$\begin{aligned}
\alpha &= \sup \{\mathrm{Gap}(\boldsymbol{x}) \mid \boldsymbol{x} \in P^n_{\mathrm{SEP}} \text{ vertex for some } n \in \mathbb{N}\} \\
&= \sup \left\{\mathrm{Gap}^+(\boldsymbol{x}) \mid \boldsymbol{x} \in P^n_{\mathrm{SEP}} \text{ vertex for some } n \in \mathbb{N}\right\} \; .
\end{aligned}$$

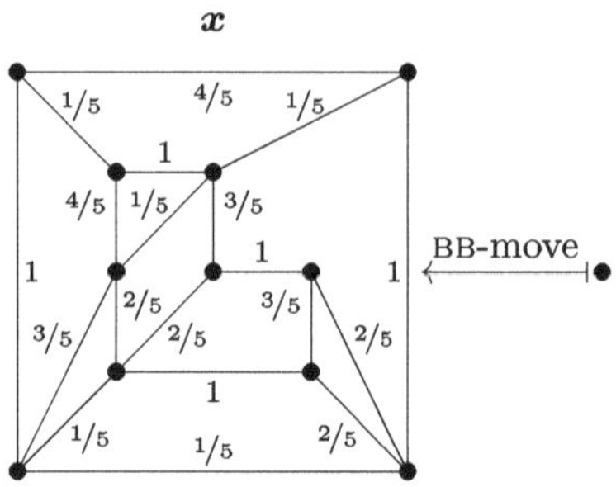

Fig. 3. Example of a vertex $\boldsymbol{x} \in P_{\text{SEP}}^{11}$ exhibiting the differences between Gap and Gap$^+$ described in Remark 2. Here, Gap$(\boldsymbol{x}) <$ Gap$^+(\boldsymbol{x})$ $(1.1091 < 1.1111)$. Moreover, a BB-move on the 1-edge pointed by the arrow yields the vertex BB$(\boldsymbol{x})$ which satisfies Gap(BB$(\boldsymbol{x})$) $<$ Gap$(\boldsymbol{x})$ $(1.1000 < 1.1091)$.

5 The Gap-Bounding algorithm

This section is entirely devoted to finding a way to bound the Gap$^+$ for all the vertices of a given family $\mathcal{F}_k$. To this end, we aim to contain the increase in Gap$^+$ originated by an iterative application of a BB-move: our goal is to provide a lower bound on OPT$^+$, which results in an upper bound on Gap$^+$, by (6).

We begin by outlining the central idea of our approach. Assume that we are given a vertex $\boldsymbol{x}_0$ alongside with an optimal solution of OPT$^+(\boldsymbol{x}_0)$ and a corresponding optimal solution $\boldsymbol{\mu}^0$ of the dual problem $\mathcal{D}\,$OPT$^+(\boldsymbol{x}_0)$. When considering a successor $\boldsymbol{x}'$ of $\boldsymbol{x}_0$, we intend to construct a feasible dual solution $\boldsymbol{\mu}'$ for $\mathcal{D}\,$OPT$^+(\boldsymbol{x}')$ starting from $\boldsymbol{\mu}^0$: succeeding in this task will directly result in a lower bound for OPT$^+(\boldsymbol{x}')$, as guaranteed by duality theory.

Indeed, for every vertex $\boldsymbol{x}$, Equations (1), (6), and duality theory yield

$$\text{Gap}(\boldsymbol{x}) \leq \text{Gap}^+(\boldsymbol{x}) = \frac{1}{\text{OPT}^+(\boldsymbol{x})} = \frac{1}{\mathcal{D}\,\text{OPT}^+(\boldsymbol{x})} \leq \frac{1}{L}\,, \tag{7}$$

where L is any lower bound on $\mathcal{D}\,$OPT$^+(\boldsymbol{x})$. In our case, L will be a bound on the objective value attained by $\boldsymbol{\mu}'$ (elaborated from $\boldsymbol{\mu}^0$). This strategy is visually represented in Fig. 4: it will be the skeleton of the Gap-Bounding algorithm, discussed in detail in Sect. 5.2.

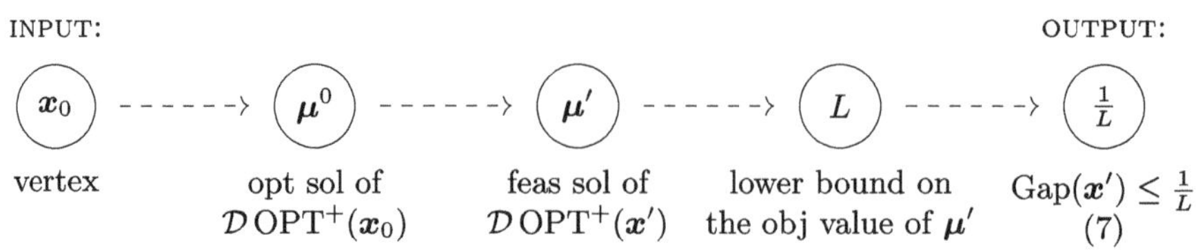

Fig. 4. Flowchart of the pipeline applied to bound Gap for a successor $\boldsymbol{x}'$ of $\boldsymbol{x}_0$.

Clearly, the crucial point in this procedure is to find a "good-enough" feasible dual solution, so that the bound obtained is meaningful (e.g., the trivial dual solution of all zeros ultimately gives no bound on Gap^+).

5.1 Dual Formulations of the OPT^+ problem

Given a vertex $\boldsymbol{x}$, the dual problem $\mathcal{D}\,\text{OPT}^+(\boldsymbol{x})$ is derived as follows.

$\mathcal{D}\,\text{OPT}^+(\boldsymbol{x})$:

$$\text{maximize} \sum_{t \text{ tour}} \mu_t \tag{8}$$

$$\text{subject to: } \sum_{k \neq i,j} (-\lambda_{ijk} + \lambda_{ikj} + \lambda_{jki}) + \sum_{t \text{ tour}} t_{ij}\mu_t \leq x_{ij} \quad \forall ij \in E_n, \tag{9}$$

$$\lambda_{ijk} \geq 0 \quad \forall ij \in E_n,\ k \neq i,j, \tag{10}$$

$$\mu_t \geq 0 \quad \forall \boldsymbol{t} \text{ tour on } K_n. \tag{11}$$

The variables λ_{ijk} correspond to the triangle inequalities of $\text{OPT}^+(\boldsymbol{x})$ and are indexed by an edge ij and a node $k \neq i,j$. Therefore, the abuse of notation $ij \equiv ji \in E$ induces also $\lambda_{ijk} \equiv \lambda_{jik}$[6]. To solve all the ambiguities, this notation may be intended as the short form for the cumbersome $\lambda_{\{i,j\},k}$.

When considering a vertex $\boldsymbol{x}$ and one of its successors $\boldsymbol{x}'$, the task of producing a feasible solution of $\mathcal{D}\,\text{OPT}^+(\boldsymbol{x}')$ starting from an optimal solution of $\mathcal{D}\,\text{OPT}^+(\boldsymbol{x})$ is anything but trivial. In what follows, we design an equivalent formulation and make use of it instead: $\mathcal{D}\,\text{OPT}^{\text{II}}$. Its main purpose is to get rid of $\boldsymbol{\lambda}$ variables: this is achieved at the price of considering walks on G_x (see Definition 1 and Remark 1) in place of tours. For a given vertex $\boldsymbol{x}$, $\mathcal{D}\,\text{OPT}^{\text{II}}(\boldsymbol{x})$ is thus defined as:

$\mathcal{D}\,\text{OPT}^{\text{II}}(\boldsymbol{x})$:

$$\text{maximize} \sum_{w \text{ walk on } G_x} \mu_w \tag{12}$$

$$\text{subject to: } \sum_{w \text{ walk on } G_x} w_{ij}\mu_w \leq x_{ij} \quad \forall ij \in E_x, \tag{13}$$

$$\mu_w \geq 0 \quad \forall \boldsymbol{w} \text{ walk on } G_x. \tag{14}$$

The equivalence of the two formulations $\mathcal{D}\,\text{OPT}^+(\boldsymbol{x})$ and $\mathcal{D}\,\text{OPT}^{\text{II}}(\boldsymbol{x})$ (that is, they have the same optimal value) is stated in the following lemma.

Lemma 4. *$\mathcal{D}\,OPT^+(\boldsymbol{x}) = \mathcal{D}\,OPT^{II}(\boldsymbol{x})$ for every vertex $\boldsymbol{x}$ of P^n_{SEP}.*

---

[6] With the symbol $\equiv$, we do not mean that two distinct variables have the same value; rather, we mean that the two notations coincide: they denote the same variable.

5.2 Bounding the Gap on successors

The goal of this section is to derive, for a generic successor $\boldsymbol{x}'$ of $\boldsymbol{x}_0$, a feasible solution of $\mathcal{D}\,\mathrm{OPT}^{\mathrm{II}}(\boldsymbol{x}')$, starting from an optimal solution of $\mathcal{D}\,\mathrm{OPT}^{\mathrm{II}}(\boldsymbol{x}_0)$; this serves as a bound on $\mathrm{Gap}^{+}(\boldsymbol{x}')$ (see Fig. 4, with $\mathcal{D}\,\mathrm{OPT}^{\mathrm{II}}$ in place of $\mathcal{D}\,\mathrm{OPT}^{+}$).

Let $\boldsymbol{x}_0$ be a vertex of P^n_{SEP} and let $\boldsymbol{\mu}^0 \in \arg\max \mathcal{D}\,\mathrm{OPT}^{\mathrm{II}}(\boldsymbol{x}_0)$. We begin by considering a 1-edge ab of $\boldsymbol{x}_0$ and applying d consecutive BB-moves (see Definition 3): we insert nodes $a_1, \dots, a_d$ and obtain, by Theorem 3, the vertices $\boldsymbol{x}_k := \mathrm{BB}(\boldsymbol{x}_{k-1},\ a_{k-1}b)$ for $k = 1, \dots, d$ (where $a_0 := a$). We aim to design a feasible solution for $\mathcal{D}\,\mathrm{OPT}^{\mathrm{II}}(\boldsymbol{x}_d)$.

An assignment $\boldsymbol{\mu}^d$ for $\mathcal{D}\,\mathrm{OPT}^{\mathrm{II}}(\boldsymbol{x}_d)$ is indexed by the walks on $G_{\boldsymbol{x}_d}$. We construct walks $\boldsymbol{w}^d$ on $G_{\boldsymbol{x}_d}$, starting from the ones on $G_{\boldsymbol{x}_0}$, and assign weights to the corresponding components of $\boldsymbol{\mu}^d$. In this perspective, we divide the walks on $G_{\boldsymbol{x}_0}$ into three families, according to how many times the edge ab is traversed: we define $\mathcal{W}^{ab}_m := \{\boldsymbol{w} \text{ walk} \mid w_{ab} = m\}$ for $m = 0, 1, 2$.

If $\boldsymbol{w}^0 \in \mathcal{W}^{ab}_0$, we construct the following $d+1$ walks on $G_{\boldsymbol{x}_d}$. For each $k = 0, \dots, d$, the walk $\boldsymbol{w}^d_k$ retraces $\boldsymbol{w}^0$ but, when it arrives at a, it deviates to pass through $a_1, a_2, \dots, a_k$ and back, and when it arrives at b, it deviates to $a_d, a_{d-1}, \dots, a_{k+1}$ and back. We set $\mu^d_{\boldsymbol{w}^d_k} := \frac{1}{d+1}\mu^0_{\boldsymbol{w}^0}$ for all $k = 0, \dots, d$.

If $\boldsymbol{w}^0 \in \mathcal{W}^{ab}_1$, we construct the walk $\boldsymbol{w}^d$ on $G_{\boldsymbol{x}_d}$ that retraces $\boldsymbol{w}^0$ but replaces ab with $aa_1, a_1a_2, \dots, a_{d-1}a_d, a_db$. We set $\mu^d_{\boldsymbol{w}^d} := \mu^0_{\boldsymbol{w}^0}$.

If $\boldsymbol{w}^0 \in \mathcal{W}^{ab}_2$, we construct the walk $\boldsymbol{w}^d$ on $G_{\boldsymbol{x}_d}$ that retraces $\boldsymbol{w}^0$ but replaces the double passage on ab by passing twice through $aa_1, a_1a_2, \dots, a_{d-1}a_d, a_db$. We set $\mu^d_{\boldsymbol{w}^d} := \mu^0_{\boldsymbol{w}^0}$.

Notice that these constructions cover all the possible walks $\boldsymbol{w}^d$ on $G_{\boldsymbol{x}_d}$, and even if they did not, we would simply set $\mu^d_{\boldsymbol{w}^d} := 0$ for the remaining cases. Figure 5 illustrates the new assignment $\boldsymbol{\mu}^d$ for $d = 2$.

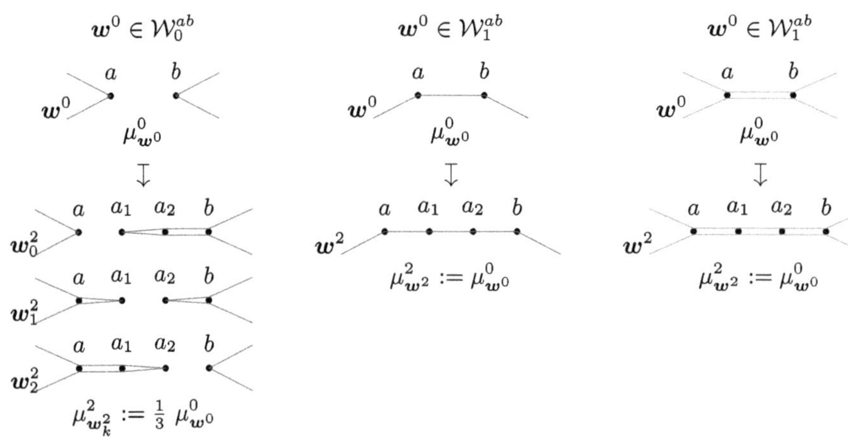

Fig. 5. Variables $\boldsymbol{\mu}^2$ for $\mathcal{D}\,\mathrm{OPT}^{\mathrm{II}}(\boldsymbol{x}_2)$ constructed from variables $\boldsymbol{\mu}^0$ for $\mathcal{D}\,\mathrm{OPT}^{\mathrm{II}}(\boldsymbol{x}_0)$.

In order to extend the construction for a generic successor of $\boldsymbol{x}_0$, we sequentially repeat the previous procedure for each 1-edge $e_1, \dots, e_p$ of $\boldsymbol{x}_0$, expanding them into 1-paths with $d_1, \dots, d_p$ internal nodes. This yields an assignment $\boldsymbol{\mu}'$ for $\mathcal{D}\,\mathrm{OPT}^{\mathrm{II}}(\boldsymbol{x}')$ such that: (i) $\boldsymbol{\mu}'$ attains the same objective value as $\boldsymbol{\mu}^0$; (ii) $\boldsymbol{\mu}'$ satisfies all the constraints (13), except those associated to the 1-edges, which nonetheless remain bounded by fixed constants. These constants are the same for all the edges in the same 1-path of $\boldsymbol{x}'$, originated from the same 1-edge e_h of $\boldsymbol{x}_0$:

$$C(\boldsymbol{x}_0, e_h) := 2 \sum_{\boldsymbol{w}^0 \in \mathcal{W}_0^{e_h}} \mu^0_{\boldsymbol{w}^0} + \sum_{\boldsymbol{w}^0 \in \mathcal{W}_1^{e_h}} \mu^0_{\boldsymbol{w}^0} + 2 \sum_{\boldsymbol{w}^0 \in \mathcal{W}_2^{e_h}} \mu^0_{\boldsymbol{w}^0} .$$

Crucially, they only depend on the initial vertex $\boldsymbol{x}_0$, not on the lengths of the 1-paths of $\boldsymbol{x}'$. Rescaling by the factor $C^*(\boldsymbol{x}_0) := \max_{h=1,\dots,p}\{C(\boldsymbol{x}_0, e_h)\}$, we thus obtain the feasible solution $\boldsymbol{\mu}^* := \frac{1}{C^*(\boldsymbol{x}_0)}\boldsymbol{\mu}'$, whose objective value accordingly becomes $\frac{1}{C^*(\boldsymbol{x}_0)}\mathcal{D}\,\mathrm{OPT}^{\mathrm{II}}(\boldsymbol{x}_0)$. This value plays the role of L in Eq. 7, yielding $\mathrm{Gap}^+(\boldsymbol{x}') \le C^*(\boldsymbol{x}_0) \cdot \frac{1}{\mathcal{D}\,\mathrm{OPT}^{\mathrm{II}}(\boldsymbol{x}_0)}$, and ultimately, by Eq. (6) and Lemma 4,

$$\mathrm{Gap}^+(\boldsymbol{x}') \le C^*(\boldsymbol{x}_0) \cdot \mathrm{Gap}^+(\boldsymbol{x}_0) .$$

Notice that we expect $C^*(\boldsymbol{x}_0)$ to be greater than or equal to 1, otherwise we would obtain $\mathrm{Gap}^+(\boldsymbol{x}_d) < \mathrm{Gap}^+(\boldsymbol{x}_0)$, contradicting Lemma 3.

We are finally able to design the Gap-Bounding (GB) algorithm as Algorithm 1; its main purpose is clarified in the next theorem.

Algorithm 1 Gap-Bounding algorithm

1: INPUT: $\boldsymbol{x} \in P^n_{\mathrm{SEP}}$ vertex
2: Solve $\mathcal{D}\,\mathrm{OPT}^{\mathrm{II}}(\boldsymbol{x})$ and retrieve an optimal assignment of variables $\{\mu_{\boldsymbol{w}}\}_{\boldsymbol{w} \text{ walk on } \boldsymbol{x}}$.
3: **for** each e 1-edge of $\boldsymbol{x}$ **do**
4: $\quad C(\boldsymbol{x}, e) \leftarrow 2 \sum_{\boldsymbol{w} \in \mathcal{W}_0^e} \mu_{\boldsymbol{w}} + \sum_{\boldsymbol{w} \in \mathcal{W}_1^e} \mu_{\boldsymbol{w}} + 2 \sum_{\boldsymbol{w} \in \mathcal{W}_2^e} \mu_{\boldsymbol{w}}$
5: **end for**
6: $C^*(\boldsymbol{x}) \leftarrow \max\{C(\boldsymbol{x}, e)\}_{e \text{ 1-edge of } \boldsymbol{x}}$
7: $\mathrm{Gap}^+(\boldsymbol{x}) \leftarrow 1/\mathcal{D}\,\mathrm{OPT}^{\mathrm{II}}(\boldsymbol{x})$
8: Return: $C^*(\boldsymbol{x}) \cdot \mathrm{Gap}^+(\boldsymbol{x})$

Theorem 4. *The Gap-Bounding algorithm, on input a vertex* $\boldsymbol{x} \in P^n_{SEP}$, *returns a value* $GB(\boldsymbol{x})$ *which is an upper bound for the gap of all the successors* $\boldsymbol{x}'$ *of* $\boldsymbol{x}$: $Gap(\boldsymbol{x}') \le GB(\boldsymbol{x})$.

Proof. This whole section is the proof that $\mathrm{Gap}^+(\boldsymbol{x}') \le C^*(\boldsymbol{x}) \cdot \mathrm{Gap}^+(\boldsymbol{x}) = \mathrm{GB}(\boldsymbol{x})$ for any successor $\boldsymbol{x}'$ of $\boldsymbol{x}$; Eq. (1) completes the argument. □

Notice that both the GB algorithm and Theorem 4 apply to all generic $\boldsymbol{x}$ vertices, without requiring that they be ancestors. The following lemma shows how to exploit this fact.

Lemma 5. *Let $\boldsymbol{x} \in P^n_{SEP}$ be a vertex and let $\boldsymbol{x}_0$ be its ancestor. Then $GB(\boldsymbol{x})$ gives a bound for the gap of all the successors $\boldsymbol{x}'$ of $\boldsymbol{x}_0$ (even for $\boldsymbol{x}'$ that are not successors of $\boldsymbol{x}$): $Gap(\boldsymbol{x}') \leq GB(\boldsymbol{x})$.*

6 Computational Results

This section reports how the execution of the Gap-bounding algorithm yields the *computer-aided* proof of our main result: Theorem 5.

Our pipeline starts by extracting the ancestors of $\mathcal{A}_k$ for $k = 3, 4, 5, 6$ from the lists of vertices already available from [1,3][7], as explained in Sect. 3. Subsequently, the Gap-Bounding algorithm is applied on each vertex $\boldsymbol{x} \in \mathcal{A}_k$.

We inform that, for some ancestors, a straightforward application of the GB algorithm returned a value higher than $4/3$. To improve the bounds returned, we exploited Lemma 5 and executed the GB algorithm on some selected successors. Even though there is no guarantee that the GB algorithm applied to successors gives strictly better results, we thought it reasonable that more "information" (more nodes, edges, and dual variables) would lead to a more detailed analysis and a tighter bound. Eventually, this intuition led us to accomplish our objective.

Theorem 5. *$Gap(\boldsymbol{x}) \leq \frac{4}{3}$ for all the vertices of $\boldsymbol{x} \in \mathcal{F}_k$ with $k = 3, 4, 5, 6$.*

We implemented the GB algorithm in Python, employing the commercial software Gurobi [12] to solve the LPs. Our code is available here. The computation is relatively lightweight and can be easily performed on a standard laptop.

7 Conclusion

In this paper, the main result of proving the $4/3$-conjecture for vertices of P^n_{SEP} with at most $n+6$ edges in their support graphs is presented. A natural continuation of this research consists of constructing $\mathcal{A}_7$ by exhaustive enumeration, to enlarge the set of ancestors on which to apply the GB algorithm and collect bounds for the whole family $\mathcal{F}_7$.

Besides this, the new *methodology* introduced is itself of independent interest, as it opens a new range of possible advancements in the field. Investigating novel transformations between vertices and analyzing their impact on the integrality gap (akin to the analysis we presented for the BB-move) may lead to interesting results on the integrality gap for diverse infinite families of vertices.

To conclude, we also mention a last research direction, motivated by an intriguing, computationally verified observation. Using data from [1], for instances with few nodes, we noticed a correlation between the number of edges

[7] Vertices data can be found at this link, last visited 04.06.2025.

in the support graphs of the vertices of P^n_{SEP} (with $6 \leq n \leq 12$) and their integrality gap. As shown in Fig. 6, for a fixed $n \leq 12$, the integrality gap is higher on vertices with few non-zero components and appears to decrease as the number of edges in the support graph increases. Although this correlation is currently supported only for small values of n, it presents a fascinating avenue for further investigation. For example, the donut instances of [5] have an integrality gap tending to $4/3$, and they do not belong to the families of instances considered in our result; nevertheless, for each donut instance, the proposed lower bound on its integrality gap is strictly smaller than the integrality gap of a vertex of $\mathcal{F}_3$ with the same number of nodes. Should one succeed in proving that, for every n, the maximum integrality gap on the polytope P^n_{SEP} is always attained on the vertices with the smallest number of edges, such a result together with the contribution of this work would be the two key ingredients to definitely resolve the $4/3$-conjecture.

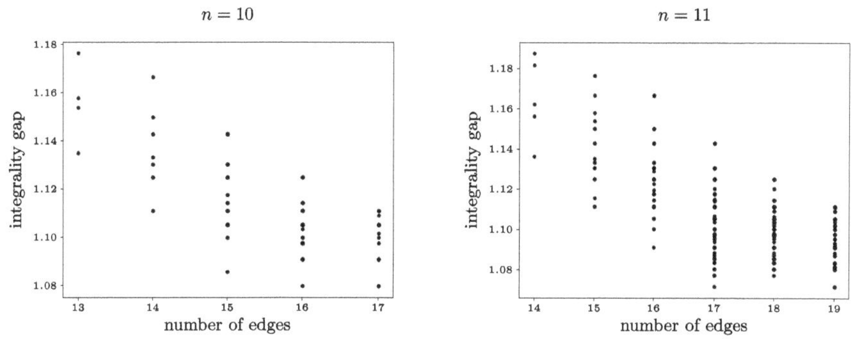

Fig. 6. Correlation between the number of edges in the support graph of a vertex and their integrality gap. Each plot collects all the vertices of P^n_{SEP} for a fixed n: $n = 10$ on the left, $n = 11$ on the right. The observed trend is the same for all small $n \leq 12$.

Acknowledgments. This work was supported by the Swiss National Science Foundation project n. 200021_212929/1 "Computational methods for integrality gaps analysis".

Disclosure of Interests. The authors have no competing interests to declare that are relevant to the content of this article.

References

1. Benoit, G., Boyd, S.: Finding the exact integrality gap for small traveling salesman problems. Math. Oper. Res. **33**, 921–931 (2008). https://doi.org/10.1287/moor.1080.0337

2. Boyd, S., Carr, R.: Finding low cost TSP and 2-matching solutions using certain half-integer subtour vertices. Disc. Optim. **8**(4), 525–539 (2011). https://doi.org/10.1016/j.disopt.2011.05.002
3. Boyd, S., Elliott-Magwood, P.: Structure of the extreme points of the subtour elimination polytope of the STSP. In: RIMS Kokyuroku Bessatsu, pp. 33–47 (2010). https://api.semanticscholar.org/CorpusID:15182527
4. Boyd, S., Pulleyblank, W.R.: Optimizing over the subtour polytope of the travelling salesman problem. Math. Program. **49**, 163–187 (1990). https://doi.org/10.1007/BF01588786
5. Boyd, S., Sebő, A.: The salesman's improved tours for fundamental classes. Math. Program. **186**(1), 289–307 (2021)
6. Boyd, S., Sitters, R., van der Ster, S., Stougie, L.: TSP on cubic and subcubic graphs. In: Günlük, O., Woeginger, G.J. (eds.) IPCO 2011. LNCS, vol. 6655, pp. 65–77. Springer, Heidelberg (2011). https://doi.org/10.1007/978-3-642-20807-2_6
7. Christofides, N.: Worst-case analysis of a new heuristic for the travelling salesman problem. Oper. Res. Forum **3** (1976)
8. Cornuejols, G., Fonlupt, J., Naddef, D.: The traveling salesman problem on a graph and some related integer polyhedra. Math. Program. **33**, 1–27 (1985). https://doi.org/10.1007/BF01582008
9. Dantzig, G., Fulkerson, R., Johnson, S.: Solution of a large-scale traveling-salesman problem. J. Oper. Res. Soc. Am. **2**(4), 393–410 (1954)
10. Goemans, M.X.: Worst-case comparison of valid inequalities for the TSP. Math. Program. **69**(1), 335–349 (1995)
11. Gupta, A., Lee, E., Li, J., Mucha, M., Newman, H., Sarkar, S.: Matroid-based TSP rounding for half-integral solutions. Math. Program. **206**(1), 541–576 (2024)
12. Gurobi Optimization, LLC: Gurobi Optimizer Reference Manual (2024). https://www.gurobi.com
13. Hougardy, S.: On the integrality ratio of the subtour LP for Euclidean TSP. Oper. Res. Lett. **42**(8), 495–499 (2014)
14. Hougardy, S., Zhong, X.: Hard to solve instances of the Euclidean traveling salesman problem. Math. Program. Comput. **13**, 51–74 (2021)
15. Jin, B., Klein, N., Williamson, D.P.: A 4/3-approximation algorithm for half-integral cycle cut instances of the TSP. Math. Program. **209** (2025). https://doi.org/10.1007/s10107-025-02193-5
16. Karlin, A.R., Klein, N., Oveis Gharan, S.: An improved approximation algorithm for TSP in the half integral case. In: Proceedings of the 52nd Annual ACM SIGACT Symposium on Theory of Computing, STOC 2020, pp. 28–39. Association for Computing Machinery (2020). https://doi.org/10.1145/3357713.3384273
17. Karlin, A.R., Klein, N., Oveis Gharan, S.: A (slightly) improved approximation algorithm for metric TSP. In: Proceedings of the 53rd Annual ACM SIGACT Symposium on Theory of Computing, STOC 2021, pp. 32–45. Association for Computing Machinery (2021). https://doi.org/10.1145/3406325.3451009
18. Karlin, A.R., Klein, N., Oveis Gharan, S.: A (slightly) improved bound on the integrality gap of the subtour LP for TSP. In: 2022 IEEE 63rd Annual Symposium on Foundations of Computer Science (FOCS), pp. 832–843. IEEE Computer Society (2022). https://doi.org/10.1109/FOCS54457.2022.00084
19. Klein, N., Taziki, M.: Dual charging for half-integral TSP (2025). arXiv preprint arXiv:2507.17999
20. Mömke, T., Svensson, O.: Removing and adding edges for the traveling salesman problem. J. ACM **63**(1), 2:1–2:28 (2016). https://doi.org/10.1145/2739008

21. Schalekamp, F., Williamson, D.P., van Zuylen, A.: 2-matchings, the traveling salesman problem, and the subtour LP: a proof of the Boyd-Carr conjecture. Math. Oper. Res. **39**(2), 403–417 (2014)
22. Vercesi, E., Gualandi, S., Mastrolilli, M., Gambardella, L.M.: On the generation of metric TSP instances with a large integrality gap by branch-and-cut. Math. Program. Comput. **15**(2), 389–416 (2023). https://doi.org/10.1007/s12532-023-00235-7
23. Villa, T., Vercesi, E., Barta, J., Mastrolilli, M.: The integrality gap of the traveling salesman problem is 4/3 if the LP solution has at most n+6 non-zero components (2025). https://arxiv.org/abs/2507.07003
24. Wolsey, L.A.: Heuristic analysis, linear programming and branch and bound. In: Rayward-Smith, V.J. (ed.) Combinatorial Optimization II, pp. 121–134. Springer, Heidelberg (1980). https://doi.org/10.1007/BFb0120913
25. Zhong, X.: Lower bounds on the integrality ratio of the subtour LP for the traveling salesman problem. Disc. Appl. Math. **365**, 109–129 (2025)

The Local Dyadic Conjecture

Mahtab Alghasi, Bertrand Guenin, and Levent Tunçel(✉)

Department of Combinatorics and Optimization, University of Waterloo,Waterloo, Canada
{malghasi,bguenin,levent.tuncel}@uwaterloo.ca

Abstract. A rational with denominator a power of two is said to be dyadic. A long standing conjecture of Paul Seymour predicts that for an ideal clutter, the dual of the set covering problem always admits an optimal solution where all variables take dyadic values.We present a local version of this conjecture, and motivate it by proving an important special case.

Keywords: Linear Programming · dyadic rationals · binary clutters · polyhedra

1 Introduction.

1.1 The Dyadic Conjecture

A family of sets $\mathcal{C}$ over finite ground set $E(\mathcal{C})$ is a *clutter* if for every $S, S' \in \mathcal{C}$ where $S \subseteq S'$ we have $S = S'$. A clutter $\mathcal{C}$ is *trivial* if $\mathcal{C} = \emptyset$ or $\mathcal{C} = \{\emptyset\}$. Throughout this paper we shall assume that all clutters are non-trivial. Given a clutter $\mathcal{C}$, $M(\mathcal{C})$ denotes the $0, 1$ matrix where the rows correspond to the characteristic vectors of members of $\mathcal{C}$.[1] Consider a clutter $\mathcal{C}$, and weights $w \in \mathbb{Z}_+^{E(\mathcal{C})}$. We consider the following primal-dual pair,

$$\min\{w^\top x : M(\mathcal{C})x \geq \mathbf{1}, x \geq \mathbf{0}\}, \tag{1}$$

$$\max\{\mathbf{1}^\top z : M(\mathcal{C})^\top y \leq w, y \geq \mathbf{0}\}. \tag{2}$$

Given a clutter $\mathcal{C}$ we associate the polyhedron,

$$P(\mathcal{C}) := \{x \geq \mathbf{0} : M(\mathcal{C})x \geq \mathbf{1}, x \geq \mathbf{0}\}.$$

The clutter $\mathcal{C}$ is *ideal* if $P(\mathcal{C})$ is integral, i.e. every vertex (also known as extreme point) of $P(\mathcal{C})$ is in the integer lattice. In 1975, Paul Seymour (see [19] 79.3e) proposed the following conjecture,

Conjecture 1. Let $\mathcal{C}$ be an ideal clutter. Then for every $w \in \mathbb{Z}_+^{E(\mathcal{C})}$ the linear program (2) has an optimal solution that is dyadic.

[1] $M(\mathcal{C})$ is only defined up to permutations.

S. S. Dey et al. (Eds.): IPCO 2026, LNCS 16588, pp. 144–157, 2026.
https://doi.org/10.1007/978-3-032-28691-8_10

Here, a rational number $\frac{r}{s}$ where $r, s \in \mathbb{Z}$ is *dyadic* if $s = 2^q$ for some $q \geq 0$ where $q \in \mathbb{Z}$. A vector is *dyadic* if every entry is dyadic.

Dyadic numbers are important for numerical computations because they have a finite binary representation, and therefore they can be represented exactly on a computer in floating-point arithmetic. Checking whether a linear program admits a dyadic optimal solution can be done in polynomial time [2]. For an implementation see [6]. For the set packing problems, integrality of the polytope, $\{x \geq \mathbf{0} : M(\mathcal{C})x \leq \mathbf{1}, x \geq \mathbf{0}\}$ implies the existence of an integral solution to the dual problem $\min\{\mathbf{1}^\top y : M(\mathcal{C})^\top y \geq w, y \geq \mathbf{0}\}$ for every $w \in \mathbb{Z}_+^{E(\mathcal{C})}$ [5]. This affirms a one-to-one correspondence between perfect matrices and perfect graphs. Conjecture 1 would establish a weak counterpart for set covering problems.

Conjecture 1 is known to hold for a number of special cases, for instance, (a) ideal clutters of odd cycles of graphs [9], (b) clutters of T-cuts of grafts [22], (c) clutters of T-joins of grafts [1], (d) clutters of dicuts of a directed graph [14], (e) clutters of dijoins of a directed graph [11] see also [4]. For (a), (b) (2) always have an $\frac{1}{2}$-integral solution. For (d) (2) always has an integral solution.

1.2 A Local Strengthening of the Dyadic Conjecture

Consider a pointed polyhedron $P \subseteq \mathbb{R}^n$. We say that a pair of vertices v^1, v^2 of P are *adjacent* if both v^1, v^2 are contained in a single face of dimension 1 of P. We say that a vertex v of a polyhedron P is an *integral corner*, if v and all vertices adjacent to v are in $\mathbb{Z}^n$. Consider now a clutter $\mathcal{C}$ and weights $w \in \mathbb{Z}_+^{E(\mathcal{C})}$. Observe that $\min\{w^\top x : x \in P(\mathcal{C})\}$ always has an optimal solution and that the set of optimal solutions is a face F of $P(\mathcal{C})$. We say that F is the *w-optimal face* of $P(\mathcal{C})$. A set $B \subseteq E(\mathcal{C})$ is a *cover* of $\mathcal{C}$ if $B \cap S \neq \emptyset$ for every $S \in \mathcal{C}$. A cover is w-minimum if it minimizes $w(B) := \sum_{j \in B} w_j$ among all covers. We say that w are *minimal* weights of $\mathcal{C}$ if $w \in \mathbb{Z}_+^{E(\mathcal{C})}$ and every $j \in \mathcal{C}$ is contained in a w-minimum cover.

We predict the following.

Conjecture 2. Let $\mathcal{C}$ be a clutter and let w be minimal weights of $\mathcal{C}$. If the w-optimal face of $P(\mathcal{C})$ contains an integral corner then (2) has an optimal solution that is dyadic.

Observe that for an integral polyhedron, every vertex is an integral corner. In particular, if $\mathcal{C}$ is ideal then every vertex of $P(\mathcal{C})$ is an integral corner. Furthermore, it is an easy exercise to show that it suffices to prove Conjecture 1 for minimal weights only. Hence, Conjecture 2 implies Conjecture 1. However, the former appears strictly stronger than the latter.

The condition that the optimal face contain an integral corner cannot be omitted. Consider for instance the case where $\mathcal{C}$ are the lines of the Fano Matroid, i.e.

$$\mathcal{C} = \{\{1,2,3\},\{1,4,5\},\{1,6,7\},\{2,4,7\},\{2,5,6\},\{3,4,6\},\{3,5,7\}\}.$$

For minimal weights $w = \mathbf{1}$ the w-optimal face contains a unique point $\frac{1}{3}\mathbf{1}$ and (2) has a unique dual solution $\frac{1}{3}\mathbf{1}$ as well.

It is also not possible to replace in Conjecture 2 the condition that "*the w-optimal face contains an integral corner*" by "*the w-optimal face contains an integral point*". To see this consider the clutter $\mathcal{C}$ with,

$$M(\mathcal{C}) = \begin{pmatrix} 0\,1\,1\,1\,1\,0\,0\,0 \\ 1\,0\,1\,1\,0\,1\,0\,0 \\ 1\,1\,0\,1\,0\,0\,1\,0 \\ 1\,1\,1\,0\,0\,0\,0\,1 \\ 0\,0\,0\,0\,1\,1\,1\,1 \end{pmatrix}.$$

Let $w = \mathbf{1}$ and let F denote the w-optimal face. It can be readily checked that (2) has a unique optimal solution $y = (\frac{1}{3}, \frac{1}{3}, \frac{1}{3}, \frac{1}{3}, \frac{2}{3})^\top$. The vertices of F are $v^1 = (1,0,0,0,1,0,0,0)$, $v^2 = (0,1,0,0,0,1,0,0,0)$, $v^3 = (0,0,1,0,0,0,1,0)$, $v^4 = (0,0,0,1,0,0,0,1)$. We claim that none of v^i, $i \in [4]$ is an integral corner. Because of symmetry it suffices to consider v^1, but v^1 is adjacent to $(\frac{1}{2}, \frac{1}{2}, \frac{1}{2}, 0, 0, 0, 0, 1)^\top$.

1.3 A Special Case of Conjecture 2

The set of all inclusion-wise minimal covers of a clutter $\mathcal{C}$ forms another clutter called the *blocker* of $\mathcal{C}$. We denote by $b(\mathcal{C})$ the blocker of $\mathcal{C}$. The blocking operation is idempotent, i.e., $b(b(\mathcal{C})) = \mathcal{C}$ [8,13]. A clutter $\mathcal{C}$ is said to be *binary* if for every $S \in \mathcal{C}$ and every $B \in b(\mathcal{C})$ we have $|S \cap B|$ odd. Cases (a)-(c) of Conjecture 1 are all instances of binary clutters. The restriction of Conjecture 1 to binary clutters remains open and is therefore of particular interest.

A main corollary of this paper is the following special case of Conjecture 2.

Theorem 1. *Let $\mathcal{C}$ be a binary clutter and let w be minimal weights of $\mathcal{C}$. If the w-optimal face of $P(\mathcal{C})$ contains a (k,ℓ)-degenerate integral corner where $k(\ell+1) \le 6$ then* (2) *has an optimal solution that is dyadic.*

A vertex v of a full-dimensional polyhedron $P = \{x : Ax \le b\} \subseteq \mathbb{R}^n$ is (k,ℓ)-*degenerate* if (i) v is contained in $n+k$ faces of dimension 1 of P, and (ii) there are $n+\ell$ tight constraints of $Ax \le b$ for v. Observe that a $(0,0)$-degenerate vertex is non-degenerate. For $\mathcal{Q}_6$, the clutter with groundset the edges of K_4 and members the triangles of K_4, every vertex of $P(\mathcal{Q}_6)$ is (k,ℓ)-degenerate, for $k,\ell \le 2$. For the case where the integral corner is non-degenerate, it can be readily shown that we in fact get an integral solution for (2) as the left-hand-side of the tight constraints must be unimodular [20].

1.4 Organization of the Paper

Section 2 presents a general result, Theorem 2, and explains how Theorem 1 follows as a corollary. Section 3 proves a result about binary clutters that might be of independent interest. In Sect. 4 we leverage these result and a general tool on dyadic linear programs to prove Theorem 2. Proofs of

- the general tool and
- the fact that (2) has a half-integer solution, when there are no more than 6 active dual variables and the clutter is binary,

are in the full version of this article.

2 The General Result

Before we can proceed, we require a number of definitions. Consider a full-dimensional polyhedron $P = \{x : Ax \leq b\}$ with a vertex v and a proper face F that contains v. Let $\bar{A}x \leq \bar{b}$ denote the implicit equalities for F. A set $\mathcal{L}$ of faces of dimension 1 containing v is *(F, v)-saturating* if no constraint of $\bar{A}x \leq \bar{b}$ is tight for every $L \in \mathcal{L}$. Observe that this is a property of the polyhedron, not of the system of linear constraints.

If F, F' are faces of P where $F' \supset F$ then F' is a *superface* of F. If in addition $\dim(F') = \dim(F) + 1$ then F' is an *immediate superface* of F. Denote by $\mathcal{L}'$ the set of faces of dimension 1 containing v and that contain a point distinct from v in some immediate superface of F. Then F is *v-compatible* if $\mathcal{L}'$ intersects every (F, v)-saturating set.

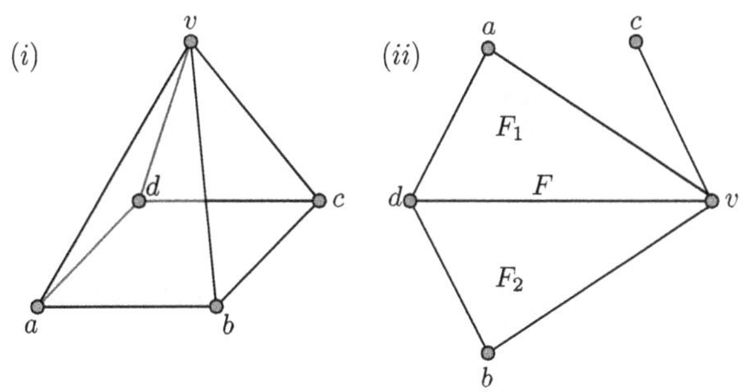

Fig. 1. Illustrating definitions.

We illustrate these definitions in Fig. 1. For both (i) and (ii) denote by L_a, L_b, L_c, L_d be the line segment between v and a, b, c, d respectively. Consider (i). The point v is $(1, 1)$-degenerate. Pick $F = \{v\}$. Then the sets $\{L_a, L_c\}$ and $\{L_b, L_d\}$ are both (F, v)-saturating and every other (F, v)-saturating set is a superset of these sets. The immediate superfaces of F are L_a, L_b, L_c and L_d. The set $\mathcal{L}'$ defined above is $\{L_a, L_b, L_c, L_d\}$. In particular, F is v-compatible. Indeed, whenever F consists of a single point v, then it is always v-compatible. Consider (ii). Pick $F = L_d$. Then the sets $\{L_a, L_b\}$ and $\{L_c\}$ are both (F, v)-saturating. The immediate superfaces of F are the triangles F_1 and F_2. The set $\mathcal{L}'$ defined above is $\{L_a, L_b\}$. Since $\{L_c\} \cap \mathcal{L}' = \emptyset$, F is not v-compatible.

Suppose P is full dimensional and determined by an irreducible system $Ax \leq b$. Given a proper face F of P, we denote by $\sigma(F, P)$ the number of tight constraints of $Ax \leq b$ for F. Note, that $\sigma(F, Q)$ is independent of the choice of $Ax \leq b$. When clear from context, we write $\sigma(F)$ for $\sigma(F, P)$.

We can now state the main result of the paper.

Theorem 2. *Let $\mathcal{C}$ be a binary clutter and let w be minimal weights of $\mathcal{C}$. Denote by $\bar{F}$ the w-optimal face of $P(\mathcal{C})$ and let v be an integral corner of $\bar{F}$. If every face $F \supseteq \bar{F}$ for which $\sigma(F) \geq 7$ is v-compatible then* (2) *has an optimal solution that is dyadic.*

We will show this implies Theorem 1. We first require.

Proposition 1. *Let P be a pointed polyhedron of dimension n and let F be a face of P of dimension d. Then there exists at least $n - d$ immediate superfaces containing F.*

Proof. We first consider the case when P is a polytope. Denote by $L(P)$ the face poset of P. It is a graded lattice with rank function $r(G) = \dim(G) + 1$ for faces G. The interval $[F, P]$ of $L(P)$ is the face lattice of a polytope P' of dimension $r(P) - r(F) - 1 = n - d - 1$ (see [23] Theorem 2.7). $F = \emptyset$ in P' and every immediate superface of F is a vertex of P'. Since $\dim(P') = n - d - 1$, P' must have at least $n - d - 1 + 1$ vertices (equality holds for simplexes). Hence, F is contained in at least $n - d$ immediate superfaces, as required.

Consider now the general case where P need not be a polytope. Pick $M \in \mathbb{Z}_+$ large enough such that every vertex of P is contained in the interior of the box $B := \{x : -M\mathbf{1} \leq x \leq M\mathbf{1}\}$. Describe P by an irreducible set of inequalities $Ax \leq b$ where A has index rows $[m]$. If G is a face of P, then there exists $I_G \subseteq [m]$ such that $G = P \cap \{x : \mathrm{row}_i(A)x = b_i, i \in I_G\}$. Moreover, if G is a face of $P \cap B$ where $G \cap F \neq \emptyset$ for a face F of P, then none of the constraints of B are tight for F as F must contain a vertex of P. Hence, there exists a bijection between faces $G \supseteq F$ of P and faces $G' \supseteq F$ of $P \cap B$. Therefore, we can count the number of superfaces of F in $P \cap B$ rather than P and use the aforementioned result for polytopes. □

Next we tie degeneracy to the number of active dual variables.

Proposition 2. *Let $P \subset \mathbb{R}^n$ be a full dimensional polyhedron. Let v be a (k, ℓ)-degenerate vertex of P where $k(\ell + 1) \leq \alpha$ for some $\alpha \in \mathbb{Z}_+$. Let $F \ni v$ be a face of P with $\sigma(F, P) > \alpha$. Then F is v-compatible.*

Proof. Let $Ax \leq b$ be an irreducible representation of A and let $\bar{A}x \leq \bar{b}$ denote the tight constraints for v. Let $\mathcal{L}$ be the set of all faces of dimension 1 of P containing v.

Claim 1. Let $A'x \leq b'$ be a subset of δ constraints of $\bar{A}x \leq \bar{b}$ and let $L \in \mathcal{L}$. Then $\geq \delta - \ell - 1$ of the constraints $A'x \leq b'$ are tight for L.

Proof of Claim. Since L is a face of dimension 1 and $P \subseteq \mathbb{R}^n$, there are $\geq n - 1$ constrains of $\bar{A}x \leq \bar{b}$ that are tight for L. As v is (k, ℓ)-degenerate, $\bar{A}x \leq \bar{b}$ has at most $n + \ell$ constraints. Hence, at most $n + \ell - (n - 1) = \ell + 1$ of the constraints of $\bar{A}x \leq \bar{b}$ are strict for L. As $A'x \leq b'$ is a subsystem of $\bar{A}x \leq \bar{b}$ the number of tight constraints for L is $\geq \delta - (\ell + 1)$.

Let $F_1, \ldots, F_p$ denote the immediate superfaces of F and let $\mathcal{F} = F \cup_{i \in [p]} F_i$.

Claim 2. At least n of the faces in $\mathcal{L}$ intersect $\mathcal{F} \setminus \{v\}$.

Proof of Claim. Let $d = \dim(F)$. Then $p \geq n - d$, by Proposition 1. Since F has dimension d, at least d of the faces in $\mathcal{L}$ intersect $F \setminus \{v\}$. For each $i \in [p]$, at least one face $L_i \in \mathcal{L}$ intersect $F_i \setminus \{v\}$. Pick distinct $i, j \in [p]$. As F_i, F_j are immediate superfaces and $F_i \neq F_i$, we have $L_i \neq L_j$. Hence, we have $d + p \geq d + (n-d) \geq n$ distinct faces of $\mathcal{L}$ that intersect $\mathcal{F} \setminus \{v\}$.

Consider now a face F and let $A'x \leq b'$ be the corresponding subset of tight constraints of $\bar{A}x \leq \bar{b}$. We may assume that there exists a saturating set $\mathcal{L}' \subseteq \mathcal{L}$ where none of the faces in $\mathcal{L}'$ intersect $\mathcal{F} \setminus \{v\}$, for otherwise F is v-compatible. Since v is (k, ℓ)-degenerate, $|\mathcal{L}| \leq n + k$. By Claim 2, n of the faces in $\mathcal{L}$ intersect $\mathcal{F} \setminus \{v\}$. It follows that $|\mathcal{L}'| \leq n + k - n \leq k$. By Claim 1, at most $k(\ell + 1)$ constraints of $A'x \leq b'$ are non-tight for $\mathcal{L}'$. Since $\mathcal{L}'$ is saturating, $k(\ell + 1) \geq \sigma(F, v)$. However, by hypothesis, $\sigma(F, v) > \alpha \geq k(\ell + 1)$, a contradiction. □

Proposition 3. *Theorem 2 implies Theorem 1.*

Proof. Let $\alpha = 6$. By hypothesis of Theorem 1, v is (k, ℓ)-degenerated for $k(\ell + 1) \leq \alpha$. It follows from Proposition 2 that every face $F \supseteq \bar{F}$ for which $\sigma(F) \geq 7$ is v-compatible. However, then Theorem 2 implies that (2) has an optimal solution that is dyadic as required. □

In the remainder of the paper we shall prove Theorem 2.

3 A Lemma on Binary Clutters

Here is the key result of the section,

Proposition 4. *Let $\mathcal{C}$ be a binary clutter and let $S \in \mathcal{C}$. Suppose that there exists $B \in b(\mathcal{C})$ where $|S \cap B| \geq 3$. Then there exists $B' \in b(\mathcal{C})$ for which $|S \cap B'| = 3$ and $S \cap B' \subseteq S \cap B$.*

Before we proceed to the proof, observe that the binary assumption cannot be omitted, even if we replace 3 with any constant bound. Namely, we have,

Remark 1. For every $n \in \mathbb{Z}_+$ there exists a clutter $\mathcal{C}$ with $|E(\mathcal{C})| = 2n$ and $S \in \mathcal{C}$, such that

- there exists $B \in b(\mathcal{C})$ with $|S \cap B| = n$, and
- for every $B \in b(\mathcal{C})$, $|S \cap B| \in \{1, n\}$.

Proof. Let $\mathcal{B}$ be the clutter with groundset $\{1, \ldots, n\} \cup \{1', \ldots, n'\}$ and members $\{1, \ldots, n\}$ and for all $i \in [n]$ members $\{i, i'\}$. Let $\mathcal{C} = b(\mathcal{B})$. Then $S = \{1, \ldots, n\} \in \mathcal{C}$ and the result follows.

If we are allowed to pick the set $S \in \mathcal{C}$ then the condition that $\mathcal{C}$ be binary is not needed.

Proposition 5. *Let $\mathcal{C}$ be a clutter. Suppose that some $S \in \mathcal{C}$ and some $B \in b(\mathcal{C})$ satisfy $|S \cap B| > 1$. Then, there exist $S' \in \mathcal{C}$ and $B' \in b(\mathcal{C})$ with $|S' \cap B'| \in \{2, 3\}$.*

Proof. If $\mathcal{C}$ is binary then the result holds by Proposition 4. Otherwise, it is proved in [21] that $\mathcal{C}$ contains as a minor, say $\mathcal{D} = \mathcal{C}/I \setminus J$, one of: $\mathcal{P}_4 = \{\{1,2\},\{2,3\},\{3,4\}\}$ or $\Delta_s = \{\{1,\ldots,s\} \cup \{0,i\} : i \in [s]\}$ where $s \geq 2$. Note that $\{2,3\} \in \mathcal{P}_4 \cap b(\mathcal{P}_4)$ and $\{0,1\} \in \Delta_s \cap b(\Delta_s)$. Hence, there exist $S' \in \mathcal{D}$ and $B' \in b(\mathcal{D})$ with $|S' \cap B'| = 2$. Then, for some $I' \subseteq I$ and $J' \subseteq J$ we have $S = S' \cup I \in \mathcal{C}$ and $B = B' \cup J \in b(\mathcal{C})$. In particular, $|S \cap B| = |S' \cap B'| = 2$.

3.1 A Restatement in Terms of Matroids

To prove Proposition 4 we shall first restate the result in the language of binary matroids. Given a matroid M with a fixed element Ω we define,

$$\mathrm{port}_\Omega(M) = \{C \setminus \Omega : C \text{ is a circuit of } M \text{ containing } \Omega\}.$$

Binary clutters are exactly the clutters that can be expressed as ports of binary matroids. Namely,

Proposition 6 ([21])**.** *Let $\mathcal{C}$ be a clutter. Then the following are equivalent,*

- *$\mathcal{C}$ is a binary clutter,*
- *for some binary matroid M and $\Omega \in E(M)$, we have $\mathcal{C} = \mathrm{port}_\Omega(M)$.*

In addition we have the following attractive relation between matroid duality and the blocking relation.

Proposition 7 ([15](2.A.1))**.** *Let M be a binary matroid and let $\Omega \in E(M)$. Then,*

$$b(\mathrm{port}_\Omega(M)) = \mathrm{port}_\Omega(M^*).$$

Here M^* denotes the dual of M. Oxley studied when a matroid contains a *quad*, namely a set of four elements that are exactly the intersection of a circuit and a cocircuit [16]. We prove the following related rooted result for binary matroids.

Proposition 8. *Let M be a binary matroid and let $\Omega \in E(M)$. Consider C and D where*

a. D is a cocircuit of M containing Ω,
b. C is a circuit of M containing Ω, and
c. $|C \cap D| \geq 4$.

Then there exists a circuit C' of M with $\Omega \in C'$ and $|C' \cap D| = 4$.

We postpone the proof until Sect. 3.2. Let us show that this implies our results on binary clutters.

Proof (Proposition 4). By Proposition 6, $\mathcal{C} = \mathrm{port}_\Omega(M^*)$ for some binary matroid M. Then $S = D \setminus \{\Omega\}$ where D is a cocircuit of M with $\Omega \in D$. By Proposition 7, $B = C \setminus \{\Omega\}$ where C is a circuit of M such that $\Omega \in C$. By Proposition 8 there exists a circuit C' of M with $\Omega \in C'$ and $|C' \cap D| = 4$. Then by Proposition 7, $B' := C' \setminus \{\Omega\}$ is a member of $b(\mathcal{C})$ where $|B' \cap S| = |(C' \cap D) \setminus \{\Omega\}| = 3$, as required.

□

3.2 The Proof of Proposition 8

We shall require a few definitions first. A *signed graph* is a pair (G, Σ) where $\Sigma \subseteq E(G)$. We say that $C \subseteq E(G)$ is a *cycle* of G if C induces an Eulerian subgraph of G. It is a *polygon* if it induces a connected subgraph where every vertex has degree two. A cycle C of G is an *even cycle* of (G, Σ) if $|C \cap \Sigma|$ is even. A set of the form $\Gamma = \Sigma \triangle \delta_G(U)$ where $\delta_G(U)$ is a cut, is a *signature* of (G, Σ). Observe that (G, Σ) and (G, Γ) have the same set of even cycles.

We denote by cycle(G) the set of all cycles of G and by ecycle(G, Σ) the set of all even cycles of (G, Σ). The set cycle(G) are exactly the cycles of the graphic matroid with representation G and we identify cycle(G) with that matroid. The set ecycle(G, Σ) are exactly the cycles of a binary matroid that is a single lift of a graphic matroid, this is known as an even-cycle matroid [10]. We identify ecycle(G, Σ) with that matroid. We use the following folklore result that says that we can represent as signed graphs, binary matroids that are graphic after contraction of a single element. This follows readily, by considering the 0, 1 representation of the binary matroid.

Remark 2 Let M be a binary matroid and let $e \in E(M)$ where e is not a loop. If $M/e =$ cycle(G) then $M =$ ecycle(G', Σ) where G' is obtained from G by adding loop e and $\Sigma \subseteq E(G)$ with $e \in \Sigma$.

Finally, we will need a characterization of the cocycles of even-cycle matroids.

Proposition 9 **([17] Remark 2.1).** *Let* (G, Σ) *be a signed graph. The cocycle of* ecycle(G, Σ) *are the cuts of* G *and the signatures of* (G, Σ).

We are now ready for the principal proof in this section.

Proof (Proposition 8). We may assume $|C \cap D| \geq 6$. It suffices to prove the following weaker statement,

There exists a circuit C' with $\Omega \in C'$ and where $4 \leq |C' \cap D| < |C \cap D|$. $(\star)$

Suppose for a contradiction that $(\star)$ does not hold for M with C and D satisfying (a)-(c). We assume that M among all such counterexamples minimizes $|E(M)|$.

Claim 3. Let $\hat{M} = M/I \setminus J$ where $\Omega \notin I \cup J \neq \emptyset$ and $I \cap J = \emptyset$. Then $\hat{C} = C \setminus I$ is not a circuit of $\hat{M}$ or $\hat{D} = D \setminus J$ is not a cocircuit of $\hat{M}$.

Proof of Claim. Suppose for a contradiction that $\hat{C}$ is a circuit of $\hat{M}$ and that $\hat{D}$ is a cocircuit of $\hat{M}$. As $\hat{C} \cap \hat{D} = C \cap D$, we have $\Omega \in \hat{C} \cap \hat{D}$, and $|\hat{C} \cap \hat{D}| \geq 6$. It follows that (a)-(c) hold for $\hat{M}$ with $\hat{C}, \hat{D}$. By our minimality assumption, there exists a circuit F of $\hat{M}$ with $\Omega \in F$ and $4 \leq |F \cap \hat{D}| < |\hat{C} \cap \hat{D}|$. By definition of minors, there exists $I' \subseteq I$ and $J' \subseteq J$ such that $C' = F \cup I'$ is a circuit of M and $D = \hat{D} \cup J'$. Then $C' \cap D = (F \cup I') \cap (\hat{D} \cup J') = F \cap \hat{D}$. It follows that $4 \leq |F \cap \hat{D}| = |C' \cap D|$ and $|C \cap D| = |\hat{C} \cap \hat{D}| > |F \cap \hat{D}| = |C' \cap D|$. However, then $(\star)$ holds for M, a contradiction.

Claim 4. $C = D$ and all elements of $E(M) \setminus C$ are in the span of C.

Proof of Claim. If there exists $e \in C \setminus D$ then let $\hat{M} = M/e$. Then $C \setminus \{e\}$ is a circuit of $\hat{M}$ and D is a cocircuit of $\hat{M}$, contradicting Claim 3. If there exists $e \in D \setminus C$ then let $M' = M \setminus e$. Then $D \setminus \{e\}$ is a cocircuit of $\hat{M}$ and C is a circuit of $\hat{M}$, contradicting Claim 3. Thus we have $C = D$. If there exists $e \notin C = D$ where e is not in the span of C, the let $\hat{M} = M/e$. Then C is a circuit of $\hat{M}$ and D is a cocircuit of $\hat{M}$, contradicting Claim 3.

Denote by $\Omega, e_1, \ldots, e_n$ the elements of $C = D$. Note, $n \geq 5$ since $|C \cap D| \geq 6$. Let $F = \{f_1, \ldots, f_m\}$ denote the elements in $E(M) \setminus C$. Pick $i \in [m]$. By Claim 4, there are exactly two circuits L, L' in $C \cup \{f_i\}$ and in addition $L \triangle L' = C$. Thus exactly one of L, L' contains Ω. Denote that circuit by C_i. If $|C_i| \geq 5$ then $4 \leq |C_i \cap D| < |C \cap D|$ and $(\star)$ holds with $C' = C_i$. Since circuits and cocircuits have even intersection in binary matroids, for every $i \in [m]$, C_i is a triangle (a circuit of cardinality three) with elements say, e_i, f_i, Ω.

Let $N = M/\Omega$. Then $C \setminus \{\Omega\}$ is a circuit of N. Moreover, for every $i \in [m]$, e_i, f_i are parallel in N. This implies that every element of f_i is paired with some element e_i, in particular, $n \geq m$. Let N' denote the matroid obtained from N by replacing every parallel class e_i, f_i by e_i. Then N' consists of a single circuit $C \setminus \{\Omega\}$, in particular, it is graphic. It follows that $N = \text{cycle}(G)$ where G is the graph obtained from the polygon $C \setminus \{\Omega\}$ by adding for each $i \in [m]$ an edge f_i parallel to e_i. (G is unique up to reordering $e_1, \ldots, e_n$.)

By Remark 2, $M = \text{ecycle}(H, \Sigma)$ where H is obtained from G by adding a loop Ω and for a suitable $\Sigma \subseteq E(H)$ with $\Omega \in \Sigma$. Proposition 9 implies that the cocircuits M are either cuts of H or signatures of (H, Σ). Since D is a cocircuit of M and Ω is a loop of H, D must be a signature of (H, Σ). Since the signatures of (H, Σ) and (H, D) are the same, we will assume that $\Sigma = D$. Suppose that $n \geq m + 2$. Then there exists $e_{m+1}, e_{m+2} \in D$ which are not parallel to edges in F in the graph H. It follows that $D' = \{e_{m+1}, e_{m+2}\}$ is a cut of H, hence, by Proposition 9, D' is a cocycle of M. However, $D' \subset D$, contradicting the fact that D is a cocircuit. Thus $n \leq m + 1$ and $m \geq n - 1 \geq 4$. In particular, there exists $f_1, f_2 \in F$ and for $i = 1, 2$, e_i, f_i are parallel in H. It follows that $C' = C \setminus \{e_1, e_2\} \cup \{f_1, f_2\}$ is a polygon of H. Moreover, $|C' \cap \Sigma|$ is even. It follows that C' is a circuit of M which satisfies $(\star)$ as required.

3.3 A Geometric Application

Given a clutter $\mathcal{C}$ and $S \in \mathcal{C}$, we denote by $\chi(S)$ the characteristic vector for S, i.e. for every $e \in E(\mathcal{C})$, $\chi(S)_e = 1$ if $e \in S$ and $\chi(S)_e = 0$ otherwise. Recall that 0, 1 vertices of $P(\mathcal{C})$ are of the form $\chi(B)$ for some $B \in b(\mathcal{C})$ ([7] Remark 1.16). In this section we describe a variant of Proposition 4 that describes where the point $\chi(B')$ can be selected on the polyhedron $P(\mathcal{C})$. Namely, the following holds.

Proposition 10. *Let $\mathcal{C}$ be a binary clutter and let $\bar{z}$ be an integral corner of $P(\mathcal{C})$. Let $S \in \mathcal{C}$ for which $\sum_{e \in S} \bar{z}_e = 1$. Suppose there exists $B \in b(\mathcal{C})$ for which*

$|S \cap B| > 1$. Then there exists $B' \in b(\mathcal{C})$ for which $|S \cap B'| = 3$ where $\chi(B')$ is adjacent to $\bar{z}$ in $P(\mathcal{C})$.

Before we proceed with the proof, recall the following result about tangent cones, see [18] Chap. 6, and also [23].

Proposition 11. *Let P be a polyhedron with a vertex $\bar{z}$. Let $L_1, \ldots, L_k$ be the faces of dimension 1 that contain $\bar{z}$ and for all $i \in [k]$ let v^i be a point in $L_i - \{\bar{z}\}$. Then $P \subseteq \bar{z} + \mathrm{cone}(v^i - \bar{z} : i \in [k])$. In particular, if $y^1, \ldots, y^k$ are the vertices that are adjacent to $\bar{z}$ then*

$$P \subseteq \bar{z} + cone\big(y^i - \bar{z} : i \in [k]\big) + \mathrm{rec}(P).$$

Here, $\mathrm{rec}(P)$ denotes the recession cone of P.

Proof. (Proposition 10). By Proposition 4 there exists $R \in b(\mathcal{C})$ for which $|S \cap R| = 3$. By Proposition 11, there exists $\lambda \in \mathbb{R}^k_+$ for which,

$$\chi(R) = \bar{z} + \sum_{i \in [k]} \lambda_i (y^i - \bar{z}). \tag{3}$$

Note, we do not need any rays of $\mathrm{rec}(P(\mathcal{C}))$ to express $\chi(R)$ in the previous equation as R is a clutter element and thus is inclusion-wise minimal. Every vertex of $P(\mathcal{C})$ is the characteristic vector of some element of the blocker ([7] Remark 1.16). Thus, for every $i \in [k]$ we have $B_i \in b(\mathcal{C})$ with $y^i = \chi(B_i)$. We also have $B_z \in b(\mathcal{B})$ with $\bar{z} = \chi(B_z)$. By construction $|S \cap B_z| = 1$ and $|S \cap B_i| \geq 1$ for all $i \in [k]$. Left multiplying (3) by $\chi(S)^\top$ yields,

$$\begin{aligned} |S \cap R| = \chi(S)^\top \chi(R) &= \chi(S)^\top \bar{z} + \sum_{i \in [k]} \lambda_i \chi(S)^\top (y^i - \bar{z}) \\ &= |S \cap B_z| + \sum_{i \in [k]} \lambda_i (|S \cap B_i| - |S \cap B_z|) \\ &= 1 + \sum_{i \in [k]} \lambda_i (|S \cap B_i| - 1). \end{aligned} \tag{4}$$

Since $|S \cap R| > 1$ it follows from (4) that for some $\kappa \in [k]$ we have $\lambda_\kappa > 0$ and $|S \cap B_\kappa| > 1$. Moreover, (3) implies that R equals the support of $\bar{z} + \sum_{i \in [k]} \lambda_i (y^i - \bar{z})$. Therefore, $R \supseteq B_\kappa \setminus B_z$ and in particular,

$$3 = |S \cap R| \geq |S \cap B_\kappa| - |S \cap B_z| = |S \cap B_\kappa| - 1.$$

Thus $|S \cap B_\kappa| \leq 4$. Since $\mathcal{C}$ is binary, we must have $|S \cap B_\kappa| \leq 3$. However, then $B' = B_\kappa$ is as required.

4 The Proof of Theorem 2

4.1 A General Tool

We shall state a result from [12] that gives a sufficient condition for a linear program to have a dyadic solution. We require a number of definitions. Consider $A \in \mathbb{Z}^{m \times n}$ and $b \in \mathbb{Z}^m$ and let P be the polyhedron $\{x : Ax \geq b\}$. Let F be a face of P. Denote by $I(F)$ the set of row indices of A corresponding to the tight constraints for F, i.e. $i \in I(F)$ if and only if $\text{row}_i(A)x = b_i$ for every $x \in F$. A face F of P is *compliant* if for every $w \in \mathbb{Z}^n$ for which F is the w-optimal face, the following linear program has an optimal solution that is dyadic,

$$\min\{b^\top y : A^\top y = w, y \geq \mathbf{0}\}. \tag{5}$$

A sequence of faces of P, $(F_0, F_1, \ldots, F_k)$ is a *face-chain rooted at* F_0 if for every $\ell \in [k]$, F_ℓ is an immediate superface of $F_{\ell-1}$. This face-chain is *well-marked* if F_k is compliant and for every $\ell \in [k]$ there exists a vector $\rho \in \mathbb{Z}^n$ and an index $\hat{\imath} \in I(F_\ell)$ such that,[2]

(g1) $\rho \in \text{aff}(F_\ell) \setminus \text{aff}(F_{\ell-1})$ and
(g2) $row_{\hat{\imath}}(A)\rho - b_{\hat{\imath}} = \pm 2^p$ for some $p \in \mathbb{Z}_+$.

Here is our key tool from [12], it is a combination of Theorem 2.2 and Proposition 3.1 in a different language but without essential changes to the proof. For the sake of completeness we will give a self-contained proof in the full version of this article.

Theorem 3. *Let $P = \{x : Ax \geq b\} \subset \mathbb{R}^n$ be a polyhedron where all entries of A, b are integer. Let $\bar{w} \in \mathbb{Z}^n$ and let $\bar{F}$ denote the $\bar{w}$-optimal face of P where $\bar{F} \cap \mathbb{Z}^n \neq \emptyset$. If there exists a well-marked face-chain rooted at $\bar{F}$, then there exists an optimal solution of (5) for $w = \bar{w}$ that is dyadic.*

4.2 The Proof

We are now ready for the proof of our key result.

Proof (Theorem 2). Let $n = |E(\mathcal{C})|$ and let $p = |\mathcal{C}|$. Let M be the matrix where the rows correspond to the characteristic vectors of $\mathcal{C}$. Let A be the matrix obtained from M by adding as additional rows the identity matrix I_n. Then A is an $m \times n$ matrix where $m = p + n$. Let b be the vector with m entries where $b_i = 1$ if $i \in [p]$ and $b_i = 0$ otherwise. Then

$$P(\mathcal{C}) = \{x : Ax \geq b\}. \tag{6}$$

Let $F_0 = \bar{F}$ denote the w-optimal face of $P(\mathcal{C})$. Let us choose $k \in \mathbb{Z}_+$ maximum for which $(F_0, F_1, \ldots, F_k)$ is a face-chain rooted at $\bar{F}$ with the property that for

[2] Note that ρ and $\hat{\imath}$ depend on ℓ.

every $\ell \in [k]$, there exists $\rho \in \mathbb{Z}^n$ and $\hat{\imath} \in I(F_\ell)$ such that properties (g1), (g2) are satisfied (such a chain always exists as we can pick $k = 0$).

If F_k is compliant then $(F_0, F_1, \ldots, F_k)$ is a well-marked face-chain and it follows from Theorem 3 that (5) has an optimal dyadic solution as required. Furthermore, we leave it as an exercise to prove that (5) has an optimal dyadic solution if and only if (2) does. Thus we will assume that F_k is not compliant, i.e., there exists $w' \in \mathbb{Z}_+^{E(\mathcal{C})}$ for which F_k is the w'-optimal face and (5) does not have an optimal dyadic solution for right-hand-side w'. Since $\bar{F} \subset F_k$, $v \in F_k$ and every w-minimal cover B is a w'-minimal cover, in particular, w' is minimal. Then by the fact[3] that (2) has a half-integer solution, when there are no more than 6 active dual variables and the clutter is binary, $\sigma(F_k) \geq 7$ and by hypothesis F_k is v-compatible.

Since w' is minimal, no constraint $x_j \geq 0$ is tight for F_k. Hence, for some $\mathcal{D} \subseteq \mathcal{C}$ the tight constraints for F_k are $x(D) = 1$ for all $D \in \mathcal{D}$. Let $\mathcal{L}$ denote the set of faces of dimension 1 containing v.

Claim 5. For every $D \in \mathcal{D}$ there exists $L \in \mathcal{L}$ and $\rho \in L \cap Z^n$ with

$$\sum_{j \in D} \rho_j \in \{2, 3\}.$$

Proof of Claim. Consider the case where there exists $B \in b(\mathcal{C})$ with $|B \cap D| > 1$. By Proposition 10 there exists $B' \in b(\mathcal{C})$ with $|D \cap B'| = 3$ and where $\rho = \chi(B')$ is adjacent to v. Let $L \in \mathcal{L}$ be the 1-dimensional face containing v and $\chi(B')$. Then L, ρ are as required. Thus we may assume that every $B \in b(\mathcal{C})$ satisfies $|B \cap D| = 1$. It can be shown that all line segments in $\mathcal{L}$ are tight for $x(D) \geq 1$ for every $D \in \mathcal{D}$. Since $P(\mathcal{C})$ is up-monotone, and by Proposition 11, some half-line $L \in \mathcal{L}$ where $L = \{v + \lambda \mathbf{1}_j : \lambda \geq 0\} \in \mathcal{L}$ is not tight for some $x(D) \geq 1$ where $D \in \mathcal{D}$. Then, $j \in D$ and $L, \rho = v + \mathbf{1}_j$ are as required. ◇

Construct $\mathcal{L}' \subseteq \mathcal{L}$ as follows: for every $D \in \mathcal{D}$, include the face L in Claim 5. Observe that by construction $\mathcal{L}'$ is (F, v)-saturating. Since F_k is v-compatible, there exists $L \in \mathcal{L}'$ for which $L \cap F_{k+1} \setminus \{v\} \neq \emptyset$ where F_{k+1} is some immediate superface of F_k. L arises in Claim 5 and consider the corresponding constraint $x(D) \geq 1$ and the corresponding $\rho \in L \cap Z^n$. Denote by $\hat{\imath}$ the index of A corresponding to constraint $x(D) \geq 1$. Then (g1), (g2) hold for $\ell = k + 1$ and $(F_0, \ldots, F_{k+1})$ contradicts our choice of k. □

5 Closing Remarks

In our proof for the binary case, we used the fact that if the number of active dual variables is bounded by 6 then the dual (2) has a $\frac{1}{2}$-integer optimal solution. Suppose that we could improve the result from 6 to some larger value α and only guarantee a dyadic optimal solution of (2). This would immediately allow us to replace 7 in Theorem 2 by $\alpha + 1$. This in turn, using Proposition 2, would allow us to replace 6 by α in Theorem 1. So, we in fact proved the following statement.

[3] a proof is in the full version of the article.

Theorem 4. *If Conjecture 2 holds for all binary instances with at most α active constraints, then it also holds for all binary instances where the corner is (k, ℓ)-degenerate for $k(\ell + 1) \leq \alpha$.*

In other words, proving the conjecture for a higher number of active constraints implies the conjecture for more degenerate corners.

Acknowledgments. We thank Ahmad Abdi, and András Sebő for many lively discussions on the topic. We are also grateful to Jim Geelen, for suggesting the proof of Proposition 8. This research was supported in part by Discovery Grants from the Natural Sciences and Engineering Research Council (NSERC) of Canada.

Declaration of Competing Interests. The authors have no competing interests to declare that are relevant to the content of this article.

References

1. Abdi, A., Cornuéjols, G., Palion, Z.: On dyadic fractional packings of T-joins. SIAM J. Disc. Math. **36**(3), 2445–2451 (2022)
2. Abdi, A., Cornuéjols, G., Guenin, B., Tunçel, L.: Dyadic linear programming and extensions. Math. Program. **213**, 473–516 (2025)
3. Abdi, A., Cornuéjols, G., Guenin, B., Tunçel, L.: Total dual dyadicness and dyadic generating sets. Math. Program. **206**, 125–143 (2024)
4. Abdi, A., Cornuéjols, G., Liu, S., Silina, O.: Strongly connected orientations and integer lattices. In: International Conference on Integer Programming and Combinatorial Optimization, pp. 1–14. Springer, Cham (2025). https://doi.org/10.1007/978-3-031-93112-3_1
5. Chvátal, V.: On certain polytopes associated with graphs. J. Comb. Theory Ser. B **18**, 138–154 (1975)
6. Cornuéjols, G., Karahalios, A., Patil, V.: Computing a Nonnegative Dyadic Solution to a System of Linear Equations. Manuscript, Tepper School of Business, Carnegie Mellon University (2025)
7. Cornuéjols, G.: Combinatorial optimization: Packing and covering. Society for Industrial and Applied Mathematics (2001)
8. Edmonds, J., Fulkerson, D.R.: Bottleneck extrema. J. Comb. Theory **8**, 299–306 (1970)
9. Geelen, J.F., Guenin, B.: Packing odd-circuits in Eulerian graphs. J. Comb. Theory B **86**, 280–295 (2002)
10. Guenin, B., Heo, C.: Recognizing even-cycle and even-cut matroids. Math. Program. **202**, 515–542 (2023)
11. B. Guenin, B., Hwang, S.: Dyadic packing of dijoins. SIAM J. Disc. Math. **39**, 593–606 (2025)
12. Guenin, B., Tunçel, L.: Generalizations of Total Dual Integrality. arXiv preprint arXiv:2503.07925 (2025)
13. Lehman, A.: A solution of the Shannon switching game. J. Soc. Ind. Appl. Math. **12**, 687–725 (1964)
14. Lucchesi, C.L., Younger, D.H.: A minimax theorem for directed graphs. J. Lond. Math. Soc. **17**(3), 369–374 (1978)

15. Novick, B., Sebő, A.: On combinatorial properties of binary spaces. In: Balas, E., Clausen, J. (eds.) IPCO 1995. LNCS, vol. 920, pp. 212–227. Springer, Heidelberg (1995). https://doi.org/10.1007/3-540-59408-6_53
16. Oxley, J.G.: On the intersections of circuits and cocircuits in matroids. Combinatorica **4**, 187–195 (1984)
17. Pivotto, I.: Even cycle and even cut matroids. Doctoral dissertation, University of Waterloo (2011)
18. Rockafellar, R.T., Wets, R.J.-B.: Variational analysis. Grundlehren Math. Wiss., 317 [Fundamental Principles of Mathematical Sciences]. Springer, Berlin (1998)
19. Schrijver, A.: Combinatorial optimization. Polyhedra and efficiency. Vol. C, vol. 24, C of Algorithms and Combinatorics. Springer, Berlin (2003). Disjoint paths, hypergraphs, Chapters 70–83
20. Sebő, A.: Personal communication (2026)
21. Seymour, P.D.: The forbidden minors of binary clutters. J. Lond. Math. Soc. **2**, 356–360 (1976)
22. Seymour, P.D.: A two-commodity cut theorem. Disc. Math. **23**, 177–181 (1978)
23. Ziegler, G. M.: Lectures on polytopes. Grad. Texts in Math., 152. Springer, New York (1995)

On Integer Programs That Look Like Paths

Marcin Briański[1], Alexandra Lassota[2], Kristýna Pekárková[3,4], Michał Pilipczuk[5], and Janina Reuter[6(✉)]

[1] Jagiellonian University, Kraków, Poland
marcin.brianski@doctoral.uj.edu.pl
[2] Eindhoven University of Technology, Eindhoven, Netherlands
a.a.lassota@tue.nl
[3] AGH University, Kraków, Poland
pekarkova@agh.edu.pl
[4] Faculty of Informatics, Masaryk University, Brno, Czech Republic
[5] Institute of Informatics, University of Warsaw, Warsaw, Poland
michal.pilipczuk@mimuw.edu.pl
[6] Kiel University, Kiel, Germany
janina.reuter@informatik.uni-kiel.de

Abstract. Solving integer programs of the form $\min\{\mathbf{x} \mid A\mathbf{x} = \mathbf{b}, \mathbf{l} \leqslant \mathbf{x} \leqslant \mathbf{u}, \mathbf{x} \in \mathbb{Z}^n\}$ is, in general, NP-hard. Hence, great effort has been put into identifying subclasses of integer programs that are solvable in polynomial or FPT time. A common scheme for many of these integer programs is a star-like structure of the constraint matrix. The arguably simplest form that is not a star is a path. We study integer programs where the constraint matrix A has such a path-like structure: every non-zero coefficient appears in at most two consecutive constraints. We prove that even if all coefficients of A are bounded by 8, deciding the feasibility of such integer programs is NP-hard via a reduction from 3-SAT. Given the existence of efficient algorithms for integer programs with star-like structures and a closely related pattern where the sum of absolute values is column-wise bounded by 2 (hence, there are at most two non-zero entries per column of size at most 2), this hardness result is surprising.

Keywords: Integer programming · Block-structured integer programs · NP-hardness

1 Introduction

We study integer programs (to which we refer to as *IP*s) of the form

$$\min\{\mathbf{x} \mid A\mathbf{x} = \mathbf{b}, \mathbf{l} \leqslant \mathbf{x} \leqslant \mathbf{u}, \mathbf{x} \in \mathbb{Z}^n\}, \tag{G}$$

with constraint matrix $A \in \mathbb{Z}^{m \times n}$, right-hand side $\mathbf{b} \in \mathbb{Z}^m$, lower and upper bounds $\mathbf{l} \in (\mathbb{Z} \cup \{-\infty\})^n$ and $\mathbf{u} \in (\mathbb{Z} \cup \{\infty\})^n$. Solving such IPs means to

S. S. Dey et al. (Eds.): IPCO 2026, LNCS 16588, pp. 158–172, 2026.
https://doi.org/10.1007/978-3-032-28691-8_11

either decide that the system is infeasible, there exists an optimal feasible solution (and give the value of it), or it is unbounded and provides a solution with an unbounded direction of improvement. Note that restricting to equality constraints does not restrict the problem, as every inequality constraint can be turned to equality by introducing slack variables (which also preserve the special structures we discuss in this paper).

Integer programming is a powerful tool that has been applied to many combinatorial problems, such as scheduling and bin packing [13, 18], graph problems [8, 9], multichoice optimization [7] and computational social choice [1, 22], among others. Unfortunately, solving integer programs is, in general, NP-hard even if $\|A\|_\infty = 1$, which is usually shown by a reduction from 3-SAT.

This motivated a flourishing line of research that aims to understand the structural properties of matrices that allow the corresponding IPs to be solved efficiently, see, e.g. [8, 13, 14, 18, 22]. Arguably, the most famous is Lenstra's 1983 algorithm, which introduced the first FPT time algorithm for constraint matrices with few columns [19]. The body of literature for the over three decades-long research and the many structures, parameters, and applications is too vast to cover here; we thus refer to [12] for an overview.

One striking property for many of these IPs is the star-likeness of the constraint matrix, which is probably best exemplified by *n-fold* and *2-stage stochastic* integer programs. Here, an integer program is of n-fold form if its constraint matrix, after removing at most k rows (*aka* constraints), consists of n independent blocks with at most k rows each. The transpose of this matrix is called 2-stage stochastic integer programs, i.e., after removal of at most k columns (*aka* variables), they are decomposed into independent blocks with at most k columns each. As proven in [15, 16], the optimization problem for n-fold and 2-stage stochastic integer programs can be solved in time $f(k, \Delta) \cdot |I|^{1+o(1)}$, where f is a computable function, Δ is the largest absolute coefficient in the constraint matrix, and $|I|$ is the total bitsize of the instance. This running time is FPT with parameters k and Δ.

The structure of n-fold and 2-stage stochastic integer programs can be generalized by allowing multiple stages of deleting rows or columns and decomposing into blocks. This gives rise to so-called *tree-fold* and *multistage stochastic* integer programs, which retain efficient solvability, see, e.g., [3, 6, 21]. The structure of tree-fold and multistage stochastic IPs can be perhaps best understood through decompositions of the associated graphs. For a matrix A, we define the *constraint graph* of A as the graph on the rows (constraints) of A in which two rows are adjacent if they simultaneously have non-zero entries in the same column (in other words, there is a variable appearing in both the corresponding constraints). The *variable graph* of A is defined symmetrically for the columns (variables) of A. Then, the *primal treedepth* of A is the treedepth of its variable graph, and the *dual treedepth* of A is the treedepth of its constraint graph. Here, the treedepth of a graph is a parameter that measures star-likeness through the maximum depth of a procedure that alternately deletes a vertex and decomposes the graph into connected components, see for example [24, Chapter 6] for a broad introduction.

As proven in [10,11,17], the optimization problem for integer programs can be solved in time $f(d, \Delta) \cdot |I|^{1+o(1)}$ for a computable function f, where d is either the primal or the dual treedepth of the constraint matrix. We refer the reader to the article of Eisenbrand, Hunkenschröder, Klein, Koutecký, Levin and Onn [6] for a comprehensive introduction to the theory of parameterized algorithms for block-structured integer programming.

It is, therefore, natural to ask to what extent the (primal or dual) treedepth of the constraint matrix captures the efficient solvability of integer programs. Several results in the literature suggest that integer programming quickly becomes NP-hard once one relaxes the assumption of bounded treedepth. For instance, Ganian and Ordyniak [10] showed that the feasibility problem for integer programs of the form $\{\mathbf{x} \mid A\mathbf{x} \leqslant \mathbf{b}\}$ is NP-hard even when all the entries of A and $\mathbf{b}$ are in $\{-2, -1, 0, 1, 2\}$ and the treewidth of the constraint graph of A is at most 4. Their reduction is based on an earlier work of Jansen and Kratsch [17], who achieved a similar result, but without obtaining the bound on the entries. We note that Ganian and Ordyniak stated their result in terms of the treewidth of the *incidence graph* of A: the bipartite graph on rows and columns of A where a row is adjacent to a column if the entry at their intersection is non-zero. As shown in [10], the treewidth of the incidence graph of the constructed matrix is at most 3, but one can also easily verify that the treewidth of the constraint graph is at most 4 (bounding the treewidth/treedepth of the constraint graph implies a bound on the treewidth of the incidence graph, but not vice versa.) Similar lower bounds were also reported in the work of Ganian, Ordyniak, and Ramanujan [11]. Later, Eiben, Ganian, Knop, Ordyniak, Pilipczuk, and Wrochna [5] showed that when the incidence graph is considered, even bounded treedepth is not enough: deciding feasibility of integer programs of the form $\{\mathbf{x} \mid A\mathbf{x} = 0, \mathbf{l} \leqslant \mathbf{x} \leqslant \mathbf{u}\}$, is NP-hard even when the entries of $A, \mathbf{l}, \mathbf{u}$ are in $\{-1, 0, 1\}$ and the incidence graph of A has treedepth 5. Klein, Polak, and Rohwedder [20] show NP-hardness for constraint matrices, where a small 3×5 block matrix is repeated in a (lower-left) triangle, i.e., the same matrix is repeated over the diagonal and below, and entries in $\{-2, -1, 0, 1\}$. Finally, Eiben, Ganian, Knop, and Ordyniak [4] showed a number of lower bounds showing that allowing A to have arbitrarily large entries quickly leads to NP-hardness, even if A has bounded treedepth.

The goal of this work is to contribute to the understanding of the complexity of integer programming by considering matrices that have the simplest possible form while still having unbounded treedepth. Since the length of the longest path (contained as a subgraph) is functionally equivalent to treedepth, the simplest possible graph with arbitrarily large treedepth is a path. Thus, our aim is to understand the complexity of integer programs whose constraint graph[1] is a path. Equivalently, these are integer programs of the form $\{\mathbf{x} \mid A\mathbf{x} = \mathbf{b}, \ell \leqslant \mathbf{x} \leqslant \mathbf{u}\}$ where every variable can appear only in at most two constraints, which

[1] One could also consider the setting when the variable graph is a path, but a quick examination of this setting reveals that it makes little sense, and in fact reduces to a subcase of the setting where the constraint graph is a path.

moreover must be consecutive. Up to permuting the columns, the constraint matrix A must have the path-like structure as depicted in Fig. 1: the non-zero entries in each row may only appear in an interval, these intervals are naturally ordered from left to right, and only the intervals in consecutive rows may overlap.

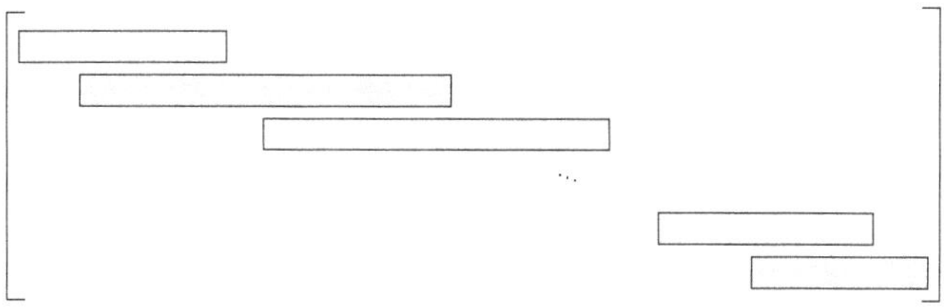

Fig. 1. A schematic depiction of a path-like constraint matrix A.

It is known that if the coefficients of the constraint matrix are in $\{-2,-1,0,1,2\}$ and the sum of absolute values is at most 2 for each column, then the problem is polynomial time solvable [25]. Indeed, the corresponding problem called (generalized) matching problem is well-studied. It captures a variety of well-known problems in P such as minimum cost flow, minimum cost (b-)matching and certain graph factor problems [25]. These structures even remain FPT time solvable parameterized by the number p of additional columns [23].

The case of *two* non-zero entries per column seems to be indeed the turning point where analyzing the computational complexity becomes intriguing. Indeed, it is easy to see that the problem is NP-hard when *three* non-zero entries per column are allowed and entries are in $\{-1,0,1,2\}$. Consider the following reduction from subset sum where n non-negative numbers $a_1,\ldots,a_n$ and a target $t \in \mathbb{Z}_{\geqslant 0}$ are given and the goal is to decide whether a subset of $\{a_1,\ldots,a_n\}$ sums up to t. Let $enc(x)$ be the binary encoding of a non-negative number x and consider the matrices

$$E = \begin{pmatrix} -1 & 2 & & & \\ & -1 & 2 & & \\ & & \ddots & \ddots & \\ & & & -1 & 2 \end{pmatrix} \quad \text{and} \quad A = \begin{pmatrix} enc(a_1) & \ldots & enc(a_n) \\ E & & \\ & \ddots & \\ & & E \end{pmatrix}.$$

The first matrix, E, is called the *encoding matrix* and is often used in lower bounds for IPs with a special structure of non-zero entries in the constraint matrix. The IP defined by the constraint matrix A, right-hand side $\mathbf{b} = (t,0,\ldots,0)$, and suitable upper and lower bounds $\mathbf{l}$ and $\mathbf{u}$ is feasible if and only if the corresponding subset sum instance is feasible. Clearly, the constraint matrix A has at most three non-zero entries per column.

In stark contrast to the above-mentioned results for an even more restricted form in terms of the placement of non-zero coefficients, we show that the feasibility problem for integer programs with path-like constraint matrices is NP-hard already when all entries are bounded by a constant.

Theorem 1. *The problem of deciding whether a given integer program* $\{\mathbf{x} \mid A\mathbf{x} = \mathbf{b}, \mathbf{l} \leqslant \mathbf{x} \leqslant \mathbf{u}\}$ *has a solution is* NP*-hard even when* A *is a path-like matrix with all entries of absolute value at most* 8*. Here, we assume that* $\mathbf{b}$ *is provided on input in binary.*

We stress that the assumption that $\mathbf{b}$ is encoded in binary is essential. Indeed, since the incidence graphs of path-like matrices have treewidth at most 2, from the result of Ganian, Ordyniak, and Ramanujan [11, Theorem 13] it follows that the feasibility of integer programs of the form discussed in Theorem 1 can be decided in time polynomial in $\|A\|_\infty + \|\mathbf{b}\|_\infty$ and the total bitsize of the input.

Theorem 1 therefore provides a strong lower bound for integer programming and intuitively shows that any deviation from constraint matrices with bounded treedepth renders the problem intractable. The proof of the theorem is significantly more elaborate than the proofs of the lower bounds from [4,5,10,11,17] mentioned above, which to a large extent relied on a direct encoding of SUBSET SUM as an integer program with a simple structure. Our reduction is from 3-SAT. Before we prove this result in Sect. 2, we give an intuitive argument and sketch of the reduction at the beginning of the section. An example is included in Appendix A.

2 Hardness

We now prove that path-like integer programs are NP-hard even when all coefficients of the constraint matrix are bounded by a constant. We prove this by reducing the 3-SAT problem to path-like IPs.

The main idea is to design constraints for each clause C_i such that the solution space of the IP represents the 3-SAT assignments that satisfy clause C_i. The first constraint introduces a variable with a solution space to represent all assignments of 3-SAT variables. Then we propagate and constrict the feasible assignments to the IP variables according to the 3-SAT clauses. Finally, the solution space of the IP variables is the set of representations for feasible assignments for 3-SAT.

We construct the matrix in such a way that every variable occurs at most twice and in consecutive constraints. We say that the first constraint for a variable *introduces* the variable, and that in a (potential) second constraint, the solution space of the variable is *forwarded* or *output*. We use a limited number of different *constraint types* to transform the solution space of a forwarded variable to a solution space for an introduced variable. Here, we consider the solution space as the set of feasible solutions for that variable considering the current and all previous constraints but ignoring all following constraints. The constraint types of the reduction are summarized in Table 1. For example, we will use the multiplication by $a \in \mathbb{Z}$ constraint type repeatedly to get a multiplication by a^n. Note that this is similar to the encoding matrix mentioned in the introduction.

A cornerstone for the reduction is a number-theoretical identity that allows us to perform operations that check satisfaction of a clause while preserving the

Table 1. Types of constraints and the operation that is thereby performed on the solution space from an output variable x and an introduced variable y that are used in the proof of Theorem 1.

Operation	Constraint
Forward	$x - y = 0$
Check $x \geqslant a$	$x - y = a$
Multiplication by $a \in \mathbb{Z}$	$ax - y = 0$
Division with remainder by $a \in \mathbb{Z}$	$x - ay - r = 0$ and $0 \leqslant r < a$
Combination of two constraints with independent variables	$\mathbf{c}_1^\intercal \mathbf{x}_1 = \mathbf{b}_1$ and $\mathbf{c}_2^\intercal \mathbf{x}_2 = \mathbf{b}_2$ with $\mathbf{0} \leqslant \mathbf{x}_1 \leqslant \mathbf{u}$ combined to $\mathbf{c}_1^\intercal \mathbf{x}_1 + a \cdot \mathbf{c}_2^\intercal \mathbf{x}_2 = \mathbf{b}_1 + a\mathbf{b}_2$ with $a > \mathbf{c}_1^\intercal \mathbf{u}$

information stored in that variable. In particular, we need to look up the assignment of specific variables, check whether they yield a true literal in the clause, and then restore the original representation of the assignment. Considering only one clause, it would be relatively easy to repeatedly divide by 2 with remainder and check whether the remainder implies a true literal. However, this procedure is in fact limited to one clause, as the information of the satisfying assignment are lost in the process.

Theorem 1. *The problem of deciding whether a given integer program* $\{\mathbf{x} \mid A\mathbf{x} = \mathbf{b}, \mathbf{l} \leqslant \mathbf{x} \leqslant \mathbf{u}\}$ *has a solution is* NP*-hard even when* A *is a path-like matrix with all entries of absolute value at most* 8*. Here, we assume that* $\mathbf{b}$ *is provided on input in binary.*

Proof. We reduce from 3-SAT, which is known to be NP-complete [2], i.e., we define a path-like matrix A, lower and upper bounds $\mathbf{l}$ and $\mathbf{u}$, and a right-hand side $\mathbf{b}$ such that $A\mathbf{x} = \mathbf{b}$ has a solution $\mathbf{l} \leqslant \mathbf{x} \leqslant \mathbf{u}$ if and only if the 3-SAT formula is satisfiable.

Let $\varphi = \bigwedge_{i=1}^{m} C_i$ be a 3-SAT formula with variables $V = \{v_1, v_2, \dots, v_n\}$. We begin by checking whether the all-zero assignment $\nu \colon V \ni v_j \mapsto 0 \in \{0, 1\}$ is a satisfying assignment. If it does satisfy φ, then the construction outputs a trivial instance where A is the zero matrix, and $\mathbf{b}, \mathbf{l}, \mathbf{u}$ are zero vectors, i.e., $[0]\ \mathbf{v} = [0], \mathbf{l} = \mathbf{u} = [0]$. Henceforth, we assume that the constant 0 assignment is not a satisfying assignment of φ.

We want to shortly describe naming conventions of the variables of the IP. Roughly speaking, variables with names f, x, y, and z will (mostly) appear in two consecutive constraints each. Hence, we use these variables to forward, evaluate, or alter information, with x, y, and z holding information on possible assignments to the 3-SAT formula and f being used as a local helper variable in the *checking clause* phase below. Respectively indexed versions of α, β, and γ are each used in exactly one constraint and thus their purpose lies in the realization of the desired operation for that constraint only.

We first construct a constraint where the set of solutions contains all the $2^n - 1$ possible variable assignments for the 3-SAT formula, except for the assignment mapping all variables to 0, as follows:

$$x_{0,0} - \alpha = 0,$$

where lower and upper bounds are $\alpha \in \{1, \ldots, 2^n - 1\}$. Bounds on $x_{0,0}$ are not required. The constraint is feasible for variable

$$x_{0,0} \in \mathcal{A}_0 = \{1, \ldots, 2^n - 1\} = \{[b_{n-1}b_{n-2} \ldots b_0]_2 \neq 0 \mid b_i \in \{0, 1\}\},$$

where $[b_{n-1}b_{n-2} \ldots b_0]_2$ represents the number $\sum_{i=0}^{n-1} b_i 2^i$.

The main part is to delete all representations of non-satisfying assignments from the set. Each element of $\mathcal{A}_i$ represents a variable assignment that satisfies all clauses C_j with $j \leqslant i$. Therefore, in order to generate the next set $\mathcal{A}_{i+1}$, we describe constraints that delete assignments that do not satisfy clause C_{i+1}.

Although it might seem natural to do the checking of the clause directly on the set $\mathcal{A}_i$, we need a trick in order to preserve the information about the assignment for checking further clauses.

Claim 1. *The following equations are true.*

1. $x = \lfloor \lfloor x \cdot 3^n / 2^n \rfloor 2^n / 3^n \rfloor + 1$ *for all* $x \in \{1, \ldots 2^n - 1\}$
2. $\{x \cdot 3^n \mod 2^n \mid x \in \{1, \ldots 2^n - 1\}\} = \{1, \ldots 2^n - 1\}$

Proof. Regarding the first equation, the inner floor operation alters the result by an amount of $0 < (3^n x/2^n - \lfloor 3^n x/2^n \rfloor) \cdot 2^n/3^n < 1$ because x is not a multiple of 2^n and of course $2^n/3^n < 1$. The outer floor operation rounds down by $2^n y/3^n - \lfloor 2^n y/3^n \rfloor < 1$ for $y = \lfloor 3^n x/2^n \rfloor$. In total, the rounding scrapes off more than 0 but less than 2, starts with an integer and ends with an integer. Thus, adding 1 after the second rounding restores x. The second equation follows directly as 2 and 3 are coprime.

□ Note that Claim 1 is the reason we excluded the all-zero assignment at the beginning of the proof since the identity of the first equation does not hold for 0. Roughly speaking, we will use the second equality in Claim 1 to extract clause satisfiability from the remainder during the division by 2^n and then restore the original set of candidate assignments using the first identity of Claim 1. For any $x \in \mathcal{A}_i$ the represented assignment are the bits of $[b_{n-1} \ldots b_0]_2 = 3^n x \mod 2^n$. The following steps recreate the identity of the first equation in Claim 1 while checking one clause for each application.

Preparation Phase. We first construct the set $\mathcal{B}_i = \{3^n \cdot a \mid a \in \mathcal{A}_i\}$. The following constraints transform variable $x_{i,0}$ with feasible assignments $\mathcal{A}_i$ to variable $y_{i,0}$ with feasible assignment $\mathcal{B}_i$

$$3x_{i,j} - x_{i,j+1} = 0 \quad \text{for all } 0 \leqslant j < n$$
$$x_{i,n} - y_{i,0} = 0.$$

Here, every constraint of the first type is a multiplication-by-3-type of constraint (see Table 1) and hence ensures that any feasible value for $x_{i,j+1}$ is three times any feasible value for $x_{i,j}$. The second constraint only adds syntactical sugar with $x_{i,n} = y_{i,0}$ to simplify variable names and readability in the next step.

Checking Clause. The next step in recreating the first equation of Claim 1 is an integer division by 2^n. We use this step to check for satisfaction of clause C_i using the remainder of the division. A remainder of $[b_{n-1}, \dots, b_0]_2$ represents the assignment $v_i \mapsto b_i$ for all 3-SAT variables $v_i \in V$. The second equation of Claim 1 implies that starting from $x_{0,0} \in \mathcal{A}_0$ all assignments except for the zero-assignment appear as a remainder in the first iteration. This phase gathers information whether $b \in \mathcal{B}_i$ represents an assignment that satisfies clause C_i. The following phase uses this information to delete all non-satisfying assignments to clause C_i.

We use an additional variable $f_{i,j}$ to remember whether the clause is satisfied. This variable is increased if a literal of the clause is true for the assignment. If $f_{i,n} \neq 0$ after the division with remainder, then the clause is satisfied. Intuitively, the goal would be to achieve a set of feasible values for $(f_{i,n}, y_{i,n})$ of

$$\mathcal{C}_i = \{(f, \lfloor b/2^n \rfloor) \mid b \in \mathcal{B}_i \text{ and } f = t_i(b)\},$$

where $t_i(b)$ is the number of true literals in clause C_i for assignment $[b_{n-1} \dots b_0]_2 = b \bmod 2^n$. However, we do not achieve this set exactly but rather a set $\hat{\mathcal{C}}_i$ such that if $[b_{n-1} \dots b_0]_2$ does not satisfy the clause, then for any $(f, \lfloor b/2^n \rfloor) \in \mathcal{C}_i$ we have $f = 0$. If otherwise $[b_{n-1} \dots b_0]_2$ satisfies the clause, then at least one $(f, \lfloor b/2^n \rfloor)$ with $f \in \{1, 2, 3\}$ is contained in $\mathcal{C}_i$. The following step filters the solution space for solutions with $f > 0$. Hence, this will result in the set of $\lfloor b/2^n \rfloor$ where $[b_{n-1} \dots b_0]_2$ satisfies clause C_i.

For each SAT variable v_j, where $1 \leqslant j \leqslant n$, we use the following types of constraints to transform feasible values for $y_{i,0}$ from $\mathcal{B}_i$ to corresponding pairs $(f_{i,n}, y_{i,n})$ in the set $\hat{\mathcal{C}}_i$ as described above:

$$\begin{aligned} f_{i,j} - f_{i,j+1} + 4y_{i,j} - 8y_{i,j+1} - 4\alpha_{i,j} &= 0 \quad \text{if } v_{j+1} \text{ is not in clause } C_i \\ f_{i,j} - f_{i,j+1} + 4y_{i,j} - 8y_{i,j+1} - 3\alpha_{i,j} &= 0 \quad \text{if } v_{j+1} \text{ is in clause } C_i \\ f_{i,j} - f_{i,j+1} + 4y_{i,j} - 8y_{i,j+1} - 5\alpha_{i,j} + \alpha'_{i,j} &= 0 \quad \text{if } \neg v_{j+1} \text{ is in clause } C_i. \end{aligned}$$

We define variable bounds $\alpha_{i,j}, \alpha'_{i,j} \in \{0, 1\}$ and $f_{i,0} = \{0\}$. Further we define for all $j \geqslant 1$ the bounds $f_{i,j} \in \{0, 1, 2, 3\}$ if all literals in clause C_i are positive literals and $f_{i,j} \in \{0, 1, 2\}$ if there is at least one negative literal.

Intuitively, the *first constraint* forwards the set of feasible values for $f_{i,j} = f_{i,j+1}$ and for every feasible value of $y_{i,j+1} = \lfloor y_{i,j}/2 \rfloor$. This is a combination of a division with remainder constraint type (for y variables) and a forward constraint type (for f variables), compare with Table 1. As the variable v_{j+1} is not in the clause, the remainder is irrelevant and gets absorbed by $\alpha_{i,j}$. The division by two, if independent from the context, could be achieved by

$$y_{i,j} - 2y_{i,j+1} - \alpha_{i,j} = 0 \quad \text{and } \alpha_{i,j} \in \{0,1\}.$$

However, we need to ensure that the feasible values for $f_{i,j}$ and $y_{i,j}$ do not interfere and create new feasible values. Hence, the multiplication of the above simple form of the division by $4 > f_{i,j+1} \in \{0,1,2,3\}$ makes sure that the feasible values transform independently as any value of $f_{i,j+1}$ can not compensate for an unwanted change in $y_{i,j+1}$. We want to make this argument precise. The constraint together with the lower and upper bounds implies

$$4y_{i,j} - 8y_{i,j+1} \in \{-3,-2,-1,0,1,2,3,4,5,6,7\} \quad \text{or}$$
$$4y_{i,j} - 8y_{i,j+1} \in \{-2,-1,0,1,2,3,4,5,6\}$$

depending on whether C_i has a negative literal. Dividing by 4 we get

$$y_{i,j} - 2y_{i,j+1} \in \{-3/4,-1/2,-1/4,0,1/4,1/2,3/4,1,5/4,3/2,7/4\} \quad \text{or}$$
$$y_{i,j} - 2y_{i,j+1} \in \{-1/2,-1/4,0,1/4,1/2,3/4,1,5/4,3/2\},$$

which implies $y_{i,j}-2y_{i,j+1} \in \{0,1\}$ in both cases since both variables are integral. The parity of the forwarded variable $y_{i,j}$ is the same as the parity of $y_{i,j}-2y_{i,j+1}$. If $y_{i,j}$ is even that means $y_{i,j} - 2y_{i,j+1} = 0$ which can only be achieved with $f_{i,j+1} = f_{i,j}$ and $\alpha_{i,j} = 0$ due to the variable bounds. If $y_{i,j}$ is odd that means $y_{i,j} - 2y_{i,j+1} = 1$ which can only be achieved with $f_{i,j+1} = f_{i,j}$ and $\alpha_{i,j} = 1$ again due to the variable bounds. In both cases it holds that $y_{i,j+1} = \lfloor y_{i,j}/2 \rfloor$ and $f_{i,j} = f_{i,j+1}$ as claimed above.

The intuitive goal for the second and third type of constraint is to increment the feasible values for $f_{i,j}$ in $f_{i,j+1}$ depending on the remainder from the division. As before, this is similar to a division with remainder constraint type and a forward constraint type. However, the variables should not evolve independently and instead the solution space for $f_{i,j+1}$ depends on the remainder of the division performed on $y_{i,j}$. In the second constraint it is incremented if the remainder is 1 and in the third constraint it *could or could not* be incremented if the remainder is 0.

For the *second constraint* the lower and upper bounds on the variables imply

$$4y_{i,j} - 8y_{i,j+1} \in \{-3,-2,-1,0,1,2,3,4,5,6\} \quad \text{or}$$
$$4y_{i,j} - 8y_{i,j+1} \in \{-2,-1,0,1,2,3,4,5\}$$

depending on whether C_i has a negative literal. Similar to above, dividing by 4 implies $y_{i,j} - 2y_{i,j+1} \in \{0,1\}$ in both cases since both variables are integral.

If $y_{i,j}$ is even, that means the literal is false. By the variable bounds this implies $y_{i,j} - 2y_{i,j+1} = 0$ with $f_{i,j+1} = f_{i,j}$ and $\alpha_{i,j} = 0$ or $f_{i,j} = 3$, $f_{i,j+1} = 0$, and $\alpha_{i,j} = 1$. Above we already showed that $f_{i,j+1} = f_{i,j}$ for the first constraint type and we will show as an invariant that the $\leqslant 3$ constraints for each literal each increase $f_{i,j+1}$ by at most one compared to $f_{i,j}$. Hence, it is not possible to have an assignment of $f_{i,j} = 3$ forwarded since $f_{i,0} = 0$. The only feasible assignment is $y_{i,j} - 2y_{i,j+1} = 0$ with $f_{i,j} = f_{i,j+1}$ and $\alpha_{i,j} = 0$ if $y_{i,j}$ is even.

If $y_{i,j}$ is odd and the literal is true that means $y_{i,j} - 2y_{i,j+1} = 1$, which can only be achieved with $f_{i,j+1} = f_{i,j} + 1$ and $\alpha_{i,j} = 1$ due to the variable bounds. In any case it holds that $y_{i,j+1} = \lfloor y_{i,j}/2 \rfloor$ and $f_{i,j+1} = f_{i,j} + \text{parity}(y_{i,j})$.

Lastly, the *third constraint* together with the variable bounds implies

$$4y_{i,j} - 8y_{i,j+1} \in \{-3, -2, -1, 0, 1, 2, 3, 4, 5, 6, 7\}.$$

Again, since $y_{i,j}$ and $y_{i,j+1}$ are integral, we get $y_{i,j} - 2y_{i,j+1} \in \{0, 1\}$, where it is 0 if $y_{i,j}$ is even and 1 otherwise.

If $y_{i,j}$ is even, which means the literal is true, we have $y_{i,j} - 2y_{i,j+1} = 0$. Within the variable bounds there are two assignments possible to satisfy the constraint. We can get $f_{i,j} - f_{i,j+1} - 5\alpha_{i,j} + \alpha'_{i,j} = 0$ by either $f_{i,j+1} = f_{i,j}$ and $\alpha_{i,j} = \alpha'_{i,j} = 0$ or $f_{i,j+1} = f_{i,j} + 1$, $\alpha_{i,j} = 0$, and $\alpha'_{i,j} = 1$ unless $f_{i,j} = 2$. Note that $f_{i,j} - f_{i,j+1} + \alpha'_{i,j} < 5$ and hence $\alpha_{i,j} = 0$. In any case a feasible assignment with $f_{i,j+1} = f_{i,j} + 1 \geqslant 0 + 1 > 0$ or $f_{i,j+1} = f_{i,j} = 2 > 0$ exists. This indicates that the literal is true and the clause satisfied by the assignment.

If $y_{i,j}$ is odd and thus $y_{i,j} - 2y_{i,j+1} = 1$, the literal is false and we need $f_{i,j} - f_{i,j+1} - 5\alpha_{i,j} + \alpha'_{i,j} = -4$ by the constraint. There are again two possibilities for this sum within the variable bounds, namely $f_{i,j+1} = f_{i,j}$ and $\alpha_{i,j} = \alpha'_{i,j} = 1$ or $f_{i,j+1} = f_{i,j} - 1$, $\alpha_{i,j} = 1$, and $\alpha'_{i,j} = 0$ unless $f_{i,j} = 0$. Hence, there is feasible assignment with $f_{i,j+1} > 0$ iff there is a feasible assignment with $f_{i,j} > 0$. In other words, a false literal does not mark the clause as satisfied. As before, we get $y_{i,j+1} = \lfloor y_{i,j}/2 \rfloor$.

Delete if Not Satisfying. Next, we delete any assignment that does not satisfy the clause C_i. We add feasible values $(f_{i,n}, y_{i,n})$ from $\hat{\mathcal{C}}_i$ to the next set $\mathcal{D}_i$ iff $f_{i,n} > 0$. We want to achieve a set of feasible solutions for $z_{i,0}$ of $\mathcal{D}_i = \{c \mid (f, c) \in \hat{\mathcal{C}}_i \text{ with } f > 0\}$. We can achieve this with the constraint

$$f_{i,n} + 4y_{i,n} - 4z_{i,0} - \beta_i = 1,$$

where $\beta_i \in \{0, 1, 2\}$. Since $\beta_i < 3$, this constraint is feasible iff $f_{i,n} > 0$ and then $y_{i,n} = z_{i,0}$. This is again a combination of constraint types with a checking type for $f_{i,n}$ and a forward type for $y_{i,n}$ to $z_{i,0}$. This transformation forwards a c iff it is feasible together with any $f > 0$, which as showed in *checking clause* is the case iff there was a true literal found and thus the clause is satisfied by the assignment. Note that the behavior is equivalent to our original (and more intuitive) goal of $\mathcal{C}_i$.

Restore Representation. For the current clause C_i, it remains to restore the original representation using the identity from Claim 1. Hence, we transform any feasible value for $z_{i,0}$ from $\mathcal{D}_i$ to a value for $x_{i+1,0}$ from

$$\mathcal{A}_{i+1} = \{\lfloor d \cdot 2^n / 3^n \rfloor + 1 \mid d \in \mathcal{D}_i\}.$$

The transformation uses similar multiplication and division constraint types as before, i.e.

$$\begin{aligned} 2z_{i,j} - z_{i,j+1} &= 0 \quad \text{for all } 0 \leqslant j < n \\ z_{i,n+j} - 3z_{i,n+j+1} - \gamma_{i,j} &= 0 \quad \text{for all } 0 \leqslant j < n \\ z_{i,2n} - x_{i+1,0} &= -1, \end{aligned}$$

with variable bounds $\gamma_{i,j} \in \{0, 1, 2\}$.

Adding the constraints for *preparations phase*, *checking clause*, *delete if not satisfying*, and *restore representation* for all clauses $C_1, \ldots, C_m$ results in a constraint matrix A with a path-like structure, right-hand side $\mathbf{b}$, and lower and upper bounds $\mathbf{l}$ and $\mathbf{u}$ such that the IP is feasible if and only if the 3-SAT formula is satisfiable. To be precise, *restore representation* is not required for the last clause C_m as it does not impact feasibility of the IP.

The set of feasible solutions for the IP at position $x_{0,0} = x_{1,0} = \ldots = x_{m,0}$ is exactly the set $\mathcal{A}_m$. Solutions of the 3-SAT formula are given by $3^n \cdot x_{m,0} \mod 2^n$ for any $x_{m,0} \in \mathcal{A}_m$. In other words, the set

$$\mathcal{S} = \{3^n \cdot a \mod 2^n \mid a \in \mathcal{A}_m\}$$

consists of the assignments $\nu\colon V \ni v_j \mapsto b_j \in \{0, 1\}$ that satisfy the 3-SAT formula φ with $[b_{n-1} \ldots b_0]_2 \in \mathcal{S}$ and any feasible solution to the path-like IP sets variable $x_{m,0}$ to an element of $\mathcal{A}_m$. Hence, this problem is hard. □

We use the construction from the above proof to additionally derive a fine-grained lower bound on the running time to solve path-like IPs assuming the Exponential Time Hypothesis (ETH).

Corollary 1. *Assuming the ETH, the problem of deciding the existence of a vector* $\mathbf{x} \in \mathbb{Z}^n$ *satisfying* $A\mathbf{x} = \mathbf{b}$, $\mathbf{l} \leqslant \mathbf{x} \leqslant \mathbf{u}$ *cannot be solved in time* $2^{\delta\sqrt{n}} \cdot \mathrm{poly}(|I|)$ *for some* $\delta > 0$, *where* $|I|$ *denotes the encoding length, even when* A *is a path-like matrix with* $\|A\|_\infty \leqslant 8$.

Proof. Consider a 3-SAT formula φ over $\tilde{n}$ variables. Using sparsification, assume that φ has $O(\tilde{n})$ clauses. The ETH states that there exists $\tilde{\delta} > 0$ such that 3-SAT can not be solved in time $2^{\tilde{\delta}\tilde{n}}$.

The transformation in the proof of Theorem 1 gives us an equivalent path-like IP with the largest entry in the coefficient matrix of $\|A\|_\infty \leqslant 8$ and $\|\mathbf{l}\|_\infty < \|\mathbf{u}\|_\infty < 6^n$. The complexity of the transformation is dominated by the variables $x_{i,j}$, $y_{i,j}$, $z_{i,j}$, $z_{i,\tilde{n}+j}$, $f_{i,j}$, $\alpha_{i,j}$, $\alpha'_{i,j}$, $\gamma_{i,j}$, where i iterates over the clauses and j over the variables (and possibly one more in both cases) of φ. Hence, the number of variables in the path-like IP is bounded by $n \in O(\tilde{n}^2)$. As every variable of the path-like IP occurs in at most two constraints, this bound carries over to the number of constraints.

Hence, path-like IPs can not be solved in time

$$2^{\delta\sqrt{n}} \cdot \text{poly}(|I|) = 2^{\tilde{\delta}\tilde{n}}$$

for some $\delta > 0$ even when the coefficient matrix has entries bounded by 8. □

3 Summary and Open Problems

We showed that the feasibility problem for integer programs with path-like constraint matrices is NP-hard already when all entries of the constraint matrix are bounded by 8. We prove this result by a reduction from 3-SAT. The reduction builds a path of constraints to repeatedly filter the set of all 3-SAT assignments for those satisfying a clause from the formula until only assignments satisfying all clauses are left. The coefficient 8 comes as a result of handling two types of information simultaneously in one constraint. This requires coefficients for one flow of information to be multiplied by the largest value for the other information in order prevent interaction between the variables of the two and their respective information held by the variables. Hence, the largest coefficient of our reduction is the largest value for one information flow 4 multiplied by the largest coefficient 2 in the other information flow. Thus, solving such IPs is harder than solving IPs where the coefficients are bounded by 1, which is known to be polynomial-time solvable (or more precisely, any IP with constraint matrix where the sum of absolute values per column is bounded by 2).

This leaves two questions: First, what is the complexity of integer programs with path-like constraint matrices with coefficients smaller than 8. Second, are integer programs with path-like constraint matrices but without upper bounds on the variables, that is, ILPs of the form $\{\mathbf{x} \mid A\mathbf{x} = \mathbf{b}, \mathbf{x} \geqslant 0\}$, NP-hard as well.

Acknowledgments. Alexandra Lassota was supported by the Dutch Research Council (NWO) under project number VI.Veni.242.293. The work of Michał Pilipczuk was supported by the project BOBR that is funded from the European Research Council (ERC) under the European Union's Horizon 2020 research and innovation programme with grant agreement No. 948057. The work of Kristýna Pekárková was supported by the project GA24-11098S of the Czech Science Foundation.

A Example for One Clause

The reduction is from 3-SAT. We want to exemplify the proof idea for checking satisfaction of one clause, which is the most involved part of the construction. Consider a clause $v_1 \vee \neg v_3$. We add the following constraints

$$\begin{aligned}
4y_0 - 3\alpha_0 - f_1 - 8y_1 &= 0\\
f_1 + 4y_1 - 4\alpha_1 - f_2 - 8y_2 &= 0\\
f_2 + 4y_2 - 5\alpha_2 + \alpha_2' - 8y_3 - f_3 &= 0\\
f_3 - \beta &= 1,
\end{aligned}$$

where variable lower and upper bounds are given by $f_1, f_2, f_3 \in \{0, 1, 2\}$ and $\alpha_0, \alpha_1, \alpha_2, \alpha_2', \beta \in \{0, 1\}$.

We want to illustrate the effects of the constraints for $y_0 \in \mathcal{B} = \{3, 4\} = \{[011]_2, [100]_2\}$. We interpret a number $[b_3 b_2 b_1]_2$ as the SAT variable assignment with $v_i \mapsto b_i$. Hence, $3 = [011]_2$ represents an assignment that satisfies $v_1 \vee \neg v_3$ and $[100]_2$ represents an assignment that does not satisfy $v_1 \vee \neg v_3$. The intended meaning is as follows. Roughly speaking, the first three constraints check whether the assignment of y_0 satisfies the clause by checking whether the respective bit yields a positive literal. For example a bit $b_1 = 1$ represents assigning $v_1 \mapsto 1$ and yields a positive literal v_1. We remember a positive literal by increasing f_i. The last constraint is feasible iff at least one true literal was found, i.e. iff the assignment satisfies $v_1 \vee \neg v_3$. We access the individual bits of y_0 using the remainder of a division by 2, compare with Table 1. In some sense all three constraints are (slight alterations) of combining a division with remainder constraint type for y with a forward constraint type for f.

With the first constraint we want to evaluate the least significant bit of the assignment for y_0. For clause $v_1 \vee \neg v_3$ this means we want to get $f_1 > 0$ iff $\text{parity}(y_0) = 1$ as this would indicate a true literal. The constraint can be restated as $4y_0 - 8y_1 = 3\alpha_0 + f_1$ and the variable bounds imply $0 \leqslant 3\alpha_0 + f_1 \leqslant 6$. If we fix y_0, there is only one feasible assignment to satisfy the constraint with $y_1 = \lfloor y_0/2 \rfloor$ and $\alpha_0 = f_1 = \text{parity}(y_0)$.

- For $y_0 = 3$ we get $y_1 = 1$ and $\alpha_0 = f_1 = \text{parity}(y_0) = 1$.
- For $y_0 = 4$ we get $y_1 = 2$ and $\alpha_0 = f_1 = \text{parity}(y_0) = 0$.

The clause $v_1 \vee \neg v_3$ does not rely on the assignment for v_2. Hence, the goal for the second constraint is to assign y_2 to y_1 divided by two, ignore the remainder, and forward the value of f_1 to f_2. This is exactly the combination of the respective constraint types as stated in Table 1. The constraint uses the forwarded variables f_1 and y_1. For a fixed y_1, a solution to $4y_1 - 8y_2 = f_2 - f_1 + 4\alpha_1$ within the variable bounds requires $y_2 = \lfloor y_1/2 \rfloor$, $f_2 = f_1$, and $\alpha_1 = \text{parity}(y_1)$.

- For $(f_1, y_1) = (1, 1)$ we get $y_1 = 0$, $f_2 = f_1 = 1$, and $\alpha_1 = \text{parity}(y_1) = 1$.
- For $(f_1, y_1) = (0, 2)$ we get $y_1 = 1$, $f_2 = f_1 = 0$, and $\alpha_1 = \text{parity}(y_1) = 0$.

The clause contains the negative literal $\neg v_3$. Hence, the third constraint aims to increase f_3 compared to f_2 iff the third bit of y_0 is 0 and hence the literal $\neg v_3$ is true. The constraint uses the forwarded variables f_2 and y_2. If $(f_2, y_2) = (1, 0)$ we get that the constraint is $\alpha_2' - 5\alpha_2 - 8y_3 - f_3 = -1$. This implies $\alpha_2 = y_3 = 0$ due to the variable bounds. The feasible solutions for $(f_2, y_2) = (1, 0)$ are thus $(f_3, y_3) \in \{(1, 0), (2, 0)\}$ since we get $f_3 = \alpha_2' + 1$ with $\alpha_2' \in \{0, 1\}$.

If $(f_2, y_2) = (0, 1)$ we get that the constraint is $\alpha_2' - 5\alpha_2 - 8y_3 - f_3 = -4$. This implies $\alpha_2 = 1$ and $y_3 = 0$ due to the variable bounds. The feasible solution for $(f_2, y_2) = (0, 1)$ is thus $(f_3, y_3) \in \{(0, 0)\}$ since we get $f_3 = \alpha_2' - 1$ with $\alpha_2' \in \{0, 1\}$ and $f_3 > 0$ by its lower bound.

Considering the first three constraints with the starting assignment of $y_0 = 3$ yielded the two feasible assignments for $f_3 \in \{1, 2\}$ and the starting assignment

$y_0 = 4$ yielded one feasible assignment for $f_3 \in \{0\}$. The fourth constraint is feasible iff $f_3 > 0$. This means, if we start with $y_0 = 3$ representing a satisfying assignment of $v_1 \vee \neg v_3$, the path-like IP is also feasible. If we start with $y_0 = 4$ representing an assignment that does not satisfy $v_1 \vee \neg v_3$, then also the path-like IP is not feasible.

References

1. Bartholdi, J., Tovey, C.A., Trick, M.A.: Voting schemes for which it can be difficult to tell who won the election. Soc. Choice Welfare **6**(2), 157–165 (1989)
2. Cook, S.A.: The complexity of theorem-proving procedures. In: ACM Symposium on Theory of Computing, pp. 151–158. ACM (1971)
3. Cslovjecsek, J., Eisenbrand, F., Pilipczuk, M., Venzin, M., Weismantel, R.: Efficient sequential and parallel algorithms for multistage stochastic integer programming using proximity. In: Proceedings of the 29th Annual European Symposium on Algorithms, ESA, vol. 204, pp. 33:1–33:14 (2021)
4. Eiben, E., Ganian, R., Knop, D., Ordyniak, S.: Unary integer linear programming with structural restrictions. In: 27th International Joint Conference on Artificial Intelligence, IJCAI 2018, pp. 1284–1290 (2018)
5. Eiben, E., Ganian, R., Knop, D., Ordyniak, S., Pilipczuk, M., Wrochna, M.: Integer programming and incidence treedepth. In: Lodi, A., Nagarajan, V. (eds.) IPCO 2019. LNCS, vol. 11480, pp. 194–204. Springer, Cham (2019). https://doi.org/10.1007/978-3-030-17953-3_15
6. Eisenbrand, F., Hunkenschröder, C., Klein, K., Koutecký, M., Levin, A., Onn, S.: An algorithmic theory of integer programming (2019)
7. Ermolieva, T., et al.: Connections between robust statistical estimation, robust decision-making with two-stage stochastic optimization, and robust machine learning problems. Cybern. Syst. Anal. **59**(3), 385–397 (2023)
8. Fellows, M.R., Lokshtanov, D., Misra, N., Rosamond, F.A., Saurabh, S.: Graph layout problems parameterized by vertex cover. In: Proceedings of the 19th International Symposium on Algorithms and Computation, ISAAC, pp. 294–305 (2008)
9. Fiala, J., Gavenčiak, T., Knop, D., Koutecký, M., Kratochvíl, J.: Parameterized complexity of distance labeling and uniform channel assignment problems. Disc. Appl. Math. **248**, 46–55 (2018)
10. Ganian, R., Ordyniak, S.: The complexity landscape of decompositional parameters for ILP. Artif. Intell. **257**, 61–71 (2018)
11. Ganian, R., Ordyniak, S., Ramanujan, M.S.: Going beyond primal treewidth for (M)ILP. In: 31st AAAI Conference on Artificial Intelligence, pp. 815–821. AAAI Press (2017)
12. Gavenčiak, T., Koutecký, M., Knop, D.: Integer programming in parameterized complexity: five miniatures. Disc. Optim. **44**(Part 1), 100596 (2022)
13. Goemans, M.X., Rothvoss, T.: Polynomiality for bin packing with a constant number of item types. J. ACM **67**(6), 38:1–38:21 (2020)
14. Gramm, J., Niedermeier, R., Rossmanith, P.: Fixed-parameter algorithms for CLOSEST STRING and related problems. Algorithmica **37**(1), 25–42 (2003)
15. Hemmecke, R., Onn, S., Romanchuk, L.: n-fold integer programming in cubic time. Math. Program. **137**(1–2), 325–341 (2013)
16. Hemmecke, R., Schultz, R.: Decomposition of test sets in stochastic integer programming. Math. Program. **94**(2–3), 323–341 (2003)

17. Jansen, B.M.P., Kratsch, S.: A structural approach to kernels for ILPs: treewidth and total unimodularity. In: Bansal, N., Finocchi, I. (eds.) ESA 2015. LNCS, vol. 9294, pp. 779–791. Springer, Heidelberg (2015). https://doi.org/10.1007/978-3-662-48350-3_65
18. Jansen, K., Klein, K., Maack, M., Rau, M.: Empowering the configuration-IP: new PTAS results for scheduling with setup times. Math. Program. **195**(1), 367–401 (2022)
19. Jr. Lenstra, H.W.: Integer programming with a fixed number of variables. Math. Oper. Res. **8**(4), 538–548 (1983)
20. Klein, K., Polak, A., Rohwedder, L.: On minimizing tardy processing time, max-min skewed convolution, and triangular structured ILPs. In: Proceedings of the 2023 ACM-SIAM Symposium on Discrete Algorithms, SODA 2023, Florence, Italy, 22–25 January 2023, pp. 2947–2960. SIAM (2023)
21. Klein, K., Reuter, J.: Collapsing the tower - on the complexity of multistage stochastic IPs. ACM Trans. Algor. **20**(3), 22 (2024)
22. Knop, D., Koutecký, M., Mnich, M.: Voting and bribing in single-exponential time. ACM Trans. Econ. Comput. **8**(3), 12:1–12:28 (2020)
23. Lassota, A., Ligthart, K.: Parameterized algorithms for matching integer programs with additional rows and columns. In: ICALP. LIPIcs, vol. 334, pp. 112:1–112:18. Schloss Dagstuhl - Leibniz-Zentrum für Informatik (2025)
24. Nešetřil, J., Ossona de Mendez, P.: Sparsity – Graphs, Structures, and Algorithms. AC, vol. 28. Springer, Heidelberg (2012). https://doi.org/10.1007/978-3-642-27875-4
25. Schrijver, A.: Combinatorial optimization: polyhedra and efficiency. Algor. Comb. **24** (2003)

Solving 4-Block Integer Linear Programs Faster Using Affine Decompositions of the Right-Hand Sides

Alexandra Lassota and Koen Ligthart(✉)

Eindhoven University of Technology, Eindhoven, Netherlands
a.a.lassota@tue.nl, k.m.ligthart@tue.nl

Abstract. We present a new and faster algorithm for the 4-block integer linear programming problem, overcoming the long-standing runtime barrier faced by previous algorithms that rely on Graver complexity or proximity bounds. The 4-block integer linear programming problem asks to compute $\min\{c_0^\top x_0 + c_1^\top x_1 + \cdots + c_n^\top x_n \mid Ax_0 + Bx_1 + \cdots + Bx_n = b_0,\ Cx_0 + Dx_i = b_i\ \forall i \in [n],\ (x_0, x_1, \ldots, x_n) \in \mathbb{Z}_{\geq 0}^{(1+n)k}\}$ for some $k \times k$ matrices A, B, C, D with coefficients bounded by $\overline{\Delta}$ in absolute value. Our algorithm runs in time $f(k, \overline{\Delta}) \cdot n^{k+\mathcal{O}(1)}$, improving upon the previous best running time of $f(k, \overline{\Delta}) \cdot n^{k^2+\mathcal{O}(1)}$ [Oertel, Paat, and Weismantel (Math. Prog. 2024), Chen, Koutecký, Xu, and Shi (ESA 2020)]. Further, we give the first algorithm that can handle large coefficients in A, B and C, that is, it has a running time that depends only polynomially on the encoding length of these coefficients. We obtain these results by extending the n-fold integer linear programming algorithm of Cslovjecsek, Koutecký, Lassota, Pilipczuk, and Polak (SODA 2024) to incorporate additional global variables x_0. The central technical result is showing that the exhaustive use of the vector rearrangement lemma of Cslovjecsek, Eisenbrand, Pilipczuk, Venzin, and Weismantel (ESA 2021) can be made *affine* by carefully guessing both the residue of the global variables modulo a large modulus and a face in a suitable hyperplane arrangement among a sufficiently small number of candidates. This facilitates a dynamic high-multiplicy encoding of a *faithfully decomposed* n-fold ILP with bounded right-hand sides, which we can solve efficiently for each such guess.

Keywords: Integer linear programming · 4-block ILPs · Fixed-parameter tractability · Polyhedral optimization

1 Introduction

We study the complexity of solving integer linear programs (ILPs) for a prominent class of ILPs that have a special block-structure in the constraint matrix, called *4-block* ILPs. The integer linear programming problem (in standard form) asks to compute an optimal solution x to $\min\{c^\top x \mid Qx = b, x \in \mathbb{Z}_{\geq 0}^n\}$, for

S. S. Dey et al. (Eds.): IPCO 2026, LNCS 16588, pp. 173–187, 2026.
https://doi.org/10.1007/978-3-032-28691-8_12

some constraint matrix $Q \in \mathbb{Z}^{m \times n}$, right-hand side vector $b \in \mathbb{Z}^m$, and objective vector $c \in \mathbb{Z}^n$. This problem is known to be NP-hard for general Q, which has catalyzed broad research to understand the (parameterized) complexity of solving ILPs with restricted classes of constraint matrices; see, e.g., [19] for a survey. Among those, we focus on *4-block* matrices, which are of the form

$$Q = \begin{pmatrix} A & B_1 & B_2 & \dots & B_n \\ C_1 & D_1 & & & \\ C_2 & & D_2 & & \\ \vdots & & & \ddots & \\ C_n & & & & D_n \end{pmatrix}$$

for blocks $A \in \mathbb{Z}^{m \times s}, B_i \in \mathbb{Z}^{m \times t}, C_i \in \mathbb{Z}^{d \times s}, D_i \in \mathbb{Z}^{d \times t}$ for $i \in [n] := \{1, 2, \dots, n\}$. Here, blocks of zeroes are omitted. The underlying 4-block ILP problem asks to solve an ILP which has a constraint matrix that is a 4-block matrix.

By splitting up x, c and b into $1 + n$ bricks matching the block-structure of a 4-block matrix Q, solving the 4-block ILP problem corresponds to finding an optimal solution to

$$\begin{aligned} \min\{c_0^\top x_0 + c_1^\top x_1 + \dots + c_n^\top x_n \mid\ & Ax_0 + B_1 x_1 + \dots + B_n x_n = b_0, \\ & C_i x_0 + D_i x_i = b_i \ \forall i \in [n], \\ & (x_0, x_1, \dots, x_n) \in \mathbb{Z}_{\geq 0}^{s+nt}\}. \end{aligned} \tag{4B}$$

We use $\overline{\Delta}$ to denote the greatest absolute value of the coefficients of Q and use $\Delta \leq \overline{\Delta}$ to denote the greatest absolute value of the coefficients in the diagonal blocks D_i. Furthermore, we let L denote the encoding length of the largest absolute value of a number in ILP (4B). We say that a 4-block problem is uniform if $B_1 = B_2 = \dots = B_n, C_1 = C_2 = \dots = C_n$ and $D_1 = D_2 = \dots = D_n$ and say that a 4-block problem is B-uniform if $B_1 = B_2 = \dots = B_n$. In the following, $\mathcal{O}_x(y)$ is used to denote $\mathcal{O}(f(x) \cdot y)$ for some (computable) function f.

It is known that 4-block ILPs can be solved in polynomial time when the dimensions and coefficients of all blocks are bounded by a constant [5,21,30], i.e., the problem is in XP when parameterized by $d, m, s, t, \overline{\Delta}$. Hemmecke, Köppe, and Weismantel [21] were the first to show this by providing an algorithm that solves a uniform 4-block ILP in slice-wise polynomial time $\mathcal{O}_{dmst\overline{\Delta}}(n^{s2^m + \mathcal{O}(1)} \cdot L^{\mathcal{O}(1)})$. Subsequent works have improved the dependence of the block dimensions on the exponent of n in the running time: Chen et al. [5] solve uniform 4-block ILPs in time $\mathcal{O}_{dmst\overline{\Delta}}(n^{sm + \mathcal{O}(1)})$ and Oertel, Paat, and Weismantel [30] solve (non-uniform) 4-block ILPs in time $\mathcal{O}_{dmst\overline{\Delta}}(n^{s \cdot \min\{s+2, m\} + \mathcal{O}(1)})$. Furthermore, attention has been given to the following special cases: Chen et al. [5] solve uniform 4-block ILPs in time $\mathcal{O}_{dmst\overline{\Delta}}(n^{s(t^2+1) + \mathcal{O}(1)})$ when A is the zero matrix and Chen, Chen, and Zhang [4] solve 4-block ILPs in time $\mathcal{O}_{dmst\overline{\Delta}}(n^{\mathcal{O}(1)})$ as long as all C_i are uniformly equal to some rank 1 matrix.

All previously mentioned algorithms share the common methodology of proving a bound u on either the proximity, which limits the distance between optimal

LP relaxation solutions and their nearest optimal integer solutions, or the Graver complexity, which represents the granularity of the lattice of integer kernel elements of the constraint matrix. These bounds are used to reduce the 4-block ILP problem to solving a polynomial number of 4-block ILPs in which the domain of x_0 is restricted to a subset of the box $[0, 2u]^s$. This makes it possible to eliminate the global variable vector x_0 by guessing its value, which is one of the $\mathcal{O}(u)^s$ integer vectors in the box.[1] Finally, the resulting 4-block ILP problems without global variables ($s = 0$), typically referred to as n-fold ILPs, can be solved efficiently in $\mathcal{O}_{dmst\overline{\Delta}}(n^{\mathcal{O}(1)})$ time [6]. Since Chen et al. [5] show that the Graver complexity (and, as an analogous construction can be used, also the proximity) of 4-block constraint matrices may be as large as $u = \Omega_{dmst\overline{\Delta}}(n^{\min\{d,m,s,t\}})$, such box guessing strategies appear to be limited to a quadratic dependence of $\Omega(s \cdot \min\{d, m, s, t\})$ of the block dimensions on the exponent of n in the running time. On the contrary, Eisenbrand and Rothvoss [14] conjecture that the 4-block ILP problem is fixed-parameter tractable (FPT).

Conjecture 1 (**[14]**). Solving 4-block ILP is FPT when parameterized by d, m, s, t, and $\overline{\Delta}$, i.e., it can be solved in time $\mathcal{O}_{dmst\overline{\Delta}}((nL)^{\mathcal{O}(1)})$.

The lower bound by Chen et al. [5] shows that in order to answer Conjecture 1 in the affirmative, different algorithmic techniques are required. In the context of Conjecture 1, Cslovjecsek et al. [8] show that uniform 4-block ILP is FPT w.r.t. $d, m, s, t, \overline{\Delta}$ if and only if uniform 4-block ILP is FPT w.r.t. d, m, s, t, Δ and $\overline{\Delta} \le n$: a proof of Conjecture 1 would imply that solving uniform 4-block ILPs with moderately large coefficients in the non-diagonal blocks is also FPT. For this reason, algorithms that are parameterized by Δ instead of $\overline{\Delta}$, whose running time grows only polynomially with L, are of interest.[2] Such results seem out of reach for a box-guessing approach as the Graver complexity and proximity of ILPs are typically at least as large as the coefficients of the constraint matrix.

Cslovjecsek et al. [8], followed by Eisenbrand and Rothvoss [14], provide FPT algorithms able to deal with large coefficients outside the diagonal blocks for two intensively studied special cases of 4-block ILPs: *n-fold* and *2-stage-stochastic* ILPs, which we discuss next. These have no global variables, i.e., $s = 0$, or no global constraints, i.e., $m = 0$, respectively. Yet, their expressiveness is sufficient to be applied in the context of graph problems [16,17], scheduling [23, 26], computational social choice [2,27], and machine learning [15]. See also [19].

Current state-of-the-art FPT algorithms for these special cases in terms of their polynomial dependence of n solve n-fold ILPs in time $\mathcal{O}_{dm\overline{\Delta}}((nt)^{1+o(1)})$ [6] and 2-stage stochastic ILPs in time $\mathcal{O}_{st\overline{\Delta}}((nd)^{1+o(1)})$ [7]. The dependence on $\overline{\Delta}$ in the running times of these algorithms is a result of the use of proximity bounds. Cslovjecsek et al. [8] avoid the use of such bounding arguments and solve

---

[1] We note that under the strong assumption that all C_i are equal to some rank 1 matrix C, the authors of [4] instead enumerate the options for Cx in a 1-dimensional box, similar to the situation where $s = 1$.

[2] Parameterization by the coefficient size Δ of the diagonal blocks D_i is known to be necessary [3,9].

B-uniform[3] n-fold ILP in time $\mathcal{O}_{dt\Delta}(m \cdot (nL)^{\mathcal{O}(1)})$ and 2-stage stochastic ILP feasibility in time $\mathcal{O}_{ds\Delta}(t \cdot (nL)^{\mathcal{O}(1)})$. Eisenbrand and Rothvoss [14] extend the former result by providing an algorithm that optimizes over a 2-stage stochastic ILP in time $\mathcal{O}_{dst\Delta}((nL)^{\mathcal{O}(1)})$ and, using similar arguments, show how to find an *almost solution* to a 4-block ILP, that violates the m global constraints by an additive error of at most $\mathcal{O}_{dt\overline{\Delta}}(m)$, in time $\mathcal{O}_{dst\Delta}(m \cdot (nL)^{\mathcal{O}(1)})$.

1.1 Contributions

We extend the B-uniform n-fold ILP algorithm from [8] to obtain an algorithm that solves B-uniform 4-block ILPs with an improved running time dependence of $s+\mathcal{O}(1)$ in the exponent of n, improving over previous quadratic dependencies. Furthermore, this algorithm is the first whose running time is polynomial in the encoding lengths of the coefficients of A, B and C_i. With this, we overcome both limitations inherent to the enumeration of vectors in an s-dimensional box with side length given by the Graver complexity or proximity.

Theorem 1. *A B-uniform 4-block ILP can be solved in time $\mathcal{O}_{dst\Delta}(n^{s+\mathcal{O}(1)} \cdot m \cdot L^{\mathcal{O}(1)})$.*

The constant hidden in the $\mathcal{O}_{dst\Delta}$ is triply exponential in d, which matches that of the n-fold algorithm by Cslovjecsek et al. [8].

The algorithm underlying Theorem 1 relies on constructing *faithful decompositions* of right-hand side vectors b_i that play a central role in the n-fold ILP algorithm of Cslovjecsek et al. [8]. In this setting, a faithful decomposition may be interpreted as a decomposition of the local bricks of an n-fold ILP into more bricks that have only a bounded number of distinct and small right-hand side vectors. We extend their approach by computing faithful decompositions in a dynamic and affine manner for varying b_i. This is made precise in Sections 2 and 4. To provide such decompositions, we exploit the constructive nature of the proof of Cslovjecsek et al. [7] of their variant of a vector rearrangement lemma by Klein [25]. The affine faithful decompositions are used to encode a domain-restricted 4-block ILP for a limited number of guesses of the domain of the global variables. The reduced dependence on the exponent of n in Theorem 1 relies on the fact that the number of faces in an arrangement of N hyperplanes in $\mathbb{R}^s$ is bounded by $\mathcal{O}(N)^s$ [10,20] and can be computed in $\mathcal{O}(N)^s$ time [10–12].

We also obtain an improved algorithm for (non-B-uniform) 4-block ILPs with bounded coefficients through a reduction to the uniform case by Dvořák et al. [9]. By additionally applying Frank and Tardos' [18] coefficient reduction technique, we obtain the slice-wise strongly polynomial-time algorithm in Theorem 2.

Theorem 2. *A 4-block ILP can be solved in time $\mathcal{O}_{dms\overline{\Delta}}(n^{s+\mathcal{O}(1)} \cdot t)$.*

[3] Note that due to the NP-hardness of the subset sum problem, the B-uniformity in the setting of large coefficients is necessary.

We present the preliminaries in Section 2. In Section 3, we discuss how affine faithful decompositions can be used to obtain an algorithm to solve 4-block ILPs. The lemma that constructs the used decompositions is proven in Section 4. All full proofs are deferred to the full version [29]; we give proof sketches instead.

2 Preliminaries

When A is a matrix or scalar and $S \subseteq \mathbb{R}^d$ is a set of vectors, we use AS to denote the image $\{A \cdot x | x \in S\}$ of S under A. Similarly, we use $S + v$ to denote the translated set $\{x + v | x \in S\}$ for an offset vector $v \in \mathbb{R}^d$. For notational convenience, we identify a set of vectors $S \subseteq \mathbb{R}^d$ with the $d \times |S|$ matrix of which the columns are the vectors in S. The *cone* generated by a set of vectors $S = \{v_1, \dots, v_n\}$ is the set of nonnegative combinations of vectors in S and the *integer cone* generated by these vectors is the set of nonnegative integral combinations. That is, $\operatorname{cone} S = \{\lambda_1 \cdot v_1 + \dots + \lambda_n \cdot v_n \mid \lambda \in \mathbb{R}^n_{\geq 0}\} = S\mathbb{R}^S_{\geq 0}$ and $\operatorname{int.cone} S = S\mathbb{Z}^S_{\geq 0}$. Here, we use K^I for sets K and I to denote the set of vectors $x \in K^I$ that are indexed by $i \in I$ and have coordinates $x_i \in K$. As shorthand, we use $\mathcal{B}^d(\Delta) = \{x \in \mathbb{Z}^d : \|x\|_\infty \leq \Delta\}$ to denote the discrete radius Δ hypercube centered at the origin. Further, $\mathcal{B}^{d\times d}_*(\Delta) = \{F \in \{-\Delta, -\Delta+1, \dots, \Delta\}^{d\times d} : \det F \neq 0\}$ denotes the set of all invertible $d \times d$ matrices F with integral entries of absolute value bounded by Δ, or equivalently, the set of all bases $F \subseteq \mathcal{B}^d(\Delta)$ of $\mathbb{R}^d$. We say that a vector $x \in \mathbb{R}^d$ is *conformal* to a vector $y \in \mathbb{R}^d$ if $x_i y_i \geq 0$ and $|x_i| \leq |y_i|$ for all $i \in [d]$. We denote this by $x \sqsubseteq y$.

We use a $(d+1)$-dimensional point $(a_h, \beta_h) \in \mathbb{R}^d \times \mathbb{R}$ to describe a *hyperplane* h that represents the set $\{x \in \mathbb{R}^d : a_h^\top x = \beta_h\}$. Note that we allow $a_h = 0$, i.e., degenerate hyperplanes that are either the whole space or the empty set. A (partially open) *polyhedron* P is defined by its relation to a finite number of hyperplanes $H_\leq, H_<$ and can be written as $P = \{x \in \mathbb{R}^d : a_h^\top x \leq \beta_h \ \forall h \in H_\leq, a_h^\top x < \beta_h \ \forall h \in H_<\}$. Similarly, a collection of hyperplanes $H \subseteq \mathbb{R}^{d+1}$ induces an *arrangement* of *faces* $\mathcal{F}_H$ each defined by a *position vector* $\bowtie$ that specifies the position of the face w.r.t. each hyperplane. Here, the (potentially empty) face $F_\bowtie$ with position vector $\bowtie \in \{<, =, >\}^H$ is the partially open polyhedron given by $F_\bowtie = \{x \in \mathbb{R}^d : a_h^\top x \bowtie_h \beta_h \ \forall h \in H\}$, and the arrangement induced by H consists of all faces $\mathcal{F}_H = \{F_\bowtie \mid \bowtie \in \{<, =, >\}^H\}$. It is known that the number of distinct faces in a hyperplane arrangement in d dimensions is bounded by $\mathcal{O}(|H|)^d$ [10,20]. Furthermore, Edelsbrunner, O'Rourke, and Seidel [11] have shown that arrangements can also be computed within $\mathcal{O}(|H|)^d$ time. See also [10,12].

A central concept in the n-fold algorithm by Cslovjecsek et al. [8] and the algorithm underlying Theorem 1 is that of *faithful decompositions* of right-hand side vectors b into smaller vectors, which captures the fact that solutions to ILPs in standard form may be decomposed as sums of solutions to ILPs with bounded right-hand sides.

Definition 1. *A faithful decomposition of $b \in \mathbb{Z}^d$ w.r.t. $D \in \mathbb{Z}^{d\times t}$ of order Ξ is a multiset of vectors $b_1, \dots, b_\ell \in \mathbb{Z}^d$ such that (i) $b = b_1 + \dots + b_\ell$; (ii) $b_i \sqsubseteq b$*

for $i \in [\ell]$; *(iii)* $\|b_i\|_\infty \leq \Xi$ *for* $i \in [\ell]$; *and (iv) any solution* $x \in \mathbb{Z}^t_{\geq 0}$ *to* $Dx = b$, *can be decomposed as* $x = x_1 + \cdots + x_\ell$ *where each* $x_i \in \mathbb{Z}^t_{\geq 0}$ *is a solution to* $Dx_i = b_i$ *for* $i \in [\ell]$;

A multiset consisting of elements in a set S may be interpreted as an integer vector $f \in \mathbb{Z}^S_{\geq 0}$ where f_s denotes the multiplicity with which $s \in S$ occurs in the multiset. For this reason, we identify multisets with integer vectors. Let $X^D_{b'} = \{x \in \mathbb{Z}^t_{\geq 0} : Dx = b'\}$ be the set of solutions for a given right-hand side b'. Translating Definition 1 reveals that a faithful decomposition of b w.r.t. D of order Ξ is an integer vector $f \in \mathbb{Z}^R$ for some $R \subseteq \mathbb{Z}^d$ such that (i) $b = \sum_{b' \in R} f_{b'} \cdot b'$; (ii) $b' \sqsubseteq b$ for $b' \in R$; (iii) $R \subseteq \mathcal{B}^d(\Xi)$; and (iv) for all $x \in X^D_b$, there exists a multiset $p_{b'} \in \mathbb{Z}^{X^D_{b'}}_{\geq 0}$ for each $b' \in R$ such that $x = \sum_{b' \in R} \sum_{x' \in X^D_{b'}} (p_{b'})_{x'} \cdot x'$ and $\sum_{x' \in X^D_{b'}} (p_{b'})_{x'} = f_{b'}$ for $b' \in R$. This high-multiplicity encoding is necessary as low-order faithful decompositions need to have a multiset cardinality that is exponential in the encoding length of b.

To prove Theorem 1, we will need Lemma 1 from [8], which shows that an arbitrarily large solution to an ILP with bounded coefficients can be decomposed into a small base solution plus small elements from the *Graver basis* of the constraint matrix. The Graver basis $\mathcal{G}(D) \subseteq \mathbb{Z}^t \setminus \{0\}$ of an integer matrix $D \in \mathbb{Z}^{d \times t}$ is the set of conformally minimal nonzero integral kernel elements of D. It is known that any Graver basis element $g \in \mathcal{G}(D)$ is bounded by $\|g\|_\infty \leq \|g\|_1 \leq \mathcal{O}(d\Delta)^d$ when $D \in \{-\Delta, -\Delta+1, \ldots, \Delta\}^{d \times t}$ [13].

Lemma 1 (Lemma 5.3 in [8],[31]**).** *Let* $D \in \{-\Delta, -\Delta+1, \ldots, \Delta\}^{d \times t}, b \in \mathcal{B}^d(\Xi)$ *and* $x \in \mathbb{Z}^t_{\geq 0}$ *be so that* $Dx = b$. *Then there is a solution* $\underline{x} \in \mathbb{Z}^t_{\geq 0}$ *to* $D\underline{x} = b$ *bounded by* $\|\underline{x}\|_\infty \leq \mathcal{O}(d(\Delta + \Xi))^d$ *and a multiset* $\mu \in \mathbb{Z}^{\mathcal{G}(D) \cap \mathbb{Z}^t_{\geq 0}}_{\geq 0}$ *of nonnegative Graver basis elements such that* $x = \underline{x} + \sum_{g \in \mathcal{G}(D) \cap \mathbb{Z}^t_{\geq 0}} \mu_g \cdot g$.

3 Algorithm

This section presents the $\mathcal{O}_{dst\Delta}(n^{s+\mathcal{O}(1)} \cdot m \cdot L^{\mathcal{O}(1)})$ time algorithm for 4-block ILPs. Before describing the algorithm in detail, we sketch how Cslovjecsek et al. [8] employ faithful decompositions to solve (the feasibility variant of) the B-uniform n-fold ILP problem and argue how this can be extended to 4-block ILPs. In essence, Cslovjecsek et al. [8] preprocess a given n-fold ILP by splitting each brick $D_i x_i = b_i$ using a faithful-decomposition $b_i = b_i^{(1)} + \cdots + b_i^{(\ell)}$ into the bricks $D_i x_i^{(1)} = b_i^{(1)}, \ldots, D_i x_i^{(\ell)} = b_i^{(\ell)}$. This results in an exponentially sized n-fold ILP which consists of only $\mathcal{O}_{dt\Delta}(1)$ different bricks with bounded right-hand sides. To exploit this latter fact, they use Lemma 1 to derive a high-multiplicity encoding of the ILP in the form of a *configuration* ILP that has only $\mathcal{O}_{dt\Delta}(1)$ integral variables and contains the high-multiplicity representations of the faithful decompositions as right-hand sides.

Now, observe that a 4-block ILP may be viewed as an "n-fold ILP" where the right-hand sides b'_i depend affinely on the global variables x_0 through $b'_i =$

$L_i(x_0) := b_i - C_i x_0$. We combine this observation with our central Lemma 2, which shows that if b_i' belongs to a fixed lattice translate $r + M\mathbb{Z}^d$ and fixed partially open polyhedron F among $\mathcal{O}_{d\Delta}(1)$ many options, there exist faithful decompositions of b_i' that depend affinely on b_i'. A sketch of the proof of Lemma 2 is postponed to Sect. 4.

Lemma 2. *Let d and $\Delta \in \mathbb{Z}_{\geq 0}$ be given. Then, there exists a modulus $M = 2^{\mathcal{O}(d\Delta)^{d^3}}$, a number $\Xi = 2^{\mathcal{O}(d\Delta)^d}$, and an algorithm that outputs: a collection H of at most $\mathcal{O}(d\Delta)^{d^4}$ hyperplanes; and an affine map $\mathcal{D}_{r,F} : \mathbb{R}^d \to \mathbb{R}^{R_{r,F}}$ with $R_{r,F} \subseteq \mathcal{B}^d(\Xi)$ and $|R_{r,F}| \leq d+1$ for every remainder vector $r \in \{0, 1, \ldots, M-1\}^d$ and face $F \in \mathcal{F}_H$ in the arrangement induced by H.*

For all $b \in (r+M\mathbb{Z}^d) \cap F$ and $D \in \{-\Delta, -\Delta+1, \ldots, \Delta\}^{d \times t}$, the image $\mathcal{D}_{r,F}(b)$ is a faithful decomposition of b w.r.t. D of order Ξ. The algorithm outputs the hyperplanes in $\mathcal{O}(d\Delta)^{d^4}$ time and constructs an affine map $\mathcal{D}_{r,F}$ for a given r and F in $\mathcal{O}(\Delta)^{d^4}$ time.

In our setting, b_i' depends affinely on x_0, meaning that we can construct high-multiplicity encoded faithful decompositions of b_i' that depend affinely on x_0. As this encoding appears in the right-hand side of the configuration ILP of Cslovjecsek et al. [8], we can replace the static faithful decompositions obtained with the $\mathcal{O}_{dt\Delta}(\log^{\mathcal{O}(1)} \|b\|_\infty)$ time algorithm by Cslovjecsek et al. [8] with affine functions of x_0 and obtain a configuration integer program, that encodes the 4-block ILP, with linear constraints. This ILP, which has $\mathcal{O}_{dt\Delta}(1) + s$ integral variables, can then be solved using Kannan's algorithm [24].

The configuration ILP can be constructed as long as the right-hand sides b_i' belong to a domain of the form $(r + M\mathbb{Z}^d) \cap F$ that is known upfront for each brick $i \in [n]$, c.f. Lemma 2. To achieve this, we simultaneously guess the domains for each brick. A key observation is that the structure of the domains enables us to guess this with significantly less than the trivial bound of $\mathcal{O}_{d\Delta}(1)^n$ guesses by "lifting" these domains to the s-dimensional space of the global variables x_0. The lattice structure $r + M\mathbb{Z}^d$ is especially convenient, as it can be guessed with a parameterized number of guesses, which has recently been exploited in 2-stage stochastic integer linear programming [8,14]. We make the guessing strategy explicit in a sketch of the proof of Theorem 1. See the full version [29] for the complete proof and running time analysis.

Sketch of the proof of Theorem 1. Let $M = 2^{\mathcal{O}(d\Delta)^{d^3}}$ and $\Xi = 2^{\mathcal{O}(d\Delta)^d}$ be the numbers defined in Lemma 2. Let $\eta = \mathcal{O}(d(\Delta + \Xi))^d = 2^{\mathcal{O}(d\Delta)^d}$ be the norm bound on $\underline{x}$ defined in Lemma 1 and let $\delta = \mathcal{O}(d\Delta)^d$ be the ∞-norm bound on Graver basis elements of any matrix $D \in \{-\Delta, -\Delta+1, \ldots, \Delta\}^{d \times t}$ from [13].

First, we apply Lemma 2 to generate a collection of hyperplanes H_i in $\mathbb{R}^d$ for each brick $i \in [n]$. These define the polyhedral regions in which the affine faithful decompositions of b_i are valid. Lift each hyperplane to the space $\mathbb{R}^s$ through $L_i^{-1}(h) := (-C_i^\top a_h, \beta_h - a_h^\top b_i)$. We consider the arrangement induced by $H = \bigcup_{i \in [n], h \in H_i} L_i^{-1}(h)$ and compute this arrangement $\mathcal{F}_H$ using the algorithm by Edelsbrunner et al. [11]. Observe that $|H| \leq n \cdot \mathcal{O}(d\Delta)^{d^4}$. Using that $|\mathcal{F}_H| \leq \mathcal{O}(|H|)^s$ [10,20], it follows that $|\mathcal{F}_H| \leq \mathcal{O}(d\Delta)^{d^4 s} \cdot n^s$.

Now, for each combination of a remainder vector $r \in \{0, 1, \ldots, M-1\}^s$ and nonempty face $F \in \mathcal{F}_H$ among the $M^s \cdot |\mathcal{F}_H| \leq 2^{\mathcal{O}(d\Delta)^{d^3} \cdot s} \cdot n^s$ options, we construct and solve a high-multiplicity encoding of the 4-block ILP restricted to $x_0 \in (r + M\mathbb{Z}^s) \cap F$ using Lemma 2. For every brick $i \in [n]$, we compute:

- the face $F_i \in \mathcal{F}_{H_i}$ that contains $L_i(F) = \{b_i - C_i x_0 \mid x_0 \in F\}$. This face has position $\bowtie_i \in \{<, =, >\}^{H_i}$ given by $(\bowtie_i)_h = \bowtie_{L_i^{-1}(h)}$ for $h \in H_i$. This follows from $a_h^\top(b_i - C_i x_0)\,(\bowtie_i)_h\,\beta_h \iff (-C_i^\top a_h)^\top x_0\,(\bowtie_i)_h\,\beta_h - a_h^\top b_i$.
- the remainder vector $r_i = (b_i - C_i r) \bmod M$ so that $L_i(x_0) \equiv r_i \pmod{M}$ when $x_0 \equiv r \pmod{M}$, performing modular arithmetic coordinate-wise.
- a dynamic faithful decomposition $\mathcal{D}_i$ using Lemma 2 for the remainder vector r_i and face F_i. That is, construct the affine map $\mathcal{D}_i \colon \mathbb{R}^d \to \mathbb{R}^{R_i}$ so that $\mathcal{D}_i(b)$ is a faithful decomposition of b using elements from $R_i \subseteq \mathcal{B}^d(\Xi)$ as long as $b \in (r_i + M\mathbb{Z}^d) \cap F_i$. By construction, this holds for $b = L_i(x_0) = b_i - C_i x_0$ when $x_0 \in (r + M\mathbb{Z}^s) \cap F$.
- the set of base solutions $\underline{X}_{b',i} = \{x \in \mathbb{Z}^t_{\geq 0} : \|x\|_\infty \leq \eta, D_i x = b'\}$ to $D_i x = b'$ for each small right-hand side vector $b' \in R_i$.
- the set of Graver basis elements $G_i := \mathcal{G}(D_i) \cap \mathbb{Z}^t_{\geq 0}$ of D_i.

We can now solve the 4-block ILP restricted to $x_0 \in (r + M\mathbb{Z}^s) \cap F$ by solving ILP (C) . Here, solutions to $Dx = b'$ are decomposed using Lemma 1.

$$
\begin{aligned}
\text{minimize} \quad & c_0^\top x_0 + \sum_{i \in [n]} c_i^\top \Big(\sum_{b' \in R_i} \sum_{x' \in \underline{X}_{b',i}} x' \cdot p^i_{b',x'} + \sum_{g \in G_i} g \cdot p^i_g \Big), & \text{(C)} \\
\text{subject to} \quad & A x_0 + B \Big(\sum_{x' \in \mathcal{B}^t(\eta) \cap \mathbb{Z}^t_{\geq 0}} x' \cdot q_{x'} + \sum_{g \in \mathcal{B}^t(\delta) \cap \mathbb{Z}^t_{\geq 0}} g \cdot q_g \Big) = b_0, & \\
& q_{x'} = \sum_{i \in [n]} \sum_{b' \in R_i : x' \in \underline{X}_{b',i}} p^i_{b',x'}, & x' \in \mathcal{B}^t(\eta) \cap \mathbb{Z}^t_{\geq 0}, \\
& q_g = \sum_{i \in [n] : g \in G_i} p^i_g, & g \in \mathcal{B}^t(\delta) \cap \mathbb{Z}^t_{\geq 0}, \\
& \sum_{x' \in \underline{X}_{b',i}} p^i_{b',x'} = (\mathcal{D}_i(b_i - C_i x_0))_{b'}, & i \in [n],\ b' \in R_i, \\
& q_{x'} \in \mathbb{Z}_{\geq 0}, & x' \in \mathcal{B}^t(\eta) \cap \mathbb{Z}^t_{\geq 0}, \\
& q_g \in \mathbb{Z}_{\geq 0}, & g \in \mathcal{B}^t(\delta) \cap \mathbb{Z}^t_{\geq 0}, \\
& p^i_{b',x'} \in \mathbb{Z}_{\geq 0}, & i \in [n],\ b' \in R_i,\ x' \in \underline{X}_{b',i}, \\
& p^i_g \in \mathbb{Z}_{\geq 0}, & i \in [n],\ g \in G_i, \\
& x_0 \in (r + M\mathbb{Z}^s) \cap F,\ x_0 \geq 0. &
\end{aligned}
$$

As observed in [8], the constraint matrix of ILP (C) contains a large totally unimodular part, showing that it can be solved efficiently as mixed integer linear program, using, for example, Kannan's algorithm [24]. A careful analysis of the dimensions and coefficients of ILP (C) shows that it can be solved in time

$$
s^{\mathcal{O}(d\Delta)^d \cdot st + 2^{\mathcal{O}(d\Delta)^d \cdot t}} \cdot m \cdot (nL)^{\mathcal{O}(1)}.
$$

By taking the optimal solution among all guesses of r and F, we can report the optimal solution to the original 4-block ILP in time

$$2^{\mathcal{O}(d\Delta)^{d^3}\cdot s}\cdot s^{\mathcal{O}(d\Delta)^d\cdot st+2^{\mathcal{O}(d\Delta)^d\cdot t}}\cdot n^s\cdot m\cdot (nL)^{\mathcal{O}(1)}.$$ ◀

When all constraint matrix coefficients are small and we parameterize by $\overline{\Delta}$, we use an adaptation of the arguments by Dvořák et al. [9] in Lemma 3, proven in the full version [29], to reduce non-uniform 4-block ILP to the B-uniform case.

Lemma 3. *4-block ILP is reducible to B-uniform 4-block ILP by changing t to $(2\overline{\Delta}+1)^{m+d}$ and additively increasing d by 1.*

Therefore, in order to show Theorem 2 it suffices to solve ILP (C) efficiently. An applicable strategy to eliminate the dependencies on L is described by Koutecký, Levin and Onn [28]: i) solve the LP relaxation with Tardos' algorithm [32], ii) apply standard proximity techniques to bound the variable domains and right-hand sides iii) use Frank and Tardos' [18] objective coefficient reduction technique. See the full version [29] for the details.

4 Dynamic Faithful Decomposition

This section provides a sketch of the proof of Lemma 2. The essence of the proof is showing that by restricting b to a suitable face F of a hyperplane arrangement and restricting it to a lattice translate $r+M\mathbb{Z}^d$, all necessary ingredients to inductively apply the proof of Theorem 4.1 of Cslovjecsek et al. [7] are known up front. For a vector $b\in(r+M\mathbb{Z}^d)\cap F$, the proof proceeds as follows:

1. We identify the collection $\mathcal{U}$ of bases $U\in\mathcal{B}_*^{d\times d}(\Delta)$ that conically generate b. This is determined by the face F.
2. We identify a linearly independent set of vectors in $\bigcap_{U\in\mathcal{U}}\operatorname{int.cone}U$ that conically generate b. This is similarly determined by F.
3. If a conic multiplier in the corresponding conic combination is large, then b faithfully decomposes into the corresponding generator w_i and $b-w_i$. We exhaustively and inductively apply this. Here, the induction base case is represented by a unique root q. The root is determined by the guessed lattice translate $r+M\mathbb{Z}^d$ and the subset of conic multipliers that are sufficiently large, which is identified through F.

We note that performing work for each lattice translate and face of an arrangement has also recently been employed in [1] to verify forall-exists statements over the integers.

As the induction step of Claim 2.6 in the proof of Lemma 2 follows the proof of Cslovjecsek et al. [7], we will also need their auxiliary lemma, Lemma 4, that shows how to find the appropriate conic generators in the second step.

Lemma 4 (Lemma 4.2 in [7]). *Let $d, \Delta \in \mathbb{Z}_{\geq 0}$ be given. Then, there exists numbers $\Psi = 2^{\mathcal{O}(d\Delta)^d}$ and $\Phi = \mathcal{O}(d\Delta)^{d^2}$ such that the following holds: for any subset of generating bases $\mathcal{U} \subseteq \mathcal{B}_*^{d\times d}(\Delta)$ for which $\bigcap_{U\in\mathcal{U}} \operatorname{cone} U \neq \emptyset$, there exist a generating set $P \subseteq \mathcal{B}^d(\Phi)$ such that $\operatorname{cone} P = \bigcap_{U\in\mathcal{U}} \operatorname{cone} U$. Furthermore, for any $v \in \mathbb{Z}^d \cap \bigcap_{U\in\mathcal{U}} \operatorname{cone} U$ it holds that $\Psi \cdot v \in \bigcap_{U\in\mathcal{U}} \operatorname{int.cone} U$.*

We can now provide a sketch of the proof of Lemma 2. The decomposition mechanism is visualized in Figure 1. See the full version [29] for a full proof and constructive algorithmic considerations.

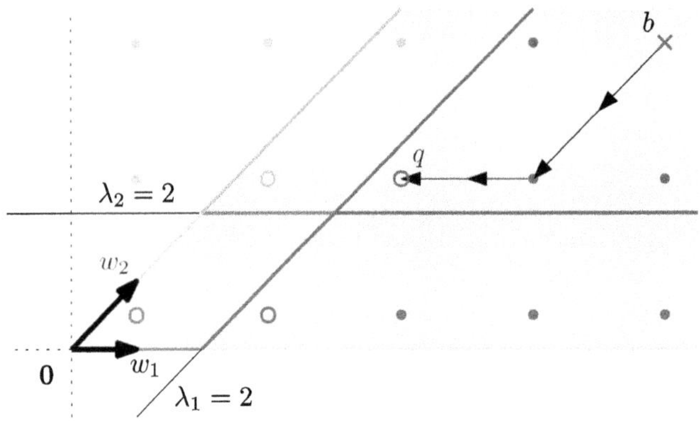

Fig. 1. Visual representation of a subregion of $\mathbb{R}^2$ in the proof of Lemma 2 for $d = 2$, $t = 2$, $\Delta = 1$ and sufficiently large multiplier $\Psi = 2$. Shown is the lattice translate $(2,1) + 4\mathbb{Z}^2$ containing the point $b = (18,9)$, which is decomposed into $q = (10,5)$, 2 times $w_1 = (2,0)$ plus 2 times $w_2 = (2,2)$.

Sketch of the proof of Lemma 2. Let $\Phi = \mathcal{O}(d\Delta)^{d^2}$ be the norm upper bound on the generating elements P defined in Lemma 4 and let $\Psi = 2^{\mathcal{O}(d\Delta)^d}$ be the large multiplier defined in Lemma 4. If necessary, we enlarge Φ such that $1 \leq \Delta \leq \Phi$.

Due to Claim 2.1, we may assume that D is full row rank and $t = \mathcal{O}(\Delta)^d$.

Claim 2.1 A faithful decomposition of b w.r.t. the matrix with column set $\mathcal{B}^d(\Delta)$ is a faithful decomposition with respect to any $D \in \{-\Delta, -\Delta+1, \ldots, \Delta\}^{d\times t}$.

For every basis $\overline{V} \in \mathcal{B}_*^{d\times d}(\Phi)$ and each $i \in [d]$ we add the two hyperplanes

$$h^0_{\overline{V}i} = ((\overline{V}^{-1}b)_i, 0) \text{ and } h^+_{\overline{V}i} = ((\overline{V}^{-1}b)_i, t\cdot\Psi)$$

to H. This satisfies $|H| = \mathcal{O}(d\Delta)^{d^4}$. Fix an arbitrary face $F \in \mathcal{F}_H$ with position vector $\bowtie \in \{<, =, >\}^H$. This information is sufficient to show Claim 2.2 and 2.3.

Claim 2.2 We can identify a set of generating bases $\mathcal{U} \subseteq \mathcal{B}_*^{d\times d}(\Delta)$ so that $F \subseteq \bigcap_{U\in\mathcal{U}} \operatorname{cone} U$. Furthermore, for all $b \in F$ and $U \in \mathcal{B}_*^{d\times d}(\Delta)$, we have that $b \in \operatorname{cone} U$ implies that $U \in \mathcal{U}$.

Proof sketch Letting $\mathcal{U}$ be the set of all $U \in \mathcal{B}_*^{d\times d}(\Delta)$ so that $\bowtie_{h^0_{Ui}} \in \{>, =\}$ for all $i \in [d]$ suffices. ◁

Claim 2.3 We can find a set of linearly independent generating vectors $V = \{v_1, \ldots, v_\ell\} \subseteq \mathcal{B}^d(\Phi)$ such that $F \subseteq \operatorname{cone} V \subseteq \bigcap_{U\in\mathcal{U}} \operatorname{cone} U$. In particular, we may write each $b \in F$ as $b = \sum_{i\in[\ell]} \mu_i \cdot v_i$ for strictly positive conic multipliers $\mu_i > 0$ that depend linearly on b through $\mu_i = (\overline{V}^{-1}b)_i$, where $\overline{V} \in \mathcal{B}_*^{d\times d}(\Phi)$.

Proof sketch This follows from Lemma 4 and Caratheodory's theorem. ◁

Let $w_i = \Psi \cdot v_i$ for $i \in [\ell]$. We inspect the scaled conic multipliers $\lambda_i := \mu_i/\Psi$ in the conic combination $b = \sum_{i\in[\ell]} \lambda_i \cdot w_i$. The next claim shows that we can pick M so that the remainder of λ_i with respect to t is fixed.

Claim 2.4 The remainder of $\alpha_i := \lambda_i \bmod t$ of the conic multipliers λ_i for $i \in [\ell]$ is constant for all $b \in (r + M\mathbb{Z}^d) \cap F$ for a fixed $r \in \{0, 1, \ldots, M-1\}^d$.

Proof sketch The proof uses the Hadamard bound and similar techniques to those used to prove Lemma 4 in [7]. In particular, M is taken sufficiently large so that the coefficients of the matrix $M \cdot \overline{V}^{-1} \cdot \frac{1}{\Psi}$ are multiples of t. ◁

Now fix r. Since we also know the position of F relative to the hyperplanes $h^+_{\overline{V}i}$, i.e., it is known whether $\mu_i = (\overline{V}^{-1}b)_i \geq t \cdot \Psi \iff \lambda_i \geq t$, we can distinguish the following two cases for every $i \in [\ell]$, uniformly over all vectors b in the domain $(r + M\mathbb{Z}^d) \cap F$: either $\lambda_i = \alpha_i$; or $\lambda_i \geq t$ and, if we define $\gamma_i = t + \alpha_i$, the difference $z_i = \lambda_i - \gamma_i \in t\mathbb{Z}_{\geq 0}$ is a nonnegative integer. Let $S \subseteq [\ell]$ be the set of directions $i \in [\ell]$ for which $\lambda_i = \alpha_i$, i.e., F is not deep in $\operatorname{cone} V$ in the direction of w_i. Let $q = \sum_{i\in S} \alpha_i \cdot w_i + \sum_{i\in[\ell]\setminus S} \gamma_i \cdot w_i$, which, as argued, only depends on r and F, but not on b. We are now ready to define the affine dynamic faithful decomposition $\mathcal{D}_{r,F} : \mathbb{R}^d \to \mathbb{R}^{R_{r,F}}$ by

$$\mathcal{D}_{r,F}(b) = 1 \cdot e_q + \sum_{i\in[\ell]\setminus S} z_i \cdot e_{w_i},$$

where $e_{b'}$ is the b'-th standard basis vector, i.e., $e_{b'}(b') = 1$ and $e_{b'}(b'') = 0$ for $b'' \neq b'$. That is, we decompose b into z_i times w_i for $i \in [\ell] \setminus S$ and one times the root q. It can be verified that the vectors in the faithful decomposition $\mathcal{D}_{r,F}$ are within $R_{r,F} = \{q, w_1, \ldots, w_\ell\} \subseteq \mathcal{B}^d(\Xi)$ for some $\Xi = 2^{\mathcal{O}(d\Delta)^d}$. Recall that each z_i is an affine function of b.

Now, fix $b \in (r + M\mathbb{Z}^d) \cap F$. It is left to show that the decomposition matches the remaining properties in Definition 1. The sign-compatibility of the decomposition follows from the membership of an appropriate orthant-generating basis in $\mathcal{U}$. We prove the fact that solutions to $Dx = b$ can be decomposed as the sum of the decomposed systems with induction on decreasing $\|z\|_1 \in \mathbb{Z}_{\geq 0}$ in Claim 2.6. For this, we need to observe that when we decrease the conic multipliers λ, the set of generating bases $\mathcal{U}$ remains the same, which is captured in Claim 2.5.

Claim 2.5 Let $\nu_i \in \mathbb{R}^\ell_{>0}$ be strictly positive multipliers and $p = \sum_{i\in[\ell]} \nu_i \cdot w_i$. Let $U \in \mathcal{B}_*^{d\times d}(\Delta)$ and suppose that $p \in \operatorname{cone} U$. Then, it holds that $U \in \mathcal{U}$.

Proof sketch This follows from the symmetry of the set $\mathcal{B}_*^{d\times d}(\Delta)$. ◁

Claim 2.6 $\mathcal{D}_{r,F}(b)$ is a faithful decomposition of b w.r.t. D.

Proof sketch We prove this by inductively showing that $\mathcal{D}_{r,F}(p)$ is a faithful decomposition of $p = \sum_{i\in S}\alpha_i\cdot w_i + \sum_{i\in[\ell]\setminus S}(\gamma_i+s_i)\cdot w_i$ w.r.t. D for all $s \in \mathbb{Z}_{\geq 0}^{[\ell]\setminus S}$. When $s = 0$, we have that $\mathcal{D}_{r,F}(p) = e_q = e_p$ and the claim holds. We proceed with the induction step. Let $i^* \in [\ell]\setminus S$ be so that $s_{i^*} \geq 1$, then the conic multiplier $\nu_{i^*} = s_{i^*} + t + \alpha_i \geq t+1$ in the conic combination $p = \sum_{i\in[\ell]}\nu_i \cdot w_i$ is sufficiently large, which, following the proof of Theorem 4.1 of [7], shows that b faithfully decomposes into $b - w_{i^*}$ plus w_{i^*}. In short, a solution $x \in \mathbb{Z}_{\geq 0}^t$ to $Dx = p$ can be written as the combination of some vertex solutions to the rational system $Dx = p, x \geq 0$. Any such vertex solution with corresponding basis U must be at least ν_{i^*} times as large as the multipliers $\tilde{y}_{i^*}$ in the conic combination $w_{i^*} = U\tilde{y}_{i^*}$. Here, Claims 2.3 and 2.5 and Lemma 4 ensure that $\tilde{y}_{i^*}$ is nonnegative and integral. Finally, an application of Caratheodory's theorem and an averaging argument combined with $\nu_{i^*} \geq t+1$ show that $\tilde{y}_{i^*}$ is a submultiset of x, which suitably decomposes the solution. ◁◀

5 Future Directions

Despite the progress obtained in this paper, the question of whether 4-block ILP is fixed-parameter tractable remains unanswered. It appears as though a different strategy must be used to obtain an FPT algorithm, if it were to exist: consider a 4-block ILP where A is the zero matrix, $C_i = -e_{j(i)}^\top$ for some standard basis vector $e_{j(i)}$, $D_i = \begin{pmatrix}1 & -1\end{pmatrix}$ and $B_i = \begin{pmatrix}a_i & 0\end{pmatrix}$ for arbitrary vectors $a_i \in \{-1,0,1\}^m$. The feasibility of such 4-block ILP is equivalent to the feasibility of

$$\left\{y_1a_1 + \cdots + y_na_n = b_0, y_i \geq \max\{0, b_i + z_{j(i)}\}\ \forall i \in [n],\ z \in \mathbb{Z}_{\geq 0}^s, y \in \mathbb{Z}^n\right\}.$$

Here, our algorithm needs to guess whether $z_{j(i)} \leq b_i$ or $z_{j(i)} < b_i$ for all $i \in [n]$, resulting in the space of the global variables being partitioned into an s-dimensional axis-aligned grid that may have up to $\Omega_s(n^s)$ many cells. We note that if the local constraints are replaced with $y_i = \max\{0, b_i + z_{j(i)}\}$, finding a solution becomes $W[1]$-hard. Roughly sketching the idea, the s-clique problem can be encoded letting each global variable z_j pick a vertex $v \in \{1, 2, \ldots, |V|\}$ to be in the clique and adding many duplicate bricks to mimic an ILP with $m = \mathcal{O}(s^2)$ constraints and unary encoded coefficients on binary variables u_{jv} indicating whether $z_j = v$, which can encode the required clique constraints as shown in [9]. In standard form, the local constraints read $y_i - s_i = b_i + z_{j(i)}$ for slack variables $s_i \in \mathbb{Z}_{\geq 0}$, which shows that a slack of s corresponds to adding the Graver basis element $g = \begin{pmatrix}1 & 1\end{pmatrix}$ a total of $p_g^i = s$ times in the decomposed ILP (C) . Therefore, we believe that a natural next question in order to resolve Conjecture 1 is to study if and how the convexity of $y_i \geq \max\{0, b_i + z_{j(i)}\}$ or the freedom of adding Graver basis elements in ILP (C) can be exploited.

Finally, we note that separable convex objective functions can be minimized using the Graver complexity and proximity bounds from [3,22,30], as described in [22]. However, it is unclear whether the algorithm underlying Theorem 1 can similarly be generalized. Therefore, it remains an interesting open question whether one can optimize such nonlinear objective functions over 4-block integer programs in time $\mathcal{O}_{dmst\overline{\Delta}}(n^{s+\mathcal{O}(1)})$.

Disclosure of Interests. The authors have no competing interests to declare that are relevant to the content of this article.

References

1. Bach, E., Eisenbrand, F., Rothvoss, T., Weismantel, R.: Forall-exist statements in pseudopolynomial time. In: Proceedings of the 2025 ACM-SIAM Symposium on Discrete Algorithms, SODA. pp. 2225–2233 (2025). https://doi.org/10.1137/1.9781611978322.73
2. Bartholdi, J., Tovey, C.A., Trick, M.A.: Voting schemes for which it can be difficult to tell who won the election. Soc. Choice Welfare **6**(2), 157–165 (1989). https://doi.org/10.1007/BF00303169
3. Chen, H., Chen, L., Zhang, G.: Block-structured integer programming: Can we parameterize without the largest coefficient? Discret. Optim. **46**, 100743 (2022). https://doi.org/10.1016/J.DISOPT.2022.100743
4. Chen, H., Chen, L., Zhang, G.: FPT algorithms for a special block-structured integer program with applications in scheduling. Math. Program. **208**(1), 463–496 (2024). https://doi.org/10.1007/S10107-023-02046-Z
5. Chen, L., Koutecký, M., Xu, L., Shi, W.: New bounds on augmenting steps of block-structured integer programs. In: Proceedings of the 28th Annual European Symposium on Algorithms, ESA. pp. 33:1–33:19 (2020). https://doi.org/10.4230/LIPICS.ESA.2020.33
6. Cslovjecsek, J., Eisenbrand, F., Hunkenschröder, C., Rohwedder, L., Weismantel, R.: Block-structured integer and linear programming in strongly polynomial and near linear time. In: Proceedings of the 2021 ACM-SIAM Symposium on Discrete Algorithms, SODA. pp. 1666–1681 (2021). https://doi.org/10.1137/1.9781611976465.101
7. Cslovjecsek, J., Eisenbrand, F., Pilipczuk, M., Venzin, M., Weismantel, R.: Efficient sequential and parallel algorithms for multistage stochastic integer programming using proximity. In: Proceedings of the 29th Annual European Symposium on Algorithms, ESA. vol. 204, pp. 33:1–33:14 (2021). https://doi.org/10.4230/LIPICS.ESA.2021.33
8. Cslovjecsek, J., Koutecký, M., Lassota, A., Pilipczuk, M., Polak, A.: Parameterized algorithms for block-structured integer programs with large entries. TheoretiCS 4, (2025). https://doi.org/10.46298/THEORETICS.25.15. previously appeared in SODA 2024
9. Dvořák, P., Eiben, E., Ganian, R., Knop, D., Ordyniak, S.: The complexity landscape of decompositional parameters for ILP: programs with few global variables and constraints. Artif. Intell. **300**, 103561 (2021). https://doi.org/10.1016/j.artint.2021.103561

10. Edelsbrunner, H.: Algorithms in Combinatorial Geometry, EATCS Monographs on Theoretical Computer Science, vol. 10. Springer (1987). https://doi.org/10.1007/978-3-642-61568-9
11. Edelsbrunner, H., O'Rourke, J., Seidel, R.: Constructing arrangements of lines and hyperplanes with applications. SIAM J. Comput. **15**(2), 341–363 (1986). https://doi.org/10.1137/0215024
12. Edelsbrunner, H., Seidel, R., Sharir, M.: On the zone theorem for hyperplane arrangements. SIAM J. Comput. **22**(2), 418–429 (1993). https://doi.org/10.1137/0222031
13. Eisenbrand, F., Hunkenschröder, C., Klein, K.: Faster algorithms for integer programs with block structure. In: Proceedings of the 45th International Colloquium on Automata, Languages, and Programming, ICALP. pp. 49:1–49:13 (2018). https://doi.org/10.4230/LIPIcs.ICALP.2018.49
14. Eisenbrand, F., Rothvoss, T.: A parameterized linear formulation of the integer hull. In: Proceedings of the 2026 ACM-SIAM Symposium on Discrete Algorithms, SODA. p. 4560–4571 (2026). https://doi.org/10.1137/1.9781611978971.164
15. Ermolieva, T., Ermoliev, Y., Havlik, P., Lessa-Derci-Augustynczik, A., Komendantova, N., Kahil, T., Balkovic, J., Skalsky, R., Folberth, C., Knopov, P.S., Wang, G.: Connections between robust statistical estimation, robust decision-making with two-stage stochastic optimization, and robust machine learning problems. Cybern. Syst. Anal. **59**(3), 385–397 (2023). https://doi.org/10.1007/s10559-023-00573-3
16. Fellows, M.R., Lokshtanov, D., Misra, N., Rosamond, F.A., Saurabh, S.: Graph layout problems parameterized by vertex cover. In: Proceedings of the 19th International Symposium on Algorithms and Computation, ISAAC. pp. 294–305 (2008). https://doi.org/10.1007/978-3-540-92182-0_28
17. Fiala, J., Gavenciak, T., Knop, D., Koutecký, M., Kratochvíl, J.: Parameterized complexity of distance labeling and uniform channel assignment problems. Discret. Appl. Math. **248**, 46–55 (2018). https://doi.org/10.1016/j.dam.2017.02.010
18. Frank, A., Tardos, É.: An application of simultaneous diophantine approximation in combinatorial optimization. Combinatorica **7**(1), 49–65 (1987). https://doi.org/10.1007/BF02579200
19. Gavenčiak, T., Koutecký, M., Knop, D.: Integer programming in parameterized complexity: Five miniatures. Discrete Optimization **44**(Part 1), 100596 (2022). https://doi.org/10.1016/j.disopt.2020.100596
20. Grünbaum, B.: Convex polytopes, Graduate Texts in Mathematics, vol. 221. Springer, second edn. (2003). https://doi.org/10.1007/978-1-4613-0019-9
21. Hemmecke, R., Köppe, M., Weismantel, R.: A polynomial-time algorithm for optimizing over N-fold 4-block decomposable integer programs. In: Proceedings of the 14th International Conference on Integer Programming and Combinatorial Optimization, IPCO. pp. 219–229 (2010). https://doi.org/10.1007/978-3-642-13036-6_17
22. Hemmecke, R., Köppe, M., Weismantel, R.: Graver basis and proximity techniques for block-structured separable convex integer minimization problems. Math. Program. , 1–18 (2013). https://doi.org/10.1007/s10107-013-0638-z
23. Jansen, K., Klein, K., Maack, M., Rau, M.: Empowering the configuration-ip: new PTAS results for scheduling with setup times. Math. Program. **195**(1), 367–401 (2022). https://doi.org/10.1007/S10107-021-01694-3
24. Kannan, R.: Minkowski's convex body theorem and integer programming. Math. Oper. Res. **12**(3), 415–440 (1987). https://doi.org/10.1287/MOOR.12.3.415
25. Klein, K.: About the complexity of two-stage stochastic ips. Math. Program. **192**(1), 319–337 (2022). https://doi.org/10.1007/S10107-021-01698-Z

26. Knop, D., Koutecký, M.: Scheduling meets n-fold integer programming. J. Sched. **21**(5), 493–503 (2018). https://doi.org/10.1007/S10951-017-0550-0
27. Knop, D., Koutecký, M., Mnich, M.: Voting and bribing in single-exponential time. ACM Transactions on Economics and Computation **8**(3), 12:1–12:28 (2020). https://doi.org/10.1145/3396855
28. Koutecký, M., Levin, A., Onn, S.: A parameterized strongly polynomial algorithm for block structured integer programs. In: Proceedings of the 45th International Colloquium on Automata, Languages, and Programming, ICALP. pp. 85:1–85:14 (2018). https://doi.org/10.4230/LIPICS.ICALP.2018.85
29. Lassota, A., Ligthart, K.: Solving 4-block integer linear programs faster using affine decompositions of the right-hand sides (2026). https://doi.org/10.48550/arXiv.2601.23083
30. Oertel, T., Paat, J., Weismantel, R.: A colorful steinitz lemma with application to block-structured integer programs. Math. Program. **204**(1), 677–702 (2024). https://doi.org/10.1007/S10107-023-01971-3
31. Pottier, L.: Minimal solutions of linear diophantine systems : bounds and algorithms. In: Proceedings of the International Conference on Rewriting Techniques and Applications, RTA. pp. 162–173 (1991). https://doi.org/10.1007/3-540-53904-_94
32. Tardos, É.: A strongly polynomial algorithm to solve combinatorial linear programs. Opererations Research **34**(2), 250–256 (1986). https://doi.org/10.1287/OPRE.34.2.250

A Threshold Phenomenon for the Shortest Lattice Vector Problem in the Infinity Norm

Stefan Kuhlmann(✉) and Robert Weismantel

Department of Mathematics, ETH Zürich, Zürich, Switzerland
{stefan.kuhlmann,robert.weismantel}@ifor.math.ethz.ch

Abstract. One important question in the theory of lattices is to detect a shortest vector: given a norm and a lattice, what is the smallest norm attained by a non-zero vector contained in the lattice? We focus on the infinity norm and work with lattices of the form $\boldsymbol{A}\mathbb{Z}^n$, where $\boldsymbol{A}$ has integer entries and is of full column rank. Finding a shortest vector is NP-hard [47]. We show that this task is fixed parameter tractable in the parameter Δ, the largest absolute value of the determinant of a full rank submatrix of $\boldsymbol{A}$. The algorithm is based on a structural result that can be interpreted as a threshold phenomenon: whenever the dimension n exceeds a certain value determined only by Δ, then a shortest lattice vector attains an infinity norm value of one. This threshold phenomenon has several applications. In particular, it reveals that integer optimal solutions lie on faces of the given polyhedron whose dimensions are bounded only in terms of Δ.

1 Introduction

A fundamental algorithmic problem in the geometry of numbers with numerous applications to other areas of mathematics and computer science is the shortest lattice vector problem. This problem is easy to state. Let $\boldsymbol{A} \in \mathbb{Z}^{m\times n}$ be a matrix of full column rank. The lattice $\boldsymbol{A}\mathbb{Z}^n$ consists of all integral combinations of the columns of $\boldsymbol{A}$, i.e., $\boldsymbol{A}\mathbb{Z}^n := \{\boldsymbol{y} \in \mathbb{Z}^m : \boldsymbol{y} = \boldsymbol{A}\boldsymbol{x} \text{ for } \boldsymbol{x} \in \mathbb{Z}^n\}$. Given such a matrix $\boldsymbol{A} \in \mathbb{Z}^{m\times n}$ of full column rank and a norm $\|\cdot\|$, a natural question is to determine a non-zero vector in the lattice $\boldsymbol{A}\mathbb{Z}^n$ that attains the smallest norm: $\min_{\boldsymbol{z}\in\mathbb{Z}^n\setminus\{\boldsymbol{0}\}} \|\boldsymbol{A}\boldsymbol{z}\|$. In this paper, the underlying norm is the infinity norm, which is given by $\|\boldsymbol{x}\|_\infty := \max\{|x_1|, |x_2|, \ldots, |x_n|\}$ for $\boldsymbol{x} = (x_1, \ldots, x_n)^\top \in \mathbb{R}^n$. Then the shortest vector problem in the infinity norm is as follows

$$\min_{\boldsymbol{z}\in\mathbb{Z}^n\setminus\{\boldsymbol{0}\}} \|\boldsymbol{A}\boldsymbol{z}\|_\infty. \tag{SVP}$$

It has been shown by van Emde Boas that (SVP) is NP-hard [47]. Hardness and algorithmic results for (SVP) and approximate versions of (SVP) play an important role in lattice-based cryptography; see [35] for a comprehensive survey on the topic. Many algorithms for solving the shortest lattice vector problem

S. S. Dey et al. (Eds.): IPCO 2026, LNCS 16588, pp. 188–203, 2026.
https://doi.org/10.1007/978-3-032-28691-8_13

exactly or approximately have been devised over the past decades. It is beyond the scope of this paper to discuss the literature in detail. Rather let us refer to several milestones and references on the topic.

It began with the remarkable work of Lenstra, Lenstra, Lovász [34], who introduced the notion of a "reduced basis". One important property of such a basis is that its first vector is approximately a shortest lattice vector with respect to the Euclidean norm. This provides a bound on the length of a shortest vector that can be utilized by an exhaustive search procedure to solve (SVP) in the Euclidean norm in running time $2^{\mathcal{O}(n^3)}$. Kannan [29,30] refined this approach and improved the running time for computing a shortest lattice vector significantly to $2^{\mathcal{O}(n \log n)}$; see also [21,25,26,38] for further modifications and improvements of Kannan's algorithm. The first randomized algorithm to compute a shortest lattice vector in single exponential time in the Euclidean norm was developed by Ajtai, Kumar, Sivakumar [5]. The authors introduced the technique of "randomized sieving". Roughly speaking, their method samples exponentially many lattice points and combines them to generate shorter vectors with positive probability; see [18,45] for excellent surveys on this topic. The randomized sieving technique has been the subject of intensive investigation. In recent years, it has been modified, generalized, improved and can be applied to arbitrary norms. It gives exact randomized algorithms for (SVP) that run in single exponential time and require single exponential space; see [3,14,36,41,44] for the fastest algorithm in the infinity norm, and [1,2,4] for state-of-the art results in the Euclidean Norm based on "discrete Gaussian sampling". Another appealing approach to solve (SVP) in the Euclidean norm was proposed by Micciancio and Voulgaris [37]. Their method is based on the "Voronoi cell" of a lattice. It provides us with the first deterministic single exponential time algorithm for (SVP) in the Euclidean norm.

The methods to tackle (SVP) discussed so far all measure complexity in terms of the dimension n. In this paper, we ask the following: can we say something more when we also fix the parameter Δ defined to be the largest absolute value among all full rank subdeterminants of the given matrix $\boldsymbol{A}$? In other words, we fix a constant Δ and work with lattices defined by Δ-modular matrices. This is made formal below.

Definition 1. *A matrix $\boldsymbol{A} \in \mathbb{Z}^{m \times n}$ of full column rank is called Δ-modular if $|\det \boldsymbol{B}| \leq \Delta$ for all full rank submatrices $\boldsymbol{B}$ of $\boldsymbol{A}$ and there exists at least one full rank submatrix $\boldsymbol{B}$ of $\boldsymbol{A}$ such that $|\det \boldsymbol{B}| = \Delta$.*

If a lattice $\boldsymbol{A}\mathbb{Z}^n$ has a basis given by a Δ-modular matrix, then this property is intrinsic to the lattice, since changing the basis of $\boldsymbol{A}\mathbb{Z}^n$ corresponds to applying a unimodular transformation from the right, see [46, Section 4], which preserves full rank subdeterminants.

It is a major open problem in integer programming whether a linear discrete optimization problem is solvable in polynomial time when the underlying matrix is Δ-modular and Δ is constant; see [10,11,16,20,32,39,40] for some partial progress in this direction. Inspired by this question, we ask ourselves whether (SVP) is solvable in polynomial time when the lattice $\boldsymbol{A}\mathbb{Z}^n$ is determined by a

Δ-modular matrix $\boldsymbol{A}$ and Δ is constant. We are not aware of any FPT algorithm for (SVP) parameterized solely by Δ prior to this work. The only available results also parameterize by m [23,24]. Our algorithmic result below is possible because of a threshold phenomenon: if the dimension n is sufficiently large, then for a Δ-modular matrix $\boldsymbol{A}$ a shortest non-zero lattice vector $\boldsymbol{Az}$ attains the smallest possible value $\|\boldsymbol{Az}\|_\infty = 1$. To make this more formal, let $\mathcal{M}_\Delta$ denote the set of Δ-modular matrices with full column rank. We consider the function

$$f(\Delta) := \sup_{\boldsymbol{A}\in\mathcal{M}_\Delta} \{n \in \mathbb{N} : \|\boldsymbol{Az}\|_\infty \geq 2 \;\forall \boldsymbol{z} \in \mathbb{Z}^n\backslash\{\boldsymbol{0}\}\}. \tag{1}$$

It turns out that one can replace the supremum in (1) by a maximum. In fact, one can formulate a quantitative upper bound on $f(\Delta)$ depending on the indicator function $\mathbf{1}_{2\mathbb{Z}}(\Delta)$, which equals 1 if Δ is even and 0 if Δ is odd.

Theorem 1. *The value $f(\Delta)$ is finite for all $\Delta \in \mathbb{N}_{\geq 1}$. In particular, one has*

$$\Delta - 1 \leq f(\Delta) \leq \left\lfloor \frac{\Delta - 1}{2} \right\rfloor \cdot (\Delta - 1) + \mathbf{1}_{2\mathbb{Z}}(\Delta).$$

For $\Delta \in \{1,2,3\}$, the upper and lower bound match. When $\Delta = 4$, the first deviation between the upper and lower bound appears with $f(\Delta) \in \{3,4\}$. Theorem 1 can also be interpreted as a variant of Minkowski's convex body theorem parameterized in Δ: Define $\mathcal{Q} := \{\boldsymbol{x} \in \mathbb{R}^n : -\mathbf{1} \leq \boldsymbol{Ax} \leq \mathbf{1}\}$. Whenever $n \geq f(\Delta) + 1$, the convex body $\mathcal{Q}$ contains a non-zero integer vector.

The proof of Theorem 1 is constructive. It gives rise to an algorithm that computes the shortest non-zero lattice vector in the lattice $\boldsymbol{A}\mathbb{Z}^n$ with respect to the infinity norm, which runs in polynomial time in m and n for fixed Δ.

Theorem 2. *Let $\Delta \in \mathbb{N}_{\geq 1}$, $\boldsymbol{A} \in \mathbb{Z}^{m\times n}$ of full column rank, and suppose that $n \geq \lfloor(\Delta-1)/2\rfloor \cdot (\Delta-1) + \mathbf{1}_{2\mathbb{Z}}(\Delta) + 1$. Then one can find $\boldsymbol{z}^* \in \mathbb{Z}^n\backslash\{\boldsymbol{0}\}$ such that $\|\boldsymbol{Az}^*\|_\infty = 1$ or return a full rank submatrix $\boldsymbol{B}$ of $\boldsymbol{A}$ with $|\det \boldsymbol{B}| > \Delta$. This can be done in $\mathcal{O}(mn^2\Delta^3)$ time.*

Theorem 2 implies an FPT algorithm for (SVP) parameterized by Δ: Given some matrix $\boldsymbol{A} \in \mathbb{Z}^{m\times n}$, one can first compute the Hermite normal form in (strongly) polynomial time, cf. [31], to verify that $\boldsymbol{A}$ has full column rank or restrict to a submatrix of full column rank. The next step is to check whether this submatrix has rank at most $\lfloor(\Delta-1)/2\rfloor \cdot (\Delta-1) + \mathbf{1}_{2\mathbb{Z}}(\Delta) + 1$ or not. In the latter case, one runs the algorithm that provides a proof of Theorem 2. Otherwise, one can use an algorithm that solves (SVP) in polynomial time when n is a constant. It will become evident in Sect. 5 that the FPT algorithm uses polynomial space. The exponents in the running time stated in Theorem 2 can be improved by incorporating fast algorithms for Gauss-Jordan elimination and matrix multiplication. In fact, Theorem 2 does not require Δ to be constant. It also applies when $\Delta \leq \sqrt{2n}$.

Theorem 1 has applications beyond algorithmic results. It can be used to derive novel results in the theory of integer programming. However, to apply

Theorem 1 to problems in integer programming requires us to assume that $f(\Delta)$ is a monotonically increasing function. It is not clear whether this is true. Hence, we replace $f(\Delta)$ by the following canonical monotonically increasing function

$$F(\Delta) := \max_{i \in [\Delta]} f(i). \tag{2}$$

If one can show that f is monotonically increasing, then one can replace $F(\Delta)$ by $f(\Delta)$ in all our subsequent results. Observe that

$$\Delta - 1 \leq F(\Delta) \leq \left\lfloor \frac{\Delta - 1}{2} \right\rfloor \cdot (\Delta - 1) + \mathbf{1}_{2\mathbb{Z}}(\Delta) \tag{3}$$

remains true. Given a polyhedron $\mathcal{P} = \{\boldsymbol{x} \in \mathbb{R}^n : \boldsymbol{A}\boldsymbol{x} \leq \boldsymbol{b}\}$, then Theorem 1 implies that the vertices of the integer hull of $\mathcal{P}$, lie on a face of $\mathcal{P}$ whose dimension is bounded by $F(\Delta)$.

Theorem 3. *Let $\boldsymbol{A} \in \mathbb{Z}^{m \times n}$ be Δ-modular, $\boldsymbol{b} \in \mathbb{Z}^m$, and $\mathcal{P} = \{\boldsymbol{x} \in \mathbb{R}^n : \boldsymbol{A}\boldsymbol{x} \leq \boldsymbol{b}\}$. Then the vertices of the convex hull of $\mathcal{P} \cap \mathbb{Z}^n$ lie on faces of $\mathcal{P}$ of dimension at most $F(\Delta)$.*

If we set $\Delta = 1$ in Theorem 3, the matrix $\boldsymbol{A}$ is unimodular and we recover that vertices of the convex hull of $\mathcal{P} \cap \mathbb{Z}^n$ coincide with vertices of $\mathcal{P}$, a fact that follows from a result by Hoffman and Kruskal [28]. When $\Delta = 2$, the vertices of the integer hull lie on faces of dimension zero or one. This also follows from a known result due to Veselov and Chirkov [48]. For $\Delta \geq 3$, upper bounds of this type were not known previously. In particular, for $\Delta = 3$, Theorem 3 ensures the existence of integer vectors on faces of dimension at most two, provided that $\mathcal{P} \cap \mathbb{Z}^n \neq \emptyset$. When $\Delta = 3$, it remains an open problem whether one can find such an integer vector or decide that no such integer vector exists in polynomial time. Theorem 3 can also be applied to obtain novel upper bounds on the number of non-zero entries of optimal solutions of integer optimization problems in standard form. This topic is discussed below.

2 Sparse Integer Optimal Solutions

The proof of Theorem 3 is based on a threshold phenomenon for the shortest lattice vector problem that is made precise in Theorem 1. This section is devoted to show that this threshold phenomenon also gives new bounds on the sparsity of integer optimal solutions for optimization problems in standard form. For this to make sense, we assume that $\boldsymbol{A} \in \mathbb{Z}^{m \times n}$ has full row rank. We call $\boldsymbol{A}$ Δ-modular if $\boldsymbol{A}^\top$ is Δ-modular according to Definition 1. Given such a matrix $\boldsymbol{A}$ and $\boldsymbol{b} \in \mathbb{Z}^m$, $\boldsymbol{c} \in \mathbb{Z}^n$, an integer optimization problem in standard form is of the form

$$\max\left\{\boldsymbol{c}^\top \boldsymbol{x} \colon \boldsymbol{A}\boldsymbol{x} = \boldsymbol{b},\, \boldsymbol{x} \geq \boldsymbol{0},\, \boldsymbol{x} \in \mathbb{Z}^n\right\}. \tag{4}$$

It is an important question to detect an optimal solution of smallest support, i.e., an optimal solution $\boldsymbol{z}^*$ with the smallest number of strictly positive values among $z_1^*, \ldots, z_n^*$.

The problem of bounding the support of solutions for integer optimization problems has been studied quite intensively in the past decade. A first bound in terms of m and determinants of $\boldsymbol{A}$ is due to Aliev et al. [8, Theorem 1]. Their result gives an upper bound on the support of an optimal integer solution of $m + \log_2(\sqrt{\det \boldsymbol{A}\boldsymbol{A}^\top})$; see also [6, Theorem 3] for a refinement of the latter statement. A problem with these bounds, for our purposes, is that the expression $\det \boldsymbol{A}\boldsymbol{A}^\top$ cannot be bounded solely by the parameter Δ, the largest full rank subdeterminant of $\boldsymbol{A}$. For instance, the matrix $\boldsymbol{T} \in \mathbb{Z}^{(n-1)\times n^2}$ obtained from deleting a row of the oriented incidence matrix of a complete graph satisfies $\det \boldsymbol{T}\boldsymbol{T}^\top = n^{n-2}$ by Kirchhoff's matrix tree theorem; see [17, Section 7.2]. However, $\boldsymbol{T}$ is totally unimodular and thus there always exist integer solutions with at most $n-1$ non-zero entries. Nevertheless, using the Cauchy-Binet formula on $\det \boldsymbol{A}\boldsymbol{A}^\top$, bounds in terms of m and Δ on the support of an optimal integer solution are available; see [33, Theorem 4] and [22, Section 3.1, Remark 5]. The latter gives an upper bound of $2.81 \cdot m + 1.18 \cdot \log_2(\Delta)$. Let us also mention that, when the columns of $\boldsymbol{A}$ span $\mathbb{R}^m$, one can improve this to $2m + \log_2(\Delta)$ [6, Theorem 2].

Below, we establish the first upper bound on the support of an optimal integer solution in form of one times m plus a function in Δ. More precisely, our bound on the support of an optimal integer solution is $m + F(\Delta)$; see (2) for the definition of $F(\Delta)$. Given that there exists an optimal solution to the LP relaxation of (4) with support at most m, the support bounds for the continuous and the integer optimal solution differ essentially by this threshold value $F(\Delta)$ and are independent of m. To the best of our knowledge, prior to the result below, upper bounds of the form m plus a function in Δ are only available when $m = 1$ [9, Theorem 1.2] or in an asymptotic setting [42]. For $\boldsymbol{x} \in \mathbb{R}^n$, we define $\operatorname{supp}(\boldsymbol{x}) := \{i \in [n] : x_i \neq 0\}$.

Theorem 4. *Let $\boldsymbol{A} \in \mathbb{Z}^{m\times n}$ have full row rank, $\boldsymbol{b} \in \mathbb{Z}^m$, and $\boldsymbol{c} \in \mathbb{R}^n$. If (4) has an optimal solution, then there exists an optimal solution $\boldsymbol{z}^*$ such that*

$$|\operatorname{supp}(\boldsymbol{z}^*)| \leq m + F(\Delta).$$

We get an explicit upper bound by applying $F(\Delta) \leq \lfloor(\Delta-1)/2\rfloor \cdot (\Delta - 1) + \mathbf{1}_{2\mathbb{Z}}(\Delta)$; cf. (3), which gives the best-known upper bound on the sparsity of optimal integer solutions when Δ is constant or small compared to m. Similar to Theorem 3, Theorem 4 extends known results for $\Delta \in \{1, 2\}$ to arbitrary values of Δ. We also provide a first non-trivial lower bound on the number of non-zero entries of integer solutions in terms of m and Δ. This bound even holds for totally Δ-modular matrices, i.e., matrices in which all $k \times k$ submatrices have a determinant bounded by Δ in absolute value for all $k \in [m]$. We omit the proof.

Proposition 1. *Let $\Delta \in \mathbb{N}_{\geq 1}$, $m = (\Delta-1)^2 + 1$, and $n = (\Delta-1)^2 + \Delta$. There exists a totally Δ-modular matrix $\boldsymbol{A} \in \mathbb{Z}^{m\times n}$ with full row rank and $\boldsymbol{b} \in \mathbb{Z}^m$*

such that

$$|\text{supp}(\boldsymbol{z}^*)| = m + \Delta - 1$$

for all optimal integer solutions $\boldsymbol{z}^*$ *of (4) with respect to all* $\mathbf{c} \in \mathbb{R}^n$*.*

For the sake of completeness, let us briefly mention that there is also a line of research investigating upper bounds on the support of optimal solutions of (4) in terms of m and $\|\boldsymbol{A}\|_\infty$, the largest entry of $\boldsymbol{A}$ in absolute value. In this regime, a first upper bound is due to Eisenbrand and Shmonin [19, Theorem 1], which has been improved to $2m \log_2(2\sqrt{m}\|\boldsymbol{A}\|_\infty)$ by Aliev et al. [8, Theorem 1]. This bound is known to be tight, up to the constant two in front of m and the logarithm [12,13]. Hence, in contrast to Theorem 4, bounds of the form m plus a function in $\|\boldsymbol{A}\|_\infty$ cannot exist.

3 Proof of the Upper Bound on $f(\Delta)$

Given a Δ-modular matrix $\boldsymbol{A}$ with sufficiently large rank, our goal is to find an integer combination of the columns with coefficients given by $\boldsymbol{z}$ such that $\|\boldsymbol{A}\boldsymbol{z}\|_\infty = 1$. One hurdle is that although all full rank subdeterminants of $\boldsymbol{A}$ are bounded by Δ, the entries of $\boldsymbol{A}$ may be arbitrarily large, and, therefore, each column of $\boldsymbol{A}$ may be arbitrarily large in the infinity norm. This makes it challenging to construct an integer vector $\boldsymbol{A}\boldsymbol{z}$ of small norm. To circumvent this issue, we apply a common normalization technique when dealing with Δ-modular matrices: we analyze subdeterminants of $\boldsymbol{A} \cdot \boldsymbol{B}^{-1} \in \mathbb{Q}^{m \times n}$, where $\boldsymbol{B}$ is some invertible full rank submatrix of $\boldsymbol{A}$. The following lemma and corollary establish that columns of $\boldsymbol{A} \cdot \boldsymbol{B}^{-1}$ have a bounded infinity norm.

To state these results, let us first explain the notation: for a matrix $\boldsymbol{A} \in \mathbb{Z}^{m \times n}$ and sets $I \subseteq [m]$ and $J \subseteq [n]$, we denote by $\boldsymbol{A}_I$ or $\boldsymbol{A}_{I,\cdot}$ the submatrix consisting of the rows indexed by I, by $\boldsymbol{A}_{\cdot,J}$ the submatrix consisting of the columns indexed by J, and by $\boldsymbol{A}_{I,J}$ the submatrix given by the rows indexed by I and columns indexed by J.

Lemma 1. *Let* $\boldsymbol{A} \in \mathbb{Z}^{m \times n}$ *have full column rank and* $\boldsymbol{B}$ *be an invertible full rank submatrix of* $\boldsymbol{A}$*. Let* $I \subseteq [m]$ *and* $J \subseteq [n]$ *with* $|I| = |J|$*. Let* $\boldsymbol{a}_{i_1}, \dots, \boldsymbol{a}_{i_{|I|}}$ *be the rows of* $\boldsymbol{A}$ *indexed by* I *and* $\boldsymbol{b}_{j_1}, \dots, \boldsymbol{b}_{j_{n-|J|}}$ *the rows of* $\boldsymbol{B}$ *not indexed by* J*. Then we have*

$$\left|\det(\boldsymbol{A} \cdot \boldsymbol{B}^{-1})_{I,J}\right| = \frac{\left|\det(\boldsymbol{a}_{i_1}, \dots, \boldsymbol{a}_{i_{|I|}}, \boldsymbol{b}_{j_1}, \dots, \boldsymbol{b}_{j_{n-|J|}})\right|}{|\det \boldsymbol{B}|}.$$

If we select the invertible submatrix $\boldsymbol{B}$ with largest determinant in absolute value in Lemma 1, we obtain the result below.

Corollary 1. *Let* $\boldsymbol{A} \in \mathbb{Z}^{m \times n}$ *be* Δ*-modular and* $\boldsymbol{B}$ *a full rank submatrix of* $\boldsymbol{A}$ *with* $|\det \boldsymbol{B}| = \Delta$*. Then* $\boldsymbol{A} \cdot \boldsymbol{B}^{-1}$ *has all its subdeterminants bounded by* 1 *in absolute value. In particular, for any column* $\boldsymbol{r}$ *of* $\boldsymbol{B}^{-1}$*, we obtain* $-\mathbf{1} \leq \boldsymbol{A}\boldsymbol{r} \leq \mathbf{1}$*.*

We only need Corollary 1 to prove Theorem 1. Lemma 1 will be used in Sect. 5.

Observe that $\boldsymbol{A} \cdot \boldsymbol{B}^{-1} \in \mathbb{Q}^{m\times n}$ may not have integer entries anymore. This does not cause difficulties since an underlying lattice structure still exists: each column of $\boldsymbol{B}^{-1}$ corresponds to an element in a finite abelian group. For the proof of Theorem 1, it is important to have enough columns such that at least Δ many of them correspond to the same element in the finite abelian group. This will determine the upper bound on the threshold dimension $f(\Delta)$, which is why we present the underlying group argument here.

Let $(G, +)$ be a finite abelian group and $I(G) := \{g \in G : g + g = 0\}$ be the set of so-called involutions of G. Observe that $0 \in I(G)$. Based on $I(G)$, we consider two subsets of $G\backslash\{0\}$, one that is contained in $G\backslash I(G)$ and the other one in $I(G)\backslash\{0\}$. For the set contained in $G\backslash I(G)$, we allow for duplicates whereas for the other set we assume that elements are pairwise disjoint. The reason for this distinction will become evident in the proof of Theorem 1, where we apply the result below when $|G| = \Delta$. Recall that $\mathbf{1}_{2\mathbb{Z}}(\Delta)$ equals 1 if Δ is even and 0 otherwise. The following result is based on the pigeonhole principle.

Lemma 2. *Let $(G, +)$ be a finite abelian group, $\{g_1, \ldots, g_k\} \subseteq G\backslash I(G)$ be a multiset, and $\{h_1, \ldots, h_l\} \subseteq I(G)\backslash\{0\}$ with $h_i \neq h_j$ for different $i, j \in [l]$. Suppose that*

$$k + l \geq \left\lfloor \frac{|G| - 1}{2} \right\rfloor \cdot (|G| - 1) + \mathbf{1}_{2\mathbb{Z}}(|G|) + 1. \tag{5}$$

Then there exist $I \subseteq [k]$ with $|I| = |G|$ and $\sigma_i \in \{-1, 1\}$ for each $i \in I$ such that

$$\sigma_i g_i = \sigma_j g_j$$

for all $i, j \in I$.

Proof (of Theorem 1). We only prove that $f(\Delta) \leq \lfloor(\Delta-1)/2\rfloor \cdot (\Delta-1) + \mathbf{1}_{2\mathbb{Z}}(\Delta)$. This implies that $f(\Delta)$ is finite for all values of Δ. Let $\boldsymbol{A} \in \mathbb{Z}^{m\times n}$ be Δ-modular and suppose that $n \geq \lfloor(\Delta-1)/2\rfloor \cdot (\Delta-1) + \mathbf{1}_{2\mathbb{Z}}(\Delta) + 1$. Our goal is to construct a non-zero integer vector $\boldsymbol{z}^*$ such that $\|\boldsymbol{A}\boldsymbol{z}^*\|_\infty = 1$. We divide the proof into three steps. In each step, we present potential candidate vectors for $\boldsymbol{z}^*$ and show that at least one of them can be chosen as $\boldsymbol{z}^*$. To do so, we need the following ingredients: Let $\boldsymbol{B}$ be an $n \times n$ submatrix of $\boldsymbol{A}$ that has determinant Δ. Set $\boldsymbol{B}^{-1} = (\boldsymbol{r}_1, \ldots, \boldsymbol{r}_n)$ and $\Lambda = \boldsymbol{B}^{-1}\mathbb{Z}^n$. The lattice Λ satisfies $\mathbb{Z}^n \subseteq \Lambda$ and $\det \Lambda = |\det \boldsymbol{B}^{-1}| = 1/\Delta$. This implies that $\Lambda/\mathbb{Z}^n$ is a finite abelian group of cardinality $(\det \Lambda)^{-1} = \Delta$. Below, we consider the set $R = \{\boldsymbol{r}_1 + \mathbb{Z}^n, \ldots, \boldsymbol{r}_n + \mathbb{Z}^n\} \subseteq \Lambda/\mathbb{Z}^n$ given by $\boldsymbol{B}^{-1}$.

Step 1 ($\boldsymbol{r}_i \in \mathbb{Z}^n$): Suppose there exists $\boldsymbol{r}_i + \mathbb{Z}^n \in R$ such that $\boldsymbol{r}_i + \mathbb{Z}^n = \mathbb{Z}^n$, that is, we have $\boldsymbol{r}_i \in \mathbb{Z}^n$ for some $i \in [n]$. In this case, we get $-\mathbf{1} \leq \boldsymbol{A}\boldsymbol{r}_i \leq \mathbf{1}$ by Corollary 1 and the theorem follows. So, for the remainder of the proof, we assume that $\boldsymbol{r}_i \notin \mathbb{Z}^n$ for all $i \in [n]$, or, equivalently, no element in R equals $\mathbb{Z}^n$.

Step 2 ($\boldsymbol{r}_i + \boldsymbol{r}_j, \boldsymbol{r}_i - \boldsymbol{r}_j \in \mathbb{Z}^n$): Suppose there exist elements in R that are not pairwise different and contained in $I(\Lambda/\mathbb{Z}^n)\backslash\{\mathbb{Z}^n\}$, e.g., there exist $i, j \in [n]$ with

$i \neq j$ such that $\boldsymbol{r}_i + \mathbb{Z}^n, \boldsymbol{r}_j + \mathbb{Z}^n \in R \cap I(\Lambda/\mathbb{Z}^n) \backslash \{\mathbb{Z}^n\}$ and $\boldsymbol{r}_i + \mathbb{Z}^n = \boldsymbol{r}_j + \mathbb{Z}^n$. Both properties imply that $\boldsymbol{r}_i + \boldsymbol{r}_j \in \mathbb{Z}^n$ and $\boldsymbol{r}_i - \boldsymbol{r}_j \in \mathbb{Z}^n$. Note that $\boldsymbol{r}_i + \boldsymbol{r}_j, \boldsymbol{r}_i - \boldsymbol{r}_j$ are non-zero because $\boldsymbol{B}^{-1}$ is invertible and since $\boldsymbol{r}_i$ and $\boldsymbol{r}_j$ are columns of $\boldsymbol{B}^{-1}$. We show that at least one of the two vectors satisfies $-\boldsymbol{1} \leq \boldsymbol{A}\boldsymbol{x} \leq \boldsymbol{1}$. Assume that $\boldsymbol{r}_i + \boldsymbol{r}_j$ does not satisfy $-\boldsymbol{1} \leq \boldsymbol{A}\boldsymbol{x} \leq \boldsymbol{1}$. So there exists a row $\boldsymbol{a}$ of $\boldsymbol{A}$ such that $|\boldsymbol{a}^\top(\boldsymbol{r}_i + \boldsymbol{r}_j)| > 1$. Let $\boldsymbol{a}^\top(\boldsymbol{r}_i + \boldsymbol{r}_j) > 1$ without loss of generality. Since $\boldsymbol{a}$ and $\boldsymbol{r}_i + \boldsymbol{r}_j$ have integer entries, the scalar product $\boldsymbol{a}^\top(\boldsymbol{r}_i + \boldsymbol{r}_j)$ is an integer and we have $\boldsymbol{a}^\top(\boldsymbol{r}_i + \boldsymbol{r}_j) \geq 2$. As $-\boldsymbol{1} \leq \boldsymbol{A}\boldsymbol{r}_i \leq \boldsymbol{1}$ and $-\boldsymbol{1} \leq \boldsymbol{A}\boldsymbol{r}_j \leq \boldsymbol{1}$ by Corollary 1, we have $\boldsymbol{a}^\top\boldsymbol{r}_i = 1$ and $\boldsymbol{a}^\top\boldsymbol{r}_j = 1$. Next, consider $\boldsymbol{r}_i - \boldsymbol{r}_j$. Suppose that there exists a row $\tilde{\boldsymbol{a}}$ of $\boldsymbol{A}$ such that $|\tilde{\boldsymbol{a}}^\top(\boldsymbol{r}_i - \boldsymbol{r}_j)| > 1$. Observe that $\tilde{\boldsymbol{a}} \neq \boldsymbol{a}$ since $\tilde{\boldsymbol{a}}^\top(\boldsymbol{r}_i - \boldsymbol{r}_j) = 0$. Following the same arguments as before, we can assume without loss of generality that $\tilde{\boldsymbol{a}}^\top(\boldsymbol{r}_i - \boldsymbol{r}_j) \geq 2$, $\tilde{\boldsymbol{a}}^\top\boldsymbol{r}_i = 1$, and $\tilde{\boldsymbol{a}}^\top\boldsymbol{r}_j = -1$. However, then the matrix $\boldsymbol{A} \cdot \boldsymbol{B}^{-1}$ contains the submatrix

$$\begin{pmatrix} \boldsymbol{a}^\top\boldsymbol{r}_i & \boldsymbol{a}^\top\boldsymbol{r}_k \\ \tilde{\boldsymbol{a}}^\top\boldsymbol{r}_i & \tilde{\boldsymbol{a}}^\top\boldsymbol{r}_k \end{pmatrix} = \begin{pmatrix} 1 & 1 \\ 1 & -1 \end{pmatrix}. \tag{6}$$

This submatrix has determinant 2 in absolute value, which contradicts that $\boldsymbol{A} \cdot \boldsymbol{B}^{-1}$ has all subdeterminants bounded by 1 as stated in Corollary 1. So we have $-\boldsymbol{1} \leq \boldsymbol{A}(\boldsymbol{r}_i - \boldsymbol{r}_j) \leq \boldsymbol{1}$ and the theorem follows. Therefore, we assume from now on that elements from R are pairwise different if they are contained in $I(\Lambda/\mathbb{Z}^n) \backslash \{\mathbb{Z}^n\}$.

Step 3 ($\boldsymbol{r}_i - \boldsymbol{r}_j \in \mathbb{Z}^n\ \forall i, j \in I$ & $\sum_{i \in I} \boldsymbol{r}_i \in \mathbb{Z}^n$, $|I| \geq \Delta$): We can exclude the cases considered in the previous two steps. Hence, the set R meets the assumptions of Lemma 2 with $n = k + l$ and $G = \Lambda/\mathbb{Z}^n$. We apply Lemma 2 and obtain $\boldsymbol{r}_1 + \mathbb{Z}^n = \ldots = \boldsymbol{r}_\Delta + \mathbb{Z}^n$, after reordering and resigning columns if necessary. This implies

1. $\boldsymbol{r}_i - \boldsymbol{r}_j \in \mathbb{Z}^n \backslash \{\boldsymbol{0}\}$ for all $i, j \in [\Delta]$ with $i \neq j$ and
2. $\boldsymbol{r}_1 + \ldots + \boldsymbol{r}_\Delta \in \mathbb{Z}^n \backslash \{\boldsymbol{0}\}$.

The vectors above are non-zero because $\boldsymbol{B}^{-1}$ is invertible. We claim that one of them satisfies $-\boldsymbol{1} \leq \boldsymbol{A}\boldsymbol{x} \leq \boldsymbol{1}$. Suppose $\boldsymbol{r}_i - \boldsymbol{r}_j$ does not satisfy $-\boldsymbol{1} \leq \boldsymbol{A}\boldsymbol{x} \leq \boldsymbol{1}$ for all $i, j \in [\Delta]$ with $i \neq j$. As before, we can assume without loss of generality that $\boldsymbol{a}^\top(\boldsymbol{r}_i - \boldsymbol{r}_j) \geq 2$, $\boldsymbol{a}^\top\boldsymbol{r}_i = 1$, and $\boldsymbol{a}^\top\boldsymbol{r}_j = -1$. We claim that $\boldsymbol{a}^\top\boldsymbol{r}_k = 0$ for all $k \in [\Delta] \backslash \{i, j\}$.

For the purpose of deriving a contradiction, suppose that there exists an index k such that $|\boldsymbol{a}^\top\boldsymbol{r}_k| > 0$. Again, without loss of generality we can assume that $\boldsymbol{a}^\top\boldsymbol{r}_k > 0$. Next, consider the integer vector $\boldsymbol{r}_i - \boldsymbol{r}_k$. Since $\boldsymbol{a}^\top(\boldsymbol{r}_i - \boldsymbol{r}_k) = 1 - \boldsymbol{a}^\top\boldsymbol{r}_k$ is an integer and $\boldsymbol{a}^\top\boldsymbol{r}_k \leq 1$ by Corollary 1, we conclude that $\boldsymbol{a}^\top(\boldsymbol{r}_i - \boldsymbol{r}_k) = 0$. This holds if and only if $\boldsymbol{a}^\top\boldsymbol{r}_k = 1$. Applying the arguments from above to $\boldsymbol{r}_i - \boldsymbol{r}_k$, we obtain another row $\tilde{\boldsymbol{a}}^\top$ of $\boldsymbol{A}$ such that $|\tilde{\boldsymbol{a}}^\top(\boldsymbol{r}_i - \boldsymbol{r}_k)| = 2$ and, without loss of generality, $\tilde{\boldsymbol{a}}^\top\boldsymbol{r}_i = 1$ and $\tilde{\boldsymbol{a}}^\top\boldsymbol{r}_k = -1$. However, then $\boldsymbol{A} \cdot \boldsymbol{B}^{-1}$ contains the submatrix

$$\begin{pmatrix} \boldsymbol{a}^\top\boldsymbol{r}_i & \boldsymbol{a}^\top\boldsymbol{r}_k \\ \tilde{\boldsymbol{a}}^\top\boldsymbol{r}_i & \tilde{\boldsymbol{a}}^\top\boldsymbol{r}_k \end{pmatrix} = \begin{pmatrix} 1 & 1 \\ 1 & -1 \end{pmatrix}, \tag{7}$$

which has determinant 2 in absolute value, a contradiction to Corollary 1. This shows the claim that $\boldsymbol{a}^\top \boldsymbol{r}_k = 0$ for all $k \in [\Delta]\backslash\{i,j\}$. In summary, we get, for each pair $i,j \in [\Delta]$ with $i \neq j$, a unique row $\boldsymbol{a}^\top$ of $\boldsymbol{A}$ such that

$$\begin{aligned} &\boldsymbol{a}^\top \boldsymbol{r}_k = 0 && \text{for all } k \in [\Delta]\backslash\{i,j\} \\ &|\boldsymbol{a}^\top(\boldsymbol{r}_k - \boldsymbol{r}_l)| \geq 2 \iff && k = i \text{ and } l = j \text{ for all } k,l \in [\Delta]. \end{aligned} \tag{8}$$

This fact is used as follows: Consider the integer vectors $\boldsymbol{r}_i - \boldsymbol{r}_{i+1}$ for $i \in [\Delta - 1]$. Suppose that none of these vectors satisfy $-\mathbf{1} \leq \boldsymbol{A}\boldsymbol{x} \leq \mathbf{1}$. It follows that, for all $i \in [\Delta - 1]$, there exists a unique row $\boldsymbol{a}_i^\top$ of $\boldsymbol{A}$ with the properties (8). This leads, up to multiplying rows with -1, to the submatrix

$$\begin{pmatrix} \boldsymbol{a}_1^\top \boldsymbol{r}_1 & \dots & \boldsymbol{a}_1^\top \boldsymbol{r}_\Delta \\ \vdots & \ddots & \vdots \\ \boldsymbol{a}_{\Delta-1}^\top \boldsymbol{r}_1 & \dots & \boldsymbol{a}_{\Delta-1}^\top \boldsymbol{r}_\Delta \end{pmatrix} = \begin{pmatrix} 1 & -1 & & & \\ & 1 & -1 & & \\ & & \ddots & \ddots & \\ & & & 1 & -1 \end{pmatrix}$$

of $\boldsymbol{A} \cdot \boldsymbol{B}^{-1}$. As a final step, we take the integer vector $\boldsymbol{r}_1 + \ldots + \boldsymbol{r}_\Delta$ into consideration. Select an arbitrary row $\boldsymbol{a}^\top$ of $\boldsymbol{A}$. We obtain a $\Delta \times \Delta$ submatrix of $\boldsymbol{A} \cdot \boldsymbol{B}^{-1}$ of the form

$$\begin{pmatrix} \boldsymbol{a}^\top \boldsymbol{r}_1 & \dots & \boldsymbol{a}^\top \boldsymbol{r}_\Delta \\ \boldsymbol{a}_1^\top \boldsymbol{r}_1 & \dots & \boldsymbol{a}_1^\top \boldsymbol{r}_\Delta \\ \vdots & \ddots & \vdots \\ \boldsymbol{a}_{\Delta-1}^\top \boldsymbol{r}_1 & \dots & \boldsymbol{a}_{\Delta-1}^\top \boldsymbol{r}_\Delta \end{pmatrix} = \underbrace{\begin{pmatrix} \boldsymbol{a}^\top \boldsymbol{r}_1 & \dots & \dots & \dots & \boldsymbol{a}^\top \boldsymbol{r}_\Delta \\ 1 & -1 & & & \\ & 1 & -1 & & \\ & & \ddots & \ddots & \\ & & & 1 & -1 \end{pmatrix}}_{=:\boldsymbol{D}}. \tag{9}$$

From Corollary 1 it follows that all subdeterminants of $\boldsymbol{A} \cdot \boldsymbol{B}^{-1}$ are at most 1 in absolute value. This applies in particular to the submatrix $\boldsymbol{D}$. We use this fact and Laplace expansion along the row given by $\boldsymbol{a}^\top$ to analyze the determinant of $\boldsymbol{D}$. For the Laplace expansion, note that, up to a sign, the uniquely scaled vector contained in the kernel of the last $n-1$ rows of $\boldsymbol{D}$ is $\mathbf{1}$. So we obtain the relation

$$1 \geq |\det \boldsymbol{D}| = \left|\boldsymbol{a}^\top \boldsymbol{r}_1 + \ldots + \boldsymbol{a}^\top \boldsymbol{r}_\Delta\right|.$$

This holds for all rows of $\boldsymbol{A}$, i.e., $-\mathbf{1} \leq \boldsymbol{A}(\boldsymbol{r}_1 + \ldots + \boldsymbol{r}_n) \leq \mathbf{1}$, and hence verifies the upper bound on $f(\Delta)$. □

4 Proofs of the Polyhedral Results

We begin with the proof of Theorem 3, which requires us to apply Theorem 1 to lower-dimensional subspaces. More precisely, we need to investigate vectors satisfying $-\mathbf{1} \leq \boldsymbol{A}\boldsymbol{x} \leq \mathbf{1}$ and $\boldsymbol{A}_I \boldsymbol{x} = \mathbf{0}$ for some $I \subseteq [m]$ such that the rows of $\boldsymbol{A}_I$ are linearly independent. To work with the appropriate determinants corresponding to such a lower-dimensional subspace defined by $\boldsymbol{A}_I\boldsymbol{x} = \boldsymbol{0}$, we introduce, for a given $I \subseteq [m]$, the refined parameter

$$\Delta_I := \max\left\{\left|\det\begin{pmatrix}\boldsymbol{A}_I\\ \boldsymbol{A}_J\end{pmatrix}\right| : J \subseteq [m],\ |J| = n - |I|\right\}.$$

Then one can show that, if the dimension of such a subspace exceeds $f(\Delta_I)$, we are guaranteed to find $\boldsymbol{z}^* \in \mathbb{Z}^n \backslash \{\boldsymbol{0}\}$ such that $-\boldsymbol{1} \leq \boldsymbol{A}\boldsymbol{z}^* \leq \boldsymbol{1}$ and $\boldsymbol{A}_I\boldsymbol{z}^* = \boldsymbol{0}$. We refer the reader to [15, Lemma 1] or [7, Lemma 6] for examples on how determinants behave when restricting to subspaces or faces of polyhedra. The key relation between the new parameter Δ_I and Δ is that $\Delta_I \leq \Delta$, by definition, and thus $f(\Delta_I) \leq F(\Delta)$ by the definition of $F(\Delta)$; cf. (2).

Proof (of Theorem 3**).** We will show the contraposition of the claim, i.e., every integer vector in $\mathcal{P} \cap \mathbb{Z}^n$ that is not contained in a face of $\mathcal{P}$ with dimension at most $F(\Delta)$ is not a vertex of the convex hull of $\mathcal{P} \cap \mathbb{Z}^n$. Let $\boldsymbol{y} \in \mathcal{P} \cap \mathbb{Z}^n$ be a vector that does not lie on a face of $\mathcal{P}$ of dimension at most $F(\Delta)$. Let $K \subseteq [m]$ be the index set for the tight inequalities of $\boldsymbol{A}\boldsymbol{y} \leq \boldsymbol{b}$, that is, $\boldsymbol{A}_K\boldsymbol{y} = \boldsymbol{b}_K$. Select a set $I \subseteq K$ such that the rows of $\boldsymbol{A}_I$ are linearly independent and $\operatorname{rank} \boldsymbol{A}_I = \operatorname{rank} \boldsymbol{A}_K$. Since $\boldsymbol{y}$ does not lie on a face of $\mathcal{P}$ of dimension at most $F(\Delta)$, we have $\operatorname{rank} \boldsymbol{A}_I \leq n - (F(\Delta) + 1)$. Consider the face $\mathcal{F} := \mathcal{P} \cap \{\boldsymbol{x} \in \mathbb{R}^n : \boldsymbol{A}_I\boldsymbol{x} = \boldsymbol{b}_I\}$ of $\mathcal{P}$. As $\operatorname{rank} \boldsymbol{A}_I \leq n - (F(\Delta)+1)$, we obtain $\dim(\mathcal{F}) \geq F(\Delta) + 1 \geq f(\Delta_I) + 1$ by definition of $F(\Delta)$. So we can use the definition of $f(\Delta_I)$, cf. (1), and that $f(\Delta_I)$ is finite due to Theorem 1: there exists $\boldsymbol{z}^* \in \mathbb{Z}^n \backslash \{\boldsymbol{0}\}$ such that $-\boldsymbol{1} \leq \boldsymbol{A}\boldsymbol{z}^* \leq \boldsymbol{1}$ and $\boldsymbol{A}_I\boldsymbol{z}^* = \boldsymbol{0}$. This implies that $\boldsymbol{y} \pm \boldsymbol{z}^* \in \mathcal{F} \cap \mathbb{Z}^n \subseteq \mathcal{P} \cap \mathbb{Z}^n$. We conclude that $1/2 \cdot (\boldsymbol{y} + \boldsymbol{z}^*) + 1/2 \cdot (\boldsymbol{y} - \boldsymbol{z}^*) = \boldsymbol{y}$. Hence, $\boldsymbol{y}$ is a convex combination of integer vectors in $\mathcal{P}$. This contradicts that $\boldsymbol{y}$ is a vertex of the integer hull. □

Our next goal is to apply Theorem 3 to obtain Theorem 4. This involves switching between inequality form and standard form while preserving determinants. To do so, we use a known result concerning orthogonal lattice bases stated below; see [43, Theorem 4.2] for a proof and [49] for an earlier reference, as cited in [22]. The result also follows from a classical identity which relates the Plücker coordinates of a Grassmannian to the Plücker coordinates of the dual Grassmannian; cf., for instance, [27, Book III, Chapter XIV, Theorem I]. In the statement, we write $\bar{I} := [n] \backslash I$ for the complement of $I \subseteq [n]$. Also, the expression $\gcd \boldsymbol{A}$ denotes the greatest common divisor of the full rank subdeterminants of $\boldsymbol{A}$. Note, for the remainder of this section, $\boldsymbol{A} \in \mathbb{Z}^{m \times n}$ is a matrix of full row rank instead of full column rank.

Lemma 3. *Let $\boldsymbol{A} \in \mathbb{Z}^{m \times n}$ have full row rank. Let $\boldsymbol{W} \in \mathbb{Z}^{n \times (n-m)}$ have full column rank such that $\boldsymbol{A} \cdot \boldsymbol{W} = \boldsymbol{0}$. Then we have*

$$1/\gcd \boldsymbol{A} \cdot |\det \boldsymbol{A}_{\cdot,I}| = 1/\gcd \boldsymbol{W} \cdot \left|\det \boldsymbol{W}_{\bar{I},\cdot}\right|$$

for all $I \subseteq [n]$ with $|I| = m$.

Let us briefly discuss this result. Denote by $\boldsymbol{a}_1^\top, \ldots, \boldsymbol{a}_m^\top$ the rows of $\boldsymbol{A}$, similarly by $\boldsymbol{w}_1, \ldots, \boldsymbol{w}_{n-m}$ the columns of $\boldsymbol{W}$, and let $L := \operatorname{lin}\{\boldsymbol{a}_1, \ldots, \boldsymbol{a}_m\}$ be the linear space spanned by $\boldsymbol{a}_1, \ldots, \boldsymbol{a}_m$. The assumption $\boldsymbol{A} \cdot \boldsymbol{W} = \mathbf{0}$ in Lemma 3 states that the linear space $\operatorname{lin}\{\boldsymbol{w}_1, \ldots, \boldsymbol{w}_{n-m}\}$ is orthogonal to L, that is, $L^\perp = \operatorname{lin}\{\boldsymbol{w}_1, \ldots, \boldsymbol{w}_{n-m}\}$. Observe that Lemma 3 applies to all bases of $L^\perp$ given by integer vectors. This gives us the freedom to choose a suitable basis. In the following proof, we select $\boldsymbol{w}_1, \ldots, \boldsymbol{w}_{n-m}$ to be a basis of the lattice $L^\perp \cap \mathbb{Z}^n$, which implies $\gcd \boldsymbol{W} = 1$ and therefore that $\boldsymbol{W}$ is a $(\Delta/\gcd \boldsymbol{A})$-modular matrix provided that $\boldsymbol{A}$ is Δ-modular.

Proof (of Theorem 4). Let $\boldsymbol{z}^*$ be a solution of (4) such that $\boldsymbol{z}^*$ is a vertex of the integer hull of $\mathcal{S} := \{\boldsymbol{x} \in \mathbb{R}^n : \boldsymbol{A}\boldsymbol{x} = \boldsymbol{b}, \boldsymbol{x} \geq \mathbf{0}\}$. Consider the polyhedron $\boldsymbol{z}^* - \mathcal{S} = \{\boldsymbol{x} \in \mathbb{R}^n : \boldsymbol{A}\boldsymbol{x} = \mathbf{0}, \boldsymbol{x} \leq \boldsymbol{z}^*\}$. Observe that $\mathbf{0}$ is a vertex of the integer hull of $\boldsymbol{z}^* - \mathcal{S}$ as $\boldsymbol{z}^*$ is a vertex of the integer hull of $\mathcal{S}$. Let L be the linear space spanned by the rows of $\boldsymbol{A}$. We apply Lemma 3: Choose $\boldsymbol{W} \in \mathbb{Z}^{n\times(n-m)}$ such that the columns of $\boldsymbol{W}$ form a basis of $L^\perp \cap \mathbb{Z}^n$. By Lemma 3, the matrix $\boldsymbol{W}$ satisfies $\gcd \boldsymbol{W} = 1$ and is a $(\Delta/\gcd \boldsymbol{A})$-modular matrix. We set $\delta := \Delta/\gcd \boldsymbol{A}$. Consider the polyhedron $\mathcal{P} := \{\boldsymbol{y} \in \mathbb{R}^{n-m} : \boldsymbol{W}\boldsymbol{y} \leq \boldsymbol{z}^*\}$. The right hand side of $\mathcal{P}$ corresponds one-to-one to vectors in $\boldsymbol{z}^* - \mathcal{S}$. More precisely, the linear map defined by $\boldsymbol{y} \mapsto \boldsymbol{W}\boldsymbol{y}$ is an isomorphism that maps $\mathcal{P}$ onto $\mathcal{S}$ and its restriction to $\mathbb{Z}^{n-m}$ maps one-to-one to $L^\perp \cap \mathbb{Z}^n$ as $\gcd \boldsymbol{W} = 1$. Therefore, $\mathbf{0} \in \mathcal{P}$ is a vertex of the integer hull of $\mathcal{P}$. From Theorem 3, it follows that $\mathbf{0}$ lies on a face of $\mathcal{P}$ of dimension at most $F(\delta)$. So there are at least $(n-m) - F(\delta)$ tight inequalities, indexed by elements in $I \subseteq [n]$, such that $\mathbf{0} = \boldsymbol{W}_I\mathbf{0} = \boldsymbol{z}_I^*$. We use this to get

$$|\operatorname{supp}(\boldsymbol{z}^*)| \leq n - |I| = n - ((n-m) - F(\delta)) = m + F(\delta) \leq m + F(\Delta)$$

as $\delta \leq \Delta$ and thus $F(\delta) \leq F(\Delta)$ by the monotonicity of $F(\Delta)$. □

5 An Algorithm for the (SVP)

This section describes how to turn the proof of Theorem 1 into an algorithm for solving (SVP). Recall that we assume $n \geq \lfloor(\Delta-1)/2\rfloor \cdot (\Delta-1) + \mathbf{1}_{2\mathbb{Z}}(\Delta) + 1$ throughout this section. We abbreviate $g(\Delta) := \lfloor(\Delta-1)/2\rfloor \cdot (\Delta-1) + \mathbf{1}_{2\mathbb{Z}}(\Delta)$.

To illustrate the idea, suppose that $\boldsymbol{B}$ is an invertible full rank submatrix of $\boldsymbol{A}$ with determinant Δ in absolute value. Let $\boldsymbol{B}^{-1} = (\boldsymbol{r}_1, \ldots, \boldsymbol{r}_n)$. We generate the following test vectors based on the three steps in the proof of Theorem 1. Step 1 implies that

1. $\boldsymbol{r}_i \in \mathbb{Z}^n \backslash \{\mathbf{0}\}$ with $i \in [n]$
   gives a shortest lattice vector $\boldsymbol{A}\boldsymbol{r}_i$. If there exist two columns $\boldsymbol{r}_i, \boldsymbol{r}_j \notin \mathbb{Z}^n$ such that
2. $\boldsymbol{r}_i + \boldsymbol{r}_j, \boldsymbol{r}_i - \boldsymbol{r}_j \in \mathbb{Z}^n \backslash \{\mathbf{0}\}$ for different $i, j \in [n]$,
   then Step 2 ensures that either $\boldsymbol{A}(\boldsymbol{r}_i - \boldsymbol{r}_j)$ or $\boldsymbol{A}(\boldsymbol{r}_i + \boldsymbol{r}_j)$ is a shortest lattice vector. Otherwise, we are in Step 3. In this case, there exist Δ columns of $\pm\boldsymbol{B}^{-1}$ that are contained in the same residue class in $\boldsymbol{B}^{-1}\mathbb{Z}^n$. Let $\boldsymbol{r}_1, \ldots, \boldsymbol{r}_\Delta$ denote these columns. Then one of the following test vectors

3. $\boldsymbol{r}_i - \boldsymbol{r}_j \in \mathbb{Z}^n \setminus \{\boldsymbol{0}\}$ for all $i, j \in [\Delta]$ with $i \neq j$ and
4. $\boldsymbol{r}_1 + \ldots + \boldsymbol{r}_\Delta \in \mathbb{Z}^n \setminus \{\boldsymbol{0}\}$

corresponds to a shortest lattice vector. The remaining issue is how to obtain such a submatrix $\boldsymbol{B}$? Unfortunately, the task of finding a submatrix $\boldsymbol{B}$ with largest subdeterminant in polynomial time for fixed Δ is a major open problem. To circumvent this difficulty, we consider a sequence of invertible submatrices $\boldsymbol{B}_{(l)}$ and generate for each $\boldsymbol{B}_{(l)}$ the four types of test vectors presented above. The key insight is that either one of these test vectors already corresponds to a shortest lattice vector or they jointly provide a certificate that there exists a full rank submatrix $\boldsymbol{B}_{(l+1)}$ of $\boldsymbol{A}$ with $\left|\det \boldsymbol{B}_{(l+1)}\right| > \left|\det \boldsymbol{B}_{(l)}\right|$. This observation gives rise to an iterative procedure. This procedure either terminates with a shortest lattice vector or it generates a submatrix whose determinant is larger than Δ in absolute value. The latter output is a contradiction if we suppose that $\boldsymbol{A}$ is Δ-modular. Hence, the procedure can be viewed as a partial recognition algorithm for testing whether $\boldsymbol{A}$ is Δ-modular.

We next introduce some notation and definitions for a lighter presentation of the algorithm. Let $n \geq g(\Delta)+1$ and $l \geq 0$ be the current iterate. Let $\boldsymbol{B}_{(l)} \in \mathbb{Z}^{n\times n}$ be an invertible matrix with $\left|\det \boldsymbol{B}_{(l)}\right| \leq \Delta$ and $\boldsymbol{B}_{(l)}^{-1} = (\boldsymbol{r}_1^{(l)}, \ldots, \boldsymbol{r}_n^{(l)})$. Suppose that $\boldsymbol{r}_i^{(l)} \notin \mathbb{Z}^n$ for all $i \in [n]$. Let

$$I_{(l)} := \left\{ (\boldsymbol{r}_i^{(l)}, \boldsymbol{r}_j^{(l)}) : \boldsymbol{r}_i^{(l)} + \boldsymbol{r}_j^{(l)}, \boldsymbol{r}_i^{(l)} - \boldsymbol{r}_j^{(l)} \in \mathbb{Z}^n \setminus \{\boldsymbol{0}\} \right\}$$

be the set of test vectors corresponding to Step 2 in the proof of Theorem 1. For Step 3, consider the set $\{\boldsymbol{h}_{j_1}^{(l)}, \ldots, \boldsymbol{h}_{j_\Delta}^{(l)}\} = H_{(l)} \subseteq \pm\{\boldsymbol{r}_1^{(l)}, \ldots, \boldsymbol{r}_n^{(l)}\}$ defined by

$$\boldsymbol{h}_{j_i}^{(l)} - \boldsymbol{h}_{j_k}^{(l)} \in \mathbb{Z}^n \text{ for all } i, k \in [\Delta] \text{ and at most } \boldsymbol{r}_i \text{ or } -\boldsymbol{r}_i \text{ is in } H_{(l)} \text{ for all } i \in [n].$$

Recall that Lemma 2 together with $n \geq g(\Delta) + 1$, $\boldsymbol{r}_i^{(l)} \notin \mathbb{Z}^n$ for all $i \in [n]$, and $I_{(l)} = \emptyset$ ensure the existence of $H_{(l)}$, though, it is in general not unique. Furthermore, let

$$J_{(l)} := \left\{ j \in [n] : \boldsymbol{r}_j^{(l)} \in H_{(l)} \text{ or } -\boldsymbol{r}_j^{(l)} \in H_{(l)} \right\} = \{j_1, \ldots, j_\Delta\}$$

be the set of indices that correspond to columns of $\pm\boldsymbol{B}_{(l)}^{-1}$ that are contained in $H_{(l)}$. Observe that $|J_{(l)}| = |H_{(l)}| = \Delta$ since $H_{(l)}$ contains at most one of $-\boldsymbol{r}_i^{(l)}$ and $\boldsymbol{r}_i^{(l)}$ for all $i \in [n]$. Finally, the set $H_{(l)}$ allows us to define a set of test vectors

$$T_{(l)} := \left\{ \boldsymbol{h} - \boldsymbol{h}' : \boldsymbol{h}, \boldsymbol{h}' \in H_{(l)} \right\} \setminus \{\boldsymbol{0}\} \cup \left\{ \sum_{k=1}^{\Delta} \boldsymbol{h}_{j_k}^{(l)} \right\}.$$

Some of the test vectors in $T_{(l)}$ are used explicitly during the algorithm. We denote them by

$$\boldsymbol{t}_k := \boldsymbol{h}_{j_k}^{(l)} - \boldsymbol{h}_{j_{k+1}}^{(l)} \text{ for } k \in [\Delta - 1] \text{ and } \boldsymbol{s} := \sum_{k=1}^{\Delta} \boldsymbol{h}_{j_k}^{(l)}.$$

Algorithm 1 Polynomial Time Algorithm for the (SVP) when $n \geq g(\Delta)+1$

Input: Full column rank matrix $\boldsymbol{A} \in \mathbb{Z}^{m\times n}$, $n \geq g(\Delta)+1$, and $\Delta \in \mathbb{N}_{\geq 1}$.
Output: Either $\boldsymbol{y} \in \boldsymbol{A}\mathbb{Z}^n$ such that $\|\boldsymbol{y}\|_\infty = 1$ or a full rank submatrix $\boldsymbol{B}$ of $\boldsymbol{A}$ with $|\det \boldsymbol{B}| > \Delta$.

1: Find some invertible full rank submatrix $\tilde{\boldsymbol{B}}$ of $\boldsymbol{A}$. Initialize $l=0$, $\boldsymbol{B}_{(0)} := \tilde{\boldsymbol{B}}$.
2: If $\left|\det \boldsymbol{B}_{(l)}\right| > \Delta$, return $\boldsymbol{B}_{(l)}$.
3: Calculate $\boldsymbol{B}_{(l)}^{-1} = (\boldsymbol{r}_1^{(l)}, \dots, \boldsymbol{r}_n^{(l)})$.
4: If $|\boldsymbol{a}_k^\top \boldsymbol{r}_j^{(l)}| > 1$ for some $k \in [m], j \in [n]$, replace the j-th row of $\boldsymbol{B}_{(l)}$ with $\boldsymbol{a}_k^\top$. Set this matrix to be $\boldsymbol{B}_{(l+1)}$, increment l and go to 2.
5: If $\boldsymbol{r}_j^{(l)} \in \mathbb{Z}^n$ for $j \in [n]$, return $\boldsymbol{y} := \boldsymbol{A}\boldsymbol{r}_j^{(l)}$.
6: Compute $I_{(l)}$ as described above.
7: If $I_{(l)} \neq \emptyset$, let $(\boldsymbol{r}_i^{(l)}, \boldsymbol{r}_j^{(l)}) \in I_{(l)}$ and set $\boldsymbol{t}_+ := \boldsymbol{r}_i^{(l)} + \boldsymbol{r}_j^{(l)}$ and $\boldsymbol{t}_- := \boldsymbol{r}_i^{(l)} - \boldsymbol{r}_j^{(l)}$. If $-\boldsymbol{1} \leq \boldsymbol{A}\boldsymbol{t}_+ \leq \boldsymbol{1}$, return $\boldsymbol{y} := \boldsymbol{A}\boldsymbol{t}_+$, if $-\boldsymbol{1} \leq \boldsymbol{A}\boldsymbol{t}_- \leq \boldsymbol{1}$, return $\boldsymbol{y} := \boldsymbol{A}\boldsymbol{t}_-$. Otherwise, collect, for $\boldsymbol{t}_+$ and $\boldsymbol{t}_-$ rows $\boldsymbol{a}_+^\top$ and $\boldsymbol{a}_-^\top$ such that $|\boldsymbol{a}_+^\top \boldsymbol{t}_+| \geq 2$ and $|\boldsymbol{a}_-^\top \boldsymbol{t}_-| \geq 2$. Replace the rows of $\boldsymbol{B}_{(l)}$ indexed by i and j with rows $\boldsymbol{a}_+^\top$ and $\boldsymbol{a}_-^\top$. Set this matrix to be $\boldsymbol{B}_{(l+1)}$, increment l and go to 2.
8: Compute $H_{(l)}$, $J_{(l)}$, and $T_{(l)}$ as described above.
9: If $-\boldsymbol{1} \leq \boldsymbol{A}\boldsymbol{t} \leq \boldsymbol{1}$ for $\boldsymbol{t} \in T_{(l)}$, return $\boldsymbol{y} := \boldsymbol{A}\boldsymbol{t}$. Otherwise, collect, for every $\boldsymbol{t} \in T_{(l)}$ a row $\boldsymbol{a}_{\boldsymbol{t}}^\top$ of $\boldsymbol{A}$ such that $|\boldsymbol{a}_{\boldsymbol{t}}^\top \boldsymbol{t}| \geq 2$.
10: If $\boldsymbol{a}_{\boldsymbol{t}_k}^\top$ does not satisfy (8) for $k \in [\Delta-1]$, compute $i \in \{j_k, j_{k+1}\}$, $j \in J_{(l)} \setminus \{j_k, j_{k+1}\}$ such that $\operatorname{sign}(\boldsymbol{a}_{\boldsymbol{t}_k}^\top \boldsymbol{h}_i^{(l)}) = \operatorname{sign}(\boldsymbol{a}_{\boldsymbol{t}_k}^\top \boldsymbol{h}_j^{(l)})$. Replace the rows of $\boldsymbol{B}_{(l)}$ indexed by i, j with the rows $\boldsymbol{a}_{\boldsymbol{t}_k}^\top$ and $\boldsymbol{a}_{\boldsymbol{h}_i^{(l)} - \boldsymbol{h}_j^{(l)}}^\top$. Set this matrix to be $\boldsymbol{B}_{(l+1)}$, increment l and go to 2.
11: Replace the rows of $\boldsymbol{B}_{(l)}$ indexed by $J_{(l)}$ with the rows $\boldsymbol{a}_{\boldsymbol{t}_1}^\top, \dots, \boldsymbol{a}_{\boldsymbol{t}_{\Delta-1}}^\top, \boldsymbol{a}_{\boldsymbol{s}}^\top$. Set this matrix to be $\boldsymbol{B}_{(l+1)}$, increment l and go to 2.

Equipped with this notation, the algorithm is given in Algorithm 1.

Proof (of Theorem 2). The correctness of Algorithm 1 follows by construction. We show that it terminates after finitely many iterations. To do so, it suffices to check that we increment index l finitely many times. In fact, one can show that $l \leq \Delta$. To reach a new increment, we have to update the matrix $\boldsymbol{B}_{(l)}$. Our claim is that $1 + \left|\det \boldsymbol{B}_{(l)}\right| \leq \left|\det \boldsymbol{B}_{(l+1)}\right|$ for all $l \geq 0$, which implies termination after finitely many iterations. We update the matrix $\boldsymbol{B}_{(l)}$ in Steps 4, 7, 10, and 11. In each of these steps we consider a submatrix of $\boldsymbol{A} \cdot \boldsymbol{B}_{(l)}^{-1}$ of the form $(\boldsymbol{A} \cdot \boldsymbol{B}_{(l)}^{-1})_{K,I}$, where $K \subseteq [m]$ is the index set of new rows of $\boldsymbol{A}$ that we add to obtain $\boldsymbol{B}_{(l+1)}$ and $I \subseteq [n]$ is the index set of rows of $\boldsymbol{B}_{(l)}$ that we replace. In each of the four steps, we claim that $1 < |\det(\boldsymbol{A} \cdot \boldsymbol{B}_{(l)}^{-1})_{K,I}|$ holds. Suppose that this is true. Then Lemma 1 implies that

$$1 < \left|\det(\boldsymbol{A} \cdot \boldsymbol{B}_{(l)}^{-1})_{K,I}\right| = \frac{\left|\det \boldsymbol{B}_{(l+1)}\right|}{\left|\det \boldsymbol{B}_{(l)}\right|}$$

and $1 + \left|\det \boldsymbol{B}_{(l)}\right| \leq \left|\det \boldsymbol{B}_{(l+1)}\right|$ follows as both determinants are integers. One can show that $1 < |\det(\boldsymbol{A} \cdot \boldsymbol{B}_{(l)}^{-1})_{K,I}|$ since such a submatrix $(\boldsymbol{A} \cdot \boldsymbol{B}_{(l)}^{-1})_{K,I}$ is either an entry of $\boldsymbol{A} \cdot \boldsymbol{B}_{(l)}^{-1}$ or of the form (6), (7), or (9). □

Acknowledgements. The authors thank the reviewers for their valuable comments, which have led to an improved upper bound on $f(\Delta)$, and Joe Paat for helpful remarks on an earlier version of this paper.

Disclosure of Interests. The authors have no competing interests to declare that are relevant to the content of this article.

References

1. Aggarwal, D., Chen, Y., Kumar, R., Shen, Y.: Improved classical and quantum algorithms for the shortest vector problem via bounded distance decoding. SIAM J. Comput. **54**(2), 233–278 (2025)
2. Aggarwal, D., Dadush, D., Regev, O., Stephens-Davidowitz, N.: Solving the shortest vector problem in 2^n time using discrete gaussian sampling: extended abstract. In: Proceedings of the Forty-Seventh Annual ACM on Symposium on Theory of Computing, STOC, pp. 733–742 (2015)
3. Aggarwal,D., Mukhopadhyay, P.: Improved algorithms for the shortest vector problem and the closest vector problem in the infinity norm. In: 29th International Symposium on Algorithms and Computation, ISAAC, vol. 123, LIPIcs, pp. 35:1–35:13 (2018)
4. Aggarwal, D., Stephens-Davidowitz, N.: Just take the average! an embarrassingly simple 2^n-time algorithm for SVP (and CVP). In: 1st Symposium on Simplicity in Algorithms, SOSA, vol. 61 of OASIcs, pp. 12:1–12:19 (2018)
5. Ajtai, M., Kumar, R., Sivakumar, D.: A sieve algorithm for the shortest lattice vector problem. In: Proceedings of the Thirty-Third Annual ACM Symposium on Theory of Computing, pp. 601–610 (2001)
6. Aliev, I., Averkov, G., De Loera, J.A., Oertel, T.: Sparse representation of vectors in lattices and semigroups. Math. Prog. Ser. B **192**, 519–546 (2022)
7. Aliev, I., Henk, M., Hogan, M., Kuhlmann, S., Oertel, T.: New bounds for the integer Carathéodory rank. SIAM J. Optim. **34**(1), 190–200 (2024)
8. Aliev, I., De Loera, J.A., Eisenbrand, F., Oertel, T., Weismantel, R.: The support of integer optimal solutions. SIAM J. Optim. **28**, 2152–2157 (2018)
9. Aliev, I., De Loera, J.A., Oertel, T., O'Neill, C.: Sparse solutions of linear Diophantine equations. SIAM J. Appl. Algebra Geom. **1**, 239–253 (2017)
10. Aprile, M., et al.: Integer programs with nearly totally unimodular matrices: the cographic case. In: Proceedings of the Annual ACM-SIAM Symposium on Discrete Algorithms (SODA), pp. 2301–2312 (2025)
11. Artmann, S., Weismantel, R., Zenklusen, R.: A strongly polynomial algorithm for bimodular integer linear programming. In: Proceedings of the 49th Annual ACM SIGACT Symposium on Theory of Computing, pp. 1206–1219 (2017)
12. Berndt, S., Jansen, K., Klein, K.-M.: New bounds for the vertices of the integer hull. In: Symposium on Simplicity in Algorithms (SOSA), pp. 25–36 (2021)
13. Berndt, S., Mnich, M., Stamm, T.: New support size bounds and proximity bounds for integer linear programming. In: International Conference on Current Trends in Theory and Practice of Computer Science, pp. 82–95. Springer (2024)

14. Blömer, J., Naewe, S.: Sampling methods for shortest vectors, closest vectors and successive minima. In: Arge, L., Cachin, C., Jurdziński, T., Tarlecki, A. (eds.) ICALP 2007. LNCS, vol. 4596, pp. 65–77. Springer, Heidelberg (2007). https://doi.org/10.1007/978-3-540-73420-8_8
15. Celaya, M., Kuhlmann, S., Paat, J., Weismantel, R.: Proximity and flatness bounds for linear integer optimization. Math. Oper. Res. **49**(4), 2446–2467 (2023)
16. Celaya, M., Kuhlmann, S., Weismantel, R.: On matrices over a polynomial ring with restricted subdeterminants. In: Integer Programming and Combinatorial Optimization, pp. 43–56 (2024)
17. Cvetković, D., Rowlinson, P., Simić, S.: An Introduction to the Theory of Graph Spectra. London Mathematical Society Student Texts, Cambridge University Press (2009)
18. Eisenbrand, F.: Integer programming and algorithmic geometry of numbers - a tutorial. In: 50 Years of Integer Programming 1958-2008 - From the Early Years to the State-of-the-Art, pp. 505–559. Springer (2010)
19. Eisenbrand, F., Shmonin, G.: Carathéodory bounds for integer cones. Oper. Res. Lett. **34**(5), 564–548 (2006)
20. Fiorini, S., Joret, G., Weltge, S., Yuditsky, Y.: Integer programs with bounded subdeterminants and two nonzeros per row. J. ACM 72(1), (2025)
21. Gama, N., Nguyen, P.Q., Regev, O.: Lattice enumeration using extreme pruning. In: Gilbert, H. (ed.) EUROCRYPT 2010. LNCS, vol. 6110, pp. 257–278. Springer, Heidelberg (2010). https://doi.org/10.1007/978-3-642-13190-5_13
22. Gribanov, D., Shumilov, I., Malyshev, D., Pardalos, P.: On Δ-modular integer linear problems in the canonical form and equivalent problems. J. Global Optim. **88**(3), 591–651 (2024)
23. Gribanov, D.V., Malyshev, D.S., Pardalos, P.M., Veselov, S.I.: FPT-algorithms for some problems related to integer programming. J. Comb. Optim. **35**(4), 1128–1146 (2018). https://doi.org/10.1007/s10878-018-0264-z
24. Gribanov, D.V., Shumilov, I.A., Malyshev, D.S.: Structured (min,+)-convolution and its applications for the shortest/closest vector and nonlinear knapsack problems. Optimiz.Lett. **18**, 73–103 (2024)
25. Hanrot, G., Stehlé, D.: Improved analysis of Kannan's shortest lattice vector algorithm. In: Advances in Cryptology - Crypto 2007. Proceedings, pp. 170–186 (2007)
26. Helfrich, B.: Algorithms to construct minkowski reduced an hermite reduced lattice bases. Theor. Comput. Sci. **41**, 125–139 (1985)
27. Hodge, W.V.D., Pedoe, D.: Methods of Algebraic Geometry. Cambridge University Press, Cambridge Mathematical Library (1994)
28. Hoffman, A.J., Kruskal, J.B.: Integral boundary points of convex polyhedra. In: Kuhn, H.W., Tucker, A.J., eds., Linear Inequalities and Related Systems, pp. 223–246 (1956)
29. Kannan, R.: Improved algorithms for integer programming and related problems. In: Proceedings of of STOC, pp. 193 – 206 (1983)
30. Kannan, R.: Minkowski's convex body theorem and integer programming. Math. Oper. Res. **12**(3), 415–440 (1987)
31. Kannan, R., Bachem, A.: Polynomial algorithms for computing the smith and hermite normal forms of an integer matrix. SIAM J. Comput. **8**(4), 499–507 (1979)
32. Kober,S.: Totally Δ-modular IPs with two non-zeros in most rows. In: Integer Programming and Combinatorial Optimization, pp. 355–370 (2025)

33. Lee, J., Paat, J., Stallknecht, I., Xu, L.: Improving proximity bounds using sparsity. In: Combinatorial Optimization: 6th International Symposium ISCO, pp. 115–127 (2020)
34. Lenstra, A.K., Lenstra, H.W., Lovász, L.: Factoring polynomials with rational coefficients. Math. Ann. **261**(4), 515–534 (1982)
35. Micciancio, D., Regev,O.: Lattice-based cryptography. In: Post-Quantum Cryptography, pp. 147–191. Springer Berlin Heidelberg (2009)
36. Micciancio, D., Voulgaris, P.: Faster exponential time algorithms for the shortest vector problem. In: Proceedings of the Twenty-First Annual ACM-SIAM Symposium on Discrete Algorithms, SODA, pp. 1468–1480 (2010)
37. Micciancio, D., Voulgaris, P.: A deterministic single exponential time algorithm for most lattice problems based on Voronoi cell computations. SIAM J. Comput. **42**(3), 1364–1391 (2013)
38. Micciancio, D., Walter, M.: Fast Lattice Point Enumeration with Minimal Overhead, pp. 276–294. In Proceedings of the Twenty-Sixth Annual ACM-SIAM Symposium on Discrete Algorithms, SODA (2015)
39. Nägele, M., Nöbel, C., Santiago, R., Zenklusen, R.: Advances on strictly Δ-Modular IPs. Math. Prog **210**, 731–760 (2025)
40. Nägele, M., Santiago, R., Zenklusen, R.: Congruency-constrained TU problems beyond the bimodular case. Math. Oper. Res. **49**(3), 1303–1348 (2023)
41. Nguyen, P., Vidick, T.: Sieve algorithms for the shortest vector problem are practical. J. Math. Crypt. **2**, 181–207 (2008)
42. Oertel, T., Paat, J., Weismantel, R.: The distributions of functions related to parametric integer optimization. SIAM J. Appl. Algeb. Geom. **4**(3), 422–440 (2020)
43. Oxley, J., Walsh, Z.: 2-modular matrices. SIAM J. Discret. Math. **36**(2), 1231–1248 (2022)
44. Pujol, X., Stehle,D.: Solving the shortest lattice vector problem in time $2^{2.465n}$. Cryptology ePrint Archive, Paper 2009/605 (2009)
45. Regev, O.: Lattices in Computer Science. Lecture Notes Tel Aviv University (2004)
46. Schrijver, A.: Theory of Linear and Integer Programming, Wiley (1986)
47. van Emde Boas, P.: Another NP-complete problem and the complexity of computing short vectors in a lattice. Technical Report 81-04, Mathematische Instituut, University of Amsterdam (1981)
48. Veselov, S.I., Chirkov, A.J.: Integer program with bimodular matrix. Discret. Optim. **6**, 220–222 (2009)
49. Veselov, S.I., Shevchenko,V.N.: Estimates of minimal distance between point of some integral lattices. Combinat. Algeb. Meth. Appl. Math., pp. 26–33 (1980)

Lattice Structure and Efficient Basis Construction for Strongly Connected Orientations

Siyue Liu(✉) and Olha Silina

Carnegie Mellon University, Pittsburgh, PA, USA
siyueliu@andrew.cmu.edu, osilina@andrew.cmu.edu

Abstract. Let $D = (V, A)$ be a digraph whose underlying undirected graph is 2-edge-connected. A subset of arcs $J \subseteq A$ is a strengthening if reversing the arcs of J makes D strongly connected. Given a family $\mathcal{F}$ of proper subsets of V, we call a strengthening J tight if after flipping J, there is exactly one arc entering U for every $U \in \mathcal{F}$. We give a polynomial-time algorithm to construct a set $\mathcal{B}$ consisting of tight strengthenings which forms an integral basis for the linear hull of tight strengthenings. That is, $\mathcal{B}$ is a linearly independent subset of tight strengthenings, and every integer vector in the linear hull of tight strengthenings can be written as an integral combination of $\mathcal{B}$. This extends the main result of Abdi, Cornuéjols, Liu, and Silina (IPCO 2025), who gave a non-constructive proof of the existence of such a basis. While the previous proof uses polyhedral theory, our proof is purely combinatorial and yields a polynomial-time algorithm. As an application of our algorithm, we show that parity-constrained tight strongly connected orientations can be solved in deterministic polynomial time. Along the way, we discover appealing connections to the theory of perfect matching lattices.

Keywords: Strongly connected orientations · Lattices · Congruency-constrained problems

1 Introduction

Let $D = (V, A)$ be a digraph whose underlying undirected graph is 2-edge-connected. For a subset of arcs $J \subseteq A$, denote by J^{-1} the arcs obtained by reversing the direction of the arcs in J. A subset of arcs $J \subseteq A$ is a *strongly connected reorientation (SCR)*, or *strengthening* if the digraph obtained from flipping the arcs in J, i.e., $D_J := \big(V, (A \setminus J) \cup J^{-1}\big)$, is strongly connected. Introduced by [43], strengthenings are important objects in digraphs [8,21,31]. In this work, we consider strengthenings that satisfy additional constraints defined by an arbitrary family $\mathcal{F} \subseteq 2^V \setminus \{\emptyset, V\}$ of proper subsets of V. Given $\mathcal{F}$, we say a strengthening J is *tight* for $(D, \mathcal{F})$ if after flipping the arcs of J, for every subset $U \in \mathcal{F}$ there is exactly one arc coming into U, i.e., $|\delta^-_{D_J}(U)| = 1$. We denote by $\mathrm{SCR}(D, \mathcal{F})$ the set of tight strengthenings of $(D, \mathcal{F})$.

S. S. Dey et al. (Eds.): IPCO 2026, LNCS 16588, pp. 204–219, 2026.
https://doi.org/10.1007/978-3-032-28691-8_14

Given a finite subset $R \subseteq \mathbb{Z}^A$, the *lattice* generated by R, denoted by lat(R), is the set of all integer linear combinations of the vectors in R. A *lattice basis for L* is a set $\mathcal{B}$ of linearly independent vectors that generates L. It is known that every lattice has a basis, but the generating set R does not necessarily contain a basis for lat(R). We study the *lattice of tight strengthenings*, i.e., the lattice generated by tight strengthenings, denoted by lat(SCR($D, \mathcal{F}$)). Denote by lin(SCR($D, \mathcal{F}$)) the linear hull of tight strengthenings. Clearly, lat(SCR($D, \mathcal{F}$)) $\subseteq$ lin(SCR($D, \mathcal{F}$)) $\cap \mathbb{Z}^A$. Given a rational linear subspace $L \subseteq \mathbb{R}^A$, an *integral basis for L* is a subset $\mathcal{B} \subseteq L \cap \mathbb{Z}^A$ of linearly independent vectors such that every vector in $L \cap \mathbb{Z}^A$ can be written as an integral linear combination of $\mathcal{B}$. Since the dimension of lin(SCR($D, \emptyset$)) is at most $|A|$, there are at most $|A|$ many sets $U \in \mathcal{F}$ for which the constraints $|\delta^-_{D_J}(U)| = 1$ are linearly independent. Thereby we may assume w.l.o.g. that $|\mathcal{F}| \leq |A|$, which implies the input size $(D, \mathcal{F})$ is bounded by the size of the digraph. Our main result is the following:

Theorem 1. *Let $D = (V, A)$ be a digraph whose underlying undirected graph is 2-edge-connected. Let $\mathcal{F} \subseteq 2^V \backslash \{\emptyset, V\}$ such that $|\mathcal{F}| \leq |A|$ and* $\gcd\{1-|\delta^-(U)| : U \in \mathcal{F}\} = 1$. *Then, we can construct an integral basis for* lin(SCR($D, \mathcal{F}$)) *consisting of tight strengthenings in polynomial time.*

This result implies two things: first, lat(SCR($D, \mathcal{F}$)) = lin(SCR($D, \mathcal{F}$)) $\cap \mathbb{Z}^A$; second, there is a basis for lat(SCR($D, \mathcal{F}$)) that consists of tight strengthenings. We remark that the gcd condition is necessary for the first thing to be true. Our work extends the previous result of Abdi, Cornuéjols and the authors [2], which proves the existence of such a basis. The previous proof is non-constructive, obtained from studying the implicit equalities and facets of the polytope of tight strengthenings. The main contribution of this paper is a new combinatorial proof that leads to a polynomial-time algorithm to construct such a basis.

Our main motivation for studying tight strengthenings is the following conjecture by Woodall [45]. In a digraph, a *dicut* is the set of arcs leaving a nonempty proper vertex subset with no incoming arcs, i.e., an arc set of the form $\delta^+(U)$ for some $\emptyset \neq U \subsetneq V$ such that $\delta^-(U) = \emptyset$. A *dijoin* is an arc subset that intersects every dicut at least once. In order to make the digraph strongly connected, at least one arc from each dicut has to be flipped, and thus, every strengthening is a dijoin. Woodall conjectured that the maximum number of disjoint dijoins is equal to the minimum size of a dicut. Schrijver [43] reformulated Woodall's conjecture in the following:

Conjecture 2 ([43]). Let $\tau \geq 2$ be an integer, and let $D = (V, A)$ be a digraph where the minimum size of a dicut is τ. Then A can be partitioned into τ strengthenings.

Observe that if such a statement were true, the only strengthenings allowed in a partition would be the ones that intersect every minimum dicut *exactly* once. Therefore, it suffices to restrict ourselves to those strengthenings, which are precisely SCR($D, \mathcal{F}$) for $\mathcal{F}$ being the family of sets $U \subseteq V$ that induce minimum dicuts $\delta^+(U)$. Theorem 1 implies a relaxation of Conjecture 2 that allows adding and subtracting tight strengthenings of $(D, \mathcal{F})$ [2]. Formally, it implies that we can find an assignment $\lambda_J \in \mathbb{Z}$ to every strengthening J that intersects every minimum dicut exactly once such that $\sum_J \lambda_J \mathbf{1}_J = \mathbf{1}_A$ and $\sum_J \lambda_J = \tau$.

As a new application, we consider the parity-constrained strongly connected orientations problem. Given a 2-edge-connected undirected graph $G = (V, E)$, its *bidirected* graph $\vec{G} = (V, E^+ \cup E^-)$ is obtained from copying every edge $e \in E$ twice and orienting them in the opposite directions, denoted as $e^+ \in E^+$ and $e^- \in E^-$. Let $\mathcal{F} \subseteq 2^V \setminus \{\emptyset, V\}$ and let $R \subseteq E^+ \cup E^-$ be a subset of arcs that are colored red. A *strongly connected orientation (SCO)* is a subset $\vec{E} \subseteq E^+ \cup E^-$ which contains exactly one of e^+, e^- for every $e \in E$ and satisfies $|\delta^-_{\vec{E}}(U)| \geq 1$ for every $\emptyset \neq U \subsetneq V$. A *tight strongly connected orientation (SCO)* of $(G, \mathcal{F})$ is an SCO $\vec{E} \subseteq E^+ \cup E^-$ such that $|\delta^-_{\vec{E}}(U)| = 1$ for every $U \in \mathcal{F}$. Our main Theorem 1 implies we can solve the following problem in deterministic polynomial time.

Parity Strongly Connected Orientation (Parity SCO). Find a tight SCO with an odd (even) number of red arcs if it exists.

We provide a deterministic polynomial-time algorithm for this problem as an application of our basis construction algorithm in Theorem 1. We will leave its proof to the full version of the paper [27].

Theorem 3. *Parity SCO can be solved in deterministic polynomial time.*

1.1 Proof Overview

We outline the proof of Theorem 1 here.

Reduction to Basic Digrafts. First, we reduce from general digraphs to *digrafts*. A digraph is *bipartite* if every node is a source or a sink.

Definition 1 (digraft). *A* digraft *is a pair (D, S^t) where $D = (V = S \dot\cup T, A)$ is a bipartite digraph with sources S and sinks T such that $|S| \leq |T|$, whose underlying undirected graph is 2-edge-connected, and $S^t \subsetneq S$ is a subset of sources with $|S^t| \leq |S| - 1$. Denote by $V^t := S^t \cup T$.*

We will be interested in dijoins satisfying degree constraints defined by $v \in V^t$. A dijoin is *tight* if it has exactly one arc incident to every $v \in V^t$. Denote by $\mathcal{J}(D, S^t)$ the set of tight dijoins of (D, S^t). A digraft is *tight dijoin-covered* if every arc $e \in A$ is contained in some tight dijoin. A dicut is *tight* if every tight dijoin intersects it exactly once. A node $v \in V$ is *tight* if every tight dijoin has degree 1 on v. Denote the set of tight nodes as V^t and the set of tight sources as S^t. Note that every $v \in V^t$ is a tight node, and thus $V^t \subseteq V^t$ and $S^t \subseteq S^t$, but there could be more vertices in $V^t \setminus V^t$ that are forced to have degree 1 for every tight dijoin. Also notice that every sink is tight by definition, so $V^t \setminus V^t$ consists solely of sources. By definition of tight nodes, $\mathcal{J}(D, S^t) = \mathcal{J}(D, S^t)$. The order in which we define tight dijoins and tight dicuts (nodes) is important. Tight dijoins are explicitly defined as degree-constrained dijoins, while tight dicuts (nodes) are implicit, dependent on tight dijoins.

The definition of digrafts can be generalized to a pair $(D, \mathcal{F})$ where $\mathcal{F}$ consists of an arbitrary family of dicut out-shores that includes $V \setminus v$ for all $v \in T$. A tight dijoin of $(D, \mathcal{F})$ is defined to be a dijoin that intersects $\delta^+(U)$ exactly once for every $U \in \mathcal{F}$. Using standard uncrossing technique, we may assume that $\mathcal{F}$ is a *cross-free* family. Our setting in Definition 1 corresponds to the case where $\mathcal{F} := \{\{v\} : v \in S^t\} \bigcup \{V \setminus v : v \in T\}$. It is proved by [2] that one can reduce basis construction for tight strengthenings of a general digraph D and family $\mathcal{F}$ to the one for tight dijoins of some digraft $(D', \mathcal{F}')$, using a bijection between $\text{SCR}(D, \mathcal{F})$ and $\mathcal{J}(D', \mathcal{F}')$ [15,43].

Further, general digrafts $(D, \mathcal{F})$ can be reduced to digrafts (D, S^t) in Definition 1 through *tight dicut decomposition*, which is described as follows. A dicut $\delta^+(U)$ is *nontrivial* if $1 < |U| < |V|-1$; otherwise, it is *trivial*. If there is some $U \in \mathcal{F}$ with $1 < |U| < |V|-1$, it induces a nontrivial tight dicut $C = \delta^+(U)$. By contracting the shores U, $V \setminus U$ while preserving the cross-free family $\mathcal{F}$ induced on the smaller digraphs, one obtains two digrafts, called *C-contractions*. When C is a nontrivial tight dicut, there is a natural way to obtain an integral basis for $\text{lin}(\mathcal{J}(D, \mathcal{F}))$ by combining the bases of the two C-contractions ([2] Lemma 10), which can be done in polynomial time. After repeatedly applying this process, we may assume $\mathcal{F}$ contains only trivial dicut out-shores. In this paper, we restrict ourselves to the setting in Definition 1 and treat the general setting $(D, \mathcal{F})$ in the full version of the paper.

A digraft (D, S^t) may still have (implicit) nontrivial tight dicuts. The next step is to further apply tight dicut decomposition on (D, S^t). To do this algorithmically, we identify two types of tight dicuts: *barrier dicuts* and *2-separation dicuts*. The terminology is inspired by matching theory [29,30]. The following theorem allows us to detect nontrivial tight dicuts efficiently.

Theorem 4. *A tight dijoin-covered digraft (D, S^t) has no nontrivial tight dicuts if and only if it does not contain any nontrivial barrier dicuts or 2-separation dicuts. Moreover, we can find a nontrivial tight dicut in polynomial time, if one exists.*

After applying tight dicut decomposition, we obtain a list of *basic digrafts*, defined as follows.

Definition 2 (basic digraft, brick, brace). *A digraft (D, S^t) is* basic *if it is tight dijoin-covered, and does not contain any nontrivial tight dicuts. A basic digraft (D, S^t) is a* brick *if $|S| < |T|$; it is a* brace *if $|S| = |T|$.*

The terminology of bricks and braces is inspired by the theory of matching lattices by Lovász [29], and Edmonds, Pulleyblank and Lovász [18]. Interestingly, in a tight dijoin-covered digraft (D, S^t) with $|S| = |T|$, a subset $J \subseteq A$ is a tight dijoin if and only if it is a perfect matching (a direct consequence of Lemma 2). So, our definition of braces is identical to the one for perfect matchings in bipartite graphs [29]. This is not the case for bricks. Nevertheless, we reuse the terminology due to its crucial role in determining the dimension of the lattice (Theorem 5). Similar to the perfect matching setting [18], we show that the results of any two tight dicut decomposition procedures on a digraft $(D, \mathcal{F})$ are the same list of bricks and braces up to the multiplicities of the arcs (Proposition 1). In particular, the number of bricks after applying tight dicut decomposition to $(D, \mathcal{F})$ is invariant, which we denote as $b(D, \mathcal{F})$. The following is our main theorem for digrafts:

Theorem 5. *Let $(D = (V, A), \mathcal{F})$ be a tight dijoin-covered digraft. Then, we can construct an integral basis $\mathcal{B}$ for $\mathrm{lin}(\mathcal{J}(D, \mathcal{F}))$ that consists of tight dijoins in polynomial time. Moreover, $|\mathcal{B}| = |A| - |V^t| - b(D, \mathcal{F}) + 2$.*

From now, we may assume the digraft is either a brick or a brace. The case of braces is identical to the setting of bipartite matchings so we omit it. We sketch the main steps in our proof on how to deal with bricks. The first step is to reduce the digraft to *robust* ones, and then to *elementary* ones. We first motivate the definition of robust digrafts and show how to reduce robust digrafts to elementary digrafts. Then, we show how to reduce non-robust bricks.

Robust Digrafts. An *edge cover* is a subset of edges that has degree at least 1 on every $v \in V$. We say an edge cover is *tight* if it has degree 1 on every $v \in V^t$. Since in a digraft every node $v \in V$ induces a dicut $\delta(v)$, every tight dijoin is a tight edge cover. We say a tight dijoin-covered digraft is *robust* if the converse also holds:

Definition 3 (robust digraft). *A digraft (D, S^t) is* robust *if it is tight dijoin-covered and every tight edge cover is a tight dijoin.*

Moving to robust digrafts allows us to simplify the problem, because a tight dijoin only needs to cover trivial dicuts as opposed to the entire family of dicuts. If a digraft is not robust, there must exist a dicut that is not intersected by some tight edge cover, which is called a *separating dicut*. A crucial step of our algorithm is to either confirm the digraft is robust or return a separating dicut:

Theorem 6. *There exists a polynomial-time algorithm that, given a digraft (D, S^t), checks whether (D, S^t) is robust. Moreover, if the answer is no, it returns a separating dicut.*

The way to construct integral basis for $\mathrm{lin}(\mathcal{J}(D, S^t))$ for a robust digraft (D, S^t) goes as follows. We show that in a robust digraft (D, S^t) with $|S^t| < |S| - 1$, the digraft $(D, S^t \cup \{v\})$ obtained from setting an arbitrary $v \in S \setminus S^t$ to be tight remains robust. Furthermore, there exists a tight dijoin J of (D, S^t) that has exactly 2 arcs incident with v. Adding J to an integral basis for $\mathrm{lin}(\mathcal{J}(D, S^t \cup \{v\}))$ gives us an integral basis for $\mathrm{lin}(\mathcal{J}(D, S^t))$. Repeatedly setting the vertices in $S \setminus S^t$ to be tight, we arrive at digrafts for which $|S^t| = |S| - 1$. This motivates the definition of *elementary digrafts*.

Definition 4 (elementary digraft). *A digraft (D, S^t) is* elementary *if it is tight dijoin-covered and $|S^t| = |S| - 1$.*

Elementary digrafts are the base case of the entire algorithm. Note that in an elementary digraft, every tight dijoin has a fixed degree on every $v \in V$.

In particular, when $|S| = |T|$, a tight dijoin is a perfect matching. We generalize the *ear decomposition* for bipartite matchings (see e.g. [39] Ch 4.1) to elementary digrafts. This gives us an algorithm to construct an integral basis for $\text{lin}(\mathcal{J}(D, S^t))$. We leave the details to the full version of the paper.

Non-robust Bricks. The most challenging part of our proof is to reduce a brick (D, S^t) if it is not robust. To this end, we will either move a non-tight vertex $u \in S \setminus S^t$ into S^t, or contract a nontrivial dicut. Both cases are equivalent to setting some non-tight dicut to be tight. In the first case, we will eventually reach elementary digrafts; in the second case, we reduce the sizes of the digrafts. Ideally, we would like to find a non-tight dicut C such that after setting C to be tight, the dimension of the lattice only drops by one; we will refer to this type of dicuts as *good* (to be defined later in Definition 6). We give a combinatorial characterization of such dicuts C in terms of the number of bricks obtained after setting C to be tight; it can also be described as maximal dicuts with respect to a certain partial order on the set of all non-tight dicuts.

We now elaborate on the algorithm of finding a good dicut. Recall that if a digraft is not robust, we can use Theorem 6 to find a separating dicut. This dicut will be a starting point of a search for a good dicut. The algorithm of finding a good dicut is a dual procedure where we keep looking for improving dicuts C in the partial order (as summarized in Theorems 11 and 12). We show that the dicuts we find are linearly independent and thus the length of the procedure is bounded by the dimension of the linear hull. The structural Theorem 4 plays a crucial role in finding improving dicuts.

Finally, after finding a good dicut C, we set C to be tight if it is trivial or contract C if it is nontrivial. We construct a basis for the tight dijoins that intersect C exactly once. This can be obtained from the basis of $\text{lin}(\mathcal{J}(D, S^t \cup \{u\})$ if C is trivial and induced by some $u \in S \setminus S^t$, or merging two bases of the C-contractions if C is nontrivial. It turns out that we only need to add one additional tight dijoin J to the basis, with the property that $|J \cap C| = 2$. The existence of such a tight dijoin follows from the *Jump-free lemma* in [2] and can be found in polynomial time. This concludes the basis construction for non-robust bricks. Figure 1 summarizes the main steps of the proof.

1.2 Related Work

Connections to Woodall's Conjecture. Schrijver [41] refuted a conjecture of Edmonds and Giles [19] stating that for every weighted digraph with weights $w \in \mathbb{Z}^A_{\geq 0}$, the maximum packing of dijoins is equal to the minimum weight of a dicut. By [43], the conjecture is equivalent to the following: taking $\mathcal{F}$ to be the out-shores of the family of minimum dicuts, every integer vector in the cone generated by $\text{SCR}(D, \mathcal{F})$ can be written as a *nonnegative* integer linear combination of tight strengthenings. Theorem 1 shows that if we relax the nonnegativity such a statement would be true. Although Edmonds-Giles conjecture has been refuted, such a statement remains interesting, since Woodall's conjecture [45] is essentially Edmonds-Giles conjecture in the unweighted digraphs, which is still

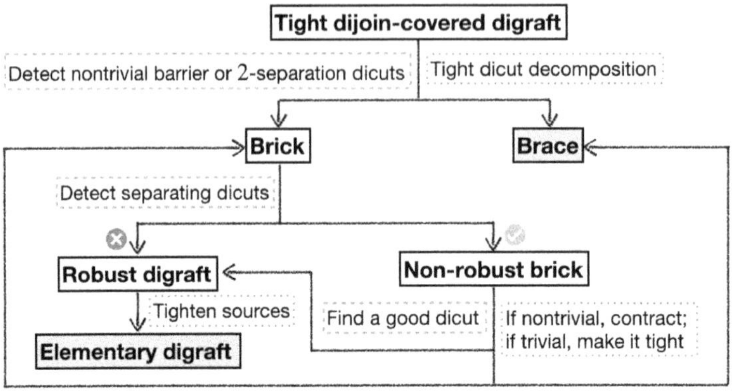

Fig. 1. The solid boxes are types of digrafts. The dashed boxes are steps involved in finding specific types of dicuts. The dotted boxes are steps executed to reduce the digrafts. The end of the procedure are elementary digrafts and braces.

widely open. Another consequence of Theorem 1 is that for every prime $p \geq 2$ and every digraph with minimum dicut size τ, there is a p-adic packing of dijoins of value τ [2], extending a result of Guenin and Hwang [25] which proves the existence of a dyadic packing of value τ. Our main Theorem 1 allows us to find such a packing in polynomial time. This can be done by first restricting ourselves to polynomially many variables as in the sparse packing procedure in [4] and then using an algorithm for p-adic programming in [1].

Comparison to Existing Proofs. Abdi, Cornuéjols, Liu and Silina [2] initiated the study of the lattice of tight strengthenings. They proved a non-constructive version of Theorem 1. Their argument can be summarized as follows. First, they use the same reduction to the digraft setting $(D, \mathcal{F})$. They closely work with the polyhedron of tight dijoins, $\mathrm{DIJ}(D, \mathcal{F}) := \{x \in \mathbb{R}^A_{\geq 0} : x(\delta^+(U)) \geq 1\ \forall \text{ dicut } U;\ x(\delta^+(U)) = 1\ \forall U \in \mathcal{F}\}$, which is given by [31]. They reduce to basic digrafts by contracting tight dicuts. This step is non-constructive. Observe that $\mathrm{DIJ}(D, \mathcal{F})$ of the basic digraft does not have implicit equalities except those defined by trivial tight dicuts induced by tight nodes. If all facet-defining inequalities (FDI) are defined by trivial dicuts or nonnegativity constraints, the digraft is robust (this is indeed equivalent to Definition 3 because the bipartite edge cover polytope is integral, see e.g. [42] Ch 19.5). Using the integer decomposition property of tight edge covers and a result of Gerards and Sebő [24], they show that an integral basis consisting of tight dijoins exists. Otherwise, there is a nontrivial dicut $C = \delta^+(U)$ that induces an FDI. They set C to be tight, which reduces the dimension of $\mathrm{lin}(\mathcal{J}(D, \mathcal{F}))$ by one. Using the Jump-free lemma, they find one additional tight dijoin J with $|J \cap C| = 2$ to add to the basis.

To make their proof constructive, the step of finding an FDI defined by a nontrivial dicut must be made algorithmic, which is highly nontrivial. For a general polyhedra, the problem of FDI recognition is D^p-hard [38]. Even for well-structured polyhedra such as perfect matching polytope, the similar question of finding an FDI defined by a nontrivial odd set is open. We discuss this in detail in the next paragraph. We overcome these difficulties by first, providing a recognition algorithm for robust digrafts and second, finding a good dicut that is, in fact, an FDI in polynomial time.

Perfect Matching Lattices. Some results of this work are reminiscent of the setting of perfect matchings. In an undirected graph $G = (V, E)$ a *perfect matching* is a set of edges incident with each $v \in V$ exactly once. A graph is *matching-covered* if every edge $e \in E$ belongs to some perfect matching. Denote the set of perfect matchings in G by $\mathcal{M}(G)$. A well-celebrated result due to Lovász [29] in the theory of matching covered graphs is that for every $x \in \text{lin}(\mathcal{M}(G)) \cap \mathbb{Z}^E$ we have $2x \in \text{lat}(\mathcal{M}(G))$. Moreover, if G has no Petersen minor, $\text{lat}(\mathcal{M}(G)) = \text{lin}(\mathcal{M}(G)) \cap \mathbb{Z}^E$. Abdi and Silina [5] recently gave a short proof of Lovász's result. In a series of follow-up works of Carvalho, Lucchesi, and Murty, it was shown that the lattice of perfect matchings $\text{lat}(\mathcal{M}(G))$ contains a basis consisting of perfect matchings [9–11,13,14] (see the excellent survey book [32]). In proving these results, the authors leverage many structural results on matching-covered graphs, such as tight cut decomposition established by Edmonds, Pulleyblank and Lovász [18]. They define the notions of *barrier cuts* and 2-*separation cuts* which are crucial for the existence of nontrivial tight odd cuts.

We give a characterization of when a digraft has a tight dijoin that has an interesting resemblance to Tutte's theorem for perfect matchings [44]. This allows us to identify similar notions of barrier dicuts and 2-separation dicuts in digrafts. However, there are important distinctions between perfect matchings and tight dijoins in digrafts. For example, a well-studied problem in matching theory is testing whether a graph is *Birkhoff-von Neumann (BvN)*, i.e., the perfect matching polytope is characterized solely by nonnegativity and degree constraints [7,12,16,17]. It is open whether recognition of BvN graphs is in $\mathcal{NP}$. In contrast, we are able to recognize robust digrafts in polynomial time, which are precisely the digrafts whose polytope of tight dijoins are solely defined by nonnegativity and degree constraints. This distinction allows us to greatly simplify the proof of Carvalho, Lucchesi, and Murty, and bypass the hardness of finding the analogous notion of good cuts in perfect matchings.

Congruency-Constrained Optimization. The literature of congruency-constrained combinatorial optimization is quite rich, e.g., [6,20,26,28,33,34,36, 37,40]. A central open question in this area is the *exact perfect matching* problem: given a graph $G = (V, E)$, a subset of edges $R \subseteq E$ that are colored red, and a number k, find a perfect matching with k red edges if it exists. A randomized polynomial algorithm exists by [35], but whether a deterministic algorithm exists is open. As a relaxation, El Maalouly, Steiner, and Wulf [33] gave an algorithm to find a perfect matching with the correct parity in deterministic polynomial time using the theory of perfect matching lattices. Tight strengthenings can be captured by submodular flows [19], whereas tight strongly connected orientations can be captured by matroid intersection [23]. Hörsch et al. [26] show that finding a common basis with an odd (even) number of red elements for two general matroids requires exponentially many independence queries. Our result on parity strongly connected orientations provides a new class of matroid intersection whose parity-constrained problem can be solved in polynomial time, generalizing the result on perfect matchings for bipartite graphs.

Organization. In Sect. 2, we give characterizations of the existence of tight dijoins and tight dicuts that will be useful in later sections. In Sect. 3, we give a structural result that allows us to recognize robust digrafts and describe how to construct bases for them. In Sect. 4, we describe basis construction for non-robust bricks. The reduction to basic digrafts, basis construction for elementary digrafts, the algorithm for parity strongly connected orientation, and all missing proofs can be found in the full version of the paper.

2 Tight Dijoins, Tight Dicuts, and Basic Digrafts

First, let us introduce the notation used throughout this work. In a digraph $D = (V, A)$, for a proper subset of nodes $\emptyset \neq X \subsetneq V$, denote by $\delta(X)$ the arcs that have exactly one end point in X. Denote by $\delta^-(X)$ and $\delta^+(X)$ the arcs entering and leaving X, resp. For a subset of arcs $J \subseteq A$, denote by $\delta_J(X) := \delta(X) \cap J$. Denote by $d_J(X) := |\delta_J(X)|$. Denote by $N(X)$ the neighbors of X in D. Denote by $\sigma_D(V \setminus X)$ the number of (weakly) connected components after deleting some subset $X \subseteq V$ from D. We omit the subscripts if the context is clear.

2.1 Tight Dijoins via Degree-Constrained Strongly Connected Orientations

First, we give a characterization of when a tight dijoin exists.

Theorem 7. *A digraft (D, S^t) has at least one tight dijoin if and only if*

$$\begin{aligned} \sigma(V \setminus X) &\leq |X|, \quad \forall \emptyset \neq X \subseteq T \\ \sigma(V \setminus X) &\leq |X|, \quad \forall \emptyset \neq X \subseteq S^t. \end{aligned} \tag{1}$$

Moreover, there is a polynomial-time algorithm that either confirms the digraft has at least one tight dijoin, or finds a set X violating (1).

This theorem essentially follows from the characterization for the existence of degree-constrained strongly connected orientations by Frank and Gyárfás [22]. This statement lies at the root of our algorithm; in particular, it will be important for us to find the set X violating this condition. The following lemma is from [4], which implies that the property of being tight dijoin-covered can also be checked in polynomial time.

Lemma 1 **([4], Theorem 3.2).** *A digraft (D, S^t) is tight dijoin-covered if and only if it has at least one tight dijoin.*

2.2 Two Types of Tight Dicuts and Basic Digraft Recognition

Consider an arbitrary tight dijoin-covered digraft (D, S^t). We begin by introducing two special classes of dicuts (Fig. 2 illustrates both classes).

Barrier Dicuts. First, if there is some $\emptyset \neq X \subseteq S^t$ or $\emptyset \neq X \subseteq T$, such that $\sigma(V \setminus X) = |X|$, then each connected component of $V \setminus X$ induces a tight dicut. Indeed, suppose $|X| = k$ and let $U_1, ..., U_k$ be the connected components of $V \setminus X$. Assume $X \subseteq T$ without loss of generality. Then, for each U_i, $\delta^-(U_i) = \emptyset$, which means $\delta^+(U_i)$ is a dicut. Thus, every dijoin J should satisfy $|J \cap \delta^+(U_i)| \geq 1$. In particular, J should have at least one arc going from U_i to X for each U_i since $U_1, ..., U_k$ are disconnected from each other. Therefore, $d_J(X) \geq k$. Moreover, every tight dijoin J has $d_J(v) = 1\ \forall v \in X$ since $X \subseteq T$. Thus, $d_J(X) = \sum_{v \in X} d_J(v) = |X| = k$. Therefore, there is exactly one arc going from X to each U_i, which implies $\delta^+(U_i)$ is a tight dicut. We call X a *barrier* and each dicut $\delta^+(U_i)$ a *barrier dicut.*

2-Separation Dicuts. Second, if there exist $u \in S^t, v \in T$ such that $\{u, v\}$ forms a 2-cut, i.e., the digraph induced by vertices $V \setminus \{u, v\}$ is disconnected, then another type of tight dicut arises. Indeed, let U and U' be unions of components of $V \setminus \{u, v\}$ such that $U \dot\cup U' = V \setminus \{u, v\}$. For every tight dijoin J, one has $1 + 1 = d_J(u) + d_J(v) = d_J(U \cup \{u\}) + d_J(U' \cup \{u\}) \geq 1 + 1$, where the inequality follows from the fact that both $\delta^+(U \cup \{u\})$ and $\delta^+(U' \cup \{u\})$ are dicuts. Therefore, $d_J(U \cup \{u\}) = d_J(U' \cup \{u\}) = 1$. This implies that both $\delta^+(U \cup \{u\})$ and $\delta^+(U' \cup \{u\})$ are tight dicuts. We call such a 2-cut $\{u, v\}$, where $u \in S^t, v \in T$, a 2-*separation.* We refer to the dicuts $\delta^+(U \cup \{u\})$ and $\delta^+(U' \cup \{u\})$ as 2-*separation dicuts.*

Fig. 2. (Left): an example of barrier dicuts. The vertices in $S \setminus S^t$ are filled solid. The vertices of the barrier are circled and the arising barrier dicuts are in dashed. (Right): an example of 2-separation dicuts. The vertices of a 2-separation are circled and the two arising tight dicuts are in dashed.

The significance of these dicuts is highlighted in Theorem 4. In the following theorem we also prove that a brick has no trivial tight dicuts other than the ones induced by $v \in V^t$, which will be useful in later sections.

Theorem 8. *A brick* (D, S^t) *has* $S^t = S^t$.

A useful fact about tight dicut decomposition is the following:

Proposition 1. *The results of any two tight dicut decomposition procedures on a digraft* $(D, \mathcal{F})$ *are the same list of bricks and braces up to the multiplicities of the arcs.*

3 Basis Construction for Robust Digrafts

Let $b : S \to \mathbb{Z}_{\geq 0}$. We say a subset of arcs $J \subseteq A$ is a *perfect b-matching* if $d_J(v) = b(v)$, $\forall v \in S$, and $d_J(v) = 1$, $\forall v \in T$. We say J is a *b-dijoin* if J is a perfect b-matching that intersects every dicut at least once. The following lemma shows that whether a subset of arcs $J \subseteq A$ is a dijoin only depends on its degree sequence. It is essentially due to an observation in [3] Sect. 4.

Lemma 2. *Given a digraft* (D, S^t) *and* $b : S \to \mathbb{Z}_{\geq 0}$, *there exists a b-dijoin if and only if*

$$\begin{aligned} &b(X) \leq |N(X)| - 1, \ \forall \emptyset \neq X \subsetneq S, \\ &b(S) = |T|. \end{aligned} \tag{2}$$

Moreover, for b satisfying (2), *every perfect b-matching is a b-dijoin.*

Below we give two characterizations of robust digrafts.

Theorem 9. *Let* (D, S^t) *be a digraft. The following are equivalent:*

(1) (D, S^t) *is a robust digraft.*
(2) For every $b : S \to \mathbb{Z}_{\geq 0}$ *such that*

$$\begin{aligned} b(v) &\geq 1, \quad \forall v \in S \\ b(v) &= 1, \quad \forall v \in S^t \\ b(S) &= |T|, \end{aligned} \tag{3}$$

there exists a b-dijoin.
(3) It holds that

$$|N(X)| - |X| \geq |T| - |S| + 1, \quad \forall \emptyset \neq X \subsetneq S, X \not\subseteq S^t \tag{4a}$$
$$|N(X)| - |X| \geq 1, \quad \forall \emptyset \neq X \subsetneq S, X \subseteq S^t, \tag{4b}$$

The condition (4) in particular leads to a polynomial-time recognition algorithm for robust digrafts (Theorem 6). It follows from Lemma 2 that Condition (4b) is needed for every tight dijoin-covered digraft. The stronger condition (4a) is more interesting. It turns out that if a set $X \in \arg\min_{\emptyset \neq X \subsetneq S, X \not\subseteq S^t} (|N(X)| - |X|)$ violates (4a), then X is a separating dicut, which can be found efficiently using submodular minimization. The characterizations also lead to a simple combinatorial way to construct a basis for robust digrafts in the following:

Theorem 10. *Let* (D, S^t) *be a robust digraft. Then, we can construct an integral basis* $\mathcal{B}$ *for* $\text{lin}(\mathcal{J}(D, \mathcal{F}))$ *that consists of tight dijoins in polynomial time. Moreover,* $|\mathcal{B}| = |A| - |V^t| + 1$.

4 Basis Construction for Non-robust Bricks

In this section, we address the most challenging step in our proof, namely, the non-robust bricks. The proof puts together everything we have established so far. Our goal is to find a good dicut to either contract if it is nontrivial or set to be tight if it is trivial. First, we introduce some necessary definitions. For a nontrivial dicut $C = \delta^+(U)$, the *C-contractions* of a digraft (D, S^t) are the two smaller digrafts (D_1, S_1^t), (D_2, S_2^t) obtained from contracting $V \setminus U$, U into $\bar{u}$, u resp., and setting the contracted vertices to be tight, that is, $S_1^t := S^t \cap U$ and $S_2^t := (S^t \cap (V \setminus U)) \cup \{u\}$. To find a good dicut, we need to find one of the following types of dicuts as an intermediate step:

Definition 5 (stable dicut, contractible dicut). *In a tight dijoin-covered digraft (D, S^t), a dicut is* stable *if it is induced by a singleton $u \in S \setminus S^t$ and $(D, S^t \cup \{u\})$ is tight dijoin-covered. A dicut C is* contractible *if it is nontrivial and both C-contractions are tight dijoin-covered.*

In particular, a good dicut has to be one of the above, but has even stronger property. We say a digraft (D, S^t) is a *near-brick* if it is tight dijoin-covered and contains exactly one brick, i.e., $b(D, S^t) = 1$.

Definition 6 (good dicut). *In a brick (D, S^t), a trivial dicut induced by node $u \in S \setminus S^t$ is* good *if $(D, (S^t)' := S^t \cup \{u\})$ is a near-brick and $(S^t)' = (S^t)'$. A nontrivial dicut C is* good *if both C-contractions (D_i, S_i^t) are near-bricks and satisfy $S_i^t = S_i^t$, $i = 1, 2$.*

We are now ready to describe the algorithm for finding a good dicut. Define a partial order over all non-tight dicuts. Let C, C' be non-tight dicuts of (D, S^t). We say C' *dominates* C if for every tight dijoin J, $|C \cap J| \geq |C' \cap J|$; we denote it by $C' \succeq C$. In this case, we say C' *strictly dominates* C if for at least one tight dijoin J, $|C \cap J| > |C' \cap J|$; we denote it by $C' \succ C$. We say C' and C are *equivalent* if they dominate each other. We say C is *maximal under* $\succeq$, or simply *maximal*, if no non-tight dicut strictly dominates C. The following theorem gives a procedure to find a contractible dicut starting from a given separating dicut.

Theorem 11. *Let (D, S^t) be a tight dijoin-covered digraft. If a separating dicut is maximal under $\succeq$, then it is contractible. Moreover, starting from an arbitrary separating dicut, one can reach a contractible dicut by a sequence of at most $|V|$ separating dicuts, each strictly dominating the previous one.*

Proof. Let $C = \delta^+(U)$ be a maximal separating dicut. Let $(D_1 = (V_1 = S_1 \cup T_1, A_1), S_1^t)$ be the C-contraction obtained from contracting $V \setminus U$ into $\bar{u}$. First, we show that (D_1, S_1^t) is tight dijoin-covered. Suppose not. Using Theorem 7, we can in polynomial time find $X \subsetneq S_1^t$ or $X \subsetneq T_1$ such that $\sigma_{D_1}(V_1 \setminus X) > |X|$. Then, $\bar{u} \in X$ since otherwise $\sigma_D(V \setminus X) > |X|$, contradicting the fact that (D, S^t) is tight dijoin-covered. In particular, we know $X \subsetneq T_1$. Let $s := |X|$ and suppose $X = \{v_1, \ldots, v_{s-1}, \bar{u}\}$. Let $k := \sigma_{D_1}(V_1 \setminus X) > s$ and $F_1, F_2, \ldots, F_k$ be the dicuts induced by the connected components of $V_1 \setminus X$ in D_1. In the original digraph D,

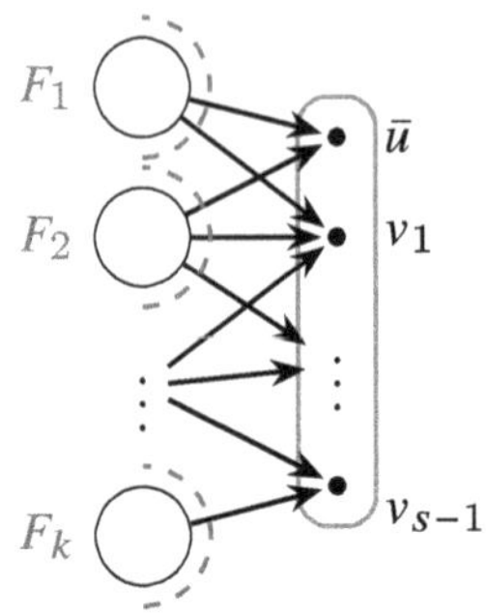

Fig. 3. (D_1, S_1^t) with a set X circled on the right-hand side; dicuts $F_1, \ldots, F_k$ are in dashed.

they are the dicuts induced by connected components of $V \setminus ((X \setminus \{\bar{u}\}) \cup (V \setminus U))$. Figure 3 illustrates the found structure.

Since C is separating, there exists a tight edge cover J such that $J \cap C = \emptyset$. Therefore,

$$\sum_{i=1}^{k} |J \cap F_i| = |J \cap C| + \sum_{i=1}^{s-1} |J \cap \delta_{\bar{D}}^-(v_i)| = s - 1 < k - 1.$$

We conclude that at least one F_j, $j \in [k]$ has $J \cap F_j = \emptyset$. Further, this F_j is nontrivial because J is an edge cover. Thus, F_j is a separating dicut as well.

Finally, we show that F_j strictly dominates C, which contradicts the fact that C is maximal under $\succeq$. Indeed, for every tight dijoin J,

$$|J \cap F_j| + (k-1) \leq \sum_{i=1}^{k} |J \cap F_i| = |J \cap C| + \sum_{i=1}^{s-1} |J \cap \delta_{\bar{D}}^-(v_i)| = |J \cap C| + (s-1) < |J \cap C| + (k-1).$$

Thus, $|J \cap F_j| < |J \cap C|$ for every tight dijoin J, which means F_j strictly dominates C. This concludes the proof that (D_1, S_1^t) is tight dijoin-covered.

The proof to show that the C-contraction obtained from contracting U is also tight dijoin-covered is analogous, and we omit it. This concludes the proof that a maximal separating dicut C is contractible. Starting from an arbitrary separating dicut C, if it is non-contractible, the above argument gives us an algorithm to find another separating dicut $F_j \succ C$ such that $|J \cap F_j| < |J \cap C|$ for every tight dijoin J. Let $C = C_1 \prec \ldots \prec C_t$ be the dicuts found by the algorithm, where C_t is a maximal separating dicut. Fix an arbitrary tight dijoin J_0, we know that $|T| = |J_0| \geq |J_0 \cap C_1| > \ldots > |J_0 \cap C_t| \geq 1$. Therefore, the process terminates in at most $|T| \leq |V|$ steps, at which point we find a contractible dicut.

Once we find a contractible dicut, the following theorem gives a procedure to find a good dicut.

Theorem 12. *Let (D, S^t) be a brick. If a stable dicut is maximal under $\succeq$, then it is good. If a contractible dicut is maximal under $\succeq$ and is not equivalent to* *a stable dicut, then it is good. Moreover, starting from an arbitrary contractible dicut, one can reach a good dicut by a sequence of at most* $|A| + |V|$ *contractible or stable dicuts, each dominating the previous one.*

The process consists of two steps. First, it finds a sequence of at most $|A|$ contractible or stable dicuts, each *strictly* dominating the previous one, after which we arrive at a maximal dicut C. If C is a stable dicut, then it is good and the algorithm terminates. If C is a maximal contractible dicut, then it goes to the second step: construct a sequence of at most $|V|$ equivalent contractible dicuts until it finds a good dicut. A similar procedure of finding a "good" cut exists in perfect matching lattices [9,14,32], but we have to overcome nontrivial technical difficulties to make it work for the digraft setting. This process allows us to finish the basis construction for bricks as follows:

Theorem 13. *Let* (D, S^t) *be a brick. Then, we can construct an integral basis* $\mathcal{B}$ *for* $\text{lin}(\mathcal{J}(D, S^t))$*. that consists of tight dijoins in polynomial time. Moreover,* $|\mathcal{B}| = |A| - |V^t| + 1$.

Acknowledgments. This work originated from a collaboration that included Ahmad Abdi and Gérard Cornuéjols and the authors are grateful for their invaluable input. We thank Karthik Chandrasekaran for suggesting the applications. We also thank the anonymous reviewers whose comments helped improve the presentation of the paper.

Disclosure of Interests. The authors have no competing interests to declare that are relevant to the content of this article.

References

1. Abdi, A., Cornuéjols, G., Guenin, B., Tunçel, L.: Dyadic linear programming and extensions. Math. Program. 1–44 (2024)
2. Abdi, A., Cornuéjols, G., Liu, S., Silina, O.: Strongly connected orientations and integer lattices. In: International Conference on Integer Programming and Combinatorial Optimization, pp. 1–14. Springer (2025)
3. Abdi, A., Cornuéjols, G., Zlatin, M.: On packing dijoins in digraphs and weighted digraphs. SIAM J. Discret. Math. **37**(4), 2417–2461 (2023)
4. Abdi, A., Cornuéjols, G., Liu, S., Silina, O.: Strongly connected orientations and integer lattices (2024). https://arxiv.org/abs/2410.13665
5. Abdi, A., Silina, O.: Integral bases, perfect matchings, and the petersen graph. arXiv preprint arXiv:2508.15602 (2025)
6. Artmann, S., Weismantel, R., Zenklusen, R.: A strongly polynomial algorithm for bimodular integer linear programming. In: Proceedings of the 49th Annual ACM SIGACT Symposium on Theory of Computing, STOC 2017, pp. 1206–1219. Association for Computing Machinery, New York, NY, USA (2017). https://doi.org/10.1145/3055399.3055473
7. Balas, E.: Integer and fractional matchings. In: North-Holland Mathematics Studies, vol. 59, pp. 1–13. Elsevier (1981)
8. Bang-Jensen, J., Gutin, G.: Theory, algorithms and applications. Monographs in Mathematics (2000)

9. Carvalho, M.H.: Decomposicao Otima em Orelhas Para Grafos Matching Covered. Ph.D. thesis, University of Campinas, Brasil, 1986. In Portuguese (1997)
10. de Carvalho, M.H., Lucchesi, C.L., Murty, U.S.R.: Optimal ear decompositions of matching covered graphs and bases for the matching lattice. J. Comb. Theory Ser. B **85**(1), 59–93 (2002). https://doi.org/10.1006/jctb.2001.2090
11. Carvalho, M.H., Lucchesi, C.L., Murty, U.S.: Ear decompositions of matching covered graphs. Combinatorica **19**(2), 151–174 (1999)
12. de Carvalho, M.H., Lucchesi, C.L., Murty, U.S.: The perfect matching polytope and solid bricks. J. Comb. Theory Ser. B **92**(2), 319–324 (2004)
13. de Carvalho, M.H., Lucchesi, C.L., Murty, U.S.R.: On a conjecture of lovász concerning bricks: I. the characteristic of a matching covered graph. J. Comb. Theory Ser. B **85**(1), 94–136 (2002). https://doi.org/10.1006/jctb.2001.2091
14. de Carvalho, M.H., Lucchesi, C.L., Murty, U.S.R.: On a conjecture of lovász concerning bricks: Ii. bricks of finite characteristic. J. Comb. Theory Ser. B **85**(1), 137–180 (2002). https://doi.org/10.1006/jctb.2001.2092
15. Cornuéjols, G., Liu, S., Ravi, R.: Approximately packing dijoins via nowhere-zero flows. Combinatorica **45**(3), 32 (2025)
16. De Carvalho, M.H., Kothari, N., Wang, X., Lin, Y.: Birkhoff-von neumann graphs that are pm-compact. SIAM J. Discret. Math. **34**(3), 1769–1790 (2020)
17. De Carvalho, M.H., Lucchesi, C.L., Murty, U.S.: How to build a brick. Discret. Math. **306**(19–20), 2383–2410 (2006)
18. Edmonds, J., Pulleyblank, W.R., Lovász, L.: Brick decompositions and the matching rank of graphs. Combinatorica **2**(3), 247–274 (1982). https://doi.org/10.1007/BF02579233
19. Edmonds, J., Giles, R.: A min-max relation for submodular functions on graphs. In: Studies in Integer Programming (Proc. Workshop, Bonn, 1975), pp. 185–204. Annals of Discrete Mathematics, vol. 1 (1977)
20. Eisenbrand, F., Rohwedder, L., Wegrzycki, K.: Sensitivity, proximity and FPT algorithms for exact matroid problems. In: 2024 IEEE 65th Annual Symposium on Foundations of Computer Science (FOCS), pp. 1610–1620. IEEE (2024)
21. Frank, A.: Connections in Combinatorial Optimization. Oxford Lecture Series in Mathematics and Its Applications. Oxford University Press (2011)
22. Frank, A., Gyárfás, A.: How to orient the edges of a graph. Combinatorics **18**, 353–362 (1978)
23. Frank, A., Tardos, É.: Matroids from crossing families. In: Finite and Infinite Sets, pp. 295–304. Elsevier (1984)
24. Gerards, A.M.H., Sebő, A.: Total dual integrality implies local strong unimodularity. Math. Program. **38**(1), 69–73 (1987). https://doi.org/10.1007/BF02591852
25. Guenin, B., Hwang, S.: Dyadic packing of dijoins. SIAM J. Discret. Math. **39**(1), 593–606 (2025)
26. Hörsch, F., Imolay, A., Mizutani, R., Oki, T., Schwarcz, T.: Problems on group-labeled matroid bases. arXiv preprint arXiv:2402.16259 (2024)
27. Liu, S., Silina, O.: Lattice structure and efficient basis construction for strongly connected orientations. arXiv preprint arXiv:2603.17424 (2026). https://arxiv.org/abs/2603.17424
28. Liu, S., Xu, C.: On the congruency-constrained matroid base. In: International Conference on Integer Programming and Combinatorial Optimization, pp. 280–293. Springer (2024)
29. Lovász, L.: Matching structure and the matching lattice. J. Comb. Theory Ser. B **43**(2), 187–222 (1987). https://doi.org/10.1016/0095-8956(87)90021-9

30. Lovász, L., Plummer, M.D.: Matching Theory, vol. 367. AMS Chelsea Publishing (2009)
31. Lucchesi, C.L., Younger, D.H.: A minimax theorem for directed graphs. J. London Math. Soc. (2) **17**(3), 369–374 (1978). https://doi.org/10.1112/jlms/s2-17.3.369
32. Lucchesi, C.L., Murty, U.S.R.: Perfect Matchings: A Theory of Matching Covered Graphs, Algorithms and Computation in Mathematics, vol. 31. Springer, Cham (2024). https://doi.org/10.1007/978-3-031-47504-7
33. Maalouly, N.E., Steiner, R., Wulf, L.: Exact matching: correct parity and FPT parameterized by independence number. In: Proceedings of the 34th International Symposium on Algorithms and Computation (ISAAC 2023). Leibniz International Proceedings in Informatics (LIPIcs), vol. 283, pp. 28:1–28:18. Schloss Dagstuhl – Leibniz-Zentrum für Informatik (2023). https://doi.org/10.4230/LIPIcs.ISAAC.2023.28
34. McCuaig, W.: Pólya's permanent problem. Electron. J. Comb. **11**(1), Research Paper R79, 83 pp. (2004). https://doi.org/10.37236/1832
35. Mulmuley, K., Vazirani, U.V., Vazirani, V.V.: Matching is as easy as matrix inversion. Combinatorica **7**(1), 105–113 (1987). https://doi.org/10.1007/BF02579224
36. Nägele, M., Santiago, R., Zenklusen, R.: Congruency-Constrained TU Problems Beyond the Bimodular Case, pp. 2743–2790. https://doi.org/10.1137/1.9781611977073.108
37. Nägele, M., Sudakov, B., Zenklusen, R.: Submodular minimization under congruency constraints. Combinatorica **39**(6), 1351–1386 (2019). https://doi.org/10.1007/s00493-019-3900-1
38. Papadimitriou, C.H., Yannakakis, M.: The complexity of facets (and some facets of complexity). J. Comput. Syst. Sci. **28**(2), 244–259 (1984). https://doi.org/10.1016/0022-0000(84)90068-7
39. Plummer, M.D., Lovász, L.: Matching Theory, vol. 29. Elsevier (1986)
40. Robertson, N., Seymour, P.D., Thomas, R.: Permanents, pfaffian orientations, and even directed circuits. Ann. Math. (2) **150**(3), 929–975 (1999). https://doi.org/10.2307/121059
41. Schrijver, A.: A counterexample to a conjecture of Edmonds and Giles. Discrete Math. **32**(2), 213–215 (1980). https://doi.org/10.1016/0012-365X(80)90057-6
42. Schrijver, A.: Combinatorial Optimization. Polyhedra and Efficiency. Springer, Heidelberg (2003)
43. Schrijver, A.: Observations on Woodall's conjecture. https://homepages.cwi.nl/~lex/files/woodall.pdf
44. Tutte, W.T.: The factorization of locally finite graphs. Can. J. Math. **2**, 44–49 (1950). https://doi.org/10.4153/CJM-1950-005-2
45. Woodall, D.: Menger and König systems. In: Alavi, Y., Lick, D. (eds.) Theory and Applications of Graphs. Lecture Notes in Mathematics, vol. 642. Springer, Heidelberg (1978)

On Lattice Diameter Segments: Algorithms and Structure

Gennadiy Averkov[1], Anouk E. Brose[2(✉)], Jesús A. De Loera[2], Gyivan Lopez-Campos[2], and Antonio J. Torres[3]

[1] Brandenburgische Technische Universität Cottbus-Senftenberg, Cottbus, Germany
averkov@b-tu.de
[2] University of California, Davis, CA 95616, USA
{aebrose,jadeloera,glopezcampos}@ucdavis.edu
[3] Smith College, Northampton, MA 01063, USA
atorres@smith.edu

Abstract. The lattice diameter of a bounded set $S \subset \mathbb{R}^d$ counts the maximal number of lattice points in a segment whose endpoints are lattice points in S. Such a segment is called a lattice diameter segment of S. This basic invariant in the geometry of numbers has several interesting applications. Our first type of contributions are algorithmic: We describe a polynomial-time algorithm that computes lattice diameter segments of polytopes in fixed dimension, with a simpler algorithm for polygons. In contrast we show hardness of computing the lattice diameters of polytopes of variable dimension and for semi-algebraic sets in dimensions three and higher.

Our second type of contributions concerns the structure of lattice diameter segments: Generalizing the classical Ehrhart function, we prove that the function that counts distinct lattice diameter segments in dilations of a lattice polytope is eventually a quasi-polynomial in the dilation factor. Finally, we prove a Borsuk-type theorem on the number of parts needed to partition a set of lattice points; such that each part has strictly smaller lattice diameter.

Keywords: Lattice diameter · Lattice polytopes · Ehrhart theory · Computational complexity · Shortest-vector problem · Borsuk's partition problem

1 Introduction

We study a fundamental geometric invariant of compact sets, called the *lattice diameter*, with a variety of applications in discrete geometry, integer optimization, and the geometry of numbers. We are not the first to study lattice diameters; this notion was introduced by Corzatt and Stolarsky [17], studied by Alarcon [1] and later by Bárány and Füredi [6]. Here are the key definitions of this paper:

S. S. Dey et al. (Eds.): IPCO 2026, LNCS 16588, pp. 220–232, 2026.
https://doi.org/10.1007/978-3-032-28691-8_15

Definition 1. *For a compact set* $S \subset \mathbb{R}^d$ *define the lattice diameter as*

$$\operatorname{diam}_{\mathbb{Z}}(S) := \max_{\mathbf{x},\mathbf{y} \in S \cap \mathbb{Z}^d} |\operatorname{conv}(\{\mathbf{x}, \mathbf{y}\}) \cap \mathbb{Z}^d| - 1.$$

If $\mathbf{x}$ *and* $\mathbf{y}$ *maximize* $\operatorname{diam}_{\mathbb{Z}}(S)$, *we call* $\operatorname{conv}(\{\mathbf{x}, \mathbf{y}\})$ *a* lattice diameter segment. *Similarly,* $\operatorname{aff}(\{\mathbf{x}, \mathbf{y}\})$ *is a* lattice diameter line, *and* $\mathbf{y} - \mathbf{x}$ *or any of its non-zero multiples is called a* lattice diameter direction.

Our paper investigates properties of the lattice diameter segments, and algorithms to compute them.

Note that the lattice diameter of S and $\operatorname{conv}(S)$ can differ when S is non-convex. When S is a convex body, then a lattice diameter line is an affine lattice line L such that $|L \cap S \cap \mathbb{Z}^d|$ is maximal among all affine lattice lines. In this case the lattice diameter of S coincides with the lattice diameter of $\operatorname{conv}(S \cap \mathbb{Z}^d)$, a lattice polytope, thus we focus on studying lattice diameters of lattice polytopes. The Euclidean diameters and lattice diameters of lattice polytopes need not correlate. For fixed lattice diameter, the Euclidean diameter can be arbitrarily large (Fig. 1a). Moreover, Euclidean diameter lines always pass through vertices [27, Lemma 1.7], while lattice diameter lines may not (Fig. 1b).

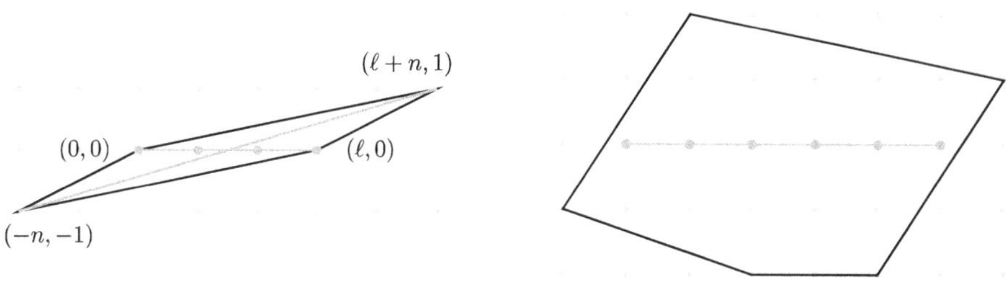

(a) Lattice polygons with fixed lattice diameter (orange) and unbounded Euclidean diameter (green).

(b) A lattice diameter segment contained in the interior of a lattice polygon.

Fig. 1. Examples of lattice diameter segments in lattice polygons.

A similar but different notion, sometimes referred to as the lattice diameter, sometimes referred to as the *rational diameter*, has also been studied [16]. It measures the longest rational segment in P with respect to its normalized lattice length. Furthermore, in [42] the lattice diameter appears as a special case of the n-th *Minkowski length* of P, $L_n(P)$, which is defined to be the largest number of lattice polytopes of positive dimension whose Minkowski sum has dimension at most n and is contained in P. It is not hard to see that $\ell_1(P)$ is the lattice diameter of P. Supronov and Supronova [42, Theorem 2.20] show that for any $n \geq 1$, the function $k \mapsto L_n(kP)$ is eventually quasi-linear. For $n = 1$ that means that $k \mapsto \operatorname{diam}_{\mathbb{Z}}(kP)$ is eventually quasi-linear.

Lattice segments and lattice diameter segments have several interesting connections. For example, in discrete tomography Gardner and Gritzmann have studied the reconstruction of lattice point sets from "X-rays", i.e., the collection of line sums over parallel lattice lines [24]. Algorithmic connections to combinatorial optimization also exist; the notion of *lattice width*, which is used in algorithms for integer programming is in a way the opposite of lattice diameter, and in dimension two it has been shown to be bounded by the lattice diameter [6]. Furthermore, understanding the polytopes that have a given polytope as a section, is directly related to intersection complexity, the dual notion of extension complexity [37]. Motivated by questions in the geometry of numbers, the volumes of sections and projections of convex bodies have been studied for many years; see the surveys [26,35]. But analogous questions for the number of lattice points inside sections have only been studied more recently [22,23,25]. Algorithmically, it is a challenge to determine a section of a convex body with maximal volume, but algorithms for hyperplane sections of convex polytopes exist [14]; however, the computational complexity of finding a section of a lattice polytope containing the most lattice points remains open. Here we solve the case of one-dimensional sections, i.e., segments.

Furthermore, classifications of convex lattice polytopes have been studied with respect to different invariants. This is of particular interest in algebraic geometry, where reflexive polytopes correspond to specific toric Fano varieties that have ties to mirror symmetry in physics [10,32]. Arnold, Bárány, Pach, Vershik, and others [2,7,8] have classified lattice polytopes by their volume while Averkov, Lagarias, Nill, Weismantel, Ziegler, and others [5,33,36] have continued the classification, based on the number of interior lattice points. We stress that classifying lattice polytopes by their lattice diameter is also possible. A simple old result by S. Rabinowitz [39] shows that if $\operatorname{diam}_{\mathbb{Z}}(P) < m$, then $|P \cap \mathbb{Z}^d| \leq m^d$. Also in more recent work, Dillon and Arun [4] asymptotically bound the number of vertices that a lattice polytope with fixed lattice diameter can have.

$$\mathbf{x} + \frac{j}{m}(\mathbf{y} - \mathbf{x}), \quad j = 0, 1, 2, \ldots$$

belong to P. The lemma is proved.

Another theme of our paper is Ehrhart theory. In its simplest version, for a d-dimensional lattice polytope P, and $k \in \mathbb{N}$, the Ehrhart counting function, $\mathrm{L}_P(k) = |kP \cap \mathbb{Z}^d|$, agrees with a polynomial in k of degree d. This foundational result extends to rational, multivariate, and weighted versions, using generating functions and featuring reciprocity theorems [11]. Here we count not individual lattice points, but subsets of them. It can be shown that there is an Ehrhart theorem for counting lattice lines with fixed direction in a rational polytope, that pass through a lattice point of the rational polytope [15]. In this paper we study the counting function of lattice diameter lines (or segments) of a lattice polygon.

A final motivation for our work is the famous Borsuk's partition problem. In 1933, Borsuk posed the following question in [13]:

Can every bounded set $S \subset \mathbb{R}^d$ be partitioned into $d+1$ sets with smaller diameter?

Here *diameter* refers to the Euclidean diameter, i.e., $\operatorname{diam}(S) := \max_{\mathbf{x},\mathbf{y}\in S} \|\mathbf{x}-\mathbf{y}\|_2$ where $\|\cdot\|_2$ is the Euclidean norm. The answer to Borsuk's question is positive for $d = 2$, see [13], and for $d = 3$, see [38], [20] and [28]. In higher dimensions it was believed to be true until Kahn and Kalai [31] constructed a counterexample for $d = 1325$ and for each $d > 2014$. Today, there are known counterexamples in dimensions 64 and higher, see [29], but the problem is still open for $4 \le d \le 63$. More generally, the problem of partitioning a set into sets with smaller diameter is studied for specific classes of bounded sets. In this paper, we study this problem for sets of lattice points with respect to the lattice diameter instead of the Euclidean diameter. We refer to this challenge as the *discrete Borsuk partition problem.*

Outline of our Contributions

In Sect. 2 we show that the lattice diameter of rational polytopes in fixed dimension can be computed in polynomial time. In contrast, in Sect. 3 we show two hardness results on computing the lattice diameter in more general settings, first in non-fixed dimensions and second for semi-algebraic sets. In Sect. 4 we extend Ehrhart Theory to counting lattice diameter segments in dilations of a polytope. Lastly, in Sect. 5 we prove the discrete version of Borsuk's partition problem. The details of our results are presented in the corresponding section. While we give ideas for some of the proofs, all the complete proofs can be found in the Appendix after the bibliography.

Notation

We write $\operatorname{span}(S)$, $\operatorname{aff}(S)$, and $\operatorname{conv}(S)$ for the linear, affine, and convex hull of a set $S \subset \mathbb{R}^d$, respectively. The segment between two points $\mathbf{x}, \mathbf{y} \in \mathbb{R}^d$ is denoted $[\mathbf{x}, \mathbf{y}] := \operatorname{conv}(\{\mathbf{x}, \mathbf{y}\})$ and when $\mathbf{x}, \mathbf{y} \in \mathbb{Z}^d$, we call this a *lattice segment.* A *line* will always mean an *affine line* and a *lattice line* is a line that passes through at least two lattice points. We say that a line of the form $\mathbf{x} + \operatorname{span}(\mathbf{u})$ *has direction* $\mathbf{u}$, and call it a $\mathbf{u}$-lattice line if $\mathbf{x}, \mathbf{u} \in \mathbb{Z}^d$, similarly we say $\mathbf{u}$-lattice segment. From now on a *diameter direction* means a lattice diameter direction. The inner product we use is the standard inner product on $\mathbb{R}^d$, denoted $\langle \cdot, \cdot \rangle$. A hyperplane is denoted $H(\mathbf{a}, \alpha) := \{\mathbf{x} \in \mathbb{R}^d : \langle \mathbf{a}, \mathbf{x} \rangle = \alpha\}$. A vector $(u_1, \ldots, u_d) \in \mathbb{Z}^d$ is called primitive if $\gcd(u_1, \ldots, u_d) = 1$, and we use the convention $\gcd(0, \ldots, 0) = 0$.

2 A Polynomial-Time Algorithm for the Lattice Diameter of Rational Polytopes in Fixed Dimension

We compute lattice diameters and their segments. It is known that when the dimension is fixed, then the number of lattice points in a rational polytope P

[9,19], as well as the lattice width of a rational polytope [41] can be computed in polynomial time, whereas when the dimension is variable, computing these invariants is #P-hard, and NP-hard, respectively. We show the same phenomenon for lattice diameters.

We can rewrite the formula for the lattice diameter in the form

$$\operatorname{diam}_{\mathbb{Z}}(P) := \max\{\gcd(\mathbf{y}-\mathbf{x})\colon \mathbf{x},\mathbf{y} \in P \cap \mathbb{Z}^d\}.$$

Using this we can encode a lattice diameter segment as a solution to an integer linear program in $3d$ variables, which yields the following result:

Theorem 1. *Let the dimension $d \in \mathbb{Z}_{\geq 1}$ be fixed. Then a lattice diameter segment of a rational polytope $P \subset \mathbb{R}^d$ can be computed in polynomial time.*

In dimension two, we give a more geometric algorithm to compute lattice diameter segments of lattice polygons that also computes *all* diameter directions. It relies on two properties: First, for every lattice diameter direction there exists a lattice diameter line in that direction, which intersects an edge e, and a vertex $\mathbf{v}$ *opposite to* e, i.e., $\mathbf{v} \notin e$ and the translate of e through $\mathbf{v}$ is tangent to P. This follows from [6, Lemma 1]. Second, for every vertex $\mathbf{v}$ of P, there exist at most three lattice diameter segments containing $\mathbf{v}$ (Lemma 8).

Algorithm 1 COMPUTING LATTICE DIAMETER SEGMENTS

Input: a lattice polygon P given by its vertices
Output: at least one lattice diameter segment for each direction in which a lattice diameter segment exists

```
Initialize two empty sets: 𝒯 for triangles, and 𝒮 for candidate segments
for all edges e of P do
    compute all opposite vertices v to e (an edge has at most two opposite vertices)
    form triangles conv({e, v}) and add them to 𝒯
end for
for all triangles in 𝒯 do
    compute up to three local lattice diameter segments of T = conv({e, v}) at v
    add these local lattice diameter segments to 𝒮
end for
return those segments in S ∈ 𝒮 that maximize |S ∩ P ∩ ℤ²|
```

We prove that Algorithm 1 runs in polynomial time in the encoding size of P.

Theorem 2. *Let P be a lattice polygon. There exists a polynomial-time algorithm that computes all diameter directions of P, and at least one representative lattice diameter segment per diameter direction.*

Remark 1. A key ingredient in Algorithm 1 to compute lattice diameters of lattice polygons is the existence of lattice diameter segments that pass through vertices. This fails in higher dimensions, as the following example shows. For $\mathbf{v}_1 = (-1,0)$, $\mathbf{v}_2 = (0,-1)$, $\mathbf{v}_3 = (1,1)$, let

$$P = \operatorname{conv}(\{(-m,\mathbf{v}_1),(-m,\mathbf{v}_2),(-m+1,\mathbf{v}_3),(m,-\mathbf{v}_1),(m,-\mathbf{v}_2),(m-1,-\mathbf{v}_3)\}),$$

where $m \geq 2$, as depicted in Fig. 2 for $m = 3$. The unique lattice diameter segment of P is the segment on the first-coordinate axis, from values $-(m-1)$ to $m-1$. It does not pass through a vertex of P, not even through a lattice point on the boundary of P.

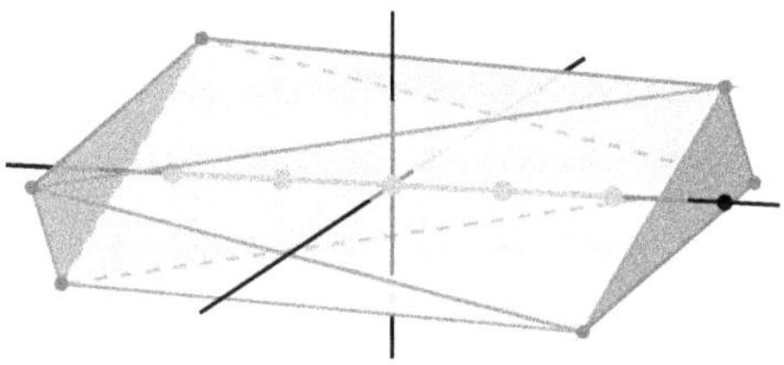

Fig. 2. A *unique* lattice diameter segment in the interior of a lattice polytope.

3 Hardness of Computing Lattice Diameters

We show two hardness results on computing lattice diameters in a more general setting.

3.1 Hardness of Computing Lattice Diameters in Non-fixed Dimension

When the dimension is non-fixed, we show that it is hard to compute the lattice diameter of centrally symmetric polytopes, even just of lattice cross-polytopes, i.e., polytopes of the form $\mathbf{B} \cdot \lozenge_d$ where $\mathbf{B} \in \mathbb{Z}^{d\times d}$ is some regular matrix and $\lozenge_d := \operatorname{conv}(\pm\mathbf{e}_1, \ldots \pm\mathbf{e}_d)$ is the standard cross-polytope. For an origin symmetric convex body K, we denote by $\|\mathbf{u}\|_K$ the norm with the unit ball being K, i.e. $\|\mathbf{u}\|_K := \min\{\lambda > 0 \colon \mathbf{u} \in \lambda K\}$. The ℓ_1 norm, which we denote by $\|\mathbf{u}\|_1$, is the norm whose unit ball is the standard cross polytope $\lozenge_d$. The *shortest-vector problem* for an origin symmetric convex body K and a regular matrix $\mathbf{A} \in \mathbb{Z}^{d\times d}$ is to determine

$$\min_{\mathbf{u}\in\mathbf{A}\cdot\mathbb{Z}^d\setminus\{\mathbf{0}\}} \|\mathbf{u}\|_K$$

and it is denoted by $\operatorname{SVP}(K, \mathbf{A})$. Regev and Rosen showed the following in [40].

Lemma 1 (Regev & Rosen, 2006). *Let* $\mathbf{A} \in \mathbb{Z}^{d\times d}$ *be a regular matrix. If* $\mathsf{NP} \not\subseteq \mathsf{RP}$*, then the shortest-vector problem*

$$\mathrm{SVP}(\Diamond_d, \mathbf{A}) = \min_{\mathbf{u}\in\mathbf{A}\cdot\mathbb{Z}^d\setminus\{\mathbf{0}\}} \|\mathbf{u}\|_1.$$

is hard to approximate within any constant factor.

A decision problem lies in RP, if there exists a polynomial time randomized algorithm that, if the correct answer is no, outputs *no* with probability 1, and if the correct answer is yes, outputs *yes* with a probability at least 1/2. See [3, Chapter 7] for more background. We would like to comment on the meaning of the hardness assertion in Lemma 1, which we have borrowed from the original source. In the study of SVP problems, one typically compares problems via randomized reductions rather than (standard) polynomial-time reductions. More technically, Regev & Rosen show that approximation of the SVP for the ℓ_1-norm within any given constant factor $c \geq 1$ is NP-hard for randomized reductions. If $\mathsf{P} = \mathsf{RP}$, which is widely assumed, then NP-hardness under randomized reductions is equivalent to the (standard) NP-hardness.

Lemma 2. *The problem of computing* $\mathrm{SVP}(\Diamond_d, \mathbf{A})$ *with a regular matrix* $\mathbf{A} \in \mathbb{Q}^{d\times d}$*, and the problem of computing* $\mathrm{SVP}(\Diamond_d, \mathbf{B}^{-1})$ *with an integer regular matrix* $\mathbf{B} \in \mathbb{Z}^{d\times d}$ *are equivalent in terms of the polynomial-time reduction.*

Next, we connect the shortest-vector problem with the problem of computing the lattice diameter.

Lemma 3. *Let* $\mathbf{B} \in \mathbb{Z}^{d\times d}$ *be a regular matrix and let* $P = \mathbf{B}\cdot\Diamond_d$*. Then*

$$\frac{2}{\mathrm{diam}_{\mathbb{Z}}(P) + 2} \leq \mathrm{SVP}(\Diamond_d, \mathbf{B}^{-1}) \leq \frac{2}{\mathrm{diam}_{\mathbb{Z}}(P)}.$$

We conclude with:

Theorem 3. *There exists a polynomial-time reduction from factor-two approximation of* SVP *with respect to the* ℓ_1*-norm for a given lattice to computation of the lattice diameter of a given origin-symmetric lattice cross-polytope.*

This implies that computation of the lattice-diameter for origin-symmetric lattice cross-polytopes is NP-hard for randomized reductions.

Proof. Consider the problem $\mathrm{SVP}(\Diamond_d, \mathbf{B}^{-1})$ for a given $\mathbf{B} \in \mathbb{Z}^{d\times d}$. The reduction maps $\mathbf{B}$ to the polytope $P = \mathbf{B}\cdot\Diamond_d$, the input for the lattice-diameter computation. For an origin-symmetric lattice cross-polytope P, by computing $\mathrm{diam}_{\mathbb{Z}}(P)$ one can also determine the value $2/(\mathrm{diam}_{\mathbb{Z}}(P)+2)$, which (by Lemma 3) approximates $\mathrm{SVP}(\Diamond_d, \mathbf{B}^{-1})$ within the factor two, because $2/\mathrm{diam}_{\mathbb{Z}}(P)$ is at most twice as large as $2/(\mathrm{diam}_{\mathbb{Z}}(P) + 2)$ in view of $\mathrm{diam}_{\mathbb{Z}}(P) \geq 2$. □

Remark 2. The same argument can also be made for hyper-parallelepipeds, relying on the hardness of approximation result for the shortest-vector problem with respect to the ℓ_∞ norm, see [21].

3.2 Hardness of Computing Lattice Diameter of Semi-algebraic Sets in Fixed-Dimension

We show that even in fixed dimension $d \in \mathbb{Z}_{\geq 3}$, computing lattice diameters is NP-hard for semi-algebraic sets that are known to contain lattice points. We rely on the following hardness result, [18, Lemma 2.1].

Lemma 4 ([18]). *The problem of minimizing a degree four polynomial over the lattice points of a polygon is* NP*-hard.*

The main idea is to construct a semi-algebraic set associated to a quartic polynomial f and a polygon R, such that the lattice diameter of K corresponds to determining the minimum of f over the lattice points of R. This yields the following result:

Theorem 4. *Let $d \in \mathbb{Z}_{\geq 3}$ be fixed. The problem of computing the lattice diameter of a compact semi-algebraic set $K \subset \mathbb{R}^d$ that satisfies $K \cap \mathbb{Z}^d \neq \emptyset$ is* NP*-hard. The problem remains* NP*-hard even when restricting to the lattice diameter in a fixed direction.*

Remark 3. We leave it as an open problem to determine if the decision version of this problem is in NP. The answer might be negative, because the similar problem of deciding whether a (not necessarily bounded) semi-algebraic set contains a lattice point is undecidable, see [18,30,34].

4 An Ehrhart-Type Function Counting Lattice Diameter Segments

Ehrhart theory studies the number of lattice points in integer dilations of a given lattice (or rational) polytope. A fundamental result by Ehrhart states that this counting function agrees with a polynomial function for lattice polytopes, and a quasi-polynomial function for rational polytopes. For details, see [11]. While counting the number of all lattice segments in a fixed direction that intersect a polytope P in a lattice point can be easily done, see [15], counting just those that are lattice diameter segments is much more difficult. Surprisingly, we have an Ehrhart-type result.

Definition 2. *For a lattice polytope P and a positive integer k, let $\mathrm{LD}_P(k)$ denote the number of lattice diameter segments of the dilated polytope kP.*

We show that for lattice polytopes P, $\mathrm{LD}_P(k)$ agrees with a quasi-polynomial function in the dilation variable k. To prove this in any dimension we use the theory of *Presburger Arithmetic*. In dimension two, we can give more insight on the structure of this quasi-polynomial.

Definition 3. *A* Presburger formula *is a first-order formula in the language of addition, evaluated over the integers. A* Presburger counting function $g_F : \mathbb{N}^n \to \mathbb{N}$ *for a Presburger formula $F(\mathbf{p}, \mathbf{s})$ is $g_F(\mathbf{s}) := \#\{\mathbf{p} \in \mathbb{N}^d \colon F(\mathbf{p}, \mathbf{s})$ holds$\}$.*

Kevin Woods proved the following lemma in [43, Theorem 1.10].

Lemma 5 (Woods, 2015). *A Presburger counting function* $g : \mathbb{N}^n \to \mathbb{Q}$*, is a piecewise quasi-polynomial.*

For $n = 1$, a function $g : \mathbb{N} \to \mathbb{Q}$ is a piecewise quasi-polynomial if and only if it *eventually* agrees with a quasi-polynomial, i.e., if there exists an integer $N \in \mathbb{N}$ such that $g : \mathbb{N}_{k \geq N} \to \mathbb{Q}$ agrees with a quasi-polynomial. Thus it suffices to express the number of lattice diameter segments in a fixed direction as a Presburger counting function. The crucial step is to show that the set of all diameter directions of dilates of a lattice polytope is finite.

Lemma 6. *Let* $P \subset \mathbb{R}^d$ *be a lattice d-polytope. Then there exists a finite set D of primitive vectors that depends only on P with the property that for each* $k \in \mathbb{Z}_{\geq 1}$ *and each* $\mathbf{x}, \mathbf{y} \in \mathbb{Z}^d \cap kP$ *satisfying* $\mathbf{y} - \mathbf{x} = \operatorname{diam}_{\mathbb{Z}}(kP)\mathbf{u}$ *for some primitive vector* $\mathbf{u}$*, one necessarily has* $\mathbf{u} \in D$.

Then, with Lemma 5 we get the following theorem.

Theorem 5. *Let* $P \subset \mathbb{R}^d$ *be a d-dimensional lattice polytope. Then the counting function* $\mathrm{LD}_P(k) : \mathbb{N} \to \mathbb{N}$ *eventually agrees with a quasi-polynomial in the dilation variable.*

For lattice polygons P, one can show that $\mathrm{LD}_P(k)$ is quasi-linear, and its period can be expressed in terms of edge slopes of the lattice polygon. We demonstrate this in the following example. Notice that the leading coefficient depends on the residue class, it is not an invariant as in the case of the Ehrhart polynomial for rational polytopes.

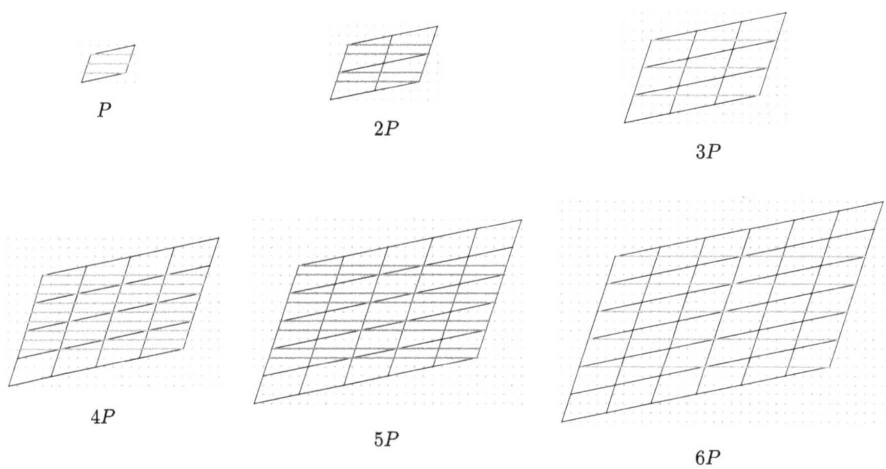

Fig. 3. Lattice diameter segments in dilations of P.

Example 1. Figure 3 depicts $P = \operatorname{conv}(\{(0,0),(1,3),(5,1),(6,4)\})$ and its dilates with their respective lattice diameter segments.

The pattern of which consecutive parallel segments are lattice diameter segments repeats modulo 3. We get

$$\mathrm{LD}_P(k) = \begin{cases} 2k+1 & \text{if } k \equiv 1 \pmod 3, \\ \frac{4}{3}k + \frac{4}{3} & \text{if } k \equiv 2 \pmod 3, \\ \frac{2}{3}k + 1 & \text{if } k \equiv 0 \pmod 3. \end{cases}$$

We conjecture that counting ℓ-dimensional slices of P that contain the most lattice points among all ℓ-dimensional slices of P have a similar behavior under dilation of P. More precisely:

Conjecture 1. Let P be a d-dimensional lattice polytope, and $2 \leq \ell \leq d-1$. Define $\mathrm{LD}_P^{\ell}(k)$ as the number of ℓ-dimensional sections of P that contain the most lattice points among all ℓ-dimensional sections of P. Then $\mathrm{LD}_P^{\ell} : \mathbb{N} \to \mathbb{N}$ eventually agrees with a quasi-polynomial function of degree $d - \ell$.

5 The Discrete Borsuk Partition Problem

In this section we continue in a more general setting, to prove the discrete version of Borsuk's parition problem.

An old lemma from [39] will be useful:

Lemma 7 (Rabinowitz, 1989). *Let $S \subset \mathbb{Z}^d$ be a bounded set and let $m \in \mathbb{N}$ satisfy $|S| > m^d$. Then* $\operatorname{conv}(S)$ *has at least $m+1$ collinear points. In other words, if* $\operatorname{diam}_{\mathbb{Z}}(S) < m$*, then $|S| \leq m^d$.*

We study lattice diameters of bounded sets $S \subset \mathbb{Z}^d$. Notice that the function $f : \mathbb{Z}^d \times \mathbb{Z}^d \to \mathbb{R}$ defined by

$$f(\mathbf{x},\mathbf{y}) = \operatorname{nvol}([\mathbf{x},\mathbf{y}]) = |[\mathbf{x},\mathbf{y}] \cap \mathbb{Z}^d| - 1$$

is a semi-metric, i.e., it is positive-definite and symmetric, so $(\mathbb{Z}^d, f)$ is a semi-metric space. Furthermore, f satisfies $f(\lambda\mathbf{x}, \lambda\mathbf{y}) = |\lambda| \cdot f(\mathbf{x},\mathbf{y})$ for all $\mathbf{x},\mathbf{y} \in \mathbb{Z}^d$ and all $\lambda \in \mathbb{Z}$. From this viewpoint, Theorem 6 is closely related to the well-known Boltyansky-Gohberg conjecture [12] which states that any bounded set S in a d-dimensional normed space can be partitioned into at most 2^d subsets, each having strictly smaller diameter.

Definition 4. *Let $S \subset \mathbb{Z}^d$ be a finite set. Define the lattice Borsuk number of S as the minimal size of a partition of S where each part has a smaller lattice diameter than* $\operatorname{diam}_{\mathbb{Z}}(S)$ *and denote it $\beta_{\mathbb{Z}}(S)$.*

In Fig. 4 we show a Borsuk partition of a set $S = P \cap \mathbb{Z}^d$, where P is a lattice polygon, into parts S_1 and S_2. Necessarily $\operatorname{diam}_{\mathbb{Z}}(S_i) < \operatorname{diam}_{\mathbb{Z}}(P)$ for $i = 1, 2$, but note that $\operatorname{diam}_{\mathbb{Z}}(\operatorname{conv}(S_1)) = \operatorname{diam}_{\mathbb{Z}}(P)$.

Define the *lattice Borsuk graph* of S, denoted $\mathsf{BG}_{\mathbb{Z}}(S)$, as the graph with vertex set S and edges $\{\mathbf{x},\mathbf{y}\} \subseteq S$ if and only if $f(\mathbf{x},\mathbf{y}) = \operatorname{diam}_{\mathbb{Z}}(S)$.

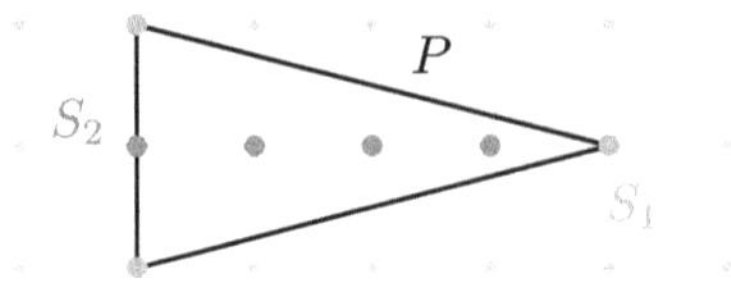

Fig. 4. A Borsuk partition of $P \cap \mathbb{Z}^d = S_1 \sqcup S_2$ where $\operatorname{diam}_{\mathbb{Z}}(S_1)$ is smaller than $\operatorname{diam}_{\mathbb{Z}}(\operatorname{conv}(S_1))$.

Remark 4. For a bounded set $S \subset \mathbb{Z}^d$, $\beta_{\mathbb{Z}}(S) = \chi(\mathsf{BG}_{\mathbb{Z}}(S))$, where χ is the *chromatic number* of $\mathsf{BG}_{\mathbb{Z}}(S)$, i.e., the minimum number of colors needed to color S so that no two adjacent vertices in S have the same color.

To find an upper bound for the Borsuk number to this problem, we characterize the lattice Borsuk graphs in the following lemma.

Lemma 8. *Let $S \subset \mathbb{Z}^d$ be a bounded set. All connected components of the lattice Borsuk graph of S have at most 2^d vertices. Furthermore, these components are cliques.*

Remark 5. In particular Lemma 8 says that the number of lattice diameter segments ending at any given point in S is at most $2^d - 1$. This generalizes what Alarcon showed in [1, Lemma 2.4] for dimension two to any dimension.

Finally, we present a solution to the discrete version of Borsuk's partition problem.

Theorem 6. *Let $S \subset \mathbb{Z}^d$ be a bounded set. Then $\beta_{\mathbb{Z}}(S) \leq 2^d$ and this bound is best possible.*

Notice that $\beta_{\mathbb{Z}}(S) = 2^d$ if and only if there exists an $\mathbf{x} \in S$ that is the endpoint of $2^d - 1$ lattice diameter segments, and $\mathsf{BG}_{\mathbb{Z}}(S)$ is isomorphic to the complete graph on 2^d vertices. Thus we conjecture the following:

Conjecture 2. Let $S \subset \mathbb{Z}^d$ be a bounded set with lattice diameter at least 2. Then $\beta_{\mathbb{Z}}(S) = 2^d$ if and only if $\operatorname{conv}(S)$ is unimodularly equivalent to a d-cube $[0, m]^d$ for some $m \in \mathbb{N}$.

In dimension two this conjecture is true and can be proven using a characterization of lattice polygons which have four diameter directions, see [6, Theorem 2(iii)].

Acknowledgments. We are grateful to Imre Bárány, Matthias Beck, Marie-Charlotte Brandenburg, Fritz Eisenbrand, Ansgar Freyer, Chiara Meroni, János Pach, and Matthias Schymura for fruitful discussions and suggestions, and to Bruce Reznick for providing important references. This research was partially supported by NSF grants 2348578 and 2434665.

Disclosure of Interests. The authors have no competing interests to declare that are relevant to the content of this article.

References

1. Alarcon II, E.G.: Convex lattice polygons. Ph.D. thesis, University of Illinois at Urbana-Champaign (1987)
2. Arnold, V.I.: Statistics of integral convex polytopes. Funkts. Anal. Pril. **14**, 1–3 (1980). (in Russian), English translation. Funct. Anal. Appl. **14**, 79–84 (1980)
3. Arora, S., Barak, B.: Computational Complexity: A Modern Approach, 1st edn. Cambridge University Press, USA (2009)
4. Arun, S., Dillon, T.: Hollow polytopes with many vertices (2025). arXiv:2504.17530
5. Averkov, G., Wagner, C., Weismantel, R.: Maximal lattice-free polyhedra: finiteness and an explicit description in dimension three. Math. Oper. Res. **36**(4), 721–742 (2011)
6. Bárány, I., Füredi, Z.: On the lattice diameter of a convex polygon. Discrete Math. **241**(1–3), 41–50 (2001)
7. Bárány, I., Pach, J.: On the number of convex lattice polygons. Comb. Probab. Comput. **1**, 295–302 (1992)
8. Bárány, I., Vershik, A.M.: On the number of convex lattice polytopes. Geom. Funct. Anal. **2**, 381–393 (1992)
9. Barvinok, A.: A polynomial time algorithm for counting integral points in polyhedra when the dimension is fixed. Math. Oper. Res. **19**, 769–779 (1994)
10. Batyrev, V.V.: On the classification of toric Fano 4-folds. J. Math. Sci. **94**(1), 1021–1050 (1999)
11. Beck, M., Robins, S.: Computing the Continuous Discretely: Integer-Point Enumeration in Polyhedra. Undergraduate Texts in Mathematics, 2nd edn. Springer (2015)
12. Boltyansky, V.G., Gohberg, I.: Results and problems in combinatorial geometry. CUP Archive (1985)
13. Borsuk, K.: Drei Sätze über die n-dimensionale Euklidische Sphäre. Fund. Math. **20**, 177–190 (1933)
14. Brandenburg, M.C., De Loera, J.A., Meroni, C.: The best ways to slice a polytope. Math. Comp. **94**, 1003–1042 (2025)
15. Brose, A.E.: The combinatorics and computation of sections of polytopes. Ph.D. thesis, UC Davis (2028). (in progress)
16. Codenotti, G., Freyer, A.: Lattice reduced and complete convex bodies. J. London Math. Soc. **110**(4), e12982 (2024)
17. Corzatt, C.E.: Some extremal problems of number theory and geometry. Ph.D. thesis, University of Illinois at Urbana-Champaign (1974)
18. De Loera, J.A., Hemmecke, R., Köppe, M., Weismantel, R.: Integer polynomial optimization in fixed dimension. Math. Oper. Res. **31**(1), 147–153 (2006). https://doi.org/10.1287/moor.1050.0169
19. De Loera, J.A., Hemmecke, R., Tauzer, J., Yoshida, R.: Effective lattice point counting in rational convex polytopes. J. Symbolic Comput. **38**(4), 1273–1302 (2004)
20. Eggleston, H.G.: Covering a three-dimensional set with sets of smaller diameter. J. Lond. Math. Soc. **s1-30**(1), 11–24 (1955)
21. van Emde-Boas, P.: Another NP-complete partition problem and the complexity of computing short vectors in a lattice. Report. Department of Mathematics, University of Amsterdam, Department, Univ. (1981). https://books.google.com/books?id=tCQiHQAACAAJ

22. Freyer, A., Henk, M.: Bounds on the lattice point enumerator via slices and projections. Discrete Comput. Geom. **67**, 895–918 (2022)
23. Freyer, A., Henk, M.: Polynomial bounds in Koldobsky's discrete slicing problem. Proc. Amer. Math. Soc. **152**(7), 3063–3074 (2024)
24. Gardner, R.J., Gritzmann, P.: Discrete tomography: determination of finite sets by X-rays. Trans. Amer. Math. Soc. **349**(6), 2271–2295 (1997)
25. Gardner, R.J., Gronchi, P., Zong, C.: Sums, projections, and sections of lattice sets, and the discrete covariogram. Discrete Comput. Geom. **34**, 391–409 (2005)
26. Giannopoulos, A., Koldobsky, A., Zvavitch, A.: Inequalities for Sections and Projections of Convex Bodies, vol. 9, pp. 223–256. De Gruyter (2023)
27. Gritzmann, P., Klee, V.: Inner and outer j-radii of convex bodies in finite-dimensional normed spaces. Discrete Comput. Geom. **7**(3), 255–280 (1992). https://doi.org/10.1007/BF02187841
28. Grünbaum, B.: A simple proof of Borsuk's conjecture in three dimensions. Math. Proc. Cambridge Philos. Soc. **53**, 776–778 (1957)
29. Jenrich, T., Brouwer, A.E.: A 64-dimensional counterexample to Borsuk's conjecture. Electron. J. Combin. **21**(1), 4–29 (2014)
30. Jeroslow, R.C.: There cannot be any algorithm for integer programming with quadratic constraints. Oper. Res. **21**(1), 221–224 (1973). https://doi.org/10.1287/opre.21.1.221
31. Kahn, J., Kalai, G.: A counterexample to Borsuk's conjecture. Bull. Amer. Math. Soc. **29**, 60–62 (1993)
32. Kasprzyk, A.: Toric Fano varieties and convex polytopes. Ph.D. thesis, University of Bath (2006)
33. Lagarias, J.C., Ziegler, G.M.: Bounds for lattice polytopes containing a fixed number of interior points in a sublattice. Canad. J. Math. **43**(5), 1022–1035 (1991)
34. Matiyasevich, Y.: Hilbert's Tenth Problem. The MIT Press, Cambridge, MA and London (1993)
35. Nayar, P., Tkocz, T.: Extremal sections and projections of certain convex bodies: a survey, vol. 9, pp. 343–390. De Gruyter (2023)
36. Nill, B., Ziegler, G.M.: Projecting lattice polytopes without interior lattice points. Math. Oper. Res. **36**(3), 462–467 (2011)
37. Padrol, A., Pfeifle, J.: Polygons as sections of higher-dimensional polytopes. Electron. J. Combin. **22**(1), P1.24 (2015). https://doi.org/10.37236/4315
38. Perkal, J.: Sur la subdivision des ensembles en parties de diamètre inférieur. Colloq. Math. **1**, 45 (1947)
39. Rabinowitz, S.: A Theorem about collinear lattice points. Util. Math. **36**, 93–95 (1989)
40. Regev, O., Rosen, R.: Lattice problems and norm embeddings. In: Proceedings of 38th ACM Symposium on Theory of Computing (STOC), pp. 447–456 (2006)
41. Sebő, A.: An Introduction to Empty Lattice Simplices. In: Integer Programming and Combinatorial Optimization (IPCO 1999). Lecture Notes in Computer Science, vol. 1610, pp. 400–414. Springer (1999)
42. Soprunov, I., Soprunova, J.: Eventual quasi-linearity of the Minkowski length. Eur. J. Combin. **58**, 107–117 (2016)
43. Woods, K.: Presburger arithmetic, rational generating functions, and quasi-polynomials. J. Symb. Log. **80**(2), 433–449 (2015)

On Circuit Diameter and Straight Line Complexity

Daniel Dadush[1], Stefan Kober[2(✉)], and Zhuan Khye Koh[3]

[1] Centrum Wiskunde & Informatica, Amsterdam, The Netherlands
dadush@cwi.nl
[2] Université Libre de Bruxelles, Brussels, Belgium
stefan.kober@ulb.be
[3] Boston University, Boston, USA
zkkoh@bu.edu

Abstract. The *circuit diameter* of a polyhedron is the maximum length (number of steps) of a shortest circuit walk between any two vertices of the polyhedron. Introduced by Borgwardt, Finhold and Hemmecke (SIDMA 2015), it is a relaxation of the *combinatorial diameter* of a polyhedron. These two notions of diameter lower bound the number of iterations taken by circuit augmentation algorithms and the simplex method respectively for solving linear programs.

Recently, an analogous lower bound for path-following interior point methods was introduced by Allamigeon, Dadush, Loho, Natura and Végh (SICOMP 2025). Termed *straight line complexity*, it refers to the minimum number of pieces of any piecewise linear curve that traverses a specified neighborhood of the central path.

We study the relationship between circuit diameter and straight line complexity. For a polyhedron $P := \{x \in \mathbb{R}^n : Ax = b, x \geq \mathbf{0}\}$, we show that its circuit diameter is up to a $\mathrm{poly}(n)$ factor upper bounded by the straight line complexity of linear programs defined over P. This yields a strongly polynomial circuit diameter bound for polyhedra with at most 2 variables per inequality. We also give a circuit augmentation algorithm with matching iteration complexity.

1 Introduction

The *simplex* method is probably the most well-known method for solving linear programs (LPs). Starting from an initial vertex, it moves from vertex to vertex along edges of the polyhedron towards an optimal vertex. This method is an example of *circuit augmentation* algorithms – a general algorithmic scheme for solving LPs. Like the simplex method, given an initial feasible point, a circuit augmentation algorithm proceeds through a sequence of points on the boundary

Part of this work was done during the Oberwolfach workshop "Combinatorial Optimization" on November 10–14, 2024. DD and ZKK were supported by European Research Council grant 805241–QIP. SK was supported by Fonds de la Recherche Scientifique–FNRS through research project BD-DELTA-3 (PDR 40028812).

S. S. Dey et al. (Eds.): IPCO 2026, LNCS 16588, pp. 233–248, 2026.
https://doi.org/10.1007/978-3-032-28691-8_16

of the polyhedron. However, instead of only moving along edges, the algorithm is allowed to travel along any direction parallel to a maximal subset of facets. So, its trajectory may pass through the interior of the polyhedron.

Interior point methods are another family of LP algorithms that move in the interior of the polyhedron. Unlike circuit augmentation algorithms, they always stay far away from the boundary. In this paper, we compare these two algorithmic frameworks via natural lower bounds that govern their performance. Ultimately, we are interested in the following question:

> *Can an algorithm that walks from boundary point to boundary point perform as well as an algorithm that moves in the 'deep' interior of the polyhedron?*

More formally, consider an LP in standard equality form

$$\min\{\langle c, x\rangle : Ax = b, x \geq \mathbf{0}\}, \tag{1}$$

where $A \in \mathbb{R}^{m\times n}$, $\mathrm{rk}(A) = m$ and $b \in \mathbb{R}^m$. Let P denote the feasible region of (1). For $W := \ker(A) \subseteq \mathbb{R}^n$, we call $h \in W \setminus \{\mathbf{0}\}$ an *elementary vector* if it is support-minimal, i.e., there is no $h' \in W \setminus \{\mathbf{0}\}$ such that $\mathrm{supp}(h') \subsetneq \mathrm{supp}(h)$. The support of an elementary vector $\mathrm{supp}(h) \subseteq [n]$ is called a *circuit*. Let $\mathcal{E}(A) \subseteq W$ and $\mathcal{C}(A) \subseteq 2^n$ denote the set of elementary vectors and circuits of A respectively.

In a circuit augmentation algorithm, the allowable movement directions are precisely $\mathcal{E}(A)$. In every iteration $t \geq 0$, the algorithm computes an elementary vector $h \in \mathcal{E}(A)$ based on a *pivot rule*, and updates the current solution $x^{(t)}$ to $x^{(t+1)} := x^{(t)} + \alpha h$, where $\alpha > 0$ is the maximal step size that preserves feasibility. The elementary vector h is called an *augmenting direction* if $\langle c, h\rangle < 0$ and $\alpha > 0$. If no augmenting direction exists, then $x^{(t)}$ is optimal and the algorithm terminates. The sequence of generated points $x^{(0)}, x^{(1)}, \ldots$ is called a *circuit walk*.

Clearly, the number of iterations taken by *any* circuit augmentation algorithm is lower bounded by the maximum length (number of steps) of a shortest circuit walk between any two vertices of P. Introduced by Borgwardt, Finhold and Hemmecke [6], this latter quantity is known as *circuit diameter*. It is a relaxation of the standard *combinatorial diameter* – the diameter of the vertex-edge graph of P. Analogously, the combinatorial diameter lower bounds the number of iterations taken by the simplex method. The best known upper bound on the combinatorial diameter in terms of m and n only is quasipolynomial [15,22,23], having the form $m^{\log O((n-m)/\log(n-m))}$.

Interior Point Methods. While circuit augmentation algorithms iterate through points on the boundary of P, interior point methods (IPMs) solve (1) by staying in the 'deep' interior of P. Path-following IPMs reach an optimal solution by following a smooth curve called the *central path*. Assuming P is bounded and has a strictly feasible solution, this is the parametric curve $x^{\mathrm{cp}} : \mathbb{R}_{>0} \to \mathbb{R}^n_{>0}$

defined by

$$x^{\mathrm{cp}}(\mu) := \arg\min_{x \in P} \left\{ \langle c, x \rangle - \mu \sum_{i=1}^{n} \log(x_i) \right\}. \tag{2}$$

The second term in (2) is called the *logarithmic barrier*, which pushes the central path away from zero on each coordinate. As $\mu \to 0$, $x^{\mathrm{cp}}(\mu)$ converges to an optimal solution of (1). Path-following IPMs maintain iterates in a certain neighborhood of the central path while geometrically decreasing μ, and hence the optimality gap. They can be implemented to run in polynomial time. Standard analyses yield a bound of $O(\sqrt{n}L)$ on the number of iterations, where L is the total encoding length of (A, b, c).

Straight Line Complexity. The trajectory of a path-following IPM is a piecewise-linear curve in a neighborhood of the central path, whose number of pieces corresponds to the number of iterations. Thus, the minimum number of pieces of *any* piecewise-linear curve in the neighborhood provides a lower bound on the number of iterations taken by the IPM. In a surprising result, Allamigeon, Benchimol, Gaubert and Joswig [1] constructed a parametric family of LPs such that for suitably large parameter values, any piecewise-linear curve in the neighborhood of the central path has exponentially many pieces. This result was later generalized to arbitrary self-concordant barriers [3].

Recently, Allamigeon, Dadush, Loho, Natura and Végh [2] complemented this lower bound by giving an IPM whose number of iterations matches it up to $\mathrm{poly}(n)$ factors. The guarantees of the IPM are stated in terms of a combinatorial proxy of the central path, which we elaborate below.

Let v^* denote the optimal value of (1). For $g \geq 0$, let

$$P_g := \{x \in P : \langle c, x \rangle \leq v^* + g\} \tag{3}$$

be the feasible sublevel set with optimality gap at most g. Assuming P_g is bounded, the *max central path* is the parametric curve $x^{\mathrm{m}} : \mathbb{R}_{\geq 0} \to \mathbb{R}^n_{\geq 0}$ defined by

$$x_i^{\mathrm{m}}(g) := \max\{x_i : x \in P_g\} \qquad \forall i \in [n]. \tag{4}$$

For each $i \in [n]$, x_i^{m} is a piecewise-linear, concave and nondecreasing function.

For $\eta \in (0, 1]$, we say that a function $f : \mathbb{R}_{\geq 0} \to \mathbb{R}^n_{\geq 0}$ lies in the *η-neighborhood* of x^{m} if $\eta x^{\mathrm{m}} \leq f \leq x^{\mathrm{m}}$. For the logarithmic barrier, its central path x^{cp} lies in the $(1/2n)$-neighborhood of x^{m}. More generally, for any ν-self-concordant barrier, its central path lies in the $\Omega(1/\nu)$-neighborhood of x^{m} [2]. For this reason, the max central path is a good proxy of the central path. Next, we need the following definition.

Definition 1. *Let $\phi : \mathbb{R}_{\geq 0} \to \mathbb{R}_{\geq 0}$ be a function and $\eta \in (0, 1]$. The* straight line complexity *of ϕ with respect to η, denoted $\mathrm{SLC}_\eta(\phi)$, is the minimum number of pieces of a continuous piecewise linear function $\psi : \mathbb{R}_{\geq 0} \to \mathbb{R}_{\geq 0}$ satisfying $\eta\phi \leq \psi \leq \phi$. For $g \geq 0$, we write $\mathrm{SLC}_\eta(\phi, g)$ to indicate the analogous quantity for approximating ϕ in the interval $[0, g]$.*

Recall that the trajectory of an IPM is a piecewise-linear curve in a neighborhood of x^{cp}. If it lies in the η-neighborhood of x^{m}, then its number of pieces is at least $\max_{i\in[n]} \mathrm{SLC}_\eta(x_i^{\mathrm{m}})$. The negative result of Allamigeon et al. [1] shows that for any $\eta \in (0,1]$, there exists an LP with n variables and $O(n)$ constraints such that $\max_{i\in[n]} \mathrm{SLC}_\eta(x_i^{\mathrm{m}}) = \Omega(2^n)$. On the positive side, Allamigeon et al. [2] gave an IPM whose number of iterations is bounded by the sum of straight line complexities of every coordinate of the max central path, i.e.,

$$O\left(\inf_{\eta\in(0,1]} \sqrt{n}\log\left(\frac{n}{\eta}\right)\sum_{i=1}^{n}\mathrm{SLC}_\eta(x_i^{\mathrm{m}})\right).$$

We remark that the neighborhood parameter η is not important, as one can show that for $0<\eta<\eta'<1$, $\mathrm{SLC}_{\eta'}(x_i^{\mathrm{m}}) = O(\log(1/\eta)/\log(1/\eta'))\mathrm{SLC}_\eta(x_i^{\mathrm{m}})$.

1.1 Our Contributions

Even though circuit diameter and straight line complexity are two geometric quantities that govern the performance of two broad classes of LP algorithms, not much is known about their relationship. From [1,15], we know that the circuit diameter is smaller than the straight line complexity in the worst case (quasipolynomial vs. exponential). However, this is unsatisfactory as it does not provide an instance-by-instance comparison between the two quantities.

Our main result shows that for any LP, the length of a shortest circuit walk from any feasible point to the optimal face can be bounded in terms of the straight line complexity.

Theorem 2. *Given a bounded LP in the form* (1)*, let F^* be its optimal face. For any feasible point x with optimality gap $g \geq 0$, the length of a shortest circuit walk from x to F^* is*

$$O\left(n^2\log(n)\sum_{i\in[n]}\mathrm{SLC}_{1/2}(x_i^{\mathrm{m}}, g)\right).$$

Our next result is an algorithmic implementation of Theorem 2. In particular, we give a circuit augmentation algorithm that solves an LP in the same number of iterations as the bound in Theorem 2. In each iteration, the algorithm calls the Ratio-Circuit oracle to obtain an augmenting direction. This oracle takes as input a matrix $A\in\mathbb{R}^{m\times n}$, costs $c\in\mathbb{R}^n$ and weights $v,w\in(\mathbb{R}_{\geq 0}\cup\infty)^n$, and returns a basic optimal solution to the system

$$\min\,\langle c,z\rangle \quad \text{s.t.} \quad Az=\mathbf{0},\quad \langle v,z^+\rangle + \langle w,z^-\rangle \leq 1, \tag{5}$$

where $(z^+)_i := \max\{z_i,0\}$ and $(z^-)_i := \max\{-z_i,0\}$. Here, we use the convention $ab=0$ if $a=\infty$ and $b=0$. If bounded, a basic optimal solution is either $\mathbf{0}$ or an elementary vector $z\in\mathcal{E}(A)$ that minimizes $\langle c,z\rangle/(\langle v,z^+\rangle+\langle w,z^-\rangle)$.

Theorem 3. *There exists a circuit augmentation algorithm such that given any bounded LP in the form* (1) *and any feasible point* x *with optimality gap* $g \geq 0$, *it returns an optimal solution using*

$$O\left(n^2 \log(n) \sum_{i\in[n]} \mathrm{SLC}_{1/2}(x_i^{\mathtt{m}}, g)\right)$$

RATIO-CIRCUIT *augmentation steps.*

Qualitatively, this signifies that circuit augmentation algorithms can achieve the convergence guarantees of path-following IPMs. In other words, moving in the 'deep' interior of the polyhedron is not more powerful than iterating along boundary points via circuits. We remark that the circuit walks in both theorems are cost-monotone.

As an immediate corollary, the circuit diameter of a polyhedron is at most the straight line complexity of its associated LPs up to poly(n) factors.

Corollary 4. *For a polyhedron* $\{x \in \mathbb{R}^n : Ax = b, x \geq \mathbf{0}\}$, *its circuit diameter is*

$$O\left(n^2 \log(n) \sup_{c\in\mathbb{R}^n} \left\{\sum_{i\in[n]} \mathrm{SLC}_{1/2}(x_i^{\mathtt{m}})\right\}\right).$$

Consequently, for a polyhedron, if the straight line complexities of its associated LPs are strongly polynomial, so is its circuit diameter. Recently, strongly polynomial bounds were obtained for the straight line complexity of LPs in the form (1) whose constraint matrix A has at most 2 non-zeroes per column [8]. In particular, $\mathrm{SLC}_{1/2}(x_i^{\mathtt{m}}) = O(mn \log n)$ for all $i \in [n]$. This class of LPs is equivalent to the minimum-cost generalized flow problem. Applying Corollary 4 yields a strongly polynomial circuit diameter bound for these polyhedra.

Corollary 5. *For a polyhedron* $\{x \in \mathbb{R}^n : Ax = b, x \geq \mathbf{0}\}$ *where* $A \in \mathbb{R}^{m\times n}$ *has at most 2 non-zeroes per column, its circuit diameter is* $O(mn^4 \log^2 n)$.

Previously, Kalai [14] gave an upper bound of $(n-m)^{\log_{11/10} 19} = (n-m)^{30.893\ldots}$ on the combinatorial diameter of generalized flow polyhedra with arc capacities, where m and n are the number of nodes and arcs respectively. With arc capacities, $\mathrm{SLC}_{1/2}(x_i^{\mathtt{m}}) = O(n^2 \log n)$ for all $i \in [n]$, so our bound in Corollary 5 becomes $O(n^5 \log^2 n)$. Further, it was shown in [2] that the straight line complexity of the primal and dual variables are essentially the same. In particular, for the dual LP of (1)

$$\max\{\langle b, y\rangle : A^\top y + s = c, s \geq \mathbf{0}\}, \tag{6}$$

we have $\mathrm{SLC}_\eta(s_i^{\mathtt{m}}) = O(\mathrm{SLC}_\eta(x_i^{\mathtt{m}}))$ for all $i \in [n]$. Hence, we also obtain a strongly polynomial circuit diameter bound for polyhedra with at most 2 variables per inequality.

Corollary 6. *For a polyhedron* $\{y \in \mathbb{R}^m : A^\top y \leq c\}$ *where* $A^\top \in \mathbb{R}^{n\times m}$ *has at most 2 non-zeroes per row, its circuit diameter is* $O(mn^4 \log^2 n)$.

Both of these bounds are an improvement over those in terms of the circuit imbalance measure [9], which could be unbounded for such polyhedra.

1.2 Technical Overview

The starting point of proving Theorem 2 is to decompose the max central path x^{m} into at most $2\sum_{i\in[n]} \mathrm{SLC}_{1/2}(x_i^{\mathrm{m}})$ *polarized segments.* A segment of the max central path $\{x^{\mathrm{m}}(g) : g \in [g_0, g_1]\}$ is *polarized* if the variable set $[n]$ can be partitioned into $B \cup N$ such that the coordinates in B are barely changing, while the coordinates in N are scaling down linearly with the optimality gap g. More precisely, for every $g \in [g_0, g_1]$, we have

$$\begin{aligned} &\frac{1}{4}x_i^{\mathrm{m}}(g_1) \leq x_i^{\mathrm{m}}(g) \leq x_i^{\mathrm{m}}(g_1) && \forall i \in B \\ &\frac{g}{g_1}x_i^{\mathrm{m}}(g_1) \leq x_i^{\mathrm{m}}(g) \leq \frac{4g}{g_1}x_i^{\mathrm{m}}(g_1) && \forall i \in N \end{aligned} \tag{7}$$

We call $[g_0, g_1]$ a *polarized interval.* Since the max central path multiplicatively approximates the central path, the latter also admits the same polarized decomposition. The connection between straight line complexity and polarization was already pointed out in the work of [2].

Given a current point $x \in P$ with optimality gap g, let $(g_0, g_1]$ be the polarized interval containing g. To finish the proof, it suffices to traverse the interval in $\mathrm{poly}(n)$ iterations, i.e., reach a point $x' \in P$ with optimality gap $g' \leq g_0$. We will take two types of circuit steps, which can be obtained using the RATIO-CIRCUIT oracle. The first type of circuit step is obtained by setting $v = \mathbf{0}$ and $w = 1/x$. First proposed by Wallacher [24] for minimum-cost flow, it decreases the optimality gap by a factor of at least $1 - 1/n$ [19]. So, if the interval is short, i.e., $g_1/g_0 \leq 2^{\mathrm{poly}(n)}$, then repeating this step for $\mathrm{poly}(n)$ iterations does the job. We remark that for flow problems, such a minimum-ratio cycle can be found in strongly polynomial time.

On the other hand, if $g_1/g_0 \gg 2^{\mathrm{poly}(n)}$, then we can no longer rely on the geometric decay of the optimality gap. Let (B, N) be the bipartition associated with this polarized interval. From (7), we know that the contribution of B to the optimality gap in this interval is negligible. So, the goal is to decrease the coordinates in N.

For simplicity of notation, we now assume that $x^{\mathrm{m}}(g_1) = \mathbf{1}_n$.[1] Unlike an IPM, a circuit step inevitably sets a coordinate to zero. So, one needs to be careful about which coordinates are zeroed out in order to avoid oscillations. To this end, we consider the *lifting operator* $\ell_N^W : \pi_N(W) \to \pi_B(W)$ for $W = \ker(A)$

$$\ell_N^W(z) = \operatorname*{arg\,min}_{(y_B, y_N)\in W} \{\|y_B\| : y_N = z\}, \tag{8}$$

[1] This is achieved by rescaling the instance: $A \to A\mathrm{diag}(x^{\mathrm{m}}(g_1))$ and $c \to \mathrm{diag}(x^{\mathrm{m}}(g_1))c$.

where π_B and π_N denote the coordinate projections onto B and N respectively. Using its singular values $\sigma_1 \geq \sigma_2 \geq \cdots \geq \sigma_{\dim(\pi_N(W))}$, we further subdivide $[g_0, g_1]$ into subintervals

$$\left[g_0, \frac{g_1}{\sigma_a}\right], \left[\frac{g_1}{\sigma_a}, \frac{g_1}{\sigma_{a+1}}\right], \ldots, \left[\frac{g_1}{\sigma_b}, g_1\right]. \tag{9}$$

Now, suppose that g lies in some long subinterval $[g_1/\sigma_j, g_1/\sigma_{j+1}]$. By (7) and (8), any step that significantly decreases the coordinates in N must be close to the singular subspace corresponding to $\sigma_{j+1}, \sigma_{j+2}, \ldots, \sigma_{\dim(\pi_N(W))}$. Otherwise, the induced change on B is too big and violates feasibility. Hence, this suggests to move as much as possible in the singular subspace to ensure significant progress.

Let $S \subseteq N$ be the set of 'small' coordinates relative to their max central path values at the end of the subinterval, i.e., $S := \{i \in N : x_i \leq 2^{\text{poly}(n)} x_i^{\text{m}}(g_1/\sigma_j)\}$. In the singular subspace, we show the existence of an elementary vector $z \in \mathcal{E}(A)$ such that 1) $\langle c, z\rangle \leq -g/(2n)$; 2) $z_S \geq 0$; and 3) z_S, z_B are 'small'. This motivates our second type of circuit step, where we choose the weights v and w to induce these properties in an optimal solution of (5). This step ensures that a coordinate in $N \setminus S$ is set to zero, while the coordinates in S stay 'small'. Thus, in $|N|$ iterations, we have $S = N$. At that point, $g/(g_1/\sigma_j) \leq 2^{\text{poly}(n)}$, so we can use the first type of circuit step to traverse the subinterval in poly(n) iterations.

The augmentation scheme described above is almost fully algorithmic, except that it uses the max central path and its decomposition into polarized segments. As our proof is robust to poly(n) approximations, it suffices to be able to approximate these additional input. A straightforward way of doing so is to simply compute the trajectory of the IPM [2]. In the full version [7], we show that one can avoid running the IPM by estimating all the required information using poly(n) calls to Ratio-Circuit together with additional linear algebraic operations. All missing proofs can also be found in the full version.

1.3 Related Work

Besides the simplex method, many network optimization algorithms can be seen as circuit augmentation algorithms. A prominent example is the Edmonds–Karp–Dinic algorithm [10,11] for maximum flow, which is an efficient implementation of the Ford-Fulkerson method [12]. Bland generalized this result by giving a circuit augmentation algorithm for LP [5,16]. Building on this work, De Loera, Hemmecke and Lee [17] analyzed different pivot rules for LP, and also extended them to the setting of integer programming via Graver basis augmentation. De Loera, Kafer and Sanità [18] studied the computational complexity of carrying out these rules, as well as their convergence on 0/1-polytopes.

Dadush, Koh, Natura and Végh [9] upper bounded the circuit diameter in terms of the circuit imbalance measure of A. They also gave circuit augmentation algorithms with similar iteration bounds. Circuit diameter bounds have been shown for some combinatorial polytopes such as dual transportation polyhedra [6], matching, travelling salesman, and fractional stable set polytopes [13].

Nöbel and Steiner proved that computing the circuit diameter is strongly NP-hard [21], which was then strengthened to a Π_2^p-completeness result by Wulf [25]. Black, Nöbel and Steiner showed that approximating the length of a shortest monotone circuit walk is already hard in dimension 2 [4]. We remark that in subsequent breakthrough work [20], Natura showed that the circuit diameter is strongly polynomial for all polyhedra.

2 Preliminaries

If not specified, we use the ℓ_2-norm, i.e., $\|\cdot\| := \|\cdot\|_2$. For $x \in \mathbb{R}^n$, we define $1/x \in (\mathbb{R} \cup \infty)^n$ as $(1/x)_i := 1/x_i$ for all $i \in [n]$, with the convention $1/0 = \infty$. We also use the convention $1/\infty = 0$.

Let v^* be the optimal value of (1), which we assume to be finite. For $x \in P$, we denote its optimality gap as $g(x) := \langle c, x \rangle - v^*$. Given an elementary vector $h \in \mathcal{E}(A) \setminus \mathbb{R}^n_{\geq 0}$, we define $\text{aug}_P(x, h) := x + \alpha h$ where $\alpha = \max\{\bar{\alpha} \in \mathbb{R} : x + \bar{\alpha} h \in P\}$. For $g \geq 0$, we define $\tilde{x}^{\text{m}}(g) := (1/n) \cdot \sum_{i \in [n]} \text{argmax}\{x_i : x \in P_g\}$ as a feasible approximation of the max central path. Note that $\tilde{x}^{\text{m}}(g) \in P$ by convexity. Moreover, a simple averaging argument yields $x^{\text{m}}(g)/n \leq \tilde{x}^{\text{m}}(g) \leq x^{\text{m}}(g)$.

2.1 The Ratio-Circuit Oracle

We will use the following oracle to perform circuit augmentations.

Oracle 2.1. Ratio-Circuit(A, c, v, w)

For a matrix $A \in \mathbb{R}^{m \times n}$, costs $c \in \mathbb{R}^n$ and weights $v, w \in (\mathbb{R}^n_{\geq 0} \cup \{\infty\})^n$, the output is a basic optimal solution to the system:

$$\min \ \langle c, z \rangle \quad \text{s.t.} \quad Az = \mathbf{0}, \quad \langle v, z^+ \rangle + \langle w, z^- \rangle \leq 1. \tag{10}$$

Recall that we use the convention $ab = 0$ if $a = \infty$ and $b = 0$. Note that (10) can be reformulated as an LP using auxiliary variables:

$$\begin{aligned} \min \ & \langle c, z^+ \rangle - \langle c, z^- \rangle \\ \text{s.t.} \ & Az^+ - Az^- = \mathbf{0}, \\ & \langle v, z^+ \rangle + \langle w, z^- \rangle \leq 1, \\ & z^+, z^- \geq \mathbf{0}. \end{aligned} \tag{11}$$

If (10) is bounded, then a basic optimal solution is either $\mathbf{0}$ or an elementary vector $z \in \mathcal{E}(A)$ that minimizes $\langle c, z \rangle / (\langle v, z^+ \rangle + \langle w, z^- \rangle)$.

The following standard lemma gives the guarantees of Wallacher's rule [24].

Lemma 7 ([9, Lemma 2.5])**.** *Given $x \in P$, let h be the optimal solution returned by* Ratio-Circuit$(A, c, \mathbf{0}, 1/x)$.

(i) If $\langle c, h \rangle = 0$, then x is optimal to (1).
(ii) If $\langle c, h \rangle < 0$, then letting $x' := \text{aug}_P(x, h)$, we have $1 \leq \alpha \leq n$ for the augmentation step size and $g(x') \leq (1 - 1/n) g(x)$.

2.2 Polarized Decomposition of the Max Central Path

We first state a basic lemma that compares any two points of the max central path. The upper bound is due to the monotonicity of x^{m}, while the lower bound follows from the concavity and nonnegativity of x^{m}.

Lemma 8. *For any $g' \geq g \geq 0$, we have $\frac{g}{g'} x^{\mathrm{m}}(g') \leq x^{\mathrm{m}}(g) \leq x^{\mathrm{m}}(g')$.*

Next, we define what it means for a max central path segment to be *polarized.*

Definition 9. *Given $\gamma \in (0,1]$ and $0 \leq g_0 < g_1$, we say that the max central path segment $\{x^{\mathrm{m}}(g) : g \in [g_0, g_1]\}$ is* γ-polarized *if there exists a partition $(B,N) = [n]$ such that for every $g \in [g_0, g_1]$,*

$$\begin{aligned} \gamma x_i^{\mathrm{m}}(g_1) \leq x_i^{\mathrm{m}}(g) &\leq x_i^{\mathrm{m}}(g_1) && \forall i \in B \\ \frac{g}{g_1} x_i^{\mathrm{m}}(g_1) \leq x_i^{\mathrm{m}}(g) &\leq \frac{g}{\gamma g_1} x_i^{\mathrm{m}}(g_1) && \forall i \in N. \end{aligned} \tag{12}$$

We say that the interval $[g_0, g_1]$ is γ-polarized, *and call (B,N) its* polarized partition.

Note that the upper bound for B and the lower bound for N always holds by Lemma 8. So, the interesting part of the definition is the lower bound for B and the upper bound for N.

The following lemma decomposes the max central path into $O(\sum_{i\in[n]} \mathrm{SLC}_\eta(x_i^{\mathrm{m}}))$ many polarized segments. We will use $\eta = 1/2$ throughout the paper.

Lemma 10 ([2, Lemma 4.5])**.** *Let $x \in P$ with optimality gap $g \geq 0$. For any $\eta \in (0,1]$, there exist points $0 = g_0 < g_1 < \cdots < g_r = g$ with $r \leq 2\sum_{i\in[n]} \mathrm{SLC}_\eta(x_i^{\mathrm{m}}, g)$ such that each interval $[g_j, g_{j+1}]$ is $(\eta/2)$-polarized.*

2.3 The Lifting Operator and Its Singular Subspaces

Recall the definition of the lifting operator in the introduction.

Definition 11 (Lifting Map and Operator). *Given a linear subspace $W \subseteq \mathbb{R}^n$ and a partition (I,J) of $[n]$, we define the* lifting map $L_I^W : \pi_I(W) \to W$ *as*

$$L_I^W(x) := \arg\min_{w\in W} \{\|w\| : w_I = x\}.$$

We define the lifting operator $\ell_I^W : \pi_I(W) \to \pi_J(W)$ *as $\ell_I^W(x) := \pi_J(L_I^W(x))$.*

As ℓ_I^W is a linear operator [2, Lemma 6.7], it has a singular value decomposition.

Definition 12 (Singular value decomposition). *Let $U \subseteq \mathbb{R}^n$ and $V \subseteq \mathbb{R}^m$ be linear subspaces. Let $T : U \to V$ be a linear operator and $\mathcal{M}(T)$ be its matrix representation. Then, T admits a* singular value decomposition *(SVD)*

$$\mathcal{M}(T) = \sum_{i=1}^{\mathrm{rk}(T)} \sigma_i(T) v_i u_i^\top,$$

where $v_1, \dots, v_{\mathrm{rk}(T)} \in V$, and $u_1, \dots, u_{\mathrm{rk}(T)} \in U$ are orthonormal vectors in their respective subspaces, and $\sigma_1(T) \geq \sigma_2(T) \geq \cdots \geq \sigma_{\mathrm{rk}(T)}(T) > 0$. We also set $\sigma_i(T) := 0$ for all $\mathrm{rk}(T) < i \leq \dim(U)$, and $\sigma_0(T) := \infty$. We denote by $\sigma_{\min}(T) := \sigma_{\dim(U)}(T)$ the smallest singular value of T.

Given an interval $I \subseteq \mathbb{R}_{\geq 0}$, we define $C_\sigma(T, I) := |\{i \in [\dim(U)] : \sigma_i(T) \in I\}|$ to count the number of singular values of T in I. Note that $\sigma_i(T) \geq \alpha$ if and only if $C_\sigma(T, [\alpha, \infty)) \geq i$.

Definition 13 (Singular subspaces). *Let $T : U \to V$ be a linear operator. A linear subspace $S \subseteq U$ is called a* singular subspace for T *if*

$$\sigma_1(T|_S) \leq \sigma_{\dim(U)-\dim(S)+1}(T).$$

3 The Circuit Diameter Bound

In this section, we give an 'existential' circuit augmentation algorithm (Algorithm 1) for solving (1), which uses the max central path and its polarized decomposition. Here, we will only show that the algorithm outputs a circuit walk whose length is at most a factor $O(n)$ larger than the bound in Theorem 2.; the stronger bound is proved in the full version.

Throughout this section, we will use the following four functions:

$$f_1(n) := 32n^3 \cdot (3n)^{5n} \qquad f_2(n) := 64n^{2.5}$$
$$f_3(n, p) := n \cdot (3n)^{5p} \qquad f_4(n) := 2n$$

The high-level description of Algorithm 1 is quite simple. In every iteration, we first identify the polarized interval $(g_j, g_{j+1}]$ containing the current optimality gap $g(x)$, along with its polarized partition (B, N). Then, we compute the lifting operator $\ell_N^{W_j}$ and its corresponding singular values $\sigma_1 \geq \sigma_2 \geq \cdots \geq \sigma_{\dim(\pi_N(W_j))}$, where $W_j := \mathrm{diag}(x^{\mathrm{m}}(g_{j+1}))^{-1} \ker(A)$.

Let $a := C_\sigma(\ell_N^{W_j}, [g_{j+1}/g_j, \infty)) + 1$ and $b := C_\sigma(\ell_N^{W_j}, [1, \infty))$. Observe that if $a \leq \dim(\pi_N(W))$, then g_{j+1}/σ_a is the first singular value breakpoint above g_j. Similarly, if $b \geq 1$, then g_{j+1}/σ_b is the last singular value breakpoint below g_{j+1}. If $a \leq b$, we further subdivide the interval $[g_j, g_{j+1}]$ into subintervals based on the singular value breakpoints, i.e.,

$$\left[g_j, \frac{g_{j+1}}{\sigma_a}\right], \left[\frac{g_{j+1}}{\sigma_a}, \frac{g_{j+1}}{\sigma_{a+1}}\right], \dots, \left[\frac{g_{j+1}}{\sigma_b}, g_{j+1}\right].$$

Algorithm 1: Existential circuit augmentation algorithm

Input : Bounded instance of (1) with constraint matrix $A \in \mathbb{R}^{m \times n}$, feasible solution $x^{(0)} \in P$ with optimality gap $g_r := g(x^{(0)})$, and polarized decomposition $[g_0, g_1] \cup [g_1, g_2] \cup \cdots \cup [g_{r-1}, g_r]$ of x^{m}

Output: Optimal solution x^*

$x \leftarrow x^{(0)}$;
while x *is not optimal* **do**
    Find $j \geq 0$ s.t. $g(x) \in (g_j, g_{j+1}]$ with associated polarized partition (B, N);
    Compute the lifting operator $\ell_N^{W_j}$ where $W_j := \mathrm{diag}(x^{\mathrm{m}}(g_{j+1}))^{-1}\mathrm{ker(A)}$;
    Compute the singular values $\sigma_1 \geq \sigma_2 \geq \cdots \geq \sigma_{\dim(\pi_N(W_j))}$ of $\ell_N^{W_j}$;
    $k \leftarrow C_\sigma(\ell_N^{W_j}, [g_{j+1}/g(x), \infty))$;
    if $k = 0$ **then** $\hat{g} \leftarrow g_j$ **else** $\hat{g} \leftarrow \max\{g_j, g_{j+1}/\sigma_k\}$ **if** $k < \dim(\pi_N(W_j))$ *and* $f_1(n) \cdot \hat{g} < g(x) \leq g_{j+1}/(f_2(n) \cdot \sigma_{k+1})$ **then**
        $\bar{x} \leftarrow \tilde{x}^{\mathrm{m}}(g(x))$;
        $y^{(0)} \leftarrow \tilde{x}^{\mathrm{m}}(n\hat{g})$;
        $x \leftarrow$ LONG-STEPS$(x, \bar{x}, y^{(0)}, N, \hat{g})$;
    $z \leftarrow$ RATIO-CIRCUIT$(A, c, \mathbf{0}, 1/x)$;
    $x \leftarrow \mathrm{aug}_P(x, z)$;
return x;

Let $k := C_\sigma(\ell_N^{W_j}, [g_{j+1}/g(x), \infty))$. If $k = 0$, then $g(x)$ lies in the subinterval with left endpoint g_j. Otherwise, $g(x)$ lies in the subinterval with left endpoint $\max\{g_j, g_{j+1}/\sigma_k\}$. We define $\hat{g}$ to be the corresponding left endpoint.

Our goal is to cross $\hat{g}$ quickly via circuit steps. If the current gap $g(x)$ is multiplicatively close to $\hat{g}$, i.e., $g(x) \leq f_1(n) \cdot \hat{g}$, then this can be achieved by repeatedly applying Wallacher's rule according to Lemma 7. It turns out that this is always the case when $k = \dim(\pi_N(W_j))$ (Lemma 17). On the other hand, if $g(x)$ is multiplicatively far away from $\hat{g}$, then we need a different circuit step. In this scenario, the algorithm first runs Wallacher's rule to move the gap away from g_{j+1}/σ_{k+1}. Once $g(x) \leq g_{j+1}/(f_2(n) \cdot \sigma_{k+1})$, then it uses the LONG-STEPS subroutine to approach $\hat{g}$.

The LONG-STEPS subroutine relies on two key points $\bar{x}, y^{(0)} \in P$. In this paper, we set $\bar{x} := \tilde{x}^{\mathrm{m}}(g(x))$ and $y^{(0)} := \tilde{x}^{\mathrm{m}}(n\hat{g})$. We remark that this choice is not used when we implement Algorithm 1 in the full version [7]. This is because we no longer have access to the max central path, so it is unclear how to compute these points. Instead, we identify the properties required by $\bar{x}$ and $y^{(0)}$ to make the proof work (which are also satisfied by our choice above), and show how to compute such points.

Let $L : \mathbb{R}_{\geq 0} \to \mathbb{R}^n$ be the ray containing the line segment $[y^{(0)}, \bar{x}]$, where $L(g)$ is the point on L whose optimality gap is equal to g. As long as $g(x)$ is multiplicatively far from $\hat{g}$, LONG-STEPS repeats the following procedure for

Algorithm 2: LONG-STEPS

Input : Feasible solution $x \in P$, points $\bar{x}, y^{(0)} \in P$, subset of coordinates $N \subseteq [n]$, target gap $\hat{g}$

Output: Feasible solution of (1)

Let L be the ray through $[y^{(0)}, \bar{x}]$, where $L(g)$ is the point on L with optimality gap g;

$p \leftarrow 0$;

while $f_1(n) \cdot \hat{g} < g(x)$ **do**

  $S_p \leftarrow \{i \in N : x_i \leq 3ny_i^{(p)}\}$;

  $y^{(p+1)} \leftarrow L(g)$ for $g \geq 0$ minimal such that $L_N(g) \geq 4ny_N^{(p)}$;

  Compute elementary vector $z \in \mathcal{E}(A)$ as a solution to (LSC);

  $x \leftarrow \mathrm{aug}_P(x, z)$;

  $p \leftarrow p + 1$;

return x;

$p \in \mathbb{Z}_{\geq 0}$, starting with $p = 0$. First, we identify the set of *small* variables in N

$$S_p := \left\{ i \in N : x_i \leq 3ny_i^{(p)} \right\}. \tag{13}$$

Next, we set $y^{(p+1)} := L(g)$, where $g \geq 0$ is the smallest value such that $L_N(g) \geq 4ny_N^{(p)}$. Then, we define the *Long Step Circuit LP* as follows

$$\begin{aligned} \min\ & \langle c, z^+ \rangle - \langle c, z^- \rangle \\ \text{s.t.}\ & Az^+ - Az^- = \mathbf{0}, \\ & \sum_{i \in S_p} \frac{z_i^+}{2y_i^{(p+1)}} + f_4(n) \sum_{i \in B} \frac{z_i^-}{x_i} + \sum_{i \in N \setminus S_p} \frac{z_i^-}{x_i} \leq 1, \\ & z^+, z^- \geq \mathbf{0},\ z_i^- = 0 \quad \forall i \in S_p. \end{aligned} \tag{LSC}$$

Note that it can be solved using the RATIO-CIRCUIT oracle. After obtaining an elementary vector $z \in \mathcal{E}(A)$ as a solution to (LSC), we augment our current point x with it. This finishes the description of an iteration of LONG-STEPS. As we will see in the next section, there are at most n such iterations.

3.1 Traversing a Long Singular Value Subinterval

In this section, we show that Algorithm 1 traverses a long singular value subinterval in $O(n^2 \log n)$ calls to RATIO-CIRCUIT. We first set up notation for the analysis. For $t \geq 0$, let $x^{(t)}$ be the iterate of Algorithm 1 after t calls to RATIO-CIRCUIT, including the ones made in LONG-STEPS. Fix $t \geq 0$, and let $(g_j, g_{j+1}]$ be the polarized interval containing $g(x^{(t)})$, with its polarized partition $(B, N) = [n]$. For simplicity of notation, we may assume that $x^{\mathrm{m}}(g_{j+1}) = \mathbf{1}$ and $g_{j+1} = 1$.[2] Denoting $\tilde{g} := g_j$, we have $g(x^{(t)}) \in [\tilde{g}, 1]$.

[2] This is achieved by rescaling the instance: $A \to A\mathrm{diag}(x^{\mathrm{m}}(g_{j+1}))$, $c \to \mathrm{diag}(x^{\mathrm{m}}(g_{j+1}))c/g_{j+1}$.

Consider the lifting operator ℓ_N^W where $W := \ker(A)$, along with its singular values $\sigma_1 \geq \sigma_2 \geq \cdots \geq \sigma_{\dim(\pi_N(W))}$. Let $k := C_\sigma(\ell_N^W, [1/g(x^{(t)}), \infty))$ and assume that $k < \dim(\pi_N(W))$; we will deal with the case $k = \dim(\pi_N(W))$ in Sect. 3.2. Then, $g(x^{(t)})$ lies in the singular value subinterval $[\hat{g}, 1/\sigma_{k+1}]$, where $\hat{g} = \max\{\tilde{g}, 1/\sigma_k\}$. We call the subinterval *long* if $f_1(n)f_2(n) \cdot \hat{g} < 1/\sigma_{k+1}$, and *short* otherwise. If the subinterval is short, then according to Lemma 7, we can traverse it in $O(n^2 \log n)$ calls to RATIO-CIRCUIT with Wallacher's rule. So, we may assume that the subinterval is long. We may also assume that $g(x^{(t)})$ lies deep in the subinterval, i.e.,

$$f_1(n) \cdot \hat{g} < g(x^{(t)}) \leq \frac{1}{f_2(n) \cdot \sigma_{k+1}}. \tag{14}$$

Noting that $[\hat{g}, f_1(n) \cdot \hat{g}]$ and $[1/(f_2(n)\sigma_{k+1}), 1/\sigma_{k+1}]$ are *short*, it suffices to show that this deep part of the subinterval $[\hat{g}, 1/\sigma_{k+1}]$ can be traversed quickly.

Recall the definition of $y^{(p)}$. The lemma below shows that for $p \in \{0, 1, \ldots, n\}$, $y^{(p)}$ lies on the line segment $[y^{(0)}, \bar{x}]$, and is therefore feasible to (1).

Lemma 14. *For every $p \in \{0, 1, \ldots, n\}$, we have $g(y^{(0)}) \leq g(y^{(p)}) \leq f_3(n, p) \cdot \hat{g}$.*

The next lemma upper bounds the optimality gap of any feasible solution to (1) in terms of its 1-norm.

Lemma 15. *For any $x \in P$, its optimality gap is $g(x) \leq 8\tilde{g}\|x_B\|_1 + 2\|x_N\|_1$.*

Since $g(x^{(t)}) \gg \tilde{g}$ and $x^{(t)} \leq x^{\mathrm{m}}(1) = \mathbf{1}$, Lemma 15 shows that the optimality gap of $x^{(t)}$ is essentially determined by $\|x_N^{(t)}\|_1$. This suggests to decrease the coordinates in N in order to approach $\tilde{g}$. Recall the definition of small coordinates (13). Once a coordinate becomes small, we want it to remain small. More formally, for $r \geq t$ and $p \in \mathbb{Z}_{\geq 0}$, let us denote $S_p^{(r)} := \{i \in N : x_i^{(r)} \leq 3ny_i^{(p)}\}$. Then, we want $S_p^{(r)} \subseteq S_{p+1}^{(r+1)}$. This sentiment is reflected in the constraint of (LSC), where any increase in the small coordinates is penalized.

The key proposition below shows that by taking circuit steps given by basic optimal solutions to (LSC), the set of small coordinates grows or we exit the deep part of the subinterval. Thus, LONG-STEPS terminates in at most $|N| \leq n$ iterations.

Proposition 16. *Let $x^{(t)}$ be an iterate of the circuit walk satisfying Condition (14). Let $\bar{x} := \tilde{x}^{\mathrm{m}}(g(x^{(t)}))$ and $y^{(0)} := \tilde{x}^{\mathrm{m}}(n\hat{g})$. For $r \geq t$, if $x^{(r)}$ satisfies Condition (14) and $S_p^{(r)} \neq N$ for some $p \in \mathbb{Z}_{\geq 0}$, then any basic optimal solution $z^* \in \mathcal{E}(A)$ to (LSC) and next iterate $x^{(r+1)} := \mathrm{aug}_P(x^{(r)}, z^*)$ satisfy*

- *$g(x^{(r+1)}) < g(x^{(r)})$ and $S_p^{(r)} \subsetneq S_{p+1}^{(r+1)}$* or
- *$g(x^{(r+1)}) \leq f_1(n) \cdot \hat{g}$.*

Fix $r \geq t$ and $p \in \mathbb{Z}_{\geq 0}$. In order to prove Proposition 16, we construct a target point y° such that $z^\circ := y^\circ - x^{(r)}$ is a good solution to (LSC). First, we shift the iterate $x^{(r)}$ by slightly pushing it towards the max central path

$$\hat{x} := \left(1 - \frac{1}{f_4(n)}\right) x^{(r)} + \frac{1}{f_4(n)} \cdot \tilde{x}^{\mathrm{m}}(g(x^{(r)})). \tag{15}$$

Then, we define the target point as

$$y^\circ := \hat{x} + (\ell_N^W(\Pi_{V_d}(y_N^{(p+1)} - \hat{x}_N)), \Pi_{V_d}(y_N^{(p+1)} - \hat{x}_N)), \tag{16}$$

where V_d is a singular subspace for ℓ_N^W of dimension $d := \dim(\pi_N(W)) - k$, and Π_{V_d} is the orthogonal projection from $\pi_N(W)$ onto V_d.

We now explain the intuition behind this construction, and defer the formal proof to the full version. As indicated in Proposition 16, the goal is to zero out a coordinate in $N \setminus S_p^{(r)}$, while keeping the coordinates in B and $S_p^{(r)}$ positive, and the coordinates in $S_p^{(r)}$ small. If we simply used $y^{(p+1)}$ as the target, the entries of $x^{(r)}_{B\cup S_p^{(r)}}$ that are larger than $y^{(p+1)}_{B\cup S_p^{(r)}}$ may be set to zero first.

To prevent the entries of $x_B^{(r)}$ from hitting zero, one could pick the target y° such that $y_B^\circ \approx x_B^{(r)}$. This can be achieved by projecting $y_N^{(p+1)} - x_N^{(r)}$ onto the singular subspace V_d, and then lifting it to the coordinates in B. However, this only yields additive closeness, which does not work for the entries of $x_B^{(r)}$ that are tiny. To fix this, we first push up these coordinates slightly using the convex combination (15).

The next consideration is to prevent the entries of $x^{(r)}_{S_p^{(r)}}$ from hitting zero. We achieve this by picking the target such that $y^\circ_{S_p^{(r)}} \geq x^{(r)}_{S_p^{(r)}}$. Since $y_i^{(p+1)} \geq x_i^{(r)} + ny_i^{(p)}$ for all $i \in S_p^{(r)}$ by definition, it suffices to show that $|y_i^\circ - y_i^{(p+1)}| \leq ny_i^{(p)}$. This can be proved using our choice of the singular subspace V_d and the point $y^{(0)}$. Finally, the same bound also ensures that the entries of $x^{(r)}_{S_p^{(r)}}$ do not increase too quickly, i.e., $S_p^{(r)} \subseteq S_{p+1}^{(r+1)}$.

3.2 Bounding the Number of Augmentation Steps

In order to bound the number of augmentation steps carried out by Algorithm 1, it remains to cross the singular value breakpoint $1/\sigma_{\dim(\pi_N(W))}$ in poly(n) iterations. We achieve this by showing that the smallest singular value of ℓ_N^W can never be too big.

Lemma 17. *Let $\tilde{g} \leq 1/8n$. Then, $\sigma_{\dim(\pi_N(W))}(\ell_N^W) \leq 2n^{3/2}$.*

We now have all the ingredients to prove our main result.

Theorem 18. *Algorithm 1 terminates in $O\left(n^3 \log n \sum_{i\in[n]} \mathrm{SLC}_{1/2}(x_i^{\mathrm{m}}, g(x^{(0)}))\right)$ Ratio-Circuit augmentation steps.*

Proof. Let $x^{(0)} \in P$ be the initial feasible point given as input. We define a potential function $\Phi : \mathbb{R}_{\geq 0} \to \mathbb{Z}_{\geq 0}$ as follows. For $g \in [0, g(x^{(0)})]$, let $(g_j, g_{j+1}]$ be the polarized interval containing g, with polarized partition (B, N). Then,

$$\Phi(g) := \begin{cases} 1 + n \cdot j + C_\sigma(\ell_N^{W_j}, [g_{j+1}/g, \infty)), & \text{if } g > 0 \\ 0, & \text{if } g = 0. \end{cases}$$

Let $x \in P$ be an iterate of Algorithm 1. First, observe that $\Phi(g(x)) \leq O(n \cdot \sum_{i\in[n]} \mathrm{SLC}_{1/2}(x_i^{\mathrm{m}}, g(x^{(0)})))$. Further, $\Phi(g(x))$ is monotonically nonincreasing as $g(x)$ decreases. Finally, $\Phi(g(x)) = 0$ if and only if $g(x) = 0$. Thus, it remains to show that Algorithm 1 decreases $\Phi(g(x))$ in $O(n^2 \log n)$ augmentation steps.

Let $(g_j, g_{j+1}]$ be the polarized interval containing $g(x)$. We denote $\sigma := \sigma(\ell_N^{W_j})$ and $k := C_\sigma(\ell_N^{W_j}, [g_{j+1}/g(x), \infty))$. As long as $k = \dim(\pi_N(W_j))$, Algorithm 1 only uses steps given by Ratio-Circuit$(A, c, \mathbf{0}, 1/x)$. By Lemma 17 and Lemma 7, $\Phi(g(x))$ decreases in $O(n \log n)$ steps. Hence, we may assume that $k < \dim(\pi_N(W_j))$, and so $\sigma_{k+1} > 0$. While $g(x) > 1/(f_2(n) \cdot \sigma_{k+1})$, Algorithm 1 again only uses steps given by Ratio-Circuit$(A, c, \mathbf{0}, 1/x)$. In $O(n \log n)$ steps, the iterate satisfies $g(x) \leq 1/(f_2(n) \cdot \sigma_{k+1})$ by Lemma 7.

If $g(x) > f_1(n) \cdot \hat{g}$, then the Long-Steps subroutine is called. We claim that $g(x) \leq f_1(n) \cdot \hat{g}$ in at most n augmentation steps via (LSC). By Lemma 16, if $g(x) > f_1(n) \cdot \hat{g}$ after each step, then the set of small coordinates S_p has grown. Thus, after $|N|$ steps, we have $S_p = N$. At this point, for every $i \in N$, we have

$$x_i \leq 3ny_i^{(p)} \leq 3nx_i^{\mathrm{m}}(f_3(n,p) \cdot \hat{g}) \leq 12nf_3(n,p) \cdot \hat{g} \leq \frac{f_1(n) \cdot \hat{g}}{2n}.$$

The second inequality is by Lemma 14, while the third inequality is due to $[\tilde{g}, 1]$ being $\frac{1}{4}$-polarized and $x^{\mathrm{m}}(1) = \mathbf{1}$. Since $x_i \leq 1$ for all $i \in B$, we have $g(x) \leq f_1(n) \cdot \hat{g}$ by Lemma 15. If x remains in the same singular value subinterval, then Algorithm 1 again only uses steps given by Ratio-Circuit$(A, c, \mathbf{0}, 1/x)$. By Lemma 7, $\Phi(g(x))$ decreases after $O(n^2 \log n)$ steps.

Disclosure of Interests. The authors have no competing interests to declare that are relevant to the content of this article.

References

1. Allamigeon, X., Benchimol, P., Gaubert, S., Joswig, M.: Log-barrier interior point methods are not strongly polynomial. SIAM J. Appl. Algebra Geom. **2**(1), 140–178 (2018)
2. Allamigeon, X., Dadush, D., Loho, G., Natura, B., Végh, L.A.: Interior point methods are not worse than simplex. SIAM J. Comput. (0), FOCS22-178 (2025)
3. Allamigeon, X., Gaubert, S., Vandame, N.: No self-concordant barrier interior point method is strongly polynomial. In: Leonardi, S., Gupta, A. (eds.) STOC 2022: 54th Annual ACM SIGACT Symposium on Theory of Computing, Rome, Italy, 20–24 June 2022, pp. 515–528. ACM (2022)

4. Black, A.E., Nöbel, C., Steiner, R.: Short circuit walks in fixed dimension. CoRR abs/2510.01916 (2025)
5. Bland, R.G.: On the generality of network flow theory (1976), presented at the ORSA/TIMS Joint National Meeting, Miami, FL
6. Borgwardt, S., Finhold, E., Hemmecke, R.: On the circuit diameter of dual transportation polyhedra. SIAM J. Discret. Math. **29**(1), 113–121 (2015)
7. Dadush, D., Kober, S., Koh, Z.K.: On circuit diameter and straight line complexity. CoRR abs/2602.05699 (2026)
8. Dadush, D., Koh, Z.K., Natura, B., Olver, N., Végh, L.A.: A strongly polynomial algorithm for linear programs with at most two nonzero entries per row or column. In: Proceedings of the 56th Annual ACM Symposium on Theory of Computing, pp. 1561–1572 (2024)
9. Dadush, D., Koh, Z.K., Natura, B., Végh, L.A.: On circuit diameter bounds via circuit imbalances. Math. Program. **206**(1), 631–662 (2024)
10. Dinic, E.A.: Algorithm for solution of a problem of maximum flow in a network with power estimation. Soviet Math. Doklady **11**(5), 1277–1280 (1970)
11. Edmonds, J., Karp, R.M.: Theoretical improvements in algorithmic efficiency for network flow problems. J. ACM **19**(2), 248–264 (1972)
12. Ford, L.R., Fulkerson, D.R.: Maximal flow through a network. Can. J. Math. **8**, 399–404 (1956)
13. Kafer, S., Pashkovich, K., Sanità, L.: On the circuit diameter of some combinatorial polytopes. SIAM J. Discret. Math. **33**(1), 1–25 (2019)
14. Kalai, G.: Upper bounds for the diameter and height of graphs of convex polyhedra. Discret. Comput. Geom. **8**, 363–372 (1992)
15. Kalai, G., Kleitman, D.J.: A quasi-polynomial bound for the diameter of graphs of polyhedra. Bull. Am. Math. Soc. **26**(2), 315–316 (1992)
16. Lee, J.: Subspaces with well-scaled frames. Linear Algebra Appl. **114**, 21–56 (1989)
17. Loera, J.A.D., Hemmecke, R., Lee, J.: On augmentation algorithms for linear and integer-linear programming: from Edmonds-Karp to bland and beyond. SIAM J. Optim. **25**(4), 2494–2511 (2015)
18. Loera, J.A.D., Kafer, S., Sanità, L.: Pivot rules for circuit-augmentation algorithms in linear optimization. SIAM J. Optim. **32**(3), 2156–2179 (2022)
19. McCormick, S.T., Shioura, A.: Minimum ratio canceling is oracle polynomial for linear programming, but not strongly polynomial, even for networks. Oper. Res. Lett. **27**(5), 199–207 (2000)
20. Natura, B.: Circuit diameter of polyhedra is strongly polynomial. CoRR abs/2602.06958 (2026)
21. Nöbel, C., Steiner, R.: Complexity of polytope diameters via perfect matchings. In: Azar, Y., Panigrahi, D. (eds.) Proceedings of the 2025 Annual ACM-SIAM Symposium on Discrete Algorithms, SODA 2025, New Orleans, LA, USA, 12–15 January 2025, pp. 2234–2251. SIAM (2025)
22. Sukegawa, N.: An asymptotically improved upper bound on the diameter of polyhedra. Discret. Comput. Geom. **62**(3), 690–699 (2019)
23. Todd, M.J.: An improved Kalai-Kleitman bound for the diameter of a polyhedron. SIAM J. Discret. Math. **28**(4), 1944–1947 (2014)
24. Wallacher, C.: A generalization of the minimum-mean cycle selection rule in cycle canceling algorithms (1989), unpublished manuscript, Institute für Angewandte Mathematik, Technische Universität Braunschweig
25. Wulf, L.: Computing the polytope diameter is even harder than NP-hard (already for perfect matchings). CoRR abs/2502.16398 (2025)

Improved Approximation Algorithms for Orthogonally Constrained Problems Using Semidefinite Optimization

Ryan Cory-Wright[1] and Jean Pauphilet[2(✉)]

[1] Imperial Business School, South Kensington Campus, London SW7 2AZ, UK
r.cory-wright@imperial.ac.uk

[2] London Business School, Regent's Park, London NW1 4SA, UK
jpauphilet@london.edu

Abstract. Building on the blueprint from Goemans and Williamson (1995) for the Max-Cut problem, we construct a polynomial-time approximation algorithm for orthogonally constrained quadratic optimization problems. First, we derive a semidefinite relaxation and propose a randomized rounding algorithm to generate feasible solutions from the relaxation. Second, we derive purely multiplicative approximation guarantees for our algorithm. When optimizing for m orthogonal vectors in dimension n, we show that our algorithm achieves a performance ratio of at least $\max\left\{\frac{2}{\pi m}, \frac{1}{\pi(\log(2m)+1)}\right\}$. Our analysis is tight in the sense that we exhibit instances where our algorithm's performance is at most $O(1/\log m)$. We also show how to compute a tighter constant for finite (n, m) by solving a univariate optimization problem, and this analysis is exact for any n when $m = 1$.

Keywords: Orthogonality constraints · semidefinite relaxation · randomized rounding · approximation algorithm

1 Introduction

Many important optimization problems, such as quadratic assignment [15], quantum nonlocality [9], and control theory [4] problems feature semi-orthogonal matrices, i.e., matrices $\boldsymbol{U} \in \mathbb{R}^{n\times m}$ where $\boldsymbol{U}^\top \boldsymbol{U} = \boldsymbol{I}_m$. Orthogonality constraints are also related to the rank of a matrix, which models a matrix's complexity in imputation [3], factor analysis [5], and multi-task regression [21] settings.

In combinatorial optimization, a major advance in the design of approximation algorithms occurred with [16], who proposed a 0.87856-approximation algorithm for Max-Cut. Their algorithm also provides a $2/\pi$-approximation for general binary quadratic optimization (BQO) problems [23], and can be extended to linearly-constrained BQO problems [7]. Conceptually, [16] established semidefinite optimization and correlated rounding at the core of approximation algorithms [see 28, 29].

S. S. Dey et al. (Eds.): IPCO 2026, LNCS 16588, pp. 249–264, 2026.
https://doi.org/10.1007/978-3-032-28691-8_17

In this work, we extend the core ideas underpinning the Goemans–Williamson algorithm to quadratic semi-orthogonal optimization problems and provide analogous constant-factor guarantees on the quality of semidefinite relaxations in the semi-orthogonal setting.

1.1 The Original Goemans–Williamson Algorithm

BQO is a canonical optimization problem [for a review of applications, see 19]. It is also an important building block for logically constrained optimization problems with quadratic objectives [see, e.g., 13]. Formally, given a matrix $\boldsymbol{Q} \succeq \boldsymbol{0}$, BQO selects a vector $\boldsymbol{z}$ in $\{-1,1\}^m$ that solves

$$\max_{\boldsymbol{z} \in \{-1,1\}^m} \sum_{i,j} Q_{i,j} z_i z_j = \max_{\boldsymbol{z} \in \{-1,1\}^m} \langle \boldsymbol{Q}, \boldsymbol{z}\boldsymbol{z}^\top \rangle, \tag{1}$$

where $\langle \cdot, \cdot \rangle$ denotes the Frobenius inner product between matrices. Problem (1) is NP-hard and often challenging to solve to optimality when $m \geq 100$ [25]. Accordingly, a popular approach for obtaining near-optimal solutions is to sample from a distribution parameterized by the solution of (1)'s convex relaxation. Specifically, we can reformulate (1) in terms of the rank-one matrix $\boldsymbol{Z} = \boldsymbol{z}\boldsymbol{z}^\top$. Then, a valid relaxation of Problem (1) is given by

$$\max_{\boldsymbol{Z} \in \mathcal{S}_+^m} \langle \boldsymbol{Q}, \boldsymbol{Z} \rangle \text{ s.t. } \mathrm{diag}(\boldsymbol{Z}) = \boldsymbol{e}, \tag{2}$$

which would be an exact reformulation with the additional (non-convex) constraint $\mathrm{rank}(\boldsymbol{Z}) = 1$. Probabilistically speaking, (2) is a device for constructing a pseudodistribution over $\boldsymbol{z} \in \{-1,1\}^m$ [14]. This suggests sampling vectors from a distribution with second moment $\boldsymbol{Z}^\star$ and rounding to restore feasibility, as proposed by [16] for Max-Cut and described in Algorithm 1. The overall idea of Algorithm 1 is that the projection step (i.e., taking the coordinate-wise sign of $\boldsymbol{y}$) partially preserves the second moment of the distribution of $\boldsymbol{y}$, $\mathbb{E}[\boldsymbol{y}\boldsymbol{y}^\top] = \boldsymbol{Z}^\star$. Precisely, we have $\mathbb{E}\left[\hat{\boldsymbol{z}}\hat{\boldsymbol{z}}^\top\right] \succeq \frac{2}{\pi}\boldsymbol{Z}^\star$ [see 7,23].

Algorithm 1. The Goemans–Williamson rounding algorithm for Problem (1)

Compute $\boldsymbol{Z}^\star$ a solution of (2)
Sample $\boldsymbol{y} \sim \mathcal{N}(\boldsymbol{0}, \boldsymbol{Z}^\star)$
Construct $\hat{\boldsymbol{z}} \in \{-1,1\}^m : \hat{z}_i := \mathrm{sign}(y_i)$
return $\hat{\boldsymbol{z}}$ a solution to Problem (1)

1.2 Orthogonally Constrained Quadratic Optimization

In this paper, we consider a family of orthogonally constrained quadratic problems that subsumes binary quadratic optimization. Formally, we search for m orthogonal vectors $\boldsymbol{u}_i \in \mathbb{R}^n$ which solve

$$\max_{\boldsymbol{u}_i \in \mathbb{R}^n,\ i \in [m]} \quad \sum_{i,j=1}^{m} \boldsymbol{u}_i^\top \boldsymbol{A}^{(i,j)} \boldsymbol{u}_j \text{ s.t. } \boldsymbol{u}_i^\top \boldsymbol{u}_j = \delta_{i,j}, \quad \forall i, j \in [m], \tag{3}$$

where $\boldsymbol{A}$ is an $nm \times nm$ positive semidefinite matrix (in short, $\boldsymbol{A} \in \mathcal{S}_+^{nm}$) with block matrices $\boldsymbol{A}^{(i,j)} \in \mathbb{R}^{n \times n}$, and $\delta_{i,j} = 1$ if $i = j$ and 0 otherwise. We require $n \geq m$. By introducing the semi-orthogonal matrix $\boldsymbol{U} \in \mathbb{R}^{n \times m}$ whose columns are the vectors $\boldsymbol{u}_i \in \mathbb{R}^n$, we can write our problem as

$$\max_{\boldsymbol{U} \in \mathbb{R}^{n \times m}} \quad \langle \boldsymbol{A}, \text{vec}(\boldsymbol{U})\text{vec}(\boldsymbol{U})^\top \rangle \quad \text{s.t.} \quad \boldsymbol{U}^\top \boldsymbol{U} = \boldsymbol{I}_m, \tag{4}$$

where the vec$(\cdot)$ operator stacks the columns of $\boldsymbol{U}$ together into a single vector.

The similarities between Problems (4) and (1) are striking: For example, we can formulate any BQO instance (1) as a special case of Problem (4). Our reduction (presented in the online appendix)[1] not only shows that Problem (4) is NP-hard [as also proved in 18, Theorem 3.1] but is also approximation-preserving. Therefore, inheriting from the inapproximability results of BQO, Problem (4) cannot be approximated in polynomial time within a factor of $2/\pi + \varepsilon$ unless P = NP [11, Theorem I.3].

In addition to its applications in clustering, quantum nonlocality, or generalized trust-region problems [and references therein, see 12], Problem (4) appears as a relevant substructure for mixed-projection formulations of rank-constrained quadratic optimization problems [6].

In this paper, inspired by the Goemans–Williamson algorithm for BQO, we develop a relax-then-round strategy with a $\Theta(1/\log m)$ multiplicative-factor performance guarantee for Problem (4).

1.3 Related Work

Our work is most closely related to [12], who develop a hierarchy of semidefinite relaxations for Problem (4). To numerically evaluate the tightness of their relaxations, they apply several 'feasible rounding procedures' but do not provide any theoretical performance guarantees. In contrast, we develop a randomized rounding procedure and show that it achieves a multiplicative factor guarantee, which is independent of the ambient dimension n and only decreases as $1/\log m$. As a non-convex quadratic optimization problem, Problem (4) can also be solved to optimality via global solvers such as `Gurobi` or `BARON`. However, the scalability of these global solvers is currently limited.

[1] Appendices are not published as part of the conference proceedings but are available with the online preprint.

Special Cases: A larger body of work considers a special case of Problem (4), where the matrix $\boldsymbol{A}$ is block-diagonal, namely $\boldsymbol{A}^{(i,j)} = \boldsymbol{0}$ if $i \neq j$. In this case, Problem (4) reduces to

$$\max_{\boldsymbol{U} \in \mathbb{R}^{n \times m}} \sum_{i \in [m]} \boldsymbol{u}_i^\top \boldsymbol{A}^{(i,i)} \boldsymbol{u}_i \quad \text{s.t.} \quad \boldsymbol{U}^\top \boldsymbol{U} = \boldsymbol{I}_m, \tag{5}$$

which is referred to as the sum of heterogeneous quadratic forms or the heterogeneous PCA problem. Indeed, when all the matrices $\boldsymbol{A}^{(i,i)}$ are equal, we recover the Principal Component Analysis (PCA) problem. [8, Section 5] solve Problem (5) in polynomial time via linear algebra techniques when the matrices $\boldsymbol{A}^{(i,i)}$ are diagonal or commute with each other. For general matrices, [15] further tailor the semidefinite relaxations of [12]. Although tighter, their relaxations are not always tight. For some instances, they even obtain optimality gaps exceeding 100%. We are not aware of any approximation algorithms with guarantees specifically for general (non-diagonal) instances of Problem (5).

Approximation Algorithms: To our knowledge, many of the existing approximation algorithms apply to optimization problems with different orthogonality structures. [10] propose an approximation algorithm for problems of the form

$$\max_{\boldsymbol{U} \in \mathbb{R}^{n \times m}} \sum_{i,j \in [m]} A_{i,j} \boldsymbol{u}_i^\top \boldsymbol{u}_j \text{ s.t. } \boldsymbol{u}_i^\top \boldsymbol{u}_i = 1 \ \forall i \in [m], \tag{6}$$

which also subsumes BQO (for $n = 1$), but does not enforce orthogonality between the columns of $\boldsymbol{U}$. They devise a relax-and-round strategy analogous to Goemans–Williamson that achieves an approximation ratio of $2/\pi + \Theta(1/n)$.

A second line of work [22,27] proposes $\Omega(1/\log(n+m))$-approximation algorithms for quadratic optimization problems over matrices $\boldsymbol{U}$ that satisfy $\boldsymbol{U}^\top \boldsymbol{U} \preceq \boldsymbol{I}_m$, i.e., whose largest singular value is at most one. This constraint does not ensure that the columns of $\boldsymbol{U}$ are orthogonal. Nonetheless, [22] shows that, in several special cases such as the orthogonal Procrustes or quadratic assignment problems (but not in the case of Problem (4)), orthogonality constraints can be relaxed into $\boldsymbol{U}^\top \boldsymbol{U} \preceq \boldsymbol{I}_m$ without loss of optimality. Our algorithm differs in that it generates matrices $\boldsymbol{U}$ with orthogonal columns, while they only guarantee $\sigma_{\max}(\boldsymbol{U}) \leq 1$. In our algorithm, however, starting from a matrix with $\sigma_{\max}(\boldsymbol{U}) \leq 1$, we show how to restore orthogonality without loss in average objective value by using a randomization trick. In addition, their performance guarantee vanishes both with n and m, while ours depends only on m.

Finally, [1] study approximation algorithms for problems of the form

$$\max_{\boldsymbol{U}_i \in \mathbb{R}^{n \times m}, i=1,\ldots,k} \sum_{i,j \in [k]} \langle \boldsymbol{A}^{(i,j)}, \boldsymbol{U}_i^\top \boldsymbol{U}_j \rangle \text{ s.t. } \boldsymbol{U}_i^\top \boldsymbol{U}_i = \boldsymbol{I}_m \ \forall i \in [k]. \tag{7}$$

Unfortunately, Problem (7) is not equivalent to (4) and [1]'s proof techniques do not extend to our case. In particular, heterogeneous PCA is a special case of our Problem (4) but cannot be cast in the form (7). There are two key differences in the objective function of (7): it involves the *inner* products between columns of *different* semi-orthogonal matrices $\boldsymbol{U}_i$, $\boldsymbol{U}_{i'}$ for $i \neq i'$. On the other hand, the objective in (4) depends on *outer* products between columns of the *same* matrix $\boldsymbol{U}$. In particular, [1] can restore feasibility for each $\boldsymbol{U}_i$, $i = 1, \ldots, k$ in (7) separately, while the columns of $\boldsymbol{U}$ in (4) need to be orthogonalized together.

Remark 1. Using a bilinearization trick analogous to that of [20, Section 5.3], we can reformulate (4) as a special case of the non-commutative Grothendieck problem, for which [20] propose a $1/(2\sqrt{2})$-factor approximation algorithm. Compared with their approach, our algorithm relies on a more compact semidefinite relaxation (involving one $nm \times nm$ semidefinite matrix vs. $2n^2 \times 2n^2$), hence is more tractable, especially when $n \gg m$. It also applies directly to real-valued variables (vs. their complex-valued rounding step) and is closer to the original Goemans–Williamson algorithm. Finally, although our performance guarantee is weaker asymptotically, it is stronger for small values of m (see Table 1).

1.4 Contributions and Structure

Our main contribution is the development of a Goemans–Williamson sampling algorithm for the class of semi-orthogonal problems (4).

In Sect. 2, we derive a semidefinite relaxation and propose a sampling procedure. We show that our algorithm achieves a purely multiplicative approximation guarantee (Theorems 1–2) for Problem (4), with a constant that scales as $1/\log m$. We also identify a class of problem instances (Proposition 2) for which our algorithm cannot achieve a performance guarantee better than $O(1/\log m)$. Notably, our approximation ratio does not depend on the ambient dimension n and depends only mildly on m. To better judge the quality of our approximations, we develop two simpler algorithms in Sect. 3: uniform sampling and a stronger benchmark inspired by PCA. We show that they achieve $1/nm$ and $1/m^2$ approximation ratios, respectively, which are dominated by our semidefinite relax-and-round procedure. Thus, we evaluate the performance of our approach numerically in Sect. 4.

1.5 Notations

We let lowercase boldfaced characters such as $\boldsymbol{x}$ denote vectors and uppercase boldfaced characters such as $\boldsymbol{X}$ denote matrices. We denote by $\boldsymbol{e}$ the vector of all ones. We let $[n]$ denote the set of running indices $\{1, ..., n\}$. The cone of $n \times n$ symmetric (resp. positive semidefinite) matrices is denoted by $\mathcal{S}^n$ (resp. $\mathcal{S}^n_+$). For a matrix $\boldsymbol{X} \in \mathbb{R}^{n \times m}$, we let $\boldsymbol{x}_i$ denote its ith column. We let $\text{vec}(\boldsymbol{X}) : \mathbb{R}^{n \times m} \to \mathbb{R}^{nm}$ denote the vectorization operator which maps matrices to vectors by stacking columns. For a matrix $\boldsymbol{W}$, we may describe it as a block matrix composed of equally sized blocks and denote the (i, i') block by $\boldsymbol{W}^{(i,i')}$. The dimension of each block will be clear from the context. We let $\mathcal{N}(\boldsymbol{0}, \boldsymbol{\Sigma})$ denote a centered multivariate normal distribution with covariance matrix $\boldsymbol{\Sigma}$. Finally, we denote $f(x) = O(g(x))$ or $g(x) = \Omega(f(x))$ when there exists a constant

$C > 0$ such that $|f(x)| \leq C|g(x)|$. The notation $f(x) = \Theta(g(x))$ means that both $f(x) = O(g(x))$ and $f(x) = \Omega(g(x))$ hold.

2 A Goemans–Williamson Approach

In this section, we propose a new Goemans–Williamson-type approach for semi-orthogonal quadratic optimization problems. We review a semidefinite relaxation for semi-orthogonal quadratic optimization in Sect. 2.1. In Sect. 2.2, we propose a randomized rounding scheme to generate feasible solutions. We derive multiplicative performance guarantees for our algorithm in Sect. 2.3, and discuss potential variants to our rounding mechanism in Sect. 2.4.

2.1 A Shor Relaxation

As reviewed in Sect. 1.3, [12, Section 2.2] derive the following semidefinite relaxation for Problem (4):

$$\max_{\boldsymbol{W} \in \mathcal{S}^{mn}_+} \langle \boldsymbol{A}, \boldsymbol{W} \rangle \quad \text{s.t.} \sum_{i \in [m]} \boldsymbol{W}^{(i,i)} \preceq \boldsymbol{I}_n, \ \operatorname{tr}\left(\boldsymbol{W}^{(j,j')}\right) = \delta_{j,j'}, \forall j, j' \in [m], \tag{8}$$

where the matrix $\boldsymbol{W}$ encodes the outer-product of $\operatorname{vec}(\boldsymbol{U})$ with itself, and the trace constraints on the blocks of $\boldsymbol{W}$ stem from the columns of $\boldsymbol{U}$ having unit norm and being pairwise orthogonal.

Similarly to the semidefinite relaxation of (1), imposing the constraint that $\boldsymbol{W}$ is rank-one in (8) would result in an exact reformulation of (4). Accordingly, the relaxation (8) is tight whenever some optimal solution is rank-one. However, the optimal solutions to (8) are often high-rank (the case $m = 1$ is one of the special cases where this semidefinite relaxation is tight). Actually, it follows from manipulating the Barvinok–Pataki bound [2,24] that there exists[2] some optimal solution to Problem (8) with rank at most $n + m$. However, not all optimal solutions obey this bound; thus, we do not use this observation in our analysis. An interesting question is how to generate a high-quality feasible solution to (4), with a provable performance guarantee, by leveraging a solution of (8), which is the focus of the rest of the section.

Note that Problem (8), which is essentially a Shor relaxation [26], corresponds to the 'DiagSum' relaxation of [12]. They also derive an even stronger relaxation, which they call a 'Kronecker' relaxation. We do not explicitly analyze their Kronecker relaxation here because it is significantly less tractable [as reported in 12, Table 1], and it would not lead to a tighter approximation guarantee for

[2] After introducing a slack matrix $\boldsymbol{S}$ to write the semidefinite inequality constraint $\boldsymbol{S} = \boldsymbol{I}_n - \sum_{i \in [m]} \boldsymbol{W}^{(i,i)}, \boldsymbol{S} \succeq \boldsymbol{0}$, we have $m(m+1)/2 + n(n+1)/2$ scalar inequalities. Thus, [24, theorem 2.2] states that there exists some optimal solution $(\boldsymbol{W}, \boldsymbol{S})$ with $\operatorname{rank}(\boldsymbol{W})(\operatorname{rank}(\boldsymbol{W}) + 1)/2 + \operatorname{rank}(\boldsymbol{S})(\operatorname{rank}(\boldsymbol{S}) + 1)/2 \leq m(m+1)/2 + n(n+1)/2$. Since $\operatorname{rank}(\boldsymbol{S}) \geq 0$, it implies $\operatorname{rank}(\boldsymbol{W}) \leq n + m$.

our algorithm in the next section[3], although it could improve the semidefinite relaxation for some specific instances.

2.2 A Sample-then-Stochastically-Project Procedure

We propose a randomized rounding scheme to generate high-quality feasible solutions to (4) from an optimal solution to (8).

Our algorithm involves three main steps: First, we solve (8) and obtain a semidefinite matrix $\boldsymbol{W}^\star$. Second, using $\boldsymbol{W}^\star$, we sample an $n \times m$ matrix $\boldsymbol{G}$ such that $\text{vec}(\boldsymbol{G})$ follows a normal distribution with mean $\mathbf{0}_{nm}$ and covariance matrix $\boldsymbol{W}^\star$. Third, from the matrix $\boldsymbol{G}$, we generate a feasible solution to (4). Specifically, we compute a singular value decomposition of $\boldsymbol{G}$, $\boldsymbol{G} = \boldsymbol{U\Sigma V}^\top$ and define $\boldsymbol{Q} := \boldsymbol{UDV}^\top$ where $\boldsymbol{D}$ is a diagonal matrix with ± 1 diagonal entries. Diagonal entries of $\boldsymbol{D}$ are sampled independently such that $\mathbb{P}(D_{i,i} = 1) = (1 + \sigma_i/\sigma_{\max})/2$, where σ_i is the ith singular value of $\boldsymbol{G}$ and $\sigma_{\max}$ is the largest singular value of $\boldsymbol{G}$. We have $\boldsymbol{Q}^\top \boldsymbol{Q} = \boldsymbol{I}_m$ because $\boldsymbol{D}^2 = \boldsymbol{I}_m$. We summarize our procedure in Algorithm 2.

In Algorithm 2, we can sample $\text{vec}(\boldsymbol{G}) \sim \mathcal{N}(\mathbf{0}_{nm}, \boldsymbol{W}^\star)$ even when $\boldsymbol{W}^\star$ is rank-deficient via the following construction—which will also be relevant for our theoretical analysis. Denoting $r = \text{rank}(\boldsymbol{W}^\star)$, we first construct a Cholesky decomposition of $\boldsymbol{W}^\star$: $\boldsymbol{W}^\star = \sum_{k\in[r]} \text{vec}(\boldsymbol{B}_k)\,\text{vec}(\boldsymbol{B}_k)^\top$ with $\boldsymbol{B}_k \in \mathbb{R}^{n\times m}$. Then, we sample $\text{vec}(\boldsymbol{G}) = \sum_{k\in[r]} \text{vec}(\boldsymbol{B}_k) z_k$ with $\boldsymbol{z} \sim \mathcal{N}(\mathbf{0}_r, \boldsymbol{I}_r)$. This procedure ensures that $\text{vec}(\boldsymbol{G}) \in \text{span}(\boldsymbol{W}^\star)$ almost surely. In particular, if our semidefinite relaxation is tight (e.g., when $m = 1$) and $\boldsymbol{W}^\star$ has rank 1, then $\boldsymbol{G}$ is optimal almost surely (our rounding is exact).

Algorithm 2. A Relax-then-project algorithm for Problem (4)

Compute $\boldsymbol{W}^\star$ a solution of (8)
Sample $\boldsymbol{G}$ according to $\text{vec}(\boldsymbol{G}) \sim \mathcal{N}(\mathbf{0}_{nm}, \boldsymbol{W}^\star)$
Compute the SVD of $\boldsymbol{G}$, $\boldsymbol{G} = \boldsymbol{U\Sigma V}^\top$
Sample $D_{i,i} = \pm 1$ independently such that $\mathbb{P}(D_{i,i} = 1) = (1 + \sigma_i/\sigma_{\max})/2$
Construct $\boldsymbol{Q} = \boldsymbol{UDV}^\top$
return the semi-orthogonal matrix $\boldsymbol{Q}$

Second, we should comment on our procedure to obtain a feasible semi-orthogonal matrix $\boldsymbol{Q}$ from $\boldsymbol{G}$. Conditioned on $\boldsymbol{G}$, we have $\mathbb{E}[\boldsymbol{D}|\boldsymbol{G}] = \boldsymbol{\Sigma}/\sigma_{\max}$, so that $\mathbb{E}[\boldsymbol{Q}|\boldsymbol{G}] = \boldsymbol{G}/\sigma_{\max}$. This observation will be crucial in our theoretical analysis (in fact, this observation is precisely the motivation for the sampling

[3] Specifically, in Proposition 2, we identify matrices $\boldsymbol{W}$ where each block matrix $\boldsymbol{W}^{(i,j)}$ is rank-one and for which we prove our algorithm cannot achieve a better approximation factor than $O(1/\log m)$. The Kronecker constraints of [12] would not rule out any of these matrices. Thus, they cannot improve the order of our approximation guarantee.

step), enabling us to relate the second moment of the distribution of $\boldsymbol{Q}$ to that of $\boldsymbol{G}$, as formally stated below.

Proposition 1. *Consider matrices* $\boldsymbol{G} \in \mathbb{R}^{n\times m}$ *and* $\boldsymbol{Q} \in \mathbb{R}^{n\times m}$ *generated according to Algorithm 2. The following holds:*

$$\mathbb{E}\left[\mathrm{vec}(\boldsymbol{Q})\mathrm{vec}(\boldsymbol{Q})^\top\right] \succeq \mathbb{E}\left[\frac{\mathrm{vec}(\boldsymbol{G})\mathrm{vec}(\boldsymbol{G})^\top}{\sigma_{\max}(\boldsymbol{G})^2}\right]. \tag{9}$$

Proof. Observe that, conditioned on $\boldsymbol{G}$, we have

$$\mathrm{Cov}(\mathrm{vec}(\boldsymbol{Q})|\boldsymbol{G}) = \mathbb{E}\left[\mathrm{vec}(\boldsymbol{Q})\mathrm{vec}(\boldsymbol{Q})^\top|\boldsymbol{G}\right] - \mathbb{E}\left[\mathrm{vec}(\boldsymbol{Q})\,|\boldsymbol{G}\right]\mathbb{E}\left[\mathrm{vec}(\boldsymbol{Q})\,|\boldsymbol{G}\right]^\top \succeq \boldsymbol{0}.$$

Since $\mathbb{E}\left[\mathrm{vec}(\boldsymbol{Q})\,|\boldsymbol{G}\right] = \mathrm{vec}(\boldsymbol{G})/\sigma_{\max}(\boldsymbol{G})$, it leads to

$$\mathbb{E}\left[\mathrm{vec}(\boldsymbol{Q})\mathrm{vec}(\boldsymbol{Q})^\top|\boldsymbol{G}\right] \succeq \frac{1}{\sigma_{\max}(\boldsymbol{G})^2}\mathrm{vec}(\boldsymbol{G})\mathrm{vec}(\boldsymbol{G})^\top.$$

Taking expectation with respect to $\boldsymbol{G}$ yields (9). □

Remark 2. For a fixed matrix $\boldsymbol{A}$, Lemma 3.5 in [20] shows how to construct in polynomial-time, for any realization of $\boldsymbol{G}$, a semi-orthogonal matrix $\widetilde{\boldsymbol{Q}}$ such that $\langle \boldsymbol{A}, \mathrm{vec}(\widetilde{\boldsymbol{Q}})\,\mathrm{vec}(\widetilde{\boldsymbol{Q}})^\top\rangle \geq \langle \boldsymbol{A}, \mathrm{vec}(\boldsymbol{G})\,\mathrm{vec}(\boldsymbol{G})^\top\rangle/\sigma_{\max}(\boldsymbol{G})^2$. In comparison, our orthogonalization satisfies the same inequality on average (instead of almost surely) while being agnostic to the objective matrix $\boldsymbol{A}$.

Similar to the original algorithm of [16], the intuition behind Algorithm 2 is that the sampled matrix $\boldsymbol{G}$ achieves an average performance equal to the relaxation value ($\mathbb{E}[\langle \boldsymbol{A}, \mathrm{vec}(\boldsymbol{G})\mathrm{vec}(\boldsymbol{G})^\top\rangle] = \langle \boldsymbol{A}, \boldsymbol{W}^\star\rangle$) and is feasible on average ($\mathbb{E}[\boldsymbol{G}^\top\boldsymbol{G}] = \boldsymbol{I}_m$). Therefore, the objective value of the feasible solution $\boldsymbol{Q}$ should not be too different from that of $\boldsymbol{G}$, as we now theoretically study.

2.3 Theoretical Analysis: Multiplicative Performance Guarantees

We analyze the performance of Algorithm 2 in the case where the objective matrix $\boldsymbol{A}$ in (4) is positive semidefinite.

Solutions $\boldsymbol{Q}$ generated by Algorithm 2 achieve an average performance of $\mathbb{E}\left[\mathrm{vec}(\boldsymbol{Q})^\top\boldsymbol{A}\,\mathrm{vec}(\boldsymbol{Q})\right] = \langle \boldsymbol{A}, \mathbb{E}\left[\mathrm{vec}(\boldsymbol{Q})\,\mathrm{vec}(\boldsymbol{Q})^\top\right]\rangle$. By Proposition 1 and the fact that $\boldsymbol{A} \succeq \boldsymbol{0}$, we have $\mathbb{E}\left[\mathrm{vec}(\boldsymbol{Q})^\top\boldsymbol{A}\,\mathrm{vec}(\boldsymbol{Q})\right] \geq \langle \boldsymbol{A}, \mathbb{E}\left[\mathrm{vec}(\boldsymbol{G})\,\mathrm{vec}(\boldsymbol{G})^\top/\sigma_{\max}(\boldsymbol{G})^2\right]\rangle$. Hence, to obtain a β-multiplicative guarantee for our algorithm, it suffices to show that $\mathbb{E}\left[\mathrm{vec}(\boldsymbol{G})\,\mathrm{vec}(\boldsymbol{G})^\top/\sigma_{\max}(\boldsymbol{G})^2\right] \succeq \beta\boldsymbol{W}^\star$.

Our first result is an analytical multiplicative performance guarantee, which asymptotically scales as $1/\log m$. It arises as a consequence of a Cauchy-Schwarz inequality and bounds on the largest singular value of $\boldsymbol{G}$. Indeed, $\sigma_{\max}(\boldsymbol{G})$ satisfies the following technical lemma (proof omitted):

Lemma 1. *Consider a random matrix* $\boldsymbol{G} \in \mathbb{R}^{n \times m}$ *sampled according to* $\mathrm{vec}(\boldsymbol{G}) \sim \mathcal{N}(\mathbf{0}, \boldsymbol{W})$*, where the matrix* $\boldsymbol{W}$ *is a feasible solution to* (8)*, and thus* $n \geq m$ *for feasibility. Then, the following inequality holds*

$$\mathbb{E}[\sigma_{\max}(\boldsymbol{G})^2] \leq \min(m, 2\log(2\,m) + 2). \tag{10}$$

From this technical lemma, we derive the following semidefinite relationship:

Theorem 1. *The matrix* $\boldsymbol{G} \in \mathbb{R}^{n \times m}$ *generated by Algorithm 2 satisfies the inequality*

$$\mathbb{E}\left[\frac{\mathrm{vec}(\boldsymbol{G})\mathrm{vec}(\boldsymbol{G})^\top}{\sigma_{\max}(\boldsymbol{G})^2}\right] \succeq \max\left(\frac{2}{\pi m}, \frac{1}{\pi(\log(2\,m)+1)}\right) \boldsymbol{W}^\star. \tag{11}$$

Proof. Consider an arbitrary unit vector $\boldsymbol{v} \in \mathbb{R}^{nm}$. From Cauchy-Schwarz, we have that for any random variables $A \geq 0, B > 0$ a.s., $\mathbb{E}[A/B] \geq \mathbb{E}[\sqrt{A}]^2/\mathbb{E}[B]$. Thus, applying this inequality to $A = (\boldsymbol{v}^\top \mathrm{vec}(\boldsymbol{G}))^2$ and $B = \sigma_{\max}(\boldsymbol{G})^2$ yields

$$\mathbb{E}[(\boldsymbol{v}^\top \mathrm{vec}(\boldsymbol{G}))^2/\sigma_{\max}^2(\boldsymbol{G})] \geq \frac{\mathbb{E}[|\boldsymbol{v}^\top \mathrm{vec}(\boldsymbol{G})|]^2}{\mathbb{E}[\sigma_{\max}(\boldsymbol{G})^2]}.$$

Since $\boldsymbol{v}^\top \mathrm{vec}(\boldsymbol{G}) \sim \mathcal{N}(\mathbf{0}, \boldsymbol{v}^\top \boldsymbol{W}^\star \boldsymbol{v})$, the numerator is equal to $\sqrt{\frac{2}{\pi}}\sqrt{\boldsymbol{v}^\top \boldsymbol{W}^\star \boldsymbol{v}}$ [e.g., 17]. For the denominator, we refer to Lemma 1, where we show that $\mathbb{E}[\sigma_{\max}(\boldsymbol{G})^2] \leq \min(m, 2\log(2m) + 2)$. Combining these two observations, we have the desired inequality. □

Theorem 1 leads to a purely multiplicative performance guarantee for Algorithm 2, $\mathbb{E}\left[\langle \boldsymbol{A}, \mathrm{vec}(\boldsymbol{Q})\mathrm{vec}(\boldsymbol{Q})^\top\rangle\right] \geq \beta\ \langle \boldsymbol{A}, \boldsymbol{W}^\star\rangle$, with $\beta = \max\left(\frac{2}{\pi m}, \frac{1}{\pi(\log(2m)+1)}\right)$. Interestingly, the multiplicative constant is independent of the ambient dimension n, but only depends on the number of vectors m. For small values of m, the $2/(\pi m)$ term dominates and drives the value of β (e.g., it equals 0.636 for $m = 1$). Asymptotically, however, our bound scales as $1/\log m$ and exhibits a very mild dependence on m.

This multiplicative guarantee stems from combining two inequalities, (9) and (11), neither of which is necessarily tight. Nonetheless, we can show that the $1/\log m$ scaling for the performance guarantee of Algorithm 2 is essentially tight, as formally stated below (full proof deferred to the appendix).

Proposition 2. *There exists a family of matrices* $\boldsymbol{W}^\star \in \mathcal{S}_+^{nm}$ *for Algorithm 2, and a constant* $C > 0$ *for which, for any* $\beta > 0$ *such that* $\mathbb{E}[\mathrm{vec}(\boldsymbol{Q})\,\mathrm{vec}(\boldsymbol{Q})^\top] \succeq \beta \boldsymbol{W}^\star$*, we must have* $\beta \leq C/\log m$*.*

Proof (sketch). Consider m (fixed) orthonormal vectors $\boldsymbol{u}_1, \ldots, \boldsymbol{u}_m$ and apply Algorithm 2 with a covariance matrix $\boldsymbol{W}$ defined as $\boldsymbol{W}^{(i,i)} := \boldsymbol{u}_i \boldsymbol{u}_i^\top$ and $\boldsymbol{W}^{(i,j)} := \alpha \boldsymbol{u}_i \boldsymbol{u}_j^\top$, for some $\alpha \in (0,1)$. The columns of the matrix $\boldsymbol{G}$ generated by Algorithm 2 are of the form $\boldsymbol{g}_i = z_i \boldsymbol{u}_i$, with $\boldsymbol{z} \sim \mathcal{N}\left(\mathbf{0}, (1-\alpha)\boldsymbol{I}_m + \alpha \boldsymbol{e}\boldsymbol{e}^\top\right)$. With this structure, the columns of the matrix $\boldsymbol{Q}$ are of the form $\boldsymbol{q}_i = d_i \boldsymbol{u}_i$, with $\mathbb{P}(d_i = 1) = 1 - \mathbb{P}(d_i = -1) = (1 + |z_i|/\sigma_{\max})/2$.

Assume that there exists $\beta > 0$ such that $\mathbb{E}[\text{vec}(\boldsymbol{Q})\,\text{vec}(\boldsymbol{Q})^\top] \succeq \beta \boldsymbol{W}$. Applied to the vector $\text{vec}(\boldsymbol{U})$, this inequality yields

$$\beta \leq \frac{1 + b_m(\alpha)(m-1)}{1+\alpha(m-1)}, \text{ with } b_m(\alpha) := \mathbb{E}\left[\frac{|z_i||z_j|}{\max_k |z_k|^2}\right].$$

In particular, one can show that there exists a constant $C_\alpha > 0$ such that $b_m(\alpha) \leq C_\alpha / \log m$, which implies $\beta = O(1/\log m)$. □

Despite its strong asymptotic behavior, for a fixed value of m, the constant in Theorem 1 can be weak (largely because of the use of the Cauchy-Schwarz inequality). To get a more accurate estimate of the performance of Algorithm 2, we now derive tighter bounds that can be computed numerically.

Theorem 2. *Let $\boldsymbol{G} \in \mathbb{R}^{n\times m}$ be a Gaussian matrix generated by Algorithm 2. Then, the matrix $\boldsymbol{G}$ satisfies the inequality:*

$$\mathbb{E}\left[\frac{\text{vec}(\boldsymbol{G})\text{vec}(\boldsymbol{G})^\top}{\sigma_{\max}(\boldsymbol{G})^2}\right] \succeq \beta_{n,m} \boldsymbol{W}^\star, \tag{12}$$

with

$$\beta_{n,m} := \min_{\lambda\in[0,1]} \int_0^\infty \left(1 + 2t\,m\frac{1-\lambda}{nm-1}\right)^{-(nm-1)/2} (1+2tm\lambda)^{-3/2} dt.$$

In particular, the constant $\beta_{n,m}$ satisfies the following properties:

(a) For any integer m, $\beta_{n,m}$ is non-increasing in n. For any integer n, $\beta_{n,m}$ is non-increasing in m.
(b) For any integer m, we have $\beta_{n,m} \to \beta_{\infty,m}$ as $n \to \infty$ with

$$\beta_{\infty,m} := \min_{\lambda\in[0,1]} \int_0^\infty e^{-t\,m(1-\lambda)} (1+2tm\lambda)^{-3/2} dt$$

(c) For $m = 1$, $\beta_{n,1}$ is optimal, i.e., there exists a covariance matrix $\boldsymbol{W}^\star$ satisfying (12) at equality.

Proof. We use the operator bound $\sigma_{\max}(\boldsymbol{G})^2 \leq \|\boldsymbol{G}\|_F^2$ to obtain

$$\mathbb{E}\left[\frac{\text{vec}(\boldsymbol{G})\text{vec}(\boldsymbol{G})^\top}{\sigma_{\max}(\boldsymbol{G})^2}\right] \succeq \int_0^\infty \mathbb{E}[\text{vec}(\boldsymbol{G})\text{vec}(\boldsymbol{G})^\top e^{-t\|\boldsymbol{G}\|_F^2}] dt.$$

Denote $r = \text{rank}(\boldsymbol{W}^\star) \leq nm$ and consider an eigenvalue decomposition of $\boldsymbol{W}^\star$, $\boldsymbol{W}^\star = \boldsymbol{H}\boldsymbol{\Lambda}\boldsymbol{H}^\top$. We have $\text{vec}(\boldsymbol{G}) = \boldsymbol{H}\boldsymbol{\Lambda}^{1/2}\boldsymbol{z}$, with $\boldsymbol{z} \sim \mathcal{N}(\boldsymbol{0}_r, \boldsymbol{I}_r)$ and thus

$$\mathbb{E}[\text{vec}(\boldsymbol{G})\text{vec}(\boldsymbol{G})^\top \exp(-t\|\boldsymbol{G}\|_F^2)] = \boldsymbol{H}\boldsymbol{\Lambda}^{1/2}\mathbb{E}[\boldsymbol{z}\boldsymbol{z}^\top \exp(-t\boldsymbol{z}^\top\boldsymbol{\Lambda}\boldsymbol{z})]\boldsymbol{\Lambda}^{1/2}\boldsymbol{H}^\top.$$

Furthermore, we can show that

$$\mathbb{E}[\boldsymbol{z}\boldsymbol{z}^\top \exp(-t\boldsymbol{z}^\top\boldsymbol{\Lambda}\boldsymbol{z})] = \frac{1}{\sqrt{\det(\boldsymbol{I}_r + 2t\boldsymbol{\Lambda})}}(\boldsymbol{I}_r + 2t\boldsymbol{\Lambda})^{-1}.$$

So, the integral is lower bounded by

$$\boldsymbol{H}\boldsymbol{\Lambda}^{1/2}\boldsymbol{B}\boldsymbol{\Lambda}^{1/2}\boldsymbol{H}^\top \text{ with } \boldsymbol{B} := \int_0^\infty \frac{1}{\sqrt{\det(\boldsymbol{I}_r + 2t\boldsymbol{\Lambda})}} (\boldsymbol{I}_r + 2t\boldsymbol{\Lambda})^{-1} dt.$$

To conclude, we show that there exists a scalar $\beta > 0$ such that $\boldsymbol{B} \succeq \beta \boldsymbol{I}_r$.

To find such β, observe that $\boldsymbol{B}$ is a diagonal matrix. Hence, it is sufficient to find a lower bound on its diagonal entries. Given the constraints on $\boldsymbol{W}^\star$, the eigenvalues $\boldsymbol{\Lambda}$ must satisfy: $\Lambda_i \geq 0$ (from $\boldsymbol{W}^\star \succeq 0$), $\sum_{i=1}^r \Lambda_i = m$ (from $\operatorname{tr}(\boldsymbol{W}^\star) = \sum_{i=1}^m \operatorname{tr}(\boldsymbol{W}^{\star(i,i)}) = m$). Hence, we can take

$$\beta = \min_{\boldsymbol{\Lambda} \in [0,m]^r \,:\, \sum_{i=1}^r \Lambda_i = m} \int_0^\infty \prod_{i=1}^r (1 + 2t\Lambda_{i'})^{-1/2} (1 + 2t\Lambda_1)^{-1} dt.$$

The function $\boldsymbol{\Lambda} \mapsto \int_0^\infty \prod_{i=1}^r (1 + 2t\Lambda_{i'})^{-1/2} (1+2t\Lambda_1)^{-1} dt$ is convex, and invariant under any permutation of the Λ_i, $i > 1$. So, by Jensen's inequality, we can restrict our attention to minimizers of the form $\Lambda_1 = \lambda$, $\Lambda_i = \dfrac{m-\lambda}{r-1}$, $i > 1$:

$$\beta = \min_{\lambda \in [0,m]} \int_0^\infty \left(1 + 2t\,\frac{m-\lambda}{r-1}\right)^{-(r-1)/2} (1 + 2t\lambda)^{-3/2} dt.$$

For a fixed value of (t, λ), the integrand is decreasing in $r \leq nm$, so

$$\beta \geq \beta_{n,m} := \min_{\lambda \in [0,m]} \int_0^\infty \left(1 + 2t\,\frac{m-\lambda}{nm-1}\right)^{-(nm-1)/2} (1 + 2t\lambda)^{-3/2} dt.$$

The change of variable $\lambda \leftarrow \lambda/m$ concludes the proof. □

Compared with Theorem 1, the value of Theorem 2 is primarily computational. By solving numerically the one-dimensional minimization problem in λ, it provides tighter estimates of the performance of our algorithm, especially for small values of m, as reported in Table 1. While the guarantee from Theorem 1 is independent of n, the constant $\beta_{n,m}$ in Theorem 2 is monotonically decreasing with n, obtaining stronger performance guarantees for finite values of n.

However, we should acknowledge that $\beta_{\infty,m}$ does not scale as $\Theta(1/\log m)$ for large values of m, and thus is asymptotically weaker than Theorem 1 (actually, we can identify a class of matrices $\boldsymbol{W}^\star$ for which $\beta_{n,m} \leq 1/m$). We can further strengthen Theorem 2 and view $\beta_{n,m}$ as a special case of an even tighter bound, which recovers the asymptotic scaling of Theorem 1. For the sake of exposition, we only present Theorem 2 in the main paper. We present and prove the more general result in the appendix.

2.4 Discussion: Algorithm Variants

Algorithm 2 can be interpreted as a two-step generalization of [16], where we sample a large multivariate normal vector $\operatorname{vec}(\boldsymbol{G}) \sim \mathcal{N}(\boldsymbol{0}, \boldsymbol{W}^\star)$ and generate

Table 1. Values of the approximation factor from Theorems 1 and 2 for some values of n and m, with $m \leq n$.

m	Theorem 1	$\beta_{n,m}$ (Theorem 2)			
		$n = 5$	$n = 10$	$n = 15$	$n = \infty$
1	0.636620	0.735264	0.706972	0.697920	0.680415
2	0.318310	0.353486	0.346734	0.344533	0.340208
3	0.212207	0.232640	0.229689	0.228720	0.226805
4	0.159155	0.173367	0.171721	0.171179	0.170104
5	0.127324	0.138164	0.137116	0.136770	0.136083
10	0.079662	—	0.068299	0.068213	0.068042
15	0.072323	—	—	0.045437	0.045361

a feasible semi-orthogonal matrix $\boldsymbol{Q}$ from $\boldsymbol{G}$. Interestingly, our algorithm introduces an additional source of randomness in the generation of $\boldsymbol{Q}$ (hence, the qualification 'two-step'), which is key for guaranteeing the relationship (9) between the second moments of $\text{vec}(\boldsymbol{Q})$ and $\text{vec}(\boldsymbol{G})$.

Alternatively, we could have taken $\boldsymbol{D} = \boldsymbol{I}_m$ in Algorithm 2, i.e., define $\boldsymbol{Q}$ as the projection (with respect to the Frobenius norm) of $\boldsymbol{G}$ onto the space of semi-orthogonal matrices. However, with this deterministic construction, Eq. (9) may not hold (we report examples of the matrix $\boldsymbol{W}^\star$ where it does not hold in Appendix B.3), and a different proof strategy would be needed.

Our rounding procedures sample $D_{i,i} \in \{\pm 1\}$ at random, in particular, without taking into account the downstream objective $\text{vec}(\boldsymbol{Q})^\top \boldsymbol{A} \, \text{vec}(\boldsymbol{Q})$. Instead, we could also optimize the diagonal entries of $\boldsymbol{D}$ to explicitly maximize the objective, by solving a binary quadratic optimization problem. Doing so would give a solution at least as good as the one obtained via a random sampling, at the expense of solving a BQO problem with m variables, which might be practically feasible for moderate values of m.

3 Benchmark: Uniform Sampling and Deflation

To appreciate the strength of our performance guarantees for Algorithm 2, we analyze the performance of two baselines.

First, a naive baseline where we draw $\boldsymbol{Q}$ uniformly from the set of semi-orthogonal matrices. Note that this is analogous to generating i.i.d. Bernoulli vectors in BQO, which achieves a $1/2$ approximation ratio in the Max-Cut case. In the appendix, we show that it provides a $1/nm$-approximation guarantee for Problem (4), and that this approximation ratio is tight (i.e., we can construct instances of arbitrary size where uniform sampling is exactly $1/nm$-suboptimal).

Second, we propose a second baseline inspired by the deflation approach for PCA. Namely, we consider a diagonal block of $\boldsymbol{A}$, $\boldsymbol{A}^{(i,i)}$, and compute its leading eigenvector. This defines the column $\boldsymbol{u}_i$. We then update (or deflate) the other diagonal blocks $\boldsymbol{A}^{(j,j)} \leftarrow (\boldsymbol{I}_n - \boldsymbol{u}_i\boldsymbol{u}_i^\top)\boldsymbol{A}^{(j,j)}(\boldsymbol{I}_n - \boldsymbol{u}_i\boldsymbol{u}_i^\top)$ and proceed with another block. We describe the procedure more formally in the appendix. We show that it provides a $1/m^2$-factor approximation.

In short, as summarized in Table 2, we show that both benchmarks achieve a significantly worse multiplicative performance guarantee than Theorem 1 for Algorithm 2.

Table 2. Summary of our guarantees for Algorithm 2 and two benchmarks.

Name	Algorithm 2	Uniform	Deflation
Ratio	$\max\left(\frac{2}{\pi m}, \frac{1}{\pi(\log(2m)+1)}\right)$	$\frac{1}{nm}$	$\frac{1}{m^2}$
Source	Theorem 1	Proposition C.1	Proposition C.2

4 Numerical Results

We numerically evaluate the performance of Algorithm 2 for semi-orthogonal quadratic optimization problems (4). For fixed (n, m), we generate a random semidefinite matrix $\boldsymbol{A} = \boldsymbol{B}\boldsymbol{B}^\top \in \mathcal{S}_+^{nm}$ where the entries of $\boldsymbol{B} \in \mathbb{R}^{nm \times 10}$ are i.i.d. standard normal random variables. We solve the Shor relaxation (8) and sample $N = 100$ feasible solutions from Algorithm 2. For comparison, we also implement the following benchmarks:

- We sample N solutions uniformly at random (Uniform).
- We sample N solutions using Algorithm 2 but generate $\boldsymbol{Q}$ by projecting the rectangular matrix $\boldsymbol{G}$ onto the set of semi-orthogonal matrices directly (Algorithm 2 with projection).
- We sample N solutions by applying the earlier deflation heuristic (Deflation).
- We follow the heuristic in [12], namely, we project the reshaped leading eigenvector of $\boldsymbol{W}^\star$ (Burer and Park).

We consider $n = 100$ and $m \in \{1, 2, 5, 10, 15, 20, 25, 30, 40, 60, 80, 100\}$. We generate five instances for each (n, m).

The left panel of Fig. 1 compares the average performance ratio for these five algorithms. Confirming our theoretical analysis, we observe that the average performance ratio degrades as m increases. We also observe that our Algorithm 2 strongly outperforms Deflation and Uniform—theoretically, Uniform and Deflation achieve a $1/nm$- and $1/m^2$-performance guarantee respectively (Proposition C.1 and C.2). A crucial step in the theoretical analysis of Algorithm 2 is the fact that we generate a feasible matrix $\boldsymbol{Q}$ from a randomly generated matrix $\boldsymbol{G}$, by randomly switching the singular values of $\boldsymbol{G}$ to ± 1. Instead, we find that using a deterministic projection (Alg. 2 with projection) leads to much stronger performance, comparable to that of the heuristic in [12]. However, as observed in

Sect. 2.4, our analysis cannot be easily generalized to such deterministic projection schemes and, to the best of our knowledge, no theoretical guarantees have been derived in the literature for those schemes.

A more relevant performance metric in practice is the performance of the best solution found (out of N), rather than the average performance. Regarding the best solution found, the right panel of Fig. 1 shows that the relative ordering of the methods remains unchanged, although the gap between methods shrinks.

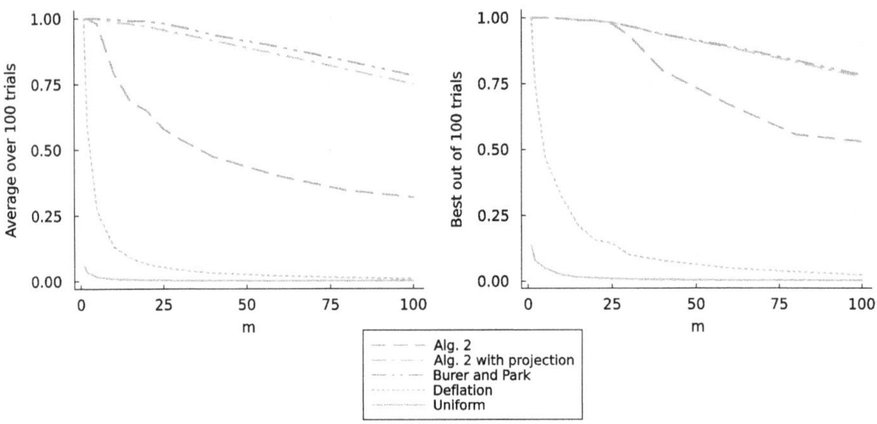

Fig. 1. Average performance ratio $\langle \boldsymbol{A}, \text{vec}(\boldsymbol{Q})\,\text{vec}(\boldsymbol{Q})^\top \rangle / \langle \boldsymbol{A}, \boldsymbol{W}^\star \rangle$ (left panel) and performance ratio $\langle \boldsymbol{A}, \text{vec}(\boldsymbol{Q})\,\text{vec}(\boldsymbol{Q})^\top \rangle / \langle \boldsymbol{A}, \boldsymbol{W}^\star \rangle$ of the best solution (right panel) over $N = 100$ samples, for different feasibility heuristics. Note that the method of [12] only returns one solution. For each value of m, results are averaged over 5 instances.

5 Conclusion

This paper proposes a new technique for relaxing and rounding quadratic optimization problems over semi-orthogonal matrices, which mirrors the blueprint of the Goemans–Williamson algorithm for BQO. Our algorithm provides a purely multiplicative performance guarantee for the orthogonally constrained quadratic optimization problem (4), which subsumes the heterogeneous PCA problem (5) among others. Future work could investigate the theoretical analysis of other sampling schemes, such as those with a deterministic projection step, or extend the approach (namely, solving a Shor relaxation followed by a sampling algorithm) to a broader class of problems, such as low-rank optimization problems.

Disclosure of Interests. The authors have no competing interests to declare that are relevant to the content of this article.

References

1. Bandeira, A.S., Kennedy, C., Singer, A.: Approximating the little Grothendieck problem over the orthogonal and unitary groups. Math. Program. **160**, 433–475 (2016)
2. Barvinok, A.: A remark on the rank of positive semidefinite matrices subject to affine constraints. Discrete Comput. Geom. **25**(1), 23–31 (2001)
3. Bell, R.M., Koren, Y.: Lessons from the Netflix prize challenge. ACM SIGKDD Explor. Newsl. **9**(2), 75–79 (2007)
4. Ben-Tal, A., Nemirovski, A.: On tractable approximations of uncertain linear matrix inequalities affected by interval uncertainty. SIAM J. Optim. **12**(3), 811–833 (2002)
5. Bertsimas, D., Copenhaver, M.S., Mazumder, R.: Certifiably optimal low rank factor analysis. J. Mach. Learn. Res. **18**(29), 1–53 (2017)
6. Bertsimas, D., Cory-Wright, R., Pauphilet, J.: Mixed-projection conic optimization: a new paradigm for modeling rank constraints. Oper. Res. **70**(6), 3321–3344 (2022)
7. Bertsimas, D., Ye, Y.: Semidefinite relaxations, multivariate normal distributions, and order statistics. In: Handbook of Combinatorial Optimization, pp. 1473–1491. Springer (1998)
8. Bolla, M., Michaletzky, G., Tusnády, G., Ziermann, M.: Extrema of sums of heterogeneous quadratic forms. Linear Algebra Appl. **269**(1–3), 331–365 (1998)
9. Briët, J., Buhrman, H., Toner, B.: A generalized Grothendieck inequality and nonlocal correlations that require high entanglement. Commun. Math. Phys. **305**(3), 827–843 (2011)
10. Briët, J., de Oliveira Filho, F.M., Vallentin, F.: The positive semidefinite Grothendieck problem with rank constraint. In: International Colloquium on Automata, Languages, and Programming, pp. 31–42. Springer (2010)
11. Briët, J., Regev, O., Saket, R.: Tight hardness of the non-commutative Grothendieck problem. In: IEEE 56th Annual Symposium on Foundations of Computer Science, pp. 1108–1122. IEEE (2015)
12. Burer, S., Park, K.: A strengthened SDP relaxation for quadratic optimization over the Stiefel manifold. J. Optim. Theory Appl. **202**(1), 320–339 (2024)
13. Dong, H., Chen, K., Linderoth, J.: Regularization vs. relaxation: a conic optimization perspective of statistical variable selection. arXiv preprint arXiv:1510.06083 (2015)
14. d'Aspremont, A., Boyd, S.: Relaxations and randomized methods for nonconvex QCQPs. EE392o Class Notes, Stanford University, vol. 1, pp. 1–16 (2003)
15. Gilman, K., Burer, S., Balzano, L.: A semidefinite relaxation for sums of heterogeneous quadratics on the Stiefel manifold. SIAM J. Matrix Anal. Appl. **46**(2), 1091–1116 (2025)
16. Goemans, M.X., Williamson, D.P.: Improved approximation algorithms for maximum cut and satisfiability problems using semidefinite programming. J. ACM (JACM) **42**(6), 1115–1145 (1995)
17. Grimmett, G., Stirzaker, D.: Probability and Random Processes. Oxford University Press (2020)
18. Lai, Z., Lim, L.H., Tang, T.: Stiefel optimization is NP-hard. Optim. Lett. (2026). https://doi.org/10.1007/s11590-025-02271-9
19. Luo, Z.Q., Ma, W.K., So, A.M.C., Ye, Y., Zhang, S.: Semidefinite relaxation of quadratic optimization problems. IEEE Signal Process. Mag. **27**(3), 20–34 (2010)

20. Naor, A., Regev, O., Vidick, T.: Efficient rounding for the noncommutative Grothendieck inequality. Theory Comput. **10**(1), 257–295 (2014)
21. Negahban, S., Wainwright, M.J.: Estimation of (near) low-rank matrices with noise and high-dimensional scaling. Ann. Stat. **39**, 1069–1097 (2011)
22. Nemirovski, A.: Sums of random symmetric matrices and quadratic optimization under orthogonality constraints. Math. Program. **109**(2–3), 283–317 (2007)
23. Nesterov, Y.: Semidefinite relaxation and nonconvex quadratic optimization. Optim. Methods Softw. **9**(1–3), 141–160 (1998)
24. Pataki, G.: On the rank of extreme matrices in semidefinite programs and the multiplicity of optimal eigenvalues. Math. Oper. Res. **23**(2), 339–358 (1998)
25. Rehfeldt, D., Koch, T., Shinano, Y.: Faster exact solution of sparse MaxCut and QUBO problems. Math. Program. Comput. **15**(3), 445–470 (2023)
26. Shor, N.Z.: Quadratic optimization problems. Sov. J. Comput. Syst. Sci. **25**, 1–11 (1987)
27. So, A.M.C.: Improved approximation bound for quadratic optimization problems with orthogonality constraints. In: Proceedings of the Twentieth Annual ACM-SIAM Symposium on Discrete Algorithms, pp. 1201–1209. SIAM (2009)
28. Williamson, D.P., Shmoys, D.B.: The Design of Approximation Algorithms. Cambridge University Press (2011)
29. Wolkowicz, H., Saigal, R., Vandenberghe, L.: Handbook of Semidefinite Programming: Theory, Algorithms, and Applications, vol. 27. Springer (1998)

Polytope Scheduling with Groups: Unified Models and Optimal Guarantees

Alexander Lindermayr[1(✉)], Zhenwei Liu[2], and Nicole Megow[3]

[1] Technische Universität Berlin, Berlin, Germany
alexander.lindermayr@tu-berlin.de
[2] Zhejiang University, Hangzhou, China
lzw98@zju.edu.cn
[3] University of Bremen, Bremen, Germany
nicole.megow@uni-bremen.de

Abstract. We propose new abstract and unified perspectives on a range of scheduling and graph coloring problems with general min-sum objectives. Specifically, we consider various problems where the objective function is the weighted sum of completion times over groups of entities (jobs, vertices, or edges), thereby generalizing two important objectives in scheduling: makespan and the sum of weighted completion times.

As one of our main results, we present a best-possible $\mathcal{O}(\log g)$-competitive algorithm in the non-clairvoyant online setting, where g denotes the size of the largest group. This is the first non-trivial competitive bound for several problems with group completion time objective, and it is an exponential improvement over previous results for non-clairvoyant coflow scheduling. For offline scheduling, we provide elegant yet powerful meta-frameworks that, in a unifying way, yield new or stronger approximation algorithms for our new abstract problems as well as for previously well-studied special cases.

1 Introduction

We introduce new unifying abstractions of machine scheduling, graph scheduling, and graph coloring problems with general min-sum objectives, and present algorithms for various information settings, both offline models and online models. More specifically, we focus on problems from these domains where the goal is to minimize the (weighted) sum of certain values C_j for certain entities $j \in J$ (jobs, vertices, or edges), such as the sum of completion times in scheduling, but also its generalization to groups: given a family $\mathcal{S}$ of subsets of entities and weights $w_S > 0$ for each $S \in \mathcal{S}$, we seek to minimize $\sum_{S \in \mathcal{S}} w_S C_S$, where $C_S = \max_{j \in S} C_j$. Notably, the latter objective generalizes both makespan—where all jobs belong to a single group—and the sum of weighted completion times—where each job is a group itself.

We establish new connections between well-studied problems and offer a unified perspective that leads to a clearer and more systematic understanding of the problems and their algorithmic solutions. This perspective allows us to transfer

S. S. Dey et al. (Eds.): IPCO 2026, LNCS 16588, pp. 265–281, 2026.
https://doi.org/10.1007/978-3-032-28691-8_18

techniques across seemingly very different problems with group completion time objectives. We establish new optimal bounds for non-clairvoyant online scheduling with group completion times, and improve previous approximation ratios for machine scheduling [9] and graph scheduling/coloring problems [10,14].

Our results cover the following well-studied problem domains. Due to space constraints, we omit precise results and refer to our full version [32] for details.

- **Machine Scheduling:** In machine scheduling problems, jobs need to be scheduled on parallel machines, where each job may have a different processing time on each machine. There are variants with and without preemption and/or migration. Well-studied objective functions are the sum of completion times [4,20,25,46], and the sum of group completion times (aka *scheduling orders* or *bag-of-tasks scheduling*) [9,19,19,31,48].
- **Polytope Scheduling:** The polytope scheduling problem generalizes preemptive unrelated machine scheduling [20] and other preemptive scheduling problems, such as multidimensional scheduling [17] and broadcast scheduling [2]. Here, the goal is to select at any time a vector in a packing polytope, which describes the amounts of progress that jobs make; see details later. While the sum of weighted completion times objective is well understood [8,24,27], we are not aware of results for group completion times.
- **Sum Multicoloring:** Given a graph with vertex weights w_v and demands p_v, the goal is to assign each vertex v a set of p_v colors (represented by $\mathbb{N}$), ensuring that no two adjacent vertices share a color. The goal is to minimize the weighted sum of the largest color assigned to a vertex over all vertices [5, 14,21]. The notion of "no preemption" means here that the colors assigned to a vertex must be consecutive. The group completion time objective (largest color per group) was studied for unit demands [10].
- **Graph Scheduling:** Given a graph, the task is to schedule edges with processing times such that the set of scheduled edges at any time forms a matching. Common objective functions include the sum of edge completion times [14,22], the sum of grouped edge completion times (called *coflow scheduling* [1,6,12,26,41]), and the sum of vertex completion times (called *data migration* [13,30,33]), where a vertex completes when all its incident edges are completed, a special case of group completion times. Graph scheduling on bipartite graphs generalizes the fundamental *open shop scheduling* problem [15,39,40].

Our abstract problem formulations are the following. In the *polytope scheduling problem with group completion times* (PSP-G), we are given n jobs $j \in J$ with processing requirements $p_j \geq 0$, and a family of groups $\mathcal{S} \subseteq 2^J$. Each group $S \in \mathcal{S}$ has associated a positive weight $w_S > 0$. At any time t, a solution must choose a *rate vector* $y(t)$ in a given downward-closed polytope $\mathcal{P} = \{y \in \mathbb{R}^n_{\geq 0} \mid B \cdot y \leq \mathbf{1}\}$ for some fixed $B = [b_{d,j}]^{D \times n} \in \mathbb{Q}^{D \times n}_{\geq 0}$. For every job j, the entry $y_j(t)$ describes how much processing job j receives at time t. Thus, its completion time is $C_j := \arg\min_{t'} \left(\int_0^{t'} y_j(t)\, \mathrm{d}t \geq p_j \right)$. The objective is to minimize the sum of weighted group completion times $\sum_{S \in \mathcal{S}} w_S C_S$. The polytope scheduling problem for the sum of weighted completion times objective (without groups) was first studied by Im, Kulkarni, and Munagala [24], and elegantly abstracts problem-specific details from various preemptive scheduling environments. PSP-G captures all the above-mentioned *preemptive* scheduling problems, for which we give formal arguments later. Note that this model does not capture time-dependent constraints like non-preemptiveness or consecutive colorings.

To address *non-preemptive* problems, we introduce a novel discrete variant of PSP-G, which we call *discrete polytope scheduling with group completion times* (DPSP-G). On a high-level, it is the discrete analogue to PSP: We are given (possibly implicitly) a subset of points $\mathcal{P}' \subseteq \mathcal{P}$ of a given downward-closed polytope $\mathcal{P} = \{y \in \mathbb{R}^n_{\geq 0} \mid B \cdot y \leq \mathbf{1}\}$ for some fixed $B = [b_{d,j}]^{D\times n} \in \mathbb{Q}^{D\times n}_{\geq 0}$. At any time t, a schedule must choose a rate vector $y(t) \in \mathcal{P}'$, describing the processing that jobs receive. Additionally, for each job j, there is a time S_j such that $y_j(t) = y_j$ for all $S_j \leq t < C_j$ and $y_j(t) = 0$ for all $t < S_j$ and $t \geq C_j$. The objective is to minimize $\sum w_S C_S$. DPSP-G captures the above-mentioned *non-preemptive* scheduling and coloring problems.

We study these problems in (i) the *non-clairvoyant online* model, where the processing requirements p_j are uncertain, and the algorithmic challenge is of information-theoretic nature, and (ii) the offline model, where all parameters of an instance are given, and the challenge is the computational complexity.

1.1 Our Results

Table 1. An overview of our new and improved results in comparison to previous work.

problem	known bound	new bound
non-clairvoyant PSP-G	-	$\mathcal{O}(\log(\max_{S\in\mathcal{S}} \lvert S\rvert))$
non-clairvoyant coflow scheduling	$\mathcal{O}(\max_{S\in\mathcal{S}} \lvert S\rvert)$ [6]	$\mathcal{O}(\log(\max_{S\in\mathcal{S}} \lvert S\rvert))$
$Q \mid r_j \mid \sum w_S C_S$	13.5 [9]	10.874
npSMC-G on interval graphs	11.273 [14]	10.874
SC-G on perfect graphs	10.874 [10]	5.437
offline PSP-G	-	$2+\varepsilon$
$R \mid r_j, \text{pmtn} \mid \sum w_S C_S$	$4+\varepsilon$ [9]	$2+\varepsilon$

We develop *general algorithmic frameworks* by framing diverse problems in machine scheduling, graph scheduling, and graph coloring as (discrete) polytope scheduling with groups (PSP-G and DPSP-G). We give both non-clairvoyant online algorithms and offline approximation algorithms; these are summarized in Table 1 with a comparison to prior work. Abstracting across previously unrelated settings and designing unified algorithms is technically challenging and constitutes a non-trivial contribution of our work. In doing so, we even improve substantially upon previous results for certain scheduling and coloring problems.

Non-clairvoyant Online Scheduling. In non-clairvoyant scheduling, an algorithm has no knowledge about the processing requirements p_j of jobs until they complete [34,45]. Moreover, it must create a schedule over time, and cannot revert decisions from previous times. We say that an online algorithm is α-competitive if it computes a solution with an objective value of at most $\alpha \cdot \text{OPT}$, where OPT denotes the optimal objective value for the instance.

Theorem 1. *There is an $\mathcal{O}(\log(\max_{S \in \mathcal{S}} |S|))$-competitive non-clairvoyant algorithm for PSP with group completion times, PSP-G.*

This bound is best-possible; we show a lower bound of $\Omega(\log \max_{S \in \mathcal{S}} |S|)$ for any maximum group size. Our result is the first non-trivial bound on the competitive ratio for non-clairvoyant PSP-G and for many of its subproblems, including unrelated machine scheduling with group completion times and data migration. For coflow scheduling, it improves exponentially upon the previous linear competitive ratio in the largest coflow (i.e., group) size [6].

Corollary 1. *There exists a single non-clairvoyant algorithm that is*

- *$\mathcal{O}(\log \rho)$-competitive for coflow scheduling, where ρ is the largest coflow size,*
- *$\mathcal{O}(\log \Delta)$-competitive for data migration, where Δ is the largest degree, and*
- *$\mathcal{O}(\log(\max_{S \in \mathcal{S}} |S|))$-competitive for unrelated machine scheduling with group completion times.*

Our algorithm is based on the Proportional Fairness (PF) algorithm, a natural and simple allocation rule from economics [28,29,36], which has recently been proven to be very powerful for PSP with the sum of weighted completion time objective [24,27]. For PSP-G, a significant challenge arises from the presence of non-trivial groups. As fairness in PF is traditionally defined over jobs rather than over groups, we cannot directly apply PF and its analysis. We overcome this challenge by introducing virtual weights over jobs and apply PF to those weights; each group distributes its weight evenly to its jobs. Surprisingly, this simple adaptation of PF achieves the best-possible competitive ratio, although the analysis becomes significantly more complicated. To compare these virtual weights with the actual instance in the analysis, we introduce a new factor-revealing LP, which was not used in previous analyses of PF [24,27].

Clairvoyant Offline Scheduling. Our second set of results addresses the offline setting, where all instance parameters are available to the algorithm. Both offline PSP-G and DPSP-G are APX-hard, even for special cases [30,39]. Therefore, we resort to approximation algorithms. We say that an algorithm is an α-approximation algorithm if it computes a solution in polynomial time with an objective value of at most $\alpha \cdot \text{OPT}$ for every instance.

For the *discrete* variant, DPSP-G, we present an algorithmic framework that reduces the weighted sum of group completion times objective to the makespan objective. It is inspired by various grouping and doubling techniques from literature [10, 14, 20, 40], but requires new ingredients to deal with general polytopes. Moreover, our framework brings these ideas under a more general umbrella, justifying their previously successful usage.

Theorem 2. *Given a polynomial-time subroutine that computes for any instance J' a schedule of makespan at most $\rho \cdot \max_{d \in [D]} \sum_{j \in J'} b_{d,j} p_j$, there is a polynomial-time $(2\rho e + \varepsilon)$-approximation for any $\varepsilon > 0$ for DPSP-G.*

By applying the framework, we obtain constant-factor approximation algorithms for minimizing the total weighted group completion time on identical machines ($P \mid r_j \mid \sum w_S C_S$), related machines ($Q \mid r_j \mid \sum w_S C_S$), even with release dates, and for non-preemptive sum multicoloring on special graphs, which means that the colors assigned to a vertex must be consecutive. We use npSMC-G to denote non-preemptive sum multicoloring with groups and SC-G for the special case with unit demands.

Theorem 3. *We can apply the algorithmic framework to obtain the following results for minimizing $\sum w_S C_S$:*

- 7.249-*approximation on identical parallel machines,* $P \mid r_j \mid \sum w_S C_S$,
- 10.874-*approximation on related machines,* $Q \mid r_j \mid \sum w_S C_S$,
- 10.874-*approximation for npSMC-G on line graphs and interval graphs, and*
- 5.437-*approximation for SC-G on perfect graphs.*

For minimizing the total weighted group completion time on related machines, we improve upon the previous approximation ratio of 13.5 [9]. For npSMC-G on interval graphs, a factor of 11.273 was previously known [14], and for SC-G on perfect graphs, a 10.874-approximation was known [10].

Finally, for offline PSP-G, we present a $(2 + \varepsilon)$-approximation algorithm.

Theorem 4. *There is a $(2 + \varepsilon)$-approximation algorithm for offline PSP-G for any $\varepsilon > 0$, even with non-uniform release dates.*

This algorithm is based on solving an interval-indexed LP relaxation of PSP-G, and randomly stretching and truncating the LP schedule [27, 39]. For our setting, we carefully combine the main ingredients of these two previous results in a non-trivial way. This ratio is best-possible up to a factor of $1 + \varepsilon$ as PSP-G contains $1 \mid \mathrm{pmtn} \mid \sum_S w_S C_S$, which has been shown to be equivalent to $1 \mid \mathrm{prec} \mid \sum_j w_j C_j$ [39] and admits a lower bound of $2 - \varepsilon$ under a stronger version of the Unique Games Conjecture [3]. Our theorem is the first result for the preemptive data migration problem, and improves over the previously known $(4 + \varepsilon)$-approximation for $R \mid r_{ij}, \mathrm{pmtn} \mid \sum w_S C_S$ [9].

Corollary 2. *There is a $(2 + \varepsilon)$-approximation algorithm for preemptive data migration and preemptive unrelated machine scheduling with group completion times, $R \mid r_{ij}, \mathrm{pmtn} \mid \sum w_S C_S$, for any $\varepsilon > 0$, even with non-uniform release dates.*

2 An Optimal Algorithm for Non-clairvoyant PSP-G

Our algorithm is inspired by the Proportional Fairness (PF) allocation rule [28, 29,35–37]. In a seminal work on PSP, Im, Kulkarni, and Munagala [24] showed that PF is $O(1)$-competitive for minimizing $\sum w_j C_j$ in PSP. At any time t, their PF implementation selects for the set of unfinished jobs $U(t)$ the rate vector $y(t) \in \mathcal{P}$ of the polytope that maximizes $\sum_{j \in U(t)} w_j \log y_j(t)$.

We propose a new variant of PF for PSP-G. For any $S \in \mathcal{S}$, let $S(t) := S \cap U(t)$ denote the set of unfinished jobs in S at time t. Further, let $\mathcal{S}(t) := \{S \in \mathcal{S} \mid S(t) \neq \emptyset\}$ denote the unfinished groups at time t. At any time t, we assign to each job a *virtual weight* in a way that each unfinished group $S(t)$ distributes its weight evenly to its unfinished jobs, i.e., $w_j(t) := \sum_{S : S \in \mathcal{S}(t), j \in S} w_S / |S(t)|$. It follows that $\sum_{j \in U(t)} w_j(t) = \sum_{S \in \mathcal{S}(t)} w_S$. We then use PF with these virtual weights to compute the rates $y_j(t)$ of the jobs in $U(t)$ as follows:

$$\max \quad \sum_{j \in U(t)} w_j(t) \cdot \log y_j(t) \tag{PF}$$

$$\text{s.t.} \quad \sum_{j \in U(t)} b_{d,j} \cdot y_j(t) \leq 1 \qquad \forall d \in [D], t \geq 0 \tag{1}$$

$$y_j(t) \geq 0 \qquad \forall j \in U(t) \tag{2}$$

We refer to [24,27] for polynomial-time implementations of PF. Let $y(t)$ denote the optimal solution to (PF) at time t.

As in the analysis in [24], our analysis uses the Lagrange multipliers $\{\eta_d \mid d \in [D]\}$ of (1) that correspond to $y(t)$. All Lagrange multipliers corresponding to (2) are zero because $y_j(t) > 0$ for each $j \in U(t)$. They satisfy the following KKT conditions:

(i) For each $j \in U(t)$, we have $\frac{w_j(t)}{y_j(t)} = \sum_{d \in [D]} b_{d,j} \eta_d(t)$.
(ii) For each $d \in [D]$, we have $\eta_d(t) \geq 0$.
(iii) For each $d \in [D]$, we have $\eta_d(t) \cdot (1 - \sum_{j \in U(t)} b_{d,j} y_j(t)) = 0$, which implies that if $\eta_d(t) > 0$ then $\sum_{j \in U(t)} b_{d,j} y_j(t) = 1$.

Using these conditions, we can relate the Lagrange multipliers to the weights.

Lemma 1. *At any time t, $\sum_{d \in [D]} \eta_d(t) = \sum_{j \in U(t)} w_j(t) = \sum_{S \in \mathcal{S}(t)} w_S$.*

To relate PF with an optimal solution to PSP-G, we use the following time-indexed LP relaxation (LP(κ)), which builds upon the mean busy time relaxation, similar to previous analyses of PF [24,27]. It incorporates an additional strengthening constraint proposed in the context of open shop scheduling [39], and has been used for related machine scheduling with groups [19]. The variables $x_{j,t}$ ($x_{S,t}$) indicate which fraction of a job (group) is done during time $(t, t+1]$.

$$
\begin{aligned}
\min \quad & \sum_{S \in \mathcal{S}} w_S C_S & & & \text{(LP}(\kappa)) \\
\text{s.t.} \quad & \sum_{t \geq 0} x_{S,t} = 1 & & \forall S \in \mathcal{S} & (3) \\
& \sum_{t'=0}^{t} x_{S,t'} \leq \sum_{t'=0}^{t} \frac{x_{j,t'}}{p_j} & & \forall S \in \mathcal{S}, j \in S, t \geq 0 & (4) \\
& C_S \geq \sum_{t \geq 0} x_{S,t} \cdot t & & \forall S \in \mathcal{S} & (5) \\
& \sum_{j \in J} b_{d,j} \cdot x_{j,t} \leq \frac{1}{\kappa} & & \forall d \in [D], t \geq 0 & (6) \\
& x_{j,t}, x_{S,t} \geq 0 & & \forall j \in J, S \in \mathcal{S}, t \geq 0 &
\end{aligned}
$$

Constraint (4) is based on precedence-constrained LP relaxations [16], ensuring that at any time t, the fraction of a group that has been done up to time t is at most that of any job in the group. For ease of exposition, we restrict the solution to run at a machine of speed $\frac{1}{\kappa}$ by strengthening the right-hand side of (6) for some fixed $\kappa \geq 1$. Since completion times scale with the machine speed, the optimal value of (LP(κ)) is at most $\kappa \cdot \text{OPT}$.

We next fit the dual of this LP, which can be written as follows.

$$
\begin{aligned}
\max \quad & \sum_{S \in \mathcal{S}} \alpha_S - \sum_{d,t} \beta_{d,t} & & & \text{(DLP}(\kappa)) \\
\text{s.t.} \quad & \alpha_S - \sum_{t' \geq t} \sum_{j \in S} \gamma_{j,S,t'} \leq t \cdot w_S & & \forall S \in \mathcal{S}, t \geq 0 & (7) \\
& \sum_{t' \geq t} \sum_{S : j \in S} \frac{\gamma_{j,S,t'}}{p_j} \leq \kappa \sum_{d \in [D]} b_{d,j} \beta_{d,t} & & \forall j \in J, t \geq 0 & (8) \\
& \gamma_{j,S,t} \geq 0 & & \forall S \in \mathcal{S}, j \in S, t \geq 0 & \\
& \beta_{d,t} \geq 0 & & \forall d \in [D], t \geq 0 &
\end{aligned}
$$

We set the dual variables such that the algorithm's objective value ALG can be bounded by the objective value of the dual. Due to the job-to-group coupling constraints (4) of the LP, we need to fit more complex dual variables than in [24,27]. For any time t, let $M(t)$ denote the *weighted median* of $\{\frac{y_j(t)}{p_j} \mid j \in U(t)\}$, where j has weight $w_j(t)$. We define a dual assignment as follows:

- For any $t \geq 0, S \in \mathcal{S}, j \in S$, if $j \in S(t)$, let $\gamma_{j,S,t} := \frac{w_S}{|S(t)|} \cdot \mathbb{1}\left[\frac{y_j(t)}{p_j} \leq M(t)\right]$; otherwise, let $\gamma_{j,S,t} := 0$.

- We set $\alpha_S := \sum_{t \geq 0} \alpha_{S,t}$, where $\alpha_{S,t} := \sum_{j \in S(t)} \gamma_{j,S,t}$.
- We set $\beta_{d,t} := \frac{1}{\kappa} \sum_{t' \geq t} \eta_d(t') \cdot M(t')$.

We first show that this dual assignment defines a feasible dual solution. Its proof hinges on the proportional distribution of group weights to jobs in the definition of γ, which has not been used in [24,27].

Lemma 2. *The assignment* $(\alpha_S, \beta_{d,t}, \gamma_{j,S,t})$ *is a feasible solution to* (DLP(κ)).

Next, we relate the dual assignment's objective value to the algorithm's objective value. Let $H_k := \sum_{i=1}^{k} \frac{1}{i}$.

Lemma 3. $\sum_{S \in \mathcal{S}} \alpha_S - \sum_{d,t} \beta_{d,t} \geq \left(\frac{1}{2} - \frac{2H_g}{\kappa}\right)\text{ALG}$ *where* $g := \max_{S \in \mathcal{S}} |S|$.

The proof of this lemma can be split into two parts. We first show $\sum_{S \in \mathcal{S}} \alpha_S \geq \frac{1}{2}\text{ALG}$. This mainly uses that at any time t at least half of the unfinished virtual weight is above the weighted median $M(t)$. Thus, we can argue that $\sum_{S \in \mathcal{S}} \alpha_{S,t} \geq \frac{1}{2} \sum_{j \in U(t)} w_j(t) = \frac{1}{2} \sum_{S \in \mathcal{S}(t)} w_S$, and conclude by integrating over time.

The proof of $\sum_{d,t} \beta_{d,t} \leq \frac{2H_g}{\kappa}\text{ALG}$ is particularly challenging and different from [24,27]. The key ingredient is to bound the following expression for any group $S \in \mathcal{S}$ and $t \geq 0$:

$$\sum_{t' \geq t} \sum_{j \in S(t')} \frac{1}{|S(t')|} \frac{y_j(t')}{p_j} \tag{9}$$

In [24,27], where each group is of size 1, this expression is clearly at most 1, because each job receives at most p_j processing. Using this fact here directly leads to an upper bound of g, and $O(g)$-competitiveness. Yet we prove the exponentially smaller bound of H_g by carefully analyzing how this expression evolves over time. By discretizing time by job completions $t_1 \leq \ldots \leq t_{|S|}$ of S, we can bound (9) by $\sum_{i=1}^{|S|} \Delta_i$ where $\Delta_i := \sum_{t'=t_{i-1}+1}^{t_i} \sum_{j \in S(t')} \frac{1}{|S(t')|} \frac{y_j(t')}{p_j}$. Since the total throughput to any prefix of ℓ jobs until time t_ℓ must be at least ℓ, that is, $\sum_{i=1}^{\ell}(|S| + 1 - i)\Delta_i \geq \ell$, and the total throughput to all jobs in S can be at most $|S|$, we can view $\sum_{i=1}^{|S|} \Delta_i$ as the objective of a *factor revealing LP* and derive that its maximum, and thus (9), is at most $H_{|S|}$.

Having Lemma 3, we set the parameter $\kappa := 8H_g$ where $g := \max_{S \in \mathcal{S}} |S|$ so that the objective value of the dual program is $\sum_{S \in \mathcal{S}} \alpha_S - \sum_{d,t} \beta_{d,t} \geq \text{ALG}/4$. Since the objective value of any feasible solution is at most κOPT, we have $\text{ALG} \leq 4(\sum_{S \in \mathcal{S}} \alpha_S - \sum_{d,t} \beta_{d,t}) \leq 32H_g \cdot \text{OPT} = \mathcal{O}(\log g) \cdot \text{OPT}$.

Theorem 1. *There is an $\mathcal{O}(\log(\max_{S\in\mathcal{S}}|S|))$-competitive non-clairvoyant algorithm for PSP with group completion times, PSP-G.*

In the full version [32] we prove a matching lower bound that builds on previous work for non-clairvoyant makespan minimization on related machines [19,45].

Theorem 5. *Every non-clairvoyant algorithm for PSP-G has a competitive ratio of at least $\Omega(\log \max_{S\in\mathcal{S}}|S|)$ for any value of* $\max_{S\in\mathcal{S}}|S|$.

Incorporating Release Times Without Performance Loss. Our algorithm can be easily extended to the online-time setting with job release times. We distinguish two models. In the first, all jobs of a group are released simultaneously, as considered previously in coflow scheduling [1,12]. In the second, jobs may arrive over time, but the group information (its weight and its group membership) is known in advance. In both models, we can obtain the same competitive ratio by solving (PF) with respect to those unfinished but released jobs.

3 Discrete PSP

We next present our results for DPSP-G. We introduce a general framework to derive polynomial-time approximation algorithms, and then apply it to various subproblems in Sects. 3.1 and 3.2 by specifying the remaining ingredients.

Our framework is based on the LP relaxation (LP(1)). As explained in Sect. 2, it relaxes PSP-G given an infinitesimal time discretization. Since every feasible schedule for DPSP-G with configurations $\mathcal{P}'$ is feasible for PSP-G with polytope $\mathcal{P} \supseteq \mathcal{P}'$, (LP(1)) is also a relaxation of DPSP-G. The size of (LP(1)) can be exponential, hence we first rewrite it into a polynomial-size interval form.

To this end, consider a large enough time horizon T, and fix constants $\delta > 0$ and $\varepsilon' > 0$. First, we slightly modify the instance by shifting the release time $r'_j := r_j + \delta$ for all j. This increases the optimal objective value by at most a factor of $1 + \delta$ assuming all jobs complete after time 1, which can be achieved w.l.o.g. by scaling. Let L be the smallest integer such that $\delta(1+\varepsilon')^L \geq T$. Note that L is polynomial in the size of the instance. For every $0 \leq i \leq L$, we define $\gamma_i := \delta(1+\varepsilon')^i$, and for every $1 \leq i \leq L$, we define the interval $I_i = (\gamma_{i-1}, \gamma_i]$. Let $|I_i|$ be the length of the ith interval, i.e., $|I_i| := \gamma_i - \gamma_{i-1}$. We consider the following *interval LP relaxation* of (LP(1)), which is analogous to (LP(1)) if we replace a time slot $(t, t+1]$ by the interval $(\gamma_{i-1}, \gamma_i]$.

$$\min \sum_{S \in \mathcal{S}} w_S C_S \tag{LP'}$$

$$\text{s.t.} \sum_{i=1}^{L} x_{S,i} \geq 1 \qquad \forall S \in \mathcal{S} \tag{10}$$

$$\sum_{i'=1}^{i} x_{S,i'} \leq \sum_{i'=1}^{i} \frac{x_{j,i'}}{p_j} |I_{i'}| \qquad \forall S \in \mathcal{S}, j \in S, 1 \leq i \leq L \tag{11}$$

$$C_S = \sum_{i=1}^{L} x_{S,i} \cdot \gamma_{i-1} \qquad \forall S \in \mathcal{S} \tag{12}$$

$$\sum_{j \in J} b_{d,j} \cdot x_{j,i} \leq 1 \qquad \forall d \in [D], 1 \leq i \leq L \tag{13}$$

$$x_{j,i} = 0 \qquad \forall j \in J, 1 \leq i \leq L, r_j > \gamma_{i-1}$$

$$x_{j,i}, x_{S,i} \geq 0 \qquad \forall j \in J, S \in \mathcal{S}, i \geq 0$$

Here, $x_{j,i}$ (resp. $x_{S,i}$) denotes the progress of job j (resp. group S) during the time interval I_i. Let $\{x_{j,i}^{\text{LP}}, x_{S,i}^{\text{LP}}, C_S^{\text{LP}}\}$ be an optimal solution to (LP'). Then, $\sum_{S \in \mathcal{S}} w_S C_S^{\text{LP}} \leq (1+\delta)\text{OPT}$. Our three-step framework is as follows:

(i) Solve (LP') and obtain an optimal solution $\{x_{j,i}^{\text{LP}}, x_{S,i}^{\text{LP}}, C_S^{\text{LP}}\}$.

(ii) Create batches of jobs according to that solution as follows. First, for every job j, define $C_j^{\text{LP}} := \sum_{i=1}^{L} x_{j,i}^{\text{LP}} |I_i| \gamma_{i-1}/p_j$. Let α be a variable uniformly drawn from $[0,1]$ and β be some parameter to be determined later. We define a partition of jobs $J_i := \{j \in J \mid \beta^{i-1+\alpha} < C_j^{\text{LP}} \leq \beta^{i+\alpha}\}$ for $i = 0, 1, \ldots, K$.

(iii) We schedule the batches $J_0, \ldots, J_K$ sequentially. That is, we start scheduling J_{i+1} after all jobs in J_i are completed. For each batch, we use a problem-specific scheduling subroutine for the following subproblem:

  Subproblem: Given a subset J' of jobs, compute a feasible non-preemptive schedule that completes all the jobs in J' within a makespan of at most $\rho \cdot \max_{d \in [D]} \sum_{j \in J'} b_{d,j} \cdot p_j$ for some $\rho \geq 1$.

Similar batching and makespan reductions have been previously used for more specific scheduling problems [14,40]. We unify these approaches by comparing the makespan to the abstract lower bound $\max_{d \in [D]} \sum_{j \in J'} b_{d,j} \cdot p_j$ that holds for all DPSP-G instances. We now explain the intuition behind the framework.

We can interpret the abstract lower bound as follows. Consider any subset J' of jobs and let $d \in [D]$. Let $y_{j,t}$ be the amount of processing that job j receives at time t. Suppose we complete J' by time T. Since each job $j \in J'$ has received p_j units of processing, we have $\int_0^T \sum_{j \in J'} b_{d,j} y_{j,t} \mathrm{d}t = \sum_{j \in J'} b_{d,j} \int_0^T y_{j,t} \mathrm{d}t \geq \sum_{j \in J'} b_{d,j} p_j$. By the polytope constraints, we have $\int_0^T \sum_{j \in J'} b_{d,j} y_{j,t} \mathrm{d}t \leq \int_0^T 1 \mathrm{d}t = T$. Hence, T must be at least $\sum_{j \in J'} b_{d,j} \cdot p_j$.

We next give an outline of the analysis of the framework. We first show that our time-indexed formulation implies inequalities $(1+\varepsilon')\sum_{j\in J'} b_{d,j}p_j C_j^{\mathrm{LP}} \geq \frac{1}{2}(\sum_{j\in J'} b_{d,j}p_j)^2$ for all $J' \subseteq J$. These can be thought of generalizations of Queyranne's subset constraints for single-machine scheduling [38,43], abstracting away the machine environment via the polytope. For a batch J_i, they imply

$$\frac{1}{2}\left(\sum_{j\in J_i} b_{d,j}p_j\right)^2 \leq (1+\varepsilon')\sum_{j\in J_i} b_{d,j}p_j C_j^{\mathrm{LP}} \leq (1+\varepsilon')\beta^{i+\alpha}\sum_{j\in J_i} b_{d,j}p_j\ ,$$

showing that our subroutine outputs a schedule for J_i with makespan at most $2(1+\varepsilon')\rho\beta^{i+\alpha}$. By the geometric partitioning of the batches, every $j \in J_i$ completes in the full schedule within time $2(1+\varepsilon')\rho\frac{\beta^{i+1+\alpha}}{\beta-1}$. Finally, using $C_S = \max_{j\in S} C_j$ and $C_j^{\mathrm{LP}} \leq C_S^{\mathrm{LP}}$, we can carefully bound the expected completion time of a group $\mathbb{E}[C_S] \leq \frac{2(1+\varepsilon')\rho\beta}{\ln\beta}C_S^{\mathrm{LP}}$. The full proof is deferred to [32].

Theorem 2. *Given a polynomial-time subroutine that computes for any instance J' a schedule of makespan at most $\rho\cdot\max_{d\in[D]}\sum_{j\in J'} b_{d,j}p_j$, there is a polynomial-time $(2\rho e+\varepsilon)$-approximation for any $\varepsilon>0$ for DPSP-G.*

3.1 Applications to Machine Scheduling

We show how to apply our framework to parallel-machine scheduling with group completion times. While the identical-machine case mainly serves to illustrate our framework, we improve upon previous results for related parallel machines.

Identical Machines. We are given m parallel identical machines, each with a speed of 1. At any time t, a schedule must satisfy $y_{j,t} \leq 1$ for any job j and $\sum_j \frac{y_{j,t}}{m} \leq 1$. These constraints form a feasible polytope for PSP. Thus, we have $\max_{d\in[D]}\sum_{j\in J'} b_{d,j}p_j \geq \max\{\max_{j\in J'} p_j, \frac{1}{m}\sum_{j\in J'} p_j\}$. To apply the framework, we can use List Scheduling [18] to compute a non-preemptive schedule for any job set J' with makespan at most $\rho\cdot\max\{\max_{j\in J'} p_j, \frac{1}{m}\sum_{j\in J'} p_j\}$ for $\rho = 2$. Hence, Theorem 2 implies a $(4e+\varepsilon)$-approximation algorithm for minimizing the total weighted group completion time on parallel identical machines. We note that a 2-approximation is known for this problem [31,48].

Uniformly Related Machines. We are given m parallel machines with speeds $s_1 \geq s_2 \geq \ldots \geq s_m$ and assume w.l.o.g. that $m = n$. At any time t, a feasible schedule satisfies for all $\{j_1, j_2, \ldots, j_\ell\} \subseteq J$ that $\sum_{q=1}^{\ell} y_{j_q,t} \leq \sum_{i=1}^{\ell} s_i$ [11,24]: These are the packing constraints of our polytope $\mathcal{P}$. They can be separated in polynomial time by enumerating ℓ and checking the first ℓ jobs with largest $y_{j,t}$. Let $J' = \{j_1, j_2, \ldots, j_k\} \subseteq J$ be any subset of jobs with $p_{j_1} \geq \ldots \geq p_{j_k}$. We have

$$\max_{d\in[D]}\sum_{j\in J'} b_{d,j}p_j = \max_{d\in[D]}\sum_{q=1}^{k} b_{d,j_q}p_{j_q} \geq \max_{1\leq\ell\leq k}\sum_{q=1}^{\ell} p_{j_q}\cdot\left(\sum_{i=1}^{\ell} s_i\right)^{-1} .$$

The last inequality uses the packing constraints on the first ℓ jobs in J' for $1 \le \ell \le k$. The polynomial-time algorithm in [23] computes a *preemptive* schedule with a makespan of at most $\max_{1\le \ell\le k} \sum_{j=1}^{\ell} p_j / \sum_{i=1}^{\ell} s_i$ for any $J' = \{j_1, j_2, \dots, j_k\} \subseteq J$. Combined with a polynomial-time conversion from a *preemptive* schedule to a *non-preemptive* schedule, which increases the makespan by a factor of at most $(2 - \frac{1}{m})$ [47], we can compute in polynomial time a non-preemptive schedule with makespan

$$2 \cdot T \le 2 \cdot \max_{1\le \ell\le k} \sum_{q=1}^{\ell} p_{j_q} \cdot \left(\sum_{i=1}^{\ell} s_i \right)^{-1} \le 2 \cdot \max_{d\in[D]} \sum_{j\in J'} b_{d,j} p_j \ ,$$

which yields the desired subroutine. Hence, Theorem 2 implies a $4e + \varepsilon \le 10.874$-approximation for $Q \mid r_j \mid \sum w_S C_S$, which improves upon the known 13.5-approximation for unrelated machines [9].

Theorem 6. *There is a* 10.874*-approximation algorithm for* $Q \mid r_j \mid \sum w_S C_S$.

3.2 Applications to Sum Multicoloring

Next, we consider the non-preemptive sum multicoloring problem with group completion times (npSMC-G), which generalizes a collection of coloring and scheduling problems, e.g., non-preemptive sum multicoloring, data migration, and coflow scheduling. We first model feasible rates as a polytope. Let $\mathcal{C}$ be the family of cliques of G. At any time, the allocation of rates to jobs should satisfy for all cliques $C \in \mathcal{C}$ that $\sum_{v\in C} y_{v,t} \le 1$. Given a subset of jobs $V' \subseteq V$, let $G[V']$ be the subgraph induced by V' and $\mathcal{C}[V']$ be the family of cliques in $G[V']$. By expressing the above constraints as (13), we derive

$$\max_{d\in[D]} \sum_{j\in J'} b_{d,j} p_j \ge \max_{C\in\mathcal{C}} \sum_{v\in C\cap V'} p_v = \max_{C\in\mathcal{C}[V']} \sum_{v\in C} p_v \ .$$

The last term is the length of the largest clique in $G[V']$. To apply our algorithmic framework, the key is to find an algorithm that computes a schedule for any given job set $V' \subseteq V$ with a makespan of $\rho \cdot \max_{C\in\mathcal{C}[V']} \sum_{v\in C} p_v$ for some $\rho \ge 1$. By plugging in different algorithms for the subproblem, we obtain approximation algorithms for various graph classes as follows.

Interval Graphs. Buchsbaum et al. [7] considered the makespan minimization variant of multicoloring and obtained a $(2+\varepsilon)$-approximation against the lower bound of the largest clique size, which implies an algorithm with $\rho = 2 + \varepsilon$. Further, if jobs' processing times are all in $\Theta(1)$, they obtained a PTAS. Applying Theorem 2, we obtain the following result.

Theorem 7. *There is a* 10.874*-approximation algorithm for npSMC-G on interval graphs. If jobs' processing times are all in* $\Theta(1)$*, the approximation ratio can be improved to* 5.437.

Perfect Graphs and Unit Processing Times. In perfect graphs, we can always find a vertex coloring in polynomial time (cf. Corollary 67.2c [42]) whose number of colors is exactly the size of the largest clique. This implies a makespan minimization algorithm with $\rho = 1$. Applying Theorem 2, we obtain the following, which improves the 10.874-approximation in [10].

Theorem 8. *There is a* 5.437*-approximation for SC-G on perfect graphs.*

Line Graphs. For simplicity, we flip the roles of vertices and edges. We say an edge is a job and no two incident edges of the same vertex can be scheduled at the same time. Hence, the length of the largest clique turns out to be the largest total incident edge lengths, i.e., $\max_{v \in V} \sum_{e \in \delta(v)} p_e$. We remark that this is equivalent to graph scheduling.

We consider the following greedy algorithm, which turns out to be a makespan minimization algorithm with $\rho = 2$: Iterate through time t and greedily schedule a job (edge) $e = (u, v)$ in the time interval $[t, t + p_e]$ whenever none of other incident edges of u or v is being scheduled at time t (consider the jobs in an arbitrary order). For the analysis, consider any edge $e = (u, v) \in E$. Before the start time t of e, either some incident edge of u or some incident edge of v is scheduled as otherwise, we would have scheduled e earlier. Hence, $C_e = t + p_e \le p(\delta(u) \setminus \{e\}) + p(\delta(v) \setminus \{e\}) + p_e \le 2 \max_{v \in V} \sum_{e \in \delta(v)} p_e$.

Applying Theorem 2, we obtain the following, which also applies to the two special cases, i.e., non-preemptive data migration and coflow scheduling with arbitrary processing times.

Theorem 9. *There is a* 10.874*-approximation algorithm for graph scheduling (or equivalently, npSMC-G on line graphs).*

We remark that for the above graph classes, the constraints $\sum_{v \in C} y_{v,t} \le 1$ for all $C \in \mathcal{C}$ can be separated in polynomial time (cf. Theorem 67.6 in [42] for interval graphs and perfect graphs and [30] for line graphs).

4 A $(2 + \varepsilon)$-Approximation for Offline PSP-G

Finally, we give a randomized $(2 + \varepsilon)$-approximation for offline PSP-G for any $\varepsilon > 0$. The idea originates from algorithms for PSP and the open shop problem [27,39,44].

Let $\delta > 0$ and $\varepsilon' > 0$. We first compute an optimal solution $\{x^{\mathrm{LP}}_{j,i}, x^{\mathrm{LP}}_{S,i}, C^{\mathrm{LP}}_S\}$ to (LP'), which is a relaxation of PSP-G. As discussed in Sect. 3, we have $\sum_{S \in \mathcal{S}} w_S C^{\mathrm{LP}}_S \le (1 + \delta)\text{OPT}$. Note that the solution $\{x^{\mathrm{LP}}_{j,i}, x^{\mathrm{LP}}_{S,i}, C^{\mathrm{LP}}_S\}$ implies a feasible preemptive schedule, which we call the LP schedule. We will derive another schedule from the LP schedule whose objective value is at most $2(1 + \varepsilon') \sum_{S \in \mathcal{S}} w_S C^{\mathrm{LP}}_S \le (2 + \varepsilon')(1 + \delta)\text{OPT}$.

Let α be a random real variable drawn from the interval $(0, 1)$ with the density function $f(\theta) = 2\theta$. We obtain a new schedule by slowing down the LP schedule by a factor of $\frac{1}{\alpha}$. That is, we map and stretch the allocation of the

LP schedule in time interval $[t, t+1]$ to time interval $[\frac{t}{\alpha}, \frac{t+1}{\alpha}]$. Now each job j receives $\frac{p_j}{\alpha}$ amount of scheduling in the stretched schedule. Then we modify the rate of every job j to 0 after the time when j has been completed. Clearly, the final schedule is feasible. The advantage of this schedule is that we can bound the (expected) completion time of any job j (resp. group S) against C_j^{LP} (resp. C_S^{LP}).

We introduce some notation for the analysis. For any job j, let C_j^α be the earliest time by which an α-fraction of j is completed in the LP schedule. Fix some $\alpha \in (0,1)$ and let C_j (resp. C_S) be the completion time of j (resp. S) in the final schedule of our algorithm. We first bound C_j against the α-point of j in the LP schedule.

Lemma 4. *For every $j \in J$, it holds $C_j \leq C_j^\alpha / \alpha$.*

Next, we bound the completion time of a group $S \in \mathcal{S}$ in the final schedule. We do this analogously to C_j^α. Consider the earliest interval i by which an α-fraction of S has been completed, i.e., $\sum_{i'=1}^{i} x_{S,i'} \geq \alpha$ and $\sum_{i'=1}^{i-1} x_{S,i'} < \alpha$. Let C_S^α be the leftmost point of the ith interval, i.e., γ_{i-1}. We obtain the following.

Lemma 5. *For any $S \in \mathcal{S}$, $C_S = \max_{j \in S} C_j \leq (1+\varepsilon')C_S^\alpha/\alpha$.*

Finally, we bound the algorithm's performance over the random choice of α.

Theorem 4. *There is a $(2+\varepsilon)$-approximation algorithm for offline PSP-G for any $\varepsilon > 0$, even with non-uniform release dates.*

Our algorithm is almost best-possible, since PSP-G contains $1 \mid \mathrm{pmtn} \mid \sum_S w_S C_S$, which has been shown to be equivalent to $1 \mid \mathrm{prec} \mid \sum_j w_j C_j$ [39] and has a lower bound of $2 - \varepsilon$ under a stronger version of the Unique Games Conjecture [3].

5 Conclusion

With polytope scheduling for group completion times, PSP-G and DPSP-G, we propose a unifying abstraction of several scheduling, coloring, and other graph problems. We present general algorithmic tools that achieve new and improved approximation factors. While we give a meta-framework for solving the new Discrete PSP, particularly for the general DPSP-G (Sect. 3), its implementation relies on problem-specific makespan minimization subroutines. It remains as an open question whether the dependence on problem-specific properties can be avoided. Further, for offline PSP-G, it is open whether the $(2+\varepsilon)$-approximation (Sect. 4) can be improved, even in the special case without groups.

Acknowledgments. This work was done while AL and ZL were affiliated with the University of Bremen. AL was supported by the "Humans on Mars Initiative", funded by the Federal State of Bremen and the University of Bremen.

Disclosure of Interests. The authors have no competing interests to declare that are relevant to the content of this article.

References

1. Ahmadi, S., Khuller, S., Purohit, M., Yang, S.: On scheduling coflows. Algorithmica **82**(12), 3604–3629 (2020). https://doi.org/10.1007/S00453-020-00741-3
2. Bansal, N., Coppersmith, D., Sviridenko, M.: Improved approximation algorithms for broadcast scheduling. SIAM J. Comput. **38**(3), 1157–1174 (2008). https://doi.org/10.1137/060674417
3. Bansal, N., Khot, S.: Optimal long code test with one free bit. In: FOCS, pp. 453–462. IEEE Comput. Soc. (2009). https://doi.org/10.1109/FOCS.2009.23
4. Bansal, N., Srinivasan, A., Svensson, O.: Lift-and-round to improve weighted completion time on unrelated machines. In: STOC, pp. 156–167. ACM (2016). https://doi.org/10.1145/2897518.2897572
5. Bar-Noy, A., Kortsarz, G.: Minimum color sum of bipartite graphs. J. Algorithms **28**(2), 339–365 (1998). https://doi.org/10.1006/JAGM.1998.0938
6. Bhimaraju, A., Nayak, D., Vaze, R.: Non-clairvoyant scheduling of coflows. In: WiOpt, pp. 81–88. IEEE (2020). https://dl.ifip.org/db/conf/wiopt/wiopt2020/32.pdf
7. Buchsbaum, A.L., Karloff, H.J., Kenyon, C., Reingold, N., Thorup, M.: OPT versus LOAD in dynamic storage allocation. SIAM J. Comput. **33**(3), 632–646 (2004). https://doi.org/10.1137/S0097539703423941
8. Chen, Q., Im, S., Petety, A.: Online scheduling via gradient descent for weighted flow time minimization. In: SODA, pp. 3802–3841. SIAM (2025). https://doi.org/10.1137/1.9781611978322.128
9. Correa, J.R., Skutella, M., Verschae, J.: The power of preemption on unrelated machines and applications to scheduling orders. Math. Oper. Res. **37**(2), 379–398 (2012). https://doi.org/10.1287/MOOR.1110.0520
10. Darbouy, S.P., Friggstad, Z.: Approximating minimum sum coloring with bundles. In: SWAT. LIPIcs, vol. 294, pp. 21:1–21:14. Schloss Dagstuhl - Leibniz-Zentrum für Informatik (2024). https://doi.org/10.4230/LIPICS.SWAT.2024.21
11. Feldman, J., Muthukrishnan, S., Nikolova, E., Pál, M.: A truthful mechanism for offline ad slot scheduling. In: SAGT. Lecture Notes in Computer Science, vol. 4997, pp. 182–193. Springer (2008).https://doi.org/10.1007/978-3-540-79309-0_17
12. Fukunaga, T.: Integrality gap of time-indexed linear programming relaxation for coflow scheduling. In: APPROX/RANDOM. LIPIcs, vol. 245, pp. 36:1–36:13. Schloss Dagstuhl - Leibniz-Zentrum für Informatik (2022). https://doi.org/10.4230/LIPICS.APPROX/RANDOM.2022.36
13. Gandhi, R., Halldórsson, M.M., Kortsarz, G., Shachnai, H.: Improved results for data migration and open shop scheduling. In: ICALP. Lecture Notes in Computer Science, vol. 3142, pp. 658–669. Springer (2004). https://doi.org/10.1007/978-3-540-27836-8_56
14. Gandhi, R., Halldórsson, M.M., Kortsarz, G., Shachnai, H.: Improved bounds for scheduling conflicting jobs with Minsum criteria. ACM Trans. Algorithms **4**(1), 11:1–11:20 (2008). https://doi.org/10.1145/1328911.1328922
15. Gandhi, R., Halldórsson, M.M., Kortsarz, G., Shachnai, H.: Corrigendum: improved results for data migration and open shop scheduling. ACM Trans. Algorithms **9**(4), 34:1–34:7 (2013). https://doi.org/10.1145/2500123
16. Garg, N., Gupta, A., Kumar, A., Singla, S.: Non-clairvoyant precedence constrained scheduling. In: ICALP. LIPIcs, vol. 132, pp. 63:1–63:14. Schloss Dagstuhl - Leibniz-Zentrum für Informatik (2019)

17. Ghodsi, A., Zaharia, M., Hindman, B., Konwinski, A., Shenker, S., Stoica, I.: Dominant resource fairness: fair allocation of multiple resource types. In: NSDI. USENIX Association (2011). https://www.usenix.org/conference/nsdi11/dominant-resource-fairness-fair-allocation-multiple-resource-types
18. Graham, R.L.: Bounds on multiprocessing timing anomalies. J. SIAM Appl. Math. **17**(2), 416–429 (1969)
19. Gupta, A., Kumar, A., Singla, S.: Bag-of-tasks scheduling on related machines. In: APPROX-RANDOM. LIPIcs, vol. 207, pp. 3:1–3:16. Schloss Dagstuhl - Leibniz-Zentrum für Informatik (2021). https://doi.org/10.4230/LIPICS.APPROX/RANDOM.2021.3
20. Hall, L.A., Schulz, A.S., Shmoys, D.B., Wein, J.: Scheduling to minimize average completion time: off-line and on-line approximation algorithms. Math. Oper. Res. **22**(3), 513–544 (1997). https://doi.org/10.1287/MOOR.22.3.513
21. Halldórsson, M.M., Kortsarz, G.: Tools for multicoloring with applications to planar graphs and partial k-trees. J. Algorithms **42**(2), 334–366 (2002). https://doi.org/10.1006/JAGM.2001.1210
22. Halldórsson, M.M., Kortsarz, G., Sviridenko, M.: Sum edge coloring of multigraphs via configuration LP. ACM Trans. Algorithms **7**(2), 22:1–22:21 (2011). https://doi.org/10.1145/1921659.1921668
23. Horvath, E.C., Lam, S., Sethi, R.: A level algorithm for preemptive scheduling. J. ACM **24**(1), 32–43 (1977). https://doi.org/10.1145/321992.321995
24. Im, S., Kulkarni, J., Munagala, K.: Competitive algorithms from competitive equilibria: non-clairvoyant scheduling under polyhedral constraints. J. ACM **65**(1), 3:1–3:33 (2018). https://doi.org/10.1145/3136754
25. Im, S., Kulkarni, J., Munagala, K., Pruhs, K.: Selfishmigrate: a scalable algorithm for non-clairvoyantly scheduling heterogeneous processors. In: FOCS, pp. 531–540. IEEE Comput. Soc. (2014). https://doi.org/10.1109/FOCS.2014.63
26. Im, S., Moseley, B., Pruhs, K., Purohit, M.: Matroid coflow scheduling. In: ICALP. LIPIcs, vol. 132, pp. 145:1–145:14. Schloss Dagstuhl - Leibniz-Zentrum für Informatik (2019). https://doi.org/10.4230/LIPICS.ICALP.2019.145
27. Jäger, S., Lindermayr, A., Megow, N.: The power of proportional fairness for non-clairvoyant scheduling under polyhedral constraints. In: SODA, pp. 3901–3930. SIAM (2025). https://doi.org/10.1137/1.9781611978322.132
28. Jain, K., Vazirani, V.V.: Eisenberg–Gale markets: algorithms and game-theoretic properties **70**(1), 84–106. https://doi.org/10.1016/j.geb.2008.11.011
29. Kaneko, M., Nakamura, K.: The Nash social welfare function **47**(2), 423–43 https://doi.org/10.2307/1914191
30. Kim, Y.A.: Data migration to minimize the total completion time. J. Algorithms **55**(1), 42–57 (2005). https://doi.org/10.1016/J.JALGOR.2004.07.009
31. Leung, J.Y.T., Li, H., Pinedo, M.: Approximation algorithms for minimizing total weighted completion time of orders on identical machines in parallel. Naval Res. Logistics (NRL) **53**(4), 243–260 (2006)
32. Lindermayr, A., Liu, Z., Megow, N.: Polytope scheduling with groups: unified models and optimal guarantees. CoRR abs/2501.17682 (2026). https://doi.org/10.48550/ARXIV.2501.17682
33. Mestre, J.: Adaptive local ratio. SIAM J. Comput. **39**(7), 3038–3057 (2010). https://doi.org/10.1137/080731712
34. Motwani, R., Phillips, S.J., Torng, E.: Non-clairvoyant scheduling. Theor. Comput. Sci. **130**(1), 17–47 (1994)
35. Moulin, H.: Fair division and collective welfare, MIT Press (2003)

36. Nash, Jr., J.F.: The bargaining problem **18**(2), 155–162. https://doi.org/10.2307/1907266
37. Panageas, I., Tröbst, T., Vazirani, V.V.: Time-efficient algorithms for Nash-bargaining-based matching market models. In: WINE. Lecture Notes in Computer Science, vol. 15534, pp. 119–136. Springer (2024). https://doi.org/10.1007/978-3-032-08560-3_7
38. Queyranne, M.: Structure of a simple scheduling polyhedron. Math. Program. **58**, 263–285 (1993). https://doi.org/10.1007/BF01581271
39. Queyranne, M., Sviridenko, M.: A (2+epsilon)-approximation algorithm for the generalized preemptive open shop problem with minsum objective. J. Algorithms **45**(2), 202–212 (2002). https://doi.org/10.1016/S0196-6774(02)00251-1
40. Queyranne, M., Sviridenko, M.: Approximation algorithms for shop scheduling problems with minsum objective. J. Sched. **5**(4), 287–305 (2002)
41. Rohwedder, L., Schnaars, L.: 3.415-approximation for coflow scheduling via iterated rounding. In: ICALP. LIPIcs, vol. 334, pp. 128:1–128:19. Schloss Dagstuhl - Leibniz-Zentrum für Informatik (2025)
42. Schrijver, A.: Combinatorial optimization: polyhedra and efficiency, vol. 24, Springer (2003)
43. Schulz, A.S.: Scheduling to minimize total weighted completion time: performance guarantees of LP-based heuristics and lower bounds. In: IPCO. Lecture Notes in Computer Science, vol. 1084, pp. 301–315. Springer (1996). https://doi.org/10.1007/3-540-61310-2_23
44. Schulz, A.S., Skutella, M.: Random-based scheduling: New approximations and LP lower bounds. RANDOM. Lecture Notes in Computer Science **1269**, 119–133. Springer (1997). https://doi.org/10.1007/3-540-63248-411
45. Shmoys, D.B., Wein, J., Williamson, D.P.: Scheduling parallel machines online. SIAM J. Comput. **24**(6), 1313–1331 (1995). https://doi.org/10.1137/S0097539793248317
46. Skutella, M.: Convex quadratic and semidefinite programming relaxations in scheduling. J. ACM **48**(2), 206–242 (2001). https://doi.org/10.1145/375827.375840
47. Woeginger, G.J.: A comment on scheduling on uniform machines under chain-type precedence constraints. Oper. Res. Lett. **26**(3), 107–109 (2000). https://doi.org/10.1016/S0167-6377(99)00076-0
48. Yang, J., Posner, M.E.: Scheduling parallel machines for the customer order problem. J. Sched. **8**(1), 49–74 (2005). https://doi.org/10.1007/S10951-005-5315-5

An Efficient Algorithm for Minimizing Ordered Norms in Fractional Load Balancing

Daniel Blankenburg[1], Antonia Ellerbrock[1](✉), Thomas Kesselheim[2], and Jens Vygen[1]

[1] Research Institute for Discrete Mathematics and Hausdorff Center for Mathematics, University of Bonn, Bonn, Germany
{blankenburg,ellerbrock,vygen}@dm.uni-bonn.de
[2] Institute of Computer Science and Hausdorff Center for Mathematics, University of Bonn, Bonn, Germany
thomas.kesselheim@uni-bonn.de

Abstract. We study the problem of minimizing an ordered norm of a load vector (indexed by a set of d resources), where a finite number n of customers c contribute to the load of each resource by choosing a solution x_c in a convex set $X_c \subseteq \mathbb{R}^d_{\geq 0}$; so we minimize $\left\|\sum_c x_c\right\|$ for some fixed ordered norm $\|\cdot\|$. We devise a randomized algorithm that computes a $(1+\varepsilon)$-approximate solution to this problem and makes, with high probability, $\mathcal{O}\left((n+d)(\varepsilon^{-2} + \log\log d)\log(n+d)\right)$ calls to oracles that minimize linear functions (with non-negative coefficients) over X_c. While this has been known for the ℓ_∞ norm via the multiplicative weights update method, existing proof techniques do not extend to arbitrary ordered norms.

Our algorithm uses a resource price mechanism that is motivated by the follow-the-regularized-leader paradigm, and is expressed by smooth approximations of ordered norms. We need and show that these have non-trivial stability properties, which may be of independent interest. For each customer, we define dynamic cost budgets, which evolve throughout the algorithm, to determine the allowed step sizes. This leads to non-uniform updates and may even reject certain oracle solutions. Using non-uniform sampling together with a martingale argument, we can guarantee sufficient expected progress in each iteration, and thus bound the total number of oracle calls with high probability.

1 Introduction

Many fractional load balancing problems can be understood as multiple customers sharing multiple resources with the goal of minimizing the maximum resource consumption. A simple example is FRACTIONAL SCHEDULING, where jobs have to be assigned to machines to minimize the makespan. Similarly, FRACTIONAL PACKING [32] can be formulated as finding a probability distribution x that minimizes the ℓ_∞ norm of Ax for some non-negative matrix A. Another

S. S. Dey et al. (Eds.): IPCO 2026, LNCS 16588, pp. 282–298, 2026.
https://doi.org/10.1007/978-3-032-28691-8_19

such problem is MAXIMUM CONCURRENT FLOW, where we have to send flows in a graph from source nodes to respective sink nodes such that all demands are fulfilled and the highest congestion of an edge is minimized. This problem can be generalized in multiple ways. For example, a highly relevant problem for VLSI routing is FRACTIONAL STEINER TREE PACKING for given terminal sets to minimize the worst congestion [19,22,31]. In all these problems, the standard goal is to minimize the maximum utilization/load/congestion of any resource, which can also be understood as minimizing the ℓ_∞ norm of the load vector.

However, minimizing the ℓ_∞ norm is not very robust. In VLSI routing, due to modeling issues, a few edges with very high congestion can often not be avoided. Therefore, design engineers typically use measures such as the average congestion of the worst $x\%$ of resources (i.e., a top-k norm) or linear combinations thereof [34]. This corresponds to an ordered norm applied to the load vector.

In this paper, we consider the problem of minimizing an ordered norm in fractional load balancing problems. We devise a fast algorithm that computes a $(1+\varepsilon)$-approximate solution for any given $\varepsilon > 0$.

1.1 Problem Statement

We assume that there is a set C of n "customers" which correspond to jobs, commodities, or requests in the underlying problem. For each customer $c \in C$, there are multiple ways to serve them. We identify these ways with their resource utilization by assuming a closed convex set $X_c \subseteq \mathbb{R}^d_{\geq 0}$ per customer, where a vector $x_c \in X_c$ denotes what fraction of each of the available d resources is used. The set of feasible solutions to our overall problem is $X = \sum_{c\in C} X_c$. Our objective is to find an $x \in X$ that (approximately) minimizes $\|x\|_\beta$, where $\|\cdot\|_\beta$ is an ordered norm, i.e.,

$$\min_{x\in X} \|x\|_\beta \quad . \tag{1}$$

An ordered norm is defined by $\|x\|_\beta = \sum_{i=1}^{d} \beta_i (x^\downarrow)_i$ for $x \in \mathbb{R}^d_{\geq 0}$, for some non-increasing weight vector $\beta \in \mathbb{R}^d_{\geq 0}$, where $(x^\downarrow)_i$ denotes the i-th largest entry in x. Recovering the ℓ_1 norm when all weights are equal and the ℓ_∞ norm when only the first weight β_1 is non-zero, ordered norms are a natural choice to interpolate between minimizing the average and the maximum resource usage.

We assume that the sets X_c are not given explicitly but by access to an oracle minimizing a linear function (with non-negative coefficients) over the fixed set X_c, an assumption that is prevalent in theory and in practice [5,21]. In the special case of FRACTIONAL PACKING, this means that A can have exponentially many columns and is not given explicitly; instead, we have an oracle that returns one of these columns (minimizing the inner product with a given price vector).

A well-known special case is the CONCURRENT FLOW problem. Here, the customers correspond to the commodities. If customer c can be served by sending one unit of flow from s to t in a unit-capacity graph, then d is the number of edges and X_c is the convex hull of incidence vectors of edge sets of s-t-paths. In this case, the linear minimization subproblem corresponds to finding a shortest s-t-path. In the MAXIMUM CONCURRENT FLOW problem, we aim at minimizing the ℓ_∞ norm over X. We will generalize this to arbitrary ordered norms; see Corollary 1.

1.2 Our Main Result

Our main result is the following. Recall that $n = |C|$ denotes the number of customers, and d denotes the number of resources (so $X = \sum_{c \in C} X_c \subseteq \mathbb{R}^d_{\geq 0}$).

Theorem 1. *There is a randomized algorithm that computes, for any given $\varepsilon > 0$, a $(1+\varepsilon)$-approximate solution to (1) and makes, with high probability, $\mathcal{O}\left((n+d)(\varepsilon^{-2} + \log\log d)\log(n+d)\right)$ calls to oracles that minimize linear functions with non-negative coefficients over X_c for $c \in C$. The additional time per oracle call is $\mathcal{O}\left(d \log d\right)$.*

This also works with τ-approximate linear minimization oracles (for any $\tau \geq 1$), and leads to a $\tau(1+\varepsilon)$-approximate solution to (1).

The dependence on ε is best possible, as shown in [25]. Further, our running time matches the best known running time for the ℓ_∞ norm [3,31] up to log-factors.

As an application, we can minimize arbitrary ordered norms rather than just the ℓ_∞ norm in the Concurrent Flow problem. The ordered norm is applied to the congestion vector of a multi-commodity flow. The congestion vector has an entry for each edge, which is the total flow along this edge divided by the edge's capacity.

Corollary 1. *Given a multi-commodity flow instance with v vertices, d edges with positive real capacities, k commodities with positive real demands, and any ordered norm $\|\cdot\|_\beta$, we can compute a multi-commodity flow with congestion vector x such that $\|x\|_\beta$ is minimum up to a factor $1+\varepsilon$, in $\mathcal{O}\left(d(d+\min\{k,v\})(\varepsilon^{-2} + \log\log d)\log^2 d\right)$ time.*

Proof. Define a customer for each vertex that is the source of at least one source-sink pair. We have $n \leq \min\{k, v\}$ customers, and we may assume $v \leq d$. For a customer (source) c, let sinks(c) denote the set of corresponding sinks; then

$$X_c = \operatorname{conv}\left\{\left(\frac{1}{u_e} \sum_{s \in \text{sinks}(c): e \in P_s} d_{cs}\right)_{e \in E} : P_s \text{ path from } c \text{ to } s \text{ for } s \in \text{sinks}(c)\right\},$$

where E is the set of edges, $u_e > 0$ is the capacity of edge e, and $d_{cs} > 0$ is the demand of the commodity $c \to s$. To minimize a linear function over X_c, we apply Dijkstra's algorithm to compute a shortest path from source c to each sink in sinks(c) in total time $\mathcal{O}\left(d \log d\right)$. The result follows from Theorem 1. □

Since our main result extends to oracles that only approximately minimize linear functions over X_c for $c \in C$ (see the full version [4]), we can also approximate ordered norms for fractional Steiner tree packings and other resource sharing problems of crucial relevance in chip design [9,10,16,31,34].

1.3 A Simple Algorithm and Its Limitations

Our algorithm is based on a well-known algorithm template, exploiting duality: Every norm can also be written as $\|x\| = \max_{y\in Y}\langle x, y\rangle$ for $x \in \mathbb{R}^d_{\geq 0}$, where the dual space Y is the set of non-negative non-dominated vectors with dual norm equal to 1. Therefore, solving problem (1) is equivalent to finding an equilibrium strategy of the minimization player in the zero-sum game $\min_{x\in X}\max_{y\in Y}\langle x, y\rangle = \max_{y\in Y}\min_{x\in X}\langle x, y\rangle$.

In order to obtain a solution to (1), one determines x and y in an iterative way, while interpreting the vector y as positive resource prices. The algorithm alternates between calls to a minimization oracle, which determines the cheapest solution x based on the current prices y, and calls to some price update mechanism, which updates y based on all previous solutions x. After a fixed number of iterations, the average of all encountered solutions x is returned. Common price update mechanisms include Mirror Descent, Follow-the-Regularized-Leader, or Follow-the-Perturbed-Leader [28,33].

In particular, in case of the ℓ_∞ norm, one can use the Multiplicative Weight Update or Hedge algorithm [2,13] to determine the prices; here, the dual space is the probability simplex. Moving to more general norms and thus dual spaces Y, one can show, using Follow-the-Regularized-Leader to update the prices, that $\mathcal{O}\left(\varepsilon^{-2}\log d\right)$ oracle calls suffice to obtain a guarantee of $\exp(\varepsilon)\,\mathrm{OPT}+\varepsilon$ if $X \subseteq [0,1]^d$, where OPT denotes the optimum solution value of (1). Similar guarantees follow with the standard Frank–Wolfe algorithm [21] with a suitable smooth approximation of the given norm (for example, the one we discuss in Sect. 3).

However, these results are not quite satisfactory for two reasons. First of all, given the additive error, the bound is only meaningful if OPT is sufficiently large. This is a strong assumption when $X \subseteq [0,1]^d$. Removing the assumption $X \subseteq [0,1]^d$ will generally increase the running time by a factor proportional to the squared ℓ_∞ diameter of X with standard techniques.

Secondly, and even more severely, in every iteration, the algorithm needs to determine an $x \in X$ that minimizes $\langle x, y\rangle$ given the current prices y. This requires n calls to single-customer oracles. Efficient algorithms, often referred to as *block-coordinate methods*, exploit the decomposable structure of X because solving such a minimization problem only over one of the sets X_c can typically be performed much faster. However, for many of these algorithms, a lower bound of $\Omega(nd)$ on the number of oracle calls can be shown. For the special case of multi-commodity flow in the ℓ_∞ case, [14] and [12] combine a width reduction technique with a round robin scheme to bypass these lower bounds and to obtain an algorithm that requires only $\tilde{\mathcal{O}}\left(\varepsilon^{-2}(n+d)\right)$ calls to single-customer oracles. [31] use a similar approach for the general ℓ_∞ case. The underlying reason for these positive results is that for the ℓ_∞ norm, prices have a simple closed-form representation, and satisfy strong stability properties [23], which do not hold for general ordered norms. While similar stability properties have been obtained for other norms, they are generally too lossy for our purpose. For example, all norm approximations given in [23] lose at least constant factors.

1.4 Our Techniques

In order to prove Theorem 1, we design a new algorithm that is inspired by the simple one sketched in the previous section but requires several new ideas.

Our price update strategy is motivated by Follow-the-Regularized-Leader with the negative entropy $F(y) = \sum_{i=1}^{d} y_i \ln y_i$ as regularizer. The prices can be expressed equivalently as the gradients of the convex conjugate of F on the dual space Y, given by $F^*(x) = \max_{y \in Y}(\langle y, x \rangle - F(y))$. This connection is known as Danskin's theorem [11]. This convex conjugate coincides with the LogSumExp function for the ℓ_∞ norm.

In this work, it is more convenient to use the language of norm approximations. Indeed, the convex conjugate of the negative entropy gives rise to a norm approximation that satisfies the properties of Theorem 2 (see Sect. 3). This result could be of independent interest in future work.

Theorem 2 (Ordered norm approximation). *For every ordered norm* $\|\cdot\|_\beta$ *on* $\mathbb{R}^d$ *and every* $\eta > 0$, *there is a differentiable convex function* Ψ *fulfilling* $\|x\|_\beta \leq \Psi(x) \leq \|x\|_\beta + \frac{\ln d}{\eta}$ *and* $\nabla\Psi(x+y) \leq \exp(\eta \|y\|_\infty) \cdot \nabla\Psi(x)$ *for all* $x, y \in \mathbb{R}^d_{\geq 0}$. *Moreover,* Ψ *and its gradient can be evaluated in* $\mathcal{O}(d \log d)$ *time.*

We prove this in Sect. 3, where we also provide our definition of a norm approximation. The property of $\nabla\Psi$ (bounded gradient increase) follows from structural properties of the generalized relative entropy projection to the dual space Y, in particular a contraction property that we prove for ordered norms. The bounds on Ψ and $\nabla\Psi$ are best possible; see full version [4].

The approximation and stability properties in Theorem 2 are similar to the ones in [23,30] but there are subtle differences. In contrast to [23], we bound the increase of the gradient while in [23] there is only a bound on the decrease. Besides, our bound on the gradient increase depends on the ℓ_∞ norm of the change, which is generally a looser bound. The norm approximations proposed in [30] require the same upper bound on the maximum gradient increase as in Theorem 2, but additionally impose a lower bound on the maximum gradient decrease, which we do not need. Further, these approximations were constructed only for ℓ_p norms.

If we had a norm approximation Ψ as in Theorem 2 (c.f. Definition 1) for more general classes of norms, which additionally satisfy the scaling assumption $\|1^d\| = 1$ and $\frac{\|\cdot\|_1}{d} \leq \|\cdot\|$, then Theorem 1 would also hold for those.

The norm approximation properties of Theorem 2 imply *weak duality* (see Lemma 2 and Corollary 2):

$$\min_{s \in X} \langle \nabla\Psi(x), s \rangle \leq \mathrm{OPT} \qquad \forall x \in \mathbb{R}^d_{\geq 0} \; . \tag{2}$$

Our core algorithm (Algorithm 1) collects, for each customer, solutions of total weight $T \in \mathbb{N}$ and takes the weighted sum. While the total weight of the solutions per customer is T, the number of oracle calls can be significantly smaller or larger than T. Since the step sizes of our algorithm are non-uniform, it is convenient to parametrize the algorithm according to the total solution weight that is collected. We interpret the evolving partial solution of our algorithm as a piecewise linear curve $S : [0, nT] \to \mathbb{R}^d_{\geq 0}$. The curve starts at $S(0) = 0$ and ends in $S(nT)$. The output of our algorithm will be $\frac{1}{T}S(nT)$. We build S by successively adding oracle solutions with a carefully chosen weight. That is, at a time point $t \in [0, nT]$, we query the oracle of a customer c with the prices $\nabla\Psi(S(t))$, and add its returned solution with weight $\xi \geq 0$, i.e., define $S(t + \xi') = S(t) + \xi' s$ for all $0 \leq \xi' \leq \xi$. We use the gradient theorem for line integrals to bound the cost of the final solution

$$\Psi(S(nT)) \;=\; \Psi(S(0)) \;+\; \int_0^{nT} \langle \nabla\Psi(S(t)), S'(t)\rangle \;\mathrm{d}t \;. \tag{3}$$

This is a continuous analogue to regret bounds of Follow-the-Regularized-Leader. The derivative $S'(t)$ is a direction of the curve S and thus always corresponds to the last added oracle solution (up to a finite number of points where S is not differentiable). We may interpret the term $\int_0^{nT} \langle \nabla\Psi(S(t)), S'(t)\rangle \;\mathrm{d}t$ as the"cost incurred along the way". Instead of the usual viewpoint, where we collect a solution s with coefficient ξ and incur a cost of $\langle y, \xi s\rangle$ for the fixed price y that was given as an input to the oracle, we pay for the solution s under the evolving costs $\nabla\Psi(S(t))$ on an interval of length ξ. To analyze these costs, the bounded increase of $\nabla\Psi$ as well as weak duality are crucial.

The algorithm for the ℓ_∞ norm by [31] proceeds in T phases, where, in every phase, oracle solutions of total weight one are collected for each customer. The analysis crucially relies on the fact that $\nabla\Psi$ remains stable for the entire phase, and the incurred cost in a phase can be bounded by approximately OPT. For general ordered norms, stability of $\nabla\Psi$ cannot be guaranteed for an entire phase, and this proof technique does not work. Thus, our algorithm does not proceed in phases, and requires a completely different proof.

Indeed, our new proof technique can be considered as one of the main technical contributions of our work. We use a *dynamic cost budget* $\sigma : [0, nT) \to [0, 1]^n$, with $\sum_{c\in C} \sigma_c(t) = 1$ at all times $t \in [0, nT)$, which represents the fraction of OPT that each customer is allowed to cost at time t. We call customer c *good* at time t if their cost is not much larger than their cost budget. Assuming that we know OPT, this means:

$$\min_{s\in X_c} \langle \nabla\Psi(S(t)), s\rangle \;\leq\; \exp(2\eta) \cdot \sigma_c(t) \cdot \mathrm{OPT} \;\;.$$

Our algorithm accepts only oracle solutions of good customers. Weak duality implies that there always exists a good customer. To ensure feasibility of the final solution $\frac{1}{T}S(nT)$, solutions of total weight T must be collected for every customer. We encode this directly in the cost budget. If, at time t, solutions of total weight T have already been collected for customer c, we have $\sigma_c(t) = 0$. We call such customers *inactive* at time t, and *active* otherwise. Only active customers can be good. With this strategy, we can bound the total cost by

$$\int_0^{nT} \langle \nabla\Psi(S(t)), S'(t)\rangle \;\mathrm{d}t \;\leq\; \exp(2\eta)\Big(\int_0^{nT} \sigma_{c(t)}(t)\;\mathrm{d}t\Big)\,\mathrm{OPT} \;,$$

where $c(t)$ is the customer whose oracle solution was chosen last before time t. We show that our dynamic cost budget satisfies $\int_0^{nT} \sigma_{c(t)}(t)\ \mathrm{d}t \le T + \frac{\ln n}{\eta}$ for every realization of our randomized algorithm, which is a key ingredient of our proof. Further, our cost budget does not decrease too rapidly: For every $\xi > 0$ such that customer c is active at time $t + \xi$, we have

$$\sigma_c(t+\xi) \ \ge\ \exp(-\eta\xi)\sigma_c(t) \ .$$

Together with the bounded gradient increase of $\nabla\Psi$ (cf. Theorem 2), this allows us to take sufficiently large steps ξ, while guaranteeing that the chosen customer remains good on the entire interval $[t, t+\xi)$. We choose the cost budget as the softmax of the remaining solution weight that needs to be collected for each active customer.

Finding a good customer may require n oracle calls in general. Another major technical contribution is the reduction of the number of oracle calls through non-uniform sampling. We emphasize that the performance guarantee of our algorithm is deterministic, and only the running time is affected by randomization. In every iteration, we sample a customer c according to the cost budget σ and take a maximal step ξ that guarantees that customer c remains good for the entire step length. By calling the oracle for c, we observe the true fraction $q_c(t) \in [0,1]$ that customer c costs relative to OPT. The step length ξ depends on the ratio $\sigma_c(t)/q_c(t)$ and a width reduction technique similar to [14]. It may happen that $\xi = 0$, but we use a martingale argument to show that, in expectation, ξ is sufficiently large.

Our core algorithm assumes OPT ≤ 1, and returns a solution of value $\varepsilon + \exp(\varepsilon)$ after $\mathcal{O}\big(\varepsilon^{-2}(n+d)\log(d+n)\big)$ oracle calls with high probability. To bound the number of oracle calls in the worst case, and to guarantee OPT ≤ 1, we design a stronger variant (see the full version [4]). This is guaranteed to terminate and either correctly decides that OPT > 1 or computes a solution of value $\varepsilon+\exp(\varepsilon)$. Overall, we apply this to scaled instances and embed this into a two-stage binary search to determine OPT approximately.

1.5 Related Work

Minimizing norms in load balancing problems has been studied extensively in the literature. Our work builds upon and brings together different lines of work, including online and offline techniques.

By formulating our problem as a zero-sum game, one can apply no-regret algorithms for online linear optimization, using price update mechanisms such as Follow-the-Regularized-Leader, Mirror Descent, or Follow-the-Perturbed-Leader. For a single customer and several specific types of norms, our problem is well-solved [17,18,26]. This line of work primarily emphasizes price updates and can be interpreted as taking a *dual* perspective. In contrast, first-order methods that operate directly on the set of feasible solutions offer a *primal* perspective. When linear minimization oracles are available, our problem can be tackled using variants of conditional gradient descent [21,27] applied to a differentiable approximation of the given norm.

Naturally, these two perspectives are closely connected: gradients of smooth norm approximations define dual vectors – a connection noted previously by many authors [1,23]. Our work is no exception; it seeks to further explore the interplay between primal and dual viewpoints, drawing inspiration from both and implicitly applying techniques from online optimization to an offline setting. A separate line of earlier work, and at the same time our original motivation for the problem, comes from multi-commodity flow [12,14] and resource sharing under the ℓ_∞ norm [15,31], which also yields width reduction strategies and methods for handling multiple customers. We generalize these techniques to ordered norms.

While minimization of ℓ_∞ and ℓ_p norms has been studied for decades, ordered norms and other (monotone symmetric) norms have only been studied quite recently. One line of research [7,8,20] considers offline *integral* load balancing problems. The techniques are quite different as they mostly focus on rounding fractional solutions of LP relaxations and only obtain constant-factor approximations. Closer in spirit are recent works on online problems [23,24] that also use norm approximations. However, the obtained approximation ratios are only poly-logarithmic.

2 Our Algorithm

In the following, we present our core algorithm (Algorithm 1) in detail. Under the assumption OPT ≤ 1, Algorithm 1 computes a solution of value at most $1+\varepsilon$, and uses, with high probability, $\mathcal{O}\left(\varepsilon^{-2}(n+d)\log(n+d)\right)$ oracle calls.

In the remainder of the section, we assume that a norm approximation Ψ as in Theorem 2 is given. As described in Sect. 1.4, we interpret our algorithm as a piecewise linear curve S from $S(0)=0$ towards $S(nT)$, such that $\bar{s}=\frac{1}{T}S(nT)\in X$ is a feasible solution and the output of our algorithm. To bound the norm of $\bar{s}$, we derive a bound on $\Psi(S(nT))$ using (3).

To ensure that a total work of exactly T is done for each customer, we consider the function $\rho:[0,nT]\to\mathbb{R}^C_{\geq 0}$ that stores the remaining work: When the total work is t, we still need to process a workload of $\rho(t)_c$ for customer c. That is, we decrease ρ_c linearly with slope $-\mathbb{1}_c$ (where $\mathbb{1}_c$ is the vector that contains a one in entry c and zeros otherwise) on the interval $[t,t+\xi]$, when we accept a solution for customer c with step size ξ at collection time t. Note that ρ is a rectilinear monotonely decreasing curve in $\mathbb{R}^C_{\geq 0}$ going from $\rho(0)=(T,\dots,T)$ to $\rho(nT)=(0,\dots,0)$.

We maintain a set of active customers $A\subseteq C$ over time. Once we have processed a total work of T for a customer, we remove this customer from the active set. Customers are sampled from the set A of active customers according to the *softmax* of ρ with smoothing parameter η, restricted to A. We denote it by $\sigma_{\eta,A} : \mathbb{R}^C_{\geq 0} \to \mathbb{R}^C_{\geq 0}$ where, for $z \in \mathbb{R}^C_{\geq 0}$,

$$\sigma_{\eta,A}(z)_c \;:=\; \begin{cases} 0 & \text{for } c \notin A \;, \\ \frac{\exp(\eta z_c)}{\sum_{a \in A} \exp(\eta z_a)} & \text{for } c \in A \;. \end{cases} \tag{4}$$

It is well-known that the softmax is the gradient of the LogSumExp function,

$$\sigma_{\eta,A}(z) \;=\; \nabla \mathrm{LSE}_{\eta,A}(z) \;, \quad \text{where} \quad \mathrm{LSE}_{\eta,A}(z) \;:=\; \frac{1}{\eta} \cdot \ln \sum_{c \in A} \exp(\eta z_c) \,. \tag{5}$$

The function $\sigma_{\eta,A}$ is not only used as a sampling distribution, but also as a cost budget (cf. Sect. 1.4) during the algorithm, which will play a crucial role in the analysis.

Algorithm 1. Core Algorithm

Input: First order oracle access to a norm approximation Ψ,
linear minimization oracles for all n customers $c \in C$,
parameters $T \in \mathbb{N}$, $\eta > 0$.

$A \leftarrow C$ ▷ Set of active customers
$\rho(0) \leftarrow (T, \ldots, T) \in \mathbb{R}^C$ ▷ Remaining work for each customer
$S(0) \leftarrow 0^d$ ▷ Collected solutions
$t \leftarrow 0$ ▷ Total collected work
while $A \neq \emptyset$ **do** ▷ Equivalent to $t < nT$
  $c \leftarrow$ Sample a customer from A with probabilities $\sigma_{\eta,A}(\rho(t))$. ▷ See (4)
  $s \leftarrow \arg\min_{x \in X_c} \langle \nabla\Psi(S(t)), x \rangle$ ▷ Query oracle with prices $\nabla\Psi(S(t))$
  $\tilde{\xi} \leftarrow \max\left\{0,\, 1 + \frac{1}{2\eta} \cdot \ln\left(\frac{\sigma_{\eta,A}(\rho(t))_c}{\langle \nabla\Psi(S(t)), s \rangle}\right)\right\}$
  $\xi \leftarrow \min\left\{\tilde{\xi}, \frac{\tilde{\xi}}{\|s\|_\infty}\right\}$ ▷ Garg–Könemann step size
  if $\xi \geq \rho_c(t)$ **then**
    $\xi \leftarrow \rho_c(t)$
    $A \leftarrow A \setminus \{c\}$
  end if
  Extend S to $[t, t+\xi]$ as a linear function from $S(t)$ to $S(t+\xi) := S(t) + \xi s$.
  Extend ρ to $[t, t+\xi]$ as a linear function from $\rho(t)$ to $\rho(t+\xi) := \rho(t) - \xi \mathbb{1}_c$.
  $t \leftarrow t + \xi$
end while
return $\bar{s} = \frac{1}{T} S(nT)$

Theorem 3 (Guarantee). *Let $\|\cdot\|_\beta$ be an ordered norm on $\mathbb{R}^d$. Let $\eta > 0$, $T \in \mathbb{N}$. Assuming that Algorithm 1 terminates, the final solution $\bar{s} = \frac{1}{T} S(nT)$ satisfies*

$$\|\bar{s}\|_\beta \;\leq\; \exp(2\eta) \cdot \left(1 + \frac{\ln n}{\eta T}\right) \;+\; \frac{\ln d}{\eta T} \;. \tag{6}$$

By choosing $\eta \in \mathcal{O}(\varepsilon)$ and $T \in \mathcal{O}\left(\varepsilon^{-2}\log(d+n)\right)$, we get $\|\bar{s}\|_\beta \leq 1+\varepsilon$.

Proof. We will drop the index η from the notation of $\sigma_{\eta,A}$ and $\mathrm{LSE}_{\eta,A}$ for this proof. Choose Ψ as in Theorem 2. First, we use the aforementioned relation to the line integral:

$$\|S(nT)\|_\beta \leq \Psi(S(nT)) = \Psi(S(0)) + \int_0^{nT} \langle \nabla\Psi(S(t)), s(t)\rangle \,\mathrm{d}t \; . \tag{7}$$

Note that $\Psi(S(0)) = \Psi(0^d) \leq \frac{\ln d}{\eta}$. Let $c(t) \in C$ be the customer corresponding to the solution $s(t)$ and $A(t)$ the corresponding set of active customers. We claim that for all $t \in [0, nT)$, the incurred cost can be upper bounded by the corresponding entry of the softmax function according to which we sample, in the following sense:

$$\langle \nabla\Psi(S(t)), s(t)\rangle \leq \exp(2\eta)\, \sigma_{A(t)}(\rho(t))_{c(t)} \; . \tag{8}$$

To prove (8), let us fix some value of t in Line 6 of Algorithm 1 for which the customer $c = c(t)$ is sampled according to $\sigma_{A(t)}(\rho(t))$ and we attain a solution $s = s(t)$ that gets added with value $\xi > 0$. We will prove (8) on the interval $[t, t+\xi)$. Then, (8) must hold on the entire interval $[0, nT)$ by the choice of t. In Line 8, $\tilde{\xi}$ is chosen such that

$$\tilde{\xi} = 1 + \frac{1}{2\eta} \cdot \ln\left(\frac{\sigma_{A(t)}(\rho(t))_c}{\langle \nabla\Psi(S(t)), s\rangle}\right) ,$$

which is equivalent to

$$\exp\bigl(2\eta\tilde{\xi}\bigr)\, \langle \nabla\Psi(S(t)), s\rangle = \exp(2\eta)\, \sigma_{A(t)}(\rho(t))_c \; . \tag{9}$$

Now let $\omega \in [0, \xi)$. For all such ω, we have $s(t+\omega) = s$ and $c(t+\omega) = c$ and $A(t+\omega) = A(t)$. Further, $\rho(t+\omega)_c = \rho(t)_c - \omega$. With this, we can derive

$$\sigma_{A(t+\omega)}(\rho(t+\omega))_{c(t+\omega)} \geq \exp(-\eta\omega)\sigma_{A(t)}(\rho(t))_c \; . \tag{10}$$

Using the bounded gradient increase of Ψ and the step size rule in Line 9, we can bound

$$\begin{aligned} \langle \nabla\Psi(S(t+\omega)), s\rangle &= \langle \nabla\Psi(S(t)+\omega s), s\rangle \\ &\leq \exp\bigl(\eta\omega \|s\|_\infty\bigr)\, \langle \nabla\Psi(S(t), s\rangle \\ &\leq \exp\bigl(\eta\tilde{\xi}\bigr)\, \langle \nabla\Psi(S(t)), s\rangle \; . \end{aligned} \tag{11}$$

Together, we have

$$\begin{aligned}
\langle \nabla \Psi(S(t+\omega)), s(t+\omega)\rangle &= \langle \nabla \Psi(S(t+\omega)), s\rangle \\
&\overset{(11)}{\leq} \exp(\eta\tilde{\xi})\, \langle \nabla \Psi(S(t)), s\rangle \\
&\overset{(9)}{=} \exp(-\eta\tilde{\xi}) \exp(2\eta)\; \sigma_{A(t)}(\rho(t))_c \\
&\overset{(\omega\leq\tilde{\xi})}{\leq} \exp(-\eta\omega) \exp(2\eta)\; \sigma_{A(t)}(\rho(t))_c \\
&\overset{(10)}{\leq} \exp(2\eta)\; \sigma_{A(t+\omega)}(\rho(t+\omega))_{c(t+\omega)} \;,
\end{aligned}$$

which proves our claim (8).

Let $a_0 = 0 < a_1 < \cdots < a_n = nT$ be the values of t at which the size of A changes. This happens only in Line 12 and we have $|A(a_i)| = n - i$. Now the approximation guarantee follows easily by comparing to an appropriate line integral of the softmax function:

$$\begin{aligned}
\int_0^{nT} \langle \nabla \Psi(S(t)), s(t)\rangle \;\mathrm{d}t &\overset{(8)}{\leq} \exp(2\eta) \int_0^{nT} \sigma_{A(t)}(\rho(t))_{c(t)} \;\mathrm{d}t \\
&= \exp(2\eta) \int_0^{nT} \left\langle \sigma_{A(t)}(\rho(t)), \mathbb{1}_{c(t)} \right\rangle \;\mathrm{d}t \\
&\overset{(5)}{=} \exp(2\eta) \int_0^{nT} -\left\langle \nabla \mathrm{LSE}_{A(t)}(\rho(t)), \rho'(t)\right\rangle \;\mathrm{d}t \\
&= \exp(2\eta) \sum_{i=0}^{n-1} \left(\int_{a_i}^{a_{i+1}} -\left\langle \nabla \mathrm{LSE}_{A(t)}(\rho(t)), \rho'(t)\right\rangle \;\mathrm{d}t \right) \\
&= \exp(2\eta) \sum_{i=0}^{n-1} \left(\mathrm{LSE}_{A(a_i)}(\rho(a_i)) - \mathrm{LSE}_{A(a_i)}(\rho(a_{i+1}))\right) \\
&\leq \exp(2\eta)\; \mathrm{LSE}_{A(a_0)}(\rho(a_0)) \\
&= \exp(2\eta)\; \mathrm{LSE}_C\,(T \cdot 1^n) \\
&= \exp(2\eta) \left(T + \tfrac{\ln n}{\eta}\right) \;. \qquad (12)
\end{aligned}$$

The last inequality follows from the fact that the LSE function increases with the dimension, i.e., $\mathrm{LSE}_{A(a_i)}(\rho(a_{i+1})) \geq \mathrm{LSE}_{A(a_{i+1})}(\rho(a_{i+1}))$. Finally, inserting (12) into (7) together with $\Psi(S(0)) \leq \frac{\ln d}{\eta}$ proves the claimed bound. □

Any order of processing customers yields the claimed approximation guarantee. However, an unfavorable order may frequently lead to tiny steps or even rejection ($\tilde{\xi} = 0$) of the oracle solution even if $\mathrm{OPT} \leq 1$. Our non-uniform sampling strategy avoids this issue. Let $\tilde{\xi}_i$ denote the value of $\tilde{\xi}$ in the i-th iteration of Algorithm 1. We show that, assuming $\mathrm{OPT} \leq 1$, the expected value of $\tilde{\xi}$, regardless of what happened before, is at least 1.

Lemma 1. *If* OPT ≤ 1, *then* $\mathbb{E}\left[\tilde{\xi}_i \mid \text{iterations } 1, \ldots, i-1\right] \geq 1$ *for all* $i \in \mathbb{N}$, *regardless of what happened in iterations* $1, \ldots, i-1$ *of Algorithm 1.*

Proof. Define $\gamma_c := \min_{s \in X_c} \langle \nabla \Psi(S(t)), s \rangle$. By weak duality (2), we have $\sum_{c \in A} \gamma_c \leq \text{OPT} \leq 1$. Therefore,

$$\begin{aligned}
\mathbb{E}\left[\tilde{\xi}_i \mid \text{iterations } 1, \ldots, i-1\right] &\geq \sum_{c \in A} \sigma_A(\rho(t))_c \left(1 + \frac{1}{2\eta} \cdot \ln \frac{\sigma_A(\rho(t))_c}{\gamma_c}\right) \\
&\geq \sum_{c \in A} \sigma_A(\rho(t))_c \left(1 + \frac{1}{2\eta} \cdot \left(1 - \frac{\gamma_c}{\sigma_A(\rho(t))_c}\right)\right) \\
&= 1 + \frac{1}{2\eta} - \frac{1}{2\eta} \sum_{c \in A} \gamma_c \\
&\geq 1 \; .
\end{aligned}$$

□

In this form, Lemma 1 is not enough to bound the number of oracle calls with high probability, but it extends to the bounded random variables $Y_i = \min\{\tilde{\xi}_i, \frac{1}{2\eta}\}$ (see full version [4]). While the step size ξ might be smaller than Y_i (due to Lines 9 and 11 of Algorithm 1), we prove that we make sufficient progress according to a potential function similar to [14]. Combining these arguments with a martingale bound, we derive that, assuming OPT ≤ 1, the number of oracle calls is in $\mathcal{O}\left(\eta^{-2}(n+d)\log(n+d)\right)$, when $T \in \mathcal{O}\left(\eta^{-2}\log(n+d)\right)$, with high probability (see full version [4]).

These are our main ingredients to prove:

Theorem 4. *Let* $\|\cdot\|_\beta$ *be an ordered norm on* $\mathbb{R}^d$. *Then there is an algorithm that, for any given instance and any* $\varepsilon \in (0,1]$, *computes a solution* $x \in X$ *with* $\|x\|_\beta \leq (1+\varepsilon) \cdot \text{OPT}$, *and makes at most* Ω^* *gradient evaluations of* Ψ, *at most* $n\Omega^*$ *oracle calls, and more than* Ω^* *oracle calls with probability at most* $(nd)^{-18}$, *where*

$$\Omega^* \in \mathcal{O}\left((n+d)\log(n+d) \cdot \left(\varepsilon^{-2} + \log\log d\right)\right) \; .$$

We refer to the full version [4] for a complete proof and an extension of Theorem 4 to approximate oracles.

Theorem 4 combined with Theorem 2 implies Theorem 1.

3 Construction of Our Norm Approximation

In this section, we prove Theorem 2 and weak duality (2). First, we define our notion of a norm approximation.

Definition 1 (Norm approximation). *Let* $\delta, \eta \in \mathbb{R}_{>0}$. *A differentiable convex function* $\Psi : \mathbb{R}^d_{\geq 0} \to \mathbb{R}_{\geq 0}$ *is a* (δ, η)-approximation *of an ordered norm* $\|\cdot\|_\beta$ *on* $\mathbb{R}^d$ *if the following two properties hold:*

1. Norm approximation *with absolute error* δ*:*

$$\forall x \in \mathbb{R}^d_{\geq 0}: \quad \|x\|_\beta \ \leq \ \Psi(x) \ \leq \ \|x\|_\beta + \delta$$

2. Bounded gradient increase *with parameter* η*:*

$$\forall x, y \in \mathbb{R}^d_{\geq 0}: \quad \nabla\Psi(x+y) \ \leq \ \exp\left(\eta \|y\|_\infty\right) \cdot \nabla\Psi(x)$$

Note that given a (δ, η)-approximation Ψ, one can construct a $\left(\frac{\delta}{a}, a\eta\right)$-approximation for every $a > 0$ by considering $x \mapsto \frac{1}{a}\Psi(ax)$. Later in this section, we show that there exists a $\left(\frac{\ln d}{a}, a\right)$-approximation for every ordered norm and every $a > 0$.

Prior to this, we prove weak duality for general norm approximations. To do so, we first show that every norm approximation Ψ is monotone and satisfies $\|\nabla\Psi\|^*_\beta \leq 1$:

Lemma 2. *Let* Ψ *be a* (δ, η)*-approximation of an ordered norm* $\|\cdot\|_\beta$ *on* $\mathbb{R}^d$*. Then,* $\nabla\Psi(x) > 0^d$ *and* $\|\nabla\Psi(x)\|^*_\beta \leq 1$ *for all* $x \in \mathbb{R}^d_{\geq 0}$*.*

Proof. To prove positivity of the gradients, assume that there exists an $x \in \mathbb{R}^d_{\geq 0}$ with $(\nabla\Psi(x))_i \leq 0$ for some $i \in \{1, \ldots, d\}$. Bounded gradient increase of Ψ implies that $(\nabla\Psi(x+K\mathbb{1}_i))_i \leq 0$ for every $K > 0$, where $\mathbb{1}_i$ is the i-th unit vector, and thus $\Psi(x+K\mathbb{1}_i) \leq \Psi(x)$. This is a contradiction to the norm approximation property of Ψ for K large enough, because $K\beta_1 \leq \Psi(x + K\mathbb{1}_i)$.

To prove that the dual norm of the gradients is bounded by 1, assume that there is an $x \in \mathbb{R}^d_{\geq 0}$ with $\|\nabla\Psi(x)\|^*_\beta = \omega > 1$, and let $y \in \mathbb{R}^d_{\geq 0}$ with $\|y\|_\beta = 1$ such that $\langle \nabla\Psi(x), y\rangle = \omega$. Let $K > \frac{\|x\|_\beta + \delta}{\omega - 1}$. By convexity and non-negativity of Ψ, we have

$$\Psi(x+Ky) \ \geq \ \Psi(x) + \langle \nabla\Psi(x), Ky\rangle \ \geq \ \langle \nabla\Psi(x), Ky\rangle \ = \ K\omega \ > \ \|x\|_\beta + K + \delta \ .$$

However, by the norm approximation property,

$$\Psi(x+Ky) \ \leq \ \|x+Ky\|_\beta + \delta \ \leq \ \|x\|_\beta + K + \delta \ ,$$

which is a contradiction. □

Positivity of the gradients is a central property. Indeed, many underlying combinatorial optimization problems, such as the Shortest Path problem, become much harder with negative prices. The bounded dual norm of the gradients implies *weak duality*:

Corollary 2 (Weak duality). *Let* Ψ *be a* (δ, η)*-approximation of an ordered norm* $\|\cdot\|_\beta$ *on* $\mathbb{R}^d$*, let* $X \subseteq \mathbb{R}^d_{\geq 0}$ *closed and* $x \in \mathbb{R}^d_{\geq 0}$*. Then,* $\min_{s \in X} \langle \nabla\Psi(x), s\rangle \leq \min_{s \in X} \|s\|_\beta$*.*

Proof. For every $y \in \mathbb{R}^d_{\geq 0}$ with $\|y\|^*_\beta \leq 1$ and every $X \subseteq \mathbb{R}^d_{\geq 0}$,

$$\min\left\{ \langle y, s\rangle : s \in X\right\} \;\leq\; \min\left\{ \|s\|_\beta : s \in X\right\} \;.$$

Let $x \in \mathbb{R}^d_{\geq 0}$. By Lemma 2, $\|\nabla\Psi(x)\|^*_\beta \leq 1$. Thus, the statement follows for $y = \nabla\Psi(x)$. □

We now turn towards the construction of our norm approximation. Observe that we can write every ordered norm as

$$\|x\|_\beta \;=\; \langle x^\downarrow, \beta\rangle \;=\; \max\left\{ \langle x, y\rangle : y \in Y\right\}$$

for $Y := \operatorname{conv}\left(\{y : y \text{ is a permutation of } \beta\}\right)$. Therefore, we call Y the *dual space* that corresponds to $\|\cdot\|_\beta$. Our norm approximation Ψ is a scaled version of the convex conjugate of the negative entropy on Y. The *negative entropy* $F : Y \to \mathbb{R}$ on Y is defined by $F(y) := \sum_{i=1}^d y_i \ln y_i$, defining $y_i \ln y_i = 0$ for $y_i = 0$. Its convex conjugate $F^* : \mathbb{R}^d_{\geq 0} \to \mathbb{R}$ is defined by

$$F^*(x) \;:=\; \max_{y\in Y}\left(\langle x, y\rangle - F(y)\right) \qquad \text{for } x \in \mathbb{R}^d_{\geq 0} \;.$$

For $\eta = 1$, the convex conjugate F^* is our norm approximation, and the general case follows by defining $\Psi(x) = \frac{1}{\eta}F^*(\eta x)$. To prove Theorem 2, we need to prove three properties about F^*. First, we need to show that F^* approximates $\|\cdot\|_\beta$ within an additive error of $\ln d$.

Lemma 3. *For all $x \in \mathbb{R}^d_{\geq 0}$, $\|x\|_\beta \leq F^*(x) \leq \|x\|_\beta + \ln d$.*

Proof. Let $x \in \mathbb{R}^d_{\geq 0}$. First, by non-positivity of F on Y,

$$F^*(x) \;\geq\; \max_{y\in Y}\langle x, y\rangle \;=\; \|x\|_\beta \;.$$

Second, using that $\max_{y\in Y}(-F(y)) = \ln d$ (attained by $y = \frac{1}{d}1^d$),

$$F^*(x) \;\leq\; \max_{y\in Y}\langle x, y\rangle \;+\; \max_{y\in Y}(-F(y)) \;=\; \|x\|_\beta \;+\; \ln d \;.$$

□

Second, we need to show that ∇F^* is stable in terms of bounded gradient increase, i.e., $\nabla F^*(x + y) \leq \exp(\|y\|_\infty)\nabla F^*(x)$ for all $x, y \in \mathbb{R}^d_{\geq 0}$. To this end, we use that ∇F^* is given by a generalized relative entropy projection to Y. The generalized relative entropy D_F is the *Bregman divergence* [6] induced by F,

$$D_F(y, p) \;:=\; \sum_{i=1}^d y_i \ln\left(\frac{y_i}{p_i}\right) + p_i - y_i \qquad \text{for } y, p \in \mathbb{R}^d_{\geq 0} \;. \tag{13}$$

For $p \in \mathbb{R}^d_{\geq 0}$, the *generalized relative entropy projection* of p to Y is $\operatorname{Proj}_Y(p) := \arg\min_{y\in Y} D_F(y, p)$. Due to strict convexity of F, $\operatorname{Proj}_Y(p)$ is unique. It is easy to see that ∇F^* is given by this projection; for all $x \in \mathbb{R}^d_{\geq 0}$,

$$\nabla F^*(x) \;=\; \operatorname{Proj}_Y\left(\exp(x_1), \ldots, \exp(x_d)\right) \;. \tag{14}$$

We prove the following non-trivial contraction property of the projection (see full version [4]), which together with (14) implies stability of ∇F^*.

Theorem 5 (Contraction property). *Let* $\xi \geq 1$ *and* $p, q \in \mathbb{R}^d_{>0}$ *with* $q \leq p \leq \xi q$. *Then,* $\mathrm{Proj}_Y(p) \leq \xi \cdot \mathrm{Proj}_Y(q)$.

Therefore, F^* satisfies the conditions of Definition 1.

Third, and last, we need to show that ∇F^* can be computed efficiently. According to (14), this is equivalent to the efficient computation of a generalized relative entropy projection to Y. An $\mathcal{O}(d \log d)$ algorithm to compute the generalized relative entropy projection to Y was given in [29].

Altogether, we get Theorem 2.

Conflicts of Interests. The authors have no competing interests to declare that are relevant to the content of this article.

References

1. Abernethy, J., Lee, C., Sinha, A., Tewari, A.: Online linear optimization via smoothing. In: Proceedings of the 27th Conference on Learning Theory, pp. 807–823 (2014)
2. Arora, S., Hazan, E., Kale, S.: The multiplicative weights update method: a meta-algorithm and applications. Theory Comput. **8**(1), 121–164 (2012)
3. Blankenburg, D.: Resource sharing revisited: local weak duality and optimal convergence. In: Proceedings of the 30th Annual European Symposium on Algorithms (2022)
4. Blankenburg, D., Ellerbrock, A., Kesselheim, T., Vygen, J.: An efficient algorithm for minimizing ordered norms in fractional load balancing (2025). https://arxiv.org/abs/2511.11237
5. Braun, G., et al.: Conditional Gradient Methods: From Core Principles to AI Applications. SIAM (2025)
6. Bregman, L.M.: The relaxation method of finding the common point of convex sets and its application to the solution of problems in convex programming. USSR Comput. Math. Math. Phys. **7**(3), 200–217 (1967)
7. Chakrabarty, D., Swamy, C.: Approximation algorithms for minimum norm and ordered optimization problems. In: Proceedings of the 51st Annual ACM SIGACT Symposium on Theory of Computing, pp. 126–137 (2019)
8. Chakrabarty, D., Swamy, C.: Simpler and better algorithms for minimum-norm load balancing. In: 27th Annual European Symposium on Algorithms. Schloss Dagstuhl – Leibniz-Zentrum für Informatik (2019)
9. Daboul, S., Hähnle, N., Held, S., Schorr, U.: Provably fast and near-optimum gate sizing. IEEE Trans. Comput. Aided Des. Integr. Circuits Syst. **37**(12), 3163–3176 (2018)
10. Daboul, S., Held, S., Natura, B., Rotter, D.: Global interconnect optimization. ACM Trans. Design Autom. Electr. Syst. **28**(5), 1–24 (2023)
11. Danskin, J.M.: The theory of max-min, with applications. SIAM J. Appl. Math. **14**(4), 641–664 (1966)
12. Fleischer, L.K.: Approximating fractional multicommodity flow independent of the number of commodities. SIAM J. Discret. Math. **13**(4), 505–520 (2000)

13. Freund, Y., Schapire, R.E.: A decision-theoretic generalization of on-line learning and an application to boosting. J. Comput. Syst. Sci. **55**(1), 119–139 (1997)
14. Garg, N., Könemann, J.: Faster and simpler algorithms for multicommodity flow and other fractional packing problems. SIAM J. Comput. **37**(2), 630–652 (2007)
15. Grigoriadis, M.D., Khachiyan, L.G.: Fast approximation schemes for convex programs with many blocks and coupling constraints. SIAM J. Optim. **4**(1), 86–107 (1994)
16. Held, S., Müller, D., Rotter, D., Scheifele, R., Traub, V., Vygen, J.: Global routing with timing constraints. IEEE Trans. Comput. Aided Des. Integr. Circuits Syst. **37**(2), 406–419 (2017)
17. Helmbold, D.P., K Warmuth, M.: Learning permutations with exponential weights. J. Mach. Learn. Res. **10**(7), 1705–1736 (2009)
18. Herbster, M., Warmuth, M.K.: Tracking the best linear predictor. J. Mach. Learn. Res. **1**, 281–309 (2001)
19. Hu, J., Sapatnekar, S.S.: A survey on multi-net global routing for integrated circuits. Integration **31**(1), 1–49 (2001)
20. Ibrahimpur, S., Swamy, C.: Minimum-norm load balancing is (almost) as easy as minimizing makespan. In: 48th International Colloquium on Automata, Languages, and Programming. Schloss Dagstuhl – Leibniz-Zentrum für Informatik (2021)
21. Jaggi, M.: Revisiting Frank-Wolfe: Projection-free sparse convex optimization. In: Proceedings of the 30th International Conference on Machine Learning, pp. 427–435 (2013)
22. Jain, K., Mahdian, M., Salavatipour, M.R.: Packing Steiner trees. In: Proceedings of the 14th Annual ACM-SIAM Symposium on Discrete Algorithms, pp. 266–274 (2003)
23. Kesselheim, T., Molinaro, M., Singla, S.: Online and bandit algorithms beyond ℓ_p norms. In: Proceedings of the 34th Annual ACM-SIAM Symposium on Discrete Algorithms. pp. 1566–1593 (2023)
24. Kesselheim, T., Molinaro, M., Singla, S.: Supermodular approximation of norms and applications. In: Proceedings of the 56th Annual ACM Symposium on Theory of Computing, pp. 1841–1852 (2024)
25. Klein, P., Young, N.E.: On the number of iterations for Dantzig-Wolfe optimization and packing-covering approximation algorithms. SIAM J. Comput. **44**(4), 1154–1172 (2015)
26. Koolen, W.M., Warmuth, M.K., Kivinen, J.: Hedging structured concepts. In: Proceedings of the 23rd Conference on Learning Theory, pp. 93–105 (2010)
27. Lacoste-Julien, S., Jaggi, M., Schmidt, M., Pletscher, P.: Block-coordinate Frank-Wolfe optimization for structural SVMs. In: Proceedings of the 30th International Conference on Machine Learning, pp. 53–61 (2013)
28. Lattimore, T., Szepesvári, C.: Bandit Algorithms, Cambridge University Press (2020)
29. Lim, C.H., Wright, S.J.: Efficient Bregman projections onto the permutahedron and related polytopes. In: Proceedings of the 19th International Conference on Artificial Intelligence and Statistics, pp. 1205–1213 (2016)
30. Molinaro, M.: Online and random-order load balancing simultaneously. In: Proceedings of the 28th Annual ACM-SIAM Symposium on Discrete Algorithms, pp. 1638–1650. SIAM (2017)
31. Müller, D., Radke, K., Vygen, J.: Faster min-max resource sharing in theory and practice. Math. Program. Comput. **3**(1), 1–35 (2011)
32. Plotkin, S.A., Shmoys, D.B., Tardos, É.: Fast approximation algorithms for fractional packing and covering problems. Math. Oper. Res. **20**(2), 257–301 (1995)

33. Shalev-Shwartz, S.: Online learning and online convex optimization. Foundations Trends Mach. Learn. **4**(2), 107–194 (2012)
34. Wei, Y., et al.: Techniques for scalable and effective routability evaluation. ACM Trans. Design Autom. Electr. Syst. **19**(2), 1–37 (2014)

Stochastic Load Balancing with Machine Reservations

David Alemán Espinosa[1], Naveen Garg[2], Sharat Ibrahimpur[3(✉)], Neil Olver[4], and Chaitanya Swamy[1]

[1] University of Waterloo, Waterloo, Canada
{dalemanespinosa,cswamy}@uwaterloo.ca
[2] Indian Institute of Technology Delhi, New Delhi, India
naveen@cse.iitd.ac.in
[3] Eidgenössische Technische Hochschule Zürich, Zürich, Switzerland
sibrahimpur@ethz.ch
[4] London School of Economics and Political Science, London, UK
n.olver@lse.ac.uk

Abstract. We introduce a novel variant of stochastic load balancing that enables a quantitative trade-off between the practical benefits of non-adaptive policies and their performance limitations. Our model describes a solution in two stages. In the first stage, given only job-size distributions, we reserve a set of at most k machines for each job (a k-reservation). In the second stage, after observing job-size realizations, we assign each job to one of its reserved machines (a consistent assignment). The goal is to minimize the expected makespan. If $k = 1$, we get the standard stochastic load balancing problem of finding a non-adaptive assignment with minimum expected makespan, and if k is equal to the number of machines, then we obtain an all-powerful omniscient optimum that can tailor the assignment arbitrarily to the job-size realizations.

We give a number of results that quantify this trade-off. Most saliently, we show that in the setting of identical machines, a 2-reservation suffices to achieve a constant-factor approximation to the omniscient optimum, establishing a "power-of-two-choices" result for stochastic load balancing. We also show that this no longer holds true in the more challenging setting of related machines. Nonetheless, we give a number of positive algorithmic results for this setting: a true $O(\log m/\log\log m)$-approximation; a bicriteria $O(1)$-approximation by reserving twice as many machines per job relative to an optimal k-reservation; and a 2-reservation whose cost is within a constant factor of the *adaptive* optimum.

Keywords: Stochastic Load Balancing · Approximation Algorithms · Two-stage Stochastic Optimization · Power of Two Choices

D. Alemán Espinosa and C. Swamy—Supported in part by C. Swamy's NSERC Discovery grant.
N. Garg—Supported by the Usha Hasteer Chair.
S. Ibrahimpur—Supported by the European Union's Horizon 2020 research and innovation programme (grant agreement no. ScaleOpt–757481), the Dutch Research Council NWO Vidi Grant 016.Vidi.189.087, and the DFG (grant no. 537750605).

S. S. Dey et al. (Eds.): IPCO 2026, LNCS 16588, pp. 299–314, 2026.
https://doi.org/10.1007/978-3-032-28691-8_20

1 Introduction

Uncertainty is an unavoidable facet of various decision environments, and a thriving area of optimization, *stochastic optimization*, deals with optimization under uncertainty. Stochastic load-balancing and scheduling problems constitute a prominent and well-studied class of stochastic-optimization problems (see, e.g., [8–10,13,14,18,19,23,25]). The typical setup in *stochastic load balancing* involves some m machines, and job sizes (a.k.a. processing times) that are random variables whose distributions are specified in the input. The goal is to come up with a job-to-machine assignment that minimizes the expected makespan, where *makespan* is the maximum load on a machine. Two ways of computing and specifying the assignment are considered in the literature, based on what information is available for the computation: a *non-adaptive* policy computes the entire job-to-machine assignment given only the distributional information and without knowing the job-size realizations, while an *adaptive* policy can make sequential decisions: it can inspect jobs sequentially and choose the assignment of the currently-inspected job based on the realizations of previously-inspected jobs.

An appealing aspect of a non-adaptive policy is that it is easy to implement and specify: since the uncertainty does not influence decisions at "run-time", i.e., when job sizes become known, there is little (logistical or computational) overhead incurred at run-time. Moreover, in certain settings, the certainty of knowing which machine will process which job can be desirable also for other compelling, non-computational reasons. For instance, consider a bank or investment firm that has to decide how to assign fund managers to cater to its clients. Here the machines correspond to fund managers, and jobs correspond to clients. The amount of time needed to serve a client is subject to uncertainty. Taking a client-centric perspective, the firm would like to assign each client to a specific, dedicated fund manager (i.e., come up with a non-adaptive policy) so as to offer a more personalized experience to the client and enhance client satisfaction, with the intent of cultivating customer loyalty and building a long-term relationship with the client. Non-adaptive policies are also often easier to design and analyze than adaptive policies, partly because an adaptive policy could lead to an exponential-size decision tree. On the flip side, being more constrained, even an optimal non-adaptive policy may incur a much larger cost than an optimal adaptive policy; that is, the non-adaptive optimum may be much larger than the adaptive optimum. Concretely, for stochastic load balancing: (1) the *adaptivity gap*, which measures the worst-case ratio of the non-adaptive and adaptive optimum, over all instances, is $\Omega(\log m/\log\log m)$ [10], even for m identical machines (see footnote 1); (2) one can devise non-adaptive policies that achieve an $O(1)$-approximation relative to the non-adaptive optimum [10,19], even on unrelated machines[1]

[1] The machine environments, identical, related, and unrelated, refer to the following. In general, the processing time of a job can depend on the machine to which it is assigned. With *unrelated* machines, the processing-time vectors $\{p_{ij}\}_{i\in[m]}$ and $\{p_{ij'}\}_{i\in[m]}$ of two jobs j and j' need not have any relationship. With *related* machines, each machine i has a speed $s_i > 0$, and we have $p_{ij} = p_j/s_i$ for every machine i and job j; with identical machines, we have $p_{ij} = p_j$ for every i, j.

(and even for more general objectives [14,24]); and (3) designing an adaptive policy that achieves a better guarantee relative to the adaptive optimum than a non-adaptive policy is open for unrelated machines, and only recently an $O(1)$-approximate adaptive policy was obtained for related machines (see footnote 1) [8].

1.1 Our Contributions

In this work, we propose and study a novel model for stochastic load balancing that allows one to quantitatively trade off the attractive features of non-adaptive policies and the associated performance loss resulting from their constrained nature. To motivate our model, let us reconsider the investment-firm example discussed above. For simplicity, suppose that the uncertainty in client service times is of the form that each client arrives with some probability, and if it arrives, then it uses up a fixed amount of service time. At the beginning of the day, the firm gets to know which clients will be arriving during the day, and has to assign these clients to fund managers. Suppose that there is no way of choosing dedicated managers for each client to obtain reasonably balanced loads for the managers.[2] As a means of alleviating this situation, the firm may consider incorporating a little flexibility in assigning clients to managers, in the following way. The firm now reserves a small dedicated *set of managers* for each client, and always assigns a client to one of the managers reserved for them. This flexibility can enable the firm to choose an assignment (meeting the above consistency constraint) at the beginning of the day (when the set of arriving clients becomes known) that better balances loads, while retaining a similar flavor of personalized client service as before if the set of dedicated managers for a client is "small"; the size of the reserved manager set is a lever that the firm can use to trade off the benefits of dedicated personalized service and the quality of load balancing.

Problem Description. We formalize the above type of setup via the following two-stage model, that we dub *two-stage stochastic load balancing with reservations* (SLBR). We have a machine set $M = [m]$ and a job set $\mathcal{J}$. In the first stage, given only the job-size distributions, we reserve, for each job j, a set $\sigma(j) \subseteq [m]$ of machines with $1 \leq |\sigma(j)| \leq k$, where k is an input parameter. In the second stage, we observe the job-size realizations, and have to assign each job j to a machine $\pi(j) \in \sigma(j)$. We refer to σ as a *k-reservation*, and say that $\pi : \mathcal{J} \to [m]$ is a σ-consistent *assignment*. The goal is to minimize the expected makespan. Throughout, we use "reservation" to denote the first-stage decision, and "assignment" to denote the assignment of jobs to machines in the second stage (consistent with the first-stage reservation).

Formally, given job-sizes $\{p_{ij}\}_{i\in[m],j\in\mathcal{J}}$ and an assignment π, the load on a machine i is $\text{load}(i;\pi,p) := \sum_{j:\pi(j)=i} p_{ij}$. The cost of a k-reservation σ for these job sizes is the minimum makespan achievable by a σ-consistent assignment:

[2] The $\Omega(\log m/\log\log m)$ adaptivity gap holds even with such weighted Bernoulli job-size distributions [10].

$$\mathrm{cost}(\sigma; p) := \min_{\substack{\sigma\text{-consistent} \\ \text{assignment } \pi}} \mathrm{makespan}(\pi; p),$$

where $\mathrm{makespan}(\pi; p) := \max_{i\in[m]} \mathrm{load}(i; \pi, p)$. In an instance of stochastic load-balancing with reservations, the vector $\{P_{ij}\}_{i\in[m]}$ of processing times of a job j on the different machines is a random vector, whose distribution is specified in the input. We often consider the setting where jobs are independent, which means that random variables P_{ij} and $P_{i'j'}$ are independent whenever $j \neq j'$. The goal is to choose a k-reservation σ to minimize $\mathrm{cost}(\sigma) := \mathbb{E}[\mathrm{cost}(\sigma; P)]$, where the expectation is taken over the realizations of the P_{ij}s.[3]

We use $\mathrm{OPT}_k(\mathcal{I})$ to denote the optimal value for instance $\mathcal{I}$; we drop $\mathcal{I}$ when this is clear from the context. When $k = 1$, the only σ-consistent assignment is σ itself, so the problem becomes that of computing the best non-adaptive assignment and OPT_1 is the non-adaptive optimum. At the other extreme, when $k = m$, taking $\sigma(j) = [m]$ for all j imposes no restrictions on the second-stage assignment; so $\mathrm{OPT}_m = \mathbb{E}[\min_{\text{assignment } \pi} \mathrm{makespan}(\pi; P)]$, which we refer to as the *omniscient optimum*. As k increases, we increase the flexibility allowed for the second-stage assignment, and so OPT_k decreases with k. (Note that with deterministic jobs, we have $\mathrm{OPT}_1 = \mathrm{OPT}_m$—we can take the optimal assignment as the 1-reservation—so the setting $k > 1$ is relevant only with stochastic jobs.)

Various interesting lines of inquiry arise in our model.

- First, there is the algorithmic question of designing good approximation algorithms for the problem. As noted above, SLBR with $k = 1$ corresponds to obtaining a good approximation relative to the non-adaptive optimum considered in prior work [10,19], so we consider $k \geq 2$.
- Another pertinent question that we explore is: *how much do we stand to lose by constraining ourselves to use* consistent *(second-stage) assignments?* Since OPT_m and OPT_k are respectively the best expected makespan one can achieve without any restrictions, and with the restriction that the assignment be consistent with a k-reservation, formally, this is asking the *analysis* question: what is the ratio $\mathrm{OPT}_k(\mathcal{I})/\mathrm{OPT}_m(\mathcal{I})$, for a specific instance $\mathcal{I}$, or the worst-case value of this ratio across all instances. We coin the term *price of k-reservation* to refer to the worst-case ratio. Note that this quantity decreases with k, and the price of m-reservation is 1 by definition.
  A low price of k-reservation for a small value of k has the appealing interpretation that a little amount of flexibility goes a long way. Instead of fixing the job-to-machine assignment beforehand, if we only require that each job be assigned to one of k preselected machines chosen beforehand for that job,

[3] This corresponds to computing a good first-stage decision. We would of course also like to compute a σ-consistent assignment in the second stage, for each realization of the P_{ij}s. This becomes an instance of *deterministic* makespan-minimization on unrelated machines, which admits well-known 2-approximation algorithms [20,28]. We therefore focus on the first-stage computation.

then one can alleviate the performance loss faced by non-adaptive assignments and get close in performance to a solution that is allowed to arbitrarily tailor the job-assignment in the second stage to the job-size realizations.

- Our two-stage reservation-based model and adaptive policies can both be viewed as enhancements of non-adaptive policies geared towards obtaining better-quality solutions, albeit of a very different nature: an adaptive policy has *partial information* about job-size realizations when assigning jobs, knowing the realizations of previously-assigned jobs, but the job-assignment is *unconstrained*; in a reservation-based algorithm, the assignment must be consistent*with the reservation*, which is chosen before any job sizes are realized, but one has *full information* about job-size realizations when determining the assignment.
  Although this is somewhat an apples-and-oranges comparison, we can ask about the relative effectiveness, or power, of these two models, by analyzing the ratio of OPT_k and the adaptive optimum, $\mathrm{OPT}_{\mathrm{AD}}$. This ratio is always bounded by the price of k-reservation; but are there settings where the price of reservation is large, yet $\mathrm{OPT}_k/\mathrm{OPT}_{\mathrm{AD}}$ is small, showing that k-reservations are nevertheless as powerful as adaptive policies, loosely speaking? If $\mathrm{OPT}_k/\mathrm{OPT}_{\mathrm{AD}}$ is small (say $O(1)$) for a small value of k (e.g., $k = 2$), then the upshot is that one can retain, to an extent, the benefits of a non-adaptive assignment, and at the same time reap the performance gains of an adaptive policy.

Our Results. We obtain especially crisp answers to the above questions for the setting of identical (a.k.a. parallel) machines where $P_{ij} = P_j$ for every machine i and job j. Strikingly, we show that 2-reservations are already powerful enough to yield an $O(1)$-approximation to OPT_m.

Theorem 1. *For any* SLBR *instance on identical machines with* $k \geq 2$, *we can efficiently compute a random* 2*-reservation* σ *such that* $\mathbb{E}[\mathrm{cost}(\sigma)] = O(\mathrm{OPT}_m)$.

Thus, we simultaneously obtain *an* $O(1)$*-approximation algorithm, for any* k, and that *the price of* k*-reservation is* $O(1)$, *even for* $k = 2$.[4] The latter statement can be viewed as a "power-of-two-choices result" for stochastic load balancing, showing that allowing even 2 preselected machines enables one to obtain substantially smaller expected makespan. The guarantee in Theorem 1 holds in fact for a random 2-reservation, that can be computed without knowing the underlying distributions, and so this guarantee holds even with *arbitrarily correlated jobs*. Note that the latter setting can model an oblivious adversary, who can choose arbitrary job-sizes but does not know the random bits used to compute the 2-reservation. Theorem 1 is proved in Sect. 4.

[4] The price of 1-reservation for m identical machines is $\Theta(\log m/\log\log m)$. The upper bound follows from reserving a single machine independently and uniformly for each job. The lower bound is realized by an instance with m^2 i.i.d. Bernoulli$(1/m)$ jobs for which OPT_m and $\mathrm{OPT}_{\mathrm{AD}}$ are both $\Theta(1)$ while OPT_1 is $\Theta(\log m/\log\log m)$.

We next consider the setting of related machines wherein each machine i has a speed $s_i > 0$, and each job j has a random size P_j, which defines the processing-time vector $\{P_{ij} = P_j/s_i\}_{i\in[m]}$ for each job j. In stark contrast with identical machines, the price of k-reservation is no longer bounded by a constant, for any constant value of k.

Theorem 2. *For every positive integer k and every $m = \exp(\Omega(k))$, there is an instance of* SLBR *on m related machines for which $\mathrm{OPT}_k/\mathrm{OPT}_m = \Omega(\sqrt[2k]{m}/k)$. In particular, for any $k = o(\log m/\log\log m)$, the price of k-reservation is unbounded for sufficiently large m.*

Complementing this, we show that the price of k-reservation is a constant already when $k \approx \log m$.

Theorem 3. *For any instance of* SLBR *on m related machines, the price of $\Omega(\log m)$-reservation is $O(1)$.*

In terms of designing approximation algorithms, related machines pose substantial technical challenges. In Sect. 2, we illustrate with a simple example how two simple strategies fail to yield constant-factor approximations for SLBR on related machines. We overcome some of these challenges by developing suitable machinery for obtaining three different flavors of approximation guarantees. First, we use a linear programming based approach combined with standard concentration inequalities to obtain a true approximation guarantee.

Theorem 4. *There is an $O(\log m/\log\log m)$-approximation algorithm for instances of* SLBR *with m related machines.*

In the second approach, we allow ourselves twice as many reservations per job and obtain a bicriteria approximation.

Theorem 5. *For any* SLBR *instance on m related machines with $k \geq 1$, we can efficiently compute a random $2k$-reservation σ with $\mathbb{E}_\sigma[\mathrm{cost}(\sigma)] = O(\mathrm{OPT}_k)$.*

Finally, we showcase the effectiveness of 2-reservations vs. adaptive policies.

Theorem 6. *For any* SLBR *instance on m related machines with $k \geq 2$, we can efficiently compute a random 2-reservation whose expected cost is within constant factors of the adaptive optimum.*

Thus, while k-reservations cannot get us close to the omniscient optimum for any constant k, even 2-reservations are as powerful as adaptive policies (modulo $O(1)$ factors). As a consequence, we obtain that for weighted Bernoulli job-size distributions—an oft-considered special case (see, e.g., [1,7,11,19]) that sometimes encapsulates the crux of the general setting [10,19]—the price of 2-reservation becomes $O(1)$, since one can show that with weighted Bernoulli distributions, the adaptive optimum is $O(\mathrm{OPT}_m)$.

Due to limited space, proofs of our results on related machines (Theorems 2, 3, 4, 5 and 6) have been deferred to the full version of this paper.

1.2 Related Work

As noted earlier, taking $k = 1$ in our model, we obtain the classical stochastic load-balancing problem of finding a non-adaptive assignment with minimum expected makespan, which is a problem with a rich history of study. In their seminal work, Kleinberg, Rabani and Tardos [19] developed an $O(1)$-approximation algorithm for this problem on identical machines. They were the first to study stochastic load balancing with arbitrary job-size distributions, whereas prior work had only considered specialized distributions (see, e.g., [4,30,31]). Recently, Gupta, Kumar, Nagarajan and Shen [10] obtained an $O(1)$-approximation for the general setting of unrelated machines, and subsequently, such results were also obtained for more-general norm-based objectives of the machine-load vector [14,16,24]. Another stream of work has focused on obtaining improved approximation guarantees for specialized distributions, such as exponential and Poisson distributions [6,9,15].

In contrast to the above line of work, which focuses on obtaining guarantees relative to the non-adaptive optimum, the study of (adaptive) policies that obtain guarantees relative to the *adaptive optimum* has received very little attention. Gupta et al. [10] showed that non-adaptive policies are quite limited in this respect by showing that the adaptivity gap is $\Omega(\log m/\log\log m)$, even for identical machines and weighted Bernoulli job-size distributions. Very recently, Eberle, Gupta, Megow, Moseley and Zhou [8] devised an $O(1)$-approximation *adaptive policy* for the setting of related machines, showing for the first time how adaptive policies can do better than non-adaptive policies. They also showed that the $\Omega(\log m/\log\log m)$ lower bound on the adaptivity gap is tight by designing a non-adaptive policy with a matching approximation guarantee, even for the setting of unrelated machines. We use some of their techniques in obtaining our approximation results for related machines. Obtaining an adaptive policy for unrelated machines that does better remains a challenging open question.

The high-level idea of our model—that providing some limited choice in deciding the job-assignment at a second stage may allow for dramatically better solutions—invites comparison to the rich stream of literature on the "power of two choices" [22]. The prototypical result in this area is the balls-and-bins result of [2], which shows that with m identical machines and m deterministic unit-size online jobs, if we choose d random machines for each job, and assign the job to the least-loaded machine among these d machines, then the expected makespan decreases from $\Theta(\log m/\log\log m)$ when $d = 1$, to $O(\log\log m/\log d)$ when $d \geq 2$. They also proved that the expected *offline* optimum is $O(1)$, i.e., if we get to see all jobs and can then assign jobs but in a way that is consistent with the machine-set chosen for each job, then the expected makespan becomes $O(1)$. The latter result has the same flavor as our result for identical machines (Theorem 1): indeed, using our terminology, they show that the cost of a random 2-reservation is $O(1)$, albeit in the extremely special case of *identical* jobs.

SLBR is an example of a two-stage stochastic optimization problem with recourse. A distinct two-stage scheduling variant is considered in [5]. Two-stage stochastic optimization problems have been extensively studied in the

Operations Research literature, from a modeling and computational viewpoint, and in the theoretical CS (TCS) literature from an approximation-algorithms viewpoint; see, e.g. the text [3], and the survey [29], which discusses work in TCS in this area. While work on stochastic load balancing has focused exclusively on independent jobs, work in two-stage stochastic optimization has also considered other, more general distribution models that incorporate arbitrary correlation effects. These include the scenario model, where the input specifies all realizations of the uncertain data, and the most-general black-box model, where one only assumes that one has sampling access to the underlying distribution; approximation algorithms have been devised even in the black-box model [12, 27]. We note that in order to obtain these positive results in the black-box model, an important property that the two-stage problem needs to satisfy is that every first-stage action has a corresponding (possibly more-expensive) second-stage action; SLBR does not satisfy this property.

See [26] for another stochastic scheduling model that considers policies with a limited amount of adaptivity.

2 Technical Overview

In this section, we give an overview of our techniques, discussing the main ideas underlying our algorithms and their analysis. A full proof of our result on identical machines can be found in Sect. 4. Due to limited space, detailed proofs of our results on related machines have been deferred to the full version of this paper.

2.1 2-Reservations vs. Omniscient Optimum for Identical Machines

Recall that for identical machines, we obtain the rather strong positive result that 2-reservations are sufficient to get within a constant factor of the omniscient optimum OPT_m, and this holds even with arbitrarily correlated job sizes. Our algorithm here is very simple: the reservation for each job is simply 2 machines chosen independently and uniformly at random. This algorithm and its analysis also lie at the heart of our results for related machines.

One of the main ingredients for analyzing this algorithm is a *density-based approximation* of the cost of a given reservation σ (see parts (i) and (ii) of Lemma 1). We can view a 2-reservation σ as specifying a graph, where we have a node for each machine, and, for each job j, we have an edge (possibly a self-loop) joining the machines in $\sigma(j)$. A σ-consistent assignment corresponds to an orientation of this graph, whose makespan, for some given job sizes $\{p_j\}_{j \in \mathcal{J}}$, is the maximum weighted in-degree of a machine node, where the weight of the edge (corresponding to) $\sigma(j)$ is p_j. We consider a *fractional* relaxation of the problem of finding an orientation to minimize the maximum weighted in-degree. This can be easily cast as a flow problem, and its dual is a cut problem that can be interpreted as finding a maximum-density node-set in this graph, where the density of a set $R \subseteq [m]$ is (total weight of edges contained in R) / $|R|$. Also, one can argue that a fractional orientation can be rounded to a σ-consistent assignment of makespan at most twice the maximum density. Thus, our goal is to find a 2-reservation σ with small maximum density.

We prove that a random 2-reservation (as described above) has small expected maximum density, for any job-sizes, where the expectation is over the random choice of the 2-reservation. It is easy to bound $\mathbb{E}[\text{density of } R]$ for any $R \subseteq [m]$; we use Chernoff bounds to show that the density of R is "close" to its expectation, followed by a union bound over all sets $R \subseteq [m]$ to obtain a bound on $\mathbb{E}[\max_{R\subseteq[m]}(\text{density of } R)]$. Although this entails taking a union bound over exponentially many node-sets, two key insights make the calculations go through: (1) if $|R|$ is small, then the expected density of R is small, and so a tail event corresponds to a large relative deviation from the expectation; (2) if $|R|$ is large, then the expected density of R is large. However, the tail event corresponds to a large absolute deviation from the expectation. In either case, we obtain a rapid decay in probability while applying Chernoff bounds, which turns out to be strong enough to weather the application of the union bound.

2.2 Price of Reservation Bounds for Related Machines

Our result in Theorem 2 is based on the intuition that any k-reservation can at best average out the load of k speed classes. So, the hard instance (shown in Table 1) is chosen such that the set of machines M can be partitioned into $k+1$ speed classes $C_1, \ldots, C_{k+1}$ such that: (a) the speed S_ℓ of class ℓ decreases by some factor Δ over the classes, where $\Delta \approx \sqrt[2k]{|M|}$; and (b) the total processing power (i.e., $|C_\ell| \cdot S_\ell$) within each class increases by a factor Δ over the classes. The job-set $\mathcal{J}$ consists of $n \gg |M|$ i.i.d. jobs. The (common) job-size distribution is such that the random variable N_ℓ denoting the number of jobs that realize a size equal to the speed of class ℓ has expected value $|C_\ell|$. The rest of the probability mass is located at size 0. We explicitly state the instance in the table below.

Table 1. Instance $\mathcal{I}$ showing unbounded gap between OPT_k and OPT_m.

(class index) ℓ	1	2	$\ldots$	$\ell-1$	ℓ	$\ell+1$	$\ldots$	$k+1$
(# class- ℓ machines) $\lvert C_\ell\rvert$	1	Δ^2	$\ldots$	$\Delta^{2\ell-4}$	$\Delta^{2\ell-2}$	$\Delta^{2\ell}$	$\ldots$	Δ^{2k}
(speed of class ℓ) S_ℓ	Δ^k	Δ^{k-1}	$\ldots$	$\Delta^{k-\ell+2}$	$\Delta^{k-\ell+1}$	$\Delta^{k-\ell}$	$\ldots$	1
(processing power in class ℓ) $\lvert C_\ell\rvert S_\ell$	Δ^k	Δ^{k+1}	$\ldots$	$\Delta^{k+\ell-2}$	$\Delta^{k+\ell-1}$	$\Delta^{k+\ell}$	$\ldots$	Δ^{2k}
$\mathbb{E}[N_\ell] = \sum_{j\in\mathcal{J}} \mathbb{P}[P_j = S_\ell] = n\mathbb{P}[P_1 = S_\ell]$	1	Δ^2	$\ldots$	$\Delta^{2\ell-4}$	$\Delta^{2\ell-2}$	$\Delta^{2\ell}$	$\ldots$	Δ^{2k}

By concentration, we can show that with high probability $N_\ell = O(|C_\ell|)$ holds for all $\ell \in [k+1]$. Thus, the omniscient optimum is $O(1)$. However, in any k-reservation, some $n/(k+1)$ jobs (call this set J') are devoid of a class-ℓ machine for some $\ell \in [k+1]$. This implies $\text{OPT}_k = \Omega(\Delta/k)$ because neither the faster speed classes nor the slower speed classes are equipped to handle the load that arises from jobs in J' in a typical realization. Faster classes lack in terms of aggregate processing power, whereas slower classes are too slow to handle a large job-size realization without the load blowing up to $\Omega(\Delta)$.

2.3 (Bicriteria) Approximation Algorithms for Related Machines

The situation for related machines is much more technically challenging. In particular, trivial extensions of our ideas from the identical-machines setting can lead to very poor approximation guarantees. For instance, suppose we have $m+1$ related machines, where the first machine has speed $\sqrt{m}$ and the remaining m machines have unit speed, and $n \gg m$ i.i.d. jobs where the size of each job is distributed as $m/n+\sqrt{m}\cdot\text{Bernoulli}(1/n)$. Observe that an $\Omega(\sqrt{m})$ loss in approximation is unavoidable in either of the following cases: (i) the 2-reservation for each job intersects at most one speed class; or (ii) the two machines are sampled proportionally to their speeds, possibly from different speed classes.

Our approximation algorithms for related machines are based on linear programming techniques. We begin with two preprocessing steps. First, we "smooth" the instance into a more convenient form in which, among other things, the machines are partitioned into classes $C_1, \ldots, C_L$ where the speed S_ℓ of class-ℓ machines decreases geometrically with ℓ while the processing power $|C_\ell| \cdot S_\ell$ increases geometrically with ℓ. This is inspired by the smoothing operation in the work of Im, Kell, Panigrahi and Shadloo [17]. It can be shown that this smoothing changes OPT_k by at most a constant factor. Second, via binary search, we suitably scale the instance so that $\text{OPT}_k = \Omega(1)$. The point of this scaling is to formulate a linear program (LP) for the scaled instance such that infeasibility of the LP certifies that OPT_k is significantly larger than 1. To simplify the exposition, we assume throughout that $\text{OPT}_k = 1$. We also restrict attention to the case $k = 2$, since the main ideas already appear in full generality in this special setting.

A natural linear relaxation allows jobs to fractionally reserve up to k machines. We relax this modeling choice even further by allowing a job to reserve up to k speed classes. Specifically, for each job $j \in \mathcal{J}$ and each pair $1 \leq \ell_1 < \ell_2 \leq L$, we introduce a variable $x_{j,\ell_1,\ell_2} \in [0,1]$, where $x_{j,\ell_1,\ell_2} = 1$ models that job j reserves the speed classes C_{ℓ_1} and C_{ℓ_2}. For each $j \in \mathcal{J}$, we impose the (assignment) constraint $\sum_{1\leq\ell_1<\ell_2\leq L} x_{j,\ell_1,\ell_2} = 1$, which models a relaxed first-stage 2-class reservation.

We next add volume constraints to control the load induced on the speed classes in the second stage. Since we assumed $\text{OPT}_k = 1$, it is useful, for each pair $1 \leq \ell_1 < \ell_2 \leq L$, to partition the realizations of P_j into three regimes: small, medium, and large, relative to the thresholds S_{ℓ_2} and S_{ℓ_1}. Thus, define $P^{\mathsf{sml}}_{j,\ell_1,\ell_2} := P_j \cdot \mathbb{1}(P_j < S_{\ell_2})$, $P^{\mathsf{med}}_{j,\ell_1,\ell_2} := P_j \cdot \mathbb{1}(S_{\ell_2} \leq P_j < S_{\ell_1})$, and $P^{\mathsf{lrg}}_{j,\ell_1,\ell_2} := P_j \cdot \mathbb{1}(P_j \geq S_{\ell_1})$. Here, $P^{\mathsf{sml}}_{j,\ell_1,\ell_2}$ represents realizations that can be processed even by a slow machine in C_{ℓ_2} without exceeding load 1. The random variable $P^{\mathsf{med}}_{j,\ell_1,\ell_2}$ represents realizations that can be handled by a machine in C_{ℓ_1}, but not by one in C_{ℓ_2}, without exceeding load 1. Finally, $P^{\mathsf{lrg}}_{j,\ell_1,\ell_2}$ captures realizations that cause load at least 1 regardless of whether the job is assigned to a machine in C_{ℓ_1} or C_{ℓ_2}.

Since we assumed $\mathrm{OPT}_k = 1$, for each $\ell \in [L]$ we include the volume constraint:

$$\sum_{j \in \mathcal{J}, 1 \le \ell' < \ell} \mathbb{E}[P^{\mathsf{sml}}_{j,\ell',\ell}] \cdot x_{j,\ell',\ell} + \sum_{j \in \mathcal{J}, \ell < \ell' \le L} \mathbb{E}[P^{\mathsf{med}}_{j,\ell,\ell'}] \cdot x_{j,\ell,\ell'} \le |C_\ell| \cdot S_\ell. \quad (1)$$

For $\ell' < \ell$, the small part $P^{\mathsf{sml}}_{j,\ell',\ell}$ could also be processed on the faster machines in $C_{\ell'}$ without exceeding load 1. However, at the cost of an $O(1)$ loss in the makespan objective, we may charge these small realizations to the slower speed class C_ℓ. This is because, after smoothing, the total processing power in class ℓ dominates that of all faster classes combined.

Finally, the large realizations $\{P^{\mathsf{lrg}}_{j,\ell_1,\ell_2}\}$ behave qualitatively differently from the small and medium parts: if the event $\{P^{\mathsf{lrg}}_{j,\ell_1,\ell_2} > 0\}$ happens, then assigning j to any machine in $C_{\ell_1} \cup C_{\ell_2}$ already incurs load at least 1. Thus, these correspond to exceptional overflow events rather than class-specific volume, and it suffices to control their total expected contribution via a single aggregate constraint. As in [10, 14], we include the following exceptional jobs constraint:

$$\sum_{j \in \mathcal{J}} \sum_{1 \le \ell_1 < \ell_2 \le L} \left(\mathbb{E}[P^{\mathsf{lrg}}_{j,\ell_1,\ell_2}] / S_{\ell_1} \right) \cdot x_{j,\ell_1,\ell_2} \le 1. \quad (2)$$

Now let $x^* \in [0,1]^{\mathcal{J} \times \binom{L}{2}}$ be a fractional solution to this feasibility LP. Observe that the coefficients in (1) are at most S_ℓ, while those in (2) are at most 1. Since $|C_\ell|$ grows geometrically with ℓ, this (assignment) LP satisfies a bounded-column-sums property. An iterative rounding result of Linhares, Olver, Swamy and Zenklusen [21] therefore yields an integral 2-class reservation ψ that approximately preserves the volume guarantees satisfied by x^*.

At this point, however, we still have reservations to two speed classes rather than to two specific machines. This is where we must give up something. A natural approach is to reserve, for each job j, one uniformly random machine from each of the two speed classes in $\psi(j)$. Due to the usual balls-and-bins behavior, this can incur an $O(\log m / \log \log m)$ loss in approximation. If we instead allow the reservation size to double, then we can define a 4-reservation by choosing *two* (uniformly) random machines from each of the two speed classes in $\psi(j)$. Each speed class can then be analyzed similarly to the identical-machines setting, yielding only a constant-factor loss relative to OPT_k.

2.4 2-Reservations Vs. Adaptive Optimum for Related Machines

Our key technical insight for proving Theorem 6 is that the adaptive optimum is lower bounded, up to constant factors, by the "cost" of an optimal 1-class reservation. Here, a 1-class reservation means that in the first stage, for each job, we must commit to (or reserve) a single speed class that will be responsible for processing whatever size this job realizes in the second stage. (The second stage assignment can choose any one machine from the reserved speed class for each job.) Thus, by sampling two machines independently and uniformly from the class that is reserved for each job, we can average-out the load within each speed class. The associated 2-reservation has an expected cost that is roughly the adaptive optimum.

3 Preliminaries

In the interest of space, we only state the tools that are needed to understand our result for identical machines. Omitted proofs are deferred to the full version. The following lemma gives bounds on the optimal makespan of an assignment that is consistent with a reservation by relating it to a certain density function.

Lemma 1. *Let $\mathcal{J}$ be a set of jobs, M be a set of related machines with positive speeds $\{s_i\}_{i \in M}$, and $\sigma : \mathcal{J} \to 2^M \setminus \{\emptyset\}$ be a reservation. For any arbitrary job sizes $p \in \mathbb{R}^{\mathcal{J}}_{\geq 0}$, define:*

$$\text{density}(p; \sigma, s) := \max_{R \subseteq M : R \neq \emptyset} \frac{\sum_{j \in \mathcal{J} : \sigma(j) \subseteq R} p_j}{\sum_{i \in R} s_i}. \tag{3}$$

The following are true:

(i) For any σ-consistent assignment $\pi : \mathcal{J} \to M$, we have $\text{makespan}(\pi; p) \geq \text{density}(p; \sigma, s)$.
(ii) There is a σ-consistent assignment $\pi : \mathcal{J} \to M$ satisfying

$$\text{makespan}(\pi; p) \leq \text{density}(p; \sigma, s) \cdot \max_{j \in \mathcal{J}} |\sigma(j)|.$$

We remark that $\text{density}(p; \sigma, s)$ is the minimum makespan that is achievable in part (i) if jobs could be fractionally assigned while still being σ-consistent. The form of the density function comes from casting this fractional makespan minimization problem as a flow problem and using the max-flow min-cut theorem.

The following result, derived by using an integral over the Chernoff upper-tail bound, will be useful to us.

Lemma 2. *Let $X_1, \ldots, X_n$ be independent $[0, 1]$-bounded random variables and $X := \sum_{i=1}^{n} X_i$ denote their sum. For any $\delta \geq \mathrm{e}^2 - 1$ and $B \geq (1 + \delta)\mathbb{E}[X]$, we have $\mathbb{E}[\max(0, X - B)] \leq (\mathrm{e}/(1 + \delta))^B$.*

4 Identical Machines

We prove Theorem 1 in this section. We show that the stated guarantee holds in fact for a random 2-reservation σ, obtained as follows: for each job $j \in \mathcal{J}$, we sample two machines $i_1(j)$ and $i_2(j)$ independently and uniformly from $[m]$, and set $\sigma(j) = \{i_1(j), i_2(j)\}$; note $|\sigma(j)| = 1$ if the same machine is sampled twice. Note that the computation of σ does not depend on the underlying job-size distributions. Consequently, the guarantee in Theorem 1 holds even with arbitrarily correlated jobs. The following claim is straightforward.

Claim. For $j \in \mathcal{J}$ and nonempty $R \subseteq [m]$ we have $\mathbb{P}[\sigma(j) \subseteq R] = |R|^2/m^2$.

We show $\mathbb{E}_\sigma[\text{cost}(\sigma)] = O(\text{OPT}_m)$ by showing the same inequality for any choice of $p \in \mathbb{R}^{\mathcal{J}}_{\geq 0}$: $\mathbb{E}_\sigma[\text{cost}(\sigma; p)] = O(\max(\max_j p_j, \sum_j p_j/m))$ holds. That is, choosing two machines independently and uniformly at random for each job is sufficient to ensure a constant-factor approximation guarantee against the optimal makespan that is achievable even for an adversarial choice of p as long as the adversary is oblivious to the choice of σ. We again remark that for the special case of deterministic unit-size jobs, a similar result was obtained by Azar, Broder, Karlin and Upfal [2] (see Lemma 6.1).

Theorem 7. *For any* $p \in \mathbb{R}^{\mathcal{J}}_{\geq 0}$, $\mathbb{E}_\sigma[\text{cost}(\sigma; p)] \leq 15 \cdot \max(\max_j p_j, \sum_j p_j/m)$.

Proof. For simplicity, we scale down the p_js by $\max(\max_j p_j, \sum_j p_j/m)$ so that $p_j \in [0, 1]$ for all j and $\sum_j p_j \leq m$ holds. By Lemma 1, we have:

$$\text{cost}(\sigma; p) = \min_{\substack{\sigma\text{-consistent} \\ \text{assignment } \pi}} \text{makespan}(\pi; p) \leq 2 \cdot \max_{R \subseteq [m]: R \neq \emptyset} \frac{\sum_{j:\sigma(j) \subseteq R} p_j}{|R|}. \tag{4}$$

We take an expectation over inequality (4) to obtain:

$$\mathbb{E}_\sigma[\text{cost}(\sigma; p)] \leq 2 \cdot \mathbb{E}_\sigma\left[\max_{R \subseteq [m]: R \neq \emptyset} \frac{\sum_j p_j \cdot \mathbb{1}(\sigma(j) \subseteq R)}{|R|}\right] \tag{5}$$

$$\leq 2\left\{e^2 + \sum_{R \subseteq [m]: R \neq \emptyset} \frac{1}{|R|} \mathbb{E}_\sigma\left[\max\left(0, \sum_j p_j \cdot \mathbb{1}(\sigma(j) \subseteq R) - e^2|R|\right)\right]\right\}, \tag{6}$$

where we use a union bound in the last inequality. Now, observe that for any nonempty $R \subseteq [m]$ the random variable $X := \sum_j p_j \cdot \mathbb{1}(\sigma(j) \subseteq R)$ is a sum of independent $[0, 1]$-bounded random variables. Note that $\mathbb{E}[X] = \left(\sum_j p_j\right) \cdot |R|^2/m^2 \leq |R|^2/m$ by Claim 4. Using Lemma 2 with $\delta = e^2 m/|R| - 1$ and $B = e^2|R|$, we get:

$$\mathbb{E}_\sigma[\max(0, X - e^2|R|)] \leq (em/|R|)^{-e^2|R|}.$$

Plugging the above bound into (6) and using the well-known bound $\binom{m}{\ell} \leq (em/\ell)^\ell$ we get:

$$\mathbb{E}_\sigma[\text{cost}(\sigma; p)] \leq 2e^2 + 2 \cdot \sum_{\ell=1}^{m} \binom{m}{\ell} \cdot \frac{1}{\ell} \cdot \left(\frac{em}{\ell}\right)^{-e^2\ell} \leq 2e^2 + 2 \cdot \sum_{\ell=1}^{m} \frac{e^{-6\ell}}{\ell} \leq 15.$$

□

The proof of our main result (Theorem 1) follows as a corollary of Theorem 7.

Proof. Since the omniscient optimum is at least $\max(\max_j P_j, \sum_j P_j/m)$, we have $\mathrm{OPT}_m \geq \mathbb{E}_P[\max(\max_j P_j, \sum_j P_j/m)]$. Thus,

$$\begin{aligned}\mathbb{E}_\sigma[\mathrm{cost}(\sigma)] = \mathbb{E}_\sigma[\mathbb{E}_P[\mathrm{cost}(\sigma;P)]] = \mathbb{E}_P[\mathbb{E}_\sigma[\mathrm{cost}(\sigma;P)]] \\ \leq \mathbb{E}_P[15 \cdot \max(\max_j P_j, \sum_j P_j/m)] \leq 15 \cdot \mathrm{OPT}_m.\end{aligned}$$

where we use the Fubini-Tonelli theorem to exchange the expectations in the second equality. □

5 Conclusions

Our work opens up the study of reservation-based algorithms for stochastic load balancing as an exciting research direction, with many interesting avenues for exploration. We mention just a few questions here, and hope that our work will stimulate further work on this topic. Our results show that for identical and related machines, even 2-reservations can offer a surprising amount of power. Obtaining a true $O(1)$-approximation for SLBR on related machines is still open. We leave the setting of unrelated machines as a concrete open question for future work. An intermediate setting is that of restricted machines, wherein each job can only be processed on some given subset of machines, but its processing time is the same on all machines in this set. It would be interesting to see if one can obtain similar guarantees here as for identical machines. In particular, can a 2-reservation yield makespan that is within an $O(1)$ factor of OPT_m, or (if not) the adaptive optimum?

Another question is whether our result for identical machines can be derandomized: can we efficiently compute a (deterministic) 2-reservation with cost $O(\mathrm{OPT}_m)$?

It would also be interesting to incorporate the concept of k-reservations into other stochastic-load-balancing models, where varying k would again allow one to interpolate between different solution concepts. For instance, while SLBR is a two-stage model with offline stochastic jobs, one can also consider the setting where jobs are assigned adaptively with partial job-realization information at hand, and one needs to be consistent with a k-reservation (chosen beforehand). This would interpolate between non-adaptive policies ($k = 1$) and adaptive policies ($k = m$).

Acknowledgments. Part of this work was conducted while Sharat was a postdoctoral researcher at the London School of Economics and the University of Bonn. He is grateful to these institutions for their support. David thanks Sepehr Assadi, and Sharat thanks Vipin Ravindran Vijayalakshmi for helpful discussions.

Disclosure of Interests. The authors have no competing interests to declare that are relevant to the content of this article.

References

1. Antoniadis, A., Hoeksma, R., Schewior, K., Uetz, M.: Stochastic scheduling with Bernoulli-type jobs through policy stratification. In: Proceedings of the 66th Symposium on Foundations of Computer Science, pp. 2446–2469 (2025). https://doi.org/10.1109/FOCS63196.2025.00128
2. Azar, Y., Broder, A.Z., Karlin, A.R., Upfal, E.: Balanced allocations. SIAM J. Comput. **29**(1), 180–200 (1999). https://doi.org/10.1137/S0097539795288490
3. Birge, J.R., Louveaux, F.: Introduction to Stochastic Programming. Springer Science & Business Media (2011)
4. Bruno, J., Downey, P., Frederickson, G.N.: Sequencing tasks with exponential service times to minimize the expected flow time or Makespan. J. ACM **28**(1), 100–113 (1981). https://doi.org/10.1145/322234.322242
5. Chen, L., Megow, N., Rischke, R., Stougie, L.: Stochastic and robust scheduling in the cloud. In: Proceedings of the 18th International Conference on Approximation Algorithms for Combinatorial Optimization Problems. vol. 40, pp. 175–186 (2015). https://doi.org/10.4230/LIPIcs.APPROX-RANDOM.2015.175
6. De, A., Khanna, S., Li, H., Nikpey, H.: An efficient PTAS for stochastic load balancing with poisson jobs. In: Proceedings of the 47th International Colloquium on Automata, Languages, and Programming, pp. 37:1–37:18 (2020). https://doi.org/10.4230/LIPIcs.ICALP.2020.37
7. Eberle, F., Fischer, F., Matuschke, J., Megow, N.: On index policies for stochastic minsum scheduling. Oper. Res. Lett. **47**(3), 213–218 (2019). https://doi.org/10.1016/j.orl.2019.03.007
8. Eberle, F., Gupta, A., Megow, N., Moseley, B., Zhou, R.: Configuration balancing for stochastic requests. Math. Program. **210**(1), 243–279 (2025). https://doi.org/10.1007/S10107-024-02132-W
9. Goel, A., Indyk, P.: Stochastic load balancing and related problems. In: Proceedings of the 40th Foundations of Computer Science, pp. 579–586 (1999). https://doi.org/10.1109/SFFCS.1999.814632
10. Gupta, A., Kumar, A., Nagarajan, V., Shen, X.: Stochastic load balancing on unrelated machines. Math. Oper. Res. **46**(1), 115–133 (2021). https://doi.org/10.1287/moor.2019.1049
11. Gupta, A., Moseley, B., Zhou, R.: Minimizing completion times for stochastic jobs via batched free times. In: Proceedings of the 34th Symposium on Discrete Algorithms, pp. 1905–1930 (2023). https://doi.org/10.1137/1.9781611977554.ch73
12. Gupta, A., Pál, M., Ravi, R., Sinha, A.: Boosted sampling: approximation algorithms for stochastic optimization. In: Proceedings of the 36th Symposium on Theory of Computing, pp. 417–426 (2004). https://doi.org/10.1145/1007352.1007419
13. Gupta, V., Moseley, B., Uetz, M., Xie, Q.: Greed works-online algorithms for unrelated machine stochastic scheduling. Math. Oper. Res. **45**(2), 497–516 (2020). https://doi.org/10.1287/moor.2019.0999
14. Ibrahimpur, S., Swamy, C.: Approximation algorithms for stochastic minimum-norm combinatorial optimization. In: Proceedings of the 61st Foundations of Computer Science, pp. 966–977 (2020). https://doi.org/10.1109/FOCS46700.2020.00094
15. Ibrahimpur, S., Swamy, C.: Minimum-norm load balancing is (almost) as easy as minimizing Makespan. In: Proceedings of the 48th International Colloquium on Automata, Languages, and Programming, pp. 81:1–81:20 (2021). https://doi.org/10.4230/LIPIcs.ICALP.2021.81

16. Ibrahimpur, S., Swamy, C.: A simple approximation algorithm for vector scheduling and applications to stochastic min-norm load balancing. In: Proceedings of the 5th Symposium on Simplicity in Algorithms, pp. 247–256 (2022). https://doi.org/10.1137/1.9781611977066.18
17. Im, S., Kell, N., Panigrahi, D., Shadloo, M.: Online load balancing on related machines. In: Proceedings of the 50th Symposium on the Theory of Computing, pp. 30–43 (2018). https://doi.org/10.1145/3188745.3188966
18. Im, S., Moseley, B., Pruhs, K.: Stochastic scheduling of heavy-tailed jobs. In: Proceedings of the 32nd Symposium on Theoretical Aspects of Computer Science, vol. 30, pp. 474–486 (2015). https://doi.org/10.4230/LIPIcs.STACS.2015.474
19. Kleinberg, J.M., Rabani, Y., Tardos, É.: Allocating bandwidth for Bursty connections. SIAM J. Comput. **30**(1), 191–217 (2000). https://doi.org/10.1137/S0097539797329142
20. Lenstra, J.K., Shmoys, D.B., Tardos, É.: Approximation algorithms for scheduling unrelated parallel machines. Math. Program. **46**, 259–271 (1990). https://doi.org/10.1007/BF01585745
21. Linhares, A., Olver, N., Swamy, C., Zenklusen, R.: Approximate multi-matroid intersection via iterative refinement. Math. Program., 397–418 (2020). https://doi.org/10.1007/s10107-020-01524-y
22. Mitzenmacher, M., Richa, A.W., Sitaraman, R.: The power of two random choices: a survey of techniques and results. In: Handbook of Randomized Computing, Volume I-II, Chapter 9. Kluwer Academic Publishers (2001). https://www.eecs.harvard.edu/%7Emichaelm/postscripts/handbook2001.pdf
23. Möhring, R.H., Schulz, A.S., Uetz, M.: Approximation in stochastic scheduling: the power of LP-based priority policies. J. ACM **46**(6), 924–942 (1999). https://doi.org/10.1145/331524.331530
24. Molinaro, M.: Stochastic ℓ_p load balancing and moment problems via the L-function method. In: Proceedings of the 30th Symposium on Discrete Algorithms, pp. 343–354 (2019). https://doi.org/10.5555/3310435.3310457
25. Pinedo, M.L.: Offline deterministic scheduling, stochastic scheduling, and online deterministic scheduling. In: Handbook of Scheduling - Algorithms, Models, and Performance Analysis, Chapter 38. Chapman and Hall/CRC (2004)
26. Sagnol, G., Schmidt genannt Waldschmidt, D.: Restricted adaptivity in stochastic scheduling. In: Proceedings of the 29th European Symposium on Algorithms, pp. 79:1–79:14 (2021). https://doi.org/10.4230/LIPIcs.ESA.2021.79
27. Shmoys, D.B., Swamy, C.: An approximation scheme for stochastic linear programming and its application to stochastic integer programs. J. ACM **53**(6), 978–1012 (2006). https://doi.org/10.1145/1217856.1217860
28. Shmoys, D.B., Tardos, E.: An approximation algorithm for the generalized assignment problem. Math. Program. **62**, 461–474 (1993). https://doi.org/10.1007/BF01585178
29. Swamy, C., Shmoys, D.B.: Approximation algorithms for 2-stage stochastic optimization problems. ACM SIGACT News **37**(1), 33–46 (2006). https://doi.org/10.1145/1122480.1122493
30. Weiss, G.: A tutorial in stochastic scheduling. In: Chretienne, P., Coffman, E., Lenstra, J., Liu, Z. (eds.) Scheduling Theory and Its Applications, pp. 33–64. Wiley, New York (1995)
31. Weiss, G., Pinedo, M.: Scheduling tasks with exponential service times on non-identical processors to minimize various cost functions. J. Appl. Probab. **17**(1), 187–202 (1980). https://doi.org/10.2307/3212936

$\{s, t\}$-Separating Principal Partition Sequence of Submodular Functions

Kristóf Bérczi[1,2,3], Karthekeyan Chandrasekaran[4], Tamás Király[1,2], and Daniel P. Szabo[2(✉)]

[1] HUN-REN-ELTE Egerváry Research Group, Budapest, Hungary
{kristof.berczi,tamas.kiraly}@ttk.elte.hu
[2] Eötvös Loránd University, Budapest, Hungary
dszabo2@wisc.edu
[3] HUN-REN Alfréd Rényi Institute of Mathematics, Budapest, Hungary
[4] Grainger College of Engineering, University of Illinois Urbana-Champaign, Urbana, USA
karthe@illinois.edu

Abstract. Narayanan showed the existence of the principal partition sequence of a submodular function, a structure with numerous applications in areas such as clustering, fast algorithms, and approximation algorithms. In this work, motivated by two applications, we develop a theory of $\{s, t\}$-separating principal partition sequence of a submodular function. We define this sequence, show its existence, and design a polynomial-time algorithm to construct it. We show two applications: (1) approximation algorithm for the $\{s, t\}$-separating submodular k-partitioning problem for monotone and posimodular functions and (2) polynomial-time algorithm for the problem of finding an orientation of a hypergraph that simultaneously has strong connectivity at least k and (s, t)-connectivity at least ℓ.

Keywords: Submodular Functions · Principal Partition Sequence · Submodular Partitioning · Hypergraph Orientation

1 Introduction

A set function $f\colon 2^V \to \mathbb{R}$ is *submodular* if it satisfies the inequality $f(A) + f(B) \geq f(A \cap B) + f(A \cup B)$ for all $A, B \subseteq V$. Submodular functions arise throughout combinatorial optimization and economics; common examples include graph and hypergraph cut function, matroid rank function, and coverage function. The *principal partition sequence* is a central tool in submodular optimization, originating from the study of principal partitions by Kishi and Kajitani [23] in the context of graph tri-partitions defined via maximally distant spanning trees. Initially viewed as decompositions of discrete systems into partially ordered components, principal partitions were later recognized to have a natural foundation in submodularity. Building on this foundation, Narayanan [30]

S. S. Dey et al. (Eds.): IPCO 2026, LNCS 16588, pp. 315–330, 2026.
https://doi.org/10.1007/978-3-032-28691-8_21

developed the theory of the principal partition sequence of submodular functions, while Fujishige [15] provided a comprehensive survey of the theory of principal partitions. Over time, the theory of principal partition and principal partition sequence has been extended well beyond graphs [35] to matrices [18,19], matroids [4,29,40], and general submodular systems [13,14,20,21,28,31], offering a lattice-theoretic decomposition framework for submodular functions.

Principal partition sequence is a well-structured sequence of partitions of the ground set minimizing $\sum_{A\in\mathcal{P}} f(A) - \lambda \cdot |\mathcal{P}|$ as λ varies from $-\infty$ to ∞ (see Section 1.2 for a formal definition). The sequence exhibits a refinement structure for submodular functions. A principal partition sequence of a submodular function given by its evaluation oracle can be found in polynomial time [24,30,31]. This is in contrast to minimizing $\sum_{A\in\mathcal{P}} f(A)$ over partitions $\mathcal{P}$ of the ground set with a given number of parts, which is NP-hard. The computational tractability has made the principal partition sequence a powerful structural tool in submodularity with many applications. Examples include Cunningham's network strength measure that is used to quantify network vulnerability [10], the realization of finite state machines [11], recursive ideal tree packing [8], approximation algorithms for graph and submodular partitioning problems [1,7,32,37], graph clustering [27,36], and dense subgraph decomposition [6,13,17].

In this paper, we extend the principal partition sequence to incorporate separation of two designated terminals – say $s,t \in V$. In particular, we show that the sequence of $\{s,t\}$-separating partitions minimizing $\sum_{A\in\mathcal{P}} f(A) - \lambda \cdot |\mathcal{P}|$ as λ varies from $-\infty$ to ∞ is well-structured and can be found in polynomial-time. The formal definition of $\{s,t\}$-separating principal partition sequence is fairly intricate, so we postpone its definition for now. Here, we discuss two concrete applications of this framework to illustrate its power.

1.1 Applications

The first application concerns approximation algorithms and the second concerns polynomial-time algorithms.

$\{s,t\}$-Separating Submodular$k$-Partition. Given a submodular function $f : 2^V \to \mathbb{R}$ via its evaluation oracle, the *submodular k-partition* problem (abbreviated Submod-k-Part) asks for a partition $\mathcal{P} = \{V_1, \dots, V_k\}$ of the ground set V into k non-empty parts that minimizes $f(\mathcal{P}) := \sum_{i=1}^{k} f(V_i)$. Many problems in combinatorial optimization can be formulated as special cases, including partitioning problems over graphs, hypergraphs, matrices, and matroids. Submod-k-Part is NP-hard [16], does not admit a $(2-\epsilon)$-approximation assuming polynomial number of function evaluation queries [38], does not admit a $n^{1/(\log\log n)^c}$-approximation for every constant c assuming the Exponential Time Hypothesis [9], and the best approximation factor that is known is $O(k)$ [34,41]. Nevertheless, constant factor approximations are known for broad subfamilies of submodular functions.

We recall that a function f is *symmetric* if $f(S) = f(V \setminus S)$ for all $S \subseteq V$, *monotone* if $f(S) \leq f(T)$ for every $S \subseteq T \subseteq V$, and *posimodu-*

lar if $f(S) + f(T) \geq f(S \setminus T) + f(T \setminus S)$ for every $S, T \subseteq V$. We observe that symmetric/monotone submodular functions are also posimodular. Prominent examples of symmetric submodular functions include graph and hypergraph cut functions, monotone submodular functions include matroid rank functions and coverage functions, and posimodular functions include positive combinations of symmetric submodular and monotone submodular functions. A well-studied special case of Submod-k-Part for symmetric submodular functions is Graph-k-Cut: the input is a graph and the goal is to find a minimum number of edges to delete so that the resulting graph has at least k components. If k is a fixed constant, then Graph-k-Cut is polynomial-time solvable [16]. For k part of input, Graph-k-Cut is NP-complete [16] and does not have a polynomial-time $(2-\epsilon)$-approximation for every constant $\epsilon > 0$ under the Small Set Expansion Hypothesis [25]. There are several approaches that achieve a 2-approximation for Graph-k-Cut [1,26,37,39]. One of these approaches is the principal partition sequence, which has been generalized to achieve a 2-approximation for posimodular submodular k-partition and a 4/3-approximation for monotone submodular k-partition [7].

Our first application of $\{s,t\}$-separating principal partition sequence is to the submodular k-partition problem with the additional constraint that the minimizing partition separates two specified terminals. We term this as $\{s,t\}$-Sep-Submod-k-Part and formally define it below. For a ground set V and a pair of terminals $s, t \in V$, a partition $\mathcal{P}$ of V is $\{s,t\}$-*separating* if $|P \cap \{s,t\}| \leq 1$ for every $P \in \mathcal{P}$.

$\{s,t\}$-SEP-SUBMOD-k-PART
Input: A submodular function $f\colon 2^V \to \mathbb{R}_{\geq 0}$ given by a value oracle, distinct elements $s, t \in V$, and $k \in \mathbb{Z}_{\geq 0}$.
Goal:

$$\min\left\{f(\mathcal{P})\colon \mathcal{P} \text{ is a } \{s,t\}\text{-separating partition of } V \text{ into } k \text{ non-empty parts}\right\}.$$

A concrete special case of $\{s,t\}$-Sep-Submod-k-Part is {s,t}-Sep-Graph-k-Cut: the input is a graph with two specified terminal vertices s, t and the goal is to find a minimum number of edges to delete so that the resulting graph has at least k components with s and t being in different components. If k is a fixed constant, $\{s,t\}$-Sep-Graph-k-Cut is solvable in polynomial time [2]. Thus, for fixed constant k, the complexity status of $\{s,t\}$-Sep-Graph-k-Cut is identical to that of Graph-k-Cut. This status for $\{s,t\}$-Sep-Graph-k-Cut for constant k inspired us to investigate $\{s,t\}$-Sep-Graph-k-Cut, and more generally, $\{s,t\}$-Sep-Submod-k-Part, when k is part of input. Is it possible to achieve the same approximation results for $\{s,t\}$-Sep-Submod-k-Part as for Submod-k-Part?

We observe that $\{s,t\}$-Sep-Submod-k-Part closely resembles Submod-k-Part, but known approaches for Submod-k-Part do not apply directly. It is easy to see that $\{s,t\}$-Sep-Submod-k-Part is a special case of matroid constrained submodular k-partition and consequently, it admits a 2-approximation for symmetric submodular functions via the Gomory-Hu tree approach [3]. In this work, we exploit $\{s,t\}$-separating principal partition sequence to design approximation algorithms for $\{s,t\}$-Sep-Submod-k-Part with an approximation guarantee that matches that of Submod-k-Part for monotone and posimodular submodular functions.

Theorem 1. *$\{s,t\}$-Sep-Submod-k-Part admits a 2-approximation for posimodular submodular functions and a 4/3-approximation for monotone submodular functions.*

We note that the guarantee for posimodular submodular functions implies the same guarantee for symmetric submodular functions. Consequently, we have a 2-approximation for $\{s,t\}$-Sep-Graph-k-Cut.

Hypergraph Orientation. In graph orientation problems, the goal is to orient a given undirected graph to obtain a directed graph with certain properties. Two fundamental properties of interest are (s,t)-connectivity and strong-connectivity. In Graph-(s,t)-ℓ-Conn-Orient, the input is an undirected graph with two specified terminals s and t and an integer $\ell \in \mathbb{Z}_{\geq 0}$. The goal is to verify if there exists an orientation of G that has ℓ arc-disjoint paths from s to t and if so, find one. By Menger's theorem, an undirected graph G has an orientation with ℓ arc-disjoint paths from s to t if and only if there exist ℓ edge-disjoint paths between s and t in the undirected graph G. Thus, Graph-(s,t)-ℓ-Conn-Orient can be solved using an application of max (s,t)-flow. In Graph-k-Conn-Orient, the input is an undirected graph and an integer $k \in \mathbb{Z}_{\geq 0}$. The goal is to verify if there exists a k-arc-connected orientation of the graph. We recall that a digraph is k-arc-connected if for every pair of distinct vertices $u, v \in V$, there exist k arc-disjoint paths from u to v. A classic result of Nash-Williams [33] shows that an undirected graph G has a k-arc-connected orientation if and only if G is $2k$-edge-connected. We recall that an undirected graph $G = (V, E)$ is $2k$-edge-connected if for every pair of distinct vertices $u, v \in V$, there exist $2k$ edge-disjoint paths between u and v. Nash-Williams' result is also constructive that leads to a polynomial time algorithm to solve Graph-k-Conn-Orient.

Frank, Király, and Király [12] showed an intriguing result concerning a generalization of these two problems that we term as Graph-$(k,(s,t),\ell)$-Conn-Orient. The input here is an undirected graph with two specified terminals s and t and integers $k, \ell \in \mathbb{Z}_{\geq 0}$ (with the interesting case being $k < \ell$). The goal is to verify if there exists an orientation $\vec{G}$ of G such that $\vec{G}$ is k-arc-connected and has ℓ arc-disjoint paths from s to t – we will call such an orientation to be $(k,(s,t),\ell)$-arc-connected orientation. They showed a complete characterization for the existence of a $(k,(s,t),\ell)$-arc-connected orientation while also giving a polynomial time algorithm to find such an orientation via Mader's splitting-off result. Our focus is on this generalized orientation problem over hypergraphs instead of graphs.

Frank, Király, and Király [12] considered all three orientation problems mentioned above in hypergraphs as opposed to graphs. We begin by discussing the two constituent problems in hypergraphs. A hypergraph $G = (V, E)$ is specified by a vertex set V and hyperedge set E where each $e \in E$ is a subset of V. An orientation $\vec{G}$ of a hypergraph $G = (V, E)$ is a directed hypergraph specified by $\vec{G} = (V, E, \text{head}: E \to V)$, where $\text{head}(e) \in e$ for every $e \in E$. In hypergraph orientation problems, the goal is to orient a given hypergraph to obtain a directed hypergraph with certain properties. For vertices a, b, an (a, b)-path in a directed hypergraph is an alternating sequence, without repetition, of vertices and hyperarcs $v_1, e_1, v_2, e_2, \ldots, v_k, e_k, v_{k+1}$ where $v_i \in e_i \setminus \text{head}(e_i)$ and $v_{i+1} = \text{head}(e_i)$. In (s,t)-ℓ-Conn-Orient, the input is a hypergraph with two specified terminals s and t and an integer $\ell \in \mathbb{Z}_{\geq 0}$. The goal is to verify if there exists an orientation of G that has ℓ hyperarc-disjoint paths from s to t and if so, then find one. Menger's theorem extends to hypergraphs and thus, resolves (s,t)-ℓ-Conn-Orient in hypergraphs. In k-Conn-Orient, the input is a hypergraph and an integer $k \in \mathbb{Z}_{\geq 0}$. The goal is to verify if there exists a k-hyperarc-connected orientation of the graph. A directed hypergraph is k-hyperarc-connected if for every pair of distinct vertices $u, v \in V$, there exist k hyperarc-disjoint paths from u to v. Frank, Király, and Király generalized Nash-Williams' result from graphs to hypergraphs by giving a complete characterization for the existence of a k-hyperarc-connected orientation of a hypergraph and moreover, their characterization is constructive, i.e., it leads to a polynomial-time algorithm to verify if a given hypergraph has a k-hyperarc-connected orientation and if so, then find one.

We now address the main application of interest to our work, namely $(k, (s,t), \ell)$-Conn-Orient that is defined as follows (with the interesting case being $k < \ell$).

> $(k, (s,t), \ell)$-Conn-Orient
> **Input:** A hypergraph $G = (V, E)$, vertices $s, t \in V$, and $k, \ell \in \mathbb{Z}_{\geq 0}$.
> **Goal:** Verify if there exists an orientation $\vec{G}$ of G such that $\vec{G}$ is k-hyperarc-connected and has ℓ hyperarc-disjoint paths from s to t.

We will call a feasible orientation for $(k, (s,t), \ell)$-Conn-Orient to be a $(k, (s,t), \ell)$-hyperarc-connected orientation. Frank, Király, and Király gave a characterization for the existence of a $(k, (s,t), \ell)$-connected orientation of a hypergraph. However, their proof of the characterization relies on certain steps which did not seem constructive. A polynomial-time algorithm to verify the characterization and solve $(k, (s,t), \ell)$-Conn-Orient was subsequently given in [22]. In this work, we exploit algorithmic aspects of $\{s,t\}$-separating principal partition sequence of a submodular function to design an alternative polynomial time algorithm for this problem, which also makes the proof of Frank, Király, and Király constructive.

Theorem 2. *There exists a polynomial time algorithm to verify whether a given hypergraph admits a $(k, (s,t), \ell)$-hyperarc-connected orientation and if so, then find one.*

We also consider optimization variants of the above problem where ℓ (or k) is given as part of input and the goal is to maximize k (or ℓ respectively) so that there exists a $(k, (s,t), \ell)$-connected orientation, present min-max relations, and show that the corresponding minimization problem is solvable in polynomial time via our results on $\{s,t\}$-separating principal partition sequence.

1.2 {s,t}-Separating Principal Partition Sequence

In this section, we present our definition and state our results concerning $\{s,t\}$-separating principal partition sequence. We begin by discussing principal partition sequence and known results to provide background and context. For a set function $f\colon 2^V \to \mathbb{R}$ and a collection $\mathcal{P}$ of subsets of V, we write $f(\mathcal{P}) := \sum_{P\in\mathcal{P}} f(P)$. A *partition* of a set $S \subseteq V$ is a collection of pairwise disjoint non-empty subsets of S whose union is S.

Principal Partition Sequence. Let $f\colon 2^V \to \mathbb{R}$ be a set function and $\mathcal{P}$ be a partition of the ground set V. We define functions $g_{f,\mathcal{P}} : \mathbb{R} \to \mathbb{R}$ and $g_f : \mathbb{R} \to \mathbb{R}$ as follows:

$$\begin{aligned} g_{f,\mathcal{P}}(\lambda) &:= f(\mathcal{P}) - \lambda \cdot |\mathcal{P}|, \\ g_f(\lambda) &:= \min\{g_{f,\mathcal{P}}(\lambda) : \mathcal{P} \text{ is a partition of } V\}. \end{aligned}$$

We drop the subscript f and simply write $g_{\mathcal{P}}$ and g, respectively, if the function f is clear from context. A *breakpoint* of a piecewise linear function is a point at which the slope of f changes, or equivalently, a point where f is continuous but not differentiable. By definition, the function g_f is piecewise linear and has at most $|V|$ slopes and hence, at most $|V|-1$ breakpoints.

Narayanan [29] showed that if $f\colon 2^V \to \mathbb{R}$ is submodular, then the partitions attaining the values of the function g_f at its breakpoints exhibit a refinement structure, which is captured by the principal partition sequence.

Definition 1 (Refinement). *Let $\mathcal{P}$ and $\mathcal{Q}$ be distinct partitions of V.*

1. *$\mathcal{Q}$ is a* refinement *of $\mathcal{P}$ if for all $B \in \mathcal{Q}$ there exists $A \in \mathcal{P}$ such that $B \subseteq A$.*
2. *$\mathcal{Q}$ is a refinement of $\mathcal{P}$* up to one set *if*
   (a) *$\mathcal{Q}$ is a refinement of $\mathcal{P}$ and*
   (b) *there exists $P \in \mathcal{P}$ such that for all $B \in \mathcal{Q}$, either $B \subsetneq P$ or $B \in \mathcal{P}$.*

See Figure 2a for an example of a refinement. We observe that if $\mathcal{P}$ and $\mathcal{Q}$ are distinct partitions and $\mathcal{Q}$ is a refinement of $\mathcal{P}$, then $|\mathcal{Q}| > |\mathcal{P}|$. Using this definition of refinements, a principal partition sequence is defined as follows.

Definition 2 (Principal Partition Sequence). *Let $f\colon 2^V \to \mathbb{R}$ be a set function. A sequence $\mathcal{P}_1, \ldots, \mathcal{P}_\ell$ of partitions of V is a* principal partition sequence *of f if there exist $\lambda_1, \ldots, \lambda_{\ell-1} \in \mathbb{R}$ such that*

P1 $\lambda_1 \le \ldots \le \lambda_{\ell-1}$,

P2 $g(\lambda) = \begin{cases} g_{\mathcal{P}_1}(\lambda) & \text{for } \lambda \in (-\infty, \lambda_1], \\ g_{\mathcal{P}_{j+1}}(\lambda) & \text{for } \lambda \in [\lambda_j, \lambda_{j+1}] \text{ and } j \in [\ell-2], \\ g_{\mathcal{P}_\ell}(\lambda) & \text{for } \lambda \in [\lambda_{\ell-1}, \infty), \end{cases}$

P3 $\mathcal{P}_1 = \{V\}$ *and* $\mathcal{P}_\ell = \{\{v\} : v \in V\}$, *and*

P4 for every $j \in [\ell-1]$, $\mathcal{P}_{j+1}$ is a refinement of $\mathcal{P}_j$ up to one set.

The values $\lambda_1, \ldots, \lambda_{\ell-1}$ *are called the* critical values associated with the principal partition sequence.

We observe that properties (P1), (P2), and (P3) in the definition of a principal partition sequence are easy to achieve for every set function. The essence of principal partition sequences lies in the fact that, for submodular functions, property (P4) can also be achieved. This was shown by Narayanan [29]. He also designed a polynomial-time algorithm to find such a sequence.

Theorem 3 (Narayanan). *Let* $f : 2^V \to \mathbb{R}$ *be a submodular function. Then, there exists a principal partition sequence of* f*. Moreover, given oracle access to a submodular function* $f\colon 2^V \to \mathbb{R}$*, there exists a polynomial-time algorithm to compute a principal partition sequence of* f *along with its associated critical values.*

Principal partition sequence resembles the notion of *principal partition* of a submodular function. It is known that the principal partition sequence of the cut function of a graph is related to the principal partition of the rank function of the graphic matroid associated with the graph. We refer the reader to [31] and [15] for more details on similarities and differences between the two notions.

$\{s,t\}$-Refinement and $\{s,t\}$-Separating Principal Partition Sequence. Let $f\colon 2^V \to \mathbb{R}$ be a submodular set function on ground set V and $s,t \in V$ be distinct elements. A partition $\mathcal{P}$ of V is an $\{s,t\}$-separating partition if $|\{s,t\} \cap P| \leq 1$ for every $P \in \mathcal{P}$. We recall that for every partition $\mathcal{P}$ of V and every $\lambda \in \mathbb{R}$, we defined $g_{f,\mathcal{P}}(\lambda) = f(\mathcal{P}) - \lambda \cdot |\mathcal{P}|$. As a natural $\{s,t\}$-separating counterpart, we define the function $g_f^{s,t}\colon \mathbb{R} \to \mathbb{R}$ by setting

$$g_f^{s,t}(\lambda) := \min\{g_{f,\mathcal{P}}(\lambda) : \mathcal{P} \text{ is an } \{s,t\}\text{-separating partition of } V\}.$$

We drop the subscript f and simply write $g^{s,t}$ if the function f is clear from context.

We observe that for every set function $f\colon 2^V \to \mathbb{R}$, one can find a sequence $\mathcal{P}_1, \ldots, \mathcal{P}_\ell$ of partitions and $\lambda_1, \ldots, \lambda_\ell \in \mathbb{R}$ satisfying properties (P1)- (P3) in Definition 2. Submodularity becomes crucial, however, in order to obtain the refinement property (P4). A key difference between principal partition and the $\{s,t\}$-separating partition sequences arises precisely in this property: minimizers for the breakpoints of $g^{s,t}$ do not necessarily satisfy the refinement property; an example is given in Figure 1.

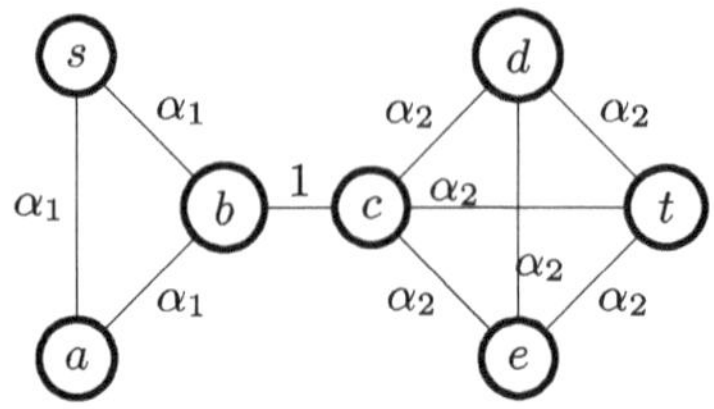

Fig. 1. An edge-weighted graph whose cut function f is such that minimizers for the breakpoints of $g_f^{s,t}$ do not necessarily satisfy the refinement property. Here $0 < \varepsilon < 1/24$ is a small constant, and $\alpha_1 = 1/2 + \varepsilon$, $\alpha_2 = 1/3 + 2\varepsilon$. The sequence of minimizers of $g_f^{s,t}$ is unique and is given by $\mathcal{P}_1 = \{\{s, a, b\}, \{t, c, d, e\}\}, \mathcal{P}_2 = \{\{s\}, \{a\}, \{b, t, c, d, e\}\}$, $\mathcal{P}_3 := \{\{s, a, b, c\}, \{d\}, \{e\}, \{t\}\}$, $\mathcal{P}_4 = \{\{s\}, \{a\}, \{b, c\}, \{d\}, \{e\}, \{t\}\}$, and $\mathcal{P}_5 = \{\{s\}, \{a\}, \{b\}, \{c\}, \{d\}, \{e\}, \{t\}\}$. The principal partition sequence, meanwhile, is $\{s, a, b, c, d, e, t\}, \mathcal{P}_3, \mathcal{P}_4, \mathcal{P}_5$.

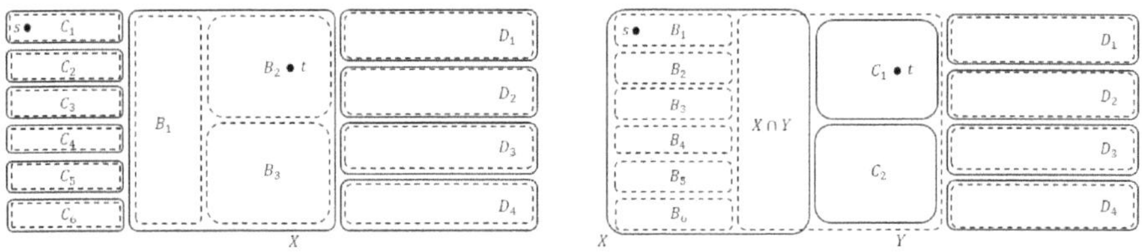

(a) An example where $\mathcal{Q}$ is a refinement of $\mathcal{P}$, where $\mathcal{P} = \{X, C_1, C_2, C_3, C_4, C_5, C_6, D_1, D_2, D_3, D_4\}$ (solid lines) and $\mathcal{Q} = \{B_1, B_2, B_3, C_1, C_2, C_3, C_4, C_5, C_6, D_1, D_2, D_3, D_4\}$ (dashed lines).

(b) An example where $\mathcal{Q}$ is an $\{s,t\}$-refinement of $\mathcal{P}$ up to X and Y, where $\mathcal{P} = \{X, C_1, C_2, C_3, C_4, C_5, C_6, D_1, D_2, D_3, D_4\}$ (solid lines) and $\mathcal{Q} = \{B_1, B_2, B_3, C_1, C_2, C_3, C_4, C_5, C_6, D_1, D_2, D_3, D_4\}$ (dashed lines).

Fig. 2. Examples of refinement and $\{s,t\}$-refinement.

Nevertheless, this sequence still exhibits a more intricate form of refinement. We now formalize this intricate form of refinement. Two sets $X \subseteq V$ and $Y \subseteq V$ are called *intersecting* if all of $X \cap Y$, $X \setminus Y$, and $Y \setminus X$ are nonempty. A pair of intersecting sets $X \subseteq V$ and $Y \subseteq V$ is called $\{s,t\}$-*uncrossable* if they do not separate s and t, meaning that either $|\{s,t\} \cap (X \setminus Y)| \neq 1$ or $|\{s,t\} \cap (Y \setminus X)| \neq 1$.

Definition 3 ($\{s,t\}$-refinements). *Let $\mathcal{P}$ and $\mathcal{Q}$ be distinct $\{s,t\}$-separating partitions of V.*

1. *$\mathcal{Q}$ is a $\{s,t\}$-*refinement of $\mathcal{P}$ along (X, Y) *if $X \in \mathcal{P} \setminus \mathcal{Q}$, $Y \in \mathcal{Q} \setminus \mathcal{P}$, and the following hold:*
   (a) X and Y are intersecting but they are not *$\{s,t\}$-uncrossable,*
   (b) for all $B \in \mathcal{Q} \setminus \{Y\}$ there exists $A \in \mathcal{P}$ such that $B \subseteq A$,
   (c) for all $A \in \mathcal{P} \setminus \{X\}$, either $A \cap Y = \emptyset$ or $A \subseteq Y$, and
   (d) $|\{A \in \mathcal{P} \colon A \subseteq Y\}| \leq |\{B \in \mathcal{Q} \colon B \subseteq X\}|$.

2. *$\mathcal{Q}$ is a* $\{s,t\}$-refinement of $\mathcal{P}$ *if there exists $X \in \mathcal{P} \setminus \mathcal{Q}$ and $Y \in \mathcal{Q} \setminus \mathcal{P}$ such that $\mathcal{Q}$ is an $\{s,t\}$-refinement of $\mathcal{P}$ along (X,Y).*
3. *$\mathcal{Q}$ is a* $\{s,t\}$-refinement of $\mathcal{P}$ up to two sets *if there exists $X \in \mathcal{P} \setminus \mathcal{Q}$ and $Y \in \mathcal{Q} \setminus \mathcal{P}$ such that*
   (a) $\mathcal{Q}$ is an $\{s,t\}$-refinement of $\mathcal{P}$ along (X,Y) and
   (b) $\{P \in \mathcal{P} \colon P \subseteq V \setminus (X \cup Y)\} \subseteq \mathcal{Q}$.

See Figures 2a and 2b for examples of refinements and $\{s,t\}$-refinements. We observe that an $\{s,t\}$-refinement allows exactly one pair of intersecting sets in $\mathcal{P} \cup \mathcal{Q}$, and these sets cannot be $\{s,t\}$-uncrossable. Moreover, if $\mathcal{Q}$ is a refinement of $\mathcal{P}$, then $|\mathcal{Q}| > |\mathcal{P}|$, whereas if $\mathcal{Q}$ is an $\{s,t\}$-refinement of $\mathcal{P}$, then $|\mathcal{Q}| \geq |\mathcal{P}|$.

Based on the notion of refinements and $\{s,t\}$-refinements, we define an $\{s,t\}$-separating principal partition sequence as follows.

Definition 4 ($\{s,t\}$-separating Principal Partition Sequence). *Let $f \colon 2^V \to \mathbb{R}$ be a set function. A sequence $\mathcal{P}_1, \ldots, \mathcal{P}_\ell$ of $\{s,t\}$-separating partitions of V is a* $\{s,t\}$-separating principal partition sequence *of f if there exist $\lambda_1, \ldots, \lambda_{\ell-1} \in \mathbb{R}$ such that*

(SP1) $\lambda_1 \leq \ldots \leq \lambda_{\ell-1}$,

(SP2) $$g^{s,t}(\lambda) = \begin{cases} g_{\mathcal{P}_1}(\lambda) & \text{for } \lambda \in (-\infty, \lambda_1], \\ g_{\mathcal{P}_{j+1}}(\lambda) & \text{for } \lambda \in [\lambda_j, \lambda_{j+1}] \text{ and } j \in [\ell-2], \\ g_{\mathcal{P}_\ell}(\lambda) & \text{for } \lambda \in [\lambda_{\ell-1}, \infty). \end{cases}$$

(SP3) $\mathcal{P}_1$ is $\{s,t\}$-separating partition with two parts such that $f(\mathcal{P}_1) \leq f(A) + f(V \setminus A)$ for every $s \in A \subseteq V - t$, and $\mathcal{P}_\ell = \{\{v\} \colon v \in V\}$,

(SP4) for every $j \in [\ell-1]$, either $\mathcal{P}_{j+1}$ is a refinement of $\mathcal{P}_j$ up to one set or $\mathcal{P}_{j+1}$ is an $\{s,t\}$-refinement of $\mathcal{P}_j$ up to two sets, and moreover, $|\mathcal{P}_j| < |\mathcal{P}_{j+1}|$.

The values $\lambda_1, \ldots, \lambda_{\ell-1}$ are called the critical values associated with the $\{s,t\}$-separating principal partition sequence.

We encourage the reader to compare and contrast Definitions 2 and 4. A principal partition sequence can be viewed as a prefix of non-$\{s,t\}$-separating partition subsequence followed by a suffix of $\{s,t\}$-separating partition subsequence. However, the suffix by itself is not necessarily an $\{s,t\}$-separating principal partition sequence. We now state our main result regarding $\{s,t\}$-separating principal partition sequence of a submodular function.

Theorem 4. *Let $f \colon 2^V \to \mathbb{R}$ be a submodular function and $s,t \in V$. Then, there exists an $\{s,t\}$-separating principal partition sequence of f. Moreover, given oracle access to f, such a sequence and its critical values can be computed in polynomial time.*

Due to space constraints, we give a short summary of the proof, and refer the reader to the full version of this work [5] for all missing proofs and further discussion.

2 Proof Outline

Here, we outline the proof of existence of an $\{s,t\}$-separating principal partition sequence. We first establish a key refinement property of $\{s,t\}$-separating partitions (Theorem 5). We say a function f is strictly submodular on intersecting pairs if $f(A) + f(B) > f(A \cap B) + f(A \cup B)$ for every $A, B \subseteq V$ for which $A \cap B, A \setminus B, B \setminus A$ are non-empty.

Theorem 5. *Let $f: 2^V \to \mathbb{R}$ be a submodular function that is strictly submodular on intersecting pairs*[1], *and let $s, t \in V$. Suppose that $\mathcal{P}$ and $\mathcal{Q}$ are distinct $\{s,t\}$-separating partitions such that at least one of the following holds:*

1. *There exists $\lambda \in \mathbb{R}$ such that*

$$\mathcal{P} \in \arg\min\{|\pi| \colon \pi \text{ is an } \{s,t\}\text{-separating partition with } g_\pi(\lambda) = g^{s,t}(\lambda)\},$$
$$\mathcal{Q} \in \arg\max\{|\pi| \colon \pi \text{ is an } \{s,t\}\text{-separating partition with } g_\pi(\lambda) = g^{s,t}(\lambda)\}.$$

2. *There exists $\alpha < \lambda$ such that $g^{s,t}(\alpha) = g_\mathcal{P}(\alpha)$ and $g^{s,t}(\lambda) = g_\mathcal{P}(\lambda) = g_\mathcal{Q}(\lambda)$.*

Then, $\mathcal{Q}$ is either a refinement or an $\{s,t\}$-refinement of $\mathcal{P}$.

We use Theorem 5 to show Lemmas 1 and 2 below. These two lemmas show the behavior of the relevant partitions at different values of λ.

Lemma 1. *Let $\lambda_a < \lambda_b$ be adjacent breakpoints of $g^{s,t}$ and let*

$$\mathcal{P}_a \in \arg\max\{|\pi| \colon \pi \text{ is an } \{s,t\}\text{-separating partition with } g_\pi(\lambda_a) = g^{s,t}(\lambda_a)\},$$
$$\mathcal{P}_b \in \arg\min\{|\pi| \colon \pi \text{ is an } \{s,t\}\text{-separating partition with } g_\pi(\lambda_b) = g^{s,t}(\lambda_b)\}.$$

Then, $g_{\mathcal{P}_a}(\lambda) = g^{s,t}(\lambda) = g_{\mathcal{P}_b}(\lambda)$ for all $\lambda \in [\lambda_a, \lambda_b]$, and $|\mathcal{P}_a| = |\mathcal{P}_b|$.

Moreover, if $g_\mathcal{P}(\lambda) = g^{s,t}(\lambda)$ for some $\lambda \in (\lambda_a, \lambda_b)$ for a partition $\mathcal{P}$, then $g_\mathcal{P}(\lambda) = g^{s,t}(\lambda)$ for every $\lambda \in [\lambda_a, \lambda_b]$, and $|\mathcal{P}| = |\mathcal{P}_a| = |\mathcal{P}_b|$.

The next lemma gives a sequence of partitions at a given breakpoint.

Lemma 2. *Let λ be a breakpoint of $g^{s,t}$ and let*

$$\mathcal{P} \in \arg\min\{|\pi| \colon \pi \text{ is an } \{s,t\}\text{-separating partition with } g_\pi(\lambda) = g^{s,t}(\lambda)\}.$$

Then, there exists a sequence $\mathcal{P} = \mathcal{R}_0, \mathcal{R}_1, \ldots, \mathcal{R}_{r+1}$ of $\{s,t\}$-separating partitions such that

1. *$\mathcal{R}_{r+1} \in \arg\max\{|\pi| \colon \pi$ is an $\{s,t\}$-separating partition with $g_\pi(\lambda) = g^{s,t}(\lambda)\}$,*
2. *$\mathcal{R}_1$ is either a refinement of $\mathcal{P}$ up to one set or an $\{s,t\}$-refinement of $\mathcal{P}$ up to two sets,*

[1] The assumption of strict submodularity is a technical detail needed in the proof of Theorem 5, and can be applied without loss of generality when constructing the sequence.

3. *$\mathcal{R}_{i+1}$ is a refinement of $\mathcal{R}_i$ up to one set for every $i \in [r]$,*
4. *$|\mathcal{R}_{i+1}| > |\mathcal{R}_i|$ for every $i \in \{0, 1, 2, \ldots, r\}$, and*
5. *$g_{\mathcal{R}_i}(\lambda) = g^{s,t}(\lambda)$ for every $i \in \{0, 1, \ldots, r+1\}$.*

Moreover, given

$$\mathcal{P} \in \arg\min\{|\pi| \colon \pi \text{ is an } \{s,t\}\text{-separating partition with } g_\pi(\lambda) = g^{s,t}(\lambda)\}, \text{ and}$$
$$\mathcal{Q} \in \arg\max\{|\pi| \colon \pi \text{ is an } \{s,t\}\text{-separating partition with } g_\pi(\lambda) = g^{s,t}(\lambda)\}$$

a sequence $\mathcal{P} = \mathcal{R}_0, \mathcal{R}_1, \ldots, \mathcal{R}_{r+1} = \mathcal{Q}$ of $\{s,t\}$-separating partitions satisfying the above five properties can be constructed in polynomial-time.

The following lemma shows what the first and last partitions will be in the sequence.

Lemma 3. *Let a, b be the leftmost and rightmost breakpoints of $g^{s,t}$ and let*

$$\mathcal{P}' \in \arg\min\{f(\pi) \colon \pi \text{ is an } \{s,t\}\text{-separating partition with } |\pi| = 2\},$$
$$\mathcal{Q}' := \{\{v\} \colon v \in V\}.$$

Then, $g^{s,t}(\lambda) = g_{\mathcal{P}'}(\lambda)$ for $\lambda \in (-\infty, a]$ and $g^{s,t}(\lambda) = g_{\mathcal{Q}'}(\lambda)$ for $\lambda \in [b, \infty)$.

We now prove the existence in Theorem 4.

Proof Proof of Existence in Theorem 4). Let $\lambda_1 < \cdots < \lambda_{\ell-1}$ be the breakpoints of $g^{s,t}$. We will use Lemmas 2 and 1 alternatively to construct the claimed sequence of partitions along with the critical value sequence. Let $\mathcal{P}_1 \in \arg\min\{|\pi| \colon \pi$ is an{s,t}-separating partition with $g_\pi(\lambda_1) = g^{s,t}(\lambda_1)\}$. By Lemma 3, we have $\mathcal{P}_1 \in \arg\min\{f(\mathcal{P}) \colon \mathcal{P}$ is an $\{s,t\}$-separating partition with $|\mathcal{P}| = 2\}$ and $g^{s,t}(\lambda) = g_{\mathcal{P}_1'}(\lambda)$ if $\lambda \in (-\infty, \lambda_1]$.

We proceed as follows for $i = 1, 2, 3, \ldots, \ell - 1$. We will inductively ensure that the $\{s,t\}$-separating partition $\mathcal{P}_i$ is such that $\mathcal{P}_i \in \arg\min\{|\pi| \colon \pi$ is an{s,t}-separating partition with $g_\pi(\lambda_i) = g^{s,t}(\lambda_i)\}$. Applying Lemma 2 at the breakpoint λ_i and the $\{s,t\}$-separating partition $\mathcal{P}_i$ gives a sequence of $\{s,t\}$-separating partitions $\mathcal{P}_i = \mathcal{R}_0^i, \mathcal{R}_1^i, \ldots, \mathcal{R}_{r_i+1}^i$ satisfying the properties guaranteed by the lemma. We set $\mathcal{P}_{i+1} := \mathcal{R}_{r_i+1}^i$.

By the first property of Lemma 2, we know that $\mathcal{P}_{i+1} = \mathcal{R}_{r_i+1}^i \in \arg\max\{|\pi| \colon \pi$ is an{s,t}-separating partition with $g_\pi(\lambda_i) = g^{s,t}(\lambda_i)\}$.

Applying Lemma 1 to adjacent breakpoints $\lambda_i < \lambda_{i+1}$ of $g^{s,t}$, we conclude that $\mathcal{P}_{i+1} \in \arg\min\{|\pi| \colon \pi$ is an{s,t}-separating partition with $g_\pi(\lambda_{i+1})$ $= g^{s,t}(\lambda_{i+1})\}$.

This implies that $\mathcal{P}_{i+1}$ satisfies the induction hypothesis for every $i \in [\ell - 2]$ and moreover, $\mathcal{P}_\ell = \{\{v\} : v \in V\}$ by Lemma 3.

With this construction, we now consider the sequence

$$\mathcal{P}_1, \mathcal{R}_1^1, \ldots, \mathcal{R}_{r_1+1}^1 = \mathcal{P}_2, \ldots, \mathcal{P}_i, \mathcal{R}_1^i, \ldots, \mathcal{R}_{r_i+1}^i = \mathcal{P}_{i+1}, \ldots, \mathcal{P}_{\ell-1}, \mathcal{R}_1^\ell, \ldots, \mathcal{P}_\ell$$

together with the critical value sequence

$$\lambda_1, \underbrace{\lambda_1, \ldots, \lambda_1}_{r_1 \text{ copies}}, \lambda_2, \ldots, \lambda_i, \underbrace{\lambda_i, \ldots, \lambda_i}_{r_i \text{ copies}}, \lambda_{i+1}, \ldots, \lambda_{\ell-1}, \underbrace{\lambda_{\ell-1}, \ldots, \lambda_{\ell-1}}_{r_{\ell-1} \text{ copies}}.$$

We claim that this sequence of partitions satisfies properties (SP1)-(SP4) of Definition 4. The critical value sequence is non-decreasing and hence, we have (SP1). We have already seen that $\mathcal{P}_1 \in \arg\min\{f(\mathcal{P})\colon \mathcal{P}$ is an $\{s,t\}$-separating partition with $|\mathcal{P}| = 2\}$ and $\mathcal{P}_\ell = \{\{v\} : v \in V\}$ and hence, we have (SP3). We now show (SP2). By Lemma 3, we have that $g^{s,t}(\lambda) = g_{\mathcal{P}_1}(\lambda)$ for $\lambda \in (-\infty, \lambda_1]$ and $g^{s,t}(\lambda) = g_{\mathcal{P}_\ell}(\lambda)$ for $\lambda \in [\lambda_{\ell-1}, \infty)$. Let $i \in [\ell - 1]$. We observe that $g^{s,t}(\lambda_i) = g_{\mathcal{R}_j^i}(\lambda_i)$ by construction of $\mathcal{R}_j^i$ via Lemma 2. It suffices to show that $g^{s,t}(\lambda) = g_{\mathcal{P}_{i+1}}(\lambda)$ for every $\lambda \in [\lambda_i, \lambda_{i+1}]$. This follows by applying Lemma 1 by considering adjacent breakpoints $\lambda_i < \lambda_{i+1}$ of $g^{s,t}$ and the partition $\mathcal{P}_{i+1} \in \arg\max\{|\pi|\colon \pi$ is an{s,t}-separating partition with $g_\pi(\lambda_i) = g^{s,t}(\lambda_i)\}$. Finally, we show (SP4). Let $i \in [\ell - 1]$. We observe that $\mathcal{R}_1^i$ is either a refinement of $\mathcal{P}_i$ or an $\{s,t\}$-refinement of $\mathcal{P}_i$ up to two sets and $|\mathcal{R}_1^i| > |\mathcal{P}_i|$ by Lemma 2. Moreover, for $j > 1$, we also have that $\mathcal{R}_j^i$ is either a refinement of $\mathcal{R}_{j-1}^i$ or an $\{s,t\}$-refinement of $\mathcal{R}_{j-1}^i$ up to two sets and $|\mathcal{R}_j^i| > |\mathcal{R}_{j-1}^i|$ by Lemma 2.

3 Approximation Algorithm for $\{s,t\}$-Sep-Submod-k-Part

In this section, we design an algorithm for $\{s,t\}$-Sep-Submod-k-Part via $\{s,t\}$-separating principal partition sequence and show that it achieves an approximation factor of 4/3 for monotone submodular functions and 2 for posimodular submodular functions. Let $f : 2^V \to \mathbb{R}$ be the input function with $s, t \in V$ and $k \in \mathbb{Z}_{\geq 2}$.

Algorithm. Our algorithm computes an $\{s,t\}$-separating principal partition sequence of f as mentioned in Theorem 4. If the sequence contains a partition with exactly k parts, the algorithm returns this partition. Otherwise, it considers the two partitions $\mathcal{P}_{i-1}$ and $\mathcal{P}_i$ in the sequence with $|\mathcal{P}_{i-1}| < k < |\mathcal{P}_i|$. By the properties of the $\{s,t\}$-separating principal partition sequence, we know that $\mathcal{P}_i$ is either a refinement of $\mathcal{P}_{i-1}$ up to one set or an $\{s,t\}$-refinement of $\mathcal{P}_{i-1}$ up to two sets. Suppose first that $\mathcal{P}_i$ is a refinement of $\mathcal{P}_{i-1}$ up to one set $X \in \mathcal{P}_{i-1}$. Then, the algorithm proceeds similarly to the k-partitioning algorithm of [7,32]: it obtains an $\{s,t\}$-separating k-partition $\mathcal{P}$ from $\mathcal{P}_{i-1}$ by replacing X with the $k - |\mathcal{P}_{i-1}|$ cheapest parts of $\mathcal{P}_i$ contained within X and an additional part containing the remainder. Now suppose that $\mathcal{P}_i$ is an $\{s,t\}$-refinement of $\mathcal{P}_{i-1}$ up to two sets, namely $X \in \mathcal{P}_{i-1} \setminus \mathcal{P}_i$ and $Y \in \mathcal{P}_i \setminus \mathcal{P}_{i-1}$. In this case, the algorithm constructs three $\{s,t\}$-separating k-partitions σ_1, σ_2, and π, and returns the one with the smallest objective value, as follows: σ_1 is obtained from $\mathcal{P}_{i-1}$ by replacing X with the $k - |\mathcal{P}_{i-1}|$ cheapest parts of $\mathcal{P}_i$ contained within X and an additional part containing the remainder; σ_2 is obtained from $\mathcal{P}_{i-1}$ by replacing X with the $k - |\mathcal{P}_{i-1}| - 1$ cheapest parts of $\mathcal{P}_i$ contained within X, a part $X \cap Y$, and an additional part containing the remainder; π is obtained from $\mathcal{P}_i$ by replacing the $|\mathcal{P}_i| - k + 1$ most expensive parts of $\mathcal{P}_i$ contained within X by their union. See full version for an example of the partitions created by the algorithm.

Since an $\{s,t\}$-separating principal partition sequence can be computed in polynomial time by Theorem 4, the above algorithm can indeed be implemented to run in polynomial time. Moreover, the algorithm returns an $\{s,t\}$-separating k-partition. The following result completes the proof of Theorem 1.

Theorem 6 *Let $\mathcal{P}^*$ denote an optimal $\{s,t\}$-separating k-partition and let $\mathcal{P}$ denote the partition returned by the above algorithm.*

1. *If f is posimodular, then $f(\mathcal{P}) \le 2f(\mathcal{P}^*)$.*
2. *If f is monotone, then $f(\mathcal{P}) \le \frac{4}{3}f(\mathcal{P}^*)$.*

4 Min-Max Relations for $(k,(s,t),\ell)$-Hyperarc-Connected Orientation Problems

In this section, we consider two optimization variants of $(k,(s,t),\ell)$-hyperarc-connected orientation problems, present min-max relations for both, and outline a polynomial-time algorithm to solve the corresponding minimization problem using our results on $\{s,t\}$-separating principal partition sequence. Our min-max relations follow immediately from the known characterization for the existence of $(k,(s,t),\ell)$-hyperarc-connected orientation [12]. Our main contribution is a polynomial-time algorithm to solve the minimization problem that arises in the min-max relation.

Firstly, we consider the problem of maximizing ℓ for a given k such that a given hypergraph G with vertices $s,t \in V$ has a $(k,(s,t),\ell)$-connected orientation and give a min-max relation. For this maximization problem, we work under the assumption that G has a k-hyperarc-connected orientation – otherwise, G has no $(k,(s,t),\ell)$-hyperarc-connected orientation for every ℓ.

Theorem 7. *Let $G=(V,E)$ be a hypergraph, $s,t \in V$, and $k \in \mathbb{Z}_{\ge 0}$ such that G has a k-hyperarc-connected-orientation. Then,*

$$\begin{aligned}&\max\{\ell\colon G \text{ has a } (k,(s,t),\ell)\text{-hyperarc-connected orientation}\}\\&\quad=\min\{|\delta_G(\mathcal{P})|-k(|\mathcal{P}|-1)\colon \mathcal{P} \text{ is an } \{s,t\}\text{-separating partition of } V\}.\end{aligned}$$

Next, we consider the problem of maximizing k for a given ℓ such that G has a $(k,(s,t),\ell)$-connected orientation and give a min-max relation. For this maximization problem, we work under the assumption that G has ℓ hyperedge-disjoint paths between s and t – otherwise, G has no $(k,(s,t),\ell)$-hyperarc-connected orientation for every k.

Theorem 8. *Let $G=(V,E)$ be a hypergraph, $s,t \in V$, and $\ell \in \mathbb{Z}_{\ge 0}$ such that G has ℓ hyperedge-disjoint paths between s and t. Then,*

$$\max\{k\colon G \text{ has a } (k,(s,t),\ell)\text{-hyperarc-connected orientation}\} = \min\{\alpha,\beta\},$$

where

$$\alpha := \min\{\lfloor\frac{|\delta_G(\mathcal{P})|}{|\mathcal{P}|}\rfloor\colon \mathcal{P} \text{ is a partition of } V\}, \text{ and}$$

$$\beta := \min\{\lfloor\frac{|\delta_G(\mathcal{P})|-\ell}{|\mathcal{P}|-1}\rfloor\colon \mathcal{P} \text{ is an } \{s,t\}\text{-separating partition of } V\}.$$

We now outline algorithms to solve the minimization problems in Theorems 7 and 8 in polynomial time. We need some background. We recall that a *directed hypergraph* $\vec{G} = (V, E, \text{head}: E \to V)$ is specified by a vertex set V, hyperedge set E where each $e \in E$ is a subset of V, and a function head: $E \to V$ with the property that head$(e) \in e$ for each $e \in E$. For a subset $U \subseteq V$, we define $\delta^{in}_{\vec{G}}(U) := \{e \in E\colon \text{ head}(e) \in U, e \backslash U \neq \emptyset\}$ and the function $d^{in}_{\vec{G}}\colon 2^V \to \mathbb{Z}$ defined by $d^{in}_{\vec{G}}(U) := |\delta^{in}_{\vec{G}}(U)|$. It is well-known that the function $d^{in}_{\vec{G}}$ is submodular.

Let $G = (V, E)$ be a hypergraph and $s, t \in V$. Let $\vec{G} = (V, E, head : E \to V)$ be an arbitrary orientation of G. We observe that for each partition $\mathcal{P}$ of V, we have that $|\delta_G(\mathcal{P})| = \sum_{U \in \mathcal{P}} d^{in}_{\vec{G}}(A) = d^{in}_{\vec{G}}(\mathcal{P})$. We define $g_{\mathcal{P}}(\lambda) := d^{in}_{\vec{G}}(\mathcal{P}) - \lambda|\mathcal{P}|$ for each partition $\mathcal{P}$ of V, $g(\lambda) := \min\{g_{\mathcal{P}}(\lambda)\colon \mathcal{P}$ is a partition of $V\}$, and $g^{s,t}(\lambda) := \min\{g_{\mathcal{P}}(\lambda)\colon \mathcal{P}$ is a $\{s,t\}$-separating partition of $V\}$.

Now, for a given $k \in \mathbb{Z}_{\geq 0}$, the minimization problem in Theorem 7 is equivalent to minimizing $g^{s,t}(k) + k$ and we note that $g^{s,t}(k)$ can be computed in polynomial time using Theorem 4. Next, for a given $\ell \in \mathbb{Z}_{\geq 0}$, we address the minimization problem in Theorem 8. Here, we note that α is the floor of the smallest number λ such that $g(\lambda) \leq 0$ and β is the floor of the smallest number λ such that $g^{s,t}(\lambda) \leq \ell - \lambda$. We can compute $g(\lambda)$ and $g^{s,t}(\lambda)$ for all λ in polynomial time using Theorems 3 and 4 respectively. Consequently, we can compute α and β in polynomial time.

5 Conclusion

Motivated by the breadth of applications of the principal partition sequence of submodular functions, we investigated the notion of $\{s,t\}$-separating principal partition sequence. We illustrated two applications of this sequence: we designed approximation algorithms for $\{s,t\}$-Sep-Submod-k-Part for monotone and posimodular submodular functions and polynomial-time algorithms for $(k, (s,t), \ell)$-Conn-Orient in hypergraphs. A natural direction for research is whether variants of principal partitioning sequence provide insights into the approximability of multiway cut. As we have seen, it does provide the current-best approximation factor for Submod-k-Part and $\{s,t\}$-Sep-Submod-k-Part for monotone and posimodular submodular functions. Another interesting direction is whether the principal partition sequence of the cut function of a given graph in near-linear time. It would also be interesting to understand other applications of the $\{s,t\}$-separating principal partition sequence. E.g., the principal partition sequence of the graph cut function is related to recursive ideal tree packing [8]; do we have a similar connection for $\{s,t\}$-separating principal partition sequence?

Acknowledgments. Karthekeyan Chandrasekaran was partially supported by NSF grant CCF-2402667. Kristóf Bérczi and Tamás Király were supported in part by the Lendület Programme of the Hungarian Academy of Sciences – grant number LP2021-1/2021, by the Ministry of Innovation and Technology of Hungary from the National Research, Development and Innovation Fund – grant numbers ELTE TKP 2021-NKTA-62, ADVANCED 150556, and ADVANCED 153096, and by the Dynasnet European Research Council Synergy project – grant number ERC-2018-SYG 810115.

Disclosure of Interests. The authors declare no competing interests.

References

1. Barahona, F.: On the k-cut problem. Oper. Res. Lett. **26**(3), 99–105 (2000)
2. Bérczi, K., Chandrasekaran, K., Király, T., Lee, E., Xu, C.: Beating the 2-approximation factor for global bicut. Math. Program. **177**(1), 291–320 (2019)
3. Bérczi, K., Chandrasekaran, K., Király, T., Szabo, D.P.: Approximating submodular matroid-constrained partitioning. arXiv preprint arXiv:2506.19507 (2025)
4. Bruno, J., Weinberg, L.: The principal minors of a matroid. Linear Algebra Appl. **4**(1), 17–54 (1971)
5. Bérczi, K., Chandrasekaran, K., Király, T., Szabo, D.P.: $\{s, t\}$-separating principal partition sequence of submodular functions (2026). https://arxiv.org/abs/2510.25664
6. Chandrasekaran, K., Chekuri, C., Kulkarni, S.: On deleting vertices to reduce density in graphs and supermodular functions. In: 52nd International Colloquium on Automata, Languages, and Programming (ICALP 2025), pp. 43:1–43:20 (2025)
7. Chandrasekaran, K., Wang, W.: Approximating submodular k-partition via principal partition sequence. SIAM J. Discret. Math. **38**, 3198–3219 (2024)
8. Chekuri, C., Quanrud, K., Xu, C.: LP relaxation and tree packing for minimum k-Cut. SIAM J. Discret. Math. **34**(2), 1334–1353 (2020)
9. Chekuri, C., Li, S.: On the hardness of approximating the k-way hypergraph cut problem. Theory Comput. **16**(1) (2020)
10. Cunningham, W.H.: Optimal attack and reinforcement of a network. J. ACM (JACM) **32**(3), 549–561 (1985)
11. Desai, M.P., Narayanan, H., Patkar, S.B.: The realization of finite state machines by decomposition and the principal lattice of partitions of a submodular function. Discret. Appl. Math. **131**(2), 299–310 (2003)
12. Frank, A., Király, T., Király, Z.: On the orientation of graphs and hypergraphs. Discret. Appl. Math. **131**(2), 385–400 (2003)
13. Fujishige, S.: Lexicographically optimal base of a polymatroid with respect to a weight vector. Math. Oper. Res. **5**(2), 186–196 (1980)
14. Fujishige, S.: Principal structures of submodular systems. Discret. Appl. Math. **2**, 77–79 (1980)
15. Fujishige, S.: Theory of principal partitions revisited. In: Research Trends in Combinatorial Optimization, pp. 127–162. Springer (2009)
16. Goldschmidt, O., Hochbaum, D.: A polynomial algorithm for the k-cut problem for fixed k. Math. Oper. Res. **19**(1), 24–37 (1994)
17. Harb, E., Quanrud, K., Chekuri, C.: Faster and scalable algorithms for densest subgraph and decomposition. Adv. Neural Inf. Process. Syst. (2022)
18. Iri, M.: A min-max theorem for the ranks and term-ranks of a class of matrices: an algebraic approach to the problem of the topological degrees of freedom of a network. Trans. Inst. Electron. Commun. Eng. Japan **51**, 180–187 (1968)
19. Iri, M.: The maximum-rank minimum-term-rank theorem for the pivotal transforms of a matrix. Linear Algebra Appl. **2**(4), 427–446 (1969)
20. Iri, M.: A review of recent work in Japan on principal partitions of matroids and their applications. Ann. N. Y. Acad. Sci. **319**(1), 306–319 (1979)
21. Iri, M.: Structural theory for the combinatorial systems characterized by submodular functions. In: Progress in Combinatorial Optimization, pp. 197–219. Academic Press, New York (1984)

22. Király, T.: Computing the minimum cut in hypergraphic matroids. Tech. rep., Technical Report QP-2009-05, Egerváry Research Group, Budapest 2009 (2009). www.cs.elte.hu/egres
23. Kishi, G., Kajitani, Y.: Maximally distant trees and principal partition of a linear graph. IEEE Trans. Circuit Theory **16**(3), 323–330 (1969)
24. Kolmogorov, V.: A faster algorithm for computing the principal sequence of partitions of a graph. Algorithmica **56**(4), 394–412 (2010)
25. Manurangsi, P.: Inapproximability of maximum biclique problems, minimum k-cut and densest at-least-k-subgraph from the small set expansion hypothesis. Algorithms **11**(1), 10 (2018)
26. Nagamochi, H., Kamidoi, Y.: Minimum cost subpartitions in graphs. Inf. Process. Lett. **102**(2), 79–84 (2007)
27. Nagano, K., Kawahara, Y., Iwata, S.: Minimum average cost clustering. Adv. Neural. Inf. Process. Syst. **23** (2010)
28. Nakamura, M.: Structural theorems for submodular functions, polymatroids and polymatroid intersections. Graphs Combin. **4**(1), 257–284 (1988)
29. Narayanan, H.: Theory of matroids and network analysis. Ph.D. Thesis, Department of Electrical Engineering, Indian Institute of Technology (1974)
30. Narayanan, H.: The principal lattice of partitions of a submodular function. Linear Algebra Appl. **144**, 179–216 (1991)
31. Narayanan, H.: Submodular functions and electrical networks, Annals of Discrete Mathematics, vol. 54. North-Holland Publishing Co., Amsterdam (1997)
32. Narayanan, H., Roy, S., Patkar, S.: Approximation algorithms for min-k-overlap problems using the principal lattice of partitions approach. J. Algorithms **21**(2), 306–330 (1996)
33. Nash-Williams, C.S.J.: On orientations, connectivity and odd-vertex-pairings in finite graphs. Can. J. Math. **12**, 555–567 (1960)
34. Okumoto, K., Fukunaga, T., Nagamochi, H.: Divide-and-conquer algorithms for partitioning hypergraphs and submodular systems. Algorithmica **62**(3), 787–806 (2012)
35. Ozawa, T.: Common trees and partition of two-graphs (in Japanese). Trans. Inst. Electron. Commun. Eng. Japan **57**(5), 383–390 (1974)
36. Patkar, S.B., Narayanan, H.: Improving graph partitions using submodular functions. Discret. Appl. Math. **131**, 535–553 (2003)
37. Ravi, R., Sinha, A.: Approximating k-cuts using network strength as a lagrangean relaxation. Eur. J. Oper. Res. **186**(1), 77–90 (2008)
38. Santiago, R.: New approximations and hardness results for submodular partitioning problems. In: Proceedings of International Workshop on Combinatorial Algorithms, pp. 516–530. IWOCA (2021)
39. Saran, H., Vazirani, V.: Finding k cuts within twice the optimal. SIAM J. Comput. **24**(1), 101–108 (1995)
40. Tomizawa, N.: Strongly irreducible matroids and principal partition of a matroid into strongly irreducible minors (in Japanese). Trans. Inst. Electron. Commun. Eng. Japan **59**, 83–91 (1976)
41. Zhao, L., Nagamochi, H., Ibaraki, T.: Greedy splitting algorithms for approximating multiway partition problems. Math. Program. **102**, 167–183 (2005)

Hedgegraph Polymatroids

Karthekeyan Chandrasekaran[1], Chandra Chekuri[1], Weihang Wang[2], and Weihao Zhu[1](✉)

[1] Grainger College of Engineering, University of Illinois, Urbana-Champaign, USA
{karthe,chekuri,weihaoz3}@illinois.edu
[2] Department of Mathematics, University of Illinois at Urbana-Champaign, Champaign, USA
ww368@cornell.edu

Abstract. Graphs and hypergraphs combine expressive modeling power with algorithmic efficiency for a wide range of applications. Hedgegraphs generalize hypergraphs further by grouping hyperedges under a color/hed-ge. This allows hedgegraphs to model dependencies between hyperedges and leads to several applications. However, it poses algorithmic challenges. In particular, the cut function of a hedgegraph is not submodular, which has been a barrier to algorithms for connectivity. In this work, we introduce two alternative partition-based measures of connectivity in hedgegraphs and study their structural and algorithmic aspects. Instead of the cut function, we investigate a polymatroid associated with hedgegraphs. The polymatroidal lens leads to new tractability results as well as insightful generalizations of classical results on graphs and hypergraphs.

Keywords: Hedgegraphs · Polymatroids · Connectivity

1 Introduction

Graphs and hypergraphs are fundamental objects in discrete and combinatorial optimization. They provide a powerful way to model many problems that arise in applications. A hypergraph $G = (V, E)$ is specified by a vertex set V and a collection E of hyperedges where each hyperedge $e \in E$ is a subset of vertices. If every hyperedge has size at most two, then the hypergraph is a graph. Hedgegraphs generalize hypergraphs further—a hedge is a collection of hyperedges in an underlying hypergraph. There are various scenarios where a collection of hyperedges (or edges in a graph) could be interdependent. For example, consider a supply-chain network where a certain collection of hyperedges could depend on the same resource and could all fail together if the resource fails; such dependencies are conveniently modeled by introducing a hedge/color that contains these hyperedges. Coudert et al. [9] introduced graphs with colored edges to model dependency between a collection of edges. The term hedgegraph for such colored graphs was coined by Ghaffari, Karger, and Panigrahi [20] who were motivated by applications in graph reliability where multiple edges can fail

S. S. Dey et al. (Eds.): IPCO 2026, LNCS 16588, pp. 331–345, 2026.
https://doi.org/10.1007/978-3-032-28691-8_22

in a coordinated manner. While the richer modeling capability of hedgegraphs naturally lends itself to applications, it also poses algorithmic and structural challenges—see [2,4,8–10,15,16,20,24,30,34]. In fact, as we will discuss later, some of the fundamental tractable problems in graphs and hypergraphs become intractable in hedgegraphs. We investigate structural and algorithmic aspects of connectivity in hedgegraphs by taking a polymatroidal viewpoint that leads to new tractability results as well as insightful connections to classical results on graphs [28,29,32] and more recent results on hypergraphs [18,19].

We formally define hedgegraphs now. A hedgegraph $G = (V, E)$ consists of a finite vertex set V and a finite set E of *hedges*. A *hyperedge* over V is a subset of vertices. Each *hedge* $e \in E$ is a collection (equivalently a set) of distinct hyperedges; by distinct we mean that a hyperedge h in a hedge e does not belong to another hedge $e' \in E$ if $e' \neq e$. If every hedge has exactly one hyperedge, then the hedgegraph is a hypergraph. Throughout this work, we will assume that each hedge is a collection of vertex-disjoint hyperedges: if we have a hedge e with hyperedges $h_1, h_2 \in e$ where $h_1 \cap h_2 \neq \emptyset$, then we remove h_1 and h_2 and add the hyperedge $h_1 \cup h_2$ to e; this does not change the connectivity measures of interest to this work. This assumption about each hedge being a vertex-disjoint collection of hyperedges is crucial for a concise description of our structural results. Moreover, it abstracts the core aspects of the hedgegraph definition and allows us to easily relate hedgegraphs to hypergraphs[1]. The representation size of a hedgegraph is $O(p)$, where $p := \sum_{e \in E} \sum_{h \in e} |h|$. See Fig. 1 for an example.

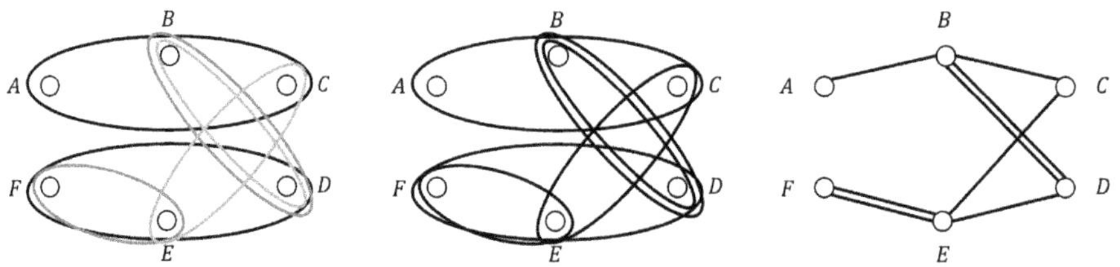

Fig. 1. Leftmost figure shows a hedgegraph with three hedges with different colors. Right pictures show an example of a hypergraph and a graph.

We now define cuts and connectivity in hedgegraphs. Let $G = (V, E)$ be a hedgegraph. For a subset $S \subseteq V$ of vertices, we define $\delta_G(S)$ to be the set of hedges $e \in E$ that cross S; a hedge e crosses S if there is a hyperedge $h \in e$ that intersects both S and $V \setminus S$. Formally,

$$\delta_G(S) := \{e \in E : \exists\ h \in e \text{ with } h \cap S \neq \emptyset, h \cap (V \setminus S) \neq \emptyset\}.$$

We define the cut function $d_G : 2^V \to \mathbb{Z}_{\geq 0}$ of the hedgegraph G as $d_G(S) := |\delta_G(S)|$ for every $S \subseteq V$. The *(hedge-)connectivity* of G is defined as $\lambda_G :=$

[1] Hedgegraphs in [20] were defined via colored edges of a graph. Our definition is equivalent to their definition for the purposes of the problems considered here.

$\min\{d_G(S) : \emptyset \neq S \subsetneq V\}$. The hedgegraph G is *connected* if $\lambda_G > 0$. We drop the subscript G from δ_G, d_G, λ_G when the hedgegraph G is clear from context. One can efficiently verify whether a given hedgegraph is connected; it reduces to testing connectivity of the underlying hypergraph.

In contrast to graph and hypergraph cut functions, the cut function of a hedgegraph is not necessarily submodular—this was pointed out in [20]. The lack of submodularity creates barriers for structure and algorithms. In particular, $\{s, t\}$-connectivity[2] and connectivity are intractable under plausible complexity-theoretic assumptions [14,24,34,35]. Despite the intractability, positive algorithmic results are known for computing connectivity—in particular, a randomized polynomial-time approximation scheme, a randomized quasi-polynomial time algorithm [16,20], a randomized polynomial-time algorithm if every hedge has a constant number of hyperedges [8], and a fixed-parameter algorithm when parameterized by the solution size [15] are known. All these algorithms are randomized and it remains open to design deterministic counterparts. Inspired by these algorithmic results, we investigate structural aspects of connectivity in hedgegraphs and some of their algorithmic consequences. We believe that our structural results will provide valuable insights for future work on hedgegraphs.

Partition-based Measures of Connectedness. Given that connectivity in hedgegraphs is intractable [24], are there other measures of connectedness that are relevant and interesting? We investigate two alternative measures of connectedness of a hedgegraph based on partitions. For ease of understanding, we define and discuss these measures for hypergraphs here and postpone the formal definition for hedgegraphs to Sect. 1.1. Let $G = (V, E)$ be a hypergraph. For a partition $\mathcal{P}$ of V, we define $\delta_G(\mathcal{P})$ to be the set of hyperedges that intersect more than one part of the partition $\mathcal{P}$, i.e.,

$$\delta_G(\mathcal{P}) := \{h \in E : \exists\ A, B \in \mathcal{P} \text{ with } h \cap A \neq \emptyset \text{ and } h \cap B \neq \emptyset\}.$$

We denote *partition connectivity* and *weak partition connectivity* of G by PC_G and WPC_G respectively, where

$$\mathrm{PC}_G := \min_{\mathcal{P} \text{ is a partition of } V} \left\lfloor \frac{|\delta_G(\mathcal{P})|}{|\mathcal{P}| - 1} \right\rfloor \text{ and} \tag{1}$$

$$\mathrm{WPC}_G := \min_{\mathcal{P} \text{ is a partition of } V} \left\lfloor \frac{\sum_{e \in E} (\# \text{ of parts of } \mathcal{P} \text{ intersecting } e - 1)}{|\mathcal{P}| - 1} \right\rfloor. \tag{2}$$

These two measures coincide in graphs while they differ in hypergraphs. Both these measures are also related to connectivity: for a hypergraph G, we have that $\mathrm{PC}_G \leq \mathrm{WPC}_G$ and $\lambda_G/2 \leq \mathrm{WPC}_G \leq \lambda_G$, i.e., weak partition connectivity is within a 2 factor of connectivity [3]. It is helpful to keep three examples in mind while understanding partition connectivity: consider the hypergraph $G_1 = (V, E_1)$ on vertex set V where $E_1 := \{V\}$, the hypergraph $G_2 = (V, E_2)$ where E_2 consists of $|V| - 1$ parallel hyperedges $e_1 = e_2 = \ldots = e_{|V|-1} = V$, and the

[2] $\{s, t\}$-connectivity in a hedgegraph is defined as $\min\{d(S) : s \in S \subseteq V - t\}$, i.e., it is the minimum number of hedges whose deletion disconnects s and t.

cycle graph C_3 on 3 vertices: we observe that $\mathrm{PC}_{G_1} = 0$, $\mathrm{WPC}_{G_1} = 1$, $\lambda_{G_1} = 1$, $\mathrm{PC}_{G_2} = 1$, $\mathrm{WPC}_{G_2} = |V|-1$, $\lambda_{G_2} = |V|-1$, $\mathrm{PC}_{C_3} = \mathrm{WPC}_{C_3} = 1$, and $\lambda_{C_3} = 2$. Tutte [32] and Nash-Williams [28] showed that partition connectivity of a graph is the maximum number of edge-disjoint spanning trees in the graph. This result has far-reaching implications in graph decomposition and orientation [17,22,23, 33]. Frank, Király, and Kriesell [19] generalized Tutte and Nash-Williams' result on partition connectivity of graphs to hypergraphs by showing a hypergraph decomposition result. Frank, Király, and Király [18] generalized Tutte and Nash-Williams' result on weak partition connectivity of graphs to hypergraphs by showing an orientation result. We observe that the connection between weak partition connectivity of hypergraphs and hypergraph orientations has recently found applications in coding theory [1,21]. In this work, we generalize and explore these two partition-based connectedness measures in the context of hedgegraphs.

Hedgegraph Polymatroid. Instead of the cut function, which lacks submodularity, we investigate an alternative function associated with a hedgegraph. We motivate this first. Several structural properties of graphs are derived via a matroidal view, in particular through the graphic matroid induced by the edges of the graph. In particular, Tutte and Nash-Williams' result on partition connectivity can be viewed as a special case of Edmonds' matroid base packing theorem applied to the graphic matroid. We recall that for a graph $G = (V, E)$, the matroid rank function $r : 2^E \to \mathbb{Z}_{\geq 0}$ is defined by $r(A) := |V| - \#\mathrm{Comps}(V, A)$ for every $A \subseteq E$, where $\#\mathrm{Comps}(V, A)$ is the number of connected components in the subgraph (V, A). We may define a similar function for hypergraphs—for a hypergraph $H = (V, E)$, consider the function $f_H : 2^E \to \mathbb{Z}_{\geq 0}$ defined as $f_H(A) := |V| - \#\mathrm{Comps}(V, A)$ for every $A \subseteq E$, where $\#\mathrm{Comps}(V, A)$ is the number of connected components in the sub-hypergraph (V, A). The function f_H is not the rank function of a matroid. Nevertheless, the function f_H is monotone and submodular, and hence, is a polymatroid. One can associate a matroid with every polymatroid over the same ground set (via Edmonds' construction of the rank function of a matroid from a polymatroid [12]) and investigate the structural properties of the resulting matroid to understand structural properties of the polymatroid. This viewpoint is implicit in [19] who study the hypergraphic matroid of Lorea [27] which is equivalent to the matroid derived from the function f_H mentioned above via Edmonds' construction of the rank function. In this work, we make this viewpoint explicit and use the polymatroid perspective to investigate partition connectivity, weak partition connectivity, and other structural aspects of hedgegraphs.

We now define the relevant polymatroid that underlies our investigation of hedgegraphs. Let $G = (V, E)$ be a hedgegraph. The number of connected components in a hedgegraph G is the number of connected components in the hypergraph (possibly multi-hypergraph) (V, H_E), where $H_E := \{h : h \in e \text{ for some } e \in E\}$ of hyperedges. For a subset $A \subseteq E$ of hedges, we define (V, A) to be the sub-hedgegraph over the vertex set V that contains all hedges in A and $\#\mathrm{Comps}(V, A)$ to denote the number of connected components in the sub-hedgegraph (V, A). We define the function $f_G : 2^E \to \mathbb{Z}_{\geq 0}$ as

$$f_G(A) := |V| - \#\text{Comps}(V, A) \ \forall \ A \subseteq E.$$

We observe that the function f_G is monotone, i.e., $f_G(A) \leq f_G(B)$ for every $A \subseteq B \subseteq E$ and submodular, i.e., $f_G(A) + f_G(B) \geq f_G(A \cap B) + f_G(A \cup B)$—see appendix for a formal proof. Hence, we denote f_G as the *hedgegraph polymatroid* associated with the hedgegraph G. We emphasize that the cut function is a set function defined over subsets of vertices while the hedgegraph polymatroid is a set function defined over subsets of hedges. If the hedgegraph is a graph, then the hedgegraph polymatroid is simply the rank function of the graphic matroid associated with the graph. The hedgegraph polymatroid is the protagonist of our work that helps generalize several aspects of graphs and hypergraphs to hedgegraphs. To understand the partition-based measures of connectedness, we rely on two notions of strengths of a polymatroid. These two notions of strengths of a polymatroid have found interesting applications recently [7,31].

1.1 Results

We define the two partition-based measures of connectedness for hedgegraphs and describe our results below.

Partition Connectivity. We define the *partition connectivity* of a hedgegraph $G = (V, E)$ as

$$\text{PC}_G := \min_{\mathcal{P} \text{ is a partition of } V} \left\lfloor \frac{|\delta(\mathcal{P})|}{|\mathcal{P}| - 1} \right\rfloor,$$

where $\delta_G(\mathcal{P}) := \{e \in E : \exists \ h \in e \text{ with } h \text{ intersecting at least two parts of } \mathcal{P}\}$. A hedgegraph G is *k-partition-connected* if $\text{PC}_G \geq k$. We show that partition connectivity of a given hedgegraph can be computed in polynomial time.

Theorem 1. *There exists a deterministic polynomial-time algorithm to compute the partition connectivity of a given hedgegraph.*

Tutte [32] and Nash-Williams [28] showed that partition connectivity in graphs measures decomposition into spanning trees: a graph is k-partition connected if and only if it has k edge-disjoint spanning trees. There is an alternative way to phrase this result that enables a generalization to hypergraphs. For this, we note that a graph is 1-partition connected if and only if it contains a spanning tree. Hence, Tutte and Nash-Williams' result is equivalent to saying that a graph is k-partition connected if and only if it contains k edge-disjoint 1-partition connected subgraphs. Frank, Király, and Kriesell [19] showed that this statement also holds for hypergraphs: a hypergraph is k-partition connected if and only if it contains k hyperedge-disjoint 1-partition connected sub-hypergraphs. We show that this statement also holds for hedgegraphs under our generalized definition of partition connectivity in hedgegraphs.

Theorem 2. *A hedgegraph is k-partition connected if and only if it contains k hedge-disjoint* 1*-partition connected sub-hedgegraphs.*

As stated earlier, a graph is 1-partition connected if and only if it contains a spanning tree. [19] characterized 1-partition connected hypergraphs via a trimming result. A *trimming of a hyperedge* h is obtained by picking a pair of vertices $u, v \in h$ and replacing the hyperedge h by the edge $\{u, v\}$. A *trimming of a hypergraph* $G = (V, H)$ is obtained by trimming each hyperedge $h \in H$. [19] showed that a hypergraph G is 1-partition connected if and only if it contains a sub-hypergraph that can be trimmed to a spanning tree. We generalize this result to hedgegraphs. A *trimming of a hedge* is obtained by picking a hyperedge $h \in e$, a pair of vertices $u, v \in h$, and replacing all hyperedges in e by the single edge $\{u, v\}$. A *trimming of a hedgegraph* $G = (V, E)$ is obtained by trimming each hedge $e \in E$. We emphasize that trimming does not allow complete deletion of hedges—in particular, every trimming of a hedgegraph with m hedges results in a graph with exactly m edges (possibly a multigraph).

Theorem 3. *A hedgegraph is 1-partition connected if and only if it contains a sub-hedgegraph that can be trimmed to a spanning tree.*

We will see a proof of Theorem 1 (i.e., an efficient algorithm to compute partition connectivity of a given hedgegraph) by reduction to submodular minimization. We mention that our techniques underlying the proof of Theorems 2 and 3 give an alternative algorithm to compute partition connectivity via matroid intersection.

Our techniques also help generalize the notion of arboricity from graphs and hypergraphs to hedgegraphs. Let $k \in \mathbb{Z}_+$. Nash-Williams [29] showed that the edge set E of a graph can be partitioned into k forests if and only if $|E[S]| \leq k(|S| - 1)$ for every $S \subseteq V$. Frank, Király, and Kriesell [19] generalized this statement to hypergraphs through the notion of acyclic-trimmability: a set of hyperedges has an *acyclic-trimming* if it has a trimming that is acyclic. They showed that the hyperedge set E of a hypergraph can be partitioned into k acyclic-trimmable subsets if and only if $|E[S]| \leq k(|S|-1)$ for every $S \subseteq V$, where $E[S]$ is the subset of hyperedges that are fully contained in S. We generalize this statement to hedgegraphs next. For a partition $\mathcal{P}$ of the vertex set V of a hedgegraph $G = (V, E)$, we define $E[\mathcal{P}] := E \setminus \delta(\mathcal{P})$. A set of hedges has an *acyclic-trimming* if it has a trimming that is acyclic. See Fig. 2 for an example.

Theorem 4. *Let $G = (V, E)$ be a hedgegraph and $k \in \mathbb{Z}_+$. Then, E can be partitioned into k acyclic-trimmable subsets if and only if $|E[\mathcal{P}]| \leq k(|V| - |\mathcal{P}|)$ for every partition $\mathcal{P}$ of V.*

If G is a hypergraph/graph, then $|E[\mathcal{P}]| \leq k(|V| - |\mathcal{P}|)$ for every partition $\mathcal{P}$ of V if and only if $|E[S]| \leq k(|S| - 1)$ for every subset $S \subseteq V$. Thus, the partition-based condition is equivalent to the subset-based condition in graphs and hypergraphs. However, this equivalence does *not* hold for hedgegraphs; see Fig. 3 for an example. We emphasize that the techniques underlying the proof of Theorem 4 also yield a polynomial-time algorithm to find the least k such that the hedge set of a given hedgegraph can be partitioned into k acyclic-trimmable

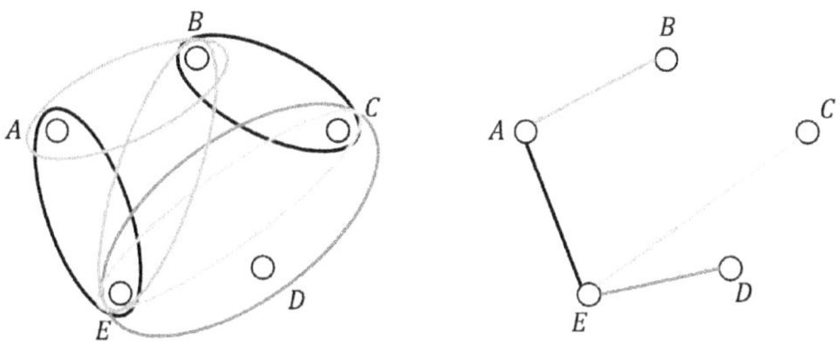

Fig. 2. Example of acyclic-trimming. A hedgegraph $G = (V, E)$ with five hedges: $e_1 := \{\{A, E\}, \{B, C\}\}$, $e_2 := \{\{A, B\}\}$, $e_3 := \{\{C, E\}\}$, $e_4 := \{\{C, D, E\}\}$, and $e_5 := \{\{B, E\}\}$. The right side shows an acyclic-trimming of $\{e_1, e_2, e_3, e_4\}$: e_1 is trimmed into $\{A, E\}$, e_2 is trimmed into $\{A, B\}$, e_3 is trimmed into $\{C, E\}$, and e_4 is trimmed into $\{D, E\}$. We may verify that $\{e_1, e_2, e_3, e_5\}$ does not have an acyclic-trimming.

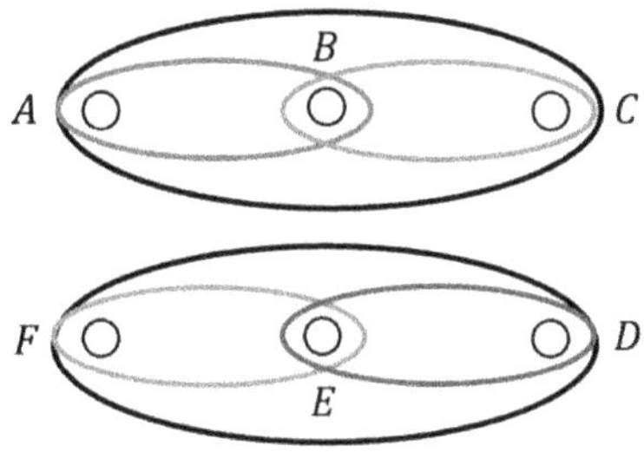

Fig. 3. A hedgegraph $G = (V, E)$ with six vertices $V := \{A, B, C, D, E, F\}$ and five hedges: $e_1 := \{\{A, B, C\}, \{D, E, F\}\}$, $e_2 := \{\{A, B\}\}$, $e_3 := \{\{B, C\}\}$, $e_4 := \{\{D, E\}\}$, and $e_5 := \{\{E, F\}\}$. This hedgegraph satisfies $|E[S]| \leq |S| - 1$ for every subset $S \subseteq V$. However, for the vertex partition $\mathcal{P} = \{\{A, B, C\}, \{D, E, F\}\}$, we have $|E[\mathcal{P}]| = 5 > 4 = |V| - |\mathcal{P}|$.

subsets. Theorems 2 and 3 also lead to a hedgegraph orientation result; see the full version of this work.

Weak Partition Connectivity. We define the weak partition connectivity of a hedgegraph $G = (V, E)$ as

$$\mathrm{WPC}_G := \min_{\mathcal{P} \text{ is a partition of } V} \left\lfloor \frac{\sum_{e \in E} (|\mathcal{P}| - \#\mathrm{Comps}(\mathcal{P}(e)))}{|\mathcal{P}| - 1} \right\rfloor,$$

where $|\mathcal{P}|$ is the number of parts in partition $\mathcal{P}$ and $\mathcal{P}(e)$ is the hedgegraph obtained from the hedgegraph $(V, \{e\})$ by contracting every part in $\mathcal{P}$ into a single vertex. See Fig. 4.

As mentioned earlier, partition connectivity and weak partition connectivity are equal in graphs but differ in hypergraphs. Our definition of weak partition connectivity in hypergraphs coincides with the definition in (2) although they might seem different on first glance. It is well-known that partition connectivity is within a 2-factor of connectivity in graphs. Moreover, weak partition connec-

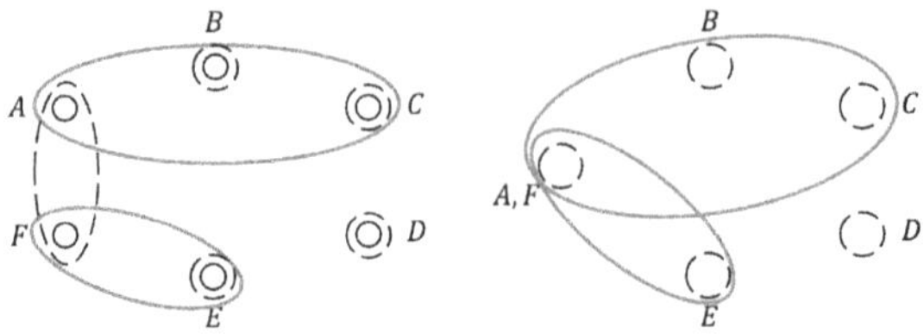

Fig. 4. Example of vertex partition $\mathcal{P}$ and hedgegraph $\mathcal{P}(e)$. The left side shows partition $\mathcal{P} := \{\{B\}, \{C\}, \{D\}, \{E\}, \{A, F\}\}$ and a hedge $e := \{\{A, B, C\}, \{E, F\}\}$. The right side shows the hedgegraph obtained from the hedgegraph $(V, \{e\})$ by contracting every part in $\mathcal{P}$ into a single vertex. We note that $|\mathcal{P}| = 5$ and $\#\text{Comps}(\mathcal{P}(e)) = 2$.

tivity is within a 2-factor of connectivity in hypergraphs [3]. We show that this fact holds in hedgegraphs for the above-mentioned definition of weak partition connectivity.

Lemma 1. *Let $G = (V, E)$ be a hedgegraph with connectivity $\lambda > 0$ and WPC_G be its weak partition connectivity. Then, $\lfloor \frac{\lambda}{2} \rfloor \leq WPC_G \leq \lambda$.*

We state two consequences of Lemma 1. The first addresses deterministic connectivity computation in hedgegraphs. Weak partition connectivity in graphs and hypergraphs can be computed in polynomial time [18]. If weak partition connectivity in hedgegraphs can be computed in deterministic polynomial time, then it immediately implies a deterministic 2-approximation for connectivity via Lemma 1. At this point, we do not know how to compute weak partition connectivity in polynomial time. However, we show that there exists a deterministic polynomial-time algorithm to compute a log-approximation to connectivity using Lemma 1 and connections to the hedgegraph polymatroid. To the best of the authors' knowledge, this is the first deterministic algorithm with a non-trivial approximation guarantee for connectivity in hedgegraphs.

Theorem 5. *There exists a deterministic polynomial-time algorithm to compute an $O(\log n)$-approximation for connectivity in hedgegraphs.*

The second consequence of Lemma 1 addresses sampling in hypergraphs and hedgegraphs. A fundamental result in graphs is that sampling every edge of an n-vertex λ-connected graph independently with probability $\Omega(\log n/\lambda)$ gives a connected graph with high probability. This result is the basis for understanding graph reliability under random edge failures [25,26]. This sampling result for graphs was first shown by Karger [25] using the fact that the number of α-approximate min-cuts in an n-vertex graph is $n^{O(\alpha)}$. It is well-known that the bound on the number of approximate min-cuts does not hold for hypergraphs—e.g., see [8]. Hence, it was unclear whether the same sampling result would hold for hypergraphs and hedgegraphs. Ghaffari, Karger, and Panigrahi [20] showed a slightly weaker bound for hedgegraphs (and hence, hypergraphs): sampling every hedge with probability $\Omega(\log^2 n/\lambda)$ gives a connected hedgegraph with high probability. We improve on their result on hedgegraphs to match the guarantee on graphs.

Theorem 6. *Let $G = (V, E)$ be an n-vertex hedgegraph with connectivity $\lambda > 0$. Let $A \subseteq E$ be a subset obtained by sampling each hedge $e \in E$ with probability at least $\frac{20 \log n}{\lambda}$. Then, the subhedgegraph (V, A) is connected with probability at least $1 - 2/n$.*

We show additional results regarding hedgegraph sparsification and hedgegraph orientations. We refer the reader to the full version [6] for these additional results, proofs of all theorems, and further discussion.

1.2 Preliminaries

A *polymatroid* $f : 2^{\mathcal{N}} \to \mathbb{Z}_{\geq 0}$ on a ground set $\mathcal{N}$ is an integer-valued monotone submodular function with $f(\emptyset) = 0$. A function $f : 2^{\mathcal{N}} \to \mathbb{Z}_{\geq 0}$ is *monotone* if $f(A) \leq f(B)$ for every $A \subseteq B$ and *submodular* if $f(A) + f(B) \geq f(A \cup B) + f(A \cap B)$ for every $A, B \subseteq \mathcal{N}$. A subset $S \subseteq \mathcal{N}$ is a *base* of the polymatroid if $f(S) = f(\mathcal{N})$. For a weight function $w : \mathcal{N} \to \mathbb{R}_{\geq 0}$ and a subset $A \subseteq \mathcal{N}$, we define $w(A) := \sum_{e \in A} w(e)$. We need two different notions of strength in a polymatroid.

Definition 1. *Let $f : 2^{\mathcal{N}} \to \mathbb{Z}_{\geq 0}$ be a polymatroid.*

1. *The* functional-strength *of f is (with the convention that $0/0 = +\infty$):*

$$k^*(f) := \min_{A \subseteq \mathcal{N}} \left\lfloor \frac{\sum_{e \in \mathcal{N}} (f(A+e) - f(A))}{f(\mathcal{N}) - f(A)} \right\rfloor.$$

2. *Let $w : \mathcal{N} \to \mathbb{R}_{\geq 0}$ be weights on the elements. The* w-strength of f *is (with the convention that $0/0 = +\infty$):*

$$\kappa_w(f) := \min_{A \subseteq \mathcal{N}} \frac{w(\mathcal{N}) - w(A)}{f(\mathcal{N}) - f(A)}.$$

For matroid rank functions f, $k^*(f) = \lfloor \kappa_{\vec{1}}(f) \rfloor$, where $\vec{1}$ is the unit weight function, and moreover, both these quantities are equal to the maximum number of disjoint bases in the matroid [11]. For polymatroids f, Călinescu, Chekuri, and Vondrák [5] showed that $k^*(f)$ is within a $O(\log f(\mathcal{N}))$-factor of the maximum number of disjoint bases and obtained the following.

Lemma 2. *[5] Given a polymatroid $f : 2^{\mathcal{N}} \to \mathbb{R}$ by its evaluation oracle, there is a deterministic polynomial time algorithm to compute an $O(\log f(\mathcal{N}))$-approximation for $k^*(f)$.*

The following lemma shows that $\kappa_w(f)$ can be computed in polynomial time.

Lemma 3. *Given a polymatroid $f : 2^{\mathcal{N}} \to \mathbb{R}$ by its evaluation oracle and weights $w : \mathcal{N} \to \mathbb{Z}_{\geq 0}$, there is a strongly polynomial-time algorithm to compute $\kappa_w(f)$.*

2 Partition Connectivity of Hedgegraphs

The following lemma relates partition connectivity of a hedgegraph to the weighted strength of the hedgegraph polymatroid. Lemmas 3 and 4 together imply that partition connectivity of a given hedgegraph can be computed in polynomial time, thereby proving Theorem 1.

Lemma 4. *Let $G = (V, E)$ be a connected hedgegraph and $f : 2^E \to \mathbb{Z}_{\geq 0}$ be the associated hedgegraph polymatroid. Then, $PC_G = \lfloor \kappa_{\vec{1}}(f) \rfloor$, where $\vec{1}$ is the unit weight function.*

Proof. We first show that $\mathrm{PC}_G \geq \lfloor \kappa_{\vec{1}}(f) \rfloor$. Let $\mathcal{P}$ be a partition of V such that $\mathrm{PC}_G = \lfloor \frac{|\delta(\mathcal{P})|}{|\mathcal{P}|-1} \rfloor$. In the subhedgegraph $(V, E \setminus \delta(\mathcal{P}))$, the number of connected components is at least $|\mathcal{P}|$, because for every hedge $e \in E \setminus \delta(\mathcal{P})$, there is no hyperedge in e intersecting different parts of the partition $\mathcal{P}$. We recall that G is connected. Hence,

$$\begin{aligned} f(E) - f(E \setminus \delta(\mathcal{P})) &= (|V| - 1) - (|V| - \#\mathrm{Comps}(V, E \setminus \delta(\mathcal{P}))) \\ &= \#\mathrm{Comps}(V, E \setminus \delta(\mathcal{P})) - 1 \geq |\mathcal{P}| - 1. \end{aligned}$$

Therefore,

$$\mathrm{PC}_G = \left\lfloor \frac{|\delta(\mathcal{P})|}{|\mathcal{P}| - 1} \right\rfloor \geq \left\lfloor \frac{|E| - |E \setminus \delta(\mathcal{P})|}{f(E) - f(E \setminus \delta(\mathcal{P}))} \right\rfloor \geq \min_{A \subseteq E} \left\lfloor \frac{|E \setminus A|}{f(E) - f(A)} \right\rfloor = \lfloor \kappa_{\vec{1}}(f) \rfloor.$$

We show that $\lfloor \kappa_{\vec{1}}(f) \rfloor \geq \mathrm{PC}_G$. Let $A \subseteq E$ be such that $\kappa_{\vec{1}}(f) = \frac{|E \setminus A|}{f(E) - f(A)}$. Let $\mathcal{P}$ be the partition of V where each part corresponds to a connected component in the subhedgegraph (V, A). Hence, $|\mathcal{P}| = \#\mathrm{Comps}(V, A)$. Hence,

$$\begin{aligned} f(E) - f(A) &= (|V| - 1) - (|V| - \#\mathrm{Comps}(V, A)) \quad \text{(since } G \text{ is connected)} \\ &= \#\mathrm{Comps}(V, A) - 1 = |\mathcal{P}| - 1. \end{aligned}$$

We also note that for every hedge $e \in \delta(\mathcal{P})$, at least one hyperedge in e intersects at least two distinct parts of the partition $\mathcal{P}$, which implies that $e \notin A$. Hence, $|\delta(\mathcal{P})| \leq |E \setminus A|$. Therefore,

$$\lfloor \kappa_{\vec{1}}(f) \rfloor = \left\lfloor \frac{|E \setminus A|}{f(E) - f(A)} \right\rfloor = \left\lfloor \frac{|E \setminus A|}{|\mathcal{P}| - 1} \right\rfloor \geq \left\lfloor \frac{|\delta(\mathcal{P})|}{|\mathcal{P}| - 1} \right\rfloor \geq \mathrm{PC}_G.$$

□

We emphasize that the hedgegraph polymatroid function value $f_G(A)$ for a given set $A \subseteq E$ can be evaluated in $O(p)$ time by computing the number of connected components of the sub-hedgegraph (V, A), where $p := \sum_{e \in E} \sum_{h \in e} |h|$ is the representation size. By Lemma 3, $\kappa_{\vec{1}}(f_G)$ can be computed in polynomial time. Together, these yield a polynomial-time algorithm for computing the partition connectivity.

3 Hedgegraph Decompositions

In this section, we sketch the proofs of Theorems 2, 3 and 4. Our proof is by associating a matroid with a hedgegraph. We recall that a *matroid* is a tuple $\mathcal{M} = (\mathcal{N}, \mathcal{I})$, where $\mathcal{N}$ is a finite set and $\mathcal{I} \subseteq 2^{\mathcal{N}}$ is a collection of independent sets satisfying two axioms: (1) if $B \in \mathcal{I}$ and $A \subseteq B$, then $A \in \mathcal{I}$ and (2) if $A, B \in \mathcal{I}$ and $|B| > |A|$, then there exists $e \in B \setminus A$ such that $A \cup \{e\} \in \mathcal{I}$. The rank function $r : 2^{\mathcal{N}} \to \mathbb{Z}_{\geq 0}$ of a matroid $\mathcal{M} = (\mathcal{N}, \mathcal{I})$ is defined by $r(U) := \max\{|A| : A \subseteq U, A \in \mathcal{I}\}$ for every $U \subseteq \mathcal{N}$.

We outline two different ways to arrive at a matroid associated with a hedgegraph. Let $G = (V, E)$ be a hedgegraph. Based on the hedgegraph polymatroid $f : 2^E \to \mathbb{Z}_{\geq 0}$, we define the function $r : 2^E \to \mathbb{Z}_{\geq 0}$ as follows:

$$r(A) := \min\{f(B) + |A \setminus B| : B \subseteq A\} \ \forall\ A \subseteq E. \tag{3}$$

It is known that the function r is the rank function of a matroid via standard results on integer-valued polymatroids [13]. Now, even though we have a matroid based on a hedgegraph, the independent sets of this matroid are not immediately clear. We show that the collection of independent sets of this matroid is

$$\mathcal{I}_G := \{F \subseteq E : \text{hedgegraph } (V, F) \text{ can be trimmed into a forest}\}. \tag{4}$$

Our proof for the above is rather long. Instead, here we first show that $\mathcal{I}_G$ corresponds to the independent sets of a matroid and subsequently, show that r is the rank function of this matroid. Due to space limitations, the second step is deferred to the full version [6].

Lemma 5. *Let $G = (V, E)$ be a hedgegraph and $\mathcal{I}_G$ be as defined in* (4)*. Then, $\mathcal{M} := (E, \mathcal{I}_G)$ is a matroid whose rank function $r : 2^E \to \mathbb{Z}_{\geq 0}$ is given by* (3)*.*

$$r(A) = \min\{f(B) + |A \setminus B| : B \subseteq A\} \ \forall\ A \subseteq E.$$

Proof. We show that $\mathcal{M} := (E, \mathcal{I}_G)$ is a matroid. The rank function requires additional arguments—see full version. It suffices to show the independent set axioms. For ease of notation, we denote a trimming of a subset $F \subseteq E$ of hedges by a function $Y_F : F \to \binom{V}{2}$ with the property that $Y_F(e) \subseteq h$ for some hyperedge $h \in e$. The first independent set axiom follows easily. We prove the second independent set axiom.

Let $A, B \in \mathcal{I}_G$ and $|A| < |B|$. Let $Y_B : B \to \binom{V}{2}$ be a trimming of hedges in B such that $(V, \cup_{e \in B} Y_B(e))$ is a forest. Since $A \in \mathcal{I}_G$, the hedgegraph (V, A) can be trimmed into a forest. Among all such trimmings, we select a trimming $Y_A : A \to \binom{V}{2}$ that maximizes $\Phi(Y_A) := |\{e \in A \cap B : Y_A(e) = Y_B(e)\}|$. If there are multiple such trimmings, then we pick an arbitrary one.

We define graphs $G_{Y_A} := (V, \cup_{e \in A} Y_A(e))$ and $G_{Y_B} := (V, \cup_{e \in B} Y_B(e))$. We note that G_{Y_A} and G_{Y_B} are forests. Hence,

$$\#\text{Comps}(G_{Y_A}) = |V| - |A| > |V| - |B| = \#\text{Comps}(G_{Y_B}),$$

where #Comps(G) is the number of connected components in graph G. Therefore, there exists an edge $\{u,v\} \in \cup_{e\in B} Y_B(e)$ such that u and v are in different components of G_{Y_A}. Let $e' \in B$ be the hedge that is trimmed to $\{u,v\}$, i.e., $Y_B(e') = \{u,v\}$. Suppose $e' \in A$. Then, $Y_A(e') \neq \{u,v\}$ since u and v are disconnected in G_{Y_A}. We note that the trimming

$$Y'_A(e) := \begin{cases} Y_A(e) \text{ if } e \in A \setminus \{e'\}, \\ \{u,v\} \text{ if } e = e', \end{cases}$$

gives a trimming such that the graph $G_{Y'_A} := (V, \cup_{e\in A} Y'_A(e))$ is a forest and $\Phi(Y'_A) > \Phi(Y_A)$, a contradiction. Hence, $e' \notin A$. This implies that $(V, A+e')$ can be trimmed into a forest via the trimming $Y : (A+e') \to \binom{V}{2}$, where

$$Y(e) = \begin{cases} Y_A(e) & \text{if } e \in A, \\ \{u,v\} & \text{otherwise.} \end{cases}$$

Therefore, $A + e' \in \mathcal{I}_G$.

□

For a hedgegraph $G = (V,E)$, we call $\mathcal{M} := (E, \mathcal{I}_G)$ as defined in Lemma 5 to be the *hedgegraph matroid.* Theorem 3 follows via the characterization of independent sets and rank function of the hedgegraph matroid given in Lemma 5. Theorem 2 and Theorem 4 follow by applying Edmonds' base packing theorem and matroid covering theorem [11] to the hedgegraph matroid, respectively.

4 Weak Partition Connectivity of Hedgegraphs

In this section, we relate weak partition connectivity and connectivity of a hedgegraph to the functional strength of the hedgegraph polymatroid. Throughout this section, let $G = (V,E)$ be a hedgegraph with connectivity $\lambda > 0$ and weak partition connectivity WPC_G. Let $f : 2^E \to \mathbb{Z}_{\geq 0}$ be the hedgegraph polymatroid associated with G.

Lemma 6. $\lfloor \lambda/2 \rfloor \leq \mathit{WPC}_G$.

Proof. Let $\mathcal{P}$ be a partition of V with $\text{WPC}_G = \lfloor \frac{\sum_{e\in E}(|\mathcal{P}| - \#\text{Comps}(\mathcal{P}(e)))}{|\mathcal{P}|-1} \rfloor$. Let $e \in E$. We recall that $\mathcal{P}(e)$ is the hedgegraph obtained from the hedgegraph $(V, \{e\})$ by contracting every part in $\mathcal{P}$ into a single vertex. We define

$$\mathcal{T}_e := \{u \in V(\mathcal{P}(e)) : u \text{ is not isolated in } \mathcal{P}(e)\}.$$

We note that $|\mathcal{P}|$ is the number of vertices in $\mathcal{P}(e)$ and #Comps($\mathcal{P}(e)$) is the number of connected components in $\mathcal{P}(e)$. Hence, we have

$$|\mathcal{P}| - \#\text{Comps}(\mathcal{P}(e)) \geq \frac{|\mathcal{T}_e|}{2},$$

since every vertex in $\mathcal{T}_e$ belongs to a connected component of size at least 2 in the hedgegraph $\mathcal{P}(e)$. This implies that

$$\text{WPC}_G = \left\lfloor \frac{\sum_{e\in E}(|\mathcal{P}| - \#\text{Comps}(\mathcal{P}(e)))}{|\mathcal{P}| - 1} \right\rfloor \geq \left\lfloor \frac{1}{2}\sum_{e\in E} \frac{|\mathcal{T}_e|}{|\mathcal{P}| - 1} \right\rfloor \geq \left\lfloor \frac{1}{2}\sum_{e\in E} \frac{|\mathcal{T}_e|}{|\mathcal{P}|} \right\rfloor .$$

For every part $S \subseteq V$ of partition $\mathcal{P}$, we note that a hedge $e \in E$ belongs to $\delta(S)$ if and only if the contracted vertex $u(S)$ corresponding to S in $\mathcal{P}(e)$ is not isolated in $\mathcal{P}(e)$, i.e., $u(S) \in \mathcal{T}_e$. Hence, we have

$$\sum_{S\in\mathcal{P}} |\delta(S)| = \sum_{S\in\mathcal{P}}\sum_{e\in E} \mathbf{1}_{[e\in\delta(S)]} = \sum_{e\in E}\sum_{S\in\mathcal{P}} \mathbf{1}_{[e\in\delta(S)]} = \sum_{e\in E}\sum_{S\in\mathcal{P}} \mathbf{1}_{[u(S)\in\mathcal{T}_e]} = \sum_{e\in E} |\mathcal{T}_e|.$$

We also have that $\sum_{S\in\mathcal{P}} |\delta(S)| \geq \lambda|\mathcal{P}|$ since $|\delta(S)| \geq \lambda$ for every $\emptyset \neq S \subsetneq V$. This implies that

$$\text{WPC}_G \geq \left\lfloor \frac{1}{2}\sum_{e\in E} \frac{|\mathcal{T}_e|}{|\mathcal{P}|} \right\rfloor = \left\lfloor \frac{1}{2}\sum_{S\in\mathcal{P}} \frac{|\delta(S)|}{|\mathcal{P}|} \right\rfloor \geq \left\lfloor \frac{1}{2}\sum_{S\in\mathcal{P}} \frac{\lambda}{|\mathcal{P}|} \right\rfloor = \left\lfloor \frac{\lambda}{2} \right\rfloor .$$

□

We now relate weak partition connectivity and connectivity to the functional strength of the hedgegraph polymatroid.

Lemma 7. *$WPC_G \leq k^*(f) \leq \lambda$.*

Lemmas 6 and 7 together imply Lemma 1. These lemmas together with Lemma 2 imply Theorem 5 as follows: Let $G = (V, E)$ be a hedgegraph and $f : 2^E \to \mathbb{R}$ be the associated hedgegraph polymatroid. Lemmas 6 and 7 imply that

$$\left\lfloor \frac{\lambda}{2} \right\rfloor \leq \text{WPC}_G \leq k^*(f) \leq \lambda.$$

By Lemma 2, there is a deterministic polynomial time algorithm to compute an $O(\log f(E)) = O(\log |V|)$-approximation for $k^*(f)$, which implies an $O(\log |V|)$-approximation for λ via the above inequalities.

5 Conclusion

We investigated structural and algorithmic aspects of hedgegraphs via the hedgegraph polymatroid. The hedgegraph polymatroid allowed us to use properties of matroids and submodularity even though the cut function of a hedgegraph is not submodular. We illustrated some of the consequences of this investigation. We mention one more polymatroidal result that leads to an interesting conclusion on hedgegraphs: there exists an $O(\log^2 |V|)$-competitive randomized algorithm to compute an online packing of maximum number of disjoint connected spanning hedgegraphs when vertices are known a priori and hedges arrive online. This is a direct consequence of applying the polymatroidal result of [7] to the hedgegraph polymatroid. We believe that the hedgegraph polymatroid is a promising tool for both algorithmic and structural aspects of hedgegraphs. We mention two interesting open questions: (1) Does there exist a deterministic approximation scheme for the connectivity of a hedgegraph? Even a constant factor deterministic approximation for hedgegraph connectivity is open. This leads to the second open question: (2) Is weak partition connectivity of a hedgegraph computable in polynomial time?

Acknowledgments. Karthekeyan Chandrasekaran, Chandra Chekuri, and Weihao Zhu were partially supported by NSF grant CCF-2402667. Part of this work was done while Weihang Wang was a PhD student at University of Illinois, Urbana-Champaign.

Disclosure of Interests. The authors have no competing interests to declare that are relevant to the content of this article.

References

1. Alrabiah, O., Guruswami, V., Li, R.: Randomly punctured Reed–Solomon codes achieve list-decoding capacity over linear-sized fields. In: Proceedings of the 56th Annual ACM Symposium on Theory of Computing, pp. 1458–1469 (2024)
2. Amburg, I., Veldt, N., Benson, A.: Clustering in graphs and hypergraphs with categorical edge labels. In: Proceedings of The Web Conference 2020, pp. 706–717 (2020)
3. Bérczi, K., Chandrasekaran, K., Király, T., Kulkarni, S.: Splitting-off in hypergraphs. In: International Colloquium on Automata, Languages and Programming (ICALP), pp. 23:1–23:20 (2024)
4. Bonchi, F., Gionis, A., Gullo, F., Tsourakakis, C.E., Ukkonen, A.: Chromatic correlation clustering. ACM Trans. Knowl. Discov. Data **9**(4) (2015)
5. Călinescu, G., Chekuri, C., Vondrák, J.: Disjoint bases in a polymatroid. Random Struct. Algorithms **35**(4), 418–430 (2009)
6. Chandrasekaran, K., Chekuri, C., Wang, W., Zhu, W.: Hedgegraph polymatroids. arXiv:2510.25043 (2025)
7. Chandrasekaran, K., Chekuri, C., Zhu, W.: Online disjoint spanning trees and polymatroid bases. In: 52nd International Colloquium on Automata, Languages, and Programming (ICALP), pp. 44:1–44:20 (2025)
8. Chandrasekaran, K., Xu, C., Yu, X.: Hypergraph k-Cut in randomized polynomial time. Math. Program. **186**, 85–113 (2021)
9. Coudert, D., Datta, P., Perennes, S., Rivano, H., Voge, M.-E.: Shared risk resource group complexity and approximability issues. Parallel Process. Lett. **17**(02), 169–184 (2007)
10. Coudert, D., Perennes, S., Rivano, H., Voge, M.-E.: Combinatorial optimization in networks with shared risk link groups. Discrete Math. Theor. Comput. Sci. **18**(3) (2016)
11. Edmonds, J.: Minimum partition of a matroid into independent sets. J. Res. Natl. Bur. Stand. Sect. B **69B**, 67–72 (1965)
12. Edmonds, J.: Submodular functions, matroids, and certain polyhedra. Comb. Struct. Appl., pp. 69–87 (1970)
13. Edmonds, J., Rota, G.-C.: Submodular set functions. In: Waterloo Combinatorics Conference. University of Waterloo Waterloo, Ontario (1966)

14. Fellows, M.R., Guo, J., Kanj, I.: The parameterized complexity of some minimum label problems. J. Comput. Syst. Sci. **76**(8), 727–740 (2010)
15. Fomin, F.V., Golovach, P.A., Korhonen, T., Lokshtanov, D., Saurabh, S.: Fixed-parameter tractability of hedge cut. In: Proceedings of the 2025 Annual ACM-SIAM Symposium on Discrete Algorithms (SODA), pp. 1402–1411 (2025)
16. Fox, K., Panigrahi, D., Zhang, F.: Minimum cut and minimum k-cut in hypergraphs via branching contractions. ACM Trans. Algorithms **19**(2) (2023)
17. Frank, A.: Connections in Combinatorial Optimization. Oxford University Press, Oxford (2011)
18. Frank, A., Király, T., Király, Z.: On the orientation of graphs and hypergraphs. Discret. Appl. Math. **131**(2), 385–400 (2003)
19. Frank, A., Király, T., Kriesell, M.: On decomposing a hypergraph into k connected sub-hypergraphs. Discret. Appl. Math. **131**(2), 373–383 (2003)
20. Ghaffari, M., Karger, D., Panigrahi, D.: Random contractions and sampling for hypergraph and hedge connectivity. In: Proceedings of the Twenty-Eighth Annual ACM-SIAM Symposium on Discrete Algorithms, pp. 1101–1114 (2017)
21. Guo, Z., Li, R., Shangguan, C., Tamo, I., Wootters, M.: Improved list-decodability and list-recoverability of Reed–Solomon codes via tree packings. SIAM J. Comput. **53**(2), 389–430 (2024)
22. Jaeger, F.: On nowhere-zero flows in multigraphs. In: Proceedings, Fifth British Combinatorial Conference, Aberdeen, pp. 373–378 (1975)
23. Jaeger, F.: Flows and generalized coloring theorems in graphs. J. Comb. Theory Ser. B **26**(2), 205–216 (1979)
24. Jaffke, L., Lima, P.T., Masařík, T., Pilipczuk, M., Souza, U.S.: A tight quasi-polynomial bound for global label min-cut. In: Proceedings of the 2023 Annual ACM-SIAM Symposium on Discrete Algorithms (SODA), pp. 290–303 (2023)
25. Karger, D.R.: Random sampling in cut, flow, and network design problems. Math. Oper. Res. **24**(2), 383–413 (1999)
26. Karger, D.R.: A randomized fully polynomial time approximation scheme for the all-terminal network reliability problem. SIAM Rev. **43**(3), 499–522 (2001)
27. Lorea, M.: Hypergraphes et matroides. Cahiers Centre Etudes Rech. Oper **17**, 289–291 (1978)
28. Nash-Williams, C.S.J.A.: Edge disjoint spanning trees of finite graphs. J. London Math. Soc. **36**, 445–450 (1961)
29. Nash-Williams, C.S.J.A.: Decomposition of Finite graphs into forests. J. London Math. Soc. **39**, 12 (1964)
30. Petruschka, A., Sapir, S., Tzalik, E.: Color Fault-Tolerant spanners. In: 15th Innovations in Theoretical Computer Science Conference (ITCS 2024), pp. 88:1–88:17 (2024)
31. Quanrud, K.: Quotient sparsification for submodular functions. In: Proceedings of the 2024 Annual ACM-SIAM Symposium on Discrete Algorithms (SODA), pp. 5209–5248. SIAM (2024)
32. Tutte, W.T.: On the problem of decomposing a graph into n connected factors. J. London Math. Soc. **36**, 221–230 (1961)
33. West, D.: Introduction to Graph Theory, vol. 2. Prentice Hall (2001)
34. Zhang, P., Cai, J.-Y., Tang, L.-Q., Zhao, W.-B.: Approximation and hardness results for label cut and related problems. J. Comb. Optim. **21**(2), 192–208 (2011)
35. Zhang, P., Fu, B.: The label cut problem with respect to path length and label frequency. Theoret. Comput. Sci. **648**, 72–83 (2016)

Efficiently Coloring the Intersection of a General Matroid and Combinatorial Matroids

Stephen Arndt[1(✉)], Benjamin Moseley[1], Kirk Pruhs[2], and Michael Zlatin[3]

[1] Carnegie Mellon University, Pittsburgh, PA, USA
{sarndt,moseleyb}@andrew.cmu.edu
[2] University of Pittsburgh, Pittsburgh, PA, USA
kirk@cs.pitt.edu
[3] Pomona College, Claremont, CA, USA
michael.zlatin@pomona.edu

Abstract. This paper shows a polynomial-time algorithm that, given a general matroid M_1 and $k-1$ partition matroids $M_2, \ldots, M_k$, produces a coloring of the intersection $M = \cap_{i=1}^{k} M_i$ using at most $1 + \sum_{i=1}^{k} (\chi(M_i) - 1)$ colors. This is the first polynomial-time $O(k)$ approximation algorithm for matroid intersection coloring where one of the matroids may be a general matroid. Leveraging the fact that most of the standard combinatorial matroids reduce to partition matroids at a loss of a factor of two in the chromatic number, this algorithm also yields a polynomial-time $O(k)$-approximation algorithm for matroid intersection coloring in the case where each of the matroids $M_2, \ldots, M_k$ are one of these standard combinatorial types. Even when $k = 2$, the previous best-known approximation ratio was $O(\log n)$ via a reduction to Set Cover.

1 Introduction

Some of the most generally applicable tools in an algorithmist's toolkit are efficient algorithms for various standard matroid-related optimization problems. In particular, a reasonable number of natural optimization problems can be viewed as set cover problems, in which the constraints can be partitioned into parts that induce a matroid [21,26]. Such problems can be viewed as matroid intersection coloring problems, defined as follows:

- The input is a collection of matroids $M_1 = (X, \mathcal{I}_1), \ldots, M_k = (X, \mathcal{I}_k)$ on common ground set X. The intersection $M = \cap_{i=1}^{k} M_i$ is defined to be the set system $\left(X, \cap_{i=1}^{k} \mathcal{I}_i\right)$.
- A feasible solution is a coloring of the elements of X such that, for every color, the set of elements assigned that color is independent in M, i.e. independent in each matroid M_i for $i \in [k]$.

S. S. Dey et al. (Eds.): IPCO 2026, LNCS 16588, pp. 346–361, 2026.
https://doi.org/10.1007/978-3-032-28691-8_23

- The objective is to *minimize the number of colors* used. For any set system M, we use $\chi(M)$ to denote the chromatic number of M, that is the minimum number of colors necessary for the existence of a feasible coloring.

There is a significant literature within the combinatorics/mathematics community that establishes existential results related in some way to matroid intersection coloring (and/or packing) [2,3,7,9,10,15,16,22]. However, most of the proofs of these results are nonconstructive, in that they are not readily adaptable to yield polynomial-time coloring algorithms. In particular, many of these results are established with topological fixed-point arguments that employ Sperner's lemma [2,3,7]. We now highlight some of these existential results that are most relevant for this paper.

Theorem 1. *[3] For k general matroids $M_1, \ldots, M_k$, it is the case that $\chi\left(\cap_{i=1}^k M_i\right) \leq (2k-1)\max_{i=1}^k \chi(M_i)$.*

For the intersection of two general matroids ($k = 2$), this can be improved to the following.

Theorem 2. *[7] For two general matroids M_1 and M_2 it is the case that $\chi(M_1 \cap M_2) \leq \chi(M_1) + \chi(M_2)$.*

The proofs of Theorems 1 and 2 both hinge on topological fixed-point arguments, deriving from an influential earlier paper by Aharoni and Berger [2] that shows that $\chi(M_1 \cap M_2) \leq 2\max\{\chi(M_1), \chi(M_2)\}$. These theorems show that in some sense the chromatic number of the intersection of k matroids grows at most linearly with k. Further, there is a lower bound that rules out the possibility of many natural types of sublinear dependence.

Theorem 3. *[3] There are infinitely many natural numbers k, such that there are $k+1$ partition matroids $M_1, \ldots, M_{k+1}$, each with chromatic number k, and where the intersection of these matroids has chromatic number k^2.*

However, the algorithmic question of producing such colorings for matroid intersection lags well behind what is known existentially. We now summarize the literature on polynomial-time algorithms for matroid intersection coloring. Note that for the duration of the paper, we allow partition matroids to have arbitrary capacity constraints on each part (sometimes called "generalized partition matroids" in the literature).

- **A single matroid.** Edmonds' algorithm optimally colors the elements in polynomial time [13,26].
- **Two partition matroids.** This case reduces to edgecoloring a bipartite graph and is solvable in polynomial time [26].
- **Two strongly-base–orderable matroids.** Davies and McDiarmid give a polynomial-time optimal algorithm for any pair of strongly-base–orderable matroids assuming oracle access to the bijective map [12]. Partition, transversal, and gammoid matroids are all strongly-base–orderable; but most notably, graphic matroids are not strongly-base–orderable [8,26].

- k **partition matroids.** A simple greedy procedure colors the intersection using $1 + \sum_{i=1}^{k}(\chi(M_i) - 1)$ colors.
- k **combinatorial matroids.** If each of the matroids are one of the standard combinatorial types (graphic, laminar, transversal, partition, or a gammoid), then there is a polynomial-time algorithm that uses $1 + \sum_{i-1}^{k}(2\chi(M_i) - 1)$ colors [10,21,23], as all of these types of combinatorial matroids can be efficiently reduced to a partition matroid at a cost of increasing the chromatic number by at most a factor of two [10,21,23]. In contrast it is known that no such reduction is possible for some binary matroids [1,24].
- **Instance optimality hardness.** It is known that there is no polynomial-time algorithm to optimally color the intersection of two general matroids under the independence oracle model [6]. Further, it is NP-hard to optimally color the intersection of a graphic matroid and a partition matroid [11,19], or to optimally color the intersection of three partition matroids [25], when these matroids are given by an explicit representation.

Bridging the constructive gap. There are no polynomial-time approximation algorithms known for matroid intersection coloring if even one of the matroids is not one of the standard combinatorial types (except via a trivial reduction to Set Cover). Our goal is to match the existential results using constructive proofs, which also yield polynomial-time algorithms to find the desired coloring. Often, exactly matching nonconstructive existential results with constructive existential results is a daunting task. Due to this, it is common to either settle for a slightly weaker result, or require slightly stronger preconditions (c.f. [5]).

In this paper, we consider matroid intersection coloring problems where all but one of the matroids is a partition matroid (or alternatively, by applying the known reductions to partition matroids, if all but one of the matroids are one of the standard combinatorial types). Besides being a logical next step towards general matroid intersection coloring, our motivation to consider such instances arises from the observation that some well-known conjectures in the combinatorics literature can be stated as conjectures about coloring the intersection of a general matroid and a collection of partition matroids. In particular, these matroid intersection coloring problems are related to Rota's Basis Conjecture [20] and to the Strong Coloring Conjecture [4]. There is significant literature around these well-known conjectures that contain existential results with nonconstructive proofs. This gives targets to shoot for when developing constructive polynomial-time algorithms.

Further, computing the maximum number of disjoint common bases between two general matroids is reducible to the special case where one matroid is general and the other is partition [17]. Thus it seems possible that such a reduction may also exist for minimum coloring, and coloring the intersection of one general matroid and one partition matroid may capture the complexity of coloring the intersection of two general matroids.

1.1 Our Results

We work in the standard model where the matroids are accessed via a polynomial-time independence oracle. The main result of this paper is:

Theorem 4. *There is a polynomial-time algorithm that, given a general matroid* $M_1 = (X, \mathcal{I}_1)$ *and* $k-1$ *partition matroids* $M_2, \ldots, M_k$*, will produce a coloring of the intersection* $M = \cap_{i=1}^{k} M_i$ *using at most* $1 + \sum_{i=1}^{k} (\chi(M_i) - 1)$ *colors.*

The upper bound in Theorem 4 matches the standard upper bound when all of the matroids are partition matroids, improves by one on the existential upper bound for two general matroids in Theorem 2, and improves by a factor of two on the existential upper bound for k general matroids from Theorem 1. Another immediate corollary of Theorem 4 is:

Corollary 5. *There is a polynomial-time algorithm that, given a general matroid* $M_1 = (X, \mathcal{I}_1)$ *and* $k-1$ *matroids* $M_2, \ldots, M_k$ *that are each of the following types: graphic, laminar, transversal, partition, or a gammoid, will produce a coloring of the intersection* $M = \cap_{i=1}^{k} M_i$ *that has worst-case approximation ratio* $2k - 1$.

Corollary 5 follows from Theorem 4 by first noting that the chromatic number of the intersection of matroids is at least the maximum chromatic number of the individual matroids and then applying the known efficient reductions of these standard combinatorial matroids to partition matroids [10,21,23]. Taking $k = 2$ in Corollary 5 yields:

Corollary 6. *There is a polynomial-time 3-approximation for matroid intersection coloring of a general matroid and a matroid which is either graphic, laminar, transversal, partition, or a gammoid.*

In the $k = 2$ case, Corollary 6 gives the first polynomial-time $O(1)$-approximation algorithm for matroid intersection coloring where one of the matroids may be a general matroid. Previously, the best-known approximation for coloring a general and partition matroid was $O(\log n)$, through a reduction to Set Cover (where the sets are given through oracle access).

The difficulty of matroid intersection coloring. A natural algorithmic design approach is to, like the standard Edmonds' algorithm for coloring one matroid [13,26], try to design an algorithm where after q iterations, it is the case that q items are feasibly colored. Thus, on each iteration, items are recolored to result in a feasible coloring of one more element. Let us first consider the problem of coloring a single matroid. The difficulty is that when an item is recolored, say item x is recolored blue for concreteness, then the feasibility constraints may require that some other item, say y, currently colored blue, be recolored to some other color. In such a case, let us say that the recoloring of x to blue was fettered. Edmonds' algorithm for coloring one matroid addresses this issue by, on each iteration, performing an augmenting sequence of recolorings, where each coloring corrects the infeasibility created by the prior fettered recoloring, and the last recoloring is unfettered.

The standard interpretation of Edmonds' algorithm is to find such an augmenting sequence by identifying a *shortest path* in a particular digraph, that we will call Edmonds' digraph [26]. However, when attempting to adapt this approach to coloring two (or more) matroids, recoloring x to blue may require that *two (or more)* elements currently colored blue be recolored. This naturally leads to considering algorithms where the augmenting path in Edmonds' algorithm is replaced by an augmenting tree of hyperedges of cardinality three, á la Haxell's hypergraph matching algorithm [18]. However, it is not at all clear how to efficiently find such an augmenting tree that would produce a feasible recoloring of one more item.[1] Rather than surmount this issue, our algorithm that establishes Theorem 4 circumvents the issue, which we now explain.

Generalized Edmonds' Algorithm. In Sect. 2 we explain one of the two main components in the design of the algorithm that establishes Theorem 4, namely a modest generalization of Edmonds' algorithm, which we call Generalized Edmonds' Algorithm. We emphasize that Generalized Edmonds' Algorithm is essentially a known algorithm and the generalization follows directly from Lemma 13.1.11 in [14]. However, the definitions and algorithm description in Sect. 2 will be important for understanding our matroid intersection coloring algorithm in Sect. 3.

One issue with building on Edmonds' algorithm to color multiple matroids, is that it is easy to construct instances where augmenting along the shortest path P in Edmonds digraph for matroid M_1 will not produce a feasible coloring with respect to the other matroids. Thus we need greater flexibility with respect to how matroid M_1 can be recolored. Upon closer examination of the standard proof of correctness for Edmonds' algorithm [26], one can observe that it is sufficient that P is what we will call a color-chordless path, which is roughly a path where there is no chord/shortcut between vertices that involve the same color. A shortest path, as it has no chords, is trivially a color-chordless path. The Generalized Edmonds' Algorithm augments along an arbitrary color-chordless path on each iteration, giving us greater flexibility when trying to adapt the algorithm to coloring multiple matroids.

Our Matroid Intersection Coloring Algorithm. In Sect. 3 we design and analyze the algorithm that establishes Theorem 4. At a high level, our algorithm finds a color-chordless path P in the Edmonds digraph for matroid M_1 that is suffix-feasible with respect to each of the partition matroids $M_2, \ldots, M_k$. The path P is suffix-feasible with respect to a matroid M_i if iteratively augmenting along P (as in Edmonds' algorithm) results in a sequence of recolorings where each recoloring is unfettered in M_i *at the time of the recoloring*. To accomplish this, the algorithm first constructs a subgraph H of the Edmonds digraph G

[1] For hypergraph matching, this was complicated but doable [5], essentially because the matroids involved were all partition matroids.

for M_1, such that in H every source-sink path is a color-chordless path in G. The algorithm then finds a source-sink path P within H, that is suffix-feasible with respect to each of the partition matroids $M_2, \ldots, M_k$, using an iterative greedy algorithm. In Sect. 4 we demonstrate one iteration of the algorithm on a particular instance.

Extensions. See https://arxiv.org/abs/2508.19473 for a full version of the paper, with applications of our algorithm to a generalization of Rota's Basis Conjecture [20] and the Strong Coloring Conjecture [4], proofs about the hardness of approximation of matroid intersection coloring, and more.

2 Generalized Edmonds' Algorithm for Coloring a Single Matroid

This section gives a generalization of Edmonds' algorithm [13,26] for coloring the elements of a single arbitrary matroid optimally that follows directly from [14]. As with the standard Edmonds' algorithm, the Generalized Edmonds' algorithm has the following input, output, and outer loop invariant.

Input: An arbitrary matroid M and target chromatic number α.

Output: An α-coloring of M if one exists.

Outer Loop Invariant: After q iterations, q elements of the ground set are feasibly colored. A coloring is *feasible* if its color classes are independent sets in the matroid.

Description of the Generalized Edmonds' Algorithm: Iteratively update the current coloring c along an arbitrary color-chordless source-sink path P in the current Edmonds digraph G.

The standard interpretation of Edmonds' algorithm [26] is an instantiation of this algorithm in which the arbitrary color-chordless source-sink path is selected to be a shortest source-sink path. To finish the description, we need to define color-chordless source-sink path. Then we remind the reader of the definition of the Edmonds digraph G and how a coloring is updated along a path (as these are the same in the standard description of Edmonds' algorithm [26]).

Definition 1 (Edmonds Digraph). *The definition of Edmonds digraph $G = (V, A)$ depends on the matroid $M = (X, \mathcal{I})$, the target chromatic number α, and the current coloring c. Let $c : X \to \{0, 1, \ldots, \alpha\}$ be the current coloring of elements, where $c(x) = 0$ if x is uncolored. Let $S_1, S_2, \ldots, S_\alpha \in \mathcal{I}$ be the color classes of c. Let $U \subseteq X$ be the set of elements not colored by c. The vertex set $V = [\alpha] \cup X$ consists of a vertex for each element of the ground set X of M and one vertex for each color. We call the vertices in $[\alpha]$ corresponding to the α available colors the* ***sources*** *and the uncolored elements in U the* ***sinks.*** *The arc set $A = \cup_{i=1}^{\alpha} A_i$ is the union of α collections of arcs, where A_i represents the feasible exchanges for color class i. In A_i there is an arc (i, x) to all elements $x \notin S_i$ with the property that $S_i \cup \{x\} \in \mathcal{I}$. Then, for all elements $x \notin S_i$ such that $S_i \cup \{x\} \notin \mathcal{I}$, A_i contains the arc (y, x) for all elements $y \in S_i$ with the property that $S_i - \{y\} \cup \{x\} \in \mathcal{I}$.*

Definition 2 (Color-Chordless Path). *Let $P = (x_1, x_2, \ldots, x_\ell)$ be a directed path in G. A* ***color-chord*** *of P is an arc $(x_j, x_k) \in A$ between two elements $x_j, x_k \in X$ where $1 \leq j \leq k-2$, and x_j is the same color as x_{k-1} in the current coloring c (note this is x_{k-1} and not x_k). P is* ***color-chordless*** *if it has no color-chords.*

Definition of Updating Along P: Let $P = (x_1, x_2, \ldots, x_\ell)$ be a source-sink path in G. The updated coloring $c\Delta P$ is computed by coloring x_2 with the color of x_1, and coloring x_i with the color of x_{i-1} in c for all $3 \leq i \leq \ell$.

We now show that upon updating the current coloring c along an arbitrary color-chordless source-sink path P, the resulting coloring $c\Delta P$ is still feasible. This follows directly from Lemma 13.1.11 in [14] but we include a full proof for completeness.

Lemma 7. *Let c be a feasible coloring of i elements in M. Let G be the Edmonds digraph of M, c. Let $P = (x_1, x_2, \ldots, x_\ell)$ be a color-chordless source-sink path in G. Then $c\Delta P$ is a feasible coloring of $q+1$ elements in M.*

Proof. The proof structure is as follows. First, we assume for sake of contradiction that $c\Delta P$ is not feasible in M, and find the first element which is not feasibly recolored. Next, we study two circuits C, C' in M containing this element. Lastly, we appeal to the Matroid Circuit Axiom to produce a third circuit C'' in M, which we use to establish the contradiction.

Note $c\Delta P$ is a coloring of $q+1$ elements, because $x_2, x_3, \ldots, x_{\ell-1}$ remain colored and x_ℓ becomes colored. Assume for the sake of contradiction that $c\Delta P$ is not feasible in M. Recolor $x_2, x_3, \ldots, x_\ell$ according to P sequentially, and let x_t be the first element for which recoloring x_t makes the coloring infeasible. Suppose x_t is recolored with color j.

Let $S_j \in \mathcal{I}$ be the jth color class in c, and $S'_j \in \mathcal{I}$ be the jth color class immediately prior to recoloring x_{t-1}. Note $t \geq 3$, because recoloring x_2 with color x_1 is a feasible recoloring by definition of G. Thus $S_j \cup \{x_t\} \notin \mathcal{I}$ because of the existence of arc (x_{t-1}, x_t) in G, so there exists a circuit $C \subseteq S_j \cup \{x_t\}$ containing x_t. Further, no elements in $C - \{x_t\}$ are recolored before x_{t-1}. This is because all elements $x \in C - \{x_t\}$ have $S_j - \{x\} \cup \{x_t\} \in \mathcal{I}$, and thus $(x, x_t) \in A$. Therefore if $x_h \in C - \{x_t\}$ for some $h \leq t-2$, then $(x_h, x_t) \in A$ would be a color-chord of P, a contradiction. Thus $C - \{x_t\} \subseteq S'_j$.

Because recoloring x_t makes the coloring infeasible, we have $S'_j - \{x_{t-1}\} \cup \{x_t\} \notin \mathcal{I}$, so there exists a circuit $C' \subseteq S'_j - \{x_{t-1}\} \cup \{x_t\}$ containing x_t. Further, $C \neq C'$ because C contains x_{t-1} but C' does not. Thus by the Matroid Circuit Axiom, there exists a circuit $C'' \subseteq (C \cup C') - \{x_t\} \subseteq S'_j$. But $S'_j \in \mathcal{I}$, a contradiction.

As with Edmonds' algorithm, the Generalized Edmonds' algorithm can be extended to compute $\chi(M)$ by using binary search to find the minimum α where the algorithm finds a feasible coloring.

3 Matroid Intersection Coloring Algorithm

In this section we prove our main result, Theorem 4, that there is an efficient algorithm for coloring the intersection of a general matroid and a collection of partition matroids. In Subsect. 3.1, we state the algorithm. In Subsect. 3.2, we prove that the algorithm produces a feasible coloring that uses at most the desired number of colors.

3.1 Algorithm Description

Our Matroid Intersection Coloring Algorithm has the following input, output, and outer loop invariant.

Input: An arbitrary matroid M_1 and partition matroids M_i for $2 \leq i \leq k$ on common ground set X.

Output: A $\left(1 + \sum_{i=1}^{k} (\chi(M_i) - 1)\right)$-coloring of $M = \cap_{i=1}^{k} M_i$.

Outer Loop Invariant: After q iterations, q elements of the ground set are feasibly colored. A coloring is *feasible* if its color classes are independent sets in all of the matroids.

Description of our Matroid Intersection Coloring Algorithm: Iteratively update the current coloring c along an arbitrary source-sink path P, in the color-chordless subgraph H of the current Edmonds digraph G of M_1, with the property that P is suffix-feasible with respect to each of the partition matroids $M_2, \ldots, M_k$.

To complete the description of our algorithm, we need to define H, what it means for a path P to be suffix-feasible with respect to a matroid M_i, and explain how to compute such a path P. It should be obvious from the definition of H how to efficiently compute H. For ease of notation, let $\alpha = \chi(M_1)$ and $B = \sum_{i=2}^{k} (\chi(M_i) - 1)$. Note that $\chi(M_1)$ can be computed using Edmonds' algorithm for single matroid coloring, and $\chi(M_i)$ for $2 \leq i \leq k$ is simply the ceiling of the maximum ratio of part size to part capacity over all parts in M_i.

Definition 3 (Color-Chordless Subgraph). *The subgraph H is computed from the Edmonds digraph $G = (V = [\alpha + B] \cup X, A)$ of M_1. The vertices in H are partitioned into layers $L_0, L_1, \ldots, L_h$, which are defined inductively as follows. The base layer is L_0, which is defined to be $[\alpha + B]$, the collection of color nodes in G. There are no arcs entering L_0. Then, L_i is defined to be the collection of vertices x not in $L_0, \ldots, L_{i-1}$ such that x has $B + 1$ incoming arcs in G from nodes of $B + 1$ distinct colors in earlier layers of the subgraph H. In this definition, the color of an element node is its color in the coloring c. Let $y_1, \ldots, y_{B+1}$ be a set of such nodes (so y_i and y_j have distinct colors for $i \neq j$) with the property that for all $i \in [B + 1]$, there does not exist a vertex z_i with the same color as y_i, such that z_i is in an earlier layer than y_i and (z_i, x) is an arc in G. Then the arcs $(y_1, x), \ldots, (y_{B+1}, x)$ are added to H. Intuitively, the construction of H prefers arcs from earlier layers.*

Definition 4 (Suffix-Feasible Path). *Let $P = (x_1, x_2, \ldots, x_\ell = u)$ be a path in the Edmonds digraph G, constructed from a feasible coloring c of matroid $M_1 = (X, \mathcal{I}_1)$, that ends at a sink, that is an uncolored vertex u. Let M_h be an arbitrary matroid.*

- *The updated coloring $c' = c\Delta P$ is defined as follows. In c', a vertex x_i, $3 \le i \le \ell$, is colored $c(x_{i-1})$. If x_1 is a color vertex, that is $x_1 \in [\alpha + B]$, then $c'(x_2) = x_1$. If x_1 is not a color vertex, that is $x_1 \notin [\alpha + B]$, then $c'(x_2) = c(x_1)$ and x_1 is uncolored in c', that is $c'(x_1) = 0$.*
- *The path P is suffix-feasible with respect to M_h if for every suffix $P_j = (x_{\ell-j}, x_{\ell-j+1}, \ldots, x_\ell = u)$ of P it is the case that $c\Delta P_j$ is a feasible coloring in M_h.*

Algorithm to Compute P: The source-sink path $P = (x_1 = r, x_2, x_3, \ldots, x_\ell = u)$ is computed iteratively. The loop invariant is that after j iterations, the suffix P_j will have been computed. Initially P_0 is an arbitrary uncolored element u. We now explain how P_j is computed from P_{j-1}. The vertex $x_{\ell-j}$ is set to an arbitrary vertex z in H such that there is an edge $(z, x_{\ell-j+1})$ in H, and the path $(z, x_{\ell-j+1}, \ldots, x_\ell)$ is suffix-feasible with respect to each of the partition matroids $M_2, \ldots, M_k$.

3.2 Algorithm Analysis

In order to prove the correctness of our algorithm, we will prove the following statements. In Lemma 8, we will show all source-sink paths in H are color-chordless paths in G. In Lemma 9, we will show that H contains an uncolored element u. In Lemma 10, we will show that during construction of the suffix-feasible path P, a vertex z always exists to extend the suffix. In Lemma 11, we will tie everything together and show that our algorithm maintains a feasible coloring c in $M_1, M_2, \ldots, M_k$.

Throughout the analysis, we fix the following objects:

- An arbitrary matroid M_1 and partition matroids M_i for $2 \le i \le k$ on common ground set X.
- An arbitrary feasible coloring c of q elements of $M = \bigcap_{i=1}^k M_i$, using $(\alpha + B)$ colors where $\alpha = \chi(M_1)$ and $B = \sum_{i=2}^k (\chi(M_i) - 1)$.
- The Edmonds digraph G of the matroid M_1 with respect to the coloring c.
- The color-chordless subgraph H in G computed by our algorithm.
- The source-sink suffix-feasible path P in H computed by our algorithm.

Lemma 8. *All source-sink paths in H are color-chordless paths in G.*

Proof. Let $P' = (x_1, x_2, \ldots, x_\ell)$ be an arbitrary source-sink path in H. Assume for the sake of contradiction that there exists a color-chord of P', that is, an arc $(x_j, x_k) \in A$ such that $1 \le j \le k - 2$ and x_j has the same color as x_{k-1} in the current coloring c. Consider the time when (x_{k-1}, x_k) was added to H. At this time, x_{k-1} had minimum level among all $x \in V(H)$ with the same color as x_{k-1} such that $(x, x_k) \in A$. But since $j \leq k-2$, the node x_j has a lower level then x_{k-1}, a contradiction. Thus P' is color-chordless.

We now state and prove Lemma 9. The central claim is that if a cut in G separates the color nodes from the uncolored elements, then it contains many arcs originating from different colors. Therefore H must contain an uncolored element u, because otherwise it could be extended by one more level.

Recall that the set of nodes $L = [\alpha + B]$ are called **sources** and the set of uncolored elements U are called **sinks**. We say that $(\overline{Y}, Y)$ is a **source-sink separating cut** if $L \subseteq V \setminus Y$ and $U \subseteq Y$. We denote by $\delta(\overline{Y}, Y)$ the set of arcs going from $\overline{Y}$ to Y, that is $\delta(\overline{Y}, Y) = \{(x, y) \in A : x \in \overline{Y}, y \in Y\}$.

Lemma 9. *Let $(\overline{Y}, Y)$ be a source-sink separating cut in G. Then there exists a subset of arcs $A' \subseteq \delta(\overline{Y}, Y)$ such that $|A'| > B \cdot |Y|$ and no two arcs in A' share both the same target node x and the same source node color j.*

Proof. We will first show that many elements in Y can change colors feasibly in M_1. We then observe that each one of these color changes induces a corresponding arc in $\delta(\overline{Y}, Y)$.

By assumption, M_1 is an α-colorable matroid. Let $T_1, T_2, \ldots, T_\alpha$ be the color classes of an α-coloring of M_1. Let $S_1, S_2, \ldots, S_{\alpha+B}$ be the color classes of our $(\alpha + B)$-coloring c. Restrict both colorings to Y, by letting $T_i' = T_i \cap Y$ for $i = 1, 2, \ldots, \alpha$ and $S_j' = S_j \cap Y$ for $j = 1, 2, \ldots, \alpha + B$.

We now loop over $j = 1, 2, \ldots, \alpha+B$, and show that many elements $x \in Y \setminus S_j'$ have $S_j' \cup \{x\} \in \mathcal{I}_1$. For each $j = 1, 2, \ldots, \alpha + B$, and for each $i = 1, 2, \ldots, \alpha$, consider S_j', T_i'. There exist at least $|T_i'| - |S_j'|$ elements $x \in T_i' \setminus S_j' \subseteq Y \setminus S_j'$ such that $S_j' \cup \{x\} \in \mathcal{I}_1$ via the matroid exchange property. Thus, for each $j = 1, 2, \ldots, \alpha + B$, the number of elements $x \in Y \setminus S_j'$ such that $S_j' \cup \{x\} \in \mathcal{I}_1$ is at least $\sum_{i=1}^{\alpha}(|T_i'| - |S_j'|)$. Summing over all $j = 1, 2, \ldots, \alpha + B$, the number of j, x pairs with this property is at least

$$
\begin{aligned}
\sum_{j=1}^{\alpha+B} \sum_{i=1}^{\alpha} (|T_i'| - |S_j'|) &= \sum_{j=1}^{\alpha+B} \sum_{i=1}^{\alpha} |T_i'| - \sum_{i=1}^{\alpha} \sum_{j=1}^{\alpha+B} |S_j'| \\
&= \sum_{j=1}^{\alpha+B} |Y| - \sum_{i=1}^{\alpha} |Y \setminus U| \\
&> \sum_{j=1}^{\alpha+B} |Y| - \sum_{i=1}^{\alpha} |Y| \\
&= (\alpha + B)|Y| - \alpha|Y| \\
&= B|Y|
\end{aligned}
$$

We now show that each j, x pair such that $x \in Y \setminus S_j'$ and $S_j' \cup \{x\} \in \mathcal{I}_1$ induces a corresponding arc in $\delta(\overline{Y}, Y)$.

Claim. For all $j = 1, 2, \ldots, \alpha + B$ and $x \in Y \setminus S'_j$, if $S'_j \cup \{x\} \in \mathcal{I}_1$, then there exists an arc $(y, x) \in \delta(\overline{Y}, Y)$ where y has color j.

Proof. If $S_j \cup \{x\} \in \mathcal{I}_1$, then $(j, x) \in \delta(\overline{Y}, Y)$. Else, $S_j \cup \{x\} \notin \mathcal{I}_1$. Thus, $S_j \cup \{x\}$ contains a unique circuit C. This circuit must contain some element $y \in C \setminus Y$, for otherwise, we would have $C \subseteq Y$, implying $C \subseteq S'_j \cup \{x\}$, a contradiction. Thus $S_j - \{y\} \cup \{x\} \in \mathcal{I}_1$, so $(y, x) \in \delta(\overline{Y}, Y)$.

Lastly, for each j, x pair such that $x \in Y \setminus S'_j$ and $S'_j \cup \{x\} \in \mathcal{I}_1$, generate a unique arc $(y, x) \in \delta(\overline{Y}, Y)$ where y has color j, and let $A' \subseteq \delta(\overline{Y}, Y)$ be the collection of these arcs. Then $|A'| > B|Y|$ and no two arcs in A' share both the same target node x and source node color j.

We now remark how Lemma 9 shows that the color-chordless subgraph H contains an uncolored element u. Assume for sake of contradiction that H does not contain an uncolored element u. Let $Y = V \setminus V(H)$. Then $(\overline{Y}, Y)$ is a source-sink separating cut in G. By Lemma 9, there must exist some node $x \in Y$ with at least $B+1$ incoming arcs from distinct colors in $V(H)$. But then H could be extended by one more level, a contradiction.

We now show that during construction of the suffix-feasible path P, a vertex z always exists to extend the suffix.

Lemma 10. *Let $P_j = (x_{\ell-j}, x_{\ell-j+1}, \ldots, x_\ell = u)$ be the suffix-feasible path with respect to each of the partition matroids $M_2, M_3, \ldots, M_k$ computed by our algorithm after j iterations of the construction of P. Then there exists a vertex z in H such that there is an edge $(z, x_{\ell-j})$ in H and $P_{j+1} = (z, x_{\ell-j}, x_{\ell-j+1}, \ldots, x_\ell = u)$ is suffix-feasible with respect to the each of the partition matroids $M_2, M_3, \ldots, M_k$.*

Proof. Let Z be the set of vertices in H with incoming arcs to $x_{\ell-j}$. Let $c' = c\Delta P_j$, and note $x_{\ell-j}$ is uncolored in c'. For each partition matroid M_i, let C_i be $x_{\ell-j}$'s part in M_i. Each C_i has at most $\chi(M_i) - 1$ color classes of c' at the capacity constraint. Thus there are at most $B = \sum_{i=2}^{k} (\chi(M_i) - 1)$ color classes of c' which are at the capacity constraint in some C_i. Because $|Z| = B+1$, there exists some vertex $z \in Z$ with a color r such that r's color class in c' is not at the capacity constraint in *any* C_i. Let $P_{j+1} = (z, x_{\ell-j}, x_{\ell-j+1}, \ldots, x_\ell = u)$. $c\Delta P_{j+1}$ is identical to $c\Delta P_j$, except that $x_{\ell-j}$ has color r and z is uncolored. By choice of z, $c\Delta P_{j+1}$ is a feasible coloring in partition matroids $M_2, M_3, \ldots, M_k$. Further, all other suffixes of $c\Delta P_{j+1}$ are suffixes of $c\Delta P_j$. Thus P_{j+1} is suffix-feasible with respect to partition matroids $M_2, M_3, \ldots, M_k$.

Lemma 11. *$c\Delta P$ is a feasible coloring of $q+1$ elements of $M = \bigcap_{i=1}^{k} M_i$ using $\alpha + B$ colors.*

Proof. P is a color-chordless path in G by Lemma 8, so $c\Delta P$ is a feasible coloring of $q+1$ elements of M_1 by Lemma 7. P is suffix-feasible with respect to partition matroids $M_2, M_3, \ldots, M_k$ via Lemma 9 and Lemma 10, so $c\Delta P$ is a feasible coloring of $q+1$ elements of $M_2, M_3, \ldots, M_k$.

4 Coloring a Matroid Intersection on an Example

Consider the following matroids M_1, M_2 and current 3-coloring of elements c in Fig. 1. M_1 is a laminar matroid with ground set $X = \{x_1, x_2, x_3, x_4, x_5, x_6\}$ and laminar family $\mathcal{F} = \{\{x_5, x_6\}, \{x_3, x_4, x_5, x_6\}, \{x_1, x_2, x_3, x_4, x_5, x_6\}\}$. The set $\{x_5, x_6\}$ has capacity 1, $\{x_3, x_4, x_5, x_6\}$ has capacity 2, and $\{x_1, x_2, x_3, x_4, x_5, x_6\}$ has capacity 3. M_2 is a partition matroid with parts $\{x_2, x_3\}, \{x_4, x_5\}$, and $\{x_1, x_6\}$, where each part has capacity 1. Color 1 is blue (vertical hatches), color 2 is red (diagonal hatches), and color 3 is green (horizontal hatches). $c(x_6) = 0, c(x_5) = 1, c(x_3) = c(x_4) = 2$, and $c(x_1) = c(x_2) = 3$. $S_1 = \{x_5\}, S_2 = \{x_3, x_4\}, S_3 = \{x_1, x_2\}$, and $U = \{x_6\}$. Note $S_1, S_2, S_3 \in \mathcal{I}_1 \cap \mathcal{I}_2$. $\chi(M_1) = \chi(M_2) = 2$, so $\alpha = 2$ and $B = 1$.

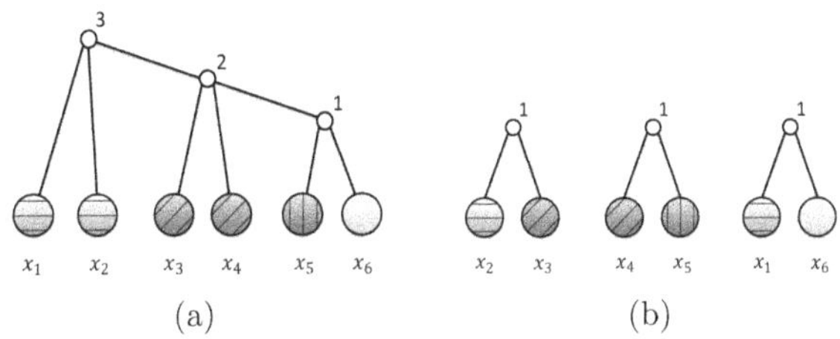

Fig. 1. (1a) is the laminar matroid M_1 and (1b) is the partition matroid M_2.

The construction of the Edmonds digraph G of M_1 is shown in Fig. 2. Note $S_1 \cup \{x_1\} = \{x_1, x_5\} \in \mathcal{I}_1$, so there is an arc $(1, x_1) \in A_1$. Note $S_2 \cup \{x_6\} = \{x_3, x_4, x_6\} \notin \mathcal{I}_1$ but $S_2 - \{x_3\} \cup \{x_6\} = \{x_4, x_6\} \in \mathcal{I}_1$, so there is an arc $(x_3, x_6) \in A_2$. Note $S_3 \cup \{x_4\} = \{x_1, x_2, x_4\} \in \mathcal{I}_1$, so there is an arc $(3, x_4) \in A_3$.

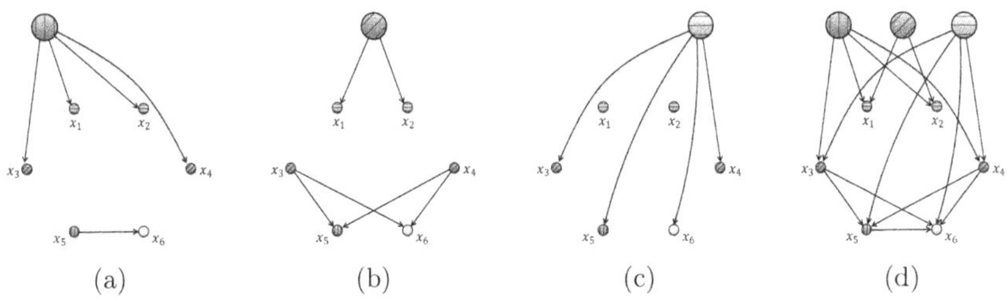

Fig. 2. (2a) is edges A_1, (2b) is edges A_2, (2c) is edges A_3, and (2d) is the Edmonds digraph G.

The construction of the color-chordless subgraph H is shown in Fig. 3. Observe that the arcs are directed down the levels, the nodes not on L_0 have in-degree 2, and the incoming arcs to nodes not on L_0 have different source colors. Observe that x_5 and x_6 are not included on L_1 because, although they have an incoming arc from the Green $= 3$ node, they do not have two incoming arcs from L_0. Note that on L_2, the incoming arcs to x_5, x_6 could be from either x_3 or x_4.

The construction of the suffix-feasible path P with respect to M_2 is shown in Fig. 4 and Fig. 5. Initially, $P_0 = x_6$. Figure 4 shows the suffix $P_1 = (x_3, x_6)$ and the feasible coloring $c\Delta P_1$ in M_2. Note that $P_1' = (3, x_6)$ is *not* suffix-feasible with respect to M_2, because $c\Delta P_1'$ is not a feasible coloring in M_2 (both x_1, x_6 are Green in $c\Delta P_1'$). Figure 5 shows the suffix $P_2 = (1, x_3, x_6)$, which is the complete suffix-feasible path P, and the feasible coloring $c\Delta P_2$ in M_2. Note that $P_2' = (3, x_3, x_6)$ is *not* suffix-feasible with respect to M_2, because $c\Delta P_2'$ is not a feasible coloring in M_2 (both x_2, x_3 are Green in $c\Delta P_2'$).

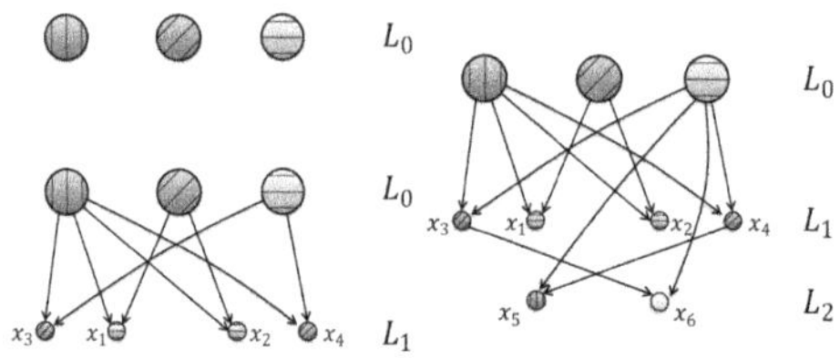

Fig. 3. Constructing the color-chordless subgraph H of G.

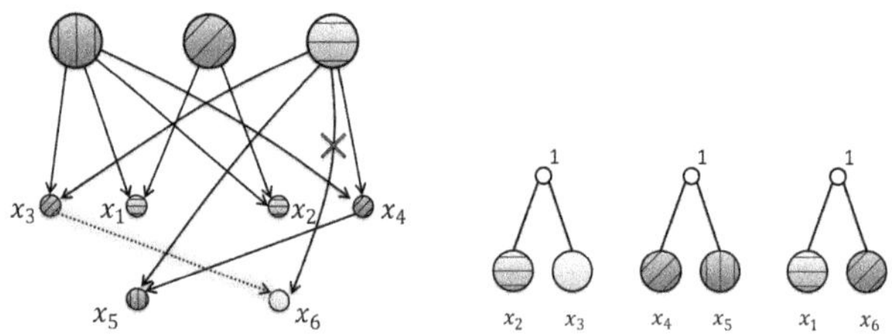

(a) (b)

Fig. 4. (4a) is the suffix P_1. (4b) is $c\Delta P_1$ on M_2.

5 Conclusion

The takeaway message from this paper is that there is a polynomial-time algorithm for matroid intersection coloring with a reasonable performance/approximation guarantee as long as all but one of the matroids are combinatorial.

As far as the algorithmics community is behind the combinatorics community in generating existential results related to matroid intersection coloring, it is even

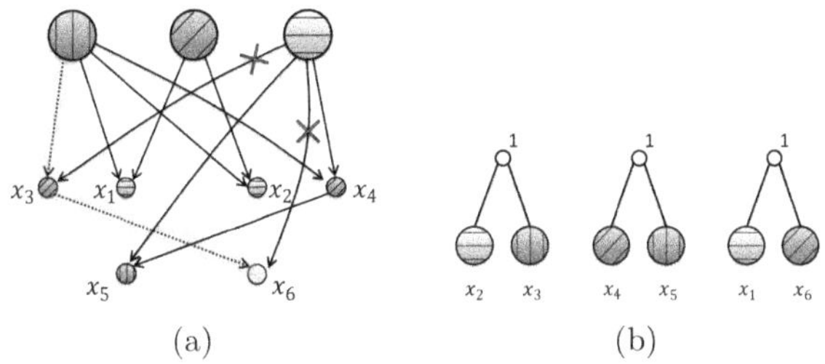

Fig. 5. (5a) is the suffix P_2 (the final path P). (5b) is $c\Delta P_2$ on M_2.

significantly more behind in generating conjectures. So to match the following conjecture from [3], that the $2k-1$ bound in Theorem 1 can be replaced by k,

Conjecture 1. [3] For any k general matroids $M_1, \ldots, M_k$ it is the case that $\chi\left(\cap_{i=1}^{k} M_i\right) \leq k \max_{i=1}^{k} \chi(M_i)$.

We also propose the following conjecture.

Conjecture 2. There exists a polynomial-time algorithm that for k arbitrary matroids $M_1, \ldots, M_k$ computes a coloring of their intersection using at most $O(k) \cdot \max_i \chi(M_i)$ colors.

A natural first step toward proving Conjecture 2 would be to prove it for the case $k = 2$. Our algorithm for matroid intersection coloring with one arbitrary matroid does not seem readily extensible to two general matroids. In particular, there is an instance of two laminar matroids and a partial coloring c, where for every source-sink path P in the Edmonds digraph G for M_1, it is the case that $c\Delta P$ is not a feasible coloring in M_2. One approach for an algorithm for two general matroids is to use an augmenting tree, instead of an augmenting path, but at least one significant difficulty with this approach is that the independence structure in general matroids is significantly more complicated than it is for partition matroids.

We should also note that as "covering" and "coloring" are essentially synonymous in this setting, our result is another example of a natural special set cover instance where better approximation is achievable than is achievable for general set cover instances, where the best achievable approximation ratio is $\Theta(\log n)$, provided P $\neq$ NP.

Acknowledgments. Benjamin Moseley is supported in part by Google Research Award, NSF grants CCF-2121744 and CCF-1845146, and ONR Grant N000142212702. Kirk Pruhs is supported by National Science Foundation grant CCF-2209654.

Disclosure of Interests. The authors have no competing interests to declare that are relevant to the content of this article.

References

1. Abdolazimi, D., Karlin, A.R., Klein, N., Gharan, S.O.: Matroid partition property and the secretary problem. In: 14th Innovations in Theoretical Computer Science Conference (ITCS 2023). Leibniz International Proceedings in Informatics (LIPIcs), vol. 251, pp. 2:1–2:9. Schloss Dagstuhl – Leibniz-Zentrum für Informatik (2023)
2. Aharoni, R., Berger, E.: The intersection of a matroid and a simplical complex. Trans. Am. Math. Soc. **358**(11), 4895–4917 (2006)
3. Aharoni, R., Berger, E., Guo, H., Kotlar, D.: Coloring, list coloring, and fractional coloring in intersections of matroids. Combinatorica **45**, 46 (2025). https://doi.org/10.1007/s00493-025-00178-8
4. Aharoni, R., Berger, E., Ziv, R.: Independent systems of representatives in weighted graphs. Combinatorica **27**, 253–267 (2007)
5. Annamalai, C.: Finding perfect matchings in bipartite hypergraphs. In: Proceedings of the Twenty-Seventh Annual ACM-SIAM Symposium on Discrete Algorithms, pp. 1814–1823. SODA '16, Society for Industrial and Applied Mathematics, USA (2016)
6. Bérczi, K., Schwarcz, T.: Complexity of packing common bases in matroids. Math. Program. **188**(1), 1–18 (2020). https://doi.org/10.1007/s10107-020-01497-y
7. Berger, E., Guo, H.: Coloring the intersection of two matroids. Proc. Am. Math. Soc. **153**(10), 4145–4154 (2025). https://doi.org/10.1090/proc/17274
8. Brualdi, R.A.: Induced matroids. In: Proceedings of the American Mathematical Society. vol. 29, pp. 213–221 (1971)
9. Bérczi, K., Schwarcz, T.: Partitioning into common independent sets via relaxing strongly base orderability. J. Comb. Theory **202**(105817) (2024)
10. Bérczi, K., Schwarcz, T., Yamaguchi, Y.: List coloring of two matroids through reduction to partition matroids. SIAM J. Discret. Math. **35**, 2192–2209 (2021)
11. Bérczi, K., Csáji, G., Király, T.: On the complexity of packing rainbow spanning trees. Discrete Math. **346**(113297) (2023)
12. Davies, J., McDiarmid, C.: Disjoint common transversals and exchange structures. J. Lond. Math. Soc. **14**, 55–62 (1976)
13. Edmonds, J.: Matroid partition. In: Mathematics of the Decision Sciences, Part 1, Lectures in Applied Mathematics, vol. 11, pp. 335–345. American Mathematical Society (1968)
14. Frank, A.: Connections in combinatorial optimization. Discrete Appl. Math. **160**(12), 1875 (2012). https://doi.org/10.1016/j.dam.2011.09.003
15. Galvin, F.: The list chromatic index of a bipartite multigraph. J. Comb. Theory **63**, 153–158 (1995)
16. Guo, H.: The list chromatic number of the intersection of two generalized partition matroids. arXiV (2024)
17. Harvey, N.J.A., Király, T., Lau, L.C.: On disjoint common bases in two matroids. SIAM J. Discret. Math. **25**(4), 1792–1803 (2011). https://doi.org/10.1137/100815232
18. Haxell, P.E.: A condition for matchability in hypergraphs. Graphs Comb. **11**(3), 245–248 (1995)
19. Hörsch, F., Kaiser, T., Kriesell, M.: Rainbow bases in matroids. SIAM J. Discret. Math. **38**(2), 1472–1491 (2024)
20. Huang, R., Rota, G.C.: On the relations of various conjectures on Latin squares and straightening coefficients. Discret. Math. **128**, 225–236 (1994)

21. Im, S., Moseley, B., Pruhs, K.: The matroid intersection cover problem. Oper. Res. Lett. **49**, 17–22 (2021)
22. Király, T., Pap, J.: On the list colouring of two matroids. Tech. Rep. EGRES Quick-Proof 2010-01, Egerváry Research Group (EGRES) (2010). https://egres.elte.hu/qp/egresqp-10-01.pdf
23. Leichter, M., Moseley, B., Pruhs, K.: An efficient reduction of a gammoid to a partition matroid. In: 29th Annual European Symposium on Algorithms (ESA 2021). Leibniz International Proceedings in Informatics (LIPIcs), vol. 204, pp. 62:1–62:13. Schloss Dagstuhl – Leibniz-Zentrum für Informatik (2021). https://doi.org/10.4230/LIPIcs.ESA.2021.62
24. Leichter, M., Moseley, B., Pruhs, K.: On the impossibility of decomposing binary matroids. Oper. Res. Lett. **50**, 623–625 (2022)
25. Obszarski, P., Jastrzębski, A.: Edge-coloring of 3-uniform hypergraphs. Discrete Appl. Math. **217** (2016). https://doi.org/10.1016/j.dam.2016.06.009
26. Schrijver, A.: Combinatorial Optimization: Polyhedra and Efficiency, Algorithms and Combinatorics, vol. 24. Springer (2003)

Polynomial Kernels with Reachability for Weighted d-Matroid Intersection

Chien-Chung Huang[1], Naonori Kakimura[2], Yusuke Kobayashi[3](✉), and Tatsuya Terao[3]

[1] CNRS, DIENS, PSL, France
Chien-Chung.Huang@ens.fr
[2] Keio Unversity, Minato, Japan
kakimura@math.keio.ac.jp
[3] Kyoto University, Kyoto, Japan
{yusuke,ttatsuya}@kurims.kyoto-u.ac.jp

Abstract. This paper studies randomized polynomial kernelization for the weighted d-matroid intersection problem. While the problem is known to have a kernel of size $O(d^{(k-1)d})$ where k is the solution size, the existence of a polynomial kernel is not known, except for the cases when either all the given matroids are partition matroids (i.e., the d-dimensional matching problem) or all the given matroids are linearly representable. The main contribution of this paper is to develop a new kernelization technique for handling general matroids. We first show that the weighted d-matroid intersection problem admits a polynomial kernel when one matroid is arbitrary and the other $d-1$ matroids are partition matroids. Interestingly, the obtained kernel has size $\tilde{O}(k^d)$, which matches the optimal bound (up to logarithmic factors) for the d-dimensional matching problem. This approach can be adapted to the case when $d-1$ matroids in the input belong to a more general class of matroids, including graphic, cographic, and transversal matroids. We also show that the problem has a kernel of pseudo-polynomial size when given $d-1$ matroids are laminar. Our technique finds a kernel such that any feasible solution of a given instance can reach a better solution in the kernel, which is sufficiently versatile to allow us to design parameterized streaming algorithms and faster EPTASs for the budgeted version of matroid intersection.

Keywords: d-matroid intersection · d-dimensional matching · kernelization · randomized algorithm

1 Introduction

In the *weighted d-matroid intersection* problem, we are given d matroids on a finite ground set E and a weight function $w : E \to \mathbb{R}_{\geq 0}$. The goal of the problem is to find a maximum-weight common independent set, that is, we aim to find a subset X of E that maximizes $\sum_{e \in X} w(e)$ subject to X is an independent

S. S. Dey et al. (Eds.): IPCO 2026, LNCS 16588, pp. 362–377, 2026.
https://doi.org/10.1007/978-3-032-28691-8_24

set for every matroid. The weighted d-matroid intersection problem is one of the most fundamental problems in combinatorial optimization, as it includes the bipartite matching problem, the arborescence problem (when $d = 2$), and the d-dimensional matching problem as a special case. The problem can be solved in polynomial time when $d \leq 2$ [1,15,16,34,39],while it is NP-hard when $d \geq 3$ (see [26]).

In this paper, we study kernelization of the weighted d-matroid intersection problem when the solution size is a parameter. In general, *kernelization* for a parameterized problem is a polynomial-time transformation that reduces any given instance to an equivalent instance of the same problem, called a *kernel*, with size and parameter bounded by a function of the parameter in the input. Kernelization plays a central role in parameterized complexity theory [23], as kernelization can be used as a preprocessing in fixed-parameter algorithms, and moreover, the existence of a kernel is equivalent to fixed-parameter tractability of a given parameterized problem.

A kernel is *polynomial* if its size is polynomially bounded by the parameter of the input. There is a line of research in parameterized complexity theory that focuses on classifying fixed-parameter tractable problems on whether they admit polynomial kernels or not. To show the existence of a polynomial kernel or to reduce the kernel size, a variety of sophisticated techniques have been developed (see e.g., [24,27,36,40]). Also, various complexity-theoretic lower bound tools for kernelization have been established [5,11,12,14,25,31,35]. See also the comprehensive textbook by Fomin et al. [23].

Recently, Huang and Ward [33] showed that the weighted d-matroid intersection problem admits a kernel of size $O(d^{(k-1)d})$, where the parameter k is the solution size. It is, however, open whether it has a polynomial kernel or not when d is a fixed constant. There are only a few special cases admitting polynomial-size kernels in the literature as below.

One of specific special cases is the d-dimensional matching problem, which is the d-matroid intersection problem in which all matroids are simple partition matroids. The problem is a special case of the d-matching problem (also known as the d-set packing problem), and the d-matching problem is known to admit a kernel of size $O(k^d)$. Thus, the d-dimensional matching problem has a kernel of size $O(k^d)$. The proof makes use of the sunflower lemma (see [23, Chapter 8]). Dell and Marx [11] showed that the kernel size $O(k^d)$ is asymptotically tight even for the d-dimensional matching problem.

Another special case is when given matroids are linearly representable. It is known that this case has a randomized FPT algorithm parameterized by d and the solution size k (see [10, Chapter 12.3.4]), which yields a kernel consisting of $O(k^d)$ elements of the ground set. The algorithm is based on the representative sets technique (see [23, Chapter 11]), specifically designed for linear matroids. In recent years, representative sets have become one of the most widely-studied tools in the parameterized algorithms community, which can effectively deal with linear matroids [21,22,38,43,45].

In the above mentioned spacial cases, a polynomial kernel is obtained via well-established kernelization techniques such as the sunflower lemma and representative sets. However, the existing techniques rely on how given matroids are represented. In fact, they fail to handle the case when even one matroid is given by an independence oracle.

1.1 Our Contribution

Our main contribution is to develop a new kernelization technique for handling general matroids. Using this technique, we design kernelization algorithms for a broader class of the weighted d-matroid intersection problem, where one matroid is arbitrary and given by an independence oracle, while the remaining $d-1$ matroids are "simple" matroids such as partition, graphic, cographic, transversal, or laminar matroids. Throughout the paper, d is regarded as a fixed constant.

The kernel that our algorithm finds satisfies an exchange property for matroids, which we call a *reachable kernel.* Intuitively, a reachable kernel satisfies the condition that any feasible solution of a given instance can reach a better solution in the kernel by exchanging elements one by one with constant probability.

Below is the formal definition of a reachable kernel.

Definition 1. *Let E be a finite set and $\mathcal{F} \subseteq 2^E$ be a family of feasible sets. Let $w\colon E \to \mathbb{R}_{\geq 0}$ be a non-negative weight function and let k be a positive integer. We say that a randomly selected set $R \subseteq E$ is a reachable kernel*with respect to $\mathcal{F}$, w, and k *if the following condition holds:*

for any $\{x_1, \ldots, x_t\} \in \mathcal{F}$ with $t \leq k$, there exist elements $y_1, \ldots, y_t \in R$ such that $w(y_i) \geq w(x_i)$ and $\{y_1, \ldots, y_i, x_{i+1}, \ldots, x_t\}$ is a set of size t in $\mathcal{F}$ for any $i \in \{1, 2, \ldots, t\}$ with probability at least $\frac{2}{3}$.[1]

When $\mathcal{F}$, w, and k are clear, we simply call it a reachable kernel.

For the problem of finding a set $X \subseteq E$ of maximum weight subject to $X \in \mathcal{F}$, by the definition, an instance restricted to the reachable kernel R has the same optimal value as the given instance when the solution size is at most k. In this sense, the set R can be regarded as a (randomized) kernel under the assumption that the membership oracle for $\mathcal{F}$ is available and it does not contribute to the input size[2].

As we will discuss later in Sect. 1.2, reachable kernels not only guarantee a reduced input instance for the purposes of FPT, but they also find additional applications in other computational models. In the following, we discuss the size of a reachable kernel that can be guaranteed in various contexts.

---

1 The value $\frac{1}{3}$ of the error probability is not essential. Indeed, by computing a reachable kernel c times and taking their union, we can reduce the error probability to less than $(1/3)^c$, increasing the output size by a factor of c.

2 The assumption is standard in the literature of kernelization for matroid-constrained problems [2].

We first consider the case when we are given one arbitrary matroid and $d-1$ simple partition matroids. Here, a matroid $\mathcal{M} = (E, \mathcal{I})$ is called a *simple partition matroid* if there exists a partition $(E^1, \ldots, E^\ell)$ of E such that $I \in \mathcal{I}$ if and only if $|I \cap E^j| \leq 1$ for any $j \in [\ell]$.

Theorem 1. *For the weighted d-matroid intersection problem such that, out of the d given matroids, $d-1$ are simple partition matroids, we can construct a reachable kernel of size*[3] $\tilde{O}(k^d)$ *in polynomial time.*

Theorem 1 implies that the d-dimensional matching problem admits a kernel of size $\tilde{O}(k^d)$, as the d-dimensional matching problem is identical with the weighted d-matroid intersection problem when all the given matroids are simple partition matroids. As mentioned before, the optimal kernel size of the d-dimensional matching problem is $O(k^d)$. Thus, our result matches the optimal bound, ignoring a logarithmic factor.

Theorem 1 can be adapted to more generalized matroids. To state our result, we introduce a notion of $g(k)$-coverable matroids, which is of independent interest. Here, $\mathrm{span}_{\mathcal{M}}(X)$ of a matroid $\mathcal{M}$ for a subset $X \subseteq E$ denotes the set of elements $e \in E$ such that adding e to X does not increase its rank.

Definition 2. *For a non-decreasing function $g : \mathbb{Z}_{\geq 0} \to \mathbb{Z}_{\geq 0}$, a matroid $\mathcal{M} = (E, \mathcal{I})$ is called* -coverable *if it satisfies the following condition: for any $X \in \mathcal{I}$ with $|X| \leq k$, there exists $F \subseteq E$ with $|F| \leq g(k)$ such that* $\mathrm{span}_{\mathcal{M}}(X) = \bigcup_{f \in F} \mathrm{span}_{\mathcal{M}}(f)$.

Roughly speaking, this definition means that any independent set X of size at most k can essentially span at most $g(k)$ elements, when we identify "parallel" elements with one element. For example, a simple partition matroid is k-coverable, while a uniform matroid of rank k on a large ground set is not $g(k)$-coverable for any function g.

Theorem 2. *For the weighted d-matroid intersection problem with d matroids $\mathcal{M}_0, \mathcal{M}_1, \ldots, \mathcal{M}_{d-1}$ where $\mathcal{M}_i$ is $g_i(k)$-coverable for each $i \in \{1, \ldots, d-1\}$, we can construct a reachable kernel of size $\tilde{O}\left(k \cdot \prod_i g_i(k)\right)$ in polynomial time.*

Since a simple partition matroid is k-coverable, Theorem 2 is a generalization of Theorem 1. Furthermore, since graphic matroids are $O(k^2)$-coverable and cographic matroids are $O(k)$-coverable as we will see in Sect. 3.3, Theorem 2 implies the following corollary.

Corollary 1. *For the weighted d-matroid intersection problem such that, out of the d given matroids, all except one are graphic (resp. cographic), we can construct a reachable kernel of size $\tilde{O}(k^{2d-1})$ (resp. $\tilde{O}(k^d)$) in polynomial time.*

Although graphic and cographic matroids are $g(k)$-coverable, it may be somehow unexpected that relatively mild generalizations of simple partition matroids,

---

[3] The notation $\tilde{O}$ hides a polylogarithmic function of k.

namely, transversal and laminar matroids are not (since they both generalize uniform matroids). To obtain the following results, significantly different techniques are required.

Theorem 3. *For the weighted d-matroid intersection problem such that, out of the d given matroids, all except one are transversal, we can construct a reachable kernel of size* $\tilde{O}(k^d)$ *in polynomial time.*

Theorem 4. *For the weighted d-matroid intersection problem such that, out of the d given matroids, all except one are laminar matroids, we can construct a reachable kernel of size* $k^{O(d \log k)}$ *in polynomial time.*

For transversal matroids, we introduce a reduction technique to transform them to simple partition matroids. On the other hand, for laminar matroids, as there are a significantly larger number of possibilities concerning how a feasible solution is "distributed" among the sets of the laminar family, we need to employ a more aggressive guessing strategy and this causes our kernel size to be quasi-polynomial.

Our kernelization technique can also be applied to certain classes of the *d-matchoid problem.* Since dealing with the general d-matchoid problem would make the discussion overly complicated, this paper focuses only on one particularly interesting example, the *weighted matching problem with a matroid constraint.* In the problem, we are given a graph $G = (V, E)$, a matroid $\mathcal{M} = (E, \mathcal{I})$ defined on the edge set of G, and a weight function $w : E \to \mathbb{R}_{\geq 0}$. The goal of the problem is to find a matching of maximum weight that is independent in $\mathcal{M}$. For this problem, we present a kernerization algorithm parameterized by the solution size k.

Theorem 5. *For the weighted matching problem with a matroid constraint, we can construct a reachable kernel of size* $\tilde{O}(k^3)$ *in polynomial time.*

The weighted matching problem with a matroid constraint is a generalization of the *rainbow matching problem* [28,29,44], which corresponds to the case where the given matroid is a simple partition matroid. The rainbow matching problem has a kernel of size $O(k^3)$ by [28, Theorem 3], which is later improved to a kernel with $O(k^2)$ vertices and $O(k^3)$ edges [29].

Finally, we discuss the case when the reachability is realized deterministically. We say that a subset $R \subseteq E$ is a deterministic reachable kernel if it satisfies the conditions in Definition 1 with probability 1.

Theorem 6. *For the weighted 2-matroid intersection problem such that the matroids are a* $g(k)$*-coverable matroid and an arbitrary matroid, we can construct a deterministic reachable kernel of size* $k^2(g(k) + 1)$ *in polynomial time.*

Proofs for Theorems 3–6 will be provided in the full version [32].

1.2 Advantages of Reachable Kernel

We here describe a few applications of a reachable kernel in different contexts.

The first application is in the design of *parameterized streaming algorithms* [7–9,18]. Suppose that we are given a feasible region $\mathcal{F} \subseteq 2^E$ over a ground set E through oracle access, and the elements of E arrive one by one. The goal is to find a set $X \subseteq E$ that maximizes $\sum_{e \in X} w(e)$ subject to $X \in \mathcal{F}$ and $|X| \leq k$. We consider the setting where $|E|$ is very large and at any moment we can keep at most $f(k)$ elements, discarding all the others. Huang and Ward [33] introduced a slight generalization of the concept of a reachable kernel, which they call a *representative set*[4]. For the d-matroid intersection problem (and, more generally, the d-matchoid problem), they proposed a streaming algorithm that always maintains a representative set whose size is bounded by an exponential function of k. In this context, we can confirm that our results can be generalized to adapt the streaming setting, resulting in streaming algorithms that use only polynomial (or quasi-polynomial) space in k for several classes of matroids.

Furthermore, our reachable kernel has applications to improving the running time of EPTASs for the budgeted version of matroid intersection [13]. Doron-Arad et al. [13] presented an EPTAS for the budgeted version, using a representative set by Huang and Ward [33]. Their EPTAS runs in $q(\varepsilon)^{O(q(\varepsilon))} \cdot \mathrm{poly}(|I|)$ time, where $q(\varepsilon) = \left\lceil \frac{1}{\varepsilon^{\varepsilon^{-1}}} \right\rceil$ and $|I|$ denotes the input size. Although this running time yields an EPTAS, the dependence on ε is extremely high. By following the argument in [13] with our reachable kernel, replaced with their representative set, we can improve the running time of the EPTAS to $2^{O\left(\varepsilon^{-2} \log \frac{1}{\varepsilon}\right)} \cdot \mathrm{poly}(|I|)$ for several classes of matroids, where the dependence on ε becomes significantly milder.

1.3 Additional Related Work

Matroids play an important role in parameterized complexity theory, as the textbooks [10,23] devote chapters to matroids. See also [2,4,19–22,30,37,38,41–43,45]. In particular, linear matroids have been actively studied in the parameterized algorithm community. For the d-matroid intersection problem when all matroids are linear and matrix representations are available, Barvinok [3] presented the first algorithm faster than naive enumeration. Marx [43] provided a fixed-parameter algorithm, and subsequent works [6,17,22] proposed faster algorithms.

Recently, there have been several studies addressing graph problems under matroid constraints, with a focus on parameterized algorithm and kernelization, such as the stable set problem [19] and various cut problems [2].

[4] Note that this notion is different from representative sets used for the linear matroid intersection.

2 Preliminaries

2.1 Basic Notation

For a positive integer t, we denote $[t] = \{1, 2, \ldots, t\}$. Throughout this paper, we often write $A + x := A \cup \{x\}$ and $A - x := A \setminus \{x\}$ for a set A and an element x.

A pair $\mathcal{M} = (E, \mathcal{I})$ of a finite set E and a non-empty set family $\mathcal{I} \subseteq 2^E$ is called a *matroid* if the following properties are satisfied.

1. If $X \in \mathcal{I}$ and $Y \subseteq X$ then $Y \in \mathcal{I}$.
2. If $X, Y \in \mathcal{I}$ and $|Y| < |X|$, then there exists $x \in X \setminus Y$ such that $Y + x \in \mathcal{I}$.

A set $X \in \mathcal{I}$ is called an *independent set.* We call a maximal independent set a *base*. For $U \subseteq E$, the *restriction* of $\mathcal{M}$ to U is the matroid $\mathcal{M}|U = (U, \mathcal{I}_U)$, where $\mathcal{I}_U = \{X \in \mathcal{I} \mid X \subseteq U\}$. Throughout the paper, we assume that every matroid has no loop, i.e., $\{x\} \in \mathcal{I}$ for every $x \in X$, since we can remove elements that do not belong to any independent set.

Let $\mathcal{M} = (E, \mathcal{I})$ be a matroid. For a subset X of E, we define the *rank* of X as $\mathrm{rank}_{\mathcal{M}}(X) = \max\{|Y| \mid Y \subseteq X, Y \in \mathcal{I}\}$. The *rank* of $\mathcal{M}$ is equal to $\mathrm{rank}_{\mathcal{M}}(E)$. For a subset X of E, the *span* of X is defined as

$$\mathrm{span}_{\mathcal{M}}(X) = \{x \in E \mid \mathrm{rank}(X) = \mathrm{rank}(X + x)\}.$$

When a given matroid $\mathcal{M}$ is clear from the context, we sometimes write $\mathrm{rank}(X)$ and $\mathrm{span}(X)$ for $\mathrm{rank}_{\mathcal{M}}(X)$ and $\mathrm{span}_{\mathcal{M}}(X)$, respectively. For an element $e \in E$, $\mathrm{span}_{\mathcal{M}}(\{e\})$ is simply denoted by $\mathrm{span}_{\mathcal{M}}(e)$. A pair of distinct elements $e, e' \in E$ are called *parallel* if $\mathrm{span}_{\mathcal{M}}(e) = \mathrm{span}_{\mathcal{M}}(e')$.

We assume that each general matroid is given by an independence oracle, while each special matroid is given in an appropriate explicit form.

2.2 Technical Tools

Let $\mathcal{F}$ be a family of subsets in E. As in Definition 1, a reachable kernel for $\mathcal{F}$ is defined by the existence of a sequence $y_1, \ldots, y_t$ of elements. We observe that a randomly selected set $R \subseteq E$ is a reachable kernel if R satisfies the following single exchangeability.

(SingleEXC) For any $X \in \mathcal{F}$ with $|X| \leq k$ and any $x \in X$, there exists $y \in R \setminus (X - x)$ such that $w(y) \geq w(x)$ and $X - x + y \in \mathcal{F}$ with probability at least $1 - \frac{1}{3k}$.

Indeed, for $i = 1, 2, \ldots, t$ in this order, by applying (SingleEXC) for $X = \{y_1, \ldots, y_{i-1}, x_i, \ldots, x_t\}$ and $x = x_i$, we can obtain y_i satisfying the conditions in Definition 1. Note that the total error probability is at most $\frac{1}{3k} \cdot t \leq \frac{1}{3}$. We remark that, when $k = 1$, (SingleEXC) obviously holds by choosing a maximum-weight element of size one in $\mathcal{F}$ as R. Thus, in what follows, we may assume that $k \geq 2$.

If $\mathcal{F}$ is a family of the independent sets of a single matroid, a simple greedy algorithm can find a maximum-weight independent set of size at most k, which satisfies (SingleEXC). As we will perform this greedy algorithm to a subset $F \subseteq E$ as a subroutine of our algorithm, we describe it formally in Algorithm 1. Here, we assume that the elements are sorted in non-increasing order of their weights, and we omit the weight function w in the input of the algorithm.

Algorithm 1: `Greedy`$(\mathcal{M}, F, k)$

Input: A matroid $\mathcal{M} = (E, \mathcal{I})$, $F \subseteq E$, and an integer k

Let $F = \{f_1, f_2, \dots, f_\ell\}$;

`// The elements in` F `are sorted in non-increasing order of the weight, i.e.,` $w(f_1) \geq w(f_2) \geq \cdots \geq w(f_\ell)$.

$I \leftarrow \emptyset$;

for $i \leftarrow 1$ **to** ℓ **do**

  if $I + f_i \in \mathcal{I}$ *and* $|I + f_i| \leq k$ **then**

    $I \leftarrow I + f_i$;

return I;

The following two lemmas are basic properties of matroids and the greedy algorithm. We provide proofs for completeness.

Lemma 1. *Let $\mathcal{M} = (E, \mathcal{I})$ be a matroid. Suppose that $T, X \in \mathcal{I}$ and $x \in X$ with $x \in \mathrm{span}_{\mathcal{M}}(T)$. Then, there exists $y \in T \setminus (X - x)$ such that $X - x + y \in \mathcal{I}$.*

Proof. Assume for contradiction that $X - x + y \notin \mathcal{I}$ for every $y \in T \setminus (X - x)$. This means that $y \in \mathrm{span}_{\mathcal{M}}(X - x)$ for all $y \in T$, and hence $T \subseteq \mathrm{span}_{\mathcal{M}}(X - x)$. By taking the span of both sides, we obtain

$$x \in \mathrm{span}_{\mathcal{M}}(T) \subseteq \mathrm{span}_{\mathcal{M}}\big(\mathrm{span}_{\mathcal{M}}(X - x)\big) = \mathrm{span}_{\mathcal{M}}(X - x).$$

Hence $x \in \mathrm{span}_{\mathcal{M}}(X - x)$, which implies that X is not independent, a contradiction. Therefore, some $y \in T$ satisfies $X - x + y \in \mathcal{I}$. □

Lemma 2. *Let $\mathcal{M} = (E, \mathcal{I})$ be a matroid with a weight function $w \colon E \to \mathbb{R}_{\geq 0}$ and let k be a positive integer. Suppose that $F \subseteq E$, $X \in \mathcal{I}$, $|X| \leq k$, and $x \in F \cap X$. Then, the output of* `Greedy`$(\mathcal{M}, F, k)$ *contains an element y such that $y \notin X - x$, $w(y) \geq w(x)$, and $X - x + y \in \mathcal{I}$.*

Proof. Let R be the output of `Greedy`$(\mathcal{M}, F, k)$. It suffices to consider the case where $x \notin R$, since otherwise $y = x$ satisfies the conditions. Since $x \notin R$ means that the greedy algorithm does not choose x, $T := \{e \in R \mid w(e) \geq w(x)\}$ satisfies that $x \in \mathrm{span}_{\mathcal{M}}(T)$ or $|T| = k$. If $x \in \mathrm{span}_{\mathcal{M}}(T)$, then Lemma 1 shows the existence of $y \in T \setminus (X - x)$ with $X - x + y \in \mathcal{I}$. If $|T| = k$, then we have $|T| > |X - x|$ and $T, X - x \in \mathcal{I}$, and hence there exists $y \in T \setminus (X - x)$ with $X - x + y \in \mathcal{I}$ by the definition of a matroid. Since $y \in T$ implies $y \in R$ and $w(y) \geq w(x)$, a desired y exists in both cases. □

For a matroid $\mathcal{M} = (E, \mathcal{I})$ and for $X \in \mathcal{I}$ and $x \in X$, we say that $Y \subseteq E$ is an *(X, x)-exchangeable set with respect to $\mathcal{M}$* if $x \in Y$ and $X - x + y \in \mathcal{I}$ for every $y \in Y$. In matroid-theoretic terms, an (X, x)-exchangeable set is a subset of a fundamental cocircuit of the matroid obtained from $\mathcal{M}$ by truncating it to rank $|X|$. In the later sections, we repeatedly use the following lemma.

Lemma 3. *Suppose we are given a matroid $\mathcal{M}_i = (E, \mathcal{I}_i)$ for $i \in \{0, 1, \ldots, d-1\}$, a weight function $w\colon E \to \mathbb{R}_{\geq 0}$, and a positive integer k. Let $\mathcal{F} = \bigcap_{i=0}^{d-1} \mathcal{I}_i$, and let $X \in \mathcal{F}$ be a feasible set with $|X| \leq k$, and $x \in X$. If $F_i \subseteq E$ is an (X, x)-exchangeable set with respect to $\mathcal{M}_i$ for any $i \in [d-1]$, then the output of* $\texttt{Greedy}(\mathcal{M}_0, \bigcap_{i\in[d-1]} F_i, k)$ *contains an element y such that $y \notin X - x$, $w(y) \geq w(x)$, and $X - x + y \in \mathcal{F}$.*

Proof. Let R be the output of $\texttt{Greedy}(\mathcal{M}_0, \bigcap_{i\in[d-1]} F_i, k)$. Since $x \in \bigcap_{i\in[d-1]} F_i$, by Lemma 2, there exists $y \in R \setminus (X - x)$ such that $w(y) \geq w(x)$ and $X - x + y \in \mathcal{I}_0$. Such an element y satisfies that $X - x + y \in \mathcal{I}_i$ for each $i \in [d-1]$, because $y \in F_i$ and F_i is an (X, x)-exchangeable set with respect to $\mathcal{M}_i$. Therefore, $X - x + y \in \bigcap_{i=0}^{d-1} \mathcal{I}_i = \mathcal{F}$, which completes the proof. □

3 Partition Matroids and Extensions

3.1 Kernelization Algorithm for Simple Partition Matroids

In this subsection, we prove Theorem 1. Recall that a matroid $\mathcal{M} = (E, \mathcal{I})$ is a *simple partition matroid* if there exists a partition $(E^1, \ldots, E^\ell)$ of E such that $\mathcal{I} = \{I \subseteq E \mid |I \cap E^j| \leq 1 \text{ for any } j \in [\ell]\}$.

Suppose that we are given d matroids $\mathcal{M}_0, \mathcal{M}_1, \ldots, \mathcal{M}_{d-1}$, where $\mathcal{M}_0 = (E, \mathcal{I}_0)$ is an arbitrary matroid and $\mathcal{M}_i = (E, \mathcal{I}_i)$ is a simple partition matroid for $i \in [d-1]$. We assume that each simple partition matroid $\mathcal{M}_i$ is represented by a partition $(E_i^1, \ldots, E_i^{\ell_i})$ of E. Define $\mathcal{F} = \bigcap_{i=0}^{d-1} \mathcal{I}_i$, which is a family of feasible sets of the problem. Our goal is to construct a reachable kernel of size $\tilde{O}(k^d)$.

We construct a reachable kernel by repeating the following procedure $T = \tilde{O}(k^{d-1})$ times, where the exact value of T will be determined later. For each $i \in [d-1]$, we sample each element $j \in [\ell_i]$ independently with probability $1/k$ to obtain a random subset $S_i \subseteq [\ell_i]$, and let $F_i = \bigcup_{j\in S_i} E_i^j$. Then we apply $\texttt{Greedy}(\mathcal{M}_0, \bigcap_{i\in[d-1]} F_i, k)$, that is, we greedily select up to k elements from $\bigcap_{i\in[d-1]} F_i$ in non-increasing order of weight, so that the selected set is independent in $\mathcal{M}_0$. We denote the resulting set by R' and add it to our reachable kernel R. The pseudocode for this algorithm is shown in Algorithm 2.

In what follows, we prove the correctness of the algorithm by showing that the obtained set R satisfies (SingleEXC). Let $X \in \mathcal{F}$ be a feasible solution of size at most k, and x be an arbitrary element in X.

Consider one round (lines 3–7) in the algorithm where we generate the sets F_i's for $i \in [d-1]$ by sampling. We say that the round is *successful* if F_i is an (X, x)-exchangeable set with respect to $\mathcal{M}_i$ (i.e., $x \in F_i$ and $X - x + y \in \mathcal{I}_i$

Algorithm 2: `PartitionKernel`$(\mathcal{M}_0, \ldots, \mathcal{M}_{d-1})$

$R \leftarrow \emptyset$;
repeat T **times**
  for $i \in [d-1]$ **do**
    Let $S_i \subseteq [\ell_i]$ be a subset obtained by independently sampling each element $j \in [\ell_i]$ with probability $\frac{1}{k}$;
    $F_i \leftarrow \bigcup_{j \in S_i} E_i^j$;
  $R' \leftarrow$ `Greedy`$(\mathcal{M}_0, \bigcap_{i \in [d-1]} F_i, k)$;
  $R \leftarrow R \cup R'$;
return R;

for any $y \in F_i$) for every $i \in [d-1]$. We now evaluate the probability that one round is successful.

For each $i \in [d-1]$, let $j_i^* \in [\ell_i]$ be the index with $x \in E_i^{j_i^*}$. Then, by the definition of a simple partition matroid, F_i is an (X, x)-exchangeable set with respect to $\mathcal{M}_i$ if and only if $j_i^* \in S_i$ and $j \notin S_i$ for any $j \in [\ell_i]$ with $(X - x) \cap E_i^j \neq \emptyset$. Hence, the probability that F_i is an (X, x)-exchangeable set with respect to $\mathcal{M}_i$ is at least

$$\frac{1}{k}\left(1 - \frac{1}{k}\right)^{|X-x|} \geq \frac{1}{k}\left(1 - \frac{1}{k}\right)^{k-1} \geq \frac{1}{ek},$$

where we use the fact that $(1 - 1/\alpha)^{\alpha - 1} \geq 1/e$ for $\alpha > 1$ and the assumption that $k \geq 2$. Therefore, the probability that one round is successful is at least $\left(\frac{1}{ek}\right)^{d-1}$. By setting $T = \lceil (ek)^{d-1} \log(3k) \rceil$, the probability that there exists no successful round among the T rounds is at most

$$\left(1 - \left(\frac{1}{ek}\right)^{d-1}\right)^T \leq \frac{1}{3k},$$

where we use $(1 - 1/\alpha)^\alpha \leq 1/e$ for $\alpha \geq 1$. Therefore, with probability at least $1 - \frac{1}{3k}$, there exists at least one successful round.

For the sets F_i's generated in a successful round, by Lemma 3, the output R' of `Greedy`$(\mathcal{M}_0, \bigcap_{i \in [d-1]} F_i, k)$ contains an element y such that $y \notin X - x$, $w(y) \geq w(x)$, and $X - x + y \in \mathcal{F}$. Since $R \subseteq E$ returned by Algorithm 2 includes R', such an element y satisfies the conditions in (SingleEXC).

Since this argument works for any $X \in \mathcal{F}$ and any $x \in X$, R satisfies (Single-EXC). The size of the set R is at most $k \cdot T = O(e^{d-1} k^d \log k)$. Since Algorithm 2 clearly runs in polynomial time, the proof of Theorem 1 is complete. □

3.2 Kernelization Algorithm for $g(k)$-coverable Matroids

In this subsection, we consider $g(k)$-coverable matroids, and prove Theorem 2. Let $\mathcal{M}_0 = (E, \mathcal{I}_0)$ be an arbitrary matroid, and $\mathcal{M}_i = (E, \mathcal{I}_i)$ be a $g_i(k)$-coverable matroid for $i \in [d-1]$. Note that $g_i(k) \geq k$ by the definition of a

$g_i(k)$-coverable matroid. We may assume that $|E| = \Omega\left(k \cdot \prod_i g_i(k)\right)$, since otherwise E is trivially a reachable kernel of size $\tilde{O}\left(k \cdot \prod_i g_i(k)\right)$.

We observe that, for $i \in [d-1]$, the ground set E is covered by the spans of singletons, that is, there exists a subset $\tilde{E}_i \subseteq E$ such that $E = \bigcup_{e \in \tilde{E}_i} \mathrm{span}_{\mathcal{M}_i}(e)$ and $\mathrm{span}_{\mathcal{M}_i}(e) \neq \mathrm{span}_{\mathcal{M}_i}(e')$ for any distinct $e, e' \in \tilde{E}_i$. Here, $\tilde{E}_i$ consists of one representative from each parallel class of $\mathcal{M}_i$. We note that $\{\mathrm{span}_{\mathcal{M}_i}(e) \mid e \in \tilde{E}_i\}$ forms a partition of E such that every independent set X contains at most one element from each part of the partition. We also note that the converse does not hold; even if a set X contains at most one element from each part of the partition, it is not necessarily independent.

Our algorithm that computes a reachable kernel for $g(k)$-coverable matroids is based on the same ideas as that for simple partition matroids in Sect. 3.1. We will use the partition $\{\mathrm{span}_{\mathcal{M}_i}(e) \mid e \in \tilde{E}_i\}$ to sample a random subset.

Algorithm 3: `CoverableKernel`$(\mathcal{M}_0, \ldots, \mathcal{M}_{d-1})$

$R \leftarrow \emptyset$;
repeat T **times**
  for $i \in [d-1]$ **do**
    Let $S_i \subseteq \tilde{E}_i$ be a subset obtained by independently sampling each element $e \in \tilde{E}_i$ with probability $\frac{1}{g_i(k)}$;
    $F_i \leftarrow \bigcup_{e \in S_i} \mathrm{span}_{\mathcal{M}_i}(e)$;
  $R' \leftarrow$ `Greedy`$(\mathcal{M}_0, \bigcap_{i \in [d-1]} F_i, k)$;
  $R \leftarrow R \cup R'$;
return R;

In our algorithm, we repeat the following procedure $T = \tilde{O}(\prod_{i=1}^{d-1} g_i(k))$ times. For each $i \in [d-1]$, we sample each element $e \in \tilde{E}_i$ independently with probability $1/g_i(k)$ to obtain a random subset $S_i \subseteq \tilde{E}_i$. Let $F_i = \bigcup_{e \in S_i} \mathrm{span}_{\mathcal{M}_i}(e)$. Then, we apply `Greedy`$(\mathcal{M}_0, \bigcap_{i \in [d-1]} F_i, k)$. We denote the resulting set by R' and add it to our reachable kernel R. The pseudocode for this algorithm is shown in Algorithm 3.

Similarly to the proof of Theorem 1, we can show that R satisfies (Single-EXC); see the full version for details.

Moreover, Algorithm 3 runs in polynomial time in $|E|$, because we assume that $|E| = \Omega\left(k \cdot \prod_i g_i(k)\right)$. Finally, the size of the set R returned by Algorithm 3 is at most $k \cdot T = O\left(4^{d-1} k \log(k) \prod_{i=1}^{d-1} g_i(k)\right)$. This completes the proof of Theorem 2. □

3.3 Examples of $g(k)$-coverable Matroids

We have already mentioned that simple partition matroids are k-coverable, which implies that Theorem 2 is an extension of Theorem 1. In this subsection, we present examples of $g(k)$-coverable matroids.

A matroid $\mathcal{M} = (E, \mathcal{I})$ is called a *graphic matroid* if there exists a graph $G = (V, E)$ such that $F \subseteq E$ is an independent set of $\mathcal{M}$ if and only if F forms a forest in G.

Lemma 4. *Graphic matroids are $O(k^2)$-coverable.*

Proof. Let $\mathcal{M} = (E, \mathcal{I})$ be a graphic matroid represented by a graph $G = (V, E)$. Fix $X \in \mathcal{I}$ with $|X| \leq k$. In a graphic matroid, an edge $e = \{u, v\} \in E$ lies in $\mathrm{span}_{\mathcal{M}}(X)$ if and only if u and v are in the same connected component of the forest (V, X). Thus, by removing parallel edges from $\mathrm{span}_{\mathcal{M}}(X)$, we obtain a set $F \subseteq \mathrm{span}_{\mathcal{M}}(X)$ such that $\mathrm{span}_{\mathcal{M}}(X) = \bigcup_{f \in F} \mathrm{span}_{\mathcal{M}}(f)$ and $|F| \leq \binom{k}{2}$. This shows that graphic matroids are $O(k^2)$-coverable. □

The dual of a graphic matroid is called a *cographic matroid*. In other words, $\mathcal{M} = (E, \mathcal{I})$ is a cographic matroid if there exists a graph $G = (V, E)$ such that $F \subseteq E$ belongs to $\mathcal{I}$ if and only if $G - F$ has the same number of connected components as G.

Lemma 5. *Cographic matroids are $O(k)$-coverable.*

Proof. Let $\mathcal{M} = (E, \mathcal{I})$ be a cographic matroid represented by a graph $G = (V, E)$. If G is not connected, we pick two vertices from different components of G and identify them, which results in another graph that also represents $\mathcal{M}$. By repeating this procedure, we may assume without loss of generality that G is connected.

If there exists an element $e \in E$ with $\{e\} \notin \mathcal{I}$, then we remove e from $\mathcal{M}$, since such an element does not affect the $O(k)$-coverability. Note that such an element e is a bridge (i.e., an edge cut of size 1) in G, and removing an element e from $\mathcal{M}$ corresponds to contracting an edge e in G. By repeating this procedure, we may assume without loss of generality that G is 2-edge-connected.

Fix $X \in \mathcal{I}$ with $|X| \leq k$. Then, $X \in \mathcal{I}$ means that X is an edge set of G such that $G - X$ is connected. We shrink each 2-edge-connected component in $G - X$ into a single vertex, which results in a tree $T = (V', E')$. Since $e \in E \setminus X$ is in $\mathrm{span}_{\mathcal{M}}(X)$ if and only if $G - X - \{e\}$ is disconnected, it holds that $E' = \mathrm{span}_{\mathcal{M}}(X) \setminus X$.

Let $\varphi \colon V \to V'$ be the mapping that represents the correspondence between V and V', that is, the 2-edge-connected component containing a vertex $v \in V$ in G is shrunk into a vertex $\varphi(v) \in V'$ in G'. Define $U \subseteq V'$ as

$$U = \{v' \in V' \mid X \text{ contains an edge incident to } \varphi^{-1}(v')\},$$

where φ^{-1} denotes the inverse image of φ. Then, $|U| \leq 2|X| \leq 2k$. Note that every leaf of T is contained in U, because G is 2-edge-connected. Let $W \subseteq V'$ be the set of vertices in T whose degree is at least 3. Since $|W|$ is at most the number of leaves in T, it holds that $|W| \leq |U| \leq 2k$.

Let $\mathcal{P}$ be the collection of inclusion wise maximal paths in T whose internal vertices are not in $U \cup W$. Since $U \cup W$ contains all the leaves and all vertices of degree at least three in T, the family $\{E(P) \mid P \in \mathcal{P}\}$ forms a partition of E', where $E(P)$ denotes the set of edges of P. Note that $|\mathcal{P}| = |U| + |W| - 1 = O(k)$.

For any $P \in \mathcal{P}$ and for any pair of distinct edges $e, e' \in E(P)$, $T - \{e, e'\}$ contains a component whose vertex set C consists only of internal vertices of P. Then, $\varphi^{-1}(C)$ and $V \setminus \varphi^{-1}(C)$ are disconnected in $G - X - \{e, e'\}$. Furthermore, since C contains no vertex in U, no edge in X is incident to $\varphi^{-1}(C)$, which implies that $\varphi^{-1}(C)$ and $V \setminus \varphi^{-1}(C)$ are also disconnected in $G - \{e, e'\}$. Therefore, $\{e, e'\} \notin \mathcal{I}$, that is, e and e' are parallel in $\mathcal{M}$. For each $P \in \mathcal{P}$, we pick one element e_P from $E(P)$, arbitrarily. Then, the above argument shows that $E(P) \subseteq \mathrm{span}_{\mathcal{M}}(e_P)$ for each $P \in \mathcal{P}$.

Let $F = \{e_P \mid P \in \mathcal{P}\}$. Then, $|F| = |\mathcal{P}| = O(k)$, and

$$\mathrm{span}_{\mathcal{M}}(X) = X \cup E' = X \cup \bigcup_{P \in \mathcal{P}} E(P) = \bigcup_{f \in X \cup F} \mathrm{span}_{\mathcal{M}}(f).$$

This shows that cographic matroid is $O(k)$-coverable. □

These lemmas show that Corollary 1 is derived from Theorem 2.

Acknowledgement. JP22H05001, JP23K21646, JP24K02901, JP24KJ1494, Japan, and JST ERATO Grant Number JPMJER2301, Japan.

Disclosure of Interests. The authors have no competing interests to declare that are relevant to the content of this article.

References

1. Aigner, M., Dowling, T.A.: Matching theory for combinatorial geometries. Trans. Am. Math. Soc. **158**(1), 231–245 (1971). https://doi.org/10.1090/S0002-9947-1971-0286689-5
2. Banik, A., Fomin, F.V., Golovach, P.A., Inamdar, T., Jana, S., Saurabh, S.: Cuts in graphs with matroid constraints. In: Proceedings of the 32nd Annual European Symposium on Algorithms (ESA 2024), vol. 308, pp. 17:1–17:15 (2024). https://doi.org/10.4230/LIPIcs.ESA.2024.17
3. Barvinok, A.I.: New algorithms for linear k-matroid intersection and matroid k-parity problems. Math. Program. **69**(1), 449–470 (1995). https://doi.org/10.1007/BF01585571
4. Bentert, M., Fomin, F.V., Golovach, P.A., Morelle, L.: Fault-tolerant matroid bases. In: Proceedings of the 33rd Annual European Symposium on Algorithms (ESA 2025), vol. 351, pp. 83:1–83:14 (2025). https://doi.org/10.4230/LIPIcs.ESA.2025.83
5. Bodlaender, H.L., Downey, R.G., Fellows, M.R., Hermelin, D.: On problems without polynomial kernels. J. Comput. Syst. Sci. **75**(8), 423–434 (2009). https://doi.org/10.1016/j.jcss.2009.04.001, preliminary version in ICALP 2008
6. Brand, C., Pratt, K.: Parameterized applications of symbolic differentiation of (totally) multilinear polynomials. In: Proceedings of the 48th International Colloquium on Automata, Languages, and Programming (ICALP 2021), pp. 38:1–38:19 (2021). https://doi.org/10.4230/LIPIcs.ICALP.2021.38

7. Chitnis, R., Cormode, G.: Towards a theory of parameterized streaming algorithms. In: Proceedings of the 14th International Symposium on Parameterized and Exact Computation (IPEC 2019). vol. 148, pp. 7:1–7:15 (2019). https://doi.org/10.4230/LIPIcs.IPEC.2019.7
8. Chitnis, R., et al.: Kernelization via sampling with applications to finding matchings and related problems in dynamic graph streams. In: Proceedings of the 27th Annual ACM-SIAM Symposium on Discrete Algorithms (SODA 2016), pp. 1326–1344. SIAM (2016). https://doi.org/10.1137/1.9781611974331.ch92
9. Chitnis, R., Cormode, G., Hajiaghayi, M.T., Monemizadeh, M.: parameterized streaming: maximal matching and vertex cover. In: Proceedings of the 26th Annual ACM-SIAM Symposium on Discrete Algorithms (SODA 2014), pp. 1234–1251. SIAM (2014). https://doi.org/10.1137/1.9781611973730.82
10. Cygan, M., et al.: Parameterized Algorithms. Springer, Cham (2015). https://doi.org/10.1007/978-3-319-21275-3
11. Dell, H., Marx, D.: Kernelization of packing problems. In: Proceedings of the 23rd annual ACM-SIAM Symposium on Discrete Algorithms (SODA 2012), pp. 68–81. SIAM (2012). https://doi.org/10.1137/1.9781611973099.6
12. Dell, H., Van Melkebeek, D.: Satisfiability allows no nontrivial sparsification unless the polynomial-time hierarchy collapses. J. ACM (JACM) **61**(4), 1–27 (2014). https://doi.org/10.1145/2629620
13. Doron-Arad, I., Kulik, A., Shachnai, H.: An EPTAS for budgeted matching and budgeted matroid intersection via representative sets. In: Proceedings of the 50th International Colloquium on Automata, Languages, and Programming (ICALP 2023), vol. 261, pp. 49:1–49:16 (2023). https://doi.org/10.4230/LIPIcs.ICALP.2023.49
14. Drucker, A.: New limits to classical and quantum instance compression. SIAM J. Comput. **44**(5), 1443–1479 (2015). https://doi.org/10.1137/130927115
15. Edmonds, J.: Matroid partition. Math. Decis. Sci. **11**, 335 (1968)
16. Edmonds, J.: Matroid intersection. In: Annals of Discrete Mathematics, vol. 4, pp. 39–49. Elsevier (1979). https://doi.org/10.1016/S0167-5060(08)70817-3
17. Eiben, E., Koana, T., Wahlström, M.: Determinantal sieving. In: Proceedings of the 25th Annual ACM-SIAM Symposium on Discrete Algorithms (SODA 2024), pp. 377–423. SIAM (2024). https://doi.org/10.1137/1.9781611977912.16
18. Fafianie, S., Kratsch, S.: Streaming kernelization. In: Proceedings of the 39th International Symposium on Mathematical Foundations of Computer Science (MFCS 2014), pp. 275–286. Springer (2014). https://doi.org/10.1007/978-3-662-44465-8_24
19. Fomin, F.V., Golovach, P.A., Korhonen, T., Saurabh, S.: Stability in graphs with matroid constraints. In: Proceedings of the 19th Scandinavian Symposium and Workshops on Algorithm Theory (SWAT 2024), vol. 294, pp. 22:1–22:16 (2024). https://doi.org/10.4230/LIPIcs.SWAT.2024.22
20. Fomin, F.V., Golovach, P.A., Lokshtanov, D., Saurabh, S., Zehavi, M.: Covering vectors by spaces in perturbed graphic matroids and their duals. In: Proceedings of the 46th International Colloquium on Automata, Languages, and Programming (ICALP 2019), vol. 132, pp. 59:1–59:13 (2019). https://doi.org/10.4230/LIPIcs.ICALP.2019.59
21. Fomin, F.V., Lokshtanov, D., Panolan, F., Saurabh, S.: Representative sets of product families. In: Proceedings of the 22nd European Symposium on Algorithms (ESA 2014), pp. 443–454. Springer (2014). https://doi.org/10.1007/978-3-662-44777-2_37

22. Fomin, F.V., Lokshtanov, D., Panolan, F., Saurabh, S.: Efficient computation of representative families with applications in parameterized and exact algorithms. J. ACM (JACM) **63**(4), 1–60 (2016). https://doi.org/10.1145/2886094
23. Fomin, F.V., Lokshtanov, D., Saurabh, S., Zehavi, M.: Kernelization: Theory of Parameterized Preprocessing. Cambridge University Press, Cambridge (2019). https://doi.org/10.1017/9781107415157
24. Fomin, F.V., Saurabh, S.: Kernelization methods for fixed-parameter tractability. In: Tractability: Practical Approaches to Hard Problems. Cambridge University Press, Cambridge (2014). https://doi.org/10.1017/CBO9781139177801.010
25. Fortnow, L., Santhanam, R.: Infeasibility of instance compression and succinct PCPs for NP. J. Comput. Syst. Sci. **77**(1), 91–106 (2011). https://doi.org/10.1016/j.jcss.2010.06.007, preliminary version in STOC 2008
26. Garey, M., Johnson, D.: Computers and Intractability: A Guide to the Theory of NP-completeness. Mathematical Sciences Series, W. H. Freeman (1979). https://doi.org/10.5555/574848
27. Guo, J., Niedermeier, R.: Invitation to data reduction and problem kernelization. ACM SIGACT News **38**(1), 31–45 (2007). https://doi.org/10.1145/1233481.1233493
28. Gupta, S., Roy, S., Saurabh, S., Zehavi, M.: Parameterized algorithms and kernels for rainbow matching. Algorithmica **81**(4), 1684–1698 (2018). https://doi.org/10.1007/s00453-018-0497-3
29. Gupta, S., Roy, S., Saurabh, S., Zehavi, M.: Quadratic vertex kernel for rainbow matching. Algorithmica **82**(4), 881–897 (2019). https://doi.org/10.1007/s00453-019-00618-0
30. Gurjar, R., Lokshtanov, D., Misra, P., Panolan, F., Saurabh, S., Zehavi, M.: Quasipolynomial-time deterministic kernelization and (gammoid) representation. In: Proceedings of the 50th International Symposium on Mathematical Foundations of Computer Science (MFCS 2025), vol. 345, pp. 54:1–54:17 (2025). https://doi.org/10.4230/LIPIcs.MFCS.2025.54
31. Hermelin, D., Wu, X.: Weak compositions and their applications to polynomial lower bounds for kernelization. In: Proceedings of the 23rd Annual ACM-SIAM Symposium on Discrete Algorithms, pp. 104–113. SIAM (2012). https://doi.org/10.1137/1.9781611973099.9
32. Huang, C.C., Kakimura, N., Kobayashi, Y., Terao, T.: Polynomial kernels with reachability for weighted d-matroid intersection (2026). https://arxiv.org/abs/2603.17345
33. Huang, C.C., Ward, J.: FPT-algorithms for the ℓ-matchoid problem with a coverage objective. SIAM J. Discret. Math. **37**(2), 1053–1078 (2023). https://doi.org/10.1137/21M1442267
34. Iri, M., Tomizawa, N.: An algorithm for finding an optimal independent assignment. J. Oper. Res. Soc. Japan **19**(1), 32–57 (1976). https://doi.org/10.15807/jorsj.19.32
35. Kenyon-Mathieu, C., Schudy, W.: How to rank with few errors. In: Proceedings of the 39th Annual ACM Symposium on Theory of Computing (STOC 2007), pp. 95–103 (2007). https://doi.org/10.1145/1250790.1250806
36. Kratsch, S.: Recent developments in kernelization: a survey. Bull. EATCS **2**(113) (2014)
37. Kratsch, S., Wahlström, M.: Compression via matroids: a randomized polynomial kernel for odd cycle transversal. ACM Trans. Algorithms (TALG) **10**(4), 1–15 (2014). https://doi.org/10.1145/2635810, preliminary version in SODA 2012

38. Kratsch, S., Wahlström, M.: Representative sets and irrelevant vertices: new tools for kernelization. J. ACM (JACM) **67**(3), 1–50 (2020). https://doi.org/10.1145/3390887, preliminary version in FOCS 2012
39. Lawler, E.L.: Matroid intersection algorithms. Math. Program. **9**(1), 31–56 (1975). https://doi.org/10.1007/BF01681329
40. Lokshtanov, D., Misra, N., Saurabh, S.: Kernelization–preprocessing with a guarantee. In: The Multivariate Algorithmic Revolution and Beyond: Essays Dedicated to Michael R. Fellows on the Occasion of His 60th Birthday, pp. 129–161. Springer, Berlin, Heidelberg (2012). https://doi.org/10.1007/978-3-642-30891-8_10
41. Lokshtanov, D., Misra, P., Panolan, F., Saurabh, S.: Deterministic truncation of linear matroids. ACM Trans. Algorithms (TALG) **14**(2), 1–20 (2018). https://doi.org/10.1145/3170444, preliminary version in ICALP 2015
42. Lokshtanov, D., Misra, P., Panolan, F., Saurabh, S., Zehavi, M.: Quasipolynomial representation of transversal matroids with applications in parameterized complexity. In: Proceedings of the 9th Innovations in Theoretical Computer Science Conference (ITCS 2018), vol. 94, pp. 32:1–32:13 (2018). https://doi.org/10.4230/LIPIcs.ITCS.2018.32
43. Marx, D.: A parameterized view on matroid optimization problems. Theor. Comput. Sci. **410**(44), 4471–4479 (2009). https://doi.org/10.1016/j.tcs.2009.07.027, preliminary version in ICALP 2006
44. Pfender, F., et al.: Complexity results for rainbow matchings. Theoret. Comput. Sci. **524**, 27–33 (2014). https://doi.org/10.1016/j.tcs.2013.12.013
45. Wahlström, M.: Representative set statements for delta-matroids and the Mader delta-matroid. In: Proceedings of the 2024 Annual ACM-SIAM Symposium on Discrete Algorithms (SODA), pp. 780–810. SIAM (2024). https://doi.org/10.1137/1.9781611977912.31

A Weighted-to-Unweighted Reduction for Matroid Intersection

Aditi Dudeja[1] and Mara Grilnberger[2(✉)]

[1] The Chinese University of Hong Kong, Shenzhen, China
aditidudeja@cuhk.edu.cn
[2] University of Salzburg, Salzburg, Austria
mara.grilnberger@plus.ac.at

Abstract. Given two matroids $\mathcal{M}_1$ and $\mathcal{M}_2$ over the same ground set, the *matroid intersection* problem is to find the maximum cardinality common independent set. In the weighted version of the problem, the goal is to find a maximum weight common independent set. It has been a matter of interest to find efficient approximation algorithms for this problem in various settings. In many of these models, there is a gap between the best known results for the unweighted and weighted versions. In this work, we address the question of closing this gap. Our main result is a *reduction* which converts any α-approximate unweighted matroid intersection algorithm into an $\alpha(1-\varepsilon)$-approximate weighted matroid intersection algorithm, while increasing the runtime of the algorithm by a $\log W$ factor, where W is the aspect ratio. Our framework is versatile and translates to settings such as streaming and one-way communication complexity where matroid intersection is well-studied. As a by-product of our techniques, we derive new results for weighted matroid intersection in these models.

1 Introduction

Matroid Intersection is a classical problem in combinatorial optimization for which faster algorithms have been a recent topic of interest. It generalizes many graph problems such as bipartite matching, colorful spanning trees, arborescences, tree packing, and disjoint spanning trees.

A matroid $\mathcal{M}$ is a tuple $(\mathcal{N}, \mathcal{I})$. Here, the groundset $\mathcal{N}$ is a finite collection of elements and $\mathcal{I} \subseteq 2^{\mathcal{N}}$ is a family of sets that are independent in $\mathcal{M}$. The collection $\mathcal{I}$ satisfies certain properties: (i) the empty set is independent (ii) every subset of an independent set is independent (**downward closure**) (iii) if $A, B \in \mathcal{I}$ such that $|A| > |B|$ then there exist $a \in A \setminus B$ such that $B \cup \{a\} \in \mathcal{I}$ (**exchange property**) The size of a maximum cardinality independent set is called the *rank* of $\mathcal{M}$. In general, the rank also extends to arbitrary $S \subseteq \mathcal{N}$

The full version is available at arXiv:2602.15702.

A. Dudeja—This work was initiated while the author was affiliated with the University of Salzburg.

S. S. Dey et al. (Eds.): IPCO 2026, LNCS 16588, pp. 378–393, 2026.
https://doi.org/10.1007/978-3-032-28691-8_25

and gives the size of the largest independent subset of S. Examples of matroids include the graphic matroid, in which all forests of a graph are independent and the linear matroid which corresponds to linear independence in a vector space.

Matroid Intersection. In the problem of matroid intersection, we are given two matroids $\mathcal{M}_1 = (\mathcal{N}, \mathcal{I}_1)$ and $\mathcal{M}_2 = (\mathcal{N}, \mathcal{I}_2)$ with ranks r_1 and r_2. The goal is to obtain the *maximum cardinality common independent set* of $\mathcal{M}_1$ and $\mathcal{M}_2$, which is the maximum-sized set independent in both $\mathcal{M}_1, \mathcal{M}_2$ (we denote its size by r). In this paper, we consider the weighted matroid intersection problem. Here, the input consists of two matroids $\mathcal{M}'_1 = (\mathcal{N}', \mathcal{I}'_1, w)$ and $\mathcal{M}'_2 = (\mathcal{N}', \mathcal{I}'_2, w)$ which are equipped with a weight function $w : \mathcal{N} \to \mathbb{R}^{>0}$ and the goal is to compute the *maximum weight common independent set* of $\mathcal{M}'_1$ and $\mathcal{M}'_2$. Note that, in general, the intersection of two matroids does not result in a matroid.

Accessing a Matroid. Since a matroid in general requires exponentially (in n) many bits to describe, in many models of interest for this paper, the algorithms for matroid intersection assume oracle access to the matroids. For example, in the *independence oracle* model, the algorithm can query if a set $S \subseteq \mathcal{N}$ is independent or not. In the *rank oracle* model, the algorithm can query the rank of a set S. In the classical setting, the efficiency of the algorithm is measured by the number of oracle queries made.

This paper concerns approximation algorithms for matroid intersection, which have been studied in the classical, parallel, streaming, and communication complexity settings. Since our reduction has implications for the streaming and communication complexity models, we give their definitions.

Communication Complexity Model. In the one-way two-party communication complexity model, Alice and Bob each have some portion of the input, and the goal is to compute some function of the entire input. All communication flows in one direction, from Alice to Bob. Understanding problems in the one-way two-party model is often seen as a first step to understanding them in more difficult models such as the streaming model, and thus the communication complexity model has received a lot of recent attention [2,3,15].

Streaming Model. The goal is to solve a problem over a stream of input while using a limited amount of memory. The algorithms may read the input once or multiple times. In the former case, they are called *one-pass* and in the latter case, they are referred to as *multi-pass*.

Steady progress [6,9,10,18] has been made in the classical setting, in the $(1-\varepsilon)$-approximation regime, culminating in the works of [7] and [27]. The former give an algorithm that computes a $(1-\varepsilon)$-approximate maximum cardinality common independent set using $O(\frac{n \log n}{\varepsilon} + \frac{r \log^3 n}{\varepsilon^5})$ independence queries. The latter gave an algorithm that computes a $(1-\varepsilon)$-approximate maximum weight common independent set using $O(\frac{n \log n}{\varepsilon} + \frac{r^{1.5}}{\varepsilon^4})$ independence queries. Beyond the classical setting, [20, 29] studied the cardinality version of the matroid intersection problem in streaming and communication complexity settings. To the best of our knowledge, there are no weighted analogs of these results.

The above discussion suggests that there is a gap between the best known results for weighted and unweighted versions of approximate matroid intersection. The question of closing this gap was first addressed by [18], who showed that one can reduce the problem of $(1-\varepsilon)$-approximate maximum weight common independent set to the problem of solving $\frac{\log r}{\varepsilon}$ instances of maximum cardinality common independent set. [11] made progress on this question in the streaming setting. They showed that any α-approximate maximum cardinality independent set algorithm can be converted into a $0.5 \cdot (1-\varepsilon) \cdot \alpha$-approximate maximum weight independent set algorithm while preserving the pass complexity of the algorithm and incurring an additional $\log n$ factor in the space complexity.

We continue this line of inquiry and obtain the following result.

Result 1 *Let $\varepsilon > 0$ and $\gamma_\varepsilon = \lceil (1/\varepsilon)^{1/\varepsilon} \rceil$. Let W be the weight ratio of the matroids. We have for query, streaming, and communication models:*

1. *Given an α-approximate unweighted matroid intersection algorithm $\mathcal{A}_u$ that makes $T_u(n, r)$ independence (resp., rank) queries, then there is an $\alpha(1-\varepsilon)$-approximate algorithm $\mathcal{A}_w$ for the weighted version that makes $O(T_u(n\gamma_\varepsilon, r\gamma_\varepsilon) \cdot \log W)$ independence (resp., rank) queries.*
2. *Given an α-approximate unweighted matroid intersection streaming algorithm $\mathcal{A}_u$ that takes space $S_u(n, r)$ and p passes, there is an $\alpha(1-\varepsilon)$-approximate streaming algorithm $\mathcal{A}_w$ for the weighted version that takes space $S_u(\gamma_\varepsilon n, \gamma_\varepsilon r)$ and p passes.*
3. *Given $\mathcal{P}_u$, an α-approximate one-way communication protocol for unweighted matroid intersection with message complexity $M_u(n, r)$, there exists an $\alpha(1-\varepsilon)$-approximate one-way communication protocol $\mathcal{P}_w$ for weighted matroid intersection with message complexity $M_u(\gamma_\varepsilon n, \gamma_\varepsilon r)$.*

In all of the above if the unweighted algorithm/protocol is deterministic, then the same is true of the weighted algorithm/protocol.

Matroid intersection captures many graph problems, and its specific instances such as bipartite b-matching have been studied in other settings such as MPC, distributed black-board model, and parallel shared-memory work-depth model. Our structural results apply to bipartite b-matching as well, and therefore improve on or match the state-of-the-art for the weighted version of this problem in these settings.[1] [20] introduced a robust sparsifier for unweighted matroid intersection. Our techniques can be used to extend this notion to the weighted problem. Lastly, our result builds on the work of [5], who gave such a reduction for the case of bipartite matching. Their reduction has subsequently been used as a subroutine in other matching algorithms [17, 21, 25]. Given the gener-

[1] We note that it is plausible that algorithms for weighted bipartite b-matching can be obtained by using other techniques. We still mention it since our result offers a unified perspective on variants of the matroid intersection problem.

ality of matroid intersection and our reduction, we hope that our techniques will also have future applications. Formal statements of Result 1 and discussions of applications can be found in the full version.

Related Work and Dependence on ε. We discuss some other works pertaining to the matroid intersection problem. [14] gave a $(0.5 - \varepsilon)$-approximate one-pass semi-streaming algorithm for weighted matroid intersection. [27,29] gave multi-pass semi-streaming algorithms for the unweighted and weighted problems, respectively. In the query model, [10] gave a scaling algorithm for weighted matroid intersection in the $(1 - \varepsilon)$-regime. Their algorithm uses unweighted $(1-\varepsilon)$-approximate matroid intersection algorithms satisfying some special properties (such as [6,12]) in each scale, incurring an overhead of $O(\varepsilon^{-2}\log(\varepsilon^{-1}))$.[2] Given the state of affairs, it is reasonable to ask if there is a cross-paradigm weighted-to-unweighted reduction for matroid intersection applicable for *arbitrary* approximation ratios with a polynomial dependence on ε^{-1}. We discuss some barriers to this in the Future Work section (Sect. 4).

Lastly, we note that while we focus on obtaining efficient approximation algorithms, there is a long line of work on the exact version of matroid intersection ([1,6,8,9,12,13,24,26]). In several instances, approximation algorithms for both the unweighted and weighted problem are used to speed up the exact algorithms. For such an approach, having a small dependence on ε in the number of queries is desirable. Our reduction on the other hand, incurs an exponential ε-dependence and is therefore not suitable for such applications.

2 Overview of Techniques

We introduce some relevant notation and tools, and then give a high-level overview of our techniques.

Notation. A matroid $\mathcal{M} = (\mathcal{N}, \mathcal{I})$ is associated with a rank function, $\text{rk} : 2^{\mathcal{N}} \to \mathbb{N}$. For $S \subseteq \mathcal{N}$, we use $\text{rk}(S)$ to denote the size of the maximal independent subset of S. The rank of a matroid $\text{rk}(\mathcal{M})$ denotes the size of the maximal independent set in $\mathcal{M}$. For a matroid intersection instance $\mathcal{M}_1, \mathcal{M}_2$, we will often use r_1, r_2 as shorthand for ranks of $\mathcal{M}_1, \mathcal{M}_2$, respectively. Moreover, let r denote the rank of the intersection. We will use OPT to denote the size of the maximum cardinality common independent set of $\mathcal{M}_1, \mathcal{M}_2$. For a set $S \subseteq \mathcal{N}$, the matroid $\mathcal{M}$ restricted to S is defined as $\mathcal{M}|S := (S, \mathcal{I}|S)$ where $\mathcal{I}|S := \{I \subseteq S | I \in \mathcal{I}\}$. We say that a set $S \subseteq \mathcal{N}$ *spans* a set $T \subseteq \mathcal{N}$ if for all $e \in T$ we have $S \cup \{e\} \notin \mathcal{I}$. A set $S \subseteq \mathcal{N}$ is a *circuit* if and only if $S \notin \mathcal{I}$ but for any $T \subsetneq S$, $T \in \mathcal{I}$.

Throughout this paper, we will use "prime" to distinguish weighted matroids from unweighted ones. An unweighted matroid will be denoted by $\mathcal{M}$, while a weighted matroid will be referred to as $\mathcal{M}'$. The same is true for associated sets (such as $\mathcal{N}'$) and functions (such as rk'). We let $W = \frac{\max_{e \in \mathcal{N}'} w(e)}{\min_{e \in \mathcal{N}'} w(e)}$.

[2] To the best of our knowledge, [10] require that the unweighted matroid intersection algorithm in each scale compute a maximal set of augmenting paths. We discuss this further in Sect. 4.

Linear Programming Relaxation. We will use duality for some of our arguments. For the unweighted problem over $\mathcal{M}_1 = (\mathcal{N}, \mathcal{I}_1)$ and $\mathcal{M}_2 = (\mathcal{N}, \mathcal{I}_2)$ we give the standard linear program relaxation. Let rk_1 and rk_2 denote the rank functions associated with $\mathcal{M}_1$ and $\mathcal{M}_2$, respectively. Let $x : \mathcal{N} \to \mathbb{R}^{\geq 0}$ and $x(S) := \sum_{e \in S} x(e)$ for all $S \subseteq \mathcal{N}$, the linear program relaxation is given by the following:

$$\begin{aligned} &\text{maximize } x(\mathcal{N}) \\ &\text{subject to } x(S) \leq \mathrm{rk}_1(S) \text{ and } x(S) \leq \mathrm{rk}_2(S) \text{ for all } S \subseteq \mathcal{N}. \end{aligned} \tag{1}$$

For the linear program relaxation of the weighted problem with weight function $w : \mathcal{N} \to \mathbb{R}^{\geq 0}$, the sum $\sum_{e \in \mathcal{N}} x(e)w(e)$ is maximized under the same constraints. Now, we give the dual linear programs for the relaxation of the matroid intersection problem as presented in Equation (1) over $\mathcal{M}_1$ and $\mathcal{M}_2$. Let $y, z : 2^{\mathcal{N}} \to \mathbb{R}^{\geq 0}$, the dual problem corresponds to:

$$\begin{aligned} &\text{minimize } \; f(y,z) = \sum_{S \subseteq \mathcal{N}} y(S)\mathrm{rk}_1(S) + z(S)\mathrm{rk}_2(S) \\ &\text{subject to } \textstyle\sum_{S : e \in S} y(S) + z(S) \geq 1 \text{ for all } e \in \mathcal{N} \end{aligned} \tag{2}$$

Let $\mathcal{M}_1' = (\mathcal{N}', \mathcal{I}_1', w)$, $\mathcal{M}_2' = (\mathcal{N}', \mathcal{I}_2', w)$ is an instance of weighted matroid intersection. Let rk_1' and rk_2' be the rank functions associated with them. Let $y', z' : 2^{\mathcal{N}} \to \mathbb{R}^{\geq 0}$, the associated dual linear program is defined as follows:

$$\begin{aligned} &\text{minimize } \; g(y',z') = \sum_{S \subseteq \mathcal{N}} y'(S)\mathrm{rk}_1'(S) + z'(S)\mathrm{rk}_2'(S) \\ &\text{subject to } \textstyle\sum_{S : e \in S} y'(S) + z'(S) \geq w(e) \text{ for all } e \in \mathcal{N}' \end{aligned} \tag{3}$$

We will use a result on the structure of duals (proved in [28, Chapter 41]).

Lemma 1 ([13]). *We say that a sequence of sets $S_1, S_2, \cdots, S_l$ form a chain if $S_1 \subsetneq \cdots \subsetneq S_l$. If w is integral then there exists optimal integral solutions y', z' for LP 3 such that* $\mathrm{supp}(y')$ *forms a chain and* $\mathrm{supp}(z')$ *forms a chain.*

Property 1 (Properties of Circuits). *A matroid can also be characterized by its circuits. The set of all circuits of $\mathcal{M}$, $C(\mathcal{M})$ satisfy the following properties.*

(a) For $C_1, C_2 \in C(\mathcal{M})$ such that $C_1 \subseteq C_2$, then $C_1 = C_2$.
(b) If $C_1, C_2 \in C(\mathcal{M})$ with $C_1 \neq C_2$ and $x \in C_1 \cap C_2$, then for all $y \in C_1 \setminus C_2$, there exist $C_y \in C(\mathcal{M})$ such that $y \in C_y$ and $C_y \subseteq C_1 \cup C_2 \setminus \{x\}$.

2.1 Our Framework

We follow the overall framework of [5], who obtained Result 1 for bipartite matching. The framework will have the following three steps.

1. **Aspect Ratio Reduction:** Suppose we are given two arbitrary matroids $\mathcal{M}_1' = (\mathcal{N}', \mathcal{I}_1', w)$ and $\mathcal{M}_2' = (\mathcal{N}', \mathcal{I}_2', w)$ with $w : \mathcal{N}' \to \mathbb{R}^{\geq 0}$. In this step we show that we can transform $\mathcal{M}_1', \mathcal{M}_2'$ into matroids whose groundset elements have weights $\{1, 2, \cdots, \gamma_\varepsilon\}$ and the maximum weight common independent set of the new matroids has weight at least $(1-\varepsilon)\textsc{opt}'$, where $\textsc{opt}'$ is the weight of the maximum weight common independent set of $\mathcal{M}_1', \mathcal{M}_2'$.

2. **Weighted-to-Unweighted Conversion:** Suppose we are given matroids $\mathcal{M}'_1 = (\mathcal{N}', \mathcal{I}'_1, w)$ and $\mathcal{M}'_2 = (\mathcal{N}', \mathcal{I}'_1, w)$ such that $w(e) \in \{1, 2, \cdots, W\}$ are integers for all $e \in \mathcal{N}'$. In this step, we will convert $\mathcal{M}'_1, \mathcal{M}'_2$ into unweighted matroids $\mathcal{M}_1 = (\mathcal{N}, \mathcal{I}_1)$ and $\mathcal{M}_2 = (\mathcal{N}, \mathcal{I}_2)$ (we call this matroid unfolding). Let OPT be the size of the maximum cardinality common independent set of $\mathcal{M}_1, \mathcal{M}_2$. This reduction ensures that the total weight of the maximum weight common independent set of $\mathcal{M}'_1$ and $\mathcal{M}'_2$, OPT$'$ equals OPT, the size of the maximum cardinality common independent set of $\mathcal{M}_1, \mathcal{M}_2$. Additionally, $|\mathcal{N}| \leq W \cdot |\mathcal{N}'|$, OPT $\leq r \cdot W$, and $\text{rk}(\mathcal{M}_i) \leq W \cdot \text{rk}'(\mathcal{M}'_i)$ for $i \in \{1, 2\}$.

Before we mention the final step, we note that despite being given weighted matroids $\mathcal{M}'_1, \mathcal{M}'_2$ with an arbitrary weight function, we can assume by Step 1 that element weights are integers in $\{1, \cdots, \gamma_\varepsilon\}$. Combining this assumption with Step 2 gives us unweighted matroids $\mathcal{M}_1, \mathcal{M}_2$ that are only a γ_ε-factor larger in size. Thus, we can apply any α-approximate matroid intersection algorithm on $\mathcal{M}_1, \mathcal{M}_2$ to obtain $I \in \mathcal{I}_1 \cap \mathcal{I}_2$ such that $|I| \geq \alpha\text{OPT} \geq \alpha(1 - \varepsilon)\text{OPT}'$. Thus, Steps 1 and 2 let us calculate the value $(1 - \varepsilon)\alpha\text{OPT}'$ efficiently.

3. **Refolding:** At this point assume by the previous discussion that we have $I \in \mathcal{I}_1 \cap \mathcal{I}_2$. We would like to "reverse" the operation of unfolding and "refold" I back in the original matroids $\mathcal{M}'_1, \mathcal{M}'_2$. The refolding process outputs $I' \subseteq \mathcal{N}'$. Note that I' may not necessarily be an independent set, but we are able to show that there exists $J' \subseteq I'$ such that $J' \in \mathcal{I}'_1 \cap \mathcal{I}'_2$ and $w(J') \geq (1 - \varepsilon) \cdot \alpha \cdot \text{OPT}'$. The key advantage is that $|I'| = O(\gamma_\varepsilon \cdot r)$. Intuitively, it seems easier to extract a maximum weight independent set from a small subset of $\mathcal{N}'$ (we do this by building on [7]).

We note that while the overall framework we follow is that of [5], implementing it requires new matroid specific arguments. Moreover, our aspect ratio reduction technique also extends to the matroid parity problem.

Roadmap. In Sects. 3.1 and 3.2 we discuss our main technical contributions, namely, matroid intersection unfolding and aspect ratio reduction. As discussed before, these steps already let us compute the value $(1 - \varepsilon)\alpha\text{OPT}'$. Moreover, obtaining an optimal solution of I' efficiently is mostly of interest in the classical settings. Thus, this proof is postponed to the full version of the paper.

3 Ingredients for Our Framework

In this section, starting with matroid intersection unfolding, we proceed to show structural results related to our framework.

3.1 Matroid Intersection Unfolding

Our starting point is the graph unfolding technique of [22], who give a weighted-to-unweighted reduction for matchings in integer-weighted graphs. Given such a graph G', they create an unweighted graph G by replacing each edge $e \in G'$ with $w(e)$-many edges in G such that $\mu(G) = \mu_w(G')$ (here $\mu(G)$ refers to the size of the maximum cardinality matching and $\mu_w(G')$ refers to the weight of the maximum weight matching of G').

Suppose we are given two matroids $\mathcal{M}'_1 = (\mathcal{N}', \mathcal{I}'_1, w)$ and $\mathcal{M}'_2 = (\mathcal{N}', \mathcal{I}'_2, w)$ with integer weights with maximum weight common independent set of weight OPT'. We create matroids $\mathcal{M}_1 = (\mathcal{N}, \mathcal{I}_1)$ and $\mathcal{M}_2 = (\mathcal{N}, \mathcal{I}_2)$ by replacing each element $e \in \mathcal{N}'$ with $w(e)$ copies of the element $\{e_1, \cdots, e_{w(e)}\}$. However, we need to impose additional structure on $\mathcal{M}_1, \mathcal{M}_2$ so that (a) $\text{OPT} = \text{OPT}'$ (b) We can implement independence (resp. rank) queries to $\mathcal{M}_1, \mathcal{M}_2$ using a small number of independence (resp. rank) queries to $\mathcal{M}'_1, \mathcal{M}'_2$. In order to achieve this, we examine the structure of the solution for the dual program of weighted matroid intersection mentioned in LP 3. To ensure $\text{OPT} \leq \text{OPT}'$, we want to give a suitable solution y, z. To this end, we can "split" the value of $y'(S')$ (and $z'(S')$) for each $S' \subseteq \mathcal{N}'$ among a collection of sets $\mathcal{S}_y$ (and $\mathcal{S}_z$), such that $\sum_{S \in \mathcal{S}_y} y(S) \leq y'(S')$ and $\text{rk}_1(S) \leq \text{rk}'_1(S')$ for all $S \in \mathcal{S}_y$ (analogous for z' and $\mathcal{M}_2$). To achieve the latter, the independence-structure of S in $\mathcal{M}_1$ should match the structure of S' in $\mathcal{M}'_1$. Simultaneously, the constraints of LP 2 need to be satisfied. We can achieve this if we can also ensure $\sum_{S \in \mathcal{S}_y} y(S) \geq y'(S')$ (similarly for z'). Intuitively, for each matroid $\mathcal{M}_1, \mathcal{M}_2$, we want to partition the new groundset $\mathcal{N}$ into W sets $N_1, N_2, \ldots, N_W$, where each set contains at most one copy of any original element in $\mathcal{N}'$. To create the new matroids $\mathcal{M}_1, \mathcal{M}_2$, the copied elements in N_i inherit the independence-structure of the corresponding original elements from $\mathcal{M}'_1, \mathcal{M}'_2$. Further, the structure given to elements in N_i should not interfere with the structure imposed upon N_j for all $j \neq i$, i.e. the circuits of the resulting matroids are contained in one set N_i for $i \in [W]$. The remaining question is how to create such partitions of $\mathcal{N}$, which inform the new independence structures given by $\mathcal{I}_1, \mathcal{I}_2$. Lemma 1 implies there exist optimal integral dual solutions y', z' that form chains. That is, let $\text{supp}(y') = \{S^1, \cdots, S^l\}$ and $\text{supp}(z') = \{T^1, \cdots, T^k\}$, then: $S^l \subsetneq \cdots \subsetneq S^1$, and $T^k \subsetneq \cdots \subsetneq T^1$. This lemma gives us some hint about the sought after partitions or rather about the right definition for $\mathcal{I}_1, \mathcal{I}_2$. For each set $S' \in \text{supp}(y')$ and each $T' \in \text{supp}(z')$, we now describe the sets in $\mathcal{S}_y$ and $\mathcal{T}_z$. For each $j \in [l]$ let $f \in \left\{\sum_{i=1}^{j-1} y'(S^i) + 1, \cdots, \sum_{i=1}^{j} y'(S^i)\right\}$ and let $S^j_f := \{e_f \mid e \in S^j\}$ and $y(S^j_f) = 1$. Similarly, for each $j \in [k]$, $f \in \left\{\sum_{i=1}^{j-1} z'(T^i) + 1, \cdots, \sum_{i=1}^{j} z'(T^i)\right\}$ let $T^j_f := \{e_{w(e)-f+1} \mid e \in T^j\}$ and $z(T^j_f) = 1$. As explained above, we need $\text{rk}_1(S^j_f) \leq \text{rk}'_1(S^j)$ and $\text{rk}_2(T^j_f) \leq \text{rk}'_2(T^j)$. For $\mathcal{M}_1$, the sets in the partition are formed by elements $\{e_j | e \in \mathcal{N}' \text{ and } j \leq w(e)\}$ for all $j \in [W]$. Similarly, $\mathcal{M}_2$-partition sets are formed by elements $\{e_{w(e)-j+1} \mid e \in \mathcal{N}' \text{ and } j \leq w(e)\}$ for all $j \in [W]$, as suggested by the dual. An illustrative example of this can be seen in Fig. 1. One can show that $f(y,z) \leq g(y',z')$ (see Lemma 4) and that the duals y, z are feasible (see Lemma 3). This motivates our Definition 1 for the matroids $\mathcal{M}_1, \mathcal{M}_2$. One of our main contributions is to show that (a),(b) can be achieved.

Definition 1 (Unfolded Matroid Intersection). *Let* $\mathcal{M}_1' = (\mathcal{N}', \mathcal{I}_1', w)$ *and* $\mathcal{M}_2' = (\mathcal{N}', \mathcal{I}_2', w)$ *be an instance of integer-weighted matroid intersection, such that for all* $e \in \mathcal{N}'$, $w(e) \in \{1, 2, \cdots, W\}$. *We create an instance* $\mathcal{M}_1 = (\mathcal{N}, \mathcal{I}_1)$ *and* $\mathcal{M}_2 = (\mathcal{N}, \mathcal{I}_2)$ *of unweighted matroid intersection as follows.*

(a) For each $e \in \mathcal{N}'$, *add* $w(e)$ *many elements,* $e_1, \cdots, e_{w(e)}$ *to* $\mathcal{N}$.
(b) Consider $I \subseteq \mathcal{N}$. *For all* $i \in [W]$, *let* $I_{1,i}' = \{e \mid e_i \in I\}$. *We say* $I \in \mathcal{I}_1$ *if and only if* $I_{1,i}' \in \mathcal{I}_1'$ *for all* $i \in [W]$.
(c) Consider $I \subseteq \mathcal{N}$. *For all* $i \in [W]$, *let* $I_{2,i}' = \{e \mid e_{w(e)-i+1} \in I\}$. *We say* $I \in \mathcal{I}_2$ *if and only if* $I_{2,i}' \in \mathcal{I}_2'$ *for all* $i \in [W]$.

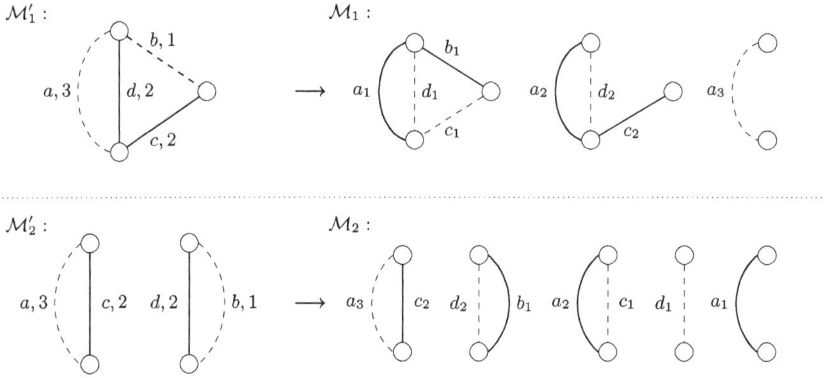

Fig. 1. An example of matroid unfolding for two graphic matroids $\mathcal{M}_1'$, $\mathcal{M}_2'$. In a graphic matroid, the elements are given by the edges of a graph and the independent sets are the forests. The labels of edges in $\mathcal{M}_1', \mathcal{M}_2'$ contain the name of the edge as well as the associated weight. The non-dashed edges show a maximum weight common independent set in $\mathcal{M}_1', \mathcal{M}_2'$ and a maximum cardinality common independent set with the same value in the unfolded matroids $\mathcal{M}_1$, $\mathcal{M}_2$. The cycles of the new graphic matroids correspond to cycles in the original instance.

We provide a proof sketch to show that $\mathcal{M}_1, \mathcal{M}_2$ are valid matroids.

Lemma 2. *Suppose* $\mathcal{M}_1' = (\mathcal{N}', \mathcal{I}_1', w)$ *and* $\mathcal{M}_2' = (\mathcal{N}', \mathcal{I}_2', w)$ *is an instance of weighted matroid intersection. Suppose that the weights* $w(e) \in \{1, 2, \cdots, W\}$ *for all* $e \in \mathcal{N}'$. *Let* $\mathcal{M}_1 = (\mathcal{N}, \mathcal{I}_1)$ *and* $\mathcal{M}_2 = (\mathcal{N}, \mathcal{I}_2)$ *be the unweighted instance obtained by applying Definition 1. Then,* $\mathcal{M}_1$ *and* $\mathcal{M}_2$ *are matroids.*

Proof (Proof Sketch). It is sufficient to verify that $\mathcal{M}_1, \mathcal{M}_2$ satisfy the three matroid properties. We show this for $\mathcal{M}_1$ and the proof for $\mathcal{M}_2$ is analogous. First, observe that the *empty set property* is satisfied because $\phi \subset \mathcal{N}$ satisfies conditions imposed by Definition 1(b). The *downward closure property* holds for $\mathcal{M}_1$ because for each $A \subseteq B \subseteq \mathcal{N}$, if we consider A_i', B_i' as in Definition 1(b), then $B_i' \in \mathcal{I}_1'$ and by downward closure of $\mathcal{M}_1'$, $A_i' \in \mathcal{I}_1'$ as well. To verify the *exchange property*, suppose $A, B \in \mathcal{I}_1$ such that $|A| > |B|$. Let $A'_i, B'_i \in \mathcal{I}'_1$ for $i \in [W]$ as in Definition 1(b). There exists $i \in [W]$ such that $|A'_i| > |B'_i|$. Since $\mathcal{M}'_1$ is a matroid and A'_i, B'_i are independent, they therefore satisfy the exchange property. Thus, one can show that A, B satisfy the exchange property. □

Oracle Access to Unfolded Matroids. We discuss how to implement queries to unfolded matroids $\mathcal{M}_1, \mathcal{M}_2$ given query access to $\mathcal{M}'_1, \mathcal{M}'_2$. Consider the independence query whether $I \in \mathcal{I}_1$. This can be implemented by checking if $I'_i \in \mathcal{I}'_1$ for $i \in [W]$ (where I'_i is as per Definition 1(b)). A similar argument holds for $\mathcal{M}_2$. We can implement the rank oracle for $\mathcal{M}_1, \mathcal{M}_2$ as follows. Let $S \subseteq \mathcal{N}$ and for $i \in [W]$, consider $S'_i \subseteq \mathcal{N}$ as per Definition 1(b). By definition of $\mathcal{M}_1$, we have, $\text{rk}_1(S) = \sum_{i \in [W]} \text{rk}'_1(S_j)$. A similar observation holds for $\mathcal{M}_2$ as well. Hence, a rank (resp., independence) query to $\mathcal{M}_i$ for $i \in \{1, 2\}$ can be implemented by W rank (resp., independence) queries to the corresponding $\mathcal{M}'_i$.

Equivalence of Original and Unfolded Matroids. We now show that the optimal solution for the weighted matroid intersection problem and the corresponding unfolded version are equal.

Theorem 1. *Let* $\mathcal{M}'_1 = (\mathcal{N}', \mathcal{I}'_1, w)$, $\mathcal{M}'_2 = (\mathcal{N}', \mathcal{I}'_2, w)$ *be an instance of integer-weighted matroid intersection with optimal value* OPT'. *Let* $\mathcal{M}_1 = (\mathcal{N}, \mathcal{I}_1)$, $\mathcal{M}_2 = (\mathcal{N}, \mathcal{I}_2)$ *be the unfolded matroids with optimal value* OPT. *Then* $\text{OPT} = \text{OPT}'$.

Showing $\text{OPT} \geq \text{OPT}'$ is relatively straightforward. Hence, our focus will be to show the converse. Towards this, we take an optimal dual solution for the weighted problem y', z' and input them to Algorithm 1. The algorithm outputs y, z, which we will show are feasible solutions to the unfolded matroid intersection problem (Lemma 3) and $f(y', z') \geq g(y, z)$ (Lemma 4). Then, by weak duality, we can conclude that $\text{OPT} \leq \text{OPT}'$. We state some observations to give intuition about Algorithm 1.

Observation 1. *For* $i \in [l]$, $f_y \in \{\sum_{j=1}^{i-1} y'(S^j) + 1, \cdots, \sum_{j=1}^{i} y'(S^j)\}$, $S^i_{f_y} = \{e_{f_y} \mid e \in S^i, w(e) \geq f_y\}$. *For* $i \in [k]$, $f_z \in \{\sum_{j=1}^{i-1} z'(T^j) + 1, \cdots, \sum_{j=1}^{i} z'(T^j)\}$, $T^i_{f_z} = \{e_{w(e)+1-f_z} \mid e \in T^i, w(e) \geq f_z\}$.

Observation 2. *At the end of the ith execution of the for loop in Algorithm 1 Line 4,* $f_y = \sum_{j=1}^{i} y'(S^j)$ *and for each* $e \in \mathcal{N}'$, *elements* $\{e_1, \cdots, e_{f_y}\} \subseteq \mathcal{N}$ *are covered by* y *and therefore,* $U^y \cap \{e_1, \cdots, e_{f_y}\} = \emptyset$. *This is because the sequence* $S^1, \cdots, S^l$ *is a chain. Similarly, at the end of the ith execution of the for loop in Algorithm 1 Line 12* $f_z = \sum_{j=1}^{i} z'(T^j)$ *and for each* $e \in \mathcal{N}'$, *elements* $\{e_{w(e)}, \cdots, e_{w(e)+1-f_z}\} \subseteq \mathcal{N}$ *are covered by* z. *Therefore,* $U^z \cap \{e_{w(e)}, \cdots, e_{w(e)+1-f_z}\} = \emptyset$. *This is because the sequence* $T^1, \cdots, T^l$ *is a chain.*

We now show feasibility of y, z output by Algorithm 1.

Lemma 3 (Feasibility of Output Duals). *Suppose* $\mathcal{M}'_1 = (\mathcal{N}', \mathcal{I}'_1, w)$ *and* $\mathcal{M}'_2 = (\mathcal{N}', \mathcal{I}'_2, w)$ *is an instance of integer-weighted matroid intersection. Let* y', z' *be optimal dual solutions, such that* $\text{supp}(y')$ *and* $\text{supp}(z')$ *are chains, input to Algorithm 1. The output* y, z *is feasible for* $\mathcal{M}_1, \mathcal{M}_2$*, the unfolded matroids.*

Proof. It is sufficient to show that for all $e \in \mathcal{N}'$, for all $i \in [w(e)]$, $e_i \notin U^z \cap U^y$. Assume to the contrary that there exists such an e_i. By Observation 2, $i \geq \sum_{j=1}^{l} y'(S^j) + 1$, and $i \leq w(e) - \sum_{j=1}^{k} z'(T^j)$. Combining this, we have, $\sum_{j=1}^{l} y'(S^j) + \sum_{j=1}^{k} z'(T^j) \leq w(e) - 1$. Since $\{S^j\}_{j\in[l]}$ and $\{T^j\}_{j\in[k]}$ form the $\text{supp}(y')$ and $\text{supp}(z')$, respectively, we can conclude that, $\sum_{S:e\in S} y'(S) + z'(S) \leq w(e) - 1$. This contradicts the feasibility of y', z'. □

Algorithm 1

Require: Integral optimal dual solutions y', z' for $\mathcal{M}'_1, \mathcal{M}'_2$,
$\text{supp}(y')$, $\text{supp}(z')$ form a chain
Ensure: Dual solutions y, z for unfolded matroids $\mathcal{M}_1, \mathcal{M}_2$.

Let $\text{supp}(y') = \{S^1, S^2, \ldots, S^l\}$, such that $S^l \subsetneq S^{l-1} \subsetneq \cdots \subsetneq S^1$.
Let $\text{supp}(z') = \{T^1, T^2, \ldots, T^k\}$, such that $T^k \subsetneq T^{k-1} \subsetneq \cdots \subsetneq T^1$.
Initialize $y \leftarrow 0$, $z \leftarrow 0$, $f_y, f_z \leftarrow 1, U^y, U^z \leftarrow \mathcal{N}$. ▷ U^y, U^z are uncovered elements.
for $i = 1, 2, \cdots, l$ **do**
  while $f_y \leq \sum_{j=1}^{i} y'(S^j)$ **do** Initialize $S^i_{f_y} \leftarrow \emptyset$. ▷ Elements covered by y.
    for $e \in S^i$ with $w(e) \geq f_y$ **do** $S^i_{f_y} \leftarrow S^i_{f_y} \cup \{e_{f_y}\}$.
    end for
    $y(S^i_{f_y}) \leftarrow 1$, $U^y \leftarrow U^y \setminus S^i_{f_y}$, $f_y \leftarrow f_y + 1$.
  end while
  $i \leftarrow i + 1$.
end for
for $i = 1, 2, \cdots, k$ **do**
  while $f_z \leq \sum_{j=1}^{i} z'(T^j)$ **do** Initialize $T^i_{f_z} \leftarrow \emptyset$. ▷ Elements covered by z.
    for $e \in T^i$ with $w(e) \geq f_z$ **do** $T^i_{f_z} \leftarrow T^i_{f_z} \cup \{e_{w(e)+1-f_z}\}$.
    end for
    $z(T^i_{f_z}) \leftarrow 1$, $U^z \leftarrow U^z \setminus T^i_{f_z}$, $f_z \leftarrow f_z + 1$.
  end while
  $i \leftarrow i + 1$.
end for
Return y, z.

We are now left with the task of showing the following lemma.

Lemma 4. *Let* $\mathcal{M}'_1 = (\mathcal{N}', \mathcal{I}'_1, w)$ *and* $\mathcal{M}'_2 = (\mathcal{N}', \mathcal{I}'_2, w)$ *be an instance of matroid intersection with rank functions* $\text{rk}'_1, \text{rk}'_2$. *Let* $\mathcal{M}_1 = (\mathcal{N}, \mathcal{I}_1)$ *and* $\mathcal{M}_2 = (\mathcal{N}, \mathcal{I}_2)$ *with rank functions* rk_1, rk_2 *be the unfolded matroids. Let* y', z' *be optimal duals for* $\mathcal{M}'_1, \mathcal{M}'_2$ *such that* $\text{supp}(y')$ *and* $\text{supp}(z')$ *each form a chain. Let* y', z' *be the input to Algorithm 1 and let* y, z *be the output. Then,* $\sum_{S\subseteq\mathcal{N}} y(S)\text{rk}_1(S) + z(S)\text{rk}_2(S) \leq \sum_{S'\subseteq\mathcal{N}'} y'(S')\text{rk}'_1(S') + z'(S')\text{rk}'_2(S')$.

Proof. Consider a set S^i in supp(y') that is assigned dual values $y'(S^i)$. The sets $S^i_{f_y}$ for $f_y \in \{\sum_{j=1}^{i-1} y'(S^j)+1, \cdots, \sum_{j=1}^{i} y'(S^j)\}$ together inherit the value from $y'(S_i)$, since $y(S^i_{f_y}) = 1$ for each such set. By Observation 1 and Definition 1(b), $\text{rk}'_1(S^i) \geq \text{rk}_1(S^i_{f_y})$ for all $f_y \in \{\sum_{j=1}^{i-1} y'(S^j)+1, \cdots, \sum_{j=1}^{i} y'(S^j)\}$. Thus,

$$\sum_{S' \subseteq \mathcal{N}'} y'(S')\text{rk}'_1(S') \overset{(i)}{=} \sum_{j=1}^{l} y'(S^i)\text{rk}'_1(S^i) \overset{(ii)}{\geq} \sum_{j=1}^{l} \sum_{f_y=\sum_{j=1}^{i-1} y'(S^j)+1}^{\sum_{j=1}^{i} y'(S^j)} \text{rk}_1(S^i_{f_y})$$

$$\overset{(iii)}{=} \sum_{j=1}^{l} \sum_{f_y=\sum_{j=1}^{i-1} y'(S^j)+1}^{\sum_{j=1}^{i} y'(S^j)} y(S^i_{f_y})\text{rk}_1(S^i_{f_y}) = \sum_{S \subseteq \mathcal{N}} y(S)\text{rk}_1(S).$$

In the above, (i) is by the fact that $\{S^1, S^2, \cdots, S^l\} = \text{supp}(y')$, (ii) is by the fact that $\text{rk}'_1(S^i) \geq \text{rk}_1(S^i_{f_y})$, and (iii) is by the fact that $y(S^i_{f_y}) = 1$ (Line 8). A similar argument gives $\sum_{S' \subseteq \mathcal{N}'} z'(S')\text{rk}'_2(S') \geq \sum_{S \subseteq \mathcal{N}} z(S)\text{rk}_2(S)$. □

We now show Theorem 1.

Proof (Theorem 1). Note that OPT $\geq$ OPT$'$ can be shown by converting a set $S' \in \mathcal{I}'_1 \cap \mathcal{I}'_2$ into a set $S \in \mathcal{I}_1 \cap \mathcal{I}_2$ such that $w(S') \leq |S|$. Let $S = \cup_{e \in S'} \{e_i \in \mathcal{N} | i \in [w(e)]\}$, using Definition 1 (b) and (c) we can see that $S \in \mathcal{I}_1 \cap \mathcal{I}_2$. To show the converse, observe that by Lemma 4, given an optimal integral dual solution y', z' for $\mathcal{M}'_1, \mathcal{M}'_2$, such that supp$(y')$ and supp(z') each form a chain, we can construct a feasible dual y, z solution for $\mathcal{M}_1, \mathcal{M}_2$ such that $\sum_{S \subseteq \mathcal{N}} y(S)\text{rk}_1(S) + z(S)\text{rk}_2(S) \leq \sum_{S' \subseteq \mathcal{N}'} y'(S')\text{rk}'_1(S') + z'(S')\text{rk}'_2(S')$. Thus, by weak duality, OPT can be upper bounded by

$$\sum_{S \subseteq \mathcal{N}} y(S)\text{rk}_1(S) + z(S)\text{rk}_2(S) \overset{(i)}{\leq} \sum_{S' \subseteq \mathcal{N}'} y'(S')\text{rk}'_1(S') + z'(S')\text{rk}'_2(S') \overset{(ii)}{=} \text{OPT}'.$$

Here, (i)is by Lemma 4, (ii) follows from strong duality. □

Finally, we note that the size and rank of two unfolded matroids $\mathcal{M}_1$, $\mathcal{M}_2$ from weighted matroids $\mathcal{M}'_1$, $\mathcal{M}'_2$ may increase by a factor W.

3.2 Aspect Ratio Reduction

In this section we first consider a special type of matroid, with elements that can be sorted into weight intervals, such that the aspect ratio within each interval is bounded by γ_ε and there is a gap of size ε^{-1} between the intervals. We show that for two such matroids, a greedy strategy for combining α-approximate maximum weight common independent sets on each interval results in an $\alpha(1-\varepsilon)$-approximation for the entire matroids. Finally, we need to be able to convert arbitrary matroids to such an instance. Our starting point is the work of [16]. At a high level, our transformation to reduce the aspect-ratio of matroids

$\mathcal{M}'_1 = (\mathcal{N}', \mathcal{I}'_1, w)$ and $\mathcal{M}'_2 = (\mathcal{N}', \mathcal{I}'_2, w)$ (where $w : \mathcal{N}' \to \mathbb{R}^{\geq 0}$) will proceed as follows. We will split $\mathcal{N}'$ into groups with geometrically increasing weights, so that within a group the ratio between weights is ε^{-1}. Then, we will delete one group for every ε^{-1} groups (we refer to these as *missing groups*) and merge consecutive groups into *weight classes*. Each weight class L_i now has a weight ratio of γ_ε. Within each class we would like to maintain for each $i \geq 0$, a common independent set of $\mathcal{M}'_1 \mid L_i$ and $\mathcal{M}'_2 \mid L_i$ called I'_i. The idea is to show that a) there exist $I' \subseteq \cup_{i\geq 0} I'_i$ such that $I' \in \mathcal{I}'_1 \cap \mathcal{I}'_2$ and $w(I') \geq (1-\varepsilon) \cdot \sum_{i \geq 0} w(I'_i)$ and b) we can extract I' from $\cup_{i \geq 0} I'_i$ efficiently.

[16] do the above procedure for the case of matchings. Let $\{M_i\}_{i\geq 0}$ be the matchings. They combine M_i's greedily (in descending order of weight classes) to obtain M. Every edge in a higher weight class can be "blamed" for excluding at most two edges from each of the lower weight classes. Since there is a gap of ε^{-1} between two consecutive weight classes, an edge e in a higher weight class only blocks out a total weight of $4\varepsilon \cdot w(e)$ from the lower weight classes, leading to a total loss of at most $4\varepsilon \cdot \sum_i w(M_i)$, and therefore, $w(M) \geq (1-\varepsilon) \cdot \sum_i w(M_i)$.

For the case of matroids we also combine I'_i's greedily (in descending order of weight classes) to obtain I'. However, we cannot hope to make a "local" charging argument as the one above. Instead, we show a structural lemma (Lemma 6) which suggests that if I' excludes l elements of I'_j, then there are at least $\frac{l}{2}$ elements of $I' \setminus \cup_{f \leq j} I'_f$ which can be "blamed" for the excluded elements. Thus, on average, each element of a higher weight class gets blamed by at most two elements from each of the lower weight classes. With this more "global" argument we are able to recover the charging argument mentioned above and conclude that $w(I') \geq (1-4\varepsilon) \cdot \sum_i w(I'_i)$. Finally, one can show by a simple pigeonhole argument that there is a choice of *missing groups* such that $\sum_i w(I'_i) \geq (1-\varepsilon) \cdot \text{OPT}'$ and therefore, $w(I') \geq (1-5\varepsilon) \cdot \text{OPT}'$.

The focus of this section will be to show aspect ratio reduction for ε^{-1}-spread matroids. These correspond to matroids containing the *missing groups* that we discussed above. We already briefly touched on how to convert arbitrary matroids into matroids containing *missing groups*. We note that the approach also extends to matroid parity.

Definition 2 (ε^{-1}-spread matroid). *Let $[l_1, r_1], \cdots, [l_k, r_k]$ be disjoint weight intervals with $l_{i+1} > r_i \varepsilon^{-1}$. Suppose $\mathcal{M}'_1 = (\mathcal{N}', \mathcal{I}'_1, w)$ and $\mathcal{M}'_2 = (\mathcal{N}', \mathcal{I}'_2, w)$ be a matroid intersection instance such that $\mathcal{N}'_i = \{e \in \mathcal{N}' \mid w(e) \in [l_i, r_i]\}$ form a partition of $\mathcal{N}'$. Then, $\mathcal{M}'_1, \mathcal{M}'_2$ are called ε^{-1}-spread matroids.*

Let $\mathcal{M}'_1, \mathcal{M}'_2$ be an instance of ε^{-1}-spread matroid intersection. For $j \in [k]$ let $L_j = [l_j, r_j]$ be the corresponding intervals and define $\mathcal{M}'|L_j := \mathcal{M}|\{e \in \mathcal{N}' \mid w(e) \in L_j\}$ for a matroid $\mathcal{M}' = (\mathcal{N}', \mathcal{I}', w)$ and a weight interval L_j. In addition, let I'_j be the optimal common independent set of $\mathcal{M}'_1|L_j$ and $\mathcal{M}'_2|L_j$. Suppose we proceed greedily to combine $\{I'_j\}_{j\geq 0}$, in descending order of intervals and only proceeding to add an element e to the combined set I_{merge} if and only if $I_{\text{merge}} \cup \{e\}$ is a common independent set of $\mathcal{M}'_1, \mathcal{M}'_2$. Now we want to bound the loss we incur due to the combining step as given in the following lemma.

Lemma 5. *Let $\mathcal{M}'_1, \mathcal{M}'_2$ be an instance of weighted matroid intersection that is ε^{-1}-spread. For $j \in [k]$, let $L_j = [l_j, r_j]$ be the corresponding intervals and I'_j be a common independent set of $\mathcal{M}'_1|L_j$ and $\mathcal{M}'_2|L_j$. Suppose I_{merge} is obtained by greedily combining I'_j for $j \in [k]$. Then, $w(I_{merge}) \geq (1-4\varepsilon)\sum_{j=1}^{k} w(I'_j)$.*

In order to show this, we will use the following structural lemma.

Lemma 6. *Let $\mathcal{M}' = (\mathcal{N}', \mathcal{I}', w)$ be a ε^{-1}-spread matroid with corresponding weight intervals L_i. Let $I' \in \mathcal{I}'$ such that $I' \subseteq \bigcup_{t \geq j} S'_t$, where S'_t is an independent set in $\mathcal{M}'|L_t$. Let further $\{e_1, \cdots e_l\} \subseteq S'_j$ be the set of elements in the independent set of level j that are spanned by I'. Let C_t be the unique circuit in $I' \cup \{e_t\}$ for every $t \in [l]$, then $\left|(\cup_{t=1}^{l} C_t) \cap (I' \setminus S'_j)\right| \geq l$.*

Proof. We will show by contradiction that if $\left|(\cup_{t=1}^{l} C_t) \cap (I' \setminus S'_j)\right| < l$ then, there must be a circuit in S'_j, thus contradicting the fact that S'_j is an independent set. To show this, we proceed by induction.

First, let us show the base case. Consider any two circuits and without loss of generality, let these be C_1 and C_2. Observe that for both C_1 and C_2, we have, $\left|C_1 \cap I' \setminus S'_j\right| \geq 1$ and $\left|C_2 \cap I' \setminus S'_j\right| \geq 1$. If this is not the case, then this would imply either $C_1 \subseteq S'_j$ or $C_2 \subseteq S'_j$ and we would have a contradiction. Assume that $\left|(C_1 \cup C_2) \cap I' \setminus S'_j\right| = 1$. Let this element be a. Thus, we have, $C_1 \cup C_2 \setminus \{a\} \subseteq S'_j$. Since $C_1 \neq C_2$ by Property 1(b), we can conclude that $C_1 \cup C_2 \setminus \{a\} \subseteq S'_j$ contains a circuit.

By inductive hypothesis, we can assume that the claim applies to any $t < l$ circuits from $\{C_1, \cdots, C_l\}$. We will now show the claim for any subset of $t+1$ circuits. Without loss of generality, let these $t+1$ circuits be $C_1, \cdots, C_{t+1}$. We additionally know that for all $s \in [t+1]$, $\left|C_s \cap I' \setminus S'_j\right| \geq 1$. Thus, if $\left|\cup_{i=1}^{t+1} C_i \cap I' \setminus S'_j\right| < t+1$, this implies that there exist an element $b \in I' \setminus S'_j$ such that at $b \in C_j$ for at least two C_j. We define $\mathcal{C}_b = \{C_i \mid b \in C_i, i \in [t+1]\}$ and without loss of generality, assume $C_1 \in \mathcal{C}_b$. For $C_t \in \mathcal{C}_b \setminus \{C_1\}$, let $C'_t \subseteq (C_t \cup C_1) \setminus \{b\}$ be a circuit containing e_t (this exists by Property 1(b)). Let this collection be called $\mathcal{C}'_b$. Observe that $C'_s \neq C'_t$ as $e_s \in C'_s$ and $e_t \in C'_t$. We now define $\mathcal{C}_{\text{new}} = (\mathcal{C} \setminus \mathcal{C}_b) \cup \mathcal{C}'_b$. Let $I'_{\text{new}} = I' \cup \{e_1\} \setminus \{b\}$ be an independent subset of $\cup_{f=j}^{k} S'_f$. Moreover, each $C_s \in \mathcal{C}_{\text{new}} \setminus \mathcal{C}_b$, is a circuit in $I'_{\text{new}} \cup \{e_s\}$ and each $C'_s \in \mathcal{C}_{\text{new}} \cap \mathcal{C}'_b$ is a circuit in $I'_{\text{new}} \cup \{e_s\}$. Thus, $\mathcal{C}_{\text{new}}$ and I'_{new} satisfy the premise of the inductive hypothesis. However, $|\mathcal{C}_{\text{new}}| = t$ and $\left|\cup_{C \in \mathcal{C}_{\text{new}}} C \cap (I'_{\text{new}} \setminus S'_j)\right| < t$, thus we can conclude by inductive hypothesis that there must be a circuit in S'_j. This contradicts the assumption that S'_j is an independent set. □

Lemma 5 follows by a charging argument as described in the beginning of the section. We conclude by noting that our aspect ratio subroutine can be implemented efficiently in classical, streaming and communication complexity settings.

4 Future Work

We obtained a reduction which converts an α-approximate unweighted matroid intersection algorithm into a weighted one with an approximation ratio of

$\alpha(1-\varepsilon)$. The reduction is versatile but incurs a loss of a factor of $\varepsilon^{-O(\varepsilon^{-1})}$ in various parameters pertaining to the models of interest. An additional feature of our reduction is its *non-adaptivity*, which means that in each weight class, one can pick an independent set arbitrarily and independently of other weight classes.[3] On one hand, non-adaptivity implies that in the streaming and one-way communication complexity settings we obtain a lossless (in terms of the number of rounds/passes) reduction. On the other hand, a result of [4] (Claim C.1) suggests that even for bipartite matching, one cannot have a weighted-to-unweighted reduction that (a) is non-adaptive, (b) works for arbitrary approximation ratios, and (c) has poly(ε^{-1}) aspect ratio. Indeed, the work of [10], which gives a scaling algorithm for weighted matroid intersection by using an unweighted algorithm in each scale, to the best of our knowledge, relies on (i) the unweighted algorithms computing a maximal set of short augmenting paths in the exchange graph, (ii) is adaptive as independent sets derived in the subsequent scales depend on previous ones, and (iii) is only applicable in $(1-\varepsilon)$-approximation ratio regime. The last two properties affect the portability of this technique to one-pass streaming and one-way communication complexity settings. This is because (ii) leads to an increase in the number of passes [19,25]. Moreover, there are results which show that it is impossible to achieve an approximation ratio of $(1-\varepsilon)$ in these models [15,23]. This leaves us with the following open question:

> Building on the techniques of [4], can we obtain a non-adaptive weighted-to-unweighted reduction for matroid intersection, with an aspect ratio of poly(ε^{-1}), in the $(1-\varepsilon)$-approximate regime?

Acknowledgements. We are grateful to Sebastian Forster for his valuable feedback and comments during this project. We thank the anonymous reviewers for their helpful comments. This project has received funding from the European Research Council (ERC) under the European Union's Horizon 2020 research and innovation programme (grant agreement No 947702). This publication has been supported by the EXDIGIT (Excellence in Digital Sciences and Interdisciplinary Technologies) project, funded by Land Salzburg under grant number 20204-WISS/263/6-6022.

Disclosure of Interests. The authors have no competing interests to declare that are relevant to the content of this article.

References

1. Aigner, M., Dowling, T.A.: Matching theory for combinatorial geometries. Trans. Am. Math. Soc. **158**(1), 231–245 (1971). https://doi.org/10.2307/1995784
2. Assadi, S., Behnezhad, S.: On the robust communication complexity of bipartite matching. Approximation, Randomization, and Combinatorial Optimization. Algorithms and Techniques (APPROX/RANDOM, (2021). https://doi.org/10.4230/LIPIcs.APPROX/RANDOM.2021.48

[3] We recall that in our reduction, we split up the elements of the matroid into different weight classes. In each weight class, we can run the unfolding and unweighted algorithms in parallel. Therefore, we refer to our reduction as a *non-adaptive* reduction from weighted to unweighted matroid intersection.

3. Assadi, S., Bernstein, A.: Towards a unified theory of sparsification for matching problems. In: 2nd Symposium on Simplicity in Algorithms (SOSA), pp. 11:1–11:20 (2019). https://doi.org/10.4230/OASIcs.SOSA.2019.11
4. Bernstein, A., Chen, J., Dudeja, A., Langley, Z., Sidford, A., Tu, T.: Matching composition and efficient weight reduction in dynamic matching. In: Proceedings of the 2025 Annual ACM-SIAM Symposium on Discrete Algorithms (SODA), pp. 2991–3028 (2025). https://doi.org/10.1137/1.9781611978322.97
5. Bernstein, A., Dudeja, A., Langley, Z.: A framework for dynamic matching in weighted graphs. In: Proceedings of the 53rd Annual ACM SIGACT Symposium on Theory of Computing (STOC), pp. 668–681 (2021). https://doi.org/10.1145/3406325.345111
6. Blikstad, J.: Breaking o(nr) for matroid intersection. In: 48th International Colloquium on Automata, Languages, and Programming (ICALP), pp. 31:1–31:17 (2021). https://doi.org/10.4230/LIPIcs.ICALP.2021.31
7. Blikstad, J., Tu, T.: Efficient matroid intersection via a batch-update auction algorithm. In: 2025 Symposium on Simplicity in Algorithms (SOSA), pp. 226–237 (2025). https://doi.org/10.1137/1.9781611978315.18
8. Blikstad, J., Van Den, Brand, J., Mukhopadhyay, S., Nanongkai, D.: Breaking the quadratic barrier for matroid intersection. In: Proceedings of the 53rd Annual ACM SIGACT Symposium on Theory of Computing (STOC), pp. 421–432 (2021). https://doi.org/10.1145/3406325.345109
9. Chakrabarty, D., Lee, Y.T., Sidford, A., Singla, S., Wong, S.C.W.: Faster matroid intersection. In: 2019 IEEE 60th Annual Symposium on Foundations of Computer Science (FOCS), pp. 1146–1168. IEEE (2019). https://doi.org/10.1109/FOCS.2019.00072
10. Chekuri, C., Quanrud, K.: A fast approximation for maximum weight matroid intersection. In: Proceedings of the Twenty-Seventh Annual ACM-SIAM Symposium on Discrete Algorithms (SODA), pp. 445–457 (2016). https://doi.org/10.1137/1.9781611974331.ch33
11. Crouch, M.S., Stubbs, D.M.: Improved streaming algorithms for weighted matching, via unweighted matching. In: Approximation, Randomization, and Combinatorial Optimization. Algorithms and Techniques (APPROX/RANDOM). LIPIcs, vol. 28, pp. 96–104 (2014). https://doi.org/10.4230/LIPIcs.APPROX-RANDOM.2014.96
12. Cunningham, W.H.: Improved bounds for matroid partition and intersection algorithms. SIAM J. Comput. **15**(4), 948–957 (1986). https://doi.org/10.1137/0215066
13. Edmonds, J.: Submodular functions, matroids, and certain polyhedra. In: Combinatorial Optimization—Eureka, You Shrink! Papers Dedicated to Jack Edmonds 5th International Workshop Aussois, France, 5–9 March 2001, pp. 11–26 (2003). https://doi.org/10.1007/3-540-36478-1
14. Garg, P., Jordan, L., Svensson, O.: Semi-streaming algorithms for submodular matroid intersection. In: Integer Programming and Combinatorial Optimization - 22nd International Conference (IPCO) (2021). https://doi.org/10.1007/978-3-030-73879-215
15. Goel, A., Kapralov, M., Khanna, S.: On the communication and streaming complexity of maximum bipartite matching. In: Proceedings of the Twenty-Third Annual ACM-SIAM Symposium on Discrete Algorithms (SODA), pp. 468–485 (2012). https://doi.org/10.1137/1.9781611973099.41
16. Gupta, M., Peng, R.: Fully dynamic (1+ e)-approximate matchings. In: 54th Annual IEEE Symposium on Foundations of Computer Science (FOCS), pp. 548–557 (2013). https://doi.org/10.1109/FOCS.2013.65

17. Hashemi, D., Wrzos-Kaminska, W.: Weighted matching in the random-order streaming and robust communication models. In: Approximation, Randomization, and Combinatorial Optimization. Algorithms and Techniques (APPROX/RANDOM), pp. 16:1–16:26 (2024). https://doi.org/10.4230/LIPIcs.APPROX/RANDOM.2024.16
18. Huang, C., Kakimura, N., Kamiyama, N.: Exact and approximation algorithms for weighted matroid intersection. In: Proceedings of the Twenty-Seventh Annual ACM-SIAM Symposium on Discrete Algorithms (SODA), pp. 430–444 (2016). https://doi.org/10.1137/1.9781611974331.ch32
19. Huang, C., Sellier, F.: Maximum weight b-matchings in random-order streams. In: 30th Annual European Symposium on Algorithms (ESA), pp. 68:1–68:14 (2022). https://doi.org/10.4230/LIPIcs.ESA.2022.68
20. Huang, C., Sellier, F.: Robust sparsification for matroid intersection with applications. In: Proceedings of the 2024 ACM-SIAM Symposium on Discrete Algorithms (SODA), pp. 2916–2940 (2024). https://doi.org/10.1137/1.9781611977912.10
21. Huang, S.E., Su, H.H.: (1-eps)-approximate maximum weighted matching in poly (1/eps, log n) time in the distributed and parallel settings. In: ACM Symposium on Principles of Distributed Computing (2023). https://doi.org/10.1145/3583668.3594570
22. Kao, M., Lam, T.W., Sung, W., Ting, H.: A decomposition theorem for maximum weight bipartite matchings. SIAM J. Comput. **31**(1), 18–26 (2001). https://doi.org/10.1137/S0097539799361208
23. Kapralov, M.: Space lower bounds for approximating maximum matching in the edge arrival model. In: Proceedings of the 2021 ACM-SIAM Symposium on Discrete Algorithms (SODA), pp. 1874–1893 (2021)
24. Lawler, E.L.: Matroid intersection algorithms. Math. Program. **9**(1), 31–56 (1975)
25. Liu, Q.C., Ke, Y., Khuller, S.: Scalable auction algorithms for bipartite maximum matching problems. In: Approximation, Randomization, and Combinatorial Optimization. Algorithms and Techniques (APPROX/RANDOM), pp. 28:1–28:24 (2023). https://doi.org/10.4230/LIPIcs.APPROX/RANDOM.2023.28
26. Nguyen, H.L.: A note on cunningham's algorithm for matroid intersection. arXiv preprint arXiv:1904.04129 (2019)
27. Quanrud, K.: Adaptive sparsification for matroid intersection. In: 51st International Colloquium on Automata, Languages, and Programming (ICALP), pp. 118:1–118:20 (2024). https://doi.org/10.4230/LIPIcs.ICALP.2024.118
28. Schrijver, A.: Combinatorial Optimization: Polyhedra and Efficiency, vol. 24. Springer (2003)
29. Terao, T.: Deterministic (2/3-ϵ)-approximation of matroid intersection using nearly-linear independence-oracle queries. CoRR abs/2410.18820 (2024). https://doi.org/10.48550/arXiv.2410.18820

Arithmetic Circuits and Neural Networks for Regular Matroids

Christoph Hertrich[1], Stefan Kober[2(✉)], and Georg Loho[3,4]

[1] University of Technology Nuremberg, Nuremberg, Germany
christoph.hertrich@utn.de
[2] Université Libre de Bruxelles, Brussels, Belgium
stefan.kober@ulb.be
[3] Freie Universität Berlin, Berlin, Germany
georg.loho@math.fu-berlin.de
[4] University of Twente, Enschede, The Netherlands

Abstract. We prove that there exist uniform $(+, \times, /)$-circuits of size $O(n^3)$ to compute the basis generating polynomial of regular matroids on n elements. By tropicalization, this implies that there exist uniform $(\max, +, -)$-circuits and ReLU neural networks of the same size for weighted basis maximization of regular matroids. As a consequence in linear programming theory, we obtain a first example where taking the *difference of two extended formulations* can be more efficient than the best known individual extended formulation for the base polytope of regular matroids of size $O(n^6)$ by Aprile and Fiorini. Such differences have recently been introduced as *virtual extended formulations*. The proof of our main result relies on a fine-tuned version of Seymour's decomposition of regular matroids which allows us to identify and maintain graphic substructures to which we can apply a *local* version of the *star-mesh transformation*.

1 Introduction

A key challenge in complexity theory is to characterize the computational power of *arithmetic circuits*, the algebraic counterparts of Boolean circuits. In such a circuit, every gate outputs an arithmetic expression, like the sum or product of its inputs. The fundamental underlying motivation for this field of research is to understand which problems one can solve efficiently with a predefined set of algrebraic operations. The most frequently studied type of arithmetic circuits is $(+, \times)$-circuits [44], as they form a very natural model to represent polynomials. However, allowing additional operations, e.g., subtraction or division, sometimes makes the model of computation exponentially more powerful [30,48]. This leads to the question of how the complexity of representing certain functions depends on the operations allowed in the circuit. To this end, Fomin, Grigoriev, and Koshevoy [18] coined the notion of *subtraction-free complexity*, which is the study of $(+, \times, /)$-circuits. These circuits are of particular interest because they can

S. S. Dey et al. (Eds.): IPCO 2026, LNCS 16588, pp. 394–409, 2026.
https://doi.org/10.1007/978-3-032-28691-8_26

be *tropicalized*: they give rise to corresponding $(\max, +, -)$-circuits computing *tropical polynomials* or *tropical rational functions* [33].

One motivation to study tropical circuits is to prove lower bounds on (pure) dynamic programming algorithms [32]. Our primary motivation, however, is the expressive power of feed-forward neural networks with ReLU activations, the most commonly used activation function in modern machine learning. Like $(\max, +, -)$-circuits, ReLU networks represent *continuous and piecewise linear* (CPWL) functions [5], which can be understood as tropical rational functions. The tropical viewpoint on neural networks was initiated by Zhang, Naitzat, and Lim [49] and Charisopoulos and Maragos [14], and was subsequently used in many theoretical works on neural networks, e.g., to prove lower bounds on the depth [22,23].

We contribute to an emerging stream of research studying ReLU networks as a kind of arithmetic circuit [9,26]. In fact, ReLU networks can be interpreted as $(\max, +, -)$-circuits with the additional power of multiplying intermediate results with real-valued constants (through the *weights* of the neural network). Thereby, using the idea of tropicalizing $(+, \times, /)$-circuits, this yields a recipe to transform subtraction-free circuits into ReLU networks.

Contribution. Our main result is the construction of efficient $(+, \times, /)$-circuits for the basis generating polynomial of regular matroids. For a matroid M with basis set $\mathcal{B}$, the *basis generating polynomial* is $f_M := \sum_{B \in \mathcal{B}} x^B := \sum_{B \in \mathcal{B}} \prod_{e \in B} x_e$.

Theorem 1. *For a regular matroid M with n elements, there is a $(+, \times, /)$-circuit of size $O(n^3)$ computing the basis generating polynomial f_M. Given an independence oracle of M, this circuit can be constructed in polynomial time.*

Theorem 1 generalizes a result by Fomin et al. [18] on subtraction-free circuits for the spanning tree polynomial. In circuit complexity, a circuit family whose members can be computed in polynomial time is called *uniform*.

Implications on Neural Networks. Using the idea of tropicalization described above, we obtain the following.

Corollary 2. *For a regular matroid M with basis set $\mathcal{B}$ and n elements, there is (i) a $(\max, +, -)$-circuit and (ii) a ReLU neural network of size $O(n^3)$ computing the tropical polynomial $\max_{B \in \mathcal{B}} \sum_{e \in B} x_e$. Given an independence oracle of M, the circuit and the neural network can be constructed in polynomial time.*

Evaluating this tropical polynomial can be seen as solving the *maximum weight basis problem* over the regular matroid, which can famously be achieved by the greedy algorithm. However, as described by Hertrich and Sering [26], finding efficient neural network representations to solve combinatorial optimization problems like this remains non-trivial, as the computational model of neural networks is missing simple algorithmic building blocks like if-branchings based on the comparison of real numbers or computing argmins over finite sets. These limitations prohibit the approach of simply implementing the greedy algorithm on a

ReLU neural network. Consequently, Corollary 2 requires a deeper understanding of the structure of regular matroids and provides new non-trivial insights on the power of ReLU neural networks as a model of computation.

Implications in Linear Programming Theory. Combining Corollary 2 with the connection between neural networks and extension complexity established in [25], we immediately obtain the following.

Corollary 3. *There exists a* virtual *extended formulation of size $O(n^3)$ of the matroid base polytope of a regular matroid on n elements.*

This is in contrast to the best known extension complexity upper bound $O(n^6)$ from [3]. A *virtual* extended formulation of a polytope P consists of extended formulations for two polytopes Q and R such that $P + Q = R$, where $P + Q$ is a Minkowski sum. This allows to solve the linear optimization problem over P by solving it over Q and R and taking the difference of the two results [25]. It is an open question whether taking a single difference of two LPs brings an advantage over solving just one LP, that is, whether virtual extended formulations can be more efficient than ordinary ones. To the best of our knowledge, our result provides the first example of virtual extended formulations that are smaller than the best known extended formulations.

Generalization to MFMC Matroids. Our results extend to the more general class of Max-Flow-Min-Cut (MFMC) matroids. This class of matroids has been introduced by Seymour [41], and MFMC matroids have been shown to have a decomposition that extends Seymour's decomposition of regular matroids [42], see [41] and [40, Corollary 12.3.22]. We formally define MFMC matroids and explain how our main result can be generalized in the full version of our paper [24].

Proof Techniques. A key ingredient to prove our main result is Maurer's generalization [38] of Kirchhoff's celebrated matrix-tree theorem [35] in order to view the basis generating polynomial of a regular matroid M as the determinant of an r-dimensional square matrix L, where r is the rank of M. Since a general determinant cannot be computed with a $(+, \times, /)$-circuit, we need to exploit the specific structure of the matrix L. We can understand the approaches by Fomin et al. [18] and Hertrich and Sering [26] for the case of spanning trees in graphs as an inductive approach to reduce the number of rows of L by one in each step. In terms of graphs, this strategy corresponds to eliminating one vertex at a time, and compensating for the lost information by introducing additional edges between neighbors of the removed vertex. This procedure is known as a *star-mesh transformation* [18]. In general regular matroids, however, the concept of a vertex does not exist, making it difficult to generalize this approach. Nevertheless, we introduce a generalization of the star-mesh transformation to $\{0, \pm 1\}$-matrices, see Definition 16. However, this generalization suffers from the fact that it does not preserve regularity, rendering the application of Maurer's theorem invalid in later iterations. In fact, for the regular matroid R_{10}, none of the possible generalized star-mesh transformations preserves regularity. We show that this is the only counterexample, in the following sense. We give a refined version of Seymour's decomposition of regular matroids, implying that for any 3-connected regular matroid M that is not R_{10}, either M or its dual has a star-mesh transformation that preserves regularity. Together with constructions of arithmetic circuits for the 1-sum and 2-sum of regular matroids, R_{10}, and all cographic matroids by dualization, this allows an inductive proof of our main result. We follow a similar strategy as Bérczi, Mátravölgyi, and Schwarcz [11], and employ Seymour's decomposition of regular matroids without explicitly dealing with the most complicated case, that is, the 3-sum. Instead, we are able to iteratively reduce the rank by 1 in any 3-connected regular matroid, see Proposition 14. It would be interesting to see whether this strategy can be employed to other problems on regular matroids, for which in many cases the 3-sum is the hardest case [3,6,7] or remains unsolved, see e.g. [4], and extensions of [11,39].

Limitations Beyond Regular Matroids. We do not expect that Theorem 1 can be generalized much beyond regular or MFMC matroids. The reason is that efficient circuits for the basis generating polynomial imply a simple algorithm to count the number of bases: simply feed the all-ones vector into the circuit. However, already for binary matroids, which would be the natural next step to generalize Theorem 1, counting the bases is #P-hard [36,45], see also [1,2] for some approximate counting results. Note that this limitation is not specific to the subtraction-free setting, as it equally holds for circuits allowing subtractions. On a more technical level, our proofs crucially rely on the generalization of the Matrix Tree Theorem to regular matroids [38]. Such a generalization is not known beyond regular matroids. Unlike Theorem 1, it may be possible that Corollary 2 could be generalized to binary matroids, but this would require entirely different proof techniques that are specific to the tropical setting. A related open problem is whether base polytopes of binary matroids always have polynomial extension complexity [3]. A first step in this direction could be to consider proper minor-closed subclasses of binary matroids, which admit strong structural results [20].

Related Work. The notion of subtraction-free circuits was coined by Fomin et al. [18], who also proved that $(+, \times, /)$-circuits of size $O(n^3)$ exist for the spanning tree generating polynomial of a graph with n vertices. This is the special case of Theorem 1 for graphic matroids. Hertrich and Sering [26] implicitly used this to conclude that ReLU networks of size $O(n^3)$ can compute the value of a minimum spanning tree from the edge weights. They also constructed polynomial-size ReLU networks for the maximum flow problem. ReLU networks as a model of computation were also studied in [27] for the knapsack problem. The book by Jukna [33] provides an in-depth treatment of tropical circuits. In particular, Jukna and Seiwert [34] proved an exponential lower bound for $(+, \times)$-circuits computing the spanning tree generating polynomial. This implies that, already in the graphic case, Theorem 1 would fail without division gates. This lower bound also translates to the tropical setting and therefore shows that subtraction gates are necessary in Corollary 2. When disallowing subtractions, the resulting model of *monotone* or the related *input-convex* ReLU networks yield a different model of computation that was investigated in [9,12,19,25]. Such networks imply the existence of (ordinary) extended formulations for the polytopes underlying the related optimization problem [25], while non-monotone networks only yield virtual extended formulations. Connections between circuit complexity and extension complexity also appeared in [17,28]. It remains an open question whether there is any class of CPWL functions that can be evaluated in polynomial time, but requires exponential-size ReLU networks.

Arora, Basu, Mianjy, and Mukherjee [5] proved that the class of functions computable by ReLU networks exactly coincides with the class of CPWL functions. We refer to [29] for a survey on polyhedral methods in deep learning. It is an open question which depth one needs for such exact representations [8,10,21–23,47]. This is in contrast to well-known universal approximation theorems [15,37], which are often restricted to a bounded domain and require very wide neural networks.

Our approach for regular matroids is based on a refinement of Seymour's decomposition of regular matroids [42]. Such refinements have been crucial in many other algorithmic and structural applications of this decomposition theorem, see, e.g., [3,11,16]. We emphasize that our refinement naturally extends text-book proofs of Seymour's decomposition and is otherwise self-contained.

2 Preliminaries

All missing proofs, as well as further formal definitions to follow the missing proofs can be found in the full version of this paper [24].

This section presents the most important preliminaries to understand our main results and the intuition behind the proof techniques.

2.1 Arithmetic Circuits and Neural Networks

An *arithmetic circuit* is a directed acyclic graph defining an arithmetic expression using 2-ary operations like max, min, $+$, $-$, $\times$, or $/$. We assume that each node (or *gate*) of the circuit has either in-degree zero, in which case it is one of n input nodes, or in-degree two, in which case it performs one of the previously mentioned arithmetic computations on the outputs of its two predecessors. Further, we assume that there is a unique output gate with out-degree zero, which defines the arithmetic expression represented by the entire circuit. This expression contains one variable for each of the n input gates and defines a function $f\colon \mathbb{R}^n \to \mathbb{R}$. We remark that allowing division gates can lead to divisions by zero when evaluating a circuit on numerical inputs. For all constructions in this paper, this is not a problem, as our statements remain correct on the level of formal rational functions. Moreover, when inputs are assumed to be nonnegative and the zero coordinates are known at the time the circuit is constructed, the construction can be modified so that the circuit can still be evaluated. We defer the details of this modification to the full version of the paper. When talking about specific types of arithmetic circuits, we usually put the allowed operations in parentheses; e.g., a $(+, \times, /)$-circuit is an arithmetic circuit where, besides input gates, there only exist addition, multiplication, and division gates. The *size* of an arithmetic circuit is the number of non-input gates.

In the context of this paper, we consider *neural networks* with *rectified linear unit* (ReLU) activations. Like arithmetic circuits, they define a computation through a directed acyclic graph with n input nodes (or *neurons*) and a particular output neuron. In contrast to arithmetic circuits, the neurons can have arbitrary large in-degree (say k) and compute a function $(z_1 \dots, z_k) \mapsto \max\{0, \sum_{i=1}^{k} w_i z_i\}$, where the z_i are the outputs of the predecessor neurons and the w_i are the *weights* of the considered neuron. Then, the whole network computes a *continuous and piecewise linear* (CPWL) function $f\colon \mathbb{R}^n \to \mathbb{R}$ defined on the input variables and parameterized by the weights of all the neurons. The *size* of such a neural network is the number of non-input neurons.

We will now make some statements to relate arithmetic circuits, their tropical counterparts, and neural networks. To this end, we define the *tropicalization* of a $(+, \times, /)$-circuit as the corresponding $(\max, +, -)$-circuit, where every $+$-gate is replaced with a max-gate, every $\times$-gate with a $+$-gate, and every $/$-gate with a $-$-gate. The function computed by a $(+, \times, /)$-circuit is a *rational function*

$$f(\mathbf{z}) = \frac{\sum_{\mathbf{i} \in \mathcal{I}_1} a_{\mathbf{i}} \mathbf{z}^{\mathbf{i}}}{\sum_{\mathbf{i} \in \mathcal{I}_2} b_{\mathbf{i}} \mathbf{z}^{\mathbf{i}}}, \tag{1}$$

that is, the quotient of two polynomials, where $\mathbf{i} = (i_1, \dots, i_n) \in \mathbb{Z}_{\geq 0}^n$ is a multi-index and $a_{\mathbf{i}}, b_{\mathbf{i}} \in \mathbb{Z}_{\geq 1}$. We highlight that by our definition, a $(+, \times, /)$-circuit, the constant coefficients $a_{\mathbf{i}}, b_{\mathbf{i}} \in \mathbb{Z}_{\geq 1}$ only arise from accumulated additions. We do not allow the multiplication with arbitrary constant values. In particular, the process of tropicalization would not yield the desired construction if we had negative coefficients.

For a rational function of the form (1), we define its *tropicalization* as the CPWL function

$$\operatorname{trop}(f)(\mathbf{z}) = \max_{\mathbf{i} \in \mathcal{I}_1} \mathbf{i}^\top \mathbf{z} - \max_{\mathbf{i} \in \mathcal{I}_2} \mathbf{i}^\top \mathbf{z}. \tag{2}$$

See, e.g., Joswig [31, Section 2.2] for more details on the process of tropicalization. Observe that our version of tropicalization does not depend on the values of $a_{\mathbf{i}}$ and $b_{\mathbf{i}}$, as long as they are non-zero. This is on purpose to induce the following behavior: $\operatorname{trop}(z) = z = \max\{z, z\} = \operatorname{trop}(z+z) = \operatorname{trop}(2z)$. In other words, one might think of this as all non-zero constants being sent to 0 by the map trop.

With these definitions we can derive the following two propositions, which immediately imply that Corollary 2 can be deduced from Theorem 1. See also [33, Corollary 6.4] and [26] for specific versions in the case of spanning trees.

Proposition 4. *If a $(+, \times, /)$-circuit computes a rational function f, then the tropicalization of this circuit computes* $\operatorname{trop}(f)$.

The proof uses the fact that trop is a semiring homomorphism between functions of the forms (1) and (2) ([24, Lemma 18]). This homomorphism is not injective. Therefore, the converse fails, that is, converting $(\max, +, -)$-circuits to $(+, \times, /)$.

Proposition 5. *If a CPWL function is computed by a* $(\max, +, -)$*-circuit of size* s*, then it can also be computed by a ReLU network of size* $3s$*.*

Proof. The circuit directly translates to a neural network, where each max-gate can be realized with three ReLU neurons using $\max\{x, y\} = \max\{0, x-y\}+y = \max\{0, x - y\} + \max\{0, y\} - \max\{0, -y\}$, see [5,23].

□

In contrast to our circuit models, neural networks do have constants encoded in their weights. Networks arising through Proposition 5 have weights in $\{0, \pm 1\}$. This is fine for proving upper bounds, but it remains unclear to what extent the use of different constants could lead to more efficient representations.

2.2 Matroid Basics

We assume the reader to be familiar with standard matroid terms. For a comprehensive overview of regular matroids, we refer to [40,46]. We lay out some matroid definitions in terms of the bases of matroids, where the definitions are necessary to follow our arguments.

Matroid Basics. Given a ground set of elements E, and a non-empty set of bases $\mathcal{B} \subseteq 2^E$ we say that $M = (E, \mathcal{B})$ is a *matroid* if for each $B_1, B_2 \in \mathcal{B}$ with $B_1 \neq B_2$ and every $b \in B_1 \backslash B_2$, there is a $b' \in B_2 \backslash B_1$ such that $B_1 \cup \{b'\} \backslash \{b\} \in \mathcal{B}$. We set $n := |E|$ and define the rank as $r := r(M) := |B|$ for any $B \in \mathcal{B}$. The rank function extends to $X \subseteq E$, i.e., $r(X) := \max\{|B \cap X| : B \in \mathcal{B}\}$. We say that a set of elements $I \subseteq E$ is *independent*, if it is a subset of a basis, and denote the set of independent sets by $\mathcal{I}(M)$. Further, we say that $C \subseteq E$ is a *circuit* of M, if $C \notin \mathcal{I}(M)$, but for all $e \in C$, $C - e \in \mathcal{I}(M)$. Note that a matroid is uniquely determined by its bases, its independent sets, or its circuits.

Given a matroid $M = (E, \mathcal{B})$, we define the *cobases* as the set of complements of bases $\mathcal{B}^* := \{E \backslash B : B \in \mathcal{B}\}$. Then the dual matroid M^* of M is defined on the same ground set E, with set of bases $\mathcal{B}^*$. This directly implies $(M^*)^* = M$. We say that $C \subseteq E$ is a cocircuit of M if C is a circuit of M^*.

Given a matroid $M = (E, \mathcal{B})$ and some element $e \subseteq E$, we define the deletion operation $M \backslash e$ as the matroid on the ground set $E \backslash \{e\}$ with bases $\{B \in \mathcal{B} : e \notin B\}$. Similarly, we define the contraction operation M/e as the matroid on the ground set $E \backslash \{e\}$ with bases $\{B \backslash \{e\} : B \in \mathcal{B}, e \in B\}$.

Representability. Given a field $\mathbb{F}$, we say that a matroid $M = (E, \mathcal{B})$ of rank r is $\mathbb{F}$*-representable*, if there is a matrix $A \in \mathbb{F}^{r \times n}$, such that a subset $E' \subseteq E$ is independent if and only if the corresponding columns are linearly independent over $\mathbb{F}$. Note that elementary row operations do not affect the independence of columns. Therefore, we may assume that any representation matrix of a matroid contains a full-rank identity matrix. Given a representation matrix of the form $A = (I_r\, C)$ of M, where I_r is the identity matrix of size r, a representation matrix of M^* is given by $(C^\intercal\, I_{n-r})$.

Regular Matroids. Recall that a matrix is *totally unimodular* (TU) if all its subdeterminants are in $\{-1, 0, 1\}$ (over $\mathbb{R}$). Let M be a binary matroid with representation matrix A over $\mathbb{F}_2$. We say that M is *regular* if the 1-entries of A can be signed such that we obtain a real totally unimodular matrix. This is equivalent to M being representable over $\mathbb{R}$ by a TU matrix.

We call a matroid $M = (E, \mathcal{B})$ *graphic* if there is a connected graph $G = (V, E)$, such that $C \subseteq E$ is a circuit of M if and only if the corresponding edges form a cycle in G. We write $M := M(G)$ and say that $M(G)$ is the graphic matroid of G. The bases of $M(G)$ map bijectively to the spanning trees of the corresponding graph G. A matroid M is said to be *cographic* if M^* is graphic. Graphic and cographic matroids are important examples of regular matroids, but not every regular matroid is of this form, see Theorem 7.

Encoding Matroids. The set of bases, independent sets, or circuits of a matroid can be exponentially large in terms of the number of elements, even for regular matroids. Hence, matroids are often given implicitly in algorithmic applications, via oracles. The most commonly used type of oracle is the *independence oracle*. Let $M = (E, \mathcal{B})$ be a matroid. Then, an independence oracle takes a set $X \subseteq E$ as its input and returns *'Yes'* if X is independent, and *'No'* otherwise. In our running time statements throughout the paper, we assume that matroids are given via independence oracles and that one oracle call takes time $O(1)$. This suffices to compute representation matrices of binary matroids and TU-representations of regular matroids in polynomial time, see [13,40,46]. We provide more details on this in the full version, leading to the following statement.

Lemma 6. *Let M be an $\mathbb{F}_2$-representable matroid given via an independence oracle, and let $D \subseteq E(M)$ be a cocircuit of M. Then we can find a representation matrix $A \in \mathbb{F}_2^{r \times n}$ of M such that $D = \operatorname{supp}(A_r)$ in polynomial time, where A_r denotes the r-th row of A.*

We remark that for general matroids, an independence oracle alone is not sufficient to test efficiently whether a given matroid is binary [43].

Connectivity. Let M be a matroid with ground set E. We say that a partition $(X, E \backslash X)$ is a k-separation of M if $r(X) + r(E \backslash X) - r(M) \leq k - 1$. For $k \geq 2$, the matroid M is called *(Tutte) k-connected*, if there is no ℓ-separation with $\ell < k$. Note that M^* is k-connected if and only M is k-connected.

1-, 2-, and Δ-sums. Let $M_1 = (E_1, \mathcal{B}_1)$ and $M_2 = (E_2, \mathcal{B}_2)$ be matroids. If $E_1 \cap E_2 = \emptyset$, then we define the 1-*sum of M_1 and M_2* as the matroid M with ground set $E := E_1 \cup E_2$ and set of bases $\mathcal{B} := \{B_1 \cup B_2 : B_1 \in \mathcal{B}_1, B_2 \in \mathcal{B}_2\}$. We write $M := M_1 \oplus_1 M_2$. If $E_1 \cap E_2 = \{e\}$, then we define the 2-*sum of M_1 and M_2 at e* as the matroid with ground set $E := E_1 \triangle E_2$ and set of bases $\mathcal{B} := \{(B_1 \cup B_2) - e : B_1 \in \mathcal{B}_1, B_2 \in \mathcal{B}_2, e \in B_1 \triangle B_2\}$. We write $M := M_1 \oplus_2 M_2$.

Let M_1 and M_2 be binary matroids, and $E_1 \cap E_2 = \{d_1, d_2, d_3\} := D$, such that D is a circuit and does not contain a cocircuit in both M_1 and M_2. Then we define the *Δ-sum of M_1 and M_2 at D* as the matroid with ground set

$E := E_1 \triangle E_2$ and set of bases $\mathcal{B}$, where $B = (B_1 \cup B_2) \setminus D \in \mathcal{B}$ if $B_1 \in \mathcal{B}_1$, $B_2 \in \mathcal{B}_2$, $B_1 \cap B_2 = \emptyset$ and either

(i) $|B_1 \cap D| = 0$ and $|B_2 \cap D| = 2$, *or*
(ii) $|B_1 \cap D| = 2$ and $|B_2 \cap D| = 0$, *or*
(iii) $B_1 \cap D = \{d_i\}$, $B_2 \cap D = \{d_j\}$, $B_1 \triangle (D - d_j) \in \mathcal{B}_1$, and $B_2 \triangle (D - d_i) \in \mathcal{B}_2$.

We write $M := M_1 \oplus_\Delta M_2$. We remark that while our definition of the 1-, 2-, and Δ-sum is non-standard, it is well-known to be equivalent to other definitions, see [3, Lemma 10] for the most general case (the Δ-sum) and [11]. Further, note that our definition of Δ-sum (following Truemper [46]) corresponds to the way that the 3-sum is defined in other contexts [3,40,42]. Finally, note that if $M := M_1 \oplus_1 M_2$, then $(E(M_1), E(M_2))$ is a 1-separation of M. Analogous statements hold for the 2-sum and the Δ-sum, provided the summands contain sufficiently many elements.

We introduce a variant of the Δ-sum, where we preserve the circuit that is used for identification. To be precise, let M_1^+ be the matroid arising from M_1 taking the parallel extension for all elements in D. Then, we define $M_1 \oplus_\Delta^+ M_2 := M_1^+ \oplus_\Delta M_2$; right from the definition, it follows that the deletion of the copied elements yields again the Δ-sum.

Finally, we state Seymour's decomposition theorem.

Theorem 7 (Seymour [42], see [46, Thm. 11.3.14]). *Every regular matroid M can be decomposed into graphic and cographic matroids and matroids isomorphic to R_{10} by repeated* 1-, 2-, *and Δ-sum decompositions.*

3 Decomposing Regular Matroids with Graphic Leafs

The crucial structural insight on regular matroids for our algorithm is Proposition 8. The statement has been derived very recently by Bérczi, Mátravölgyi, and Schwarcz [11] to solve two conjectures for regular matroids. Their proof is based on a global statement for the refined decomposition of regular matroids due to Aprile and Fiorini [3]. In the full version [24], we give a new short proof of Proposition 8 that is self-contained outside of text-book proofs of Seymour's decomposition [40,46] and relies only on local information of the decomposition.

Proposition 8 (Bérczi, Mátravölgyi, and Schwarcz [11, Proposition 5.7]). *Let M be a* 3*-connected regular matroid, such that M is not graphic, cographic or isomorphic to R_{10}. Then there are* 3*-connected regular matroids M_1 and M_2, such that $|E(M_2)| \geq 9$, M_2 is graphic and $M_1 \oplus_\Delta M_2 \in \{M, M^*\}$.*

Our proof of Proposition 8 builds on an inductive variant of the 3-connected case of Seymour's decomposition, see [46, Lemma 11.3.18]. In a nutshell, every 3-connected regular matroid that is not isomorphic to R_{10} can be written as a Δ-sum of a graphic or cographic matroid M_1 with a regular matroid M_2. If M_1 is graphic, then our proof is done. Otherwise, we inductively apply the variant of Seymour's decomposition to the dual matroid. Note that the Δ-sum $M_1 \oplus_\Delta M_2$ is not stable under dualization, since the dual of a circuit may not be a circuit, that is, the circuit that is used for identification cannot be used anymore. Instead, the matroids M_1 and M_2 have to be modified in order to replace this circuit by a cocircuit. This operation is called a Δ-Y-exchange. We complete the proof by analyzing minors of graphs and regular matroids under Δ-Y-exchanges. This analysis guarantees that the number of elements shrinks in each induction step.

The following lemma shows roughly speaking that cocircuits induced by vertex cuts of graphic matroids are preserved as cocircuits under a Δ-sum. More precisely, if M is a graphic matroid, C a cocircuit induced by a vertex cut of M, and D a triangle in M such that $C \cap D = \emptyset$, then C is a cocircuit of $M \oplus_\Delta N$, where N is a regular matroid and the Δ-sum is performed on D. Together with Proposition 8, this implies that for any 3-connected regular matroid M that is not R_{10}, there is a cocircuit in M or M^* that 'locally behaves' like a vertex of a graph.

Lemma 9. *Let M_1 be a regular matroid, $\ell \in \mathbb{Z}_{\geq 4}$ and $M := M_1 \oplus_\Delta^+ M(K_\ell)$, where $V(K_\ell) := \{v_i : i \in [\ell]\}$ and the Δ-sum is performed on the triangle $\{v_1v_2, v_1v_3, v_2v_3\}$ of K_ℓ. Then, $\{e \in E(K_\ell) : v_\ell \in e\}$ is a cocircuit of M.*

It suffices to prove the statement for the complete graph K_ℓ. Indeed, Lemma 9 naturally extends to sufficiently connected graphic matroids, and any graph can be completed to K_ℓ by adding the missing edges and assigning them weight 0.

4 Reduction to the 3-Connected Case

Before we demonstrate how we can use our technical insights on regular matroids to handle the 3-connected case, in this section, we demonstrate how to deal with 1- and 2- sums in our inductive proof of Theorem 1.

Proposition 10. *Let M_1 and M_2 be matroids and suppose that f_{M_1} and f_{M_2} can be represented by $(+, \times, /)$-circuits of size s_1 and s_2, respectively. Then, the basis generating polynomial f_M of $M = M_1 \oplus_1 M_2$ can be represented by a $(+, \times, /)$-circuit of size $s_1 + s_2 + 1$.*

Proof. By the definition of the 1-sum, we obtain $f_M = f_{M_1} \cdot f_{M_2}$. Hence, we can simply combine the two circuits with one additional multiplication gate.

□

Proposition 11. *Let M_1 and M_2 be matroids with $\{d\} = E(M_1) \cap E(M_2)$. Suppose that f_{M_1}, $f_{M_2 \backslash d}$, and $f_{M_2/d}$ can be represented by $(+, \times, /)$-circuits of size s_1, $s_2^{\backslash}$, and $s_2^{/}$, respectively. Then, the basis generating polynomial f_M of $M = M_1 \oplus_2 M_2$ can be represented by a $(+, \times, /)$-circuit of size $s_1 + s_2^{\backslash} + s_2^{/} + 2$.*

Proposition 11 can be proved by grouping the bases of M into two parts, depending on which side contains the gluing element d. This allows to write f_M in terms of f_{M_1}, $f_{M_2 \backslash d}$, $f_{M_2/d}$, and two additional arithmetic operations.

5 Handling the 3-Connected Case

Proposition 12. *Let $M = M_1 \oplus_\Delta M_2$ with M_1 being a regular matroid and M_2 being a 3-connected graphic matroid with $n_2 := |E(M_2)| \geq 9$. Assume f_{M_1} can be represented by a $(+, \times, /)$-circuit of size s_1. Then, f_M can be represented by a $(+, \times, /)$-circuit of size $s_1 + n_2^3/2$. Moreover, this circuit can be constructed in polynomial time using an independence oracle of M.*

Once we have Proposition 12, we can prove Theorem 1 by induction. We need one little additional helpful lemma to control dualization.

Lemma 13. *Let M be a matroid on n elements and assume f_M can be represented by a $(+, \times, /)$-circuit of size s. Then, f_{M^*} can be represented by a $(+, \times, /)$-circuit of size $s + 2n$.*

Proof. Observe that $f_{M^*}(x) = x^E \cdot f_M((1/x_e)_{e \in E})$. This expression involves n additional multiplications and n additional divisions compared to f_M, implying the statement.

□

Proof (Proof sketch of Theorem 1). We prove by induction on the number n of elements of the matroid M that f_M can be represented by a $(+, \times, /)$-circuit of size $g(n) := n^3$. If M is graphic, the statement was proven by Fomin et al. [18]. For the cographic case, we additionally apply Lemma 13 to reduce to the graphic case. Finally, R_{10} has rank 5 and a constant number of bases, so a naive implementation yields a constant size circuit. If M is a 1-sum or 2-sum, we apply Proposition 10 and Proposition 11 to recursively obtain a circuit of size $O(n^3)$.

If none of the cases considered so far applies, then the assumption of Proposition 8 is fulfilled. Thus, we can write either M or M^* as $M_1 \oplus_\Delta M_2$ with M_1 being a regular matroid and M_2 being a 3-connected graphic matroid with $n_2 := |E(M_2)| \geq 9$. By induction, Proposition 12, and Lemma 13, we again obtain a circuit of size $O(n^3)$. We remark that all steps are constructive and can be performed by a standard implementation in polynomial time.

□

5.1 A Generalized Star-Mesh Transformation

To prove Proposition 12, we first focus on the case that M_2 is the matroid corresponding to the complete graph. We show that we can "remove one vertex" from the complete graph by performing an operation known as the *star-mesh transformation*. The important observation is that we can do this even though M is not graphic globally, by exploiting that M_2 is graphic. We first state Proposition 14. Next, we outline how to use Proposition 14 in order to complete the proof of Proposition 12. The remainder of the section is devoted to the proof of Proposition 14.

Proposition 14. *Let M_1 be a regular matroid and $\ell \in \mathbb{Z}_{\geq 4}$. Let $M := M_1 \oplus_\Delta^+ M(K_\ell)$ and $M' := M_1 \oplus_\Delta^+ M(K_{\ell-1})$, where the respective Δ-sum is performed on the triangle $\{v_1v_2, v_1v_3, v_2v_3\}$. Let z_{ij} be the variable of $f_M(z)$ corresponding to the edge v_iv_j of K_ℓ and let z'_{ij} be the variable of $f_{M'}(z')$ corresponding to the edge v_iv_j of $K_{\ell-1}$. Then, setting $z'_{ij} := z_{ij} + (z_{i\ell}z_{j\ell})/y$ for $1 \leq i < j \leq \ell - 1$ with $y := \sum_{k\in[\ell-1]} z_{k\ell}$, we have that $f_M(z) = yf_{M'}(z')$.*

Proof (Proof of Proposition 12*).* Observe that in order to bound the complexity of f_M for $M_1 \oplus_\Delta M_2$, we can instead bound the complexity of f_{M^+} with $M^+ = M_1 \oplus_\Delta^+ M_2$, as we obtain f_M by just plugging in 0 for the three additional variables.

First, assume that $M_2 = M(K_\ell)$. We show by induction on $\ell \geq 3$ that f_{M^+} and thus f_M can be represented with size $s_1 + \ell^3/2$. For $\ell = 3$, observe that $M_1 \oplus_\Delta^+ M(K_3)$ is isomorphic to M_1. Thus, the induction start follows. Now assume that $\ell > 3$. We apply Proposition 14 once to M^+. For calculating y we need $\ell - 2$ addition gates, for calculating all the z'-variables, we need three gates each, amounting to $3(\ell-1)(\ell-2)/2$ many gates, and then we need one further gate to multiply y with $f_{M'}(z')$. Thus, by induction, we can compute f_M with

$$s_1 + (\ell-1)^3/2 + (\ell-2) + 3(\ell-1)(\ell-2)/2 + 1 \leq s_1 + \ell^3/2$$

many gates, finishing the induction.

Now we complete the proof by considering $M_2 = M(G)$ for an arbitrary graph G with ℓ many vertices and n_2 many edges. By simply plugging in 0 for all missing edges, we can complete G to the complete graph K_ℓ. As M_2 is 3-connected, we have $n_2 \geq \ell$. Thus, the statement follows. Clearly, the above construction can be performed in polynomial time. Further, we can also efficiently find the relevant representation matrices by Lemma 6.

□

To derive the representation arising from the star-mesh transformation, we use the following generalization of Kirchhoffs celebrated matrix-tree theorem on the number of spanning trees of a connected graph [35].

Theorem 15 (Maurer [38, Thm. 3]). *Let M be a regular matroid and let $A \in \{0, \pm 1\}^{r\times n}$ be a totally unimodular matrix representing M. Let $X := \operatorname{diag}(x_e : e \in E(M))$ and $L := AXA^\intercal$. Then, $f_M = \det(L)$.*

Definition 16. *Let $A \in \{0, \pm 1\}^{r\times n}$ be a matrix. Let $N_1 := \operatorname{supp}(A_r)$ and $N_0 := [n]\backslash N_1$, where A_r denotes the r-th row of A. We define a matrix $A' \in \mathbb{R}^{(r-1)\times n'}$, where $n' := |N_0| + \binom{|N_1|}{2}$ as the matrix obtained by performing a* star-mesh transformation *on A with respect to r. Specifically, we index the columns of A' by $N_0 \cup \binom{N_1}{2}$ and define $A'_{i,j} := A_{i,j}$ if $i \in [r-1]$, $j \in N_0$ and $A'_{i,(j,k)} := A_{i,j} - (A_{r,j}A_{r,k}A_{i,k})$ for $i \in [r-1]$ and $j, k \in N_1$ with $j < k$.*

We note that our definition of star-mesh transformation for matrices generalizes the well-known star-mesh transformation for graphs (see for instance [18]) in the following sense: If G is a graph and $A \in \{0, \pm 1\}^{V(G)\times E(G)}$ is the vertex-edge incidence matrix of G, where each edge is arbitrarily directed, then A' is the (directed) vertex-edge incidence matrix of a graph G', where some vertex $v \in V(G)$ has been removed along with its incident edges, and has been replaced by a complete graph on the neighborhood of v. In this sense, the star-mesh transformation generalizes the notion of Y-Δ-exchanges, with equivalence for vertices of degree 3.

Our final ingredient to prove Proposition 14 is that our version of the star-mesh transformation can actually be applied to reduce the number of rows of the matrix L in Theorem 15 by one. This is captured by the following proposition.

Proposition 17. *Let $A \in \{0, \pm 1\}^{r\times n}$ be a matrix and let $X = \mathrm{diag}(z_i : i \in [n])$ be a diagonal matrix of rational functions. Further, let $N_1 := \mathrm{supp}(A_r)$ and $N_0 := [n] \backslash N_1$. Then, $\det(AXA^\intercal) = y \cdot \det(A'X'A'^\intercal)$, where $y := \sum_{i\in N_1} z_i$, the matrix $A' \in \mathbb{R}^{(r-1)\times n'}$ is obtained by performing a star-mesh transformation on A with respect to r, and $X' \in \mathbb{R}(x)^{n'\times n'}$ is a diagonal matrix of rational functions such that $X'_{i,i} := z_i$ for $i \in N_0$ and $X'_{(\alpha,\beta),(\alpha,\beta)} := z_\alpha z_\beta / y$ for $\alpha, \beta \in N_1$ with $\alpha < \beta$.*

Proof (Proof sketch of Proposition 14*).* In a first step, we establish that there is a representation matrix $A \in \mathbb{R}^{r\times n}$ of M, such that A is totally unimodular and the set of edges $\{v_i v_\ell : i \in [\ell - 1]\}$ corresponds to the support of the r-th row of A. In addition, we can efficiently compute A. Secondly, we show that if A' is obtained by performing a star-mesh transformation on A with respect to r, then A' is a representation of M' (up to parallel elements) and in particular still totally unimodular. We remove parallel elements and update the rational functions in order to obtain the claimed result, using Theorem 15 and Proposition 17.

□

Acknowledgments. We thank Matthias Walter for discussions about matroid decompositions. We thank Steven Noble for clarifications on the hardness of counting bases. We thank Samuel Fiorini for initiating the group retreat in Wissant in June 2024, during which this project gained momentum. Part of this work was completed while Christoph Hertrich was affiliated with Université Libre de Bruxelles, Belgium, and received support by the European Union's Horizon Europe research and innovation program under the Marie Skłodowska-Curie grant agreement No 101153187—NeurExCo. Stefan Kober acknowledges funding from *Fonds de la Recherche Scientifique* - FNRS through research project BD-DELTA-3 (PDR 40028812).

Disclosure of Interests. The authors have no competing interests to declare that are relevant to the content of this article.

References

1. Anari, N., Gharan, S.O., Vinzant, C.: Log-concave polynomials, entropy, and a deterministic approximation algorithm for counting bases of matroids. In: 2018 IEEE 59th Annual Symposium on Foundations of Computer Science (FOCS), pp. 35–46. IEEE (2018)

2. Anari, N., Liu, K., Gharan, S.O., Vinzant, C.: Log-concave polynomials II: high-dimensional walks and an FPRAS for counting bases of a matroid. In: Proceedings of the 51st Annual ACM SIGACT Symposium on Theory of Computing, pp. 1–12 (2019)
3. Aprile, M., Fiorini, S.: Regular matroids have polynomial extension complexity. Math. Oper. Res. **47**(1), 540–559 (2022)
4. Aprile, M., et al.: Integer programs with nearly totally unimodular matrices: the cographic case. In: Proceedings of the 2025 Annual ACM-SIAM Symposium on Discrete Algorithms (SODA), pp. 2301–2312. SIAM (2025)
5. Arora, R., Basu, A., Mianjy, P., Mukherjee, A.: Understanding deep neural networks with rectified linear units. In: International Conference on Learning Representations (2018)
6. Artmann, S.: Optimization of bimodular integer programs and feasibility for three-modular base block IPs. Ph.D. thesis, ETH Zurich (2020)
7. Artmann, S., Weismantel, R., Zenklusen, R.: A strongly polynomial algorithm for bimodular integer linear programming. In: Proceedings of the 49th Annual ACM SIGACT Symposium on Theory of Computing, pp. 1206–1219 (2017)
8. Averkov, G., Hojny, C., Merkert, M.: On the expressiveness of rational ReLU neural networks with bounded depth. In: The Thirteenth Internatinal Conference on Learning Representations (2025)
9. Bakaev, E., Brunck, F., Hertrich, C., Reichman, D., Yehudayoff, A.: On the depth of monotone ReLU neural networks and ICNNs. arXiv preprint arXiv:2505.06169 (2025)
10. Bakaev, E., Brunck, F., Hertrich, C., Stade, J., Yehudayoff, A.: Better neural network expressivity: subdividing the simplex. arXiv preprint arXiv:2505.14338 (2025)
11. Bérczi, K., Mátravölgyi, B., Schwarcz, T.: Reconfiguration of basis pairs in regular matroids. In: Proceedings of the 56th Annual ACM Symposium on Theory of Computing, pp. 1653–1664 (2024)
12. Brandenburg, M.C., Grillo, M., Hertrich, C.: Decomposition polyhedra of piecewise linear functions. In: The Thirteenth International Conference on Learning Representations (2025)
13. Camion, P.: Matrices totalement unimodulaires et problemes combinatoires. Ph.D. thesis, Université Libre de Bruxelles (1964)
14. Charisopoulos, V., Maragos, P.: A tropical approach to neural networks with piecewise linear activations. arXiv preprint arXiv:1805.08749 (2018)
15. Cybenko, G.: Approximation by superpositions of a sigmoidal function. Math. Control Signals Syst. **2**(4), 303–314 (1989)
16. Dinitz, M., Kortsarz, G.: Matroid secretary for regular and decomposable matroids. SIAM J. Comput. **43**(5), 1807–1830 (2014)
17. Fiorini, S., Huynh, T., Weltge, S.: Strengthening convex relaxations of 0/1-sets using Boolean formulas. Math. Program. **190**(1), 467–482 (2021)
18. Fomin, S., Grigoriev, D., Koshevoy, G.: Subtraction-free complexity, cluster transformations, and spanning trees. Found. Comput. Math. **16**, 1–31 (2016)
19. Gagneux, A., Massias, M., Soubies, E., Gribonval, R.: Convexity in ReLU neural networks: beyond ICNNs? arXiv preprint arXiv:2501.03017 (2025)
20. Geelen, J., Gerards, B., Whittle, G.: The highly connected matroids in minor-closed classes. Ann. Comb. **19**(1), 107–123 (2015)
21. Grillo, M., Hertrich, C., Loho, G.: Depth-bounds for neural networks via the braid arrangement. In: Conference on Neural Information Processing Systems (NeurIPS) (2025)

22. Haase, C.A., Hertrich, C., Loho, G.: Lower bounds on the depth of integral ReLU neural networks via lattice polytopes. In: The Eleventh International Conference on Learning Representations (2023)
23. Hertrich, C., Basu, A., Di Summa, M., Skutella, M.: Towards lower bounds on the depth of ReLU neural networks. SIAM J. Discret. Math. **37**(2), 997–1029 (2023)
24. Hertrich, C., Kober, S., Loho, G.: Arithmetic circuits and neural networks for regular matroids. arXiv preprint arXiv:2511.02406 (2025)
25. Hertrich, C., Loho, G.: Neural networks and (virtual) extended formulations. arXiv preprint arXiv:2411.03006 (2024)
26. Hertrich, C., Sering, L.: ReLU neural networks of polynomial size for exact maximum flow computation. Math. Program. 1–30 (2024)
27. Hertrich, C., Skutella, M.: Provably good solutions to the knapsack problem via neural networks of bounded size. INFORMS J. Comput. **35**(5), 1079–1097 (2023)
28. Hrubeš, P., Yehudayoff, A.: Shadows of newton polytopes. Israel J. Math. **256**(1), 311–343 (2023)
29. Huchette, J., Muñoz, G., Serra, T., Tsay, C.: When deep learning meets polyhedral theory: a survey. arXiv preprint arXiv:2305.00241 (2023)
30. Jerrum, M., Snir, M.: Some exact complexity results for straight-line computations over semirings. J. ACM (JACM) **29**(3), 874–897 (1982)
31. Joswig, M.: Essentials of tropical combinatorics, vol. 219. American Mathematical Society (2021)
32. Jukna, S.: Lower bounds for tropical circuits and dynamic programs. Theory Comput. Syst. **57**(1), 160–194 (2015)
33. Jukna, S.: Tropical Circuit Complexity: Limits of Pure Dynamic Programming. Springer, Heidelberg (2023)
34. Jukna, S., Seiwert, H.: Greedy can beat pure dynamic programming. Inf. Process. Lett. **142**, 90–95 (2019)
35. Kirchhoff, G.: Ueber die Auflösung der Gleichungen, auf welche man bei der Untersuchung der linearen Vertheilung galvanischer Ströme geführt wird. Ann. Phys. **148**(12), 497–508 (1847)
36. Knapp, C., Noble, S.: The complexity of the greedoid tutte polynomial. Electron. J. Comb. **32**(3), #P3.3 (2025)
37. Leshno, M., Lin, V.Y., Pinkus, A., Schocken, S.: Multilayer feedforward networks with a nonpolynomial activation function can approximate any function. Neural Netw. **6**(6), 861–867 (1993)
38. Maurer, S.B.: Matrix generalizations of some theorems on trees, cycles and cocycles in graphs. SIAM J. Appl. Math. **30**(1), 143–148 (1976)
39. Nägele, M., Nöbel, C., Santiago, R., Zenklusen, R.: Advances on strictly Δ-modular IPs. Math. Program. **210**(1), 731–760 (2025)
40. Oxley, J.G.: Matroid theory, vol. 3. Oxford University Press, Oxford (2006)
41. Seymour, P.D.: The matroids with the max-flow min-cut property. J. Comb. Theory, Ser. B **23**(2–3), 189–222 (1977)
42. Seymour, P.D.: Decomposition of regular matroids. J. Comb. Theory, Ser. B **28**(3), 305–359 (1980)
43. Seymour, P.D.: Recognizing graphic matroids. Combinatorica **1**(1), 75–78 (1981)
44. Shpilka, A., Yehudayoff, A.: Arithmetic circuits: a survey of recent results and open questions. Found. Trends® Theor. Comput. Sci. **5**(3–4), 207–388 (2010)
45. Snook, M.: Counting bases of representable matroids. Electron. J. Comb. P41–P41 (2012)
46. Truemper, K.: Matroid decomposition, vol. 6, Citeseer (1992)

47. Valerdi, J.L.: On minimal depth in neural networks. arXiv preprint arXiv:2402.15315 (2024)
48. Valiant, L.G.: Negation can be exponentially powerful. In: Proceedings of the Eleventh Annual ACM Symposium on Theory of Computing, pp. 189–196 (1979)
49. Zhang, L., Naitzat, G., Lim, L.H.: Tropical geometry of deep neural networks. In: International Conference on Machine Learning, pp. 5824–5832. PMLR (2018)

Bayesian Probing on Graphs

Anupam Gupta[1], Benjamin Moseley[2], and Rudy Zhou[3(✉)]

[1] New York University, New York, NY, USA
anupam.g@nyu.edu
[2] Carnegie Mellon University, Pittsburgh, PA, USA
moseleyb@andrew.cmu.edu
[3] Microsoft, Washington, DC, USA
rudyzhou@microsoft.com

Abstract. We introduce a stochastic probing problem with correlated items. In our model, which we call *Bayesian Probing*, the correlations are modeled by an underlying graph G. Each vertex is independently active with a known probability. Each item corresponds to an edge in the graph. Probing an edge has some cost, gives some reward if both endpoints are active, and also reveals the state of its endpoints. Hence a probe induces a Bayesian update on the remaining edges. The goal is to adaptively probe items/edges subject to a knapsack constraint to maximize the expected total reward obtained from the probed edges.

Bayesian Probing generalizes stochastic knapsack and stochastic probing by allowing correlations between items. Moreover, it gives a tractable model for the *Bayesian Active Search* problem, a popular problem considered in the machine learning community.

In this paper, we design efficient approximation algorithms for Bayesian Probing:

- We give asymptotically tight bounds on the *adaptivity gap*: we design a non-adaptive $O(d_{\max} \cdot S)$-approximation algorithm, where $d_{\max}$ is the maximum vertex degree of G and S is the ratio of the maximum and minimum edge costs. We also show a matching $\Omega(d_{\max} \cdot S)$ lower bound for non-adaptive policies.
- We design an adaptive policy that achieves a $O(dens(G) \cdot S^2)$-approximation, where $dens(G)$ is the density of the densest subgraph of G.

These results give the first efficient approximation algorithms for Bayesian Active Search, for a class of practically-relevant prior distributions.

Keywords: Approximation Algorithm · Stochastic Optimization · Probing Problem

1 Introduction

Active search is a fundamental problem in applications like fraud detection and drug discovery [10], where the objective is to identify elements of a target class

S. S. Dey et al. (Eds.): IPCO 2026, LNCS 16588, pp. 410–425, 2026.
https://doi.org/10.1007/978-3-032-28691-8_27

within a large search space under a limited budget. In this paradigm, items within the search space – such as financial transactions or chemical compounds – are often interrelated; transactions may share metadata, or chemical compounds may have common substructures. Consequently, evaluating a single item can provide information about others, a process often modeled via a Bayesian update on the likelihood of other items belonging to the target class. Bayesian optimization provides a popular framework for sequentially selecting items to evaluate, leveraging the outcomes of past evaluations to inform future choices. See the book by Garnett [9] and the work of Jiang et al. [20,21] for an overview.

More formally, in the *Active Search problem* [11], as input we are given a finite set of elements X and budget $k \in \mathbb{N}$. We assume there is an *unknown* subset $T \subset X$ of *targets* but with known distribution $T \sim \mathcal{D}$. Note that in general, there may be correlations among different elements being in the target set. For each $x \in X$, we do not know if $x \in T$ initially, but we can request a binary test for x, which returns either "$x \in T$" or "$x \notin T$". The goal is to adaptively choose a set of at most k tests, say $P \subset X$ to maximize the expected total number of targets found: $\mathbb{E}[\sum_{x \in P} \mathbb{1}(x \in T)]$.

Prior work has focused on computing the optimal policy for a given objective function – typically maximizing the number of items found in the class as stated above – but unfortunately the optimal policy requires exponential space and time in general. Further, for general correlation models, no polynomial time algorithm can have better than a polylogartihmic approximation ratio for this objective [21]. While this lower bound is known, there has been no progress on upper bounds on the approximation ratio. Instead, researchers have focused on heuristic methods with no performance guarantees [21,28].

The main motivation of this paper is to give a *concrete correlation model* for Bayesian active search that enables us to efficiently compute *provably near-optimal* policies. This work gives the foundation for algorithms with provable guarantees for Bayesian active search problem by leveraging techniques from stochastic combinatorial optimization.

Bayesian Probing Problem. Motivated by Bayesian active search, we introduce the *Bayesian probing problem* (BAYESPROBE): We are given as input an undirected graph $G = (V, E)$[1], where each vertex $v \in V$ is independently *active* with probability $p_v \in [0, 1]$. The probabilities are *known* to the algorithm apriori, but the active vertices are not. We let X_v be the indicator random variable for vertex v being active. Each edge $uv \in E$ has a random *reward* $R_{uv} = w_{uv} \cdot X_u X_v$, where $w_{uv} \geq 0$ and $s_{uv} > 0$ are the *weight* and *size* of this edge, respectively.

We can access the graph only through *edge probes*. We are given a budget $B > 0$. Probing edge e uses s_e of the budget and gives us reward R_e. Further, we learn the outcome of both endpoint vertices, X_u and X_v. We allow probes to be chosen adaptively, so the subsequent probes can depend on the information

[1] We allow parallel edges, but not self loops; given a graph with self loops, for each such edge we create a dummy vertex with probability 1 and make one endpoint of the self loop this dummy vertex.

learned from prior ones. The goal is to adaptively probe edges of total size at most B to maximize the expected reward obtained from the probed edges – precisely:

$$\mathbb{E}[\sum_{e \in P} R_e] = \mathbb{E}[\sum_{uv \in P} w_{uv} \cdot X_u X_v],$$

where P is the adaptively-chosen probe set.

We note that BayesProbe captures a version of Bayesian active search with a particular correlation model and Bayesian update. The edges of G correspond to the items to be tested, and probing an edge corresponds to testing whether that item belongs to the target set. In our model, each item's status is determined by two "features" – its endpoints (e.g. each chemical compound may have two ingredients in a drug discovery setting). An item is in the target set if and only if both its corresponding vertices are active. Since edges can share endpoints, probing one edge reveals information about its vertices, thereby updating the expected reward of all other edges incident to those same vertices.

For the rest of this paper, we focus on designing algorithms for BayesProbe. One can compute the optimal adaptive policy – which can be represented by a decision tree – via a dynamic program, but unfortunately computing this policy and its representation may require exponential time/space. Thus, we focus on efficient *approximation algorithms*. For any $\alpha \geq 1$, we say an algorithm, Alg, is an *α-approximation* for BayesProbe if for any instance $\mathcal{I}$, we have $\frac{\mathbb{E}[\text{Opt}(\mathcal{I})]}{\mathbb{E}[\text{Alg}(\mathcal{I})]} \leq \alpha$, where Opt is the reward of the optimal *adaptive* policy.

1.1 Applications and Related Work

Beyond the Bayesian active search problem described above, BayesProbe is related to many other directions in stochastic combinatorial optimization.

Stochastic Probing. Bayesian probing is closely connected with stochastic probing. In stochastic probing, we are given a collection of elements E, such that each element is active *independently* with known probability. We can adaptively probe elements to learn if they are active or not. Further, we are given a set function $f : 2^E \rightarrow \mathbb{R}$. The goal of stochastic probing is to adaptively probe elements, say $P \subset E$, and choose a subset $Q \subset P$ of the probe set to maximize $\mathbb{E}[f(Q \cap A)]$, where $A \subset E$ is the random subset of active elements. Stochastic probing has been studied for different f (linear, submodular, XOS) and constraints on the P and chosen subset Q [15–17].

The main distinction is that in BayesProbe, we remove the assumption that the elements are active independently. There has been some prior work on probing problems with correlated elements. However, the results rely on strong assumptions on the correlation model and function f. Most relevant to our work is the work of [12], where they introduce a property called *adaptive submodularity* (Definition 2), which enables simple greedy-type $O(1)$-approximations for a variety of stochastic probing problems. In the full version of this paper, we show that BayesProbe does *not* satisfy adaptive submodularity, so our work is not subsumed by theirs.

Stochastic Knapsack. Stochastic knapsack is a well-studied generalization of the classic knapsack problem. In stochastic knapsack, we are given a deterministic knapsack budget, say B, but the items are stochastic. There are $O(1)$-approximations known for multiple variants of the problem, the simplest being that each item has a random size with known distribution but deterministic reward [7]. This has been generalized to random sizes with random rewards such that the size and reward distributions of an item may be arbitrarily correlated [14], and even further such that each item is a Markov chain [25]. However, in all of these variations, it is assumed that the size/reward distributions are independent *across items*.

In contrast, BAYESPROBE is another variation of stochastic knapsack where instead the sizes are deterministic, but the rewards are random and correlated with each other. To the best of our knowledge, we give the first positive results for stochastic knapsack with correlations across items for any correlation model.

Algorithmic Site Percolation. Site percolation is a model of random graphs studied in probabilistic combinatorics. Given a graph G and probability parameter $p \in [0, 1]$, let $G[V_p]$ be the random induced subgraph on $V_p \subset V(G)$, where each vertex $v \in V(G)$ is in V_p independently with probability p.

Researchers have focused on proving threshold results for graph properties of the percolated graph, $G[V_p]$, such as for what range of p does $G[V_p]$ contains a linear-sized component [13,22] or properties of particular families of graphs such as the grid graph [26]. Further, percolation theory has been applied to model and predict the spread of diseases [6] and the extinction of species [2].

Given a graph G and probability parameter p, consider the Bayesian probing instance where $p_v = p$ for all $v \in V(G)$. The edges which give non-zero reward are exactly those that are in the percolated graph $G[V_p]$. Thus, BAYESPROBE captures an algorithmic question on the percolated graph: How much of the percolated graph can we discover given a budget of edge queries?

This question is of particular relevance to the above applications, where it is expensive to query the underlying graph, which involves expensive diagnoses or field work, respectively.

Stochastic Optimization with Correlations. Another common theme throughout stochastic combinatorial optimization is the independence assumption: We typically assume that the jobs to be scheduled, items to be packed, or elements to be probed are independent of one another. There has been recent interest in relaxing this assumption by allowing correlations in problems such as Pandora's box [4], mechanism design [3], and prophet inequality [24]. Our work on BAYESPROBE adds stochastic knapsack and probing to this list.

1.2 Our Results

Our main question is how well can efficiently-computable policies approximate the optimal adaptive policy for BAYESPROBE?

First, we consider the simplest class of policies: *non-adaptive policies*. A non-adaptive policy fixes a probe set $P \subset E$ up-front without any knowledge of the probe outcomes. In other words, a non-adaptive policy performs no Bayesian updates. Our first result gives an upper bound on the approximation ratio achieved by non-adaptive policies.

Theorem 1. *There exists a poly-time algorithm for* BayesProbe *that outputs a non-adaptive policy that $O(d_{max} \cdot S)$-approximates the optimal adaptive policy, where d_{max} is the max degree of G and S is the ratio between the maximum and minimum edge sizes in G.*

Further, we show that our upper bound is asympotically best possible by lower bounding the *adaptivity gap* of BayesProbe. We define the *adaptivity gap* for BayesProbe by $\max_{\mathcal{I}} \frac{\mathbb{E}[\text{Opt}_A(\mathcal{I})]}{\mathbb{E}[\text{Opt}_{NA}(\mathcal{I})]}$, where Opt_A and Opt_{NA} are the reward obtained by the optimal adaptive and non-adaptive policies, respectively. The maximum is taken over all instances $\mathcal{I}$ of BayesProbe.

Lemma 1. *The adaptivity gap of* BayesProbe *is $\Omega(d_{max} \cdot S)$.*

In terms of the size of the graph G itself, we have a different adaptivity gap lower bound.

Lemma 2. *The adaptivity gap of* BayesProbe *is $\Omega(|V|^2)$ or $\Omega(|E|)$.*

Thus, to improve over our first algorithm, we need to use adaptivity. We consider a class of adaptive policies that have two phases: an exploration phase and an exploitation phase. Roughly speaking, in the exploration phase we first use half of the overall budget to learn information (which vertices are active or not). Then in the exploitation phase, we use the remaining half of the budget to gain reward conditioned on what we learned. See Definition 1 for a formal definition of such policies. We show that this allows us to improve d_{max}-dependence for non-adaptive policies to $dens(G) = \max_{H \subset V(G)} \frac{|E[H]|}{|H|}$ – the density of the densest subgraph of G.

Theorem 2. *There exists a poly-time algorithm for* BayesProbe *that computes an adaptive policy that $O(dens(G) \cdot S^2)$-approximates the optimal adaptive policy, where $dens(G) = \max_{H \subset V(G)} \frac{|E[H]|}{|H|}$ is the density of the densest subgraph of G and S is the ratio between the maximum and minimum edge sizes in G.*

One special case of particular interest is Bayesian active search, where every edge has size 1. In this case, we obtain a $O(d_{max})$-approximation by a non-adaptive policy, but a $O(dens(G))$-approximation with an *efficiently computable* adaptive policy. These are the first provable approximation guarantees for Bayesian active search, and we hope that our work showcases the potential of applying algorithmic and modeling techniques from stochastic combinatorial optimization to problems in Bayesian machine learning.

1.3 Technical Overview

We highlight our main ideas with a running example.

Example 1. Let $d \in \mathbb{N}$. Then let G be a collection d stars, each with d leaves. The center of each star is active with probability $\frac{1}{d}$ and each leaf with probability 1. Further, each edge weight and size is 1, and our budget is d. See Fig. 1. This completes the description of the Bayesian probing instance.

Thus, our goal in this instance is to probe at most d edges to maximize the expected number of active edges (those with both endpoints active) probed.

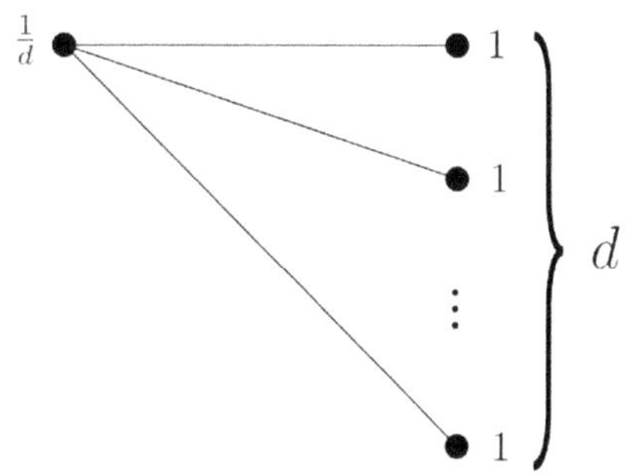

Fig. 1. A single star in G. Vertices are labeled with their probabilities. All edges have $w_e = 1$ and $s_e = 1$.

First, consider probing any edge of a star. Doing so, we learn the realized value of its center. If its center is active, then we now know all edges in this star are active and will give reward 1. Otherwise, every such edge gives reward 0. In contrast, prior to probing this edge, every edge of this star gives expected reward $\frac{1}{d}$.

Interestingly, such stars correspond exactly to *clumps* in the Bayesian active search literature. In the language of Bayesian active search, a clump is a subcollection of items such that either all items in the clump are active or all are inactive (with some probability). Collections of clumps were introduced in [11] as a hard instance for Bayesian active search. Thus, the algorithms we will design for BayesProbe can provably perform well on hard benchmarks for Bayesian active search.

Going back to our particular instance in Example 1, an adaptive policy on this instance can probe one edge per star until it finds a star with an active center. Then, it uses its remaining budget to probe edges in this particular star. The adaptive policy finds an active center within its first $\frac{d}{2}$ probes with at least constant probability, and – conditioned on this – probes at least $\frac{d}{2}$ edges that each give reward 1. This policy has expected reward $\Omega(d)$.

On the other hand, a non-adaptive policy can only fix a set of d edges to probe in advance, each of which gives expected reward $\frac{1}{d}$. Thus, any non-adaptive policy has expected reward at most 1, so the adaptivity gap of this example is $\Omega(d) = \Omega(d_{max})$.

This example highlights the key difference between BAYESPROBE and other stochastic combinatorial optimization problems: In BAYESPROBE *the reward obtained from an edge is not independent of the decision to probe that edge.* A common approach in other stochastic combinatorial optimization problems is to define a linear programming (LP) relaxation of the problem and show that the marginals of the optimal adaptive policy certify that the relaxation is a good one. Let's try this approach for Bayesian probing. The natural LP relaxation for BAYESPROBE is exactly the knapsack LP

$$\mathrm{LP}(G,B) = \max_x \{ \sum_{e \in E(G)} \mathbb{E}[R_e] \cdot x_e \mid \sum_{e \in E(G)} s_e \cdot x_e \leq B, \quad x_e = 0 \text{ if } s_e > B, \quad 0 \leq x \leq 1 \}. \quad (1)$$

Note that $\mathrm{LP}(G,B)$ exactly models the problem of finding the optimal *non-adaptive* policy. Throughout this paper, we use $\mathrm{LP}(G,B)$ to denote the LP itself and its optimal objective value. Similarly, we use $\mathrm{OPT}(G,B)$ to denote the optimal adaptive policy and its reward. Consider setting $x_e = \Pr(e \in Q)$, where the event $e \in Q$ is with respect to the optimal adaptive policy $\mathrm{OPT}(G,B)$. This solution is feasible for $\mathrm{LP}(G,B)$ and achieves objective value $\sum_{e \in E(G)} \mathbb{E}[R_e] \cdot \Pr(e \in Q)$. On the instance in Example 1, this is $\sum_{e \in E(G)} \frac{1}{d} \cdot \Pr(e \in Q) \leq 1$, whereas $\mathbb{E}[\mathrm{OPT}(G,B)] = \Omega(d)$, so this LP is *not* a good benchmark. The issue is that $\mathbb{E}[R_e] \cdot \Pr(e \in Q) \neq \mathbb{E}[R_e \cdot \mathbb{1}_{e \in Q}]$ since for many edges, the optimal adaptive policy queries those edges only if it *knows* that it will give reward 1.

To overcome this issue, instead of strengthening the LP by adding more constraints – perhaps surprisingly – we actually *simplify* the LP by restricting it to a simpler subgraph where the LP is a good benchmark. This is in contrast to prior work on generalizations of stochastic knapsack, where a stronger time-indexed LP is used [14,25]. Suppose that G is a matching (no two edges share an endpoint). Now probing an edge gives no information about other edges, so the event $\{e \in Q\}$ is independent of the reward R_e. This allows us to carry out the idealized calculation above to obtain $LP(G,B) \geq \sum_{e \in E(G)} \mathbb{E}[R_e] \cdot \Pr(e \in Q) = \sum_{e \in E(G)} \mathbb{E}[R_e \cdot \mathbb{1}_{e \in Q}] = \mathbb{E}[\mathrm{OPT}(G,B)]$.

Our main technical tool is a *decomposition lemma* that allows us to decompose the input graph G into simpler graphs – say, matchings – and relate $\mathrm{OPT}(G,B)$ to OPT on these simpler graphs (up to a multiplicative violation of the budget).

Lemma 3 (Decomposition Lemma). *Given graph G, let $G_1, \ldots, G_\ell$ be graphs on the same vertex set as G such that $E(G_1) \cup \cdots \cup E(G_\ell) = E(G)$. Then we have $\mathbb{E}[\mathrm{OPT}(G_1, 2S \cdot B)] + \cdots + \mathbb{E}[\mathrm{OPT}(G_\ell, 2S \cdot B)] \geq \mathbb{E}[\mathrm{OPT}(G,B)]$ for any budget B.*

Thus, to design our non-adaptive policy, we decompose the input graph G into $O(d_{max})$-many matchings. On some such matching, say $M \subset G$, by averaging we have $LP(M, 2S \cdot B) \geq \mathbb{E}[\mathrm{OPT}(M, 2S \cdot B)] = \Omega(\frac{1}{d_{max}}) \cdot \mathbb{E}[\mathrm{OPT}(G,B)]$. We use $LP(M, 2S \cdot B)$ (up to a budget rescaling argument) as the benchmark for our algorithm, which will be a standard knapsack LP rounding algorithm. This gives the $O(S \cdot d_{max})$-approximation guaranteed by Theorem 1.

How can we improve on this with an adaptive policy? Let's go back to Example 1, and slightly modify the adaptive policy that achieves expected reward $\Omega(1)$: First, we pick $\frac{d}{2}$ arbitrary stars and probe one edge per star. Then, *conditioned on the outcomes of those probes*, we non-adaptively probe the $\frac{d}{2}$ edges with the largest expected reward. As before, with at least constant probability, in the first phase, we find an edge that gives reward 1 and thus a center that is active. In this case, in the second phase, we are guaranteed to probe $\frac{d}{2}$ edges that each give reward 1. We define *explore-exploit (EE)* policies to capture this type of policy.

Definition 1 (Explore-Exploit (EE) Policy). *For any $B > 0$, an explore-exploit policy with budget B on graph $G = (V, E)$ has two phases, an exploration phase and exploitation phase.*

- ***(Exploration)*** *First, the policy has budget $\frac{B}{2}$ for adaptive* vertex probes. *In each vertex probe the policy chooses a vertex $v \in V$, and learns the realized value of X_v (obtaining no reward). We say the policy* explores *vertex v. The cost of this probe is $s_v = \min_{uv \in E} s_{uv}$.*
- ***(Exploitation)*** *After the policy has explored $P_V \subset V$ and learned the values of X_v for all $v \in P_V$ – conditioned on these outcomes – the policy* non-adaptively *probes an* exploitation set *$P_E \subset E$ of size at most $\frac{B}{2}$ (with respect to the edge sizes s_e for $e \in E$).*

The total reward of such a policy is the rewards of all edges probes in the exploitation phase, $\mathbb{E}[\sum_{e \in P_E} R_e]$.

It turns out that EE policies are sufficient[2] to obtain a $O(S^2 \cdot dens(G))$-approximation for BayesProbe, as in Example 1. To achieve this, we again simplify our LP benchmark to make it more amenable to analysis. Instead of restricting to a matching as in the non-adaptive case, now we restrict to a collection of stars and condition on the outcome of the exploration phase.

In a collection of stars, conditioning on the exploration phase has a nice structure. Consider exploring the center vertex of some star. If it is active, then the rewards of all incident edges are now independent (since their rewards are completely determined by their leaf values). If it is inactive, then all incident edges give reward 0 and can be ignored.

Using this idea, an analogous calculation to the matching case shows that the knapsack LP conditioned on the exploration phase gives a good LP relaxation for the reward obtained in the subsequent exploitation phase. Thus, we view the exploration phase as adaptively exploring vertices to maximize the knapsack LP value conditioned on the outcomes of the explored vertices. Since vertices are active independently, the upshot of this view is that we can leverage techniques for stochastic probing with *independent* items to efficiently find a good exploration set.

To summarize, in a collection of stars, our adaptive EE policy will use the first half of the budget to maximize a conditional knapsack LP benchmark.

[2] up to the standard knapsack issue of needing to take the single best item sometimes.

Then we will round the resulting knapsack LP to pick the edges to exploit with the remaining half of the budget. To lift the above algorithm on collections of stars to general graphs, we again use the decomposition lemma to decompose an arbitrary graph G into $O(dens(G))$-many collections of stars.

2 Non-adaptive Policies

Our goals in this section are to show that the adaptivity gap for BAYESPROBE is $\Theta(d_{max} \cdot S)$ and to give a non-adaptive algorithm that achieves this approximation ratio. We first give our algorithm (Theorem 1) and then a matching lower bound on the adaptivity gap (Lemma 1). We defer all proofs to the full version of this paper.

2.1 Non-adaptive Algorithm

First, we observe that the problem of finding a non-adaptive policy that approximates the optimal *non-adaptive* policy is just the classic knapsack problem: Our goal is to choose a subset of edges Q of total size at most B to maximize $\sum_{e \in Q} \mathbb{E}[R_e]$. For this problem a fully polynomial time approximation scheme (FPTAS) is known, so we can obtain a $(1+\epsilon)$-approximate solution in time $poly(n, \frac{1}{\epsilon})$ [19,23]. However, it will be convenient for our analysis here and when we later design adaptive policies to use the standard 2-approximate greedy algorithm.

NONADAPT: Given input graph G to BAYESPROBE with budget $B > 0$: Pick the best of the following two solutions:

- Sort the edges in decreasing order of *value density*: $\frac{\mathbb{E}[R_e]}{s_e}$. Probe edges in this order until we use up the budget.
- Probe the single edge with largest expected reward.

It is known that this algorithm has the following LP-based performance guarantee, see e.g. Exercise 3.1 in [29]:

Proposition 1. *Given input graph G and budget $B > 0$,* NONADAPT *outputs a solution with reward satisfying* NONADAPT$(G, B) \geq \frac{1}{2} \cdot \text{LP}(G, B)$.

Suppose we run NONADAPT to output a non-adaptive policy that obtains expected reward at least $\frac{1}{2}\text{LP}(G, B)$. However, as we saw in the technical overview (Sect. 1.3), $\text{LP}(G, B)$ is *not* a good relaxation for $\mathbb{E}[\text{OPT}(G, B)]$ in general, but it is a good relaxation if G is a matching. Our strategy to overcome this is to choose an appropriate matching, say $M \subset G$, so we have $\text{LP}(G, B) \geq \text{LP}(M, B) \geq \mathbb{E}[\text{OPT}(M, B)]$. It remains to relate $\mathbb{E}[\text{OPT}(M, B)]$ to $\mathbb{E}[\text{OPT}(G, B)]$.

The challenge here is that the edges in $E(G) \backslash E(M)$ not only allow $\text{OPT}(G, B)$ to obtain more reward, but also more information, which allows $\text{OPT}(G, B)$ to better exploit the edges in $E(M)$. To bound this discrepancy in both reward and information, we use our decomposition lemma.

Lemma 3 (Decomposition Lemma). *Given graph G, let $G_1, \ldots, G_\ell$ be graphs on the same vertex set as G such that $E(G_1) \cup \cdots \cup E(G_\ell) = E(G)$. Then we have $\mathbb{E}[\text{OPT}(G_1, 2S \cdot B)] + \cdots + \mathbb{E}[\text{OPT}(G_\ell, 2S \cdot B)] \geq \mathbb{E}[\text{OPT}(G, B)]$ for any budget B.*

This allows us, by averaging, to choose the matching M such that $\mathbb{E}[\text{OPT}(M, 2S \cdot B)] = \Omega(\frac{1}{d_{max}}) \cdot \mathbb{E}[\text{OPT}(G, B)]$. Then to complete the proof, we argue that re-scaling down the budget by $2S$ scales the LP solution value by the same factor. This gives us our main guarantee about NonAdapt, restated below.

Theorem 1. *There exists a poly-time algorithm for* BayesProbe *that outputs a non-adaptive policy that $O(d_{max} \cdot S)$-approximates the optimal adaptive policy, where d_{max} is the max degree of G and S is the ratio between the maximum and minimum edge sizes in G.*

2.2 Adaptivity Gap Lower Bound

Our lower bound construction that achieves Lemma 1 is a slight modification of Example 1.

Lemma 1. *The adaptivity gap of* BayesProbe *is $\Omega(d_{max} \cdot S)$.*

In that example, each component was a star, and further each edge size and weight was 1. This gave an adaptivity gap lower bound of $\Omega(d_{max})$. To obtain the stronger lower bound of $\Omega(d_{max} \cdot S)$, we make one edge on each star have 0 weight but very small size. Such an edge is useless for a non-adaptive policy since it gives no reward, but it is useful for an adaptive policy to learn if the center is active or not.

3 Adaptive Policies

In the previous section, we showed that the adaptivity gap for this problem is $\Theta(d_{max} \cdot S)$. Now, we show how to leverage adaptivity to go beyond this lower bound to achieve a $O(dens(G) \cdot S^2)$ by combining a non-adaptive policy with a class of limited-adaptivity policies called explore-exploit (EE) policies. We recall our main theorem and the definition of EE policies here.

Theorem 2. *There exists a poly-time algorithm for* BayesProbe *that computes an adaptive policy that $O(dens(G) \cdot S^2)$-approximates the optimal adaptive policy, where $dens(G) = \max_{H \subset V(G)} \frac{|E[H]|}{|H|}$ is the density of the densest subgraph of G and S is the ratio between the maximum and minimum edge sizes in G.*

Definition 1 (Explore-Exploit (EE) Policy). *For any $B > 0$, an explore-exploit policy with budget B on graph $G = (V, E)$ has two phases, an exploration phase and exploitation phase.*

- ***(Exploration)*** *First, the policy has budget* $\frac{B}{2}$ *for adaptive* vertex probes. *In each vertex probe the policy chooses a vertex* $v \in V$*, and learns the realized value of* X_v *(obtaining no reward). We say the policy* explores *vertex* v*. The cost of this probe is* $s_v = \min_{uv \in E} s_{uv}$.
- ***(Exploitation)*** *After the policy has explored* $P_V \subset V$ *and learned the values of* X_v *for all* $v \in P_V$ *– conditioned on these outcomes – the policy* non-adaptively *probes an* exploitation set $P_E \subset E$ *of size at most* $\frac{B}{2}$ *(with respect to the edge sizes* s_e *for* $e \in E$*).*

The total reward of such a policy is the rewards of all edges probes in the exploitation phase, $\mathbb{E}[\sum_{e \in P_E} R_e]$.

In the rest of this section, we sketch the algorithm and proof of Theorem 2. Again, we defer all proofs to the full version of this paper. Our proof strategy will be to use the decomposition lemma again to decompose the input graph into simpler subgraphs, but this time the subgraphs will be *collections of stars*. Precisely, we say G is a collection of stars with center set C if we can write $G = \cup_{c \in C} G_c$, where G_c is a star graph with center c (so we index each star by its center vertex) and leaf set L_c. Thus, each G_c consists of its center c connected individually to each of its leaves $\ell \in L_c$. We will show later that we can decompose any graph G into $O(dens(G))$-many collections of stars.

Thus, it remains to design a good EE policy for a collection of stars. To achieve this, we leverage some structural properties of EE policies on collections of stars. Our main structural property is the following lemma. We define $\text{OPT}^C_{\text{EE}}(G, B)$ to be the optimal EE policy (and its reward) on a *collection of stars* G with budget B *that only explores the centers of stars.*

Lemma 4. *For any collection of stars* G *with center set* C *and budget* $B > 0$*, we have* $\mathbb{E}[\text{OPT}^C_{\text{EE}}(G, 8B)] \geq \mathbb{E}[\text{OPT}(G, B)]$.

The above lemma allows us to reduce the exploration phase to another probing problem with *independent* items (AdaptSubmodKnapsack, Definition 3). Roughly, this is because exploring the center of a star induces a Bayesian update only on the edges of that star, so we can view each star as an independent item to be probed. In particular, we show that our objective function is *adaptive submodular.*

3.1 Adaptive Submodularity

Suppose we have a ground set Y of elements such that each element $y \in Y$ is in one of two possible states: inactive or active. Further, we have some probability distribution over the subset of active elements, say $\mathcal{A}$ on 2^Y,

Now let $f : 2^Y \times 2^Y$ be a function. We interpret the first argument as the choice of our algorithm and the second as the realization of the active elements. A partial realization of a subset $Y' \subset Y$ is a subset $A_{Y'} \subset Y'$, which specifies the active elements in Y'. For subsets $Y_1 \subset Y_2 \subset Y$ with partial realizations

A_{Y_1} and A_{Y_2}, respectively, we say these partial realizations are *consistent* if $A_{Y_2} \cap Y_1 = A_{Y_1}$ (i.e. if they agree on the active elements in the smaller subset).

Then we define, the marginal gain of adding element $y \in Y$ to a subset $Y' \subset Y$ of already-probed elements with partial realization $A_{Y'}$ by

$$\textsc{Gain}(y, Y' \mid A_{Y'}) = \mathbb{E}_{A \sim \mathcal{A}}[f(Y' \cup \{y\}, A) \mid A \cap Y' = A_{Y'}] - \mathbb{E}_{A \sim \mathcal{A}}[f(Y', A) \mid A \cap Y' = A_{Y'}].$$

We are now ready to define adaptive submodularity.

Definition 2 (Adaptive Submodularity). *The function* $f : 2^Y \times 2^Y \to \mathbb{R}_{\geq 0}$ *is adaptive submodular (with respect to distribution* $\mathcal{A}$ *on* 2^Y*) if for any subsets* $Y_1 \subset Y_2 \subset Y$, $y \notin Y_2$, *and consistent partial realizations* A_{Y_1} *and* A_{Y_2} *of* Y_1 *and* Y_2, *respectively, we have*

$$\textsc{Gain}(y, Y_1 \mid A_{Y_1}) \geq \textsc{Gain}(y, Y_2 \mid A_{Y_2}).$$

We note that one can define adaptive submodularity for more general state spaces (i.e. not only elements being active or inactive), but we chose to specialize our definition to prevent excess notation. See [12] for a more general definition.

Adaptive submodularity is a useful property, which enables greedy-type algorithms to achieve good performance guarantees for a variety of optimization problems [5,8,12,18,27]. Of particular interest to us is the problem of maximizing an adaptive submodular function subject to a knapsack constraint, AdaptSubmodKnapsack.

Definition 3 (AdaptSubmodKnapsack)**.** *In an instance of* AdaptSubmodKnapsack, *we are given an adaptive submodular function* $f : 2^Y \times 2^Y \to \mathbb{R}_{\geq 0}$ *with respect to distribution* $\mathcal{A}$ *on* 2^Y, *element costs* $c_y \geq 0$ *for* $y \in Y$, *and a budget* $B \geq 0$. *The goal is to adaptive probe elements* $P \subset Y$ *of total cost at most* B *to maximize* $\mathbb{E}_{A \sim \mathcal{A}, P}[f(P, A)]$.

This problem admits a $O(1)$-approximation by a natural adaptive greedy algorithm with respect to value density (marginal gain of adding an item divided by its cost) along with an edge case of picking the single best item. This is given in the next proposition due to [1].

Proposition 2. *Given an oracle that computes* $\textsc{Gain}(\cdot, \cdot \mid \cdot)$ *(as defined in Definition 2) for any inputs in polynomial time, there exists a polynomial time algorithm for* AdaptSubmodKnapsack *that adaptively probes* $P \subset Y$ *of total cost at most* B *such that* $\mathbb{E}[f(P, A)] \geq \frac{1}{9} \cdot \mathbb{E}[f(P^*, A)]$, *where* P^* *is the optimal adaptive policy.*

Thus, we can reduce the problem of finding a good exploration set to an instance of AdaptSubmodKnapsack with objective function $f(P, A) = \mathrm{LP}(G \mid P \cap A, B/2)$, where the latter term is our *conditional LP benchmark* defined as follows. For any collection of stars $G = \bigcup_{c \in C} G_c$ and center set $C' \subset C$, we define $G \mid C'$ to be the subgraph $\bigcup_{c \in C'} G_c$, where we update the probabilities $p_c \to 1$ for every $c \in C'$. In other words, we condition on C' being the set of active centers and only consider those stars.

In particular, the natural knapsack LP on instance $G \mid C'$ is given by

$$\mathrm{LP}(G \mid C', B) = \max_x \{ \sum_{c\ell \in E(G|C')} \mathbb{E}[w_{c\ell} \cdot X_\ell] \cdot x_{c\ell} \mid \sum_{e \in E(G|C')} s_e \cdot x_e \leq B, \quad x_e = 0 \text{ if } s_e > B, \quad 0 \leq x \leq 1 \}. \tag{2}$$

It turns out that this LP is a good relaxation of the reward obtained in the exploitation phase after conditioning on the outcome of the exploration phase.

3.2 Algorithm for Collections of Stars

Morally, our algorithm to find a good EE policy for collections of stars will use the algorithm guaranteed by Proposition 2 for AdaptSubmodKnapsack to find a good set of centers to explore. Then, conditioned on the outcome of the exploration phase, we run NonAdapt to choose the exploitation set. This is given in the next algorithm.

EEStar: Given input collection of stars G and budget $B > 0$:

1. **(Exploration)** Run the algorithm guaranteed by Proposition 2 on function $f(P, A) = \mathrm{LP}(G \mid P \cap A, B/2)$ with budget $\frac{B}{2}$ to adaptive explore centers, say $Q \subset C$.
2. **(Exploitation)** With probability $\frac{1}{2}$ each, run one of the two following algorithms:
   - Exploit the edges output by NonAdapt$(G \mid Q \cap A, B/2)$. where $A \subset C$ is the realized set of active centers.
   - Exploit the edges output by NonAdapt$(G, B/2)$.

To obtain the final algorithm guaranteed by Theorem 2, we need two additional ingredients:

- We give an algorithm to decompose an input graph G into $O(dens(G))$-many collections of stars.
- Then we randomize between running EEStar on such a collection of stars or another trivial algorithm to handle the typical knapsack edge case of picking a single valuable item/edge

The above is a sketch of our final algorithm AdaptGeneral, which gives the desired $O(dens(G) \cdot S^2)$-approximation.

4 Conclusion and Future Work

We introduced BayesProbe, one of the first stochastic combinatorial optimization models that incorporates correlated rewards, while still admitting provably efficient approximation algorithms. Our non-adaptive and adaptive algorithms give the first poly-time guarantees for Bayesian active search (and stochastic probing and knapsack with correlations across items). These results show that realistic dependencies among items need not preclude strong worst-case performance. Further, our correlation model is rich enough to capture typical bad examples for Bayesian active search as well as practically-relevant prior distributions, such as those arising in site percolation.

Our work gives the foundation for how techniques from stochastic combinatorial optimization can have a large impact in Bayesian machine learning problems, and it opens up many directions for future work:

- **Hypergraphs** Extending our model from graphs to hypergraphs would allow us to capture higher-order correlations common in biological and social networks.
- **Different reward functions** In BayesProbe, one can view the reward of an edge as the (weighted) AND of its endpoints, but it would be interesting and practically relevant to consider a generalization where the reward of a (hyper-)edge is some Boolean formula applied to its endpoints. For example, threshold functions (at least t endpoints are active) would be an interesting generalization.
- **Different probes** Further, in our model, when we probe an edge we learn the outcomes of both of its endpoints. One could consider a more restricted setting where an edge probe only reveals the realized edge reward (in BayesProbe, this means one would only learn the AND of the endpoints, rather than both endpoint values). This would introduce much more complicated Bayesian updates.

We hope the techniques developed here – particularly the decomposition lemma and our structural properties of EE policies – serve as building blocks for tackling richer correlated stochastic optimization problems.

Acknowledgments. We thank Viswanath Nagarajan for helpful discussions leading to the connection with adaptive submodularity, and an anonymous reviewer for pointing out the alternative adaptivity gap lower bound in Lemma 2. AG's research was supported in part by NSF awards CCF-2422926 and CCF-2608359. Benjamin Moseley was supported in part by ONR grant N000142212702.

Disclosure of Interests. The authors have no competing interests to declare that are relevant to the content of this article.

References

1. Amanatidis, G., Fusco, F., Lazos, P., Leonardi, S., Reiffenhäuser, R.: Fast adaptive non-monotone submodular maximization subject to a knapsack constraint. Adv. Neural. Inf. Process. Syst. **33**, 16903–16915 (2020)
2. Boswell, G.P., Britton, N.F., Franks, N.R.: Habitat fragmentation, percolation theory and the conservation of a keystone species. Proc. Royal Soc. London. Ser. B Biol. Sci. **265**(1409), 1921–1925 (1998)
3. Cai, Y., Oikonomou, A.: On simple mechanisms for dependent items. In: Proceedings of the 22nd ACM Conference on Economics and Computation, pp. 242–262(2021)

4. Chawla, S., Gergatsouli, E., Teng, Y., Tzamos, C., Zhang, R.: Pandora's box with-correlations: Learning and approximation. In: 2020 IEEE 61st Annual Symposium on Foundations of Computer Science (FOCS), pp. 1214–1225. IEEE (2020)
5. Cui, Y., Nagarajan, V.: Minimum cost adaptive submodular cover. In: Symposium on Simplicity in Algorithms (SOSA), pp. 12–27. SIAM (2023)
6. Davis, S., Trapman, P., Leirs, H., Begon, M., Heesterbeek, J.: The abundance threshold for plague as a critical percolation phenomenon. Nature **454**(7204), 634–637 (2008)
7. Dean, B.C., Goemans, M.X., Vondrák, J.: Approximating the stochastic knapsack problem: the benefit of adaptivity. Math. Oper. Res. **33**(4), 945–964 (2008)
8. Esfandiari, H., Karbasi, A., Mirrokni, V.: Adaptivity in adaptive submodularity. In: Conference on Learning Theory, pp. 1823–1846. PMLR (2021)
9. Garnett, R.: Bayesian Optimization. Cambridge University Press (2023)
10. Garnett, R., Gärtner, T., Vogt, M., Bajorath, J.: Introducing the "active search" method for iterative virtual screening. J. Comput. Aided Mol. Des. **29**(4), 305–314 (2015)
11. Garnett, R., Krishnamurthy, Y., Xiong, X., Schneider, J.G., Mann, R.P.: Bayesian optimal active search and surveying. In: Proceedings of the 29th International Conference on Machine Learning, ICML 2012, Edinburgh, Scotland, UK, June 26 - July 1, 2012. icml.cc / Omnipress (2012). http://icml.cc/2012/papers/618.pdf
12. Golovin, D., Krause, A.: Adaptive submodularity: theory and applications in active learning and stochastic optimization. J. Artif. Intell. Res. **42**, 427–486 (2011). https://doi.org/10.1613/JAIR.3278
13. Grimmett, G.R., Stacey, A.M.: Critical probabilities for site and bond percolation models. Ann. Probab. **26**(4), 1788–1812 (1998). http://www.jstor.org/stable/2652790
14. Gupta, A., Krishnaswamy, R., Molinaro, M., Ravi, R.: Approximation algorithms for correlated knapsacks and non-martingale bandits. In: Ostrovsky, R. (ed.) IEEE 52nd Annual Symposium on Foundations of Computer Science, FOCS 2011, Palm Springs, CA, USA, October 22–25, 2011, pp. 827–836. IEEE Computer Society (2011). https://doi.org/10.1109/FOCS.2011.48
15. Gupta, A., Nagarajan, V.: A Stochastic Probing Problem with Applications. In: Goemans, M., Correa, J. (eds.) IPCO 2013. LNCS, vol. 7801, pp. 205–216. Springer, Heidelberg (2013). https://doi.org/10.1007/978-3-642-36694-9_18
16. Gupta, A., Nagarajan, V., Singla, S.: Algorithms and adaptivity gaps for stochastic probing. In: Krauthgamer, R. (ed.) Proceedings of the Twenty-Seventh Annual ACM-SIAM Symposium on Discrete Algorithms, SODA 2016, Arlington, VA, USA, January 10–12, 2016, pp. 1731–1747. SIAM (2016). https://doi.org/10.1137/1.9781611974331.CH120
17. Gupta, A., Nagarajan, V., Singla, S.: Adaptivity gaps for stochastic probing: Submodular and XOS functions. In: Klein, P.N. (ed.) Proceedings of the Twenty-Eighth Annual ACM-SIAM Symposium on Discrete Algorithms, SODA 2017, Barcelona, Spain, Hotel Porta Fira, January 16–19, pp. 1688–1702. SIAM (2017). https://doi.org/10.1137/1.9781611974782.111
18. Hellerstein, L., Kletenik, D., Parthasarathy, S.: A tight bound for stochastic submodular cover. J. Artif. Intell. Res. **71**, 347–370 (2021)
19. Ibarra, O.H., Kim, C.E.: Fast approximation algorithms for the knapsack and sum of subset problems. J. ACM **22**(4), 463–468 (1975)

20. Jiang, S., Malkomes, G., Abbott, M., Moseley, B., Garnett, R.: Efficient nonmyopic batch active search. In: Bengio, S., Wallach, H.M., Larochelle, H., Grauman, K., Cesa-Bianchi, N., Garnett, R. (eds.) Advances in Neural Information Processing Systems 31: Annual Conference on Neural Information Processing Systems 2018, NeurIPS 2018, December 3-8, 2018, Montréal, Canada, pp. 1107–1117 (2018). https://proceedings.neurips.cc/paper/2018/hash/a7aeed74714116f3b292a982238f83d2-Abstract.html
21. Jiang, S., Malkomes, G., Converse, G., Shofner, A., Moseley, B., Garnett, R.: Efficient nonmyopic active search. In: Precup, D., Teh, Y.W. (eds.) Proceedings of the 34th International Conference on Machine Learning, ICML 2017, Sydney, NSW, Australia, 6–11 August 2017. Proceedings of Machine Learning Research, vol. 70, pp. 1714–1723. PMLR (2017). http://proceedings.mlr.press/v70/jiang17d.html
22. Krivelevich, M.: The phase transition in site percolation on pseudo-random graphs. Electron. J. Comb. **22**, P00 (2015)
23. Lawler, E.L.: Fast approximation algorithms for knapsack problems. Math. Oper. Res. **4**(4), 339–356 (1979)
24. Livanos, V., Patton, K., Singla, S.: Improved mechanisms and prophet inequalities for graphical dependencies. In: Proceedings of the 25th ACM Conference on Economics and Computation, pp. 782–805 (2024)
25. Ma, W.: Improvements and generalizations of stochastic knapsack and markovian bandits approximation algorithms. Math. Oper. Res. **43**(3), 789–812 (2018)
26. Reidys, C.M.: Large components in random induced subgraphs of n-cubes. Discret. Math. **309**(10), 3113–3124 (2009)
27. Tang, S.: Beyond pointwise submodularity: non-monotone adaptive submodular maximization subject to knapsack and k-system constraints. Theoret. Comput. Sci. **936**, 139–147 (2022)
28. Wang, X., Garnett, R., Schneider, J.G.: Active search on graphs. In: Dhillon, I.S., et al. (eds.) The 19th ACM SIGKDD International Conference on Knowledge Discovery and Data Mining, KDD 2013, Chicago, IL, USA, August 11–14, 2013, pp. 731–738. ACM (2013). https://doi.org/10.1145/2487575.2487605
29. Williamson, D.P., Shmoys, D.B.: The design of approximation algorithms. Cambridgeuniversity Press (2011)

Complexity of Bilevel Linear Programming with a Single Upper-Level Variable

Nagisa Sugishita[1(✉)] and Margarida Carvalho[2]

[1] Département de sciences de la décision, HEC Montréal, Montreal, Canada
nagisa.sugishita@hec.ca
[2] Département d'informatique et de recherche opérationnelle, Université de Montréal, Montreal, Canada

Abstract. Bilevel linear programming (LP) is one of the simplest classes of bilevel optimization problems, yet it is known to be NP-hard in general. Specifically, determining whether the optimal objective value of a bilevel LP is at least as good as a given threshold, a standard decision version of the problem, is NP-complete. However, this decision problem becomes tractable when either the number of lower-level variables or the number of lower-level constraints is fixed, which prompts the question: What if restrictions are placed on the upper-level problem? In this paper, we address this gap by showing that the decision version of bilevel LP remains NP-complete even when there is only a single upper-level variable, no upper-level constraints (apart from the constraint enforcing optimality of the lower-level decision) and all variables are bounded between 0 and 1. This result implies that fixing the number of variables or constraints in the upper-level problem alone does not lead to tractability in general. On the positive side, we show that there is a polynomial-time algorithm that finds a local optimal solution of such a rational bilevel LP instance. We also demonstrate that many combinatorial optimization problems, such as the knapsack problem and the traveling salesman problem, can be written as such a bilevel LP instance.

Keywords: Bilevel Programming · Computational Complexity · Lexicographic Linear Programming

1 Introduction

Bilevel programming is a class of hierarchical optimization problems in which a *leader* makes a decision first, and a *follower* subsequently optimizes their own objective subject to the leader's choice. In this paper, we focus on the computational complexity of bilevel linear programming (LP), a setting where both the upper- and lower-level problems are LPs. Unless stated otherwise, we consider the optimistic variant: when multiple optimal solutions exist for the follower, the one most favorable to the leader is chosen. Even this simplest form of bilevel programming poses serious challenges in theory and practice. A standard way

S. S. Dey et al. (Eds.): IPCO 2026, LNCS 16588, pp. 426–440, 2026.
https://doi.org/10.1007/978-3-032-28691-8_28

to analyze the complexity of optimization problems is to consider their decision versions, specifically, determining whether the optimal objective value is equal to or less than a given threshold in a minimization problem. Jeroslow [9] was the first to show that the decision version of bilevel LP is NP-hard. An alternative proof was given by Ben-Ayed and Blair [2]. Later, Hansen *et al.* [6] showed that the problem remains NP-hard even when restricted to the min-max case. More recently, Buchheim [3] proved its inclusion in NP, thereby establishing its NP-completeness. Other decision problems have also been studied in the context of bilevel LP: Rodrigues *et al.* [17] showed that checking whether the problem is unbounded is NP-complete, and Vicente *et al.* [22] showed that recognizing local optimality of a given point is NP-hard. Recently, Prokopyev and Ralphs [16] studied the complexity of the search of a local optimal solution.

Given these negative complexity results, a natural line of research is to identify conditions under which bilevel LP becomes tractable. Such conditions can guide the design of efficient algorithms for practical instances. For example, in the context of mixed-integer LP, a classical result by Lenstra Jr [12] establishes that problems with a fixed number of integer variables are solvable in polynomial time. Similar results have been derived for bilevel LP. Deng [5] demonstrated that bilevel LP is solvable in polynomial time when the number of lower-level variables is fixed. Likewise, the arguments in [1] and [3] imply polynomial-time solvability when the number of lower-level constraints is fixed. It was also shown that the toll pricing problem can be solved in polynomial time when the number of upper-level variables is fixed, enabling efficient algorithmic strategies [4]. Related arguments and connections to a pessimistic variant of bilevel LP were very recently discussed in [10], released after the initial submission of this manuscript.

In this paper, we examine the decision version of bilevel LP under a particularly restricted setting: where there is only a single upper-level variable. We show that even under this restricted structure, the decision problem remains NP-complete. In particular, we prove a stronger result: bilevel LP is NP-complete even when there is only one upper-level variable and no upper-level constraints (other than the constraint enforcing the optimality of the lower-level variable with respect to the lower-level problem), and all variables are constrained to lie in the unit cube (i.e., when the lower-level constraints explicitly include bounds of 0 and 1 on each variable). Our proof also establishes the NP-hardness of the decision version of the pessimistic formulation of bilevel LP. See the discussion following the proof of Theorem 1 (page 11) for details, including the definition of the pessimistic formulation. We also prove that given a candidate solution, the problem of deciding whether it is optimal is also coNP-complete. On the positive side, we show that a locally optimal solution can be computed in polynomial time. Furthermore, we show that many combinatorial problems, such as the knapsack problem and the traveling salesman problem, can be reformulated as bilevel LPs of this form, while preserving their optimal solutions.

Paper Structure. This paper is organized as follows. Section 2 formally introduces the class of bilevel LP instances, its decision version and its hardness, the main result of this paper. Section 3 presents auxiliary results that will be used to prove the main result in Sect. 4. Section 5 discusses how combinatorial optimization problems can be transformed to bilevel LPs with a single upper-level variable, using the technique developed in the preceding sections. Section 6 is concerned with the complexity of searching for a globally and locally optimal solution.

Conventions. We denote the binary encoding size of a rational number p and a rational vector v by $\langle p \rangle$ and $\langle v \rangle$, respectively. Throughout the paper, we refer to the binary encoding size simply as the size. For any sequence $a_1, \ldots, a_n$, we write $\langle a_1, \ldots, a_n \rangle$ to denote the sum $\langle a_1 \rangle + \cdots + \langle a_n \rangle$. We follow the encoding conventions described in [18]. In particular, the size of a rational number $x = p/q$, where p and q are relatively prime integers, is given by $1 + \lceil \log_2(|p| + 1) \rceil + \lceil \log_2(|q| + 1) \rceil$. We say that a vector, matrix, or optimization instance is *rational* if and only if all its data are rational numbers. Given an optimization instance P, we use $v(P)$, $\mathcal{F}(P)$, and $\mathcal{S}(P)$ to denote the optimal objective value, the set of feasible solutions, and the set of optimal solutions, respectively. We use $\mathbf{0}$ (respectively, $\mathbf{1}$) to denote the all-zeros (respectively, all-ones) vector, with the dimension clear from the context.

2 Main Result

In this paper, we study the hardness of rational bilevel LPs with a single upper-level variable. In fact, we consider a more restrictive class where the upper-level problem has no constraints and all the variables are bounded to be within the unit cube. Let $\text{BLP}_{\text{SINGLE}}$ denote the set of such bilevel LP instances. An instance in $\text{BLP}_{\text{SINGLE}}$ can be written as:

$$\begin{aligned} \min_{x_1, x_2} \; & c_{21}^\top x_1 + c_{22} x_2 && (\text{P}_{\text{SINGLE}}) \\ \text{s.t. } & x_1 \in \operatorname*{argmin}_{x_1'} \{ c_{11}^\top x_1' : A_{11} x_1' + A_{12} x_2 \geq b_1, \mathbf{0} \leq x_1' \leq \mathbf{1}, 0 \leq x_2 \leq 1 \}, \end{aligned}$$

where $c_{11}, c_{21} \in \mathbb{Q}^n$, $c_{22} \in \mathbb{Q}$, $A_{11} \in \mathbb{Q}^{m \times n}$, and $A_{12}, b_1 \in \mathbb{Q}^m$. A subtle remark is in order. Note that the lower-level constraints impose the bounds on the upper-level variable x_2. If the leader chooses x_2 outside these bounds, the lower-level problem becomes infeasible, consequently resulting into a bilevel infeasible solution to (P_{SINGLE}). In other words, although the upper-level problem has no variable bounds on x_2, effective bounds are implicitly enforced by the lower-level constraints.

Consider the following decision problem:

Given a rational instance (P_{SINGLE}) and a rational number k, is the optimal objective value less than or equal to k? ($\text{VAL}_{\text{SINGLE}}$)

We will prove the following result.

Theorem 1. *The decision problem* $\text{VAL}_{\text{SINGLE}}$ *is NP-complete.*

In Sect. 4, we prove Theorem 1. To this end, we provide a reduction from 3-SAT to $\text{VAL}_{\text{SINGLE}}$. That is, given a Boolean formula F in 3-conjunctive normal form (3CNF), we construct a $\text{BLP}_{\text{SINGLE}}$ instance such that the optimal objective value is -1 if F is satisfiable and 0 otherwise.

To simplify the exposition, we divide the reduction into several steps. First, we construct a bilevel LP instance with multiple upper-level variables and upper-level constraints (instance $\text{P}_{3\text{SAT}}(F)$). We then reduce the problem to a bilevel LP with a single upper-level variable (instance $\text{P}_{\text{weight}}(F)$). Lastly, we eliminate the upper-level constraints via a penalty method (instance $\text{P}(F)$). The final instance is a bilevel LP with a single upper-level variable and no upper-level constraints. We will show that the optimal objective value is preserved throughout these transformations.

To develop our argument, we need to analyse general bilevel LPs, i.e., instances with multiple variables and constraints in the upper-level problem. Thus, in Sect. 3, we provide some auxiliary results to study general bilevel LPs.

3 Auxiliary Results

This section presents auxiliary results that will be used in Sect. 4 to prove Theorem 1. Section 3.1 provides an estimate on the size of an optimal solution to a bilevel LP. To the best of our knowledge, such bounds have not been explicitly discussed in the literature. Section 3.2 presents a method for reformulating upper-level constraints as lower-level constraints.

3.1 Estimate of Optimal Solution Size

We begin by presenting a bound on the size of an optimal solution to an LP, which will be used to estimate the size of solutions for bilevel LPs.

Lemma 1. *Let $A, B \in \mathbb{Q}^{m \times n}$, $a, b \in \mathbb{Q}^m$, and $c \in \mathbb{Q}^n$. Let σ denote the maximum size of the entries in A, B, a, and b. Consider the LP*

$$\min_x \left\{c^\top x : Ax \geq a, Bx > b\right\}.$$

If an optimal solution exists, then there exists a rational optimal solution $\bar{x}$ of size bounded by a polynomial in n and σ. That is, there exists a polynomial f_{sol} such that $\langle \bar{x} \rangle \leq f_{\text{sol}}(n, \sigma)$.

Note that the optimization instance in Lemma 1 includes strict inequalities. Nonetheless, we refer to such instances as LPs throughout this paper. Strict inequalities are used later in the proof of Lemma 3. For the proof, see Corollary 10.2 in [18] (the corollary only discusses weak inequalities, but the extension to strict inequalities is straightforward). We are now ready to state an analogous result for bilevel LPs.

Lemma 2. *Let $A_{11} \in \mathbb{Q}^{m_1 \times n_1}, A_{12} \in \mathbb{Q}^{m_1 \times n_2}, A_{21} \in \mathbb{Q}^{m_2 \times n_1}, A_{22} \in \mathbb{Q}^{m_2 \times n_2}$, $b_1 \in \mathbb{Q}^{m_1}, b_2 \in \mathbb{Q}^{m_2}$, $c_{11}, c_{21} \in \mathbb{Q}^{n_1}$ and $c_{22} \in \mathbb{Q}^{n_2}$. Let $n = n_1 + n_2$ and σ denote* *the maximum size of the entries in* $A_{11}, A_{12}, A_{21}, A_{22}$, b_1, *and* b_2. *Consider the bilevel LP*

$$\begin{aligned} \min_{x_1,x_2}\ & c_{21}^\top x_1 + c_{22}^\top x_2 \\ s.t.\ & A_{21}x_1 + A_{22}x_2 \geq b_2, \\ & x_1 \in \operatorname*{argmin}_{x_1'}\{c_{11}^\top x_1' : A_{11}x_1' + A_{12}x_2 \geq b_1\}. \end{aligned} \tag{1}$$

If an optimal solution exists, then there exists a rational optimal solution $(\bar{x}_1, \bar{x}_2)$ *such that* $\langle \bar{x}_1, \bar{x}_2 \rangle \leq f_{\mathrm{sol}}(n, \sigma)$, *where* f_{sol} *is the polynomial defined in Lemma 1.*

The proof is given in the full version of the paper [21]. We note that this is closely related to the result of [3], where it is shown that the decision version of bilevel LP belongs to NP. Their argument can be used to establish that, whenever the optimal objective value is below a given threshold, there exists a rational feasible solution of polynomial encoding length with objective value below the threshold. In contrast, Lemma 2 guarantees the existence of a rational optimal solution of polynomial encoding length, provided that an optimal solution exists.

3.2 Reformulation of Upper-Level Constraints

If the upper-level problem has a constraint that does not involve any lower-level variables, it can be "moved" to the lower-level problem without changing the feasible solution set. However, if an upper-level constraint involves lower-level variables, special care must be taken. In this section, we present our approach for removing such constraints by introducing a penalty term in the objective. Let

$$\begin{aligned} \min_{x_1,x_2}\ & c_{21}^\top x_1 + c_{22}^\top x_2 \\ \text{s.t.}\ & A_{21}x_1 + A_{22}x_2 \geq b_2, \\ & \bar{A}_{21}x_1 + \bar{A}_{22}x_2 \geq \bar{b}_2, \\ & x_1 \in \operatorname*{argmin}_{x_1'}\{c_{11}^\top x_1' : A_{11}x_1' + A_{12}x_2 \geq b_1\}, \end{aligned} \tag{Q_{blp}}$$

where $A_{11} \in \mathbb{Q}^{m_1\times n_1}, A_{12} \in \mathbb{Q}^{m_1\times n_2}, A_{21} \in \mathbb{Q}^{m_2\times n_1}, A_{22} \in \mathbb{Q}^{m_2\times n_2}, \bar{A}_{21} \in \mathbb{Q}^{m_2'\times n_1}, \bar{A}_{22} \in \mathbb{Q}^{m_2'\times n_2}$, $b_1 \in \mathbb{Q}^{m_1}, b_2 \in \mathbb{Q}^{m_2}, \bar{b}_2 \in \mathbb{Q}^{m_2'}$, $c_{11}, c_{21} \in \mathbb{Q}^{n_1}$ and $c_{22} \in \mathbb{Q}^{n_2}$. We are interested in "replacing" constraints $\bar{A}_{21}x_1 + \bar{A}_{22}x_2 \geq \bar{b}_2$ in (Q_{blp}) with a penalty term in the objective with a help of a single variable $e \in \mathbb{R}$ and a real number M:

$$\begin{aligned} \min_{x_1,x_2,e}\ & c_{21}^\top x_1 + c_{22}^\top x_2 + Me \\ \text{s.t.}\ & A_{21}x_1 + A_{22}x_2 \geq b_2, \\ & (x_1, e) \in \operatorname*{argmin}_{x_1',e'} \left\{ c_{11}^\top x_1' : \begin{array}{c} A_{11}x_1' + A_{12}x_2 \geq b_1, \\ \bar{A}_{21}x_1' + \bar{A}_{22}x_2 + e'\mathbf{1} \geq \bar{b}_2, \\ e' \in I_e \end{array} \right\}, \end{aligned} \tag{$Q_{\mathrm{blp}}'(M)$}$$

where I_e is $[0,1]$ or $[0,\infty)$. The following result holds for both choices of I_e. The former is more convenient for our purposes, as we would like all variables to be constrained within $[0,1]$, whereas the latter is more natural in applications.

Lemma 3. *Let I_e is $[0,1]$ or $[0,\infty)$, and suppose (Q_{blp}) and $(Q'_{\text{blp}}(M))$ have an optimal solution for any $M > 0$. Let $n = n_1 + n_2$, σ denote the maximum size of entries in A_{11}, A_{12}, A_{21}, A_{22}, $\bar{A}_{21}$, $\bar{A}_{22}$, b_1, b_2, $\bar{b}_2$, and $c_\infty = \max\{\|c_{21}\|_\infty, \|c_{22}\|_\infty, 1\}$. For any $M \geq 3nc_\infty 4^{f_{\text{sol}}(n+1,\sigma)}$, the following relations hold:*

(i) $v(Q_{\text{blp}}) = v(Q'_{\text{blp}}(M))$;
(ii) $\mathcal{S}(Q'_{\text{blp}}(M)) = \{(x_1, x_2, 0) : (x_1, x_2) \in \mathcal{S}(Q_{\text{blp}})\}$.

A related penalty-based reformulation appears in [7], although it is unclear whether their transformation can be performed efficiently. Recent works [8,11] investigate this question. In the full version of the paper [21], we provide an alternative and simpler proof of Lemma 3.

4 Proof of Theorem 1

We begin by constructing a bilevel LP instance with multiple upper-level variables. Let F be a Boolean formula in 3CNF with n variables $u_1, \ldots, u_n$ and p clauses. Following [13,15], F can be represented by a system of linear inequalities over binary variables: $A_F x \geq \mathbf{1} - a_F$, $x \in \{0,1\}^n$, where $A_F \in \mathbb{Z}^{p \times n}$, $a_F \in \mathbb{Z}^p$. A satisfying truth assignment corresponds exactly to a feasible solution, where $u_i =$ true corresponds to $x_i = 1$. The inequalities are constructed clause by clause. For a clause $(l_1 \vee l_2 \vee l_3)$, we add the inequality

$$\sum_{i:u_i \in \{l_1,l_2,l_3\}} x_i + \sum_{i:\neg u_i \in \{l_1,l_2,l_3\}} (1 - x_i) \geq 1.$$

For example, $F = (u_1 \vee \neg u_2) \wedge (\neg u_1 \vee \neg u_2 \vee u_3)$ corresponds to

$$x_1 + (1 - x_2) \geq 1, \quad (1 - x_1) + (1 - x_2) + x_3 \geq 1, \quad x \in \{0,1\}^3.$$

In the following, we say that a binary vector satisfies F if and only if the corresponding truth assignment satisfies F.

Now, consider the following bilevel LP instance:

$$\min_{x,y,w} \; 2\sum_{i=1}^{n} (y - 2w_i) - \frac{2}{n}\sum_{i=1}^{n} w_i \qquad (\text{P}_{3\text{SAT}}(F))$$

$$\text{s.t. } A_F x \geq \left(\frac{3}{2} - \frac{1}{2}y\right)\mathbf{1} - a_F, \qquad (2)$$

$$\frac{1}{2} - \frac{1}{2}y \leq x_i \leq \frac{1}{2} + \frac{1}{2}y, \quad \forall i = 1, \ldots, n, \qquad (3)$$

$$0 \leq x_i \leq 1, \; 0 \leq y \leq 1, \quad \forall i = 1, \ldots, n,$$

$$w \in \operatorname*{argmin}_{w'} \left\{ \sum_{i=1}^{n} w'_i : -w'_i \leq x_i - 1/2 \leq w'_i, 0 \leq w'_i \leq 1, \; i = 1, \ldots, n \right\}.$$

The lower-level problem measures the closeness of the upper-level decision variable x to the nearest integers by setting $w_i = |x_i - 1/2|$, which is maximized when $x_i \in \{0, 1\}$. This instance is obtained from the one studied by Vicente *et al.* [22] (see also [17]). Instance ($\mathrm{P}_{3\mathrm{SAT}}(F)$) differs from their instance in that we removed some redundant variables from the lower-level problem for simplification and added variable bounds $0 \leq x_i \leq 1$, $0 \leq y \leq 1$ and $0 \leq w_i \leq 1$ for each $i = 1, \ldots, n$. Without the variable bounds, the instance is unbounded if and only if F is satisfiable. In our case (i.e., with the variable bounds), the optimal objective value of ($\mathrm{P}_{3\mathrm{SAT}}(F)$) is -1 if F is satisfiable and 0 otherwise, as stated below.

Lemma 4.

(i) Suppose F is unsatisfiable. Then, $x_i = 1/2$ for $i = 1, \ldots, n$, $y = 0$, and $w = \mathbf{0}$ is optimal for ($\mathrm{P}_{3\mathrm{SAT}}(F)$) and $v(\mathrm{P}_{3\mathrm{SAT}}(F)) = 0$.
(ii) Suppose F is satisfiable and let μ be a binary vector satisfying F. Then, $x = \mu$, $y = 1$, and $w = \mathbf{0}$ is optimal for ($\mathrm{P}_{3\mathrm{SAT}}(F)$) and $v(\mathrm{P}_{3\mathrm{SAT}}(F)) = -1$.

See the full version of the paper [21] for the proof.

To simplify the notation in the following argument, we write $z = (x, y) \in \mathbb{R}^{n+1}$ and the constraints (2) and (3) as $B_F z \geq b_F$. Furthermore, we introduce $f \in \mathbb{R}^{n+1}$ such that $f_i = |z_i - 1/2|$ for $i = 1, \ldots, n + 1$. With this notation, ($\mathrm{P}_{3\mathrm{SAT}}(F)$) can be written as

$$\begin{aligned} \min_{z,f}\ & 2\sum_{i=1}^{n}(z_{n+1} - 2f_i) - \frac{2}{n}\sum_{i=1}^{n} f_i \qquad (\mathrm{P}_{\mathrm{mult}}(F)) \\ \text{s.t. }& B_F z \geq b_F,\ 0 \leq z_i \leq 1,\ \forall i = 1, \ldots, n+1, \\ & f \in \operatorname*{argmin}_{f'} \left\{ \sum_{i=1}^{n+1} f'_i : -f'_i \leq z_i - 1/2 \leq f'_i, 0 \leq f'_i \leq 1,\ i = 1, \ldots, n+1 \right\}. \end{aligned}$$

The formulation ($\mathrm{P}_{\mathrm{mult}}(F)$) differs from ($\mathrm{P}_{3\mathrm{SAT}}(F)$) by the inclusion of an additional variable, f_{n+1}. Nonetheless, both problems attain the same optimal objective value. The additional variable is introduced purely for notational convenience in the analysis that follows.

Now, we use a lexicographic LP to replace the upper-level variables z in ($\mathrm{P}_{\mathrm{mult}}(F)$) with a single upper-level variable θ. Consider

$$\begin{aligned} \min_{\theta,z,f,s,t,u}\ & 2\sum_{i=1}^{n}(z_{n+1} - 2f_i) - \frac{2}{n}\sum_{i=1}^{n} f_i \qquad (\mathrm{P}_{\mathrm{lex}}(F)) \\ \text{s.t. }& B_F z \geq b_F, 0 \leq \theta \leq 1, \\ & (z, f, s, t, u) \in \mathcal{S}(\mathrm{P}^{\mathrm{low}}_{\mathrm{lex}}(n+1, \theta)), \end{aligned}$$

where

$$
\begin{aligned}
\operatorname{lexmin}_{z,f,s,t,u}\ & \left\{ s_{n+1}+t_{n+1},\ \ldots,\ s_1+t_1,\ \sum_{i=1}^{n+1} f_i \right\} && (\mathrm{P}^{\mathrm{low}}_{\mathrm{lex}}(n+1,\theta))\\
\text{s.t. } & u_{n+1}=\theta, \\
& s_i \geq (3/2)u_i - 1/2, \qquad t_i \geq (3/2)u_i - 1, && \forall i=1,\ldots,n+1,\\
& z_i = 2(s_i - t_i), \qquad -f_i \leq z_i - 1/2 \leq f_i, && \forall i=1,\ldots,n+1,\\
& u_{i-1} = 3u_i - 2z_i, && \forall i=2,\ldots,n+1,\\
& 0 \leq z_i, f_i, s_i, t_i, u_i \leq 1, && \forall i=1,\ldots,n+1,
\end{aligned}
$$

with $\theta \in \mathbb{R}$, $z, f, s, t, u \in \mathbb{R}^{n+1}$. Here, lexmin denotes the lexicographic minimization: the first objective is given highest priority, and later objectives are considered only when earlier ones are equal. In $(\mathrm{P}_{\mathrm{lex}}(F))$, z denotes a lower-level variable. We intentionally use the same symbol z, for reasons that will be clarified later. For each $\theta \in [0,1]$, $(\mathrm{P}^{\mathrm{low}}_{\mathrm{lex}}(n+1,\theta))$ has a unique solution, and can be solved analytically, as stated below.

Lemma 5.

1. *For any $\theta \in [0,1]$, it holds that*

$$
\mathcal{S}(\mathrm{P}^{\mathrm{low}}_{\mathrm{lex}}(n+1,\theta)) = \left\{ \begin{pmatrix} z\\ f\\ s\\ t\\ u \end{pmatrix} : \begin{array}{lll} u_{n+1}=\theta, & & \\ s_i = [(3/2)u_i - 1/2]^+, & t_i = [(3/2)u_i - 1]^+, & \forall i=1,\ldots,n+1,\\ z_i = f^z(u_i), & f_i = |z_i - 1/2|, & \forall i=1,\ldots,n+1,\\ u_{i-1} = f^u(u_i), & & \forall i=2,\ldots,n+1, \end{array} \right\},
$$

where $[a]^+ = \max\{a, 0\}$, and

$$
f^z(\theta') = \begin{cases} 0, & 0 \leq \theta' < 1/3,\\ 3\theta' - 1, & 1/3 \leq \theta' < 2/3,\\ 1, & 2/3 \leq \theta' \leq 1, \end{cases} \qquad f^u(\theta') = \begin{cases} 3\theta', & 0 \leq \theta' < 1/3,\\ -3\theta' + 2, & 1/3 \leq \theta' < 2/3,\\ 3\theta' - 2, & 2/3 \leq \theta' \leq 1. \end{cases}
$$

2. *Let $\theta = 1/6$ and $(z, f, s, t, u) \in \mathcal{S}(\mathrm{P}^{\mathrm{low}}_{\mathrm{lex}}(n+1,\theta))$. Then, $z_{n+1} = 0$ and $z_i = 1/2$ for $i = 1, \ldots, n$.*
3. *For any binary vector $\mu \in \{0,1\}^n$, let $\theta = (2/3)(1 + \sum_{i=1}^{n} \mu_i/3^{n+1-i})$ and $(z, f, s, t, u) \in \mathcal{S}(\mathrm{P}^{\mathrm{low}}_{\mathrm{lex}}(n+1,\theta))$. Then, $z_{n+1} = 1$ and $z_i = \mu_i$ for $i = 1, \ldots, n$.*

The proof is given in the full version of the paper [21]. For each θ, the lower-level problem $(\mathrm{P}^{\mathrm{low}}_{\mathrm{lex}}(n+1,\theta))$ has a unique optimal solution. Plots illustrating the relationship between u, z and θ are provided in Fig. 1. As θ varies from 0 to 1, each component of the vector z takes values between 0 and 1. Collectively, the vector z traces all the vertices of the $(n+1)$-dimensional unit cube, as well as the point $z = (x, y)$, where $x_i = 1/2$ for all $i = 1, \ldots, n$ and $y = 0$. Notably,

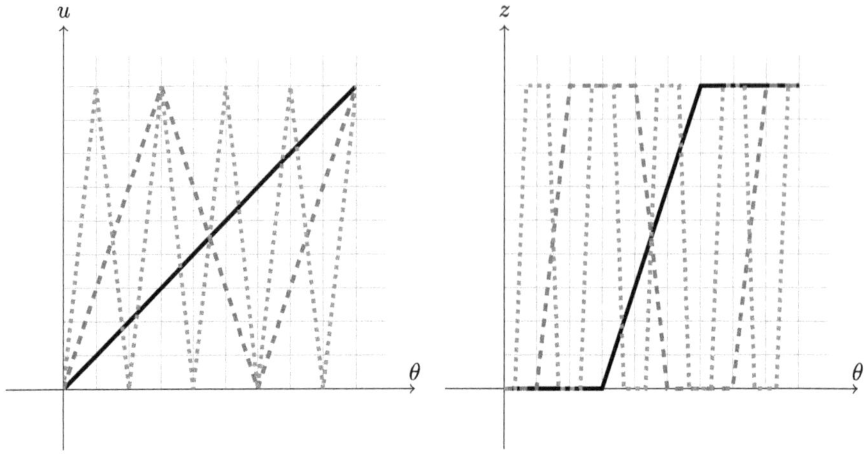

Fig. 1. Plot of u and z in $\mathcal{S}(\mathrm{P}_{\mathrm{lex}}^{\mathrm{low}}(n+1,\theta)))$. The solid black lines represent u_3 and z_3, the dashed blue lines represent u_2 and z_2, and the dotted red lines represent u_1 and z_1. (Color figure online)

as θ increases from 0 to 1, z visits all candidate solutions that may be optimal for $(\mathrm{P}_{\mathrm{mult}}(F))$, as established in Lemma 4. Therefore, despite $(\mathrm{P}_{\mathrm{lex}}(F))$ involving only a single upper-level variable, it is capable of "searching" for the solution to $(\mathrm{P}_{\mathrm{mult}}(F))$ by varying θ. Formally, we establish the following result.

Lemma 6. *It holds that* $v(\mathrm{P}_{\mathrm{mult}}(F)) = v(\mathrm{P}_{\mathrm{lex}}(F))$.

Proof. We first show that $v(\mathrm{P}_{\mathrm{mult}}(F)) \geq v(\mathrm{P}_{\mathrm{lex}}(F))$. Suppose F is satisfiable, and let μ be a binary vector satisfying F, $\theta = (2/3)(1+\sum_{i=1}^{n} \mu_i/3^{n+1-i})$ and $(z, f, s, t, u) \in \mathcal{S}(\mathrm{P}_{\mathrm{lex}}^{\mathrm{low}}(n+1,\theta))$. By Lemma 5 (iii), we have $z_i = \mu_i$ for $i = 1,\ldots,n$, $z_{n+1} = 1$, $f_i = |z_i - 1/2| = 1/2$ for $i = 1,\ldots,n+1$. Therefore, $B_F z \geq b_F$, and (θ, z, f, s, t, u) is feasible for $(\mathrm{P}_{\mathrm{lex}}(F))$ and

$$v(\mathrm{P}_{\mathrm{lex}}(F)) \leq 2\sum_{i=1}^{n}(z_{n+1} - 2f_i) - \frac{2}{n}\sum_{i=1}^{n} f_i = -1 = v(\mathrm{P}_{\mathrm{mult}}(F)).$$

If F is unsatisfiable, we can show $v(\mathrm{P}_{\mathrm{lex}}(F)) \leq 0 = v(\mathrm{P}_{\mathrm{mult}}(F))$ with $\theta = 1/6$.

Next, we show that $v(\mathrm{P}_{\mathrm{mult}}(F)) \leq v(\mathrm{P}_{\mathrm{lex}}(F))$. To this end, we prove that the objective value of any feasible solution to $(\mathrm{P}_{\mathrm{lex}}(F))$ is larger than or equal to $v(\mathrm{P}_{\mathrm{mult}}(F))$. Let (θ, z, f, s, t, u) be any feasible solution to $(\mathrm{P}_{\mathrm{lex}}(F))$. Then, we have $0 \leq z_i \leq 1$, $f_i = |z_i - 1/2|$ for $i = 1,\ldots,n+1$, and $B_F z \geq b_F$. Therefore, (z, f) is feasible for $(\mathrm{P}_{\mathrm{mult}}(F))$, and hence

$$2\sum_{i=1}^{n}(z_{n+1} - 2f_i) - \frac{2}{n}\sum_{i=1}^{n} f_i \geq v(\mathrm{P}_{\mathrm{mult}}(F)).$$

Since (θ, z, f, s, t, u) was an arbitrary feasible solution to $(\mathrm{P}_{\mathrm{lex}}(F))$, it follows that $v(\mathrm{P}_{\mathrm{mult}}(F)) \leq v(\mathrm{P}_{\mathrm{lex}}(F))$. □

Next, we aim to reformulate the lexicographic LP ($\mathrm{P}_{\mathrm{lex}}^{\mathrm{low}}(n+1,\theta)$) as a single-objective LP. There is a substantial body of literature on transforming lexicographic LPs into LPs. For example, Sherali [19] and Sherali and Soyster [20] explored approaches based on weighted sums of the objectives. However, these techniques are difficult to apply to ($\mathrm{P}_{\mathrm{lex}}^{\mathrm{low}}(n+1,\theta)$): Since the number of objectives increases with n, it is nontrivial to design appropriate weights of polynomial size to combine the objectives. Nevertheless, since ($\mathrm{P}_{\mathrm{lex}}^{\mathrm{low}}(n+1,\theta)$) has a special structure, we can derive an equivalent LP of polynomial size using a tailored argument. Consider

$$\begin{aligned}\min_{\theta,z,f,s,t,u}\ & 2\sum_{i=1}^{n}(z_{n+1}-2f_i)-\frac{2}{n}\sum_{i=1}^{n}f_i \qquad (\mathrm{P}_{\mathrm{weight}}(F))\\ \text{s.t. } & B_F z\geq b_F, 0\leq\theta\leq 1,\\ & (z,f,s,t,u)\in\mathcal{S}(\mathrm{P}_{\mathrm{weight}}^{\mathrm{low}}(n+1,\theta)),\end{aligned}$$

where

$$\begin{aligned}\min_{z,f,s,t,u}\ & \sum_{i=1}^{n+1}16^i(s_i+t_i)+\sum_{i=1}^{n+1}\frac{1}{4^{n-i+1}}f_i \qquad (\mathrm{P}_{\mathrm{weight}}^{\mathrm{low}}(n+1,\theta))\\ \text{s.t. } & (z,f,s,t,u)\in\mathcal{F}(\mathrm{P}_{\mathrm{lex}}^{\mathrm{low}}(n+1,\theta)).\end{aligned}$$

We show that $v(\mathrm{P}_{\mathrm{lex}}(F))=v(\mathrm{P}_{\mathrm{weight}}(F))$.

Lemma 7.

(i) For any $n\geq 1$ and $\theta\in[0,1]$, $\mathcal{S}(\mathrm{P}_{\mathrm{lex}}^{\mathrm{low}}(n+1,\theta))=\mathcal{S}(\mathrm{P}_{\mathrm{weight}}^{\mathrm{low}}(n+1,\theta))$.
(ii) It holds that $v(\mathrm{P}_{\mathrm{lex}}(F))=v(\mathrm{P}_{\mathrm{weight}}(F))$.

See the full version of the paper [21] for the proof. As the last step, we remove $B_F z\geq b_F$ from the upper-level problem using Lemma 3 and obtain

$$\begin{aligned}\min_{\theta,z,f,s,t,u,e}\ & 2\sum_{i=1}^{n}(z_{n+1}-2f_i)-\frac{2}{n}\sum_{i=1}^{n}f_i+Me \qquad (\mathrm{P}(F))\\ \text{s.t. } & (z,f,s,t,u,e)\in\mathcal{S}(\mathrm{P}^{\mathrm{low}}(F,\theta)),\end{aligned}$$

where

$$\begin{aligned}\min_{z,f,s,t,u,e}\ & \sum_{i=1}^{n+1}16^i(s_i+t_i)+\sum_{i=1}^{n+1}\frac{1}{4^{n-i+1}}f_i \qquad (\mathrm{P}^{\mathrm{low}}(F,\theta))\\ \text{s.t. } & B_F z+e\mathbf{1}\geq b_F, 0\leq\theta\leq 1, 0\leq e\leq 1,\\ & (z,f,s,t,u)\in\mathcal{F}(\mathrm{P}_{\mathrm{lex}}^{\mathrm{low}}(n+1,\theta)),\end{aligned}$$

with $M=3(5n+7)4^{f_{\mathrm{sol}}(5n+3,\langle 6\rangle)+1}$. We note that the penalty term e can be bounded from above (see Lemma 3).

Lemma 8. *It holds that* $v(\mathrm{P}_{\text{weight}}(F)) = v(\mathrm{P}(F))$.

The proof is in the full version of the paper [21]. Note that the size of $(\mathrm{P}(F))$ is polynomial in n.

Lemma 9.

(i) Suppose F *is satisfiable. Then,* $v(\mathrm{P}(F)) = -1$.
(ii) Suppose F *is unsatisfiable. Then,* $v(\mathrm{P}(F)) = 0$.

Proof. It follows from Lemmata 4–8. □

Theorem 1 follows immediately from the above argument.

Proof (Proof of Theorem 1*).* The inclusion in NP follows from [3]. By Lemma 9, the above transformation gives a logarithmic-space reduction [14] from 3-SAT to $\text{VAL}_{\text{SINGLE}}$. Since 3-SAT is NP-complete, $\text{VAL}_{\text{SINGLE}}$ is NP-hard. □

A few remarks are in order. First, by adding 1 (or, equivalently, by adding a variable fixed to 1) to the objective of $(\mathrm{P}(F))$, one can obtain an instance whose optimal objective value is 0 if F is satisfiable and 1 otherwise. Thus, even approximate versions of $\text{VAL}_{\text{SINGLE}}$ are NP-hard (See Corollary 4.6 in Jeroslow [9]). Second, instance $(\mathrm{P}(F))$ contains constants of exponential magnitude. Consequently, a bilevel LP with a single variable and no upper-level constraints may not be strongly NP-hard. In particular, such instances may admit pseudo-polynomial algorithms. Third, our proof of NP-hardness for the decision version of the optimistic formulation of bilevel LP also carries over to the pessimistic formulation. Recall that in the pessimistic formulation the follower, when faced with multiple optimal solutions, selects the one least favorable to the leader. Since the lower-level problem in our construction has a unique optimal solution, the optimal objective value of the pessimistic variant of $(\mathrm{P}(F))$ coincides with that of the optimistic variant. Fourth, the argument of Basu *et al.* [1] shows that the feasible set of a bilevel LP is the union of at most exponentially many polyhedra. We observe that instance $(\mathrm{P}(F))$, a bilevel LP with a single upper-level variable and no upper-level constraints, requires exponentially many polyhedra to describe its feasible set. Formally, we have the following result, proven in the full version of the paper [21].

Proposition 1. *Let* q *be the smallest integer such that there exists polyhedra* $Q_1, \dots, Q_q$ *satisfying* $\mathcal{F}(\mathrm{P}(F)) = \cup_{j=1\dots,q} Q_j$. *Then,* $q \geq 2^n$. *In other words, any representation of* $\mathcal{F}(\mathrm{P}(F))$ *as a union of polyhedra requires exponentially many polyhedra in* n.

5 Reformulation of Combinatorial Optimization Problem

In this section, we briefly present a reformulation of combinatorial optimization problems into instances of $\text{BLP}_{\text{SINGLE}}$. Given a NP-hard combinatorial optimization problem, one can construct a reduction from its decision version to the decision version of $\text{BLP}_{\text{SINGLE}}$. However, by exploiting the techniques developed in Sect. 4, one can in fact design a transformation that preserves optimal solutions, in the sense that an optimal solution to the original combinatorial problem can be constructed from one for the resulting $\text{BLP}_{\text{SINGLE}}$ instance, and vice versa. Consider the following 0–1 integer LP:

$$\min_z \{c^\top z : Az \geq a, z \in \{0,1\}^r\}, \tag{C}$$

where $c \in \mathbb{Q}^r$, $A \in \mathbb{Q}^{m \times r}$, and $a \in \mathbb{Q}^m$. We assume that (C) has at least one feasible solution, which holds when it is a 0–1 integer LP formulation of the knapsack or traveling salesman problem over a fully-connected graph. By appropriate scaling if necessary, we may assume $a_j \in [-1, 1]$ for all $j = 1, \ldots, m$. We first replace the binary variables using the lexicographic LP ($\text{P}^{\text{low}}_{\text{lex}}(n+1, \theta)$) or its LP equivalent ($\text{P}^{\text{low}}_{\text{weight}}(n+1, \theta)$) (Lemma 7 (i) is stated for $r \geq 2$ but it is straightforward to show that it also holds for $r = 1$):

$$\begin{aligned} \min_{\theta, z, f, s, t, u} \quad & c^\top z & (\text{C}_{\text{mult}}) \\ \text{s.t.} \quad & Az \geq a, 0 \leq \theta \leq 1, f_i \geq 1/2, \ \forall i = 1, \ldots, n, & (4) \\ & (z, f, s, t, u) \in \mathcal{S}(\text{P}^{\text{low}}_{\text{weight}}(n+1, \theta)), \end{aligned}$$

where $\theta \in \mathbb{R}$, and $z, f, s, t, u \in \mathbb{R}^r$. For each $i = 1, \ldots, r$, the optimality of the lower-level problem implies $f_i = |z_i - 1/2|$, thus the condition $f_i \geq 1/2$ enforces θ to be such that $z_i \in \{0, 1\}$. Next, we replace the constraints (4) with a penalty term added to the objective:

$$\min_{\theta, z, f, s, t, u, e} \{c^\top z + Me : (z, f, s, t, u, e) \in \mathcal{S}(\text{C}^{\text{low}}_{\text{single}}(\theta))\}, \tag{C_{single}}$$

where

$$\begin{aligned} \min_{z, f, s, t, u, e} \quad & \sum_{i=1}^{r} 16^i (s_i + t_i) + \sum_{i=1}^{r} \frac{1}{4^{r-i}} f_i & (\text{C}^{\text{low}}_{\text{single}}(\theta)) \\ \text{s.t.} \quad & Az + e\mathbf{1} \geq a, f_i + e \geq 1/2, \ \forall i = 1, \ldots, r, \\ & 0 \leq \theta \leq 1, 0 \leq e \leq 1, \\ & (z, f, s, t, u) \in \mathcal{F}(\text{P}^{\text{low}}_{\text{lex}}(n+1, \theta)). \end{aligned}$$

and M is a penalty coefficient. It is straightforward to verify that both instances (C_{mult}) and (C_{single}) admit optimal solutions for any value of M. For example, (C_{single}) has a feasible solution given by $\theta = 0$, $z = f = s = t = u = \mathbf{0}$, and $e = 1$. Moreover, (C_{single}) cannot be unbounded since all the variables are bounded, thus guaranteeing the existence of an optimal solution. Therefore, by invoking Lemma 3, one can compute a penalty coefficient M of polynomial size (in the size of (C)) such that $v(\text{C}) = v(\text{C}_{\text{single}})$. In fact, it is straightforward to see the following: $\mathcal{S}(\text{C}_{\text{single}}) = \{(\theta, z, f, s, t, u, e) \in \mathcal{F}(\text{C}_{\text{single}}) : z \in \mathcal{S}(\text{C})\}$.

6 Complexity of Local Search

To further understand the computational complexity of $\text{BLP}_{\text{SINGLE}}$, in the remainder of the paper, we study the complexity of searching for a solution. We begin by discussing the coNP-completeness of deciding whether a given point is a globally optimal solution. We then turn our attention to the problem of finding a locally optimal solution. We show that it is possible to compute a locally optimal solution in polynomial time.

Consider the following decision problem:

Given a rational instance (P_{SINGLE}) and a rational point (x_1, x_2), is it an optimal solution? ($\text{RECOG}^{\text{GLOBAL}}_{\text{SINGLE}}$)

Using instance ($\text{P}(F)$), we obtain the following result, which is proven in the full version of the paper [21].

Proposition 2. *The decision problem* $\text{RECOG}^{\text{GLOBAL}}_{\text{SINGLE}}$ *is coNP-complete.*

Proposition 2 shows that merely deciding whether a given point is an optimal solution is coNP-complete. Therefore, searching for a globally optimal solution is likely to be computationally intractable. A natural question is whether replacing global optimality with local optimality leads to a tractable problem.

We say that $(\bar{x}_1, \bar{x}_2) \in \mathbb{R}^{n+1}$ is a *locally optimal solution* to (P_{SINGLE}) if $(\bar{x}_1, \bar{x}_2) \in \mathcal{F}(\text{P}_{\text{SINGLE}})$ and there exists $\epsilon > 0$ such that

$$c_{21}^\top \bar{x}_1 + c_{22}\bar{x}_2 \leq c_{21}^\top x_1 + c_{22}x_2, \quad \forall (x_1, x_2) \in \mathcal{F}(\text{P}_{\text{SINGLE}}) \cap \mathcal{B}_\epsilon(\bar{x}_1, \bar{x}_2),$$

where $\mathcal{B}_\epsilon(\bar{x}_1, \bar{x}_2) = \{(x_1, x_2) : \|(x_1, x_2) - (\bar{x}_1, \bar{x}_2)\|_\infty \leq \epsilon\}$. The following result holds.

Theorem 2. *There exists a polynomial-time algorithm that finds a locally optimal solution to a rational instance of* (P_{SINGLE}).

See the full version of the paper [21] for the proof. We show that if there is a local optimal solution, there is a rational one of polynomial size. Thus, we apply a binary search on the upper-level variable x_2 to find a local optimal solution. This result contrasts with a well-known result by Vicente *et al.* [22], which shows that recognizing local optimality is NP-hard in the general case. Thus, Theorem 2 suggests a fundamental difference between general bilevel LPs and $\text{BLP}_{\text{SINGLE}}$.

Acknowledgments. This study was funded by the FRQ-IVADO Research Chair in Data Science for Combinatorial Game Theory.

Disclosure of Interests. The authors have no competing interests to declare that are relevant to the content of this article.

References

1. Basu, A., Ryan, C.T., Sankaranarayanan, S.: Mixed-integer bilevel representability. Math. Program. **185**(1–2), 163–197 (2021). https://doi.org/10.1007/s10107-019-01424-w
2. Ben-Ayed, O., Blair, C.E.: Computational difficulties of bilevel linear programming. Oper. Res. **38**(3), 556–560 (1990). https://doi.org/10.1287/opre.38.3.556
3. Buchheim, C.: Bilevel linear optimization belongs to NP and admits polynomial-size KKT-based reformulations. Oper. Res. Lett. **51**(6), 618–622 (2023). https://doi.org/10.1016/j.orl.2023.10.006
4. Bui, Q.M., Carvalho, M., Neto, J.: Asymmetry in the complexity of the multi-commodity network pricing problem. Math. Program. **208**(1), 425–461 (2024). https://doi.org/10.1007/s10107-023-02043-2
5. Deng, X.: Complexity issues in bilevel linear programming. Multilevel optimization: Algorithms and Applications, pp. 149–164 (1998). https://doi.org/10.1007/978-1-4613-0307-7_6
6. Hansen, P., Jaumard, B., Savard, G.: New branch-and-bound rules for linear bilevel programming. SIAM J. Sci. Stat. Comput. **13**(5), 1194–1217 (1992). https://doi.org/10.1137/0913069
7. Henke, D., Lefebvre, H., Schmidt, M., Thürauf, J.: On coupling constraints in linear bilevel optimization. Optim. Lett. **19**(3), 689–697 (2025). https://doi.org/10.1007/s11590-024-02156-3
8. Henke, D., Lefebvre, H., Schmidt, M., Thürauf, J.: On coupling constraints in pessimistic linear bilevel optimization. arXiv (2025). https://arxiv.org/abs/2503.01563
9. Jeroslow, R.G.: The polynomial hierarchy and a simple model for competitive analysis. Math. Program. **32**(2), 146–164 (1985). https://doi.org/10.1007/BF01586088
10. Ketkov, S.S., Prokopyev, O.A.: A note on the complexity of bilevel linear programs in fixed dimensions. arXiv (2025). https://arxiv.org/abs/2511.15592
11. Lefebvre, H., Schmidt, M.: Exact augmented Lagrangian duality for nonconvex mixed-integer nonlinear optimization. Optimization Online (2024). https://optimization-online.org/?p=27046
12. Lenstra, H.W., Jr.: Integer programming with a fixed number of variables. Math. Oper. Res. **8**(4), 538–548 (1983)
13. Marcotte, P., Savard, G.: Bilevel programming: A combinatorial perspective. In: Graph Theory and Combinatorial Optimization, pp. 191–217. Springer, Cham (2005). https://doi.org/10.1007/0-387-25592-3_7
14. Papadimitriou, C.H.: Computational Complexity. Addison-Wesley (1993)
15. Pardalos, P.M., Schnitger, G.: Checking local optimality in constrained quadratic programming is NP-hard. Oper. Res. Lett. **7**(1), 33–35 (1988). https://doi.org/10.1016/0167-6377(88)90049-1
16. Prokopyev, O.A., Ralphs, T.K.: On the complexity of finding locally optimal solutions in bilevel linear optimization. Oper. Res. **0**(0) (2026). https://doi.org/10.1287/opre.2024.1411
17. Rodrigues, B., Carvalho, M., Anjos, M.F., Sugishita, N.: Unboundedness in bilevel optimization. Math. Program. (2026). https://doi.org/10.1007/s10107-026-02344-2
18. Schrijver, A.: Theory of Linear and Integer Programming. Wiley-Interscience series in discrete mathematics and optimization, Wiley, Chichester (1998)

19. Sherali, H.D.: Equivalent weights for lexicographic multi-objective programs: characterizations and computations. Eur. J. Oper. Res. **11**(4), 367–379 (1982). https://doi.org/10.1016/0377-2217(82)90202-8
20. Sherali, H.D., Soyster, A.L.: Preemptive and nonpreemptive multi-objective programming: relationship and counterexamples. J. Optim. Theory Appl. **39**(2), 173–186 (1983). https://doi.org/10.1007/BF00934527
21. Sugishita, N., Carvalho, M.: Complexity of bilevel linear programming with a single upper-level variable. arXiv (2025). https://arxiv.org/abs/2510.21126
22. Vicente, L.N., Savard, G., Júdice, J.: Descent approaches for quadratic bilevel programming. J. Optim. Theory Appl. **81**(2), 379–399 (1994). https://doi.org/10.1007/BF02191670

A Framework for Handling and Exploiting Symmetry in Benders Decomposition

Christopher Hojny and Cédric Roy(✉)

Eindhoven University of Technology, Eindhoven, The Netherlands
{c.hojny,c.j.roy}@tue.nl

Abstract. Benders decomposition (BD) is a framework for solving optimization problems by removing some variables and modeling their contribution to the original problem via so-called Benders cuts. While many advanced optimization techniques can be applied in a BD framework, one central technique has not been applied systematically in BD: symmetry handling. The main reason for this is that Benders cuts are not known explicitly but only generated via a separation oracle.

In this work, we close this gap by providing a framework of symmetry detection within the BD framework. To this end, we introduce a tailored family of graphs that capture the symmetry information of both the Benders master problem and the Benders oracles. Once symmetries of these graphs are known, which can be found by established techniques, classical symmetry handling approaches become available to accelerate BD. We complement these approaches by devising techniques for the separation and aggregation of symmetric Benders cuts by means of tailored separation routines and extended formulations. Both substantially reduce the number of executions of the separation oracles. In a numerical study, we show the effect of both symmetry handling and cut aggregation for bin packing and scheduling problems.

Keywords: Logic Based Benders Cuts · Symmetry Exploitation · Extended Formulation

1 Introduction

We consider mixed-integer programs (MIP)

$$\min\{c^\top x + d^\top y : Ax \leq b,\ Cx + Dy \leq f,\ x \in K(p,n),\ y \in K(q,m)\}, \tag{1}$$

where A, C, D are matrices and b, c, d, f are vectors of suitable dimensions, and $K(r,s) := \mathbb{Z}^r \times \mathbb{R}^{s-r}$ for integers $0 \leq r \leq s$. In many applications like healthcare delivery [14], sports scheduling [35], facility location [11], or packing optimization [21], the matrix D has a block structure. That is, when the x-variables are fixed, the MIP decomposes into several independent smaller MIPs.

Benders Decomposition (BD) [4] exploits this structure by splitting the MIP into a master and subproblem. The master problem can be considered an abstract

S. S. Dey et al. (Eds.): IPCO 2026, LNCS 16588, pp. 441–456, 2026.
https://doi.org/10.1007/978-3-032-28691-8_29

problem (2), where $\varphi(x) = \min\{d^\top y : Dy \leq f - Cx,\ y \in K(q, m)\}$ models the optimal objective value of a y-solution for a given x-solution. Variable z then models the contribution of the y-variables to the objective and is linked to the x-variables by (2c).

$$\min_{\substack{x \in K(p,n)\\ z \in \mathbb{R}}} \quad c^\top x + z \tag{2a}$$
$$Ax \leq b \tag{2b}$$
$$\varphi(x) \leq z \tag{2c}$$

$$\min_{\substack{x \in K(p,n)\\ z \in \mathbb{R}}} \quad c^\top x + z \tag{3a}$$
$$Ax \leq b \tag{3b}$$
$$\alpha^\top x + \beta \leq z, \quad (\alpha, \beta) \in \mathcal{I} \tag{3c}$$

Since φ does not admit a closed algebraic description, (2) cannot be solved by standard techniques. Instead, (2c) is linearized by a family of linear constraints (3c), where $\mathcal{I}$ is some (usually finite) subset of $\mathbb{R}^n \times \mathbb{R}$. To solve the original MIP, BD removes (3c) from the master problem and adds, if needed, violated inequalities by calling a separation oracle. The original BD [4] showed how to implement a separation oracle for (3c) when $\varphi(x)$ is a linear program; logic-based BD extends these ideas to general MIPs [17,18]. In particular, when D has a block structure, each block can have an independent oracle, which can reduce the running time of each oracle substantially.

Once a separation oracle has been implemented, Problem (3) can be solved by standard branch-and-cut algorithms. BD therefore also benefits from sophisticated MIP techniques (presolving, cutting planes, etc.), which are readily available in MIP software, with one important exception: symmetry handling. The reason is that MIP software needs to have access to the full problem (3) to detect symmetries, whereas Inequalities (3c) are only available through their oracles. Many researchers [1,2,9,12] therefore add some symmetry handling or symmetry breaking inequalities by hand, but do not make use of advanced techniques such as orbital branching [27] or orbitopal fixing [5,10,19] to handle symmetries. This motivates the following two questions: (Q1) *Can we develop a framework for detecting symmetries in BD?* (Q2) *Can we exploit symmetries in the generation of Benders cuts?* Our main contributions are:

1. We present a practical framework for detecting symmetries in BD by using the recently introduced concept of symmetry detection graphs (SDG) [15] to provide the master problem (3) symmetry information of the Benders oracles. Detecting automorphisms of SDGs then allows to compute symmetries of (3) and to apply established symmetry handling methods. (Sect. 2)
2. We leverage symmetry information to enhance the separation of Benders cuts: Let π be a symmetry of (3). If $\alpha^\top x + \beta \leq z$ is a Benders cut (3c), Sect. 3 shows that also $\pi^{-1}(\alpha)^\top x + \beta \leq z$ is a Benders cut. A single Benders cut thus gives access to an entire family of symmetric Benders cuts. For actions of the symmetric group, which arise in many applications, we show that this family of symmetric cuts can be separated very efficiently, thus avoiding to call

potentially expensive separation oracles. Moreover, by adding linearly many auxiliary variables and constraints we can expresses all equivalent Benders cuts at once. That is, separating symmetric cuts becomes obsolete once one member of the family is known. (Sects. 3 and 4)

3. In a numerical study, we demonstrate that exploiting symmetry in BD can reduce the running time of BD by orders of magnitude. (Sect. 5)

2 Benders Symmetry

In this section, we introduce our framework for detecting symmetries in BD. We start by defining our notion of symmetries. For a set A, let $\mathcal{S}_A$ be the set of all permutations of A, the so-called *symmetric group* of A. If $A = [n] := \{1, \dots, n\}$ for some positive integer n, we write $\mathcal{S}_n$ instead of $\mathcal{S}_{[n]}$. For $\pi \in \mathcal{S}_n$ and $x \in \mathbb{R}^n$, let $\pi(x) = (x_{\pi^{-1}(1)}, \dots, x_{\pi^{-1}(n)})$, i.e., π acts on x by permuting its coordinates. Following the MIP literature [22,25], a *symmetry* of $\min\{c^\top x : Ax \leq b, x \in K(p, n)\}$ is a permutation $\pi \in \mathcal{S}_n$ such that $c^\top x = c^\top \pi(x)$ for all $x \in \mathbb{R}^n$, and x is feasible if and only if $\pi(x)$ is feasible. As deciding if $\pi \in \mathcal{S}_n$ is a symmetry is NP-hard [25], one usually considers symmetries that keep a specific MIP formulation invariant. Assuming A has m rows, a permutation $\pi \in \mathcal{S}_n$ is a *formulation symmetry* if there is $\rho \in \mathcal{S}_m$ such that $\pi^{-1}(c) = c$, $\rho(b) = b$, and $A_{\rho(i),\pi^{-1}(j)} = A_{i,j}$ for all $(i, j) \in [n] \times [m]$.

To detect formulation symmetries, one usually translates the problem into the language of graph automorphisms. To this end, one constructs a colored graph $G = (V, E)$ with the following properties: (i) there exists an injective map of the MIP's variables to a subset V' of V, the *variable nodes*; (ii) for every color-preserving automorphism $\pi \colon V \to V$ of G, the restriction of π to V' corresponds to a symmetry of the MIP. We refer to such a graph as a *symmetry detection graph (SDG)*. For MIPs, [31] suggests a concrete construction of an SDG by introducing a node for each constraint i and variable j. These nodes are connected by an edge whose color corresponds to $A_{i,j}$. The constraint nodes i are assigned a color corresponding to their right-hand side b_i. Similarly, the variable nodes j are assigned a color encoding their objective coefficient c_j and their type (integer or continuous). In the context of BD, however, this method is not applicable as the list of constraints (3c) if often too large or not explicitly given.

Recently, [15] introduced a framework for constructing an SDG for a problem by combing SDGs of its building blocks. A central concept for this construction are anchors. An SDG $G = (V, E)$ with variable nodes V' is called *anchored*, if there exists a node $a \in V \setminus V'$, the *anchor*, such that, for every $v \in V$, there exists an a-v-path p in G such that no interior point of p belongs to V'.

Theorem 1 (cf. [15]). *Let $P = \min\{c^\top x : Ax \leq b,\ Cx \leq d,\ x \in K(p, n)\}$ be a MIP. Let G_1 and G_2 be anchored SDGs of $\min\{c^\top x : Ax \leq b,\ x \in K(p, n)\}$ and $\min\{c^\top x : Cx \leq d,\ x \in K(p, n)\}$, respectively. Then, an SDG for P is given by the disjoint union of G_1 and G_2, and identifying the variable nodes of G_1 and G_2 with each other.*

One can thus find an SDG for (2) by merging anchored SDGs for (2b) and (2c) (or (3b) and (3c)). This theorem forms the core of our framework for detecting symmetries in BD, which we illustrate for two different cases of BD.

Classical BD. For illustration purposes, consider the linear program (LP)

$$\begin{aligned} \min\ & 5x_1+5x_2+5y_1+5y_2+3z_1+3z_2 \\ & x_1 + x_2 = 1, \quad (4a) \\ & x_1 + y_1 + 2z_1 = 2, \quad (4b) \\ & x_2 + y_2 + 2z_2 = 2. \quad (4c) \end{aligned}$$

If we keep only the x-variables in the master problem, the master problem is given by $\min\{5x_1 + 5x_2 + w_1 + w_2 : x_1 + x_2 = 1,\ \varphi(x_1) \le w_1,\ \varphi(x_2) \le w_2\}$, where $\varphi(x_i) = \min\{5y_i + 3z_i : y_i + 2z_i = 2 - x_i\}$ and $i \in \{1,2\}$. By Theorem 1, an SDG for the BD master problem is given by merging anchored SDGs for $\min\{5x_1+5x_2+w_1+w_2 : x_1+x_2 = 1\}$ and $\min\{5x_1+5x_2+w_1+w_2 : \varphi(x_i) \le w_i\}$ for $i \in \{1,2\}$, see Fig. 1 for an illustration. Here, the dashed arcs correspond to the identification of variable nodes and anchors. Since (3c) is derived from (2c), the SDG constructed this way is also an SDG for (3).

Of course, this construction can be generalized to the classical BD framework, in which the cut generation subproblem is an LP: the SDG for (2c) can be chosen as the anchored SDG for the subproblem (using the construction from [31]).

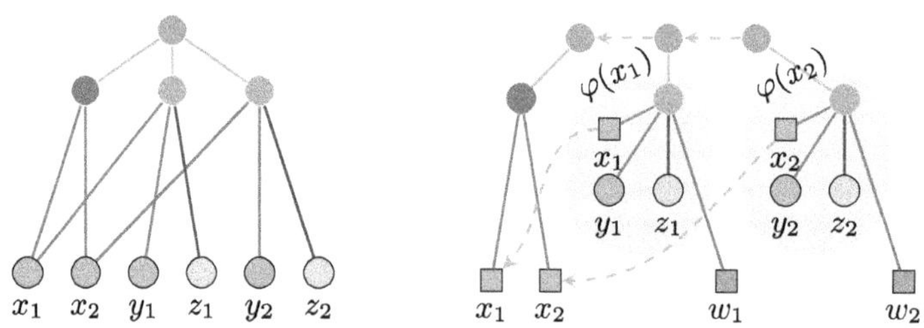

Fig. 1. Anchored symmetry detection graph for both formulations of (4). On the left, SDG of the original formulation. On the right, example on how the SDG of the BD can be recovered. We can join The SDG of the subproblems $\varphi(x_i)$, in the pink areas, to the master problem by identifying the nodes x_i. Variables of the master problem are depicted as squares nodes. (Color figure online)

No-Good Cuts. Theorem 1 can also be used to find an SDG for a Benders master problem if the subproblem is not an LP. Note that we then refer to them as *Logic Based Benders Decompositions* (LBBD) [18]. These often appear for example in assignment problems, where we are given a set N of items and a set M of slots, and the task is to assign each item to exactly one slot while minimizing some cost $c \in \mathbb{R}^{M \times N}$. The Benders master problem with cuts (3c) can then be formulated as

$$\min_{x \in \{0,1\}^{M \times N}} c^\top x : \quad \sum_{i \in M} x_{ij} = 1, \qquad j \in N, \tag{5a}$$

$$\sum_{j \in C} x_{ij} \le |C| - 1, \qquad (C, i) \in \mathcal{I}. \tag{5b}$$

The variable x_{ij} indicates whether item j is assigned to slot i, and (5a) ensures that each item j is assigned to exactly one slot. Constraint (5b) contains an inequality for every tuple (C, i), where $C \subseteq N$ is a set of items that cannot simultaneously be assigned to slot $i \in M$. These inequalities are called *no-good cuts*. Depending on the application, the set $\mathcal{I}$ is defined differently.

One example of the assignment problem is the Multiple Knapsack problem (MKP). Here, each item $j \in N$ has a weight a_j and each slot $i \in M$ has a capacity β_i. Given a cost c_{ij} for assigning item j to slot i, the task is to solve

$$\min_{x \in \{0,1\}^{M \times N}} \Big\{ c^\top x : \sum_{i \in M} x_{ij} = 1 \text{ for all } j \in N \text{ and } \sum_{j \in N} a_j x_{ij} \le \beta_i \text{ for all } i \in M \Big\}.$$

In an LBBD (5) of the MKP, one initially discards all no-good cuts, i.e., $\mathcal{I} = \emptyset$. The no-good cuts are then generated by repeatedly solving (5) and checking if the obtained solution $\hat{x}$ corresponds to a solution of the MKP. That is, the Benders oracles test if the sum of weights $\sum_{j \in N} a_j \hat{x}_{ij}$ exceeds the capacity β_i for some $i \in M$. In this case, the no-good cut (C, i) for $C = \{j \in N : \hat{x}_{ij} = 1\}$ is added to $\mathcal{I}$; otherwise, the LBBD terminates and returns $\hat{x}$ as optimal solution.

As the Benders subproblem decomposes into the feasibility problems $P_i(x) = \min\{0 : \sum_{j \in N} a_j x_j \le \beta_i\}$ for $i \in M$, an SDG for (5) is given by merging an SDG for $\min\{c^\top x : x \text{ satisfies (5a)}, x \in \{0,1\}^{M \times N}\}$ and SDGs for $P_i(x)$, $i \in M$. The SDG for $P_i(x)$ has a single constraint node connected to all variables nodes for x_{ij}, $j \in N$, with edge weight a_j.

3 Symmetry Exploitation

Once symmetries of (3) are known, a solver can make use of built-in symmetry handling methods. Next to solving (3) by branch-and-bound, a computational bottleneck is the separation of Benders cuts (3c). We therefore present two ways to exploit symmetries to generate Benders cuts. The main idea is that, whenever $\pi \in \mathcal{S}_n$ is a symmetry of (3) and $(\alpha, \beta) \in \mathcal{I}$ is a Benders cut, $\pi^{-1}(\alpha)^\top x + \beta \le z$ is also a Benders cut. Indeed, for every feasible (x, z) of (3) and symmetry π, we have $z \ge \alpha^\top \pi(x) + \beta = \pi^{-1}(\alpha)^\top x + \beta$, where the last equality holds since permutations are orthogonal maps. Every Benders cut (α, β) thus gives rise to an entire family of symmetric Benders cuts $\{(\pi^{-1}(\alpha), \beta) : \pi \text{ symmetry of (3)}\}$. One can thus potentially avoid solving an expensive separation problem for (3c) by first separating symmetric Benders cuts of already separated cuts.

In general, this problem is difficult, but for the MKP, separating symmetric Benders cuts is easy by Proposition 1 below: items j with identical weight a_j form groups $\mathcal{G}$ of variables that can be exchanged arbitrarily, i.e., $\mathcal{S}_G$ acts on group $G \in \mathcal{G}$ of identical items. These symmetries arise in many applications and separating symmetric Benders cuts can greatly improve the performance of BD, see Sect. 5. For some applications, this idea has been applied before [8,13], but it has not been described from a general perspective.

Proposition 1. *Let Π be a symmetry group of (3) and $(\alpha, \beta) \in \mathcal{I}$ a Benders cut. If Π is isomorphic to the Cartesian product $\bigotimes_{G\in\mathcal{G}} \mathcal{S}_G$ for some partition $\mathcal{G}$ of $[n]$, then $\{(\pi^{-1}(\alpha), \beta) : \pi \in \Pi\} \subseteq \mathcal{I}$ can be separated in $O(n\log(n))$ time.*

We refer to $\{(\pi^{-1}(\alpha), \beta) : \pi \in \Pi\}$ as a *symmetric cut pool* derived from (α, β).

Proof. Let $(\hat{x}, \hat{z}) \in \mathbb{R}^{n+1}$. Solving the separation problem is equivalent to deciding if $\max_{\pi\in\Pi} \pi^{-1}(\alpha)^\top \hat{x} > \hat{z} - \beta$. Since Π is isomorphic to $\bigotimes_{G\in\mathcal{G}} \mathcal{S}_G$, finding a maximizer π is the same as finding a permutation that matches the k-th largest value of $\{\alpha_j : j \in G\}$ to the k-th largest value of $\{\hat{x}_j : j \in G\}$ for all $G \in \mathcal{G}$, $k \in [|G|]$, which can be done by sorting both α and $\hat{x}$ in $\mathcal{O}(n\log(n))$ time. □

Note that the requirement of Π being a product of full symmetric groups is, in practice, not very restrictive, because these groups arise very frequently [28].

Proposition 1 allows to possibly accelerate the separation of (3c). A natural question is if the separation of symmetric Benders cuts can be avoided at all. If all x-variables are binary and the symmetries have the structure as in Proposition 1, this question can be answered affirmatively. Extending ideas of [30], we introduce for each group $G \in \mathcal{G}$ variables $\zeta_G^k \in \{0, 1\}$ and constraints that enforce $\zeta_G^k = 1$ if and only if $\sum_{j\in G} x_j \geq k$, enforced via the two following constraint sets:

$$\sum_{j\in G} x_j \leq k - 1 + (|G| - (k-1))\zeta_G^k, \qquad G \in \mathcal{G},\ k \in [|G|], \tag{6}$$

$$k \cdot \zeta_G^k \leq \sum_{j\in G} x_j, \qquad G \in \mathcal{G},\ k \in [|G|]. \tag{7}$$

The variables $\zeta_G^1, \ldots, \zeta_G^{|G|}$ can thus be considered a non-increasing reordering of $\{x_j : j \in G\}$. By denoting the k-th largest value in $\{\alpha_j : j \in G\}$ by α_G^k, the most violated cut constructed in the proof of Proposition 1 can be phrased in terms of the ζ-variables as

$$\sum_{G\in\mathcal{G}} \sum_{k=1}^{|G|} \alpha_G^k \zeta_G^k + \beta \leq z. \tag{8}$$

We thus immediately obtain the following theorem.

Theorem 2. *Assume all x-variables in (3) are binary. Let $(\alpha, \beta) \in \mathcal{I}$ and symmetry group Π adhere to the conditions in Proposition 1. Then, $(\hat{x}, \hat{z}) \in \{0, 1\}^n \times \mathbb{R}$ satisfies all inequalities in $\{(\pi^{-1}(\alpha), \beta) : \pi \in \Pi\}$ if and only if $(\hat{x}, \hat{\zeta}, \hat{z})$ in the extended model satisfies (8).*

Theorem 2 shows that the extended model (6)–(8) allows to avoid the separation of symmetric Benders cuts. The LP relaxation of the extended model is, in general, weaker than the one of (3) though. Indeed, one can show that every feasible $(x, z) \in [0,1]^{n+1}$ induces a feasible $(x, z, \zeta) \in [0,1]^{2n+1}$. The next example, however, shows that there might exist some (x, z, ζ) with projection to (x, z) not always satisfying $\alpha^\top x + \beta \le z$.

Example 1. Let $n = 3$ and denote the variable $x = (x_1, y_1, y_2)$ with both y-entries symmetric. The no-good cuts $x_1 + y_1 \le 1$ and $x_1 + y_2 \le 1$ are represented by $\zeta_x^1 + \zeta_y^1 \le 1$. However, the solution (x, ζ) with $(x_1, y_1, y_2) = (1 - \varepsilon, 2\varepsilon, 0)$ and $(\zeta_x^1, \zeta_y^1, \zeta_y^2) = (1 - \varepsilon, \varepsilon, \varepsilon)$ is feasible in the relaxation of the extended formulation for any $\varepsilon \in [0,1]$ but $x_1 + y_1 \le 1$ is not satisfied if $\varepsilon > 0$.

Note that another, arguably more natural extended formulation could have used integer variables $\xi_G \in \mathbb{Z}_{\ge 0}$ with $\xi_G = \sum_{j \in G} x_j$. It is however unclear how they can be applied in a cut. In fact, since the convex hull of all feasible (x, ξ) can contain infeasible integer points (see Fig. 2), they would likely require additional auxiliary variables which would then be equivalent to the ζ variable.

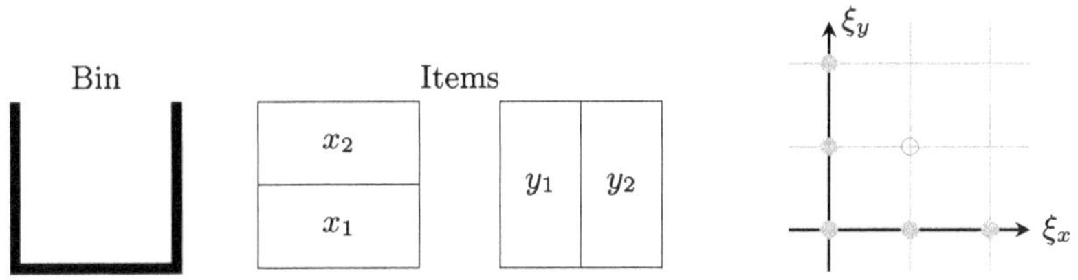

Fig. 2. Example of rectangle packing (left) where an infeasible solution for the ξ variable (right) cannot be separated. In particular, it is possible to fit up to two rectangles of the same group in the square bin, represented by the green dots. However, mixing is impossible without rotating the items (i.e. $(1, 1)$ is not a feasible point).

Application to No-Good Cuts. We illustrate both techniques for the MKP. Let $\mathcal{G}$ be a partition of the set of items N such that, for every $G \in \mathcal{G}$ and $j_1, j_2 \in G$, we have both $a_{j_1} = a_{j_2}$ and $c_{i,j_1} = c_{i,j_2}$ for all $i \in M$, i.e., items j_1 and j_2 have the same weight and objective value. Then, a symmetry group of the MKP is $\Pi = \bigotimes_{G \in \mathcal{G}} \mathcal{S}_G$, where $\pi \in \Pi$ acts on a solution x of (5) by permuting the indices of items, i.e., $\pi(x)_{i,j} = x_{i,\pi^{-1}(j)}$. Note that Prop. 1 is not directly applicable, because this action of Π reorders entire columns of the solution matrix x rather than individual entries. Nevertheless, the arguments of Prop. 1 apply analogously since the Benders cuts (5b) only contain variables from a single row i of x. In general, separating an entire family of symmetric cuts reduces to $\max\left\{\alpha^\top \pi(\hat{x}) : \pi \in \Pi\right\}$ where knowledge on the structure of both Π and α can help, but more general groups are out of the scope of this work.

Let $(C, i) \in \mathcal{I}$ be an index of a cut from (5b). Then, the family of symmetric cuts is $\{(\pi^{-1}(C), i) : \pi \in \Pi\}$, and it is uniquely determined by the number of elements selected per group G. We can thus characterize the entire family by a representative vector $R(C) \in \mathbb{Z}_+^{\mathcal{G}}$, where $R_G(C) = |C \cap G|$ for all $G \in \mathcal{G}$. The cuts of the family derived from (C, i) are then given by $\sum_{G \in \mathcal{G}} \sum_{j \in G \cap C'} x_{ij} \leq |C'| - 1$ for all $C' \subseteq N$ with $R(C') = R(C)$.

In practice, to find a violated cut for a solution $\hat{x}$, one can iterate through all $i \in M$, compute $N_i = \{j \in N : \hat{x}_{ij} = 1\}$, and check if there is (C, i) in the list of known cuts with $R(N_i) = R(C)$. If such a pair exists, we can immediately derive a symmetric Benders cut separating $\hat{x}$ without solving the subproblem. This idea can be enhanced by observing that, if $(C, i) \in \mathcal{I}$, also $(C', i) \in \mathcal{I}$ for all C' with $C \subseteq C' \subseteq N$. Hence, instead of checking $R(N_i) = R(C)$, we can check whether there is $(C, i) \in \mathcal{I}$ with $R(N_i) \geq R(C)$ to find a violated symmetric cut.

By introducing ζ-variables for each slot $i \in M$, Inequality (8) for a no-good cut (C, i) can be easily phrased as $\sum_{G \in \mathcal{G}} \sum_{k=1}^{|G \cap C|} \zeta_{iG}^k \leq |C| - 1$. Since this inequality is violated if and only if $\zeta_{iG}^{|G \cap C|} = 1$ for all $G \in \mathcal{G}$ (recall $\zeta_{iG}^k = 1$ if and only if $\sum_{j \in G} x_{ij} \geq k$), the inequality can be simplified to $\sum_{G \in \mathcal{G}} \zeta_{iG}^{|C \cap G|} \leq |\mathcal{G}| - 1$.

4 Applications

We applied our techniques to three problems with varying difficulty of the cut generation subproblem: sphere packing (SP), rectangle packing (RP), and machine scheduling (MS). In the following, we discuss how the ideas explained in Sects. 2 and 3 can be realized for these examples.

2D Bin Packing. Given a set of items N and bins B, the bin packing problem is to find the least number of bins that can hold all the items. We assume that all bins $i \in B$ are rectangles of width W_i and height H_i, and consider two different types of items: spheres and rectangles. For both variants, we use the same master problem (9). Note that, as this is an assignment problem, its formulation stems from the generic one in (5).

$$
\begin{aligned}
\min \sum_{i \in B} u_i & & & \\
\sum_{i \in B} x_{ij} &= 1, & j &\in N, & \text{(9a)}\\
\sum_{j \in C} x_{ij} &\leq |C| - 1, & (C, i) &\in \mathcal{I}, & \text{(9b)}\\
\sum_{j \in N} a_j x_{ij} &\leq W_i H_i u_i, & i &\in B, & \text{(9c)}\\
x_{ij} &\leq u_i, & j &\in N,\ i \in B. & \text{(9d)}
\end{aligned}
$$

In addition to the constraints from (5), the master problem adds auxiliary variables $u_i \in \{0, 1\}$, $i \in B$, indicating whether the corresponding bin i has assigned items. The coefficients a_j, $j \in N$, correspond to the area of item j. Thus, Equation (9c) strengthens the master problem (5) by excluding all solutions that assign a single bin $i \in B$ items whose total area exceeds the bin's area $W_i H_i$.

Given a solution $\hat{x}$, the Benders subproblem is to decide, for each bin $i \in B$, if the items $\{j \in N : \hat{x}_{ij} = 1\}$ fit into the bin. For SP, this problem can be solved by deciding feasibility of the following system, cf. [20]:

$$(y_j^{(1)} - y_k^{(2)})^2 + (y_j^{(2)} - y_k^{(2)})^2 \geq (\hat{x}_{ij} + \hat{x}_{ik} - 1)(r_j + r_k)^2, \quad j, k \in N,\ j \neq k, \tag{10a}$$

$$r_j \leq y_j^{(1)} \leq W_i - r_j, \qquad j \in N, \tag{10b}$$

$$r_j \leq y_j^{(2)} \leq H_i - r_j, \qquad j \in N. \tag{10c}$$

The variable pair $(y_j^{(1)}, y_j^{(2)})$ acts as the position of the center of sphere j in the bin. Observe that Equation (10a) enforces that two spheres do not overlap only if they are both assigned to bin i ($\hat{x}_{ij} = \hat{x}_{ik} = 1$). To build an SDG for Subproblem (10), one can either use rules for general MINLPs, see [22], or capture the symmetries by a more compact tailored graph. To this end, recall that the SDG only needs to capture the symmetries of the x-variables, because the $y^{(1)}$- and $y^{(2)}$-variables are not present in the master problem. More precisely, we define the node set

$$V = \{a\} \cup \bigcup_{j \in N} \{v_j, r_j\},$$

where a is the anchor of the graph, v_j is a variable node representing variable x_{ij}, and r_j represents the radius of sphere j. The anchor a receives a color that uniquely characterizes the dimensions (W_i, H_i) of the respective bin i, the variable nodes are colored according to their objective coefficient and type in the master problem, and the node r_j receives a unique color that indicates it as radius-node. Moreover, we define the edge set

$$E = \bigcup_{j \in N} \{\{a, r_j\}, \{r_j, v_j\}\},$$

where $\{a, r_j\}$ receives a color corresponding to the radius r_j, and edge $\{r_j, v_j\}$ remains uncolored. Due to this construction, an automorphism of G can only exchange variable nodes v_j and $v_{j'}$ if $r_j = r_{j'}$. Thus, the SDG G captures the symmetries of the subproblem.

The subproblem of RP can be modeled as a MIP, cf. [29], and an SDG can either be derived from this MIP, or via a similar construction as for SP.

Machine Scheduling With Setup Times. We consider a machine scheduling problem with setup times between different jobs. We are given a set of machines M, set of jobs N, processing time p_{ij} for job j on machine i, and setup times s_{ijk} for processing job k directly after job j on machine i that satisfy the triangle inequality $s_{ijk} + s_{ikl} \geq s_{ijl}$. The objective is to find the minimum required time to have all jobs processed, called the *makespan*, given that each machine can only process one job at a time. Using the LBBD approach of [34], we add to (5) a variable $T \geq 0$ representing the makespan, and set the objective to minimize it. To link T with the remaining variables, [34] introduces Benders cuts

$$T \geq T_i(C) - \sum_{j \in C} (1 - x_{ij})\theta_{ij}, \quad (C, i) \in \mathcal{I}, \tag{11}$$

where $C \subseteq N$, $i \in M$, and $T_i(C)$ denotes the minimum makespan for processing all jobs in C on machine i. The coefficients $\theta_{ij} := p_{ij} + \max\{s_{ikj} : k \in C\}$ represent an upper bound on the time saved when not assigning job j to that machine. Next to (11), they strengthen the model by introducing auxiliary continuous variables to the master problem. Since the details are not relevant for the following discussion, we refer to [34] for details.

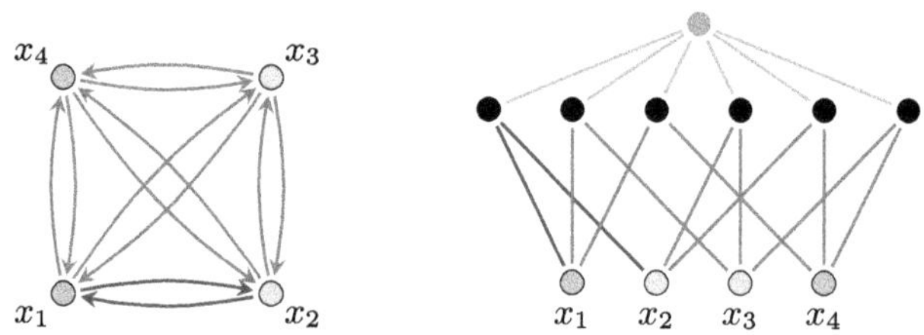

Fig. 3. Example of compact SDG for TSP. Left: exemplary instance of the MS subproblem with four selected jobs; node colors represent processing times, and arc colors the setup times. Right: corresponding SDG. One only needs to add a dummy node for each job pair and the grey anchor node, where $x_1, \ldots, x_4$ are variable nodes.

Given a solution $\hat{x} \in \{0,1\}^{M \times N}$, the subproblem is to determine, for all $i \in M$, the value $T_i(N_i)$, where $N_i = \{j \in N : \hat{x}_{ij} = 1\}$. Note that for any subset of jobs C, $T_i(C) = \sum_{j \in C} p_{ij} + S_i(C)$, where $S_i(C)$ is the minimum total setup time of jobs in C on machine i. Since the setup times depend on the ordering, finding $S_i(C)$ can be reduced to solving a traveling salesperson problem (TSP), see [34]. Note that the classical subtour elimination formulation of the TSP has exponentially many constraints. To detect symmetries, standard approaches list all these constraints explicitly, which is impractical. Theorem 1, however, allows to define a compact SDG for the subproblem, see Fig. 3 for an illustration.

Cut Pool and Extended Formulation. Since solving the TSP subproblem can be costly, the symmetric cut pool can potentially save a lot of time. When scanning the cut pool, we need to decide if the set of jobs C used in a previously derived cut (C, i) is symmetric to the current assignment $N_i = \{j \in N : \hat{x}_{ij} = 1\}$, $i \in M$. To this end, we define for each machine $i \in M$ an equivalence relation $\sim_i$ defining the groups of identical jobs. We say that two jobs $j_1, j_2 \in N$ satisfy

$$j_1 \sim_i j_2 \iff \begin{cases} p_{ij_1} = p_{ij_2},\ s_{ij_1j_2} = s_{ij_2j_1} \\ s_{ij_1k} = s_{ij_2k}, \text{ for all } k \in N \setminus \{j_1, j_2\}, \\ s_{ikj_1} = s_{ikj_2}, \text{ for all } k \in N \setminus \{j_1, j_2\}. \end{cases}$$

This relation induces job groups $G \in \mathcal{G}_i$ that depend on machine $i \in M$, and again, we can define a representative vector $R(N_i) \in \mathbb{Z}_+^{\mathcal{G}_i}$, $R_G(N_i) = |N_i \cap G|$, to decide whether a symmetric violated cut exists in the pool. Moreover, by introducing ζ-variables, all cuts symmetric to (11), can be represented by

$$T \geq T_i(C) - \sum_{G \in \mathcal{G}_i} \sum_{h=1}^{|G \cap C|} (1 - \zeta_{iG}^h)\theta_{iG}, \quad (C, i) \in \mathcal{I} \tag{12}$$

where $\theta_{iG} := p_{iG} + \max\{s_{iHG} : H \in \mathcal{G}_i, \ H \cap C \neq \emptyset\}$, $p_{iG} = p_{ij}$ if $j \in G$, $s_{iGH} = s_{ijk}$ if $j \in G$ and $k \in H$ for groups $G, H \in \mathcal{G}_i$.

5 Numerical Results

Section 3 has introduced different means to exploit symmetry in BD. This section's goal is to empirically evaluate the impact of these methods. We are particularly interested in the impact of state-of-the-art symmetry handling methods when applied within BD, and if the cut pool or extended formulation has a stronger impact on the performance of BD. The former has not been investigated yet because symmetry detection for BD was not available; regarding the latter, although cut pools and extended formulations have been discussed before [8,13,30], to the best of our knowledge, they have not been compared with each other.

Implementation Details. We have implemented LBBD schemes for the RP, SP, and MS problem using the solver SCIP [6]. For each problem, we implemented a so-called constraint handler (CH) to separate Benders cuts. To enable automatic symmetry detection for BD in SCIP, we implemented callbacks of the respective CHs that add symmetry information of the Benders oracles to SCIP's SDG. Our source code is available at GitHub[1] and zenodo [16].

Symmetric cut pools are implemented by maintaining a list of abstract Benders cuts that have already been separated. An *abstract cut* does not make use of explicit variables (items/jobs in our applications), but only counts how many variables of which symmetry class $G \in \mathcal{G}$ have been involved in the cut via representative vectors $R(C)$. Abstract cuts thus correspond to families of cuts. Whenever a solution needs to be separated, we iterate through the abstract cuts in our list and for each check whether a member of its family of cuts is violated. Only when no violated cut is found in the list, the separation oracles are triggered to possibly generate a new cut, which is then stored in our list of cuts.

Extended formulations are implemented by adding the auxiliary ζ-variables to the initial Benders master problem. The separation of cuts then follows the same framework as for the cut pool, but cuts are added in terms of the ζ-variables. For both the cut pool and extended formulation, we only separate integer solutions to avoid too many calls of the separation oracles.

[1] https://github.com/Cedric-Roy/Benders-Symmetry-Supplements.

Settings. We compare the performance of a default implementation of LBBD, not exploiting symmetry information (setting `plain`) with LBBD making use of symmetric cut pools (`pool`) and extended formulations. For extended formulations, we investigate the effect of adding Benders cuts as constraints to the problem (`EFcons`) or as rows to the LP relaxation (`EFrow`). Constraints remain in the problem, whereas rows can be removed again. Using rows, the LP relaxation has potentially less constraints and a weaker bound, but can be solved potentially faster. We also investigate the effect of enabling/disabling SCIP's built-in default symmetry handling methods.

All experiments use a time limit of 2 h. Mean numbers are reported in shifted geometric mean $\prod_{i=1}^{n}(t_i + s)^{1/n} - s$ for numbers $t_1, \dots, t_n$. We use $s = 10$ for nodes of the branch-and-bound tree, and $s = 1$ for all remaining quantities.

Hardware and Software Specifications. All experiments have been conducted on a Linux Cluster with Intel Xeon E5-1620 v4 3.5 GHz quad core processors and 32 GB memory. The code was executed single-threaded with a time limit of 2 h. We use SCIP 9.2.3 [6] as branch-and-bound framework; LP relaxations are solved using SoPlex 7.1.5 and nonlinear problems are solved by Ipopt 3.14.20 [36]. Symmetries of SDGs are detected by sassy 1.1 [3] and Nauty 2.8.8 [26].

Numerical Results. We have randomly generated 80 RP, 80 SP, and 120 MS instances with many symmetries, see [16]. The results are summarized in Table 1, where column "#solved" reports on the number of solved instances, "time" and "cut-time" give the mean solving time and time needed to separate Benders cuts, respectively; "#sepa" is the mean number of solutions that have been separated.

Table 1. Statistics on different approaches for solving RP, SP, and MS problems.

setting	with symmetry handling				without symmetry handling			
	#solved	time	cut-time	#sepa	#solved	time	cut-time	#sepa
RP								
plain	24	2678.0	2613.6	894.4	24	2704.8	2642.1	1022.0
pool	35	1159.9	395.0	2600.1	26	1960.4	498.5	6981.4
EFrow	42	1059.5	935.7	979.1	34	1451.0	1286.6	1480.8
EFcons	59	232.9	118.4	103.7	41	793.3	233.7	199.0
SP								
plain	46	287.4	287.1	8.8	42	357.9	357.8	11.6
pool	57	109.2	109.0	10.1	55	127.8	113.3	24.3
EFrow	55	202.7	202.4	7.1	46	250.0	249.8	9.0
EFcons	55	173.4	173.1	5.7	53	187.4	187.0	6.6
MS								
plain	7	6801.9	6274.9	230070.3	8	6790.8	6263.8	229783.5
pool	45	3318.7	29.7	2018389.5	46	3286.2	30.2	2039326.9
EFrow	84	561.7	6.4	143804.3	83	588.6	6.8	158353.7
EFcons	120	29.9	1.0	126.7	119	30.5	1.1	145.5

In the introduction, we have posed two questions, which we now answer in turn. Question (Q1) concerned whether symmetries can be automatically detected when using BD. Using the framework of SDGs, this is indeed possible and Table 1 shows that we can greatly benefit from handling symmetries in BD when using SCIP's default symmetry handling methods. For the RP problem, we can solve considerably more instances when handling symmetries and substantially reduce the running time (between 27 % and 71 % for `pool`, `EFrow`, and `EFcons`). Using `plain`, however, the effect is moderate. This indicates that symmetry handling only becomes effective when it is combined with exploiting symmetries in generating Benders cuts.

The effect of symmetry handling for the SP problem is less pronounced with running time improvements between 8 % and 19 %. To explain this behavior, we compared the size of the branch-and-bound trees for `pool`, `EFrow`, and `EFcons` for the instances that could be solved by all three methods within the time limit. The mean number of nodes in these trees is 11.5–16.5, i.e., the effect of symmetry handling is minor due to the rather small trees.

Finally, symmetry handling seems to have almost no effect on the MS problem, which can be explained by the symmetry handling methods used by SCIP. Since all master problem variables for the RP and SP problems are binary, SCIP handles symmetries by the methods orbital fixing [27,28] and lexicographic reduction [10]. The MS problem, however, has predominantly continuous variables in the master problem, and SCIP decides that it is more important to handle symmetries on the continuous variables using SST cuts [23,32]. SST cuts handle only a small amount of symmetries and their structure is comparable to inequalities that one would add by hand. We therefore also tested whether the more sophisticated techniques lexicographic reduction and orbital reduction [10] yield a performance improvement for MS problems. Indeed, using these techniques, `plain` solves 61 more instances and reduces the running time by more than 92 %, and also `pool`, `EFrow`, and `EFcons` substantially improve their performance. In particular, the most competitive setting `EFcons` reduces its running time by another 78 %. We thus conclude that automatic symmetry detection and handling in BD is very important, because it allows solvers to make use of sophisticated (already built-in) symmetry handling techniques that users can not easily implement on their own.

Question (Q2) asked whether the approach using a cut pool or extended formulation is better suited to exploit symmetries in BD. In the following discussion, we focus on the results for BD that handles symmetries of the master problem. From Table 1 it is apparent that both approaches substantially improve `plain` BD, reducing the mean running times for RP, SP, and MS up to 91 %, 62 %, and 99 %, respectively. In particular, while `plain` could only solve 7 of the 120 MS instances, `EFcons` solves all instances.

For the RP and MS problem, `EFcons` clearly dominates `pool` and `EFrow` when separating solutions. A possible explanation is that adding Benders cuts as constraints in an extended formulation guarantees that no symmetric solutions can be computed anymore. The search space for `EFcons` is thus considerably smaller than for `EFrow` and `pool`. This hypothesis can be confirmed by comparing the number of nodes in the branch-and-bound trees of the instances that are solved by all three settings: For the three test sets, the mean trees for `EFcons` are 30–95 % smaller than for `EFrow` (resp. 22–99 % for `pool`).

For SP problems, however, `EFcons` is considerably slower than `pool`, which has two explanations. On the one hand, many of our instances admit an optimal solution that uses two or three bins. That is, a few Benders cuts are sufficient to achieve a matching dual bound. On the other hand, we observed that the time needed by `EFcons` to find an optimal solution is much higher than for `pool`. We explain the latter by the expensive separation problem of Benders cuts, which requires to solve a sphere packing problem. Since, as argued above, `EFcons` will never encounter symmetric infeasible solutions, every separation problem is time consuming. Setting `EFrow` and `pool`, however, can explore symmetric parts of the B&B tree, which arguably increases the chance that heuristics find good feasible solution, while symmetric solutions can be separated easily by making use of a symmetric cut pool. Table 1 confirms that the mean time needed to separate a cut using `pool` and `EFrow` is much smaller than for `EFcons` when comparing the time spent for cut separation and the number of separated solutions.

To answer (Q2), the results show that problems in which a small number of Benders cuts suffices to derive a good dual bound benefit from `pool`; problems that require a lot of Benders cuts to prove optimality benefit from `EFcons` (e.g., `EFcons` reduces the number of separated solutions for MS from more than two millions to less than 200). This suggests that a hybrid strategy could be beneficial that first uses the `pool` approach to explore the branch-and-bound tree to find good solutions, and then switches to `EFcons` to keep the tree small.

Conclusion. We have addressed the topic of detecting and exploiting symmetries in BD. While automatic symmetry handling became a powerful component of modern MIP solvers [28], BD cannot benefit from these techniques because MIP solvers are not aware of the symmetries of the cuts that are separated. We demonstrated that symmetry detection for BD is, in principle, possible by creating an SDG that captures the symmetries of Benders cuts, and that it can be easily implemented within the solver SCIP. This allowed us to leverage state-of-the-art symmetry handling methods for MIP to BD to achieve substantial performance improvements. Moreover, we systematically compared (to the best of our knowledge) for the first time the cut pool and extended formulation approach to enhance the separation of symmetric Benders cuts. Based on our numerical study, we have seen that neither approach is dominant and suggested a hybrid approach that aims to create synergies between these approaches.

These findings provide fundamental insights into how symmetries should be exploited in BD. This is particularly relevant for automatic BD schemes for generating Benders cuts that exist, e.g., for the solvers CPLEX [7], SAS [33], and SCIP [24]. Our long term goal is to incorporate our findings into such an automatic BD scheme, including a fully automatic symmetry detection algorithm for generic BD and different strategies for exploiting symmetries: next to the separation of Benders cuts, we aim to use symmetry information for strengthening Benders cuts by means of lifting and sparsification.

Acknowledgments. We want to thank three anonymous reviewers for their valuable feedback. This article is part of the project "Local Symmetries for Global Success" with project number OCENW.M.21.299, which is financed by the Dutch Research Council (NWO).

Disclosure of interest. The authors have no competing interest to declare that are relevant to the content of this article.

References

1. Adulyasak, Y., Cordeau, J.F., Jans, R.: Benders decomposition for production routing under demand uncertainty. Oper. Res. **63**(4), 851–867 (2015)
2. Akpinar, S., Elmi, A., Bektaş, T.: Combinatorial Benders cuts for assembly line balancing problems with setups. Eur. J. Oper. Res. **259**(2), 527–537 (2017)
3. Anders, M., Schweitzer, P., Stieß, J.: Engineering a preprocessor for symmetry detection (2023). https://doi.org/10.48550/arXiv.2302.06351
4. Benders, J.F.: Partitioning procedures for solving mixed-variables programming problems. Numer. Math. **4**, 238–252 (1962)
5. Bendotti, P., Fouilhoux, P., Rottner, C.: Orbitopal fixing for the full (sub-)orbitope and application to the unit commitment problem. Math. Program. **186**, 337–372 (2021). https://doi.org/10.1007/s10107-019-01457-1
6. Bolusani, S., et al.: The SCIP optimization suite 9.0 (2024)
7. Bonami, P., Salvagnin, D., Tramontani, A.: Implementing automatic benders decomposition in a modern MIP solver. In: Bienstock, D., Zambelli, G. (eds.) Integer Programming and Combinatorial Optimization, pp. 78–90 (2020)
8. Daryalal, M., Pouya, H., DeSantis, M.A.: Network migration problem: a hybrid logic-based Benders decomposition approach. INFORMS J. Comput. **35**(3), 593–613 (2023). https://doi.org/10.1287/ijoc.2023.1280
9. Delorme, M., Iori, M., Martello, S.: Logic based Benders' decomposition for orthogonal stock cutting problems. Comput. Oper. Res. **78**, 290–298 (2017). https://doi.org/10.1016/j.cor.2016.09.009
10. van Doornmalen, J., Hojny, C.: A unified framework for symmetry handling. Math. Program. **212**, 217–271 (2025)
11. Fazel-Zarandi, M.M., Beck, J.C.: Using logic-based Benders decomposition to solve the capacity- and distance-constrained plant location problem. INFORMS J. Comput. **24**(3), 387–398 (2012)
12. Guo, C., Bodur, M., Aleman, D.M., Urbach, D.R.: Logic-based Benders decomposition and binary decision diagram based approaches for stochastic distributed operating room scheduling. INFORMS J. Comput. **33**(4), 1551–1569 (2021). https://doi.org/10.1287/ijoc.2020.1036
13. Han, P., et al.: Integrated optimization of train makeup problem and resource scheduling in railway marshalling yards: a hybrid MILP-CP approach with logic-based Benders decomposition. Transp. Res. Part B: Methodol. **200**, 103306 (2025). https://doi.org/10.1016/j.trb.2025.103306
14. Heching, A., Hooker, J.N., Kimura, R.: A logic-based Benders approach to home healthcare delivery. Transp. Sci. **53**(2), 510–522 (2019)
15. Hojny, C.: Detecting and handling reflection symmetries in mixed-integer (nonlinear) programming and beyond. Math. Program. Comput. 1–48 (2025)

16. Hojny, C., Roy, C.: Supplementary material for the article "a framework for handling and exploiting symmetry in benders decomposition" (2026). https://doi.org/10.5281/zenodo.19068205
17. Hooker, J.: Logic-Based Benders Decomposition. Synthesis Lectures on Operations Research and Applications, 1st edn. Springer, Cham (2024)
18. Hooker, J.N., Ottosson, G.: Logic-based Benders decomposition. Math. Program. **96**(1), 33–60 (2003)
19. Kaibel, V., Peinhardt, M., Pfetsch, M.E.: Orbitopal fixing. Discret. Optim. **8**(4), 595–610 (2011). https://doi.org/10.1016/j.disopt.2011.07.001
20. Khajavirad, A.: The circle packing problem: a theoretical comparison of various convexification techniques. Oper. Res. Lett. **57**, 107197 (2024)
21. Lehmann, A., Kleiminger, W., Invernizzi, H., Gautschi, A.: Accelerated benders decomposition for variable-height transport packaging optimisation. arXiv preprint arXiv:2308.01104 (2023)
22. Liberti, L.: Reformulations in mathematical programming: automatic symmetry detection and exploitation. Math. Program. **131**(1), 273–304 (2012)
23. Liberti, L., Ostrowski, J.: Stabilizer-based symmetry breaking constraints for mathematical programs. J. Global Optim. **60**, 183–194 (2014)
24. Maher, S.J.: Implementing the branch-and-cut approach for a general purpose Benders' decomposition framework. Eur. J. Oper. Res. **290**(2), 479–498 (2021). https://doi.org/10.1016/j.ejor.2020.08.037
25. Margot, F.: Symmetry in integer linear programming. 50 years of integer programming 1958–2008: from the early years to the state-of-the-art, pp. 647–686 (2009)
26. McKay, B.D., Piperno, A.: Practical graph isomorphism, II. J. Symb. Comput. **60**, 94–112 (2014). https://doi.org/10.1016/j.jsc.2013.09.003
27. Ostrowski, J., Linderoth, J., Rossi, F., Smriglio, S.: Orbital branching. Math. Program. **126**(1), 147–178 (2011). https://doi.org/10.1007/s10107-009-0273-x
28. Pfetsch, M.E., Rehn, T.: A computational comparison of symmetry handling methods for mixed integer programs. Math. Program. Comput. **11**(1), 37–93 (2019). https://doi.org/10.1007/s12532-018-0140-y
29. Pisinger, D., Sigurd, M.: Using decomposition techniques and constraint programming for solving the two-dimensional bin-packing problem. INFORMS J. Comput. **19**(1), 36–51 (2007)
30. Riise, A., Mannino, C., Lamorgese, L.: Recursive logic-based Benders' decomposition for multi-mode outpatient scheduling. Eur. J. Oper. Res. **255**(3), 719–728 (2016)
31. Salvagnin, D.: A dominance procedure for integer programming. Master's thesis, University of Padova, Padova, Italy (2005)
32. Salvagnin, D.: Symmetry breaking inequalities from the Schreier-sims table. In: van Hoeve, W.-J. (ed.) CPAIOR 2018. LNCS, vol. 10848, pp. 521–529. Springer, Cham (2018). https://doi.org/10.1007/978-3-319-93031-2_37
33. SAS Institute Inc.: SAS Programming Documentation. https://documentation.sas.com/doc/en/pgmsascdc/v_068/casmopt/casmopt_benders_overview.htm
34. Tran, T.T., Araujo, A., Beck, J.C.: Decomposition methods for the parallel machine scheduling problem with setups. INFORMS J. Comput. **28**(1), 83–95 (2016)
35. Van Bulck, D., Goossens, D.: A traditional Benders' approach to sports timetabling. Eur. J. Oper. Res. **307**(2), 813–826 (2023)
36. Wächter, A., Biegler, L.T.: On the implementation of an interior-point filter line-search algorithm for large-scale nonlinear programming. Math. Program. **106**, 25–57 (2006). https://doi.org/10.1007/s10107-004-0559-y

Structure-Preserving Symmetry Presolving for Mixed-Binary Linear Problems

Annika Jäger(✉) and Marc E. Pfetsch

TU Darmstadt, Darmstadt, Germany
{jaeger,pfetsch}@opt.tu-darmstadt.de

Abstract. This paper investigates a presolving method for handling symmetries in mixed-binary programs, based on inequalities computed from so-called Schreier-Sims tables. We show that an iterative application of this method together with merging variables will produce an instance for which the symmetry group is trivial. We then prove that the problem structure can be preserved for certain monotone binary problems, e.g., packing and covering integer programs, maximum stable set (or clique) problems and set covering. The result is then a changed instance of the same type such that existing black-box solvers can directly be applied. We showcase our approach on stable set problems via computational experiments. For a mixed set of graphs, the presolving method significantly outperforms a black-box branch-and-cut method, while being close to the performance of state-of-the-art symmetry handling methods. For Johnson and coding theory graphs, we significantly outperform the stand-alone solver with and without its integrated symmetry handling. A similar speed-up can be observed when running the maximum clique solver CliSAT on the Johnson graphs.

Keywords: Symmetry handling · Schreier-Sims table · Stable set problem

1 Introduction

Symmetry handling for mixed-integer programs is one important component of modern branch-and-cut solvers and allows for reducing the search space if symmetries are present. Over the last two decades, an abundance of different methods have been developed, from isomorphism pruning [20–22], fundamental domains [6,29,30], orbitopes [3,14,15], orbital branching [18,23–25], Schreier-Sims inequalities [27], to newer methods [7,9]; see also the overview article [19] and the computational evaluation in [26]. These methods are usually divided into the ones that add symmetry handling inequalities and the ones that handle symmetries during the branch-and-bound search.

A common property of these methods is that they often do not exploit the particular problem structure beyond the symmetry group. In fact, combining symmetry handling inequalities with problem structure often leads to a very

S. S. Dey et al. (Eds.): IPCO 2026, LNCS 16588, pp. 457–473, 2026.
https://doi.org/10.1007/978-3-032-28691-8_30

complicated polyhedral description, see, e.g., [9,12], with the exception of particular structures like packing/partitioning orbitopes [15]. Moreover, depending on the method, more or less work has to be invested into the implementation and checking whether a combination of methods is compatible with each other, or could lead to the elimination of all optimal solutions.

In this paper, we propose a way to deal with these two challenges for certain monotone mixed-binary problems. We build on the inequalities based on Schreier-Sims tables, which have been proposed by Liberti and Ostrowski [17] and Salvagnin [27]. They iteratively add ordering constraints like $x_i \geq x_j$ for symmetric variables x_i, x_j and proceed with the subgroup that stabilizes the previously used variables; see Sect. 2.1 for a detailed description. We propose a presolving method that iteratively computes symmetries and adds these inequalities until no (formulation) symmetry is left. Indeed, we show in Sect. 2.1 that this process terminates if we further merge variables for which equality holds. It works well for problems (or solution methods) for which adding $x_i \geq x_j$ does not change the structure like SAT. This, however, changes the structure for monotone problems. We therefore develop a general presolving method, consisting of several steps, which adds inequalities as well as fixes and deletes variables. We prove in Sect. 3 that iteratively applying this method will terminate with a trivial symmetry group.

Then, one key contribution is the observation that for several problem classes this procedure does not change the problem structure (Sect. 4). This has the advantage that all problem specific techniques can be applied. We consider packing and covering integer programs (Sect. 4.1), the maximum stable set problem (Sect. 4.2) and the set cover problem (Sect. 4.3) as examples. For instance, for stable set problems, the procedure will delete some nodes in the graph (including incident edges) and add edges resulting in a graph with trivial automorphism group; this case is motivated by the observations in [10]. The resulting stable set problem can then be solved with any suitable solver. As such, one can use black-box solvers on the presolved graphs.

We also use the stable set problem as an example to highlight the computational performance. We test an implementation of the presolving method together with a branch-and-cut solver on a mixed testset of graphs collected from many sources, on Johnson graphs, and graphs arising from coding theory (Sect. 5). We show that the performance of the presolving step plus "black-box" branch-and-cut solver is significantly better than applying the solver without symmetry handling on all testsets. On the mixed graphs, its performance is a bit worse than the one of the solver applying state-of-the-art symmetry handling, as implemented in SCIP [8]. For Johnson and especially coding theory graphs, the presolving method is vastly superior by factors around 1.3 to 5.5. We also observe a strong increase of performance of the maximum clique solver CliSAT [28] when applied on the presolved and complemented Johnson graphs.

2 Preliminaries

In this paper, we consider the following mixed-binary linear problems:

$$\max \{c^\top x : Ax \le b,\ x_i \in \{0,1\} \text{ for } i \in I,\ x \in [0,1]^n\}, \tag{P}$$

where $A \in \mathbb{R}^{m\times n}$, $b \in \mathbb{R}^m$, $c \in \mathbb{R}^n$ and $I \subseteq [n] := \{1,\ldots,n\}$. The kth row of A is denoted by a_k.

We consider the formulation symmetry of the problem (P) as defined in [26], which we will simply refer to as *symmetry* in the following. The symmetry group Γ of (P) is a subgroup of the full symmetric group $\mathcal{S}_n$ of $[n]$ acting on $x \in \mathbb{R}^n$ as $\gamma(x) := (x_{\gamma^{-1}(1)},\ldots,x_{\gamma^{-1}(n)})$. The group Γ consists of all permutations γ for which there exists a permutation $\delta \in \mathcal{S}_m$ such that $\gamma(c) = c$, $\delta(b) = b$ and $A_{\delta(i),\gamma(j)} = A_{i,j}$ for all $i \in [m]$, $j \in [n]$. We call two variables x_i, x_j *symmetric* if there exists $\gamma \in \Gamma$ with $\gamma(i) = j$ and call the set of variable indices

$$\mathrm{orb}(i,\Gamma) := \{j \in [n] : \exists \gamma \in \Gamma \text{ s.t. } \gamma(i) = j\}$$

the *orbit* of i. The set of permutations in Γ that keep each element of a set $S \subseteq [n]$ invariant, i.e.,

$$\mathrm{stab}(\Gamma,S) := \{\gamma \in \Gamma : \gamma(i) = i \text{ for all } i \in S\}$$

is the *pointwise stabilizer* of S.

2.1 Schreier-Sims Table Inequalities

One way of handling symmetries in (P) is adding *Symmetry Handling Inequalities* (SHI). One particular case of SHIs are generated using the following *Schreier-Sims Table (SST)* algorithm, see, e.g., [20], that stores a sequence of stabilized indices in L:

(A0) initialize $\Gamma' = \Gamma$ and $L = \emptyset$;
(A1) choose $\ell \in [n] \setminus L$ and compute $O_\ell = \mathrm{orb}(\ell,\Gamma')$;
(A2) append ℓ to L and update Γ' to $\mathrm{stab}(\Gamma',L)$;
(A3) repeat steps A1 and A2 until $\Gamma' = \{\mathrm{id}_{[n]}\}$.

The SHIs obtained from this algorithm are the so-called *SST-cuts* $x_\ell \ge x_f$ for all $\ell \in L$ and all $f \in O_\ell \setminus \{\ell\}$, where ℓ is called *leader* and f *follower* of ℓ. Liberti and Ostrowski [17] and Salvagnin [27] showed the validity of these cuts as SHIs. For binary variables, an SST-cut implies that whenever a follower is chosen, i.e., $x_f = 1$, its leader has to be chosen as well. We usually refer to a single SST-cut by a pair (ℓ, f) with $\ell \in L$ and $f \in O_\ell \setminus \{\ell\}$ which we call *SST-pair*. Moreover, $S(L) := \{(\ell, f) : \ell \in L,\ f \in O_\ell \setminus \{\ell\}\}$ denotes the set of all SST-pairs, which we just denote by S if the corresponding L is clear. As discussed in [10], the set of SST-pairs can be computed in polynomial time.

Salvagnin [27] showed that adding all SST-cuts to a mixed-integer program (MIP) completely breaks the symmetry of the initial formulation. Nevertheless, he also demonstrated that this can produce new symmetries. A straightforward approach to obtain a problem formulation without symmetry is to iteratively compute symmetry and add SST-cuts. This approach does not seem to have been investigated so far. We show that it terminates if we additionally merge variables for which equality holds. By *merging* variables x_i and x_j we mean fixing $x_j = 0$, updating the objective value of x_i to $c_i + c_j$ and updating the ith column of A, from $A_{\star,i}$, to $A_{\star,i} + A_{\star,j}$.

Lemma 1. *Consider a MIP on variables $x_1, \dots, x_n$ and iteratively compute the formulation symmetry of the problem and add all computed SST-cuts. Further, whenever there is a constraint $x_i \geq x_j$ to be added and $x_i \leq x_j$ is already part of the problem formulation, the variables are merged. This procedure terminates after finitely many iterations with a problem with trivial symmetry group.*

Proof. Assume this is false. There are only finitely many different SST-cuts that could possibly be added and variables that can be merged. Thus, there exists an iteration at which the problem still has non-trivial symmetry, all associated SST-cuts are already part of the problem formulation and no variable merging is applicable. More precisely, there exists a symmetry γ of the problem at the current iteration with $\gamma(i) \neq i$ for some $i \in [n]$ and all inequalities $x_i \geq x_{\gamma^k(j)}$ with $\gamma^k(i) \neq i$ are already part of the problem formulation. Since each orbit is cyclic under a single element, there exists some $p \in \mathbb{N}$ such that $\gamma^p(i) = i$, which also means that $\gamma^{p-1}(i) \neq i$ because $\gamma(i) \neq i$. Thus, $x_i \geq x_{\gamma^{p-1}(j)}$ is part of the problem formulation. Since γ is a symmetry, the constraint $x_{\gamma(i)} \geq x_i$ obtained by applying γ is also part of the problem formulation. But this contradicts that we computed the SST-pair $(i, \gamma(i))$ and no merging is applicable. □

Therefore, iteratively adding SST-cuts and merging is a presolving method that removes all non-trivial symmetry from a MIP. This works especially for problem classes that allow the addition of such constraints and merging without structural change like the SAT problem (where $x_i \geq x_j$ can be expressed as $\overline{x}_j \vee x_i$). For SAT, the use of SST clauses and the addition of binary symmetry breaking clauses have been investigated in [1] and [5], respectively, where they were not added iteratively but as part of an extensive procedure to reduce symmetry.

3 Presolving for Mixed-Binary Linear Problems

The central step in the presolving routine for (P) is adding constraints that enforce the implication of the SST-pair. In particular, for an SST-pair (i, j) over binary variables, we want to add constraints such that $x_j = 1$ implies $x_i = 1$. To achieve this goal, we add constraints that cause all constraints containing x_i (resp. x_j) to be fulfilled independently of the value of x_i (resp. x_j) if x_j (resp. x_i) is fixed to 0 (resp. 1). Then to obtain optimality, the variable has to be 0 or 1 depending on the signs of c_i and c_j. Throughout this section, we want to argue that certain modifications of problem (P) cannot cut off all optimal solutions.

We then say that the operation is *valid* for (P). We further say that some constraint holds for (P) if it is satisfied by all feasible solutions. For $N := [n] \setminus \{i, j\}$ and $x \in [0,1]^n$, we use the notation

$$\langle a_k, x \rangle_N := \sum_{p \in N} a_{k,p} x_p.$$

Definition 1. *The* ***addition operation*** *is:*

- *Let a_k be a row of A such that $a_{k,i} > 0$. If it does not already exist in* (P), *we add the constraint*

$$\langle a_k, x \rangle_N + (a_{k,i} + a_{k,j}) x_j \leq b_k. \qquad (\text{AO}^+(k,i,j))$$

- *Let a_k be a row of A such that $a_{k,j} < 0$. If it does not already exist in* (P), *we add the constraint*

$$\langle a_k, x \rangle_N + (a_{k,i} + a_{k,j}) x_i \leq b_k. \qquad (\text{AO}^-(k,i,j))$$

Note that $a_{k,j} = 0$ is possible for $(\text{AO}^+(k,i,j))$ *and $a_{k,i} = 0$ for* $(\text{AO}^-(k,i,j))$.

Lemma 2. *Let $(i,j) \in S$. Then* $(\text{AO}^+(k,i,j))$ *and* $(\text{AO}^-(k,i,j))$ *are valid for* (P) *if they are added by an addition operation.*

Proof. We have to check that the constraints added by the addition operation are satisfied by an optimal solution $x^\star$ of (P) that satisfies the SST-cut $x_i^\star \geq x_j^\star$ associated to the SST-pair $(i,j) \in S$. This implies

$$\begin{aligned} (a_{k,i} + a_{k,j}) x_j^\star &\leq a_{k,i} x_i^\star + a_{k,j} x_j^\star && \text{if } a_{k,i} > 0, \\ (a_{k,i} + a_{k,j}) x_i^\star &\leq a_{k,i} x_i^\star + a_{k,j} x_j^\star && \text{if } a_{k,j} < 0. \end{aligned}$$

Hence, the validity of the constraint $a_k x \leq b_k$ guarantees the validity of the constraints $(\text{AO}^+(k,i,j))$ and $(\text{AO}^-(k,i,j))$ added by addition operations. □

We call adding $(\text{AO}^+(k,i,j))$ and $(\text{AO}^-(k,i,j))$ to (P) by addition operations for $(i,j) \in S$ *applying addition operations regarding S*. Next, we want to show that the operation indeed yields the desired implication.

Lemma 3. *Let x_i and x_j be variables in* (P) *such that for all rows a_k of A with $a_{k,i} > 0$ the constraint* $(\text{AO}^+(k,i,j))$ *holds for* (P). *If $c_i > 0$, for every optimal solution $x^\star$, one has $x_i^\star \geq x_j^\star$. In particular, if the variables are binary, $x_i^\star = 0$ implies $x_j^\star = 0$.*

Proof. Assume $x^\star$ is an optimal solution of (P) with $x_i^\star < x_j^\star$. Define $\tilde{x}_p = x_p^\star$ for all $p \in [n] \setminus \{i\}$ and $\tilde{x}_i = x_j^\star$. Then $\tilde{x}$ is feasible for (P): Since $x^\star$ is feasible, we only have to check feasibility for constraints $a_k x \le b_k$ with $a_{k,i} > 0$. Thus,

$$a_k \tilde{x} = \langle a_k, x^\star \rangle_N + (a_{k,i} + a_{k,j}) x_j^\star \le b_k,$$

where we used that ($\mathrm{AO}^+(k,i,j)$) holds for (P). Hence, $\tilde{x}$ is feasible for (P). Moreover, $c^\top \tilde{x} > c^\top x^\star$, since $c_i > 0$. But this contradicts optimality of $x^\star$. □

Lemma 4. *Let x_i and x_j be variables in* (P) *such that for all rows a_k of A with $a_{k,j} < 0$ the constraint* ($\mathrm{AO}^-(k,i,j)$) *holds for* (P). *If $c_j < 0$, for every optimal solution $x^\star$, one has $x_i^\star \ge x_j^\star$. In particular, if the variables are binary, $x_i^\star = 0$ implies $x_j^\star = 0$.*

Proof. This can be shown analogously to the proof of Lemma 3, this time defining $\tilde{x}$ by $\tilde{x}_p = x_p^\star$ for all $p \in [n] \setminus \{j\}$ and $\tilde{x}_j = x_i^\star$. □

We next merge variables for which no further constraints can be added.

Definition 2. *Let x_i and x_j be variables in* (P). *Further, let either $c \ge 0$ and the constraints* ($\mathrm{AO}^+(k,i,j)$) *with $a_{k,i} > 0$ and* ($\mathrm{AO}^+(k,j,i)$) *with $a_{k,j} > 0$ for all $k \in [m]$ hold in* (P) *or $c \le 0$ and the constraints* ($\mathrm{AO}^-(k,i,j)$) *with $a_{k,j} < 0$ and* ($\mathrm{AO}^-(k,j,i)$) *with $a_{k,i} < 0$ for all $k \in [m]$ hold in* (P). *Then the* ***merge operation*** *merges variables x_i and x_j.*

Lemma 5. *The merge operation is a valid presolving step for* (P).

Proof. Let $c \ge 0$ and, for all $k \in [m]$, let the constraints ($\mathrm{AO}^+(k,i,j)$) and ($\mathrm{AO}^+(k,j,i)$) with $a_{k,i} > 0$ and $a_{k,j} > 0$, respectively, hold in (P). Then if c_i, $c_j > 0$, by Lemma 3 all optimal solutions $x^\star$ satisfy $x_i^\star \ge x_j^\star$ as well as $x_i^\star \le x_j^\star$, implying that $x_i^\star = x_j^\star$. We claim that if $c_i = 0$ or $c_j = 0$, there still exists at least one optimal solution $x^\star$ for which this equality holds. Indeed, we can construct a feasible solution $\tilde{x}$ like we did in the proof of Lemma 3 with $\tilde{x}_i = \tilde{x}_j$ and $\tilde{x} \ge x^\star$. Thus, $c^\top \tilde{x} \ge c^\top x^\star$, showing optimality of $\tilde{x}$. Therefore, we have $x_i = x_j$ for at least one optimal solution of (P).

If $c \le 0$, we can argue analogously and generally get $x_i^\star = x_j^\star$ for at least one optimal solution $x^\star$ of (P). Therefore, we can replace every x_j in the problem formulation by x_i, which agrees with the definition of the merge operation. □

Remark 1. Note that all constraints changed by the merge operation are of types we could possibly add by an addition operation. For instance, for a constraint $a_k x \le b_k$ with $x_i = x_j$ we get

$$a_k x = \langle a_k, x \rangle_N + a_{k,i} x_i + a_{k,j} x_i = \langle a_k, x \rangle_N + (a_{k,i} + a_{k,j}) x_i \le b_k,$$

which agrees with ($\mathrm{AO}^+(k,j,i)$) if $a_{k,j} > 0$, ($\mathrm{AO}^-(k,i,j)$) if $a_{k,j} < 0$ and the initial constraint if $a_{k,j} = 0$.

Proposition 1. *If no addition operation regarding S or merge operation is applicable for* (P), *the symmetry group acting on the problem variables is trivial.*

Proof. Assume the symmetry group is non-trivial. Then there exists $(i, j) \in S$ with corresponding automorphism $\gamma \in \Gamma$. Since the addition operation does not add any constraints to (P), we know that for all $k \in [m]$ with $a_{k,i} > 0$ or $a_{k,j} < 0$, the constraints (AO$^+(k, i, j)$) and (AO$^-(k, i, j)$) exist in (P), respectively.

We claim that for a row a_k with $a_{k,j} > 0$, the constraint (AO$^+(k, j, i)$), i.e., $\langle a_k, x\rangle_N + (a_{k,i} + a_{k,j})x_i \leq b_k$, holds for all feasible x of (P). Since constraint $a_k x = \langle a_k, x\rangle_N + a_{k,i}x_i + a_{k,j}x_j \leq b_k$ is in (P), there exits some row $\ell \in [m]$ such that $a_\ell x = \langle a_k, \gamma(x)\rangle_N + a_{k,i}x_{\gamma^{-1}(i)} + a_{k,j}x_i \leq b_\ell = b_k$ due to formulation symmetry. Since $a_{k,j} > 0$ and $(i, \gamma^{-1}(i)) \in S$, we have by assumption that the constraint $\langle a_k, \gamma(x)\rangle_N + (a_{k,i} + a_{k,j})x_{\gamma^{-1}(i)} \leq b_k$ is in the formulation of (P), because it corresponds to (AO$^+(\ell, i, \gamma^{-1}(i))$). Since γ^{-1} is a symmetry, we obtain that the constraint $\langle a_k, x\rangle_N + (a_{k,i} + a_{k,j})x_i \leq b_k$ is in (P) which is the desired constraint.

We can analogously show that the constraints of the type (AO$^-(k, j, i)$) hold in (P). Thus, the claim is true and the merge operation is applicable. But this contradicts the initial assumption. Hence, there cannot exist a non-trivial symmetry. □

Note that for the above proof we only need to merge symmetric variables. Nevertheless, the definition is more general in order to increase the chances to fix variables and the efficiency of presolving.

It is possible that the addition operation adds constraints that enforce the fixing of a certain binary variable. Since variable fixing is usually more efficient in presolving than the addition of constraints, we want to fix the variable right away instead of adding constraints that imply the fixing. This approach aims not only to handle symmetry, but also to make use of it as soon as possible.

For $i \in [n]$ we define $S^i := \{p \in [n] : (i, p) \in S\}$, the set of followers of i in S and $S_i := \{p \in [n] : (p, i) \in S\}$, the set of leaders having follower i and denote by $\overline{S}_i$ its complement $[n] \setminus S_i$. Further, we define $\mathcal{A}^i_k := \{p \in [n] \setminus \{i\} : a_{k,p} < 0\}$ and denote by $a_k(I)$ the sum $\sum_{i \in I} a_{k,i}$.

Definition 3. *Let a_k be a row in A such that*

$$a_k(S_j) + a_{k,j} + a_k(\overline{S}_j \cap \mathcal{A}^j_k) > b_k. \qquad (\mathrm{DO}(k,j))$$

*Then, if x_j is a binary variable, we fix it to 0 and call it **deletion operation**. Let a_k be a row in A such that*

$$a_k(\overline{S^i} \cap \mathcal{A}^i_k) > b_k. \qquad (\mathrm{FO}(k,i))$$

*Then, if x_i is a binary variable, we fix it to 1 and call it **fixation operation**.*

Lemma 6. *The deletion operation is a valid presolving step for* (P).

Proof. Assume (DO(k, j)) holds for some $k \in [m]$. Then fixing $x_j = 1$ implies $x_p = 1$ for all $p \in S_j$. Therefore, we have for feasible $\tilde{x}$ with $\tilde{x}_j = 1$ respecting the SST-cuts in S that

$$a_k\tilde{x} = a_k(S_j) + a_{k,j} + \sum_{p \in \overline{S}_j \setminus \{j\}} a_{k,p}\,\tilde{x}_p \geq a_k(S_j) + a_{k,j} + a_k(\overline{S}_j \cap \mathcal{A}^j_k) > b_k$$

contradicting that $a_k\tilde{x} \leq b_k$. Hence $\tilde{x}_j$ has to be 0. □

Lemma 7. *The fixation operation is a valid presolving step for* (P).

Proof. Assume (DO(k,j)) holds for some $k \in [m]$. Then fixing $x_i = 0$ implies that $x_p = 0$ for all $p \in S^i$. Therefore, we have for feasible $\tilde{x}$ with $\tilde{x}_i = 0$ respecting the SST-cuts in S that

$$a_k \tilde{x} = \sum_{p \in \overline{S^i} \setminus \{i\}} a_{k,p}\, \tilde{x}_p \geq a_k(\overline{S^i} \cap \mathcal{A}_k^i) > b_k$$

contradicting that $a_k \tilde{x} \leq b_k$. Hence $\tilde{x}_i$ has to be 1. □

We propose the following presolving routine:

Definition 4 (Presolving routine).

1. Merging: *Apply merge operations exhaustively.*
2. *Compute symmetry group of Problem* (P) *and corresponding SST-pairs.*
3. Deletion & Fixation*: Apply deletion and fixation operations exhaustively.*
4. Addition: *Apply addition operations regarding S exhaustively.*

Proposition 2. *After finitely many applications of the presolving routine of Definition 4, the symmetry group of problem* (P) *becomes trivial.*

Proof. Assume this is not true. By Proposition 1, this means that there can be infinitely many applications of the presolving routine in which the problem is modified. Clearly, there can only be finitely many presolving rounds in which merge, deletion or fixation presolving is active, since all of them reduce the number of variables of the problem. So we need to show that the number of constraints that can possibly be added by addition operations is also bounded. But this is clear, since all added constraints use a constraint of the current problem formulation and update coefficients by sums of the initial coefficients that only appear once in these sums. Hence, there are only finitely many possibilities for a constraint that can be added by an addition operation. □

Unfortunately the argument in the proof just leads to an exponential upper bound. Anyhow, the hope is that it is not tight and not all possible constraints do actually have to get generated. In fact, in the computational results for the stable set problem presented in Sect. 4.2, only few rounds of the presolving routine are needed to handle the symmetry completely.

4 Structure Preserving Presolving

In this section, we discuss examples where the general presolving method of Sect. 3 leaves the particular structure invariant.

4.1 Packing and Covering Integer Programs

Let $I = [n]$. A *Packing Integer Program* arises from Problem (P) if A, b and c are nonnegative. A *Covering Integer Program* arises if A, b and c are nonpositive, which can be reformulated as a minimization problem over $-Ax \geq -b$. In both cases, the presolving routine keeps the structure invariant, i.e., the coefficients stay nonnegative and nonpositive, respectively. Then, for instance, approximation algorithms can be applied, see, e.g., [16].

4.2 Presolving for the Stable Set Problem

Let $G = (V, E)$ be an undirected simple graph and $c \in \mathbb{R}_+^V$ be nonnegative node weights. The maximum (weight) stable set problem is

$$\max \{c^\top x \,:\, x_u + x_v \leq 1 \quad \forall \{u, v\} \in E,\ x \in \{0,1\}^V\}. \tag{MSSP}$$

A special feature of the stable set problems is that the formulation symmetry of the problem coincides with the overall symmetry of the problem. Some of the following steps have been investigated in [10], but without finiteness proof. For a node $u \in V$ we define the neighborhood of u by $N(u) := \{v \in V \,:\, \{u, v\} \in E\}$.

Lemma 8. *If the presolving routine in Definition 4 is applied on* (MSSP)*, the presolving steps can be translated for the stable set setting in the following way:*

- Merge: *Let u, $v \in V$ such that $N(u) = N(v)$, in particular $\{u, v\} \notin E$. Fix $x_v = 0$, delete all edges incident to v from E and update c_u to $c_u + c_v$.*
- Deletion: *Let $\{u, v\} \in E$ and $(u, v) \in S$. Fix variable $x_v = 0$.*
- Fixation: *Not applicable.*
- Addition: *Let $\{u, v\} \in E$ and $(v, w) \in S$. Then the edge $\{u, w\}$ is added to E if it is not already part of G.*

Proof. Since A is nonnegative for the stable set problem, the condition for the deletion operation reduces to $a_k(S_v) + a_{k,v} > 1$. So if the variables for both members of an SST-pair are contained in the same constraint, which means there is an edge between leader and follower in E, this condition is fulfilled and the follower variable is fixed to 0. The condition of the fixation operation reduces to $0 > 1$ and is therefore never fulfilled.

Let $(u, v) \in S$. Again by the nonnegativity of A, the addition operation only adds constraints of the form (AO$^+(k, u, v)$) with $a_{k,u} > 0$. Every row k with $a_{k,u} > 0$ corresponds to an edge $\{u, w\} \in E$. If $w \neq v$ and therefore $a_{k,v} = 0$, the constraint (AO$^+(k, u, v)$) just replaces x_u by x_v, i.e., it corresponds to the edge $\{u, \ell\}$. If $w = v$, the variables for both members of the SST-pair (u, v) are contained in the constraint corresponding to row k. Thus, x_v would have been removed by the deletion operation.

Since $c \geq 0$, the merge condition is that for u, $v \in V$, the constraints (AO$^+(k, u, v)$) with $a_{k,u} > 0$ and (AO$^+(k, v, u)$) with $a_{k,v} > 0$ hold in (P) for all $k \in [m]$. First assume that $\{u,v\} \in E$. Then the constraint ($\text{AO}^+(k,u,v)$) is $2x_v \leq 1$. This constraint cannot be part of the initial formulation of (MSSP), since G is assumed to be a simple graph, or implied by edge constraints or added by an addition operation. Thus, the merge condition is never met if the involved nodes are adjacent. For non-adjacent nodes u and v, the constraints ($\text{AO}^+(k,u,v)$) with $a_{k,u} > 0$ correspond to edges between v and neighbors of u (see translation of the addition operation). More precisely, the merge condition that all constraints ($\text{AO}^+(k,u,v)$) with $a_{k,u} > 0$ for all $k \in [m]$ hold in (MSSP) means that all neighbors of u are also neighbors of v. Here we use again that all edge inequalities that hold for the stable set problem need to be part of the problem formulation. Since this condition also has to hold if u and v switch places, we obtain the condition $N(u) = N(v)$. Thus, $a_{k,u}$ is updated by a merge operation if $a_{k,v} > 0$ which means that $a_{k,v} = 1$. This implies that $a_{k,u} = 0$, because u and v are non-adjacent. So the update changes an edge $\{v,w\}$ to an edge $\{u,w\}$. But this edge already has to exist because $N(u) = N(v)$. Thus, it is valid to just delete the edge $\{v,w\}$. □

No constraint with a coefficient larger than 1 can be added by an addition operation, because of the prior deletion operation. Therefore, the addition operation only adds constraints corresponding to edges. Hence, we obtain an upper bound of $O(n^2)$ on the number of presolving rounds for (MSSP) derived from the maximal number of non-parallel edges that can be added to and nodes that can be deleted from a simple graph. The next section will include an investigation of the gap between this naive upper bound and practical results.

Remark 2. As a side note, the *maximum k-colorable subgraph problem* is a special stable set problem [12] arising from the product of a graph with a complete graph with k nodes. Here, the presolving routine alters the problem structure to some extent – the added constraints might run across different color classes.

4.3 Presolving for the Set Cover Problem

Let U be a set of nodes and let P be a family of sets $\sigma \subseteq U$. The weighted set cover problem with set weights $c \in \mathbb{R}^P_+$ is defined by

$$\min \{c^\top x : \sum_{u \in \sigma \in P} x_\sigma \geq 1 \quad \forall u \in U,\ x \in \{0,1\}^P\}. \tag{SSCP}$$

The presolving steps of Definition 4 can be translated to the set cover setting in the following way:

- *Addition:* Let $(\sigma,\tau) \in S$ and let $u \in \tau$. Add a new node $\tilde{u}$ to U and add it to σ as well as all sets in $P \setminus \{\tau\}$ that contain u.
- *Merge:* Let σ, $\tau \in P$ such that for all $u \in \sigma$ with $u \notin \tau$ there exists $u' \in \tau$ such that $\{\rho \in P : u \in \rho\} \setminus \{\sigma\} = \{\rho \in S : u' \in \rho\} \setminus \{\tau\}$ and the other way around. Then fix $x_\tau = 0$, delete all nodes $u \in \tau$ from U and update c_σ to $c_\sigma + c_\tau$.

- *Deletion:* Not applicable.
- *Fixation:* Let $u \in U$ such that $u \in \sigma$ and $(\sigma, \tau) \in S$ for all τ with $u \in \tau, \tau \neq \sigma$. Then the variable x_v is fixed to 0.

As in the case of the stable set problem, the addition operation can possibly add constraints with coefficients larger than 1. But since all variables are binary in (SSCP), the constraints $2x_\sigma + \sum x_\tau \geq 1$ and $x_\sigma + \sum x_\tau \geq 1$ are equivalent. Therefore it is equally valid to add the second constraint and the problem structure remains intact after applying the presolving routine.

5 Computational Results

We use a developer version of SCIP 10.0.2 (GitHash: 36ba424) and CPLEX 12.10 as LP solver. The experiments were run on a Linux cluster with 3.5 GHz Intel Xeon E5-1620 Quad-Core CPUs, having 32 GB main memory and 10 MB cache each. All computations were run single-threaded and with a time limit of two hours.

We demonstrate the effect of our presolving method on the example of the stable set problem. To this end, we use the branch-and-cut solver BACS (https://github.com/dopt-TUDa/bacs), because it allows to apply the state-of-the-art symmetry handling methods of the framework SCIP [8]. The BACS implementation features a clique separator based on a greedy algorithm and, if unsuccessful, a combinatorial algorithm for the maximal weight clique problem implemented in SCIP. All specialized presolving/propagation methods of BACS are turned off.

Our presolving routine (Definition 4, Lemma 8) is performed on the original graph before the optimization process starts. After the symmetry is computed in the second step of the presolving routine, BACS first checks whether there are symmetric components and possibly merges them. Since nodes that get merged during this operation do not satisfy the merge condition (otherwise they would have been merged in the first step of the presolving routine), we do not count them towards the merge operation even if the action applied on them is the same. The presolving routine is run iteratively until no further modifications of the graph are made, which means that no symmetry remains by Proposition 2. Instead of computing stabilizers for the leaders and followers, we filter the generators and therefore only operate on a subgroup of the actual symmetry group of the current graph for performance reasons.

To detect automorphisms of graphs, we apply bliss [13] and sassy [2]. To highlight the effect on the dual bound, we initialized the runs with a cutoff using the best primal value. All primal heuristics are turned off. We use a stringent leader choice as motivated by [10], which roughly means that we prioritize choosing leaders from maximum orbits as long as they fulfill some structure-preserving condition.

The computational experiments were carried out on three different testsets: the *mixed-testset*, which is a collection of 73 instances from various sources, the *Johnson-testset*, which includes 62 Johnson graphs of adequate complexity, and the *binary-code-testset*, containing 23 graphs processing Hamming distances between binary code words also of adequate complexity. The Johnson- as well as the binary-code-testset and an explanation of their compositions is available on github (https://github.com/dopt-TUDa/SymGraphs). The mixed-testset is based on Table 1 of [10] which contains a collection of sources of interesting instances for the stable set or maximum clique problem. The instances generated for the maximum clique problem were complemented before we started the computations. Then we ran BACS in default-mode with disabled symmetry handling of SCIP on all of the instances listed in the table. We removed all instances with trivial symmetry and instances that appeared too easy in being solvable in the root node or in less than 10 s. All remaining 73 instances form the mixed-testset.

Table 1. Average indicators of the presolving routine on different testsets

Testset	#merge	#del	#add	time	$\|L\|$	$\|S(L)\|$
Mixed	1032.8	160.6	18 755.5	5.18	31.1	575.9
Johnson	2.5	270.0	38 428.5	0.32	7.8	2973.2
Binary-code	2.5	441.7	42 022.4	0.33	15.1	2443.7

Table 1 summarizes the structural effect that the presolving routine has on the instances of the three different testsets. More specifically, the arithmetic mean of the number of variables deleted or edges added by the presolving routine over all instances in each testset is displayed. The same information is given for the individual instances in Tables 3, 4 and 5 of [11]. Further, the average of the presolving time as well as the total number of leaders $|L|$ and followers $|S(L)|$ computed is listed in the table.

Each operation of the presolving routine becomes active, while the presolving time is small compared to the overall solving time (see Table 2). While for the mixed-testset the largest number of leaders is used on average, the number of followers and therefore the number of SST-pairs strongly increases from the mixed- to the Johnson- and binary-code-testset. Also the increase of deletion and addition operations is comparable to the increase of SST-pairs. This effect cannot be observed for the merge operation due to two outliers in the mixed-testset. The maximal number of iterations of the presolving routine is 6, meeting the expectation that the upper bound of $O(n^2)$ is far from being tight (see Tables 3, 4 and 5 of [11]).

Table 2 summarizes the overall solution performance after the presolving routine was iteratively applied until no symmetry was left (setting *iter presol*) and compares this to only applying one round of the presolving routine (setting *presol*) as well as default settings with SCIP symmetry handling (setting *base-sym*) and without SCIP symmetry handling (setting *base*). For all three testsets, the number of instances solved to optimality within two hours, the shifted geometric means over all instances of the solution time and the number of nodes in

Table 2. Comparison of presolving performance on different testsets

Setting	Mixed (73)			Johnson (62)			Binary-code (23)		
	#opt	time	#nodes	#opt	time	#nodes	#opt	time	#nodes
base	44	458.75	5716.0	29	830.92	4256.5	11	1038.98	3637.9
base-sym	53	161.78	1431.5	42	123.20	590.0	15	165.45	668.4
presol	50	259.99	3323.9	41	96.79	968.8	16	36.27	479.5
iter presol	50	261.27	3197.3	41	88.41	858.8	16	29.29	369.8

Table 3. Comparison of CliSAT performance with and without presolving

name	with presolving		without presolving			
	nodes	time	nodes	time	nodes ratio	time ratio
Johnson_15_5_4	10 436 800	3600.1	9 797 200	3600.1	0.94	1.00
Johnson_14_3_1	453	0.5	363 956	4.3	803.43	8.60
Johnson_14_4_0	1	0.5	47	0.5	47.00	1.00
Johnson_14_4_1	162	0.5	1087	0.5	6.71	1.00
Johnson_14_4_3	56 576 000	3600.0	51 291 700	3600.0	0.91	1.00
Johnson_12_5_0	1	0.5	11 922 800	3600.1	Large	Large
Johnson_13_4_3	84 417 000	3600.0	79 088 000	3600.0	0.94	1.00
Johnson_14_6_1	285 600	3600.1	1 480 400	3600.1	5.18	1.00
Johnson_19_3_0	1	0.5	1	0.5	1.00	1.00
Johnson_20_4_3	6 513 700	3600.1	7 823 300	3600.1	1.20	1.00
Johnson_11_4_0	1	0.5	179	0.5	179.00	1.00
Johnson_15_4_0	1	0.5	27	0.5	27.00	1.00
Johnson_17_4_3	19 262 900	3600.0	15 911 200	3600.0	0.83	1.00
Johnson_13_3_1	98	0.5	25 972	0.5	265.02	1.00
Johnson_14_6_5	7 917 300	3600.1	9 736 000	3600.0	1.23	1.00
Johnson_17_3_2	119 486 500	3600.0	113 778 500	3600.0	0.95	1.00
Johnson_15_4_3	40 313 300	3600.0	37 396 000	3600.0	0.93	1.00
Johnson_12_5_1	1	0.5	20 767 200	3600.1	Large	Large
Johnson_16_6_5	1 998 600	3600.1	2 340 200	3600.2	1.17	1.00
Johnson_16_5_4	6 304 300	3600.1	6 097 300	3600.1	0.97	1.00
Johnson_14_5_2	58 920 694	2117.1	75 657 900	3600.0	1.28	1.70
Johnson_13_5_1	154	0.5	14 312 507	2247.6	Large	Large
Johnson_18_4_3	13 977 400	3600.1	12 211 900	3600.1	0.87	1.00
Johnson_18_3_2	109 292 600	3600.0	98 890 700	3600.0	0.90	1.00
Johnson_11_4_3	115 265 515	2623.6	162 187 300	3600.0	1.41	1.37

continued

Table 3. continued

name	with presolving		without presolving			
	nodes	time	nodes	time	nodes ratio	time ratio
Johnson_16_3_1	388	0.5	2 759 036	62.4	Large	124.80
Johnson_12_4_2	21	0.5	1 803 397	57.6	Large	115.20
Johnson_17_3_0	1	0.5	1	0.5	1.00	1.00
Johnson_20_3_0	1	0.5	1	0.5	1.00	1.00
Johnson_13_5_2	3290	0.5	1 196 258	106.2	363.60	212.40
Johnson_13_5_0	1	0.5	2 716 200	3600.1	Large	Large
Johnson_11_4_1	50	0.5	64	0.5	1.28	1.00
Johnson_11_3_2	193	0.5	2 431 438	28.6	Large	57.20
Johnson_15_6_5	4 699 200	3600.1	3 655 600	3600.0	0.78	1.00
Johnson_14_3_0	1	0.5	1	0.5	1.00	1.00
Johnson_10_4_0	1	0.5	51	0.5	51.00	1.00
Johnson_20_3_2	82 258 400	3600.0	71 574 000	3600.0	0.87	1.00
Johnson_10_4_2	37	0.5	2975	0.5	80.41	1.00
Johnson_17_4_0	1	0.5	14	0.5	14.00	1.00
Johnson_12_5_4	62 911 700	3600.0	59 947 200	3600.0	0.95	1.00
Johnson_11_4_2	1	0.5	12 290	0.5	Large	1.00
Johnson_15_3_1	452	0.5	1 770 076	26.3	Large	52.60
Johnson_12_4_1	1	0.5	93	0.5	93.00	1.00
Johnson_15_4_1	183	0.5	4766	1.4	26.04	2.80
Johnson_13_4_0	1	0.5	40	0.5	40.00	1.00
Johnson_16_3_2	166 534 900	3600.0	144 259 000	3600.0	0.87	1.00
Johnson_19_4_3	9 750 100	3600.1	8 805 400	3600.1	0.90	1.00
Johnson_13_5_4	36 340 300	3600.0	33 686 100	3600.0	0.93	1.00
Johnson_12_3_1	2	0.5	1178	0.5	589.00	1.00
Johnson_10_4_1	1	0.5	99	0.5	99.00	1.00
Johnson_14_5_4	17 648 900	3600.0	18 491 400	3600.0	1.04	1.00
Johnson_17_5_4	3 412 100	3600.1	3 925 200	3600.1	1.15	1.00
Johnson_16_4_1	171	0.5	21 337	8.2	124.78	16.40
Johnson_12_5_2	1	0.5	4659	0.5	Large	1.00
Johnson_14_5_0	1	0.5	1 059 900	3600.5	Large	Large
Johnson_18_5_4	2 146 400	3600.1	2 156 900	3600.1	1.00	1.00
Johnson_11_3_1	3	0.5	1052	0.5	350.67	1.00
Johnson_12_4_3	115 465 900	3600.0	106 809 100	3600.0	0.93	1.00
Johnson_14_5_1	1	0.5	11 121 400	3600.1	Large	Large

the branch-and-bound tree are listed. For detailed information on the solution times see again Tables 3, 4 and 5 of [11].

For the mixed-testset, the (iter) presol setting solves six more instances than the base setting with a solving time reduction of around 43 %. However, base-sym dominates presol in solving three more instances while taking about 37 % less time. For the other two testsets, however, the performance of (iter) presol is clearly superior to both base and base-sym with around 21–97 % reduction of solving time even though one more Johnson graph is solved in the base-sym setting compared to (iter) presol. The comparison of iter presol and presol shows that there is rather small computational benefit in iterating the presolving routine and thus, handle symmetry completely. However, the speedup effect seems to correlate with the number of presolving rounds needed (displayed in Tables 3, 4 and 5 of [11]) since iter presol always performs a final check that no symmetry is left, while this computation is never done in the presol setting. Thus, for instances where only one presolcving round is necessary, clearly presol has to be a bit faster than iter presol.

We further ran computations on the presolved instances with the combinatorial maximum clique problem solver CliSAT [4,28]. Since the stable set problem is equivalent to the maximum clique problem on the complemented graph, we can apply CliSAT on the complemented graphs of our testset. The mixed-testset is not suitable for CliSAT, because it contains weighted graphs and CliSAT does not allow weighted instances. Therefore, we exemplarily only performed computations on the Johnson-testset. That CliSAT does not support weighted instances is also an issue if the presolving applies merge operations, since they modify node weights. To avoid this, we ran the presolving routine without the merge operation. The results of Table 4 of [11] show that only a fraction of the Johnson graphs are influenced by the merge operation and therefore we expect the loss caused by omitting this operation being small.

For all CliSAT computations we used the same machine as the above with a time limit of one hour. Table 3 contains information about the solving performance of CliSAT for each instance in the Johnson-testset. Due to the three graphs `Johnson_14/15/16_6_0` being very sparse, their complements are too large for CliSAT to properly run on these instances. Therefore they are not listed in the results table. The table shows the number of nodes generated and the solving time in seconds first on instances where presolving was applied prior to the CliSAT run and next for instances that were solely complemented. The last two lines contain the ratios of nodes and time, where the numbers without presolving are divided by the numbers with presolving. Note that the time for presolving and complementing is not included in the results.

From Table 3, we conclude that after presolving, CliSAT can solve 7 more instances in around one third of the time compared to CliSAT without presolving. Note that only a binary executable of CliSAT is available, such that it can be seen as a true black-box solver for us.

Disclosure of Interests. The authors have no competing interests to declare that are relevant to the content of this article.

References

1. Anders, M., Brenner, S., Rattan, G.: Satsuma: structure-based symmetry breaking in SAT. In: Chakraborty, S., Jiang, J.H.R. (eds.) 27th International Conference on Theory and Applications of Satisfiability Testing (SAT 2024). Leibniz International Proceedings in Informatics (LIPIcs), vol. 305, pp. 4:1–4:23. Schloss Dagstuhl – Leibniz-Zentrum für Informatik, Dagstuhl, Germany (2024). https://doi.org/10.4230/LIPIcs.SAT.2024.4
2. Anders, M., Schweitzer, P., Stieß, J.: Engineering a preprocessor for symmetry detection. In: Georgiadis, L. (ed.) 21st International Symposium on Experimental Algorithms (SEA 2023). Leibniz International Proceedings in Informatics (LIPIcs), vol. 265, pp. 1:1–1:21. Schloss Dagstuhl – Leibniz-Zentrum für Informatik, Dagstuhl, Germany (2023). https://doi.org/10.4230/LIPIcs.SEA.2023.1
3. Bendotti, P., Fouilhoux, P., Rottner, C.: Orbitopal fixing for the full (sub-)orbitope and application to the unit commitment problem. Math. Program. **186**, 337–372 (2021). https://doi.org/10.1007/s10107-019-01457-1
4. Instances for CliSAT – an efficient state-of-the-art exact algorithm for the maximum clique problem. https://github.com/psanse/CliSAT
5. Devriendt, J., Bogaerts, B., Bruynooghe, M., Denecker, M.: Improved static symmetry breaking for SAT. In: Creignou, N., Le Berre, D. (eds.) SAT 2016. LNCS, vol. 9710, pp. 104–122. Springer, Cham (2016). https://doi.org/10.1007/978-3-319-40970-2_8
6. Friedman, E.J.: Fundamental domains for integer programs with symmetries. In: Dress, A., Xu, Y., Zhu, B. (eds.) COCOA 2007. LNCS, vol. 4616, pp. 146–153. Springer, Heidelberg (2007). https://doi.org/10.1007/978-3-540-73556-4_17
7. Hojny, C.: Packing, partitioning, and covering symresacks. Discret. Appl. Math. **283**, 689–717 (2020). https://doi.org/10.1016/j.dam.2020.03.002
8. Hojny, C., et al.: The SCIP Optimization Suite 10.0. Technical report, Optimization Online (2025). https://optimization-online.org/2025/11/the-scip-optimization-suite-10-0/
9. Hojny, C., Pfetsch, M.E.: Polytopes associated with symmetry handling. Math. Program. **175**, 197–240 (2019). https://doi.org/10.1007/s10107-018-1239-7
10. Hojny, C., Pfetsch, M.E., Verschae, J.: The impact of symmetry handling for the stable set problem via Schreier-Sims cuts. Technical report 2311.06085, arXiv (2023). https://arxiv.org/abs/2311.06085
11. Jäger, A., Pfetsch, M.E.: Structure-preserving symmetry presolving for mixed-binary linear problems. Preprint, Optimization Online (2026). https://optimization-online.org/?p=32590
12. Januschowski, T., Pfetsch, M.E.: Branch-cut-and-propagate for the maximum k-colorable subgraph problem with symmetry. In: Achterberg, T., Beck, J.C. (eds.) CPAIOR 2011. LNCS, vol. 6697, pp. 99–116. Springer, Heidelberg (2011). https://doi.org/10.1007/978-3-642-21311-3_11
13. Junttila, T., Kaski, P.: bliss: a tool for computing automorphism groups and canonical labelings of graphs (2012). https://users.aalto.fi/~tjunttil/bliss/
14. Kaibel, V., Peinhardt, M., Pfetsch, M.E.: Orbitopal fixing. Discret. Optim. **8**(4), 595–610 (2011). https://doi.org/10.1016/j.disopt.2011.07.001

15. Kaibel, V., Pfetsch, M.E.: Packing and partitioning orbitopes. Math. Program. **114**(1), 1–36 (2008). https://doi.org/10.1007/s10107-006-0081-5
16. Kolliopoulos, S.G., Young, N.E.: Approximation algorithms for covering/packing integer programs. J. Comput. Syst. Sci. **71**(4), 495–505 (2005). https://doi.org/10.1016/j.jcss.2005.05.002
17. Liberti, L., Ostrowski, J.: Stabilizer-based symmetry breaking constraints for mathematical programs. J. Global Optim. **60**(2), 183–194 (2013). https://doi.org/10.1007/s10898-013-0106-6
18. Linderoth, J., Núñez Ares, J., Ostrowski, J., Rossi, F., Smriglio, S.: Orbital conflict: cutting planes for symmetric integer programs. INFORMS J. Optim. **3**(2), 139–153 (2021). https://doi.org/10.1287/ijoo.2019.0044
19. Margot, F.: Symmetry in integer linear programming. In: Jünger, M., et al. (eds.) 50 Years of Integer Programming, pp. 647–686. Springer (2010)
20. Margot, F.: Pruning by isomorphism in branch-and-cut. Math. Program. **94**(1), 71–90 (2002). https://doi.org/10.1007/s10107-002-0358-2
21. Margot, F.: Exploiting orbits in symmetric ILP. Math. Program. **98**(1–3), 3–21 (2003). https://doi.org/10.1007/s10107-003-0394-6
22. Margot, F.: Small covering designs by branch-and-cut. Math. Program. **94**(2), 207–220 (2003). https://doi.org/10.1007/s10107-002-0316-z
23. Ostrowski, J.: Symmetry in Integer Programming. Ph.D. thesis, Lehigh University (2008)
24. Ostrowski, J., Anjos, M.F., Vannelli, A.: Modified orbital branching for structured symmetry with an application to unit commitment. Math. Program. **150**(1), 99–129 (2014). https://doi.org/10.1007/s10107-014-0812-y
25. Ostrowski, J., Linderoth, J., Rossi, F., Smriglio, S.: Orbital branching. Math. Program. **126**(1), 147–178 (2011). https://doi.org/10.1007/s10107-009-0273-x
26. Pfetsch, M.E., Rehn, T.: A computational comparison of symmetry handling methods for mixed integer programs. Math. Program. Comput. **11**(1), 37–93 (2018). https://doi.org/10.1007/s12532-018-0140-y
27. Salvagnin, D.: Symmetry breaking inequalities from the schreier-sims table. In: van Hoeve, W.-J. (ed.) CPAIOR 2018. LNCS, vol. 10848, pp. 521–529. Springer, Cham (2018). https://doi.org/10.1007/978-3-319-93031-2_37
28. San Segundo, P., Furini, F., Álvarez, D., Pardalos, P.M.: CliSAT: a new exact algorithm for hard maximum clique problems. Eur. J. Oper. Res. **307**(3), 1008–1025 (2023). https://doi.org/10.1016/j.ejor.2022.10.028
29. Verschae, J., Villagra, M., von Niederhäusern, L.: On the geometry of symmetry breaking inequalities. In: Singh, M., Williamson, D.P. (eds.) Integer Programming and Combinatorial Optimization, pp. 73–88. Springer, Cham (2021)
30. Verschae, J., Villagra, M., von Niederhäusern, L.: On the geometry of symmetry breaking inequalities. Math. Program. **197**(2), 693–719 (2023). https://doi.org/10.1007/s10107-022-01819-2

The Strength of Root Cuts in an Extended Abstract Branch-and-Cut Model

Boyang Han and Aleksandr M. Kazachkov(✉)

Center for Applied Optimization, Department of Industrial and Systems Engineering, University of Florida, Gainesville, FL 32611, USA
{bhan1,akazachkov}@ufl.edu

Abstract. The branch-and-cut paradigm has led to effective solvers for a wide range of mathematical optimization problems. Despite the extensive empirical success and significant theoretical work on pure branching and pure cutting, little is theoretically known about their interaction. In this paper, we study an extended abstract branch-and-cut model and investigate tradeoffs between branching and cutting within the framework. We prove that there always exists a minimal-time tree following a simple root-cut structure, where all cuts are placed before the first branching node, for certain families of settings. For others, although the root-cut structure may be suboptimal and differ substantially from the minimal-time tree, it nevertheless provides a reliable approximation that does not deviate arbitrarily from the actual minimal-time tree. In addition, we identify conditions under which minimal-time trees necessarily contain local cuts, i.e., cuts added after some branching decisions, under a single-cut-per-path restriction. Lastly, simulations indicate that trees restricted to only root cuts significantly outperform our proved worst-case approximation guarantee.

Keywords: mixed-integer optimization · branch and cut · root cuts · local cuts

1 Introduction

Branch-and-cut (B&C), a hybrid version of the branch-and-bound (B&B) [15] and cutting plane (CP) methods [10–12], is a powerful tree-based framework that is a key ingredient of mathematical optimization solvers. Each node of the tree corresponds to a relaxation of the original optimization problem that is refined by either valid inequalities (*cuts*) or splitting the search space (*branching*). Solvers determine *where* in the tree to add rounds of cuts based on empirical guidelines and typically focus on the *root node* of the tree. In this paper, we provide complementary theory on cut placement with provable guarantees on the computational cost of the resulting tree, within an extension of the *abstract B&C model* by Kazachkov, Le Bodic, and Sankaranarayanan [13,14].

Our model assumes that key solution parameters are known: the bound that needs to be proved (e.g., an optimal solution is provided, but we need to certify

S. S. Dey et al. (Eds.): IPCO 2026, LNCS 16588, pp. 474–490, 2026.
https://doi.org/10.1007/978-3-032-28691-8_31

this optimality via the tree), the evolution of the bound over iterated branching and cutting operations, and the approximate time to solve each node of the tree. Our perspective (in which such information might be available) is a *post-optimality analysis* of the quality of the best-possible B&C tree for a problem, which can provide beneficial insights when solving the same problem repeatedly.

In the abstract B&C model by Kazachkov, Le Bodic, and Sankaranarayanan [14], branching yields fixed (constant) bound improvements, while cut nodes have diminishing strength across rounds and may increase node processing time. Under a symmetric branching assumption, they prove that a *root-cut* structure, where all cut nodes are before the first branch node, is sufficient: there always exists a minimal-time root-cut tree, for any function defining node solving time.

We generalize this model to accommodate broader branching and cutting scenarios and show that a root-cut tree solving in minimal (or nearly minimal) time continues to exist. Even when the root-cut structure is suboptimal, i.e., when any minimal-time tree requires *local* cuts (deeper in the tree), we derive an upper bound on how much its processing time can deviate from the minimal-time tree, potentially offering guidance (or validation) for the root-cut dominant approach of solvers. Our contributions are summarized next.

- ▷ Section 3.1: We prove that, under symmetric branching, a minimal-time root-cut tree always exists, even when simultaneously considering multiple cut families, multiple branching families, or node-dependent cut generation times.
- ▷ Section 3.2: We prescribe how to best add a sequence of weak and strong cuts, when the weak cuts cause a relatively smaller increase in node solving time.
- ▷ Section 3.3: We demonstrate a family of asymmetric branching settings where a simple root-cut structure achieves near-optimal performance.
- ▷ Section 4.1: We show by example that the minimal-time tree can be structurally complex for asymmetric settings: a local cut may appear at any depth.
- ▷ Section 4.2: We bound the approximation error of the best root-cut structure with respect to the minimal-time tree.
- ▷ Section 4.3: We characterize conditions that allow the construction of minimal-time trees featuring local cuts under a single-cut-per-path assumption.

Recent theory has been developed to analyze the strength of the pure B&B process [5–7,16], pure CP [9], and comparing B&C, B&B, and CP [1,2]. Basu, Conforti, Di Summa, and Jiang [1] show that B&C can be strictly superior to B&B and CP. In contrast, and counterintuitively, Shah, Dey, and Molinaro [19] prove that adding a single cut can make B&B exponentially worse. Other papers adopt a learning-based perspective on cutting decisions. Berthold, Francobaldi, and Hendel [3] propose methods for learning to use local cuts effectively within the B&C framework. Cheng and Basu [4] establish generalization guarantees for robustness of learned B&C policies across problem instances. However, in all this research, the explicit tradeoffs and interplay between branching and cutting decisions remain poorly understood. Our high-level motivation is to systematically study the interaction between branching and cutting within a B&C framework in a stylized (but already nontrivial to analyze) model, through which we can progress towards theoretical guidance for cut placement and selection in solvers.

2 Preliminaries: Notation and Definitions

A typical B&C approach involves solving a sequence of tightened subproblems in a tree structure. For example, in mixed-integer optimization, the B&C process starts from the *root node*, a convex relaxation of the original problem. The optimal value to the root node provides a dual bound on the original problem. From there, the problem formulation (and dual bound) will be strengthened by either: (1) partitioning the feasible space and creating two subproblems, or (2) adding a round of valid cuts that are violated by points in the current relaxation but satisfied by all feasible mixed-integer solutions. The B&C tree is built by recursively selecting between these two operations until the tree *proves a (dual) bound* matching the primal optimal value (we assume this is given as Z).

We largely adopt the notions modeling a B&C tree and its processing time from Kazachkov, Le Bodic, and Sankaranarayanan [14, Definitions 1–5], including representing the bound improvement from a cut node with c, and from a branch node by constants ℓ (to the left) and r (to the right). We assume that $\ell \leq r$ (without loss of generality) and that cut and branch nodes are derived from the same information (e.g., disjunction), so $c \leq \ell \leq r$.

Kazachkov, Le Bodic, and Sankaranarayanan [14, Definition 2] introduce a *time function* to model that cuts make formulations more complex, thus increasing the processing time of subsequent subproblems. In practice, solvers can choose among *different families of cuts* that have heterogeneous properties, and *cut generation time* also affects cut selection. For simplicity, we assume that the bound improvement from cut family i only depends on the number of cuts previously added from that family and is given by a *(worsening) cut function* $c_i \in \mathcal{F}_{\searrow}$, where $\mathcal{F}_{\searrow} : \mathbb{Z}_{\geq 0} \to \mathbb{R}_{\geq 0}$ denotes the set of monotone nonincreasing functions and $c_i(k)$ represents the bound improvement by the kth round of cuts.

We introduce two extended models of solving time: a *simple* time function in which solving time of a node in the tree only depends on the number of cuts previously generated from each family, and a *generalized* time function that also accounts for the time to compute cuts.

Let $\mathcal{V}_T$ denote the nodes of a B&C tree T. In our model, each node in the tree has a type $i \in \mathcal{I} := \{0, 1, \ldots, m\}$, where type 0 represents a branch node, and type $i \in [m] := \{1, \ldots, m\}$ represents a cut node from family i. We assume that no cut family strictly dominates another (i.e., when a cut family yields a better bound improvement, it causes slower relaxations). Let $\boldsymbol{h}(v) \in \mathbb{Z}^{\mathcal{I}}_{\geq 0}$ be the *history vector*, where $\boldsymbol{h}(v)_i$ is the number of nodes of type i applied before a node $v \in \mathcal{V}_T$. Let $e_i \in \mathbb{R}^{\mathcal{I}}$ denote the unit vector in direction i.

Definition 1 (Simple time function). Given cut functions $\{c_i\}_{i=1}^m$, a function $\mathrm{w} : \mathbb{Z}^{\mathcal{I}}_{\geq 0} \to [1, \infty)$ is a *simple time function* if it is elementwise nondecreasing, $\mathrm{w}(\mathbf{0}) = 1$, and if, for all history vectors $\boldsymbol{h} \in \mathbb{Z}^{\mathcal{I}}_{\geq 0}$, when a cut from family i is stronger than one from family j, i.e., $c_i(\boldsymbol{h}_i) \geq c_j(\boldsymbol{h}_j)$, then $\mathrm{w}(\boldsymbol{h}+e_i) \geq \mathrm{w}(\boldsymbol{h}+e_j)$.

Though Definition 1 allows dependence on the number of branch nodes, in this work, we assume that if $\boldsymbol{h}_i = \boldsymbol{h}'_i$ for all $i \in \mathcal{I} \setminus \{0\}$, then $\mathrm{w}(\boldsymbol{h}) = \mathrm{w}(\boldsymbol{h}')$. As a result, for simplicity, we omit index 0 for a history vector input to w.

Definition 2 (Separation time function). A family of functions $\{r_i : i \in \mathcal{I}\}$ with $r_i : \mathbb{Z}_{\geq 0} \to \mathbb{R}_{\geq 0}$ for all $i \in \mathcal{I}$, is a *separation time function family* if $r_0 \equiv 0$ and $r_i(h_i)$, $i \in [m]$, represents the time to generate one cut from family i given h_i prior cuts from that same family.

Definition 3 (Generalized time function). Given a simple time function w and a separation time function family $\{r_i\}_{i=0}^m$, a function $\widetilde{\mathrm{w}} : \mathbb{Z}_{\geq 0}^{\mathcal{I}} \times \mathcal{I} \to [1, \infty)$ is a *generalized time function* if for all history vectors $\boldsymbol{h}$ and node types $i \in \mathcal{I}$, it holds that $\widetilde{\mathrm{w}}(\boldsymbol{h}, i) = \mathrm{w}(\boldsymbol{h}) + r_i(\boldsymbol{h}_i)$. Given a B&C tree T and a node $v \in \mathcal{V}_T$ of type $i \in \mathcal{I}$ having a history vector $\boldsymbol{h}(v)$, we will also use $\widetilde{\mathrm{w}}(v) := \widetilde{\mathrm{w}}(\boldsymbol{h}(v), i)$.

Optimizing tree size, i.e., $\mathrm{w} \equiv 1$, is studied by Le Bodic and Nemhauser [16] (for B&B) and Kazachkov, Le Bodic, and Sankaranarayanan [14] (for B&C). However, a smaller branch-and-bound tree does not guarantee a better solving time. Example 1 illustrates this phenomenon. In this work, we focus on optimizing *tree time* $\tau_{\widetilde{\mathrm{w}}}(T) := \sum_{v \in \mathcal{V}_T} \widetilde{\mathrm{w}}(v)$ with respect to nontrivial time functions ($\widetilde{\mathrm{w}} \not\equiv 1$). The resulting optimal structure is referred to as a *minimal-time tree.*

Example 1. Figure 1 shows two B&C trees that prove bound $Z = 9$. In the left panel, the tree has 10 nodes and solves in tree time $1 + 9\mathrm{w}(1) = 12.7$. The tree in the right panel has 11 nodes, but solves more quickly, with a tree time of $7 + 4\mathrm{w}(1) = 12.2$, and is in fact the unique minimal-time tree with respect to w.

Our focus is on the *structure* of the tree, by which we mean the relative placement of cut and branch nodes. To this end, we distinguish between *root cut* nodes (those applied in the tree before any branch node) and *local cut* nodes (applied after the first branch node). Define a *leaf cut* node as a cut node without a branch node descendant (i.e., part of a path ending in a *leaf node*), and an *internal cut* node as a local (not at root) cut that is not a leaf cut. We define a *branch-then-cut* node to be a branch node with at least one child being an internal cut. In the right panel of Fig. 1, the root is a branch-then-cut node because its child is an internal cut, while the node labeled 7 is a leaf cut.

3 Root-Cut Structure Can Be Sufficient

In this section, we focus on general settings, including both symmetric and asymmetric, where it is (nearly) sufficient to only consider root cuts to attain a minimal-time tree. We also prove how to best add a sequence of weak and strong cuts, when weak cuts cause comparatively smaller increase in node solving time.

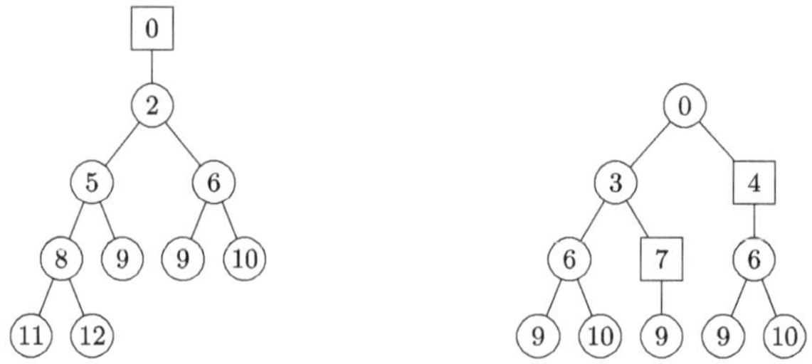

Fig. 1. Two B&C trees proving $Z = 9$, with $c = 2$, $\ell = 3, r = 4$, $\mathrm{w}(1) = 1.3$, and with $\mathrm{w}(k) = M$ for $k \geq 2$ and sufficiently large $M \in \mathbb{R}_{\geq 0}$.

3.1 Symmetric Branching

While the assumption of symmetric branching (when $\ell = r$) is restrictive in general, we initiate our study with this setting as it is a natural theoretical starting point. Basu, Conforti, Di Summa, and Jiang [1] (see also [14, Example 12]) provide a concrete construction demonstrating that adding cuts can lead to an exponential improvement in the B&C tree under a symmetric branching framework. Further, many combinatorial optimization problems, such as scheduling and set covering, have inherent algebraic or structural symmetries [17].

We prove, in Theorem 1, that under symmetric branching, a minimal-time tree exists with all cuts at the root node, even with multiple cut families and accounting for cut generation time. This generalizes a previous result [14, Theorem 23] that assumes only one cut family and constant cut strength.

Theorem 1. *Let* $(\boldsymbol{c}, \ell, r) \in \mathcal{F}^m_{\searrow} \times \mathbb{R}_{\geq 0} \times \mathbb{R}_{\geq 0}$ *such that* $\ell = r$*. For any target bound* Z *and generalized time function* $\widetilde{\mathrm{w}}$*, there exists a minimal-time B&C tree proving bound* Z *with only root cuts.*

Proof. Suppose, for the sake of contradiction, that there exists a target bound Z and generalized time function $\widetilde{\mathrm{w}}$ such that any minimal-time tree proving bound Z contains local cuts. Among those minimal-time trees, let T be one that maximizes the depth of any branch node whose child is a cut node. The symmetry assumption ensures that we can assume T is structurally symmetric, meaning all nodes at any given depth share the same type [14, Lemma 25]. Let u denote a branch node at this maximal depth whose children are both cut nodes.

Let T_L denote the subtree rooted at the left child of u. By our choice of u, T_L consists of a path of k cut nodes $v_1, \ldots, v_k$ at its root, followed by a branch (or leaf) node w, whose children (if they exist) are both identical pure branching trees, denoted as T^*. Let $\boldsymbol{h}(v)$ be the history vector at $v \in \mathcal{V}_T$ *with respect to* T.

We consider replacing T_L in T by a modified tree T'_L, where the path of root cuts in T_L is shifted down a "level". Denote by u' the root (branch) node of T'_L. When $T^* \neq \emptyset$, u' has (identical in both branches) a child cut node v'_1, part of a path of k cut nodes, and let w' be the child of v'_k. When $T^* = \emptyset$ (w is a leaf node), then each child of root u' in T'_L is a path of $k - 1$ cut nodes ending with a leaf node that we will call v'_k (in which case, unlike v_k, v'_k is not a cut node and has no cut generation cost). Note that T'_L proves a bound at least as good as T_L. Further, since the solving time for a node does not change based on the number of previous branch nodes, we define $\boldsymbol{h}'$ as a convenient modified history vector for T'_L, where $\boldsymbol{h}'(u') = \boldsymbol{h}(u)$, $\boldsymbol{h}'(v'_i) = \boldsymbol{h}(v_i)$ for $i \in [k]$, and $\boldsymbol{h}'(x) = \boldsymbol{h}(w)$ for x in the subtree rooted at w'. The (sub)tree times for T_L and T'_L are:

$$\tau_{\widetilde{\mathrm{w}}}(T_L) = \sum_{i=1}^{k} \widetilde{\mathrm{w}}(v_i) + \widetilde{\mathrm{w}}(w) + 2\mathrm{w}(\boldsymbol{h}(w))|T^*| = \sum_{i=1}^{k} \widetilde{\mathrm{w}}(v_i) + \widetilde{\mathrm{w}}(w) + 2\widetilde{\mathrm{w}}(w)|T^*|,$$

$$\begin{aligned}\tau_{\widetilde{\mathrm{w}}}(T'_L) &= \widetilde{\mathrm{w}}(u') + 2\left(\sum_{i=1}^{k} \widetilde{\mathrm{w}}(v'_i) + \mathrm{w}(\boldsymbol{h}'(w'))|T^*|\right) \\ &\leq \widetilde{\mathrm{w}}(u) + 2\sum_{i=1}^{k} \widetilde{\mathrm{w}}(v_i) + 2\widetilde{\mathrm{w}}(w)|T^*| \qquad \text{(as } \widetilde{\mathrm{w}}(v'_k) \leq \widetilde{\mathrm{w}}(v_k)).\end{aligned}$$

As T is a minimal-time tree having the deepest branch node with cut node children, we have that $\tau_{\widetilde{\mathrm{w}}}(T_L) < \tau_{\widetilde{\mathrm{w}}}(T'_L)$, from which we derive

$$\widetilde{\mathrm{w}}(w) < \widetilde{\mathrm{w}}(u) + \sum_{i=1}^{k} \widetilde{\mathrm{w}}(v_i). \tag{1}$$

Next, we consider shifting the cut node children of u up a "level". Let T_u be the subtree rooted at u defined as before, and T''_u the tree that starts with a path of k cut nodes $v''_1, \dots, v''_k$ followed by a branch node u'', where each child of u'' is the subtree T_w of T rooted at w. Observe that

$$\begin{aligned}\tau_{\widetilde{\mathrm{w}}}(T_u) &= \widetilde{\mathrm{w}}(u) + 2\tau_{\widetilde{\mathrm{w}}}(T_L) = \widetilde{\mathrm{w}}(u) + 2\sum_{i=1}^{k} \widetilde{\mathrm{w}}(v_i) + 2\tau_{\widetilde{\mathrm{w}}}(T_w), \quad \text{and} \\ \tau_{\widetilde{\mathrm{w}}}(T''_u) &= \sum_{i=1}^{k} \widetilde{\mathrm{w}}(v''_i) + \widetilde{\mathrm{w}}(u'') + 2\tau_{\widetilde{\mathrm{w}}}(T_w) = \sum_{i=1}^{k} \widetilde{\mathrm{w}}(v_i) + \widetilde{\mathrm{w}}(w) + 2\tau_{\widetilde{\mathrm{w}}}(T_w).\end{aligned}$$

Then $\tau_{\widetilde{\mathrm{w}}}(T''_u) - \tau_{\widetilde{\mathrm{w}}}(T_u) = \widetilde{\mathrm{w}}(w) - \left(\widetilde{\mathrm{w}}(u) + \sum_{i=1}^{k} \widetilde{\mathrm{w}}(v_i)\right)$ is negative by (1). Thus, if shifting the path of cuts down is strictly worse, then shifting the cuts up strictly decreases tree time, contradicting the assumption that T is minimal time. □

We emphasize that symmetry plays a crucial role in this proof. In particular, symmetry of a tree (which holds under symmetric branching) will force cut nodes to move up, regardless of how the time function is defined. On the other hand, the proof does not rely on branching strength: the result holds even with multiple symmetric branch variable candidates, i.e., possibly different values throughout the tree but always equal left/right bound improvements.

3.2 Weak-Sparse Versus Strong-Dense Cut Families

We now consider how cut nodes should be placed inside the minimal-time tree when there are families of weak and strong (constant-strength) cuts, in which weaker cuts cause a relatively smaller increase in node processing time.

Theorem 2. *There exists a minimal-time tree with respect to generalized time function* $\widetilde{\mathrm{w}}$*, such that for any path of cuts in the tree, cut nodes from weaker families come before cut nodes from stronger families.*

Proof (Sketch). Given a sequence of cuts, performing a lifting action that moves the cut node from the weakest family to the front of the sequence can only improve the total processing time. Notably, even when considering cut generation time, the total generation time remains invariant during the reordering procedure. In other words, cut placement is determined by w, rather than by $\widetilde{\mathrm{w}}$. □

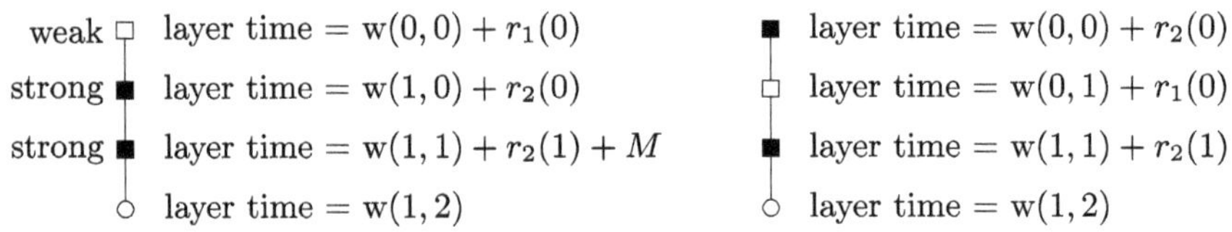

Fig. 2. Under more general separation time functions, weaker cuts may not come before stronger ones. Here, generating two consecutive rounds from the same family incurs a large cost M. Hence, the alternating structure in the right panel is cheaper than the weak-then-strong pattern in the left panel.

The full proof of Theorem 2 is deferred to Appendix B of the extended paper. We emphasize that, under our current setting, cut generation time affects the optimal number of cut nodes but not where cuts are placed in a minimal-time tree. The crucial assumption is that separation time depends only on the *number* of previous cuts from the same family. Figure 2 depicts one example where this assumption is relaxed, which heavily penalizes consecutive nodes belonging to the same cut family; the resulting optimal pattern favors alternating between cut families over the weak-then-strong pattern.

3.3 Asymmetric Branching

Having analyzed symmetric branching with multiple cut families in the previous subsection, we now turn to the single cut-family, single branching-family B&C model with constant but nonidentical bound improvements ($\ell \neq r$).

As established in Sect. 3.1, symmetry in a tree compels cut nodes to shift upward toward the root. Consequently and intuitively, under an asymmetric branching regime, where symmetry is not always preserved, one might wonder whether for any asymmetric setting, there always exists some choices of target bound and time function that force the inclusion of local cuts within any minimal-time tree. We prove that this is not universally true: there exists an asymmetric setting ($c : \ell : r = 1 : 2 : 4$) in which a minimal-time tree without any internal cuts always exists, regardless of the specific form of the time function. Proving this involves several intermediate results that we present first.

First, akin to Theorem 1, we show that symmetry in a subtree under asymmetric branching also shifts cut nodes upward. For simplicity, define a *branch-then-bothcut* node to be a branch node with both children being cut nodes.

Theorem 3. *Given a setting* $(c, \ell, r) \in \mathbb{R}^3_{\geq 0}$*, for any target bound* Z *and simple time function* w*, there always exists a minimal-time B&C tree proving bound* Z *without any branch-then-bothcut node in the tree.*

The proof is similar to that of Theorem 1 and appears in full in Appendix C of the extended paper. The next corollary follows directly.

Corollary 4. *For every* $(c, \ell, r) \in \mathbb{R}^3_{\geq 0}$ *and simple time function* w*, a minimal-time tree with the fewest cut nodes will have no branch-then-bothcut node.*

Corollary 4 rules out branch nodes where both children are cut nodes. Nevertheless, it remains unclear whether branch nodes with exactly one child as a cut node may be necessary. More importantly, if such branch nodes do appear in a minimal-time tree, it is not immediately clear whether the deepest such node has a left-child cut or a right-child cut. We address this question (for nonleaf cut nodes) in Lemma 5, whose proof is in Appendix D of the extended paper.

Lemma 5. *When* $r = \beta\ell$ *for some* $\beta \in \mathbb{Z}_{\geq 1}$*, among all minimal-time trees with the least number of cut nodes, there is a minimal-time tree where the deepest branch-then-cut node (when one exists) is a branch-then-leftcut node.*

We emphasize that, unlike symmetric branching, where symmetry is maintained throughout the tree, any node-shifting operation can significantly alter the tree structure under an asymmetric branching setting. Therefore, it is crucial to understand how tree structure changes as the target bound varies. Lemma 6 reveals the pattern under this special setting. For a given ℓ, r, and *even* target bound Z, define $p(Z)$ to be the function that calculates size of the smallest B&B tree that proves bound Z, and define $k(Z)$ to be the function that returns the number of leaf nodes proving exactly bound Z inside this tree.

Lemma 6. *When* $\ell = 2$ *and* $r = 4$*, for any even target bound* Z*,*

1) $k(Z{+}4) = k(Z) + k(Z{+}2)$,
2) $p(Z{+}2) = p(Z) + 2k(Z)$, *and*
3) $p(Z{+}4) = 1 + p(Z) + p(Z{+}2)$.

Lemma 7 states that a substructure consisting of two consecutive leaf cuts (two cuts followed by a leaf node) is dominated by a branch node (3 nodes total).

Lemma 7. *For any* $(c, \ell, r) \in \mathbb{R}^3_{\geq 0}$ *with* $c : \ell : r = 1 : 2 : 4$*, simple time function* w*, and target bound* Z*, every minimal-time tree proving bound* Z *with the fewest number of cut nodes contains at most one leaf cut node (at each leaf).*

Finally, we prove Theorem 8, the main result of this subsection, by integrating these intermediate results, that for (c, ℓ, r) satisfying a ratio $1:2:4$, there exists a minimal-time tree where all cuts are either at the root or at the leaves, with at most one cut node at each leaf node. Due to length constraints, we present only the key idea behind the proof, while the visualization and detailed discussion are in Appendix E of the extended paper.

Theorem 8. *For $(c, \ell, r) \in \mathbb{R}^3_{\geq 0}$ with $c : \ell : r = 1 : 2 : 4$, every target bound Z, and any simple time function* w*, there exists a minimal-time tree that does not contain any internal cuts.*

Proof (Sketch). Suppose there exists a time function w such that every minimal-time tree proving a given target bound has an internal cut. Consider, among the minimal-time trees that have the fewest cut nodes, one that has the deepest branch-then-cut node (call this node u). Let T be the subtree rooted at u. By Lemma 5, we assume u is a branch-then-leftcut node, i.e., the left branch T_ℓ begins with κ cut nodes followed by a pure branching tree (possibly with leaf cuts), while the right branch T_r is a pure branching tree (possibly with leaf cuts). Let T_1 and T_2 be the left and right branches of the root node of T_r.

We construct a modified tree T' by shifting the cut nodes down *two* levels among the leftmost path in T_ℓ while keeping T_r unchanged. In T', for the left child of u, we insert a right branch identical to T_1 (since it is obtained by one left and one right branch operation). For the left descendant two layers below u in T', the left branch is the path of κ cut nodes, and for the right branch, we insert T_2 (since branching twice left and once right is equivalent to two right branches when $r = 2\ell$).

By calculating the difference in tree time between T and T', we derive bounds on the time function, which are then used to construct another valid tree, denoted T'', by lifting the sequence of internal cuts, to derive a contradiction. Depictions of T, T', and T'' appear in Fig. 9 in Appendix E of the extended paper. □

Note that the results derived for the single cut-family, single branching-family model may not easily extend to multiple-family cases. For instance, while internal cuts are unnecessary under the individual settings $(c_1, \ell_1, r_1) = (1, 2, 4)$ and $(c_2, \ell_2, r_2) = (1, 3, 3)$, the existence of an internal-cuts-free minimal-time tree is no longer guaranteed when the two settings are combined.

4 Local Cuts Can Be Necessary

In this section, we explore general asymmetric branching settings with a single constant-strength cut family in which a root-cut framework is no longer sufficient. In other words, internal cuts become essential for achieving a minimal-time structure. This phenomenon captured by our model aligns with practical experience in modern solvers, where cuts generated beyond the root node play a critical role in reducing solving time and improving performance [3].

4.1 Minimal-Time Trees Can Be Complex

Unlike the symmetric branching setting or the special case analyzed in Sect. 3.3, where a minimal-time tree exists with cuts only in one of two locations (at the root or at a leaf), general asymmetric branching settings together with special choices of time functions can yield far more complex optimal structures, making prescribing best cut locations nontrivial. We construct an intuitive example in Lemma 9 to illustrate the complexity of minimal-time trees under asymmetric branching settings. We use $T(\ell, r, Z)$ to represent the minimal-size pure branching tree proving bound Z using branching variable (ℓ, r).

Lemma 9. *Let* $(c, \ell, r) = (2, 3, 4)$. *For any depth* $d \in \mathbb{Z}_{\geq 0}$, *there exists a simple time function* w *such that any minimal-time tree proving bound* $Z := 3d + 5$ *will have a cut node at depth d along the leftmost root-to-leaf path.*

Proof. Set $\mathrm{w}(1) = 2 - |T(\ell, r, Z')|^{-1}$ for $Z' := 3d + 9$ and for all $k \geq 2$, $\mathrm{w}(k) = M$ for a sufficiently large M. This time function ensures that in a minimal-time tree, a leaf-cut structure dominates a root-cut structure (with the exception of the base case, where a branch-then-bothcut node is replaced by a root cut). See Fig. 3 for intuition that, when $d = 1$, a minimal-time tree has a cut node immediately after the root.

We prove the lemma by induction on d. When $d = 0$, one can verify that the minimal time tree is a cut node followed by a branch node. Assume inductively that for any d, a minimal-time tree with respect to w proving bound $3d + 5$, a cut node lies at depth d along the leftmost path. Now consider depth $d + 1$ and a minimal-time tree that proves target bound $3d + 8$. By [14, Lemma 24], the left branch of the minimal-time tree must be a minimal-time tree proving bound $3d + 5$ with respect to the same time function w. The desired conclusion then follows directly from the induction hypothesis. □

Fig. 3. Two B&C trees proving $Z = 8$, with $c = 2, \ell = 3, r = 4$. The tree in the right panel dominates the one on the left whenever $\mathrm{w}(1) < 2$, as a consequence of the induced symmetry.

Lemma 9 presents an asymmetric setting in which the minimal-time structure differs significantly from a simple root-cut configuration. Notably, although the internal cut node may appear at an arbitrary absolute depth, it consistently lies two levels above the bottom of the tree. In other words, the placement of cut nodes is not entirely arbitrary in this setting.

4.2 Root-Cut Approximation for Minimal-Time Tree

Having established that the minimal-time tree can exhibit a complex structure, a natural question arises: how close is it, in terms of total tree processing time, to a root-cut structured tree? Theorem 11 proves that the tree time of the best root-cut structured tree has bounded deviation from that of a minimal-time tree. We need the following lemma for an intermediate derivation.

Lemma 10. *For any $(c, \ell, r) \in \mathbb{R}^3_{\geq 0}$, let T be any tree and let k be the maximum number of cut nodes along any root-leaf path in T. Let T' be the tree obtained by removing all cut nodes from T and adding a path of k root cuts. Then $|T| \geq |T'|$.*

Lemma 10 restates [14, Lemma 7], that moving cut nodes to the root does not increase the total size of the tree. With this, we bound the root-cut approximation for a minimal-time tree.

Theorem 11. *For $(c, \ell, r) \in \mathbb{R}^3_{\geq 0}$, simple time function w, and bound Z, let T^* be a minimal-time tree proving Z, k^* the maximum number of cuts along any root-leaf path, and T the best root-cut tree proving Z. Then $\tau_{\mathrm{w}}(T) \leq \mathrm{w}(k^*)\tau_{\mathrm{w}}(T^*)$.*

Proof. Given any setting $(Z, c, \ell, r; \mathrm{w})$, let $\tau_{\text{root}}(k)$ and $t_{\text{root}}(k)$ denote the tree time and size of the root-cut tree that has k root cuts and proves Z. Let k' be the number of cuts in the minimal-time root-cut structured tree. Denote by τ^* and $|T^*|$ the time and size of T^*, respectively. Then, we have:

$$\tau_{\text{root}}(k') \leq \tau_{\text{root}}(k^*) \leq \mathrm{w}(k^*)t_{\text{root}}(k^*) \overset{\text{Lemma 10}}{\leq} \mathrm{w}(k^*)|T^*| \leq \mathrm{w}(k^*)\tau^*. \qquad \square$$

Theorem 11 provides a straightforward upper bound on the error when approximating the actual minimal-time tree with the best root-cut structure. One may observe that this bound is tight if and only if the time function is trivial, i.e., $\mathrm{w} \equiv 1$, since the last inequality relies on using the size of the minimal-time tree as a proxy for its processing time. We revisit this via simulation in Sect. 4.4.

Interestingly, for the construction in Lemma 9, Theorem 11 reveals that the best root-cut structure is at most twice as slow as the optimal one, despite the significant differences in their structures.

4.3 Adding at Most One Cut Node Per Root-to-Leaf Path

Typically, solvers limit the number of cut rounds, as successive rounds often yield diminishing bound improvements [8] while substantially increasing problem size and solution time. Moreover, excessive cuts can harm numerical stability and lead to ill-conditioned subproblems that degrade solver performance [18,20].

In this subsection, we investigate necessary and sufficient conditions under which internal cuts become required for a minimal-time tree. We restrict our focus to the *one-cut* setting, defined as a configuration where at most one round of cuts is added in each root-leaf path. This assumption simplifies the analysis while preserving key structural insights.

Note that if $\mathrm{w}(1) > 2$ under the one-cut setting, no cut nodes would appear in any minimal-time tree, as branch nodes provide both better bound improvements and smaller processing time when $\mathrm{w}(1)$ is large. Meanwhile, when $\mathrm{w}(1) = 1$, there always exists a minimal-time root-cut tree. Accordingly, we focus our analysis on $\mathrm{w}(1) \in (1, 2)$. Under this assumption, by Theorem 3 (for simple time functions), a branch node in a minimal-time tree will not have two cut node children.

Other than the time function, the relationship among Z, c, ℓ, and r also determines whether cut nodes will help. For instance, if $c = 1$ and all other parameters are even numbers, cut nodes would not appear in any minimal-time tree, for any value of $\mathrm{w}(1)$. In Lemma 12, we show that one sufficient condition for cuts to be strictly beneficial is if the pure branching tree has a node with at most c bound left to prove. The proof is in Appendix F of the extended paper.

Lemma 12. *Given a one-cut setting* $(Z, c, \ell, r) \in \mathbb{Z}^4_{\geq 0}$ *and* $\mathrm{w}(1) \in (1, 2)$, *every minimal-time tree proving* Z *has a cut node if and only if there exists* $k_1 \in [\lceil \frac{Z}{\ell} \rceil - 1] \cup \{0\}$, $k_2 \in [\lceil \frac{Z}{r} \rceil - 1] \cup \{0\}$ *such that* $k_1\ell + k_2 r \in \{Z-c, Z-c+1, \ldots, Z-2, Z-1\}$.

When the pure branching tree contains a node with remaining bound at most c, the one-cut setting admits exactly two canonical structures: the *one-root-cut* tree, which places the cut at the root and then branches optimally to prove $Z - c$; and the minimal leaf-cut tree, which branches until the remaining bound first enters $\{1, \ldots, c\}$ and places a cut there. These two structures are uniquely determined by (Z, c, ℓ, r) and serve as the extreme cases—root-cut is best when $\mathrm{w}(1)$ is small (cheap cuts), leaf-cut is best when $\mathrm{w}(1)$ is large (expensive cuts).

We next prove that Lemma 12 constrains the *relative* placement of cuts in sibling subtrees. Concretely, in a one-cut setting, if one child of a branch-then-cut node begins with an internal cut (a one-root-cut subtree), the other child will also contain a node where a cut saves time—so the sibling cannot be cut-free. Lemma 13 formalizes this, with proof in Appendix G of the extended paper.

Lemma 13. *Given a one-cut setting* $(Z, c, \ell, r) \in \mathbb{Z}^4_{\geq 0}$ *and* $\mathrm{w}(1) \in (1, 2)$, *a minimal-time tree cannot contain a branch-then-cut node where one child is a pure branching subtree.*

Specifically, for the deepest branch-then-cut node (which has an internal cut child) in a minimal-time tree under a one-cut setting, one branch has a root cut and the other must have leaf cuts. Consequently, the remaining bound to be proven at this node must be at least $r + c$. This because the right branch subtree must contain at least one cut node, one branch node, and one leaf node.

We now turn our attention to characterizing the structure of a minimal-time tree and, subsequently, the placement of the cut nodes. Lemma 14 characterizes how the structure of a minimal-time tree evolves as $\mathrm{w}(1)$ grows. The detailed proof is provided in Appendix H of the extended paper.

Lemma 14. *Given a one-cut setting* $(Z, c, \ell, r) \in \mathbb{Z}^4_{\geq 0}$ *and a simple time function* w*, suppose the condition in Lemma 12 is satisfied. As* w(1) *increases from* 1 *to* 2*, the minimal-time tree* T *undergoes an ordered, monotonic structural shift in the following sequence: one-root-cut* $\rightarrow$ *internal-cut* $\rightarrow$ *leaf-cut.* T *may skip one or more stages depending on the specific parameter setting* (Z, c, ℓ, r).

Recall in the base case of Lemma 9, the one-root-cut structure remains minimal for all $\mathrm{w}(1) \in (1, 2)$. Let $\alpha(Z, c, \ell, r)$ denote the value of w(1) at which the one-root-cut and minimal leaf-cut trees proving Z require the same processing time. Lemma 14 suggests that this threshold plays a crucial role in determining the architecture of the minimal-time tree. Corollary 15 follows immediately.

Corollary 15. *Given a one-cut setting* $(Z, c, \ell, r) \in \mathbb{Z}^4_{\geq 0}$*, if there exists an interval* $[a, b] \subseteq (1, 2)$ *for* w(1) *within which every minimal-time tree proving bound* Z *is internal-cut structured, then* $\alpha(Z, c, \ell, r) \in [a, b]$.

Building on Lemma 14 and Corollary 15, we describe the structural transition of the minimal-time tree as w(1) increases. In a setting where all one-root-cut, internal-cut, leaf-cut structured minimal-time trees are achieved for different choices of $\mathrm{w}(1) \in (1, 2)$, there exist $\epsilon_1, \epsilon_2 \in \mathbb{R}_{\geq 0}$ such that every minimal-time tree begins with a root-cut structure at $\mathrm{w}(1) = 1$ and remains unchanged for $\mathrm{w}(1) \in (1, \alpha(Z, c, \ell, r) - \epsilon_1]$. As w(1) increases from $\alpha(Z, c, \ell, r) - \epsilon_1$ to $\alpha(Z, c, \ell, r) + \epsilon_2$, every minimal-time tree transitions to an internal-cut structure, and further shifts to a leaf-cut structure once $\mathrm{w}(1) > \alpha(Z, c, \ell, r) + \epsilon_2$.

Finally, by aligning Lemmas 12 to 14 and Corollary 15, we can determine whether it is possible to construct such ϵ_1, ϵ_2, i.e., an interval for time function w(1) within which the minimal-time tree is internal-cut structured.

Theorem 16. *Given a one-cut setting* $(Z, c, \ell, r) \in \mathbb{Z}^4_{\geq 0}$ *with* $\alpha_T := \alpha(Z, c, \ell, r) \in (1, 2)$*, there exists* $[a, b] \subseteq (1, 2)$ *such that* $\mathrm{w}(1) \in [a, b]$ *guarantees an internal-cut structured minimal-time tree if and only if there is a* $k_1 \in [\lceil \frac{Z}{\ell} \rceil - 1] \cup \{0\}$, $k_2 \in [\lceil \frac{Z}{r} \rceil - 1] \cup \{0\}$ *such that* $\min\{\alpha_\ell, \alpha_r\} \leq \alpha_T \leq \max\{\alpha_\ell, \alpha_r\}$*, where* $Z' := Z - k_1\ell - k_2 r > r + c$, $\alpha_\ell := \alpha(Z' - \ell, c, \ell, r)$*, and* $\alpha_r := \alpha(Z' - r, c, \ell, r)$.

Proof. If such an interval exists, Lemma 14 implies that there exists a minimal-time internal-cut tree T with respect to $\mathrm{w}(1) = \alpha_T$. Let u be the deepest branch-then-cut node from T, reached by taking k_1 left branches and k_2 right branches from the root. As a branch-and-cut node is followed by an *internal* cut, $k_1\ell + k_2 r < Z - r - c$. Let T_ℓ and T_r denote the left and right branch of u. By [14, Lemma 24], T_ℓ and T_r are minimal with respect to the same w. From Lemma 13, either T_ℓ or T_r is root-cut structured, while the other contains leaf cuts under $\mathrm{w}(1) = \alpha_T$. Suppose the left branch T_ℓ has a cut at its root. Then $\alpha_\ell \geq \alpha_T$, as otherwise the one-root-cut structure is dominated by a leaf-cut structure under α_T. Since now T_r has only leaf cuts, we must have $\alpha_r \leq \alpha_T$, as otherwise the leaf-cut structure would be dominated by a one-root-cut structure. An analogous argument holds if T_ℓ has only leaf cuts. Hence, $\min\{\alpha_\ell, \alpha_r\} \leq \alpha_T \leq \max\{\alpha_\ell, \alpha_r\}$.

Now consider the converse. Let T be a minimal-time tree with respect to $\mathrm{w}(1) = \alpha_T$ that begins with a branch node at the root. Such a tree is guaranteed to exist, as the one-root-cut tree and the minimal leaf-cut tree are identical in tree time when $\mathrm{w}(1) = \alpha_T$. We show that T is either an internal-cut structure itself, or a leaf-cut structure from which an alternative internal-cut tree T' achieving the same processing time can be constructed. Let u be a node reached by taking any k_1 left branches and k_2 right branches from the root of T. If there is a cut along the path from the root to u, then that cut node must have branching descendants in the one-cut setting, as $k_1\ell + k_2 r < Z - c$, completing the proof.

Otherwise, consider the subtree from T rooted at u, denoting its left and right branches as T_ℓ and T_r. Without loss of generality, we can further assume $\alpha_r \leq \alpha_\ell$. Since $\alpha_T \leq \alpha_\ell$, Lemma 14 implies that T_ℓ is either a one-root-cut or internal-cut structure under $\mathrm{w}(1) = \alpha_T$, or it could be a leaf-cut structure in the limiting case where $\alpha_T = \alpha_\ell$. Similarly, T_r is either leaf-cut or internal-cut structured as $\alpha_T \geq \alpha_r$ (or root-cut structured when equality holds). Since $\mathrm{w}(1) < 2$, we know T_ℓ and T_r are not both one-root-cut trees. The proof is complete if one of T_ℓ or T_r is root-cut or internal-cut structured, as that makes T an internal-cut tree. The remaining case is when both are leaf-cut structures, which occurs when $\alpha_T = \alpha_\ell$. This means there exists a one-root-cut tree T'_ℓ with the same processing time as the leaf-cut tree T_ℓ. Replacing T_ℓ by T'_ℓ, the assumed inequality $Z - k_1\ell - k_2 r > r + c$ ensures that the root cut of T'_ℓ has a branch node descendant: it is an internal cut in T. Thus, there exists an interval $a \leq \alpha_T \leq b$ for which a minimal-time tree featuring internal cuts is guaranteed.
□

4.4 Empirical Evaluation of the Root-Cut Approximation

Following Sect. 4.3, we continue to restrict attention to settings that allow at most one cut per path. To assess the strength of the root-cut structure, we conduct a series of simulation experiments designed to empirically examine the performance of the bound established in Theorem 11.

We generate 1,000 synthetic instances parameterized by (Z, c, ℓ, r) that satisfy the condition of Lemma 12. For each instance, we construct both the minimal root-cut tree and the actual minimal-time tree, then compute their total processing times, by solving a mixed-integer linear program (with formulation specified in Appendix I of the extended paper) as $\mathrm{w}(1)$ varies. The ratio between these two quantities serves as a measure of approximation quality, which we further compare against the proposed theoretical upper bound $\mathrm{w}(1)$.

Across all 1,000 configurations, we identify the 10 parameter settings with the largest approximation errors. Following Lemma 14, the approximation ratio is expected to exhibit a stable–increase–decrease pattern as $\mathrm{w}(1)$ gradually increases, reaching its maximum at $\mathrm{w}(1) = \frac{T(\ell,r,Z)-1}{T(\ell,r,Z-c)}$, that is, when the minimal pure branching structure and the root-cut structure have identical processing time. This theoretical pattern is fully consistent with our simulation results.

As shown in Fig. 4, our factor bound provides a conservative estimate of the actual ratio. Empirically, the deviation between the root-cut approximation and the minimal-time tree is generally modest, remaining below 1.2 even in the worst cases. This indicates that the root-cut structure provides a robust and practically effective approximation to the actual minimal-time tree in our model.

Lastly, as w(1) grows, by Lemma 14, the minimal-time structure shifts from root-cut to internal-cut to leaf-cut. Intuitively, cuts get pushed down in a minimal-time tree. *Does this shift progress one layer at a time?* The answer is *not always.* The shallowest cut among all minimal trees may jump from the root to a much deeper layer as w(1) grows (see Fig. 11 in Appendix J of the extended paper.

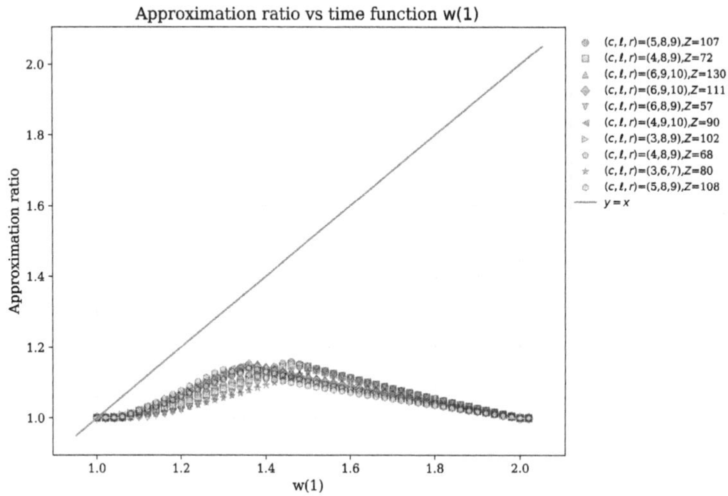

Fig. 4. 10 parameter settings with the largest root-cut approximation error, with the red line representing the proposed factor bound. (Color figure online)

Acknowledgments. This material is based upon work supported by the Air Force Office of Scientific Research under award number FA9550-23-1-0340.

Disclosure of Interests. The authors have no competing interests to declare.

References

1. Basu, A., Conforti, M., Di Summa, M., Jiang, H.: Complexity of branch-and-bound and cutting planes in mixed-integer optimization–II. Combinatorica **42**, 971–996 (2022). https://doi.org/10.1007/s00493-022-4884-7
2. Basu, A., Conforti, M., Di Summa, M., Jiang, H.: Complexity of branch-and-bound and cutting planes in mixed-integer optimization. Math. Program. **198**(1), 787–810 (2023). https://doi.org/10.1007/s10107-022-01789-5

3. Berthold, T., Francobaldi, M., Hendel, G.: Learning to use local cuts. Math. Program. Comput. **17**(3), 437–450 (2025). https://doi.org/10.1007/s12532-025-00278-y
4. Cheng, H., Basu, A.: Generalization guarantees for learning score-based branch-and-cut policies in integer programming. In: The Thirty-Ninth Annual Conference on Neural Information Processing Systems (2025). https://openreview.net/forum?id=v5ru9MGjsW
5. Dadush, D., Tiwari, S.: On the complexity of branching proofs. In: Proceedings of the 35th Computational Complexity Conference. CCC 2020. Schloss Dagstuhl–Leibniz-Zentrum fuer Informatik, Dagstuhl, DEU (2020). https://doi.org/10.4230/LIPIcs.CCC.2020.34
6. Dey, S.S., Dubey, Y., Molinaro, M.: Lower bounds on the size of general branch-and-bound trees. Math. Program. **198**(1), 539–559 (2023). https://doi.org/10.1007/s10107-022-01781-z
7. Dey, S.S., Dubey, Y., Molinaro, M., Shah, P.: A theoretical and computational analysis of full strong-branching. Math. Program. **205**(1), 303–336 (2024). https://doi.org/10.1007/s10107-023-01977-x
8. Dey, S.S., Kazachkov, A., Lodi, A., Munoz, G.: Cutting plane generation through sparse principal component analysis. SIAM J. Optim. **32**(2), 1319–1343 (2022). https://doi.org/10.1137/21M1399956
9. Dey, S.S., Molinaro, M.: Theoretical challenges towards cutting-plane selection. Math. Program. **170**(1), 237–266 (2018). https://doi.org/10.1007/s10107-018-1302-4
10. Gomory, R.E.: Outline of an algorithm for integer solution to linear programs. Bull. Am. Math. Soc. **64**(5), 275–278 (1958). https://doi.org/10.1090/S0002-9904-1958-10224-4
11. Gomory, R.E.: Solving linear programming problems in integers. In: Bellman, R., Hall, M. (eds.) Combinatorial Analysis, Proceedings of Symposia in Applied Mathematics, vol. 10, pp. 211–215. Providence, RI (1960)
12. Gomory, R.E.: An algorithm for integer solutions to linear programs. Recent Adv. Math. Program. **64**, 260–302 (1963)
13. Kazachkov, A.M., Le Bodic, P., Sankaranarayanan, S.: An abstract model for branch-and-cut. In: Aardal, K., Sanità, L. (eds.) Integer Programming and Combinatorial Optimization. Lecture Notes in Computer Science, pp. 333–346. Springer, Cham (2022). https://doi.org/10.1007/978-3-031-06901-7_25
14. Kazachkov, A.M., Le Bodic, P., Sankaranarayanan, S.: An abstract model for branch and cut. Math. Program. **206**(1), 175–202 (2024). https://doi.org/10.1007/s10107-023-01991-z
15. Land, A.H., Doig, A.G.: An automatic method of solving discrete programming problems. Econometrica **28**, 497–520 (1960). https://doi.org/10.2307/1910129
16. Le Bodic, P., Nemhauser, G.: An abstract model for branching and its application to mixed integer programming. Math. Program. **166**(1–2), 369–405 (2017). https://doi.org/10.1007/s10107-016-1101-8
17. Margot, F.: Symmetry in integer linear programming. In: Jünger, M., Liebling, T.M., Naddef, D., Nemhauser, G.L., Pulleyblank, W.R., Reinelt, G., Rinaldi, G., Wolsey, L.A. (eds.) 50 Years of Integer Programming 1958-2008, pp. 647–686. Springer, Heidelberg (2010). https://doi.org/10.1007/978-3-540-68279-0_17
18. Miltenberger, M., Ralphs, T., Steffy, D.E.: Exploring the numerics of branch-and-cut for mixed integer linear optimization. In: Kliewer, N., Ehmke, J.F., Borndörfer, R. (eds.) Operations Research Proceedings 2017. ORP, pp. 151–157. Springer, Cham (2018). https://doi.org/10.1007/978-3-319-89920-6_21

19. Shah, P., Dey, S.S.,Molinaro, M.: Non-monotonicity of branching rules with respect to linear relaxations. INFORMS J. Comput. **38**(1), 53–66 (2025). https://doi.org/10.1287/ijoc.2024.0709
20. Zanette, A., Fischetti, M., Balas, E.: Lexicography and degeneracy: can a pure cutting plane algorithm work? Math. Program. **130**(1, Ser. A), 153–176 (2011). https://doi.org/10.1007/s10107-009-0335-0

Linear Programming Hierarchies Collapse Under Symmetry

Yuri Faenza[1], Victor Verdugo[2,3], José Verschae[2], and Matías Villagra[1(✉)]

[1] Columbia University, IEOR, New York, USA
{yf2414,mjv2153}@columbia.edu

[2] Institute for Mathematical and Computational Engineering, Pontificia Universidad Católica de Chile, Santiago, Chile
{victor.verdugo,jverschae}@uc.cl

[3] Department of Industrial and Systems Engineering, Pontificia Universidad Católica de Chile, Santiago, Chile

Abstract. The presence of symmetries is one of the central structural features that make some integer programs challenging for state-of-the-art solvers. In this work, we study the efficacy of Linear Programming (LP) hierarchies in the presence of symmetries. Our main theorem unveils a connection between the algebraic structure of these relaxations and the geometry of the initial integer-empty polytope: We show that under $(k+1)$-transitive symmetries–a measure of the underlying symmetry in the problem–the corresponding relaxation at level k of the hierarchy is non-empty if and only if the initial polytope intersects all $(n-k)$-dimensional faces of the hypercube. In particular, the hierarchies of Sherali-Adams, Lovász-Schrijver, and the Lift-and-Project closure are equally effective at detecting integer emptiness. Our result provides a unifying, group-theoretic characterization of the poor performance of LP-based hierarchies, and offers a simple procedure for proving lower bounds on the integrality gaps of symmetric polytopes under these hierarchies.

Keywords: Linear Programming Hierarchies · Symmetry · k-transitivity · Integer Programming

1 Introduction

Understanding what makes an Integer Program (IP) challenging for state-of-the-art algorithms is a fundamental step toward developing more effective solvers. While it is reasonable to assume that a larger number of variables and constraints would slow down any algorithm for solving an IP, there are other, more subtle features of IPs that are problematic due to the techniques we currently employ. One such feature is *symmetry*, informally defined as any transformation of a mathematical object that leaves its fundamental properties unchanged.

Mathematicians usually welcome a high degree of symmetry, as it often implies strong properties of the objects under consideration. For instance, systems of linear equalities $Ax = b$ can be solved faster in practice when $A = A^\top$,

S. S. Dey et al. (Eds.): IPCO 2026, LNCS 16588, pp. 491–506, 2026.
https://doi.org/10.1007/978-3-032-28691-8_32

because factorizations of A are guaranteed to exist in this case, see, e.g., [6]. In contrast, symmetry is usually problematic in Integer Programming. Even small-sized symmetric problems are often hard to solve [14,31,34–36]. In his survey on the topic [31], Margot states that *the trouble comes from the fact that many sub-problems in the enumeration tree are isomorphic, forcing a wasteful duplication of effort*–but as we will see, this is not the only problem caused by symmetry.

Understanding how to tackle symmetric IPs is essential, as some degree of symmetry is prevalent in both theory and practice. The structure of the problem itself sometimes causes symmetry – for instance, if we wish to schedule jobs on identical machines, or if we solve a combinatorial optimization problem on a highly symmetric graph. Whichever the cause, more than 36% of the instances from MIPLIB 2010 admit nontrivial symmetries; see the work by Pfetsch and Rehn [36] for details. Moreover, symmetries have played an important role for theoretical work, both from a polyhedral perspective (e.g., [3,40]), as a devise to construct lower bounds of relaxations (e.g., [18,28]), and a way of reducing the dimensionality of convex relaxations (e.g., [15,34]), to mention a few.

Our Contributions. In this work, we investigate how symmetry can harm state-of-the-art techniques for solving Integer Programs. Our work distinguishes itself from the literature with respect to two features. First, unlike the most common approach of studying the effect of symmetry on creating many "equivalent" integer solutions, we investigate the relationship between symmetry and the performance of *linear programming hierarchies*, which are well-known techniques that iteratively improve LP relaxations of an integer program. Second, while most research has focused on specific classes of problems such as, e.g., matchings [22,32], knapsack [18,21], max-cut and sparsest-cut [7,19,28], scheduling [25,39], vertex-cover [16], and planted clique [2,33], with the notable exception of the work by Kurpisz et al. [24], our assumptions only rely on a well-established group-theoretic parametrization of the symmetry of the underlying polytope, namely *k-transitivity.* Our main result is the following.

Theorem 1. *Let $k, n \in \mathbb{N}$ such that $k < n$, and let $P \subseteq [0,1]^n$ be a $(k+1)$-transitive integer-empty polytope. Then the following statements are equivalent:*

(A) P intersects all of the $(n-k)$-dimensional faces of the hypercube $[0,1]^n$;
(B) $\mathrm{SA}^k(P) \neq \emptyset$;
(C) $\mathrm{LS}^k(P) \neq \emptyset$;
(D) $\mathrm{LS}_0^k(P) \neq \emptyset$.

We defer to Sect. 3 the definition of the hierarchies SA^k, LS^k, and LS_0^k. A polytope P is G-invariant for a permutation group G over $[n] := \{1, \ldots, n\}$, if under the action of the elements of G, given by $\pi x := (x_{\pi^{-1}(1)}, \ldots, x_{\pi^{-1}(n)})$, we have that $P = \pi(P)$ for all $\pi \in G$. A group G is k-transitive if it can map any ordered sequence of k distinct indices in $[n]$ to any other ordered sequence of k distinct indices in $[n]$ by some element in G. A polytope is k-transitive if it is G-invariant for a k-transitive group G; observe that $(k+1)$-transitivity implies k-transitivity. Thus, the larger the k, the *more symmetric* P is.

A high-level description of Theorem 1 is as follows. Assume that the polytope $P \subseteq [0,1]^n$ is integer-empty and "symmetric enough", i.e., $(k+1)$-transitive. Theorem 1 implies that we can decide whether k rounds of any of the three hierarchies certify that P is integer-empty just by looking at the intersection of P with the faces of the hypercube $[0,1]^n$. In particular, one of the hierarchies gives us such a certificate if and only if all of them do. It is well-known that, for any $k \in \mathbb{N}$ and polytope $P \subseteq [0,1]^n$, $\mathrm{SA}^k(P) \subseteq \mathrm{LS}^k(P) \subseteq \mathrm{LS}_0^k(P)$, and in general they can be significantly different.

In contrast, in our setting, we observe a *hierarchy collapse*: the three methods are equally effective in determining whether the polytope is integer-empty. This result may explain why the integer-emptiness of symmetric polytopes is hard to prove for LP-based hierarchies: the strongest of the three operators does not outperform the weakest one. Moreover, if the other features stay constant, a "more symmetric" polytope (i.e., k-transitive for larger k) will "resist" these operators for at least as many rounds as a "less symmetric" polytope.

Although the concept of k-transitivity is well-known in group theory–see, e.g., [10, Sec. 7]–we believe that using it as a way of measuring the level of symmetry of a polytope–especially in the context of Integer Programming–is one of the contributions of this work. We briefly discuss in Example 2 how our approach can be employed to obtain bounds on the number of rounds the hierarchies need to improve the bound given by an LP relaxation.

Organization of the Paper. In Sect. 2, we discuss some applications and consequences of Theorem 1. In Sect. 3 we introduce the Lift-and-Project operators that will be used throughout the paper, along with definitions and results on permutation groups that are needed for Theorem 1. In Sect. 4 we prove Theorem 1, after introducing additional tools from polyhedral and group theory. The implication (A)⇒(B) of Theorem 1 is the technical core of this work, thus it is the one we dedicate most space to. Section 5 concludes with a discussion and open questions. Proofs missing from the current extended abstract are deferred to the journal version of the paper.

Further Related Work. Linear and semidefinite hierarchies offer systematic procedures for strengthening an initial relaxation of an integer program [26,29,37,38]; see, e.g., [27] for a detailed comparison. They are used as an automatic way to generate stronger relaxations, while keeping efficiency at low levels of the hierarchy. However, their success in improving integrality gaps is limited, and symmetries partially explain this phenomenon. Machine scheduling showcases this behavior: while Sherali-Adams and Lasserre are not able to close the gap after linearly many rounds over the symmetric assignment relaxation, they yield approximation schemes after symmetry-breaking [25,39]. Kurpisz et al. [24] study the impact of symmetries for Lasserre and show a sufficient condition to prove rank lower bounds whenever the solutions and constraints are invariant under every permutation of $[n]$.

Several approaches for handling symmetries in IPs are found in the literature, including symmetry-breaking procedures in the formulation or in the algorithms employed; see, e.g., Pfetsch and Rehn for a detailed exposition [36]. Towards the goal of providing a unified perspective, van Doornmalen and Hojny [11] recently introduced a framework that captures many of the existing symmetry-handling methods. In the context of LP hierarchies, Ostrowski [34] explored the impact of isomorphism pruning for symmetric Sherali-Adams relaxations.

2 Examples and Applications

Example 1 (3-dimensional example). We first illustrate our theorem on the case of a 3-dimensional polytope. Let $\mathrm{conv}(S)$ denote the convex hull of a set $S \subseteq \mathbb{R}^n$. In Fig. 1 we consider the polytope $P = \mathrm{conv}(\{\pi(0, 1/2, 1) : \pi \in \mathcal{S}_3\})$, that is, the convex hull of all vectors obtained by permuting the coordinates of $(0, 1/2, 1)$ in every possible way. It is easy to see that P is 3-transitive by construction (and hence 2-transitive). In the left figure, one can see the original polytope. Observe that P intersects every 2-dimensional face of $[0,1]^3$. On the right, we depict $\mathrm{SA}^1(P)$, $\mathrm{LS}^1(P)$, and $\mathrm{LS}_0^1(P)$. As predicted by Theorem 1, when taking $k = 1$ these three polytopes are non-empty; however they are not necessarily equal as $\mathrm{SA}^1(P) = \mathrm{LS}^1(P) \subsetneq \mathrm{LS}_0^1(P)$. On the other hand, since P does not intersect every 1-dimensional face (e.g., the face defined by $x_1 = 1$ and $x_2 = 1$ does not intersect P), Theorem 1 implies that $\mathrm{SA}^2(P) = \mathrm{LS}^2(P) = \mathrm{LS}_0^2(P) = \emptyset$.

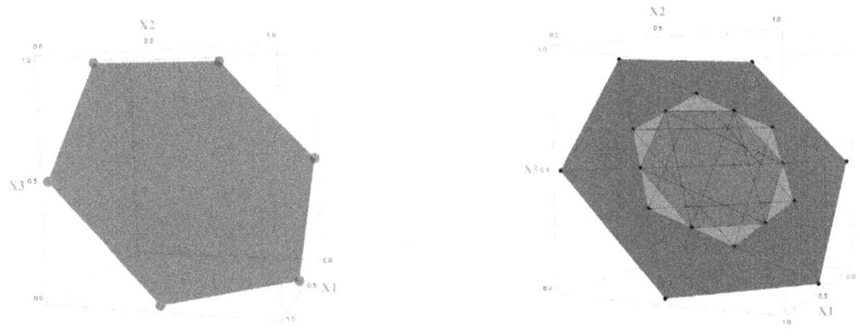

Fig. 1. A depiction of $P = \mathrm{conv}(\{\pi(0, 1/2, 1) : \pi \in \mathcal{S}_3\})$ (left), $\mathrm{SA}^1(P) = \mathrm{LS}^1(P)$ (right, violet), and $\mathrm{LS}_0^1(P)$ (right, gray).

Example 2 (Steiner triple cover and LP bounds). Starting with the work of Fulkerson, Nemhauser, and Trotter [14], Steiner Triple Covering Problems (STSs) have been studied computationally by many researchers in Integer Programming [12,30,34,35]. Most of the work in the area has focused on two classes of STSs, whose number of variables is a power of 3, or 5 times a power of 3. For $n \in \mathbb{N}$, we let STS_n be STSs instances with 3^n variables defined as follows. STS_1 is defined as the following set covering instance

$$\min\{\mathbf{1}^T x : x_1 + x_2 + x_3 \geq 1, \; x \in \{0,1\}^3\}.$$

For $n \in \mathbb{N}_{\geq 2}$, we define STS_n as a function of STS_{n-1} as follows. We let A_{n-1} be the constraint matrix for STS_{n-1}. We let I_{n-1} be the identity matrix with 3^{n-1} rows and columns. We let B_n be the matrix with 3^n columns and whose rows are all and only the vectors that satisfy the following: (a) three entries i_0, i_1, i_2 are equal to 1, and the remaining entries are equal to 0; (b) $3^j + 1 \leq i_j \leq 3^{j+1}$ for $j = 0, 1, 2$; (c) $i_j \mod 3^n$ assume three different values for $j \in \{0, 1, 2\}$. STS_n is then defined as the following set covering problem (for the sake of readability, we omit the 0s):

$$\begin{aligned} &\min \mathbf{1}^T x \\ &\text{s.t.} \quad \begin{pmatrix} A_{n-1} & & \\ & A_{n-1} & \\ & & A_{n-1} \\ I_{n-1} & I_{n-1} & I_{n-1} \\ & B_n & \end{pmatrix} x \geq \mathbf{1} \\ &\qquad\qquad\qquad\qquad\qquad x \in \{0,1\}^{3^n}. \end{aligned}$$

For instance, STS_2 is given by the following binary integer program.

$$\begin{array}{lllllllllll} \min x_1 & +x_2 & +x_3 & +x_4 & +x_5 & +x_6 & +x_7 & +x_8 & +x_9 & \\ x_1 & +x_2 & +x_3 & & & & & & & \geq 1 \\ & & & x_4 & +x_5 & +x_6 & & & & \geq 1 \\ & & & & & & x_7 & +x_8 & +x_9 & \geq 1 \\ x_1 & & & +x_4 & & & +x_7 & & & \geq 1 \\ & x_2 & & & +x_5 & & & +x_8 & & \geq 1 \\ & & x_3 & & & +x_6 & & & +x_9 & \geq 1 \\ x_1 & & & & +x_5 & & & & +x_9 & \geq 1 \\ x_1 & & & & & +x_6 & & +x_8 & & \geq 1 \\ & x_2 & & +x_4 & & & & & +x_9 & \geq 1 \\ & x_2 & & & & +x_6 & +x_7 & & & \geq 1 \\ & & x_3 & +x_4 & & & & +x_8 & & \geq 1 \\ & & x_3 & & +x_5 & & +x_7 & & & \geq 1 \\ & & & & & & & & & x \in \{0,1\}^9. \end{array}$$

An optimal solution to this instance is, e.g., $x^* = (1,1,0,1,0,1,0,1,0)$ with objective function value 5.

For $n \in \mathbb{N}_{\geq 2}$, the feasible region of the LP relaxation of STS_n is known to be 2-transitive [20] (it is also easy to argue that it is not 3-transitive). Consider adding a non-optimal objective function cut to the LP relaxation of STS_n. Since the cut is parallel to the objective and does not attain the optimal value, the optimal face of the resulting LP relaxation, denoted by P, is 2-transitive and integer-empty. Therefore, P satisfies the hypothesis of Theorem 1 with $k = 1$. It is straightforward to verify that P satisfies property (A) in Theorem 1. Thus, one round of any of the three hierarchies considered in this paper does not improve the bound provided by the LP relaxation with the added cut, i.e., to certify that P is integer-empty.

STS_n illustrates how Theorem 1 can be employed to give a bound on the number of rounds we need to apply the hierarchies for, in order to improve the bound given by some LP relaxation of a (symmetric) integer program.

Example 3 (Parity and Knapsack-Cover). Understanding the strength of hierarchies to prove approximate bounds has attracted significant attention. In particular, constructing lower bounds is a technically challenging but important task in order to understand the limitation of such techniques. The Sum-of-Squares (SoS) hierarchy, a stronger construction than SA, LS, and LS_0, has been studied extensively in this setting. The starting point of several of these constructions [23,28,33] was given by Grigoriev [18], who showed that $k = \lfloor n/2 \rfloor$ rounds of the SoS hierarchy are needed to prove that $P = \{x \in [0,1]^n : \sum_{i=1}^n x_i = n/2\}$ is integer-empty when n is odd. Noting that P is $\mathcal{S}_n$-invariant, and thus n-transitive, our main theorem implies directly that the (weaker) Sherali-Adams hierarchy needs exactly $k+1 = \lfloor n/2 \rfloor + 1$ rounds to prove that P is integer-empty. Although this is not a new result, it gives a very simple proof for this fact and exemplifies the usefulness of our techniques: note that it suffices to find a vector in the intersection of P with each $(n-k)$-dimensional face of $[0,1]^n$, a straightforward exercise for this polytope. Moreover, we note that the construction of a feasible point of $\mathrm{SA}^k(P)$ we use in the proof of (A)⇒(B) in Theorem 1 is a strict generalization of the ideas of Grigoriev [18] to full-dimensional polytopes.

Another application considers the Knapsack-Cover problem (also known as Min-Knapsack). While the basic linear relaxation of this problem has an unbounded integrality gap, the introduction of knapsack-cover inequalities reduces this gap to 2. Despite efforts to tighten this relaxation using hierarchies like Sherali-Adams, Lovász-Schrijver, and SoS, the integrality gap remains at 2 even after several rounds. We show that our construction generalizes the approach by Bienstock and Zuckerberg [4,5], offering a simpler method to generate feasible vectors in the Sherali-Adams hierarchy. For a specific instance, the knapsack-cover linear relaxation yields

$$P = \min\{\textstyle\sum_{i=1}^n x_i : \sum_{i=1}^n x_i \geq 1 + \frac{1}{n-1}, \sum_{j \in [n] \setminus \{i\}} x_j \geq 1, x \in [0,1]^n\}.$$

Our techniques from Sect. 4 can also be used to devise a direct method for constructing a solution in $x \in \mathrm{SA}^k(P)$ with value $\sum_{i=1}^n x_i = 1 + O(k/n)$, while the optimal integral solution has value 2, which implies an integrality gap of 2 as long as k is sublinear, $k = o(n)$.

We emphasize that in both constructions our explicit solution is guaranteed to belong to the Sherali-Adams hierarchy; we leave as an open problem if the same solution belongs to the tighter SoS hierarchy at the same level.

Example 4 (Cropped cube). The cropped cubed is a polytope obtained by cropping each vertex of the 0–1 hypercube, given by

$$P^{cc} := \left\{x \in [0,1]^n \ : \ \textstyle\sum_{i \in I} x_i + \sum_{i \notin I}(1 - x_i) \geq 1/2, \ \text{ for all } I \subseteq [n]\right\}.$$

This polytope is integer-empty and n-transitive. Goemans and Tunçel [17] showed that it resists $n-1$ rounds of the Lovász-Schrijver operator, meaning that $\mathrm{LS}^{n-1}(P^{cc}) \neq \emptyset$. Later, Laurent [28] extended this results, showing that even $\mathrm{SA}^{n-1}(P^{cc}) \neq \emptyset$. Theorem 1 implies directly these results: we only need to observe that P^{cc} intersects every $n-1$ dimensional face of $[0,1]^n$. Namely, we need to verify that for any two disjoint sets $J_0, J_1 \subseteq [n]$ with $|J_0 \cup J_1| = n-1$, there is a vector $x \in P^{cc}$ with $x_i = 0$ if $i \in J_0$ and $x_i = 1$ for $i \in J_1$. To do so, simply define the only free variable as $x_i = 1/2$, for $i \in [n] \setminus (J_0 \cup J_1)$.

Discussion on the Hypotheses of Theorem 1. We illustrate here how Theorem 1 is tight in several ways. First, we claim that we cannot weaken the $(k+1)$-transitivity assumption to k-transitivity. Consider the polytope

$$P := \mathrm{conv}\left(\{\pi(1,1/2,0), \pi(1,1,1/10) \,:\, \pi \in C_3\}\right),$$

where C_3 is the cyclic group acting on the set $[3]$. This polytope is 1-transitive but not 2-transitive. A routine computation shows that $\mathrm{SA}(P) = \emptyset$ while using the definition of LS_0, one can check that $\mathrm{LS}_0(P) \neq \emptyset$, see Fig. 2. Thus, (B) does not hold, while (D) holds.

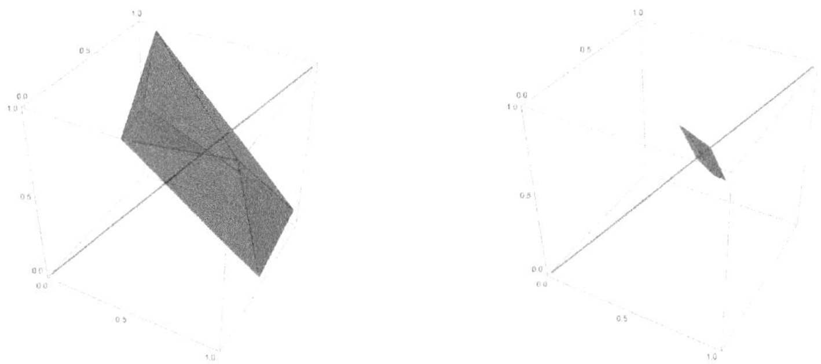

Fig. 2. On the left, the polytope $P = \mathrm{conv}\left(\{\pi(1,1/2,0), \pi(1,1,1/10) \,:\, \pi \in C_3\}\right)$. On the right, $\mathrm{LS}_0(P)$. The line (red) denotes the symmetry axis spanned by $(1,1,1)$.

Next, observe that we cannot drop the integer-emptyness assumption, since trivially (B), (C), (D) hold for any polytope with an integer point, irrespectively of whether (A) holds.

Last, recall that we already argued that $\mathrm{SA}^1(P)$ and $\mathrm{LS}_0^1(P)$ may differ (while both being non-empty), even if the hypothesis of the theorems hold for $k=1$; see Example 1.

3 Preliminaries

For $i \in \mathbb{N}$ we denote $\{1, \ldots, i\}$ by $[i]$ and $[0] := \emptyset$. For $i, j \in \mathbb{Z}_{\geq 0}$ and $x \in \mathbb{R}^n$ we write $\binom{i}{j} := 0$ and $\prod_j^i x_j := 1$ if $j > i$. Moreover, we define $\binom{i}{r} := 0$ and $\prod_j^r x_j := 1$ if $r \in \mathbb{Z}_{<0}$. We denote by $\mathbf{1}$ and $\mathbf{0}$ the all-ones vector and the all-zeros vector, respectively. Moreover, we write $\binom{n}{\leq i} := \sum_{j=0}^{i} \binom{n}{j}$. For $\emptyset \neq S \subseteq [n]$ and subsets $J_0, J_1 \subseteq S$, we say that (J_0, J_1) is a 2-partition of S if $J_0 \cup J_1 = S$ and $J_0 \cap J_1 = \emptyset$, and we denote it as $J_0 \sqcup J_1 = S$. For any vectors $y \in \mathbb{R}^{\binom{n}{\leq j}}$ and $z \in \mathbb{R}^{\binom{n}{\leq i}}$, let (y, z) denote their vertical concatenation, i.e., $(y, z) := \binom{y}{z}$. Whenever $I = \{j\}$, we simply write $y_I = y_j$. Throughout this section, we let $n \in \mathbb{N}$ and $P := \{x \in \mathbb{R}^n : a_i^\top x \geq b_i, i \in [m]\} \subseteq [0,1]^n$. For technical reasons, we will assume that the description of P contains explicitly the linear inequalities $\mathbf{0} \leq x \leq \mathbf{1}$.

3.1 Linear Programming Hierarchies

We now present the Linear Programming hierarchies of interest for this work. For the polytope P as above, let

$$P^{i,1} := P \cap \{x \in \mathbb{R}^n : x_i = 1\} \quad \text{and} \quad P^{i,0} := P \cap \{x \in \mathbb{R}^n : x_i = 0\} \text{ for } i \in [n].$$

We denote by P_i the *Balas-Céria-Cornuéjols operator* [1] by the variable x_i, i.e., $P_i(P) := \text{conv}(P^{i,1} \cup P^{i,0})$. Hence the *Lift-and-Project closure* of P can be written as $\text{LS}_0(P) := \bigcap_{i \in [n]} P_i(P)$. We will employ a useful characterization of the Lift-and-Project closure (see, e.g., [17,29]):

$$\begin{aligned}\text{LS}_0(P) := \{x \in \mathbb{R}^n :\, & \exists Y \in \mathbb{R}^{(n+1)\times(n+1)}, Ye_i, Y(e_0 - e_i) \in K(P), \forall i \in [n], \\ & Ye_0 = Y^\top e_0 = \text{diag}(Y) = (1, x)^\top\},\end{aligned}$$

where the matrices Y have columns and rows indexed by $\{0, 1, \ldots, n\}$, e_i denotes the i-th canonical vector, and $K(P) := \{(\lambda, \lambda x)^\top : \lambda \in \mathbb{R}_+, x \in P\}$. We denote by LS_0^k the k iterative applications of LS_0. That is, $\text{LS}_0^2(P) = \text{LS}_0(\text{LS}_0(P))$, $\text{LS}_0^3(P) = \text{LS}_0(\text{LS}_0^2(P))$, etc. The *Lovász-Schrijver* operator, LS, further requires the lifting matrices Y to be symmetric, and we similarly denote repeated applications of LS by LS^k. Following [13], we define the *Sherali-Adams level k* relaxation [38] $\text{M}^k(P)$ of P as:

(i) **Extend:** For every row $a_i^\top x \geq b_i$, with $i \in [m]$, every set $S \subseteq [n]$, $|S| = k$ and every 2-partition $J_0 \sqcup J_1 = S$, introduce a new constraint

$$\prod_{i \in J_1} x_i \prod_{j \in J_0} (1 - x_j) \cdot (a_i^\top x - b_i \geq 0). \tag{1}$$

(ii) **Linearize:** multilinearize each of the constraints in the previous step by replacing every monomial $\prod_{i \in I} x_i$ with a new variable y_I, define $y_\emptyset := 1$, and set $x_i^2 = x_i$

$$\text{M}^k(P) := \left\{ y \in \mathbb{R}^{\binom{n}{\leq k+1}} : \text{multilinearized inequalities (1)} \right\}.$$

We denote by $\mathrm{SA}^k(P)$ the (coordinate) projection of $\mathrm{M}^k(P)$ onto the space of original variables, i.e.,

$$\mathrm{SA}^k(P) := \{x \in \mathbb{R}^n : \text{there exists } y \in \mathrm{M}^k(P),\ y_{\{i\}} = x_i,\ i \in [n]\}.$$

Let P_I be the integer hull of P. It is well known that for all $k \in [n]$ we have $P_I \subseteq \mathrm{SA}^k(P) \subseteq \mathrm{LS}^k(P) \subseteq \mathrm{LS}_0^k(P) \subseteq P$. Moreover, $P_I = \mathrm{LS}_0^n(P)$; see, e.g., [8,27].

Fix $d \leq k \leq n$ and a polytope $P \subseteq [0,1]^n$. The Sherali-Adams hierarchy has the following *locally consistent distribution property*: for every point $y \in \mathrm{M}^k(P)$ and every set of d coordinates $S \subseteq [n]$, we can write y as convex combination of certain points z^{J_0,J_1} (for any $J_0 \sqcup J_1 = S$) that belong to $\mathrm{M}^{k-d}(P)$ and are integer on coordinates from S. The Sherali-Adams hierarchy shares this property with any stronger hierarchy, such as SoS. However, it can be shown that this property *characterizes* the Sherali-Adams hierarchy (see, e.g., [13]): y belongs to $\mathrm{M}^k(P)$ if and only if the vectors z^{J_0,J_1} belong to $\mathrm{M}^{k-d}(P)$. To make this property formal, we begin by introducing some notation.

Let $d,k \in \mathbb{Z}_{\geq 0}$ such that $d \leq k$ and $y \in [0,1]^{\binom{n}{\leq k+1}}$. For $S \subseteq [n]$ of cardinality d and a 2-partition $J_0 \sqcup J_1 = S$, we define:

$$y_I^{J_0,J_1} := \sum_{H \subseteq J_0} (-1)^{|H|} y_{J_1 \cup I \cup H}, \qquad z^{J_0,J_1}(y)_I := \begin{cases} \frac{y_I^{J_0,J_1}}{y_\emptyset^{J_0,J_1}} & \text{if } y_\emptyset^{J_0,J_1} > 0, \\ y_I^{J_0,J_1} & \text{if } y_\emptyset^{J_0,J_1} = 0, \end{cases} \tag{2}$$

for every $I \subseteq [n]$ with $|I| \leq k-d+1$.

Proposition 1 *Let $P \subseteq [0,1]^n$, $d,k \in \mathbb{Z}_{\geq 0}$ such that $d \leq k$ and $y \in [0,1]^{\binom{n}{\leq k+1}}$. Then $y \in \mathrm{M}^k(P)$ if and only if for every $S \subseteq [n]$ of cardinality d:*

(i) For every 2-partition $J_0 \sqcup J_1 = S$, we have $z^{J_0,J_1}(y) \in \mathrm{M}^{k-d}(P)$; and
(ii) The point y restricted to its first $\binom{n}{\leq k-d+1}$ coordinates can be written as follows

$$y_I = \sum_{J_0,J_1 \subseteq [n]\,:\, J_0 \sqcup J_1 = S} y_\emptyset^{J_0,J_1} z_I^{J_0,J_1}(y), \tag{3}$$

for every $I \subseteq [n]$ with $|I| \leq k-d+1$.

It is worth emphasizing that in the context of (3), due to (2), it holds that, for any S, $\sum_{J_0 \sqcup J_1 = S} y_\emptyset^{J_0,J_1} = 1$, and thus the restriction of the vector y to $\mathbb{R}^{\binom{n}{\leq k-d+1}}$ is a convex combination of the vectors $\{z^{J_0,J_1}\}_{J_0 \sqcup J_1 = S}$. Moreover, the vectors $\{z^{J_0,J_1}\}_{J_0 \sqcup J_1 = S}$ are integer in the coordinates indexed by S. Throughout, expressions of the form $J_0 \sqcup J_1 = S$ in indices range all over $J_0, J_1 \subseteq [n]$ satisfying $J_0 \sqcup J_1 = S$.

3.2 Symmetries

Let $\mathcal{S}_n := \{\text{bijection } \pi : [n] \to [n]\}$ denote the set of all permutations on $[n]$. For $\pi, \pi' \in \mathcal{S}_n$, we denote their (function) composition $\pi' \circ \pi$ as $\pi'\pi$, and by π^{-1}

the inverse function of π. We say that a subset $G \subseteq \mathcal{S}_n$ is a *(permutation) group* if G is nonempty and it is closed under compositions and inverses. We write $H \leq G \leq \mathcal{S}_n$ to denote that H, G are permutation groups with $H \subseteq G$. The *set of (permutation) symmetries acting on* P is defined as $G(P) := \{\pi \in \mathcal{S}_n : \pi x \in P \text{ for all } x \in P\}$, where $\pi x := (x_{\pi^{-1}(1)}, x_{\pi^{-1}(2)}, \ldots, x_{\pi^{-1}(n)})$. The following fact is well-known and easy to check.

Lemma 1 *$G(P)$ is a permutation group.*

The group $G(P)$ contains (possibly strictly) the set of permutation symmetries of any linear formulation for P. We refer to [31,36] for details.

Let $k \in [n]$ and $G \leq \mathcal{S}_n$. G is *k-transitive* if, for every pair of (ordered) k-tuples t^1 and t^2 of $[n]$, where the elements of each k-tuple are all-different, there exists a permutation $\pi \in G$ such that $\pi(t^1_j) = t^2_j$ for $j \in [k]$. We say that P is *G-invariant* if $G \leq G(P)$ and it is k-transitive if moreover G is k-transitive.

Given a subset $S \subseteq [n]$ and $G \leq \mathcal{S}_n$, we call the subgroup of G which fixes pointwise each element of S the *pointwise stabilizer of* S and we denote it by $G_S := \{\pi \in G : \pi(i) = i, \forall \pi \in G, \forall i \in S\}$. Given any $x \in \mathbb{R}^n$, we define the *orbit of x under G* as $\text{orb}_G(x) := \{\pi x : \pi \in G\}$ where $\pi x := (x_{\pi^{-1}(i)})_{i \in [n]}$. Moreover, we define the *Reynolds operator* [9] $\beta_G : \mathbb{R}^n \to \mathbb{R}^n$ as

$$\beta_G(x) := \frac{1}{|G|} \sum_{\pi \in G} \pi x$$

for $x \in \mathbb{R}^n$. If $x \in P$ for some polytope P and $G \leq G(P)$, then the convexity of P implies that $\beta_G(x) \in P$ for every $x \in P$.

Proposition 2 *Let $P \subseteq [0,1]^n$ be a polytope, $G \leq G(P)$, and S a subset of $[n]$ of cardinality at most k, for some $k \in \mathbb{Z}_{\geq 0}$. Assume that G is $(k+1)$-transitive. Then for any $x \in P$, the point $\beta_{G_S}(x)$ belongs to P. Moreover, for $i \in [n]$, it has entries:*

$$\beta_{G_S}(x)_i = \begin{cases} x_i & \text{if} i \in S, \\ \frac{\sum_{j \in [n] \setminus S} x_j}{n - |S|} & \text{if} i \notin S. \end{cases}$$

In all the definitions given above, we omit the subscript if the group is clear from the context. For $G \leq \mathcal{S}_n$ we say that G is *transitive on* $T \subseteq [n]$ if for every $i, j \in T$, there exists $\pi \in G$ such that $\pi(i) = \pi(j)$. Thus, in particular, G is 1-transitive if and only if it is transitive on $[n]$. The proof of the following fact can be found in [10, Section 7.1].

Proposition 3 *A permutation group $G \leq \mathcal{S}_n$ is $(k+1)$-transitive if and only if G is k-transitive and for each $S \subseteq [n]$, with $|S| = k$, we have that G_S is transitive on $[n] \setminus S$.*

We can naturally extend the *permutation group action* of $\mathcal{S}_n$ over singletons $\{i\} \subseteq [n]$ to arbitrary subsets $S \subseteq [n]$ as: $\pi(S) := \{\pi(i) : i \in S\}$ for all $\pi \in \mathcal{S}_n$. Hence we can define the following *action* on $\mathbb{R}^{\binom{n}{\leq k+1}}$:

$$(\pi x)_S := x_{\pi^{-1}(S)} \tag{4}$$

for every $x \in \mathbb{R}^{\binom{n}{\leq k+1}}$, $S \subseteq [n]$ and $\pi \in \mathcal{S}_n$. See [10] for further background on permutation groups and their actions. The following is an observation that is implicit in the literature, see e.g. Ostrowski [34].

Proposition 4 *Let $G \leq G(P)$ and $k \in \mathbb{N}$. Then $\mathrm{M}^k(P)$ is G-invariant.*

A similar statement holds for LS_0 and LS.

Proposition 5 *Let $G \leq G(P)$, $k \in \mathbb{N}$. Then $\mathrm{LS}_0^k(P)$ and $\mathrm{LS}^k(P)$ are G-invariant.*

4 Proof of Theorem 1

The most complex part of the proof is the (A)⇒(B) implication, for which we give below a high-level description, and then some details. We then briefly discuss the other implications.

4.1 Proof of (A)⇒(B)

A High-Level Description of the Proof. Assume that (A)holds. Let P be a polytope with $G \leq G(P)$ and G $(k+1)$-transitive. We apply Proposition 1 with $d = k$. Thus, we want to exhibit a point $\overline{y}$ in $\mathrm{M}^k(P)$, and show that, for every $S \subseteq [n]$ with $|S| = k$, it can be supported by the points $z^{J_0,J_1}(\overline{y}) \in \mathrm{M}^0(P)$ as in Proposition 1. However, our construction will proceed in reverse: we build the supporting points first and then show that they satisfy (2) and (3) for some $\overline{y}$. In particular, we perform the following steps.

1. Construction of Points from $\mathrm{M}^0(P)$. By leveraging the $(k+1)$-transitivity of P, for every $S \subseteq [n]$ with $|S| = k$ and $J_0 \sqcup J_1 = S$, we first construct "highly symmetric" points, i.e., points $x^{J_0,J_1} \in P$ such that, for any $S' \subseteq [n]$ with $|S'| = k$ and $J_0' \sqcup J_1'$ with $|J_1| = |J_1'|$, we have $x^{J_0,J_1} = \pi x^{J_0',J_1'}$ for some $\pi \in G$. By definition, $(1, x^{J_0,J_1}) \in \mathrm{M}^0(P)$.

2. System of Constraints Defining $\overline{y}$. Note that any point $\overline{y}$ that satisfies (2) and (3) with respect to our constructed vectors $z^{J_0,J_1} := (1, x^{J_0,J_1})$ is guaranteed to belong to $\mathrm{M}^k(P)$. Imposing these conditions leads to a system of equations whose variables are the coordinates of $\overline{y}$. This system appears to be highly complex and non-tractable. However, because of the symmetric structure of z^{J_0,J_1}, we look for solutions $\overline{y}$ that are highly symmetric. Indeed, for every $I, J \subseteq [n]$ with $|I| = |J| \leq k+1$ and every $J_0 \sqcup J_1 = S$, $J_0' \sqcup J_1' = S'$ with $|J_1| = |J_1'|$ and $|S| = |S'| = k$, we impose the following conditions:

(C-1) $\overline{y}_I = \overline{y}_J =: \overline{\gamma}_{|I|}$;
(C-2) $\overline{y}_\emptyset^{J_0,J_1} = \overline{y}_\emptyset^{J_0',J_1'} =: \overline{\lambda}_{|J_1|}$.

Conditions (C-1) and (C-2) above lead to a (much smaller) system in the variables γ and λ, which have $k+1$ components each. Thus, we are left to show that this system has a solution. This is the most technical part of the proof, divided into two steps.

3. Solution to the System Defining $\overline{\gamma}$. We first consider a subsystem of linear equalities in the variables γ only, and show that the corresponding constraint matrix is invertible.

4. Solution to the General System. Based on the previous step, we give an explicit formula for $\overline{\gamma}$ and $\overline{\lambda}$, and proceed to show that they satisfy all constraints of the system in γ and λ defined in Step 2.

We now present full details on Step 1. above, and the main lemma leading to Step 2.

1. Construction of Points from $\mathrm{M}^0(P)$. For $\ell \in \{0, \dots, k\}$, define

$$P_\ell := P \cap \{x \in [0,1]^n \,:\, x_i = 1 \,\text{for all}\, i \in [\ell], x_j = 0 \,\text{for all}\, j \in [k] \setminus [\ell]\},$$

and observe that it is non-empty because of Assumption (A). Thus, let $x^\ell \in P_\ell$. Recall that $G_{[k]}$ is the pointwise stabilizer of $[k]$. Observe that P_ℓ is $G_{[k]}$-invariant. Moreover, since G is $(k+1)$-transitive, by Proposition 3, $G_{[k]}$ is transitive on $[n] \setminus [k]$. Therefore, by applying Proposition 2 to x^ℓ with $G_{[k]}$ and P_ℓ we obtain

$$\overline{x}^\ell := \beta_{G_{[k]}}(x^\ell) = (\underbrace{1, \dots, 1}_{\ell}, \underbrace{0, \dots, 0}_{k-\ell}, \underbrace{\Delta_\ell, \dots, \Delta_\ell}_{n-k})^T \in P_\ell, \tag{5}$$

for some $\Delta_\ell \in [0,1]$. Define $\Delta := (\Delta_0, \dots, \Delta_k) \in [0,1]^{k+1}$ and observe that $\Delta \in (0,1)^{k+1}$ since P is integer-empty.

Next, let $S \subseteq [n]$ with $|S| = k$ and consider an arbitrary 2-partition $J_0 \sqcup J_1 = S$ such that $|J_1| = \ell$. By definition, the point $\overline{x}^\ell$ has ℓ ones in entries indexed by $[\ell]$ and zeros in entries indexed by $[k] \setminus [\ell]$. Since G is $(k+1)$-transitive and $G \leq G(P)$, there exists some permutation $\pi \in G$ and some point $\overline{x}^{J_0,J_1} \in P$ such that π maps: all of the ones in positions $[\ell]$ of $\overline{x}^\ell$ to entries indexed by J_1 of $\overline{x}^{J_0,J_1}$; all of the zeros in positions $[k] \setminus [\ell]$ to entries indexed by J_0; and the fractional Δ_ℓ's in positions $[n] \setminus [k]$ to entries indexed by $[n] \setminus S$. In other words,

$$\overline{x}_i^{J_0,J_1} := (\pi \overline{x}^\ell)_i = \begin{cases} 0 & \text{if} \, i \in J_0, \\ 1 & \text{if} \, i \in J_1, \\ \Delta_\ell & \text{if} \, i \in [n] \setminus S. \end{cases} \tag{6}$$

By definition of $\mathrm{M}^0(P)$, we immediately deduce the following.

Lemma 2 *For each* $S \subseteq [n]$ *with* $|S| = k$ *and* 2*-partition* $J_0 \sqcup J_1 = S$*, we have* $(1, \overline{x}^{J_0,J_1}) \in \mathrm{M}^0(P)$.

2. System of Constraints Defining $\overline{y}$. We give next sufficient conditions for the existence of $\overline{y} \in \mathrm{M}^d(P)$. For $t \in \mathbb{N}$ and $\Theta = (\Theta_0, \dots, \Theta_t) \in \mathbb{R}^{t+1}$, we let $A^t_\Theta \in \mathbb{R}^{(t+1)\times(t+1)}$ with rows indices in $\{0, \dots, t\}$ and column indices in $\{1, \dots, t+1\}$ be defined as follows:

$$(A^t_\Theta)_{\ell,r} := \begin{cases} (-1)^{\ell+r-1}\left[\binom{t-\ell}{r-\ell-1} + \binom{t-\ell}{r-\ell}\Theta_\ell\right] & \text{if}\, \ell < r, \\ -\Theta_\ell & \text{if}\, \ell = r, \\ 0 & \text{if}\, \ell > r. \end{cases} \tag{7}$$

Lemma 3 *Assume that the following system in the variables $\gamma_0, \dots, \gamma_{k+1} \in \mathbb{R}$, $\lambda_0, \dots, \lambda_k \in \mathbb{R}$*

$$A^k_\Delta \gamma = \Delta_0(1, 0, \dots, 0), \tag{8a}$$

$$1 = \sum_{\ell=0}^{k} \binom{k}{\ell} \lambda_\ell, \tag{8b}$$

$$\gamma_1 = \sum_{\ell=1}^{k} \binom{k-1}{\ell-1} \lambda_\ell, \tag{8c}$$

$$\gamma_1 = \sum_{\ell=0}^{k} \binom{k}{\ell} \lambda_\ell \Delta_\ell, \tag{8d}$$

$$\lambda_\ell = \sum_{r=0}^{k-\ell} (-1)^r \binom{k-\ell}{r} \gamma_{r+\ell} \quad \forall \ell \in \{0, 1, \dots, k\}, \tag{8e}$$

$$\mathbf{0} \le \lambda, \gamma \le \mathbf{1}, \quad \gamma_0 = 1, \tag{8f}$$

has a solution that is strictly positive in all components, where we abbreviate $\gamma = (\gamma_1, \dots, \gamma_{k+1})$. Then, $\overline{y} \in \mathrm{M}^k(P)$, where

$$\overline{y}_I = \overline{\gamma}_{|I|} \quad \textit{for all}\, I \subseteq [n], |I| \le k+1 \tag{9}$$

and $(\overline{\lambda}, \overline{\gamma})$ is a solution to the system (8).

4.2 Proofs of the Remaining Statements from Theorem 1

We note that (B) implies (C) and (C) implies (D) are well-known facts, see e.g. [27]. We show that (D) implies (A). Suppose therefore that $\mathrm{LS}^k_0(P)$ is nonempty and let G be a $(k+1)$-transitive subgroup of $G(P)$. We make use of the following lemma.

Lemma 4 *Let $j \in [k]$ and $i \in [n]$. Suppose that P is $(k+1)$-transitive and that there exists $x \in \mathrm{LS}^{k-j+1}_0(P)$ such that $x_i \in (0,1)$. Then for every subset $J \subseteq [n]$ with $i \in J$ and $|J| = j$, there exists $x^{i,1}, x^{i,0} \in \mathrm{LS}^{k-j}_0(P)$ such that: a) $x^{i,1}_i = 1$ and $x^{i,0}_i = 0$; b) $x^{i,1}_\ell = x^{i,1}_r$ and $x^{i,0}_\ell = x^{i,0}_r$ for all $\ell, r \in [n] \setminus J$.*

Let $x \in \mathrm{LS}_0^k(P)$ and consider $\overline{x} := \beta_G(x)$. Since P is integer-empty, we deduce that $\overline{x}$ is fractional in all components. By Proposition 5, $\mathrm{LS}_0^k(P)$ is G-invariant, hence $\overline{x} \in \mathrm{LS}_0^k(P)$. Pick $i = 1$ and apply Lemma 4 as to obtain points $x^{1,1}$ and $x^{1,0} \in \mathrm{LS}_0^{k-1}(P)$. Let G_1 be the subgroup of G that fixes the first coordinate. Using again the fact that P is integer-empty, we deduce that all coordinates of $\overline{x}^{1,1} := \beta_{G_1}(x^{1,,1})$ are fractional, except the first that is equal to 1. Similarly, all coordinates of $\overline{x}^{1,0} := \beta_{G_1}(x^{1,0})$ are fractional, except the first that is equal to 0. Since $\mathrm{LS}^{k-1}(P)$ is G-invariant by Proposition 5, we have $\overline{x}^{1,0}, \overline{x}^{1,1} \in \mathrm{LS}^{k-1}(P)$. Iterating, for all $j \in [k]$ and vectors $q^j \in \{0,1\}^j$ we obtain points $\overline{x}^{q^2}, \overline{x}^{q^3} \ldots, \overline{x}^{q^k}$ such that

- $\overline{x}^{q^j} \in \mathrm{LS}_0^{k-j}(P)$;
- $\overline{x}_\ell^{q^j} = q_\ell^j$ for all $\ell \in [j]$.

In particular, for all binary k-tuples q, we obtain a point $\overline{x}^q \in \mathrm{LS}_0^0(P) = P$ with $\overline{x}_\ell^q = q_\ell$ for all $\ell \in [k]$.

Now fix a $(n-k)$-dimensional face F of $[0,1]^n$, defined as:

$$F = \{x \in [0,1]^n : x_i = 1\, \forall i \in S_1;\ x_i = 0\, \forall i \in S_0\}.$$

Let $q \in \{0,1\}^k$ such that $q_1 = q_2 = \cdots = q_{|S_1|} = 1$ and $q_{|S_1|+1} = q_{|S_1|+2} = \cdots = q_k = 0$. Since G is $(k+1)$-transitive, there exists $\pi \in G$ that maps $\overline{x}^q$ to a point of $\pi\overline{x}^q \in F$. In particular, $\pi\overline{x}^q \in F \cap P$, as required.

5 Conclusion

In this paper, we investigate bounds on the number of iterations that LP-based hierarchies need to certify integer-emptiness of a polytope P contained in the binary cube. Roughly speaking, the "more symmetric" P is, the larger is the number of rounds for which the hierarchies "collapse" – that is, they can either all certify integer-emptiness, or none can. An important feature of our work is that we do not assume any particular structure of the polytopes, beyond the assumption on symmetry. We illustrate how certain results from the literature and new insights can be deduce from our main result. It is a natural question to investigate the behavior under various (partial) symmetry of LP hierarchies and other automatic approaches, such as the SoS (Lasserre) hierarchy, as well as the Split and Chvàtal-Gomory closures. Finally, while our results suggest that some of the current methods for tightening LP relaxations behave poorly for symmetric instances, it is a relevant task to develop new techniques that are enhanced–or at least not negatively affected–by symmetry.

Acknowledgments.. Yuri Faenza and Matías Villagra acknowledge support from the AFOSR grant FA9550-23-1-0697 and wish to thank Warren Adams for useful discussions on hierarchies. Victor Verdugo acknowledges support from ANID Chile, grant FONDECYT 1241846.

Disclosure of Interests. The authors have no competing interests to declare that are relevant to the content of this article.

References

1. Balas, E., Ceria, S., Cornuéjols, G.: A lift-and-project cutting plane algorithm for mixed 0–1 programs. Math. Program. **58**(1), 295–324 (1993)
2. Barak, B., Hopkins, S., Kelner, J., Kothari, P.K., Moitra, A., Potechin, A.: A nearly tight sum-of-squares lower bound for the planted clique problem. SIAM J. Comput. **48**(2), 687–735 (2019)
3. Ben-Tal, A., Nemirovski, A.: On polyhedral approximations of the second-order cone. Math. Oper. Res. **26**(2), 193–205 (2001)
4. Bienstock, D., Zuckerberg, M.: Subset algebra lift operators for 0–1 integer programming. SIAM J. Optim. **15**(1), 63–95 (2004)
5. Bienstock, D., Zuckerberg, M.: Approximate fixed-rank closures of covering problems. Math. Program. **105**(1), 9–27 (2006)
6. Bunch, J.R., Kaufman, L.: Some stable methods for calculating inertia and solving symmetric linear systems. Math. Comput. **31**(137), 163–179 (1977)
7. Cohen-Addad, V., Mömke, T., Verdugo, V.: A 2-approximation for the bounded treewidth sparsest cut problem in FPT time. Math. Program. **206**(1), 479–495 (2024)
8. Cook, W., Dash, S.: On the matrix-cut rank of polyhedra. Math. Oper. Res. **26**(1), 19–30 (2001)
9. Derksen, H., Kemper, G.: Computational Invariant Theory. Encyclopaedia of Mathematical Sciences, vol. 130. Springer, Berlin (2002)
10. Dixon, J., Mortimer, B.: Permutation Groups. Springer, Cham (1996)
11. van Doornmalen, J., Hojny, C.: A unified framework for symmetry handling. Math. Program. **212**(1), 217–271 (2025)
12. Feo, T.A., Resende, M.G.: A probabilistic heuristic for a computationally difficult set covering problem. Oper. Res. Lett. **8**(2), 67–71 (1989)
13. Fleming, N., Kothari, P., Pitassi, T.: Semialgebraic proofs and efficient algorithm design. Found. Trends Theor. Comput. Sci. **14**(1–2), 1–221 (2019)
14. Fulkerson, D., Nemhauser, G.L., Trotter, L.E.: Two computationally difficult set covering problems that arise in computing the 1-width of incidence matrices of steiner triple systems. In: Approaches to Integer Programming, pp. 72–81. Springer (2009)
15. Gatermann, K., Parrilo, P.A.: Symmetry groups, semidefinite programs, and sums of squares. J. Pure Appl. Algebra **192**(1–3), 95–128 (2004)
16. Georgiou, K., Magen, A., Pitassi, T., Tourlakis, I.: Integrality gaps of 2-o(1) for vertex cover SDPs in the Lovász-Schrijver hierarchy. SIAM J. Comput. **39**(8), 3553–3570 (2010)
17. Goemans, M., Tunçel, L.: When does the positive semidefiniteness constraint help in lifting procedures. Math. Oper. Res. **26**(4), 769–815 (2001)
18. Grigoriev, D.: Complexity of positivstellensatz proofs for the knapsack. Comput. Complex. **10**, 139–154 (2001)
19. Gupta, A., Talwar, K., Witmer, D.: Sparsest cut on bounded treewidth graphs: algorithms and hardness results. In: ACM Symposium on Theory of Computing (STOC), pp. 281–290 (2013)
20. Hall, M.: Automorphisms of steiner triple systems. IBM J. Res. Dev. **4**(5), 460–472 (1960)
21. Karlin, A.R., Mathieu, C., Nguyen, C.T.: Integrality gaps of linear and semidefinite programming relaxations for knapsack. In: Integer Programming and Combinatorial Optimization (IPCO), pp. 301–314 (2011)

22. Kurpisz, A., Leppänen, S., Mastrolilli, M.: Sum-of-squares rank upper bounds for matching problems. J. Comb. Optim. **36**(3), 831–844 (2018)
23. Kurpisz, A., Leppänen, S., Mastrolilli, M.: Sum-of-squares hierarchy lower bounds for symmetric formulations. In: Integer Programming and Combinatorial Optimization (IPCO). Lecture Notes in Computer Science, pp. 362–374 (2016)
24. Kurpisz, A., Leppänen, S., Mastrolilli, M.: Sum-of-squares hierarchy lower bounds for symmetric formulations. Math. Program. **182**(1), 369–397 (2020)
25. Kurpisz, A., Mastrolilli, M., Mathieu, C., Mömke, T., Verdugo, V., Wiese, A.: Semidefinite and linear programming integrality gaps for scheduling identical machines. Math. Program. **172**(1), 231–248 (2018)
26. Lasserre, J.B.: Global optimization with polynomials and the problem of moments. SIAM J. Optim. **11**(3), 796–817 (2001)
27. Laurent, M.: A comparison of the Sherali-Adams, Lovász-Schrijver, and Lasserre relaxations for 0–1 programming. Math. Oper. Res. **28**(3), 470–496 (2003)
28. Laurent, M.: Lower bound for the number of iterations in semidefinite hierarchies for the cut polytope. Math. Oper. Res. **28**(4), 871–883 (2003)
29. Lovász, L., Schrijver, A.: Cones of matrices and set-functions and 0–1 optimization. SIAM J. Optim. **1**(2), 166–190 (1991)
30. Mannino, C., Sassano, A.: Solving hard set covering problems. Oper. Res. Lett. **18**(1), 1–6 (1995)
31. Margot, F.: Symmetry in integer linear programming. In: 50 Years of Integer Programming 1958-2008, pp. 647–686. Springer, Heidelberg (2010)
32. Mathieu, C., Sinclair, A.: Sherali-Adams relaxations of the matching polytope. In: ACM Symposium on Theory of Computing (STOC), pp. 293–302 (2009)
33. Meka, R., Potechin, A., Wigderson, A.: Sum-of-squares lower bounds for Planted Clique. In: ACM Symposium on Theory of Computing (STOC), pp. 87–96 (2015)
34. Ostrowski, J.: Using symmetry to optimize over the Sherali-Adams relaxation. Math. Program. Comput. **6**(4), 405–428 (2014)
35. Ostrowski, J., Linderoth, J., Rossi, F., Smriglio, S.: Solving large steiner triple covering problems. Oper. Res. Lett. **39**(2), 127–131 (2011)
36. Pfetsch, M., Rehn, T.: A computational comparison of symmetry handling methods for mixed integer programs. Math. Program. Comput. **11**(1), 37–93 (2019)
37. Sherali, H.D., Adams, W.P.: A hierarchy of relaxations between the continuous and convex hull representations for zero-one programming problems. SIAM J. Discret. Math. **3**(3), 411–430 (1990)
38. Sherali, H., Adams, W.: A hierarchy of relaxations between the continuous and convex hull representations for zero one programming problems. SIAM J. Discret. Math. **3**(3), 411–430 (1990)
39. Verdugo, V., Verschae, J., Wiese, A.: Breaking symmetries to rescue sum of squares in the case of makespan scheduling. Math. Program. (1), 583–618 (2020). https://doi.org/10.1007/s10107-020-01511-3
40. Yannakakis, M.: Expressing combinatorial optimization problems by linear programs. J. Comput. Syst. Sci. **43**, 441–466 (1991)

Closing the Gap: Efficient Algorithms for Discrete Wasserstein Barycenters

Jiaqi Wang and Weijun Xie(✉)

H. Milton Stewart School of Industrial and Systems Engineering, Georgia Institute of Technology, Atlanta, GA, USA
{jwang3737,wxie}@gatech.edu

Abstract. The Wasserstein barycenter problem seeks a probability measure that minimizes the weighted average of the Wasserstein distances to a given collection of probability measures. We study the discrete setting, in which each measure has finite support—a regime that frequently arises in machine learning and operations research. The discrete Wasserstein barycenter problem is known to be NP-hard, which motivates us to study approximation algorithms with provable guarantees. The best-known algorithm to date achieves an approximation ratio of two. We close this gap by developing a polynomial-time approximation scheme (PTAS) for the discrete Wasserstein barycenter problem that generalizes and improves upon the 2-approximation method. In addition, for the special case of equally weighted measures, we obtain a strictly tighter approximation guarantee. Numerical experiments show that the proposed algorithms are computationally efficient and produce near-optimal barycenter solutions.

Keywords: Wasserstein barycenter · Optimal transport · Candidate support reduction · Polynomial-time approximation scheme · Sampling algorithm

1 Introduction

A Wasserstein barycenter, also known as the "center of mass" in Wasserstein space, is the probability measure that minimizes the average optimal-transport cost using the type-2 Wasserstein metric for a given collection of measures. The type-2 Wasserstein distance between $\mathbb{P}, \mathbb{Q} \in \mathcal{P}_2(\mathbb{R}^d)$ is

$$W_2(\mathbb{P}, \mathbb{Q}) := \left(\inf_{\Pi \in \mathcal{M}(\mathbb{P}, \mathbb{Q})} \int_{\mathbb{R}^d \times \mathbb{R}^d} \|\boldsymbol{x} - \boldsymbol{y}\|_2^2 \, d\Pi(\boldsymbol{x}, \boldsymbol{y}) \right)^{1/2},$$

where $\mathcal{M}(\mathbb{P}, \mathbb{Q})$ denotes the set of joint distributions with marginals $\mathbb{P}$ and $\mathbb{Q}$.

Let $\{\mathbb{P}_i\}_{i \in [k]}$ be discrete marginal probability measures with $[k] := \{1, \ldots, k\}$, each supported at most on n points, denoted by $\{\Xi_i\}_{i \in [k]} \subseteq \mathbb{R}^d$. For notational convenience, for each discrete probability measure $\mathbb{P}_i$ with $i \in [k]$, we let $\hat{\boldsymbol{x}}_{ij}$

S. S. Dey et al. (Eds.): IPCO 2026, LNCS 16588, pp. 507–520, 2026.
https://doi.org/10.1007/978-3-032-28691-8_33

denote the j-th support point of $\mathbb{P}_i$ with mass $\hat{p}_{ij}$ for any $j \in [|\Xi_i|]$. Given weights $\boldsymbol{\lambda} = (\lambda_1, \ldots, \lambda_k) \in \Delta_k := \{\boldsymbol{\lambda} \in \mathbb{R}_+^k : \sum_{i\in[k]} \lambda_i = 1\}$, the discrete Wasserstein barycenter can be formulated as

$$v^* = \inf_{\mathbb{P}\in\mathcal{P}(\mathbb{R}^d)} \sum_{i\in[k]} \lambda_i W_2^2(\mathbb{P}, \mathbb{P}_i), \tag{WBCenter}$$

where $\mathcal{P}(\mathbb{R}^d)$ is the set of Borel probability measures on $\mathbb{R}^d$ with finite second moment. The Wasserstein barycenter problems are fundamental and have been widely applied to clustering [33], regression [7], dictionary learning [27], texture mixing [8], image morphing [28], medical imaging [19], facial detection [31], time series modeling [14], and distributionally robust optimization [22].

1.1 Literature Review

The notion of the Wasserstein barycenter was introduced in [1], which also connected it to the multi-marginal optimal transport (MOT) problem. The Wasserstein barycenter problem has also been extended to different notions (see, e.g., [6,18,29]). For discrete probability measures with finite support, [4] established foundational structural properties—including discrete support, sparsity, and a non–mass-splitting property of the optimal barycenter—and proposed a linear programming (LP) formulation via MOT. However, the resulting LP formulation has $O(n^k)$ variables and constraints, which are exponential in the number of marginals k, and thus becomes computationally expensive as n and k grow. For the *sparse* Wasserstein barycenter problem (i.e., when the optimal barycenter has sparse support), the work [11] proved NP-hardness even in dimension $d = 2$ with only $k = 3$ measures. For the general barycenter problem, the paper [3] showed that there is no polynomial-time randomized algorithm with probability at least 2/3 computing the Wasserstein barycenter problem to ϵ additive accuracy even for uniform weights unless NP$\subsetneq$BPP. However, this result does not contradict ours since our approximation is multiplicative. These hardness results motivate the development of efficient approximation algorithms. However, existing work is limited. The work [9] proposed a 2-approximation that takes the union of the input supports as a candidate support, and the work [25] gave another 2-approximation based on a sequence of two-marginal optimal transport plans. In this paper, we improve the factor-2 guarantee. In particular, we develop a polynomial-time approximation scheme (PTAS) for the Wasserstein barycenter problem. That is, for any $\alpha \in (0, 1]$, our algorithm computes a $(1 + \alpha)$-approximate Wasserstein barycenter in time polynomial in $(nk)^{1/\alpha}$ and d.

Despite NP-hardness, there is a line of work on computing Wasserstein barycenters to optimality. In fixed dimension d, the work [2] showed that one can compute an exact type-p Wasserstein barycenter for $p \in \{1, 2\}$ in time polynomial in n, k, and $\log U$ (where U is the upper bound of the input bit-length), or an approximate barycenter with additive error $\epsilon > 0$ in time polynomial in n, k, and $\log(1/\epsilon)$, by solving an exponential-size LP via an efficient separation oracle based on power diagrams. Beyond the fixed-dimension method, the work [10] proposed strengthened LP formulations that exploit the non-mass-splitting structure of optimal barycenters. Their models first encoded that each barycenter support point aggregates mass from a single tuple of input support points, then aggregated transport variables at the level of such tuples, therefore shrinking the variable set relative to the naïve MOT LP formulation, and finally incorporated preprocessing/pruning rules and structural bounds on the barycenter support size, which together reduce problem size substantially in practice. Building on these structural ideas, the work [12] developed a column-generation framework whose restricted master problem uses a compact LP, while the pricing subproblem searches for new barycenter atoms (i.e., support tuples) with negative reduced costs. They design efficient pricing routines with bounding and heuristics, stabilization, and warm starts to generate exact solutions. However, none of these approaches yields a polynomial-time algorithm in the general input parameters, and they remain challenging to apply at scale when n, k, and d are large.

A related line of work considers a simpler variant in which the candidate support of the barycenter is fixed *a priori*, reducing the problem to a linear program. This fixed-support formulation can still be computationally challenging when the given support is large. To address this, the work [16] introduced entropic regularization and developed subgradient-based schemes; the work [5] proposed a tractable, parallelizable iterative Bregman projection (Sinkhorn) method that achieves scalability at the cost of an entropic bias; and the work [20] provided a debiased variant that removes this regularization bias while preserving convergence rates comparable to the Sinkhorn approach. Other directions include an interior-point method [17], a fast deterministic Bregman-projection algorithm with improved complexity guarantees [23], and a symmetric Gauss–Seidel-based ADMM from a dual perspective [32] for the given-support barycenter LP formulation. However, none of these methods can be applied to the Wasserstein barycenter problems studied in this paper.

For the general Wasserstein barycenter problem, the work [15] proposed a stochastic scheme that updates the barycenter support by sampling subgradients from the input measures. The work [26] developed a Frank–Wolfe optimization strategy for estimating barycenters with respect to the Sinkhorn divergence and established convergence guarantees. The work [24] formulated the free-support barycenter problem as an integer program and designed a federated algorithm via a dual subgradient method. The work [21] analyzed the complexity of iterative Bregman projections and accelerated gradient methods for the entropically regularized barycenter objective. Despite these advances, none of the above methods yield a polynomial-time algorithm in the general input parameters (n, k, d) (and, when applicable, $(nk)^{1/\alpha}$ and d).

1.2 Summary of Contributions

In this paper, we develop a unified framework of approximation algorithms for the discrete Wasserstein barycenter problem (WBCenter). Our approach is based on systematic candidate-support reduction and yields provable guarantees that improve upon the best-known results in the literature. The main contributions are summarized below.

(i) We establish the first polynomial-time approximation scheme (PTAS) for the Wasserstein barycenter problem. In particular, we propose a family of algorithms parameterized by α that achieve a $(1+\alpha)$-approximation guarantee with running time polynomial in $(nk)^{1/\alpha}$ and d.
(ii) We present both randomized (sampling-based) and deterministic variants, with the same approximation guarantee. In the special case of equally weighted input measures, we further refine the construction to obtain an improved approximation ratio.
(iii) We evaluate the proposed algorithms on both synthetic and real datasets, including instances with many input measures and instances with a large support size. The results demonstrate that our methods are computationally practical and produce high-quality barycenters in challenging regimes.

The remainder of the paper is organized as follows. In Sect. 2, we present preliminary results that form the basis of our approximation framework. In Sect. 3, we develop a subset sampling algorithm and its deterministic counterpart, and analyze their computational complexity and approximation guarantees. In Sect. 4, we show that, under uniform weights, the approximation ratio can be further improved via a modified sampling procedure and its deterministic counterpart. Section 5 reports numerical results, and Sect. 6 concludes the paper.

Notation. Let $[k] := \{1, \ldots, k\}$, $\|\cdot\|_2$ denote the ℓ_2-norm, $\delta_{(\cdot)}$ denote the Dirac measure, $\mathbb{I}_{(\cdot)}$ denote the indicator function, $(\mathbb{R}^n_+)^{\otimes k}$ denote the k-fold tensor product of the nonnegative orthant $\mathbb{R}^d_+$, $\mathbb{P}(S)$ denote a probability measure with support S, $O(\cdot)$ denote the standard big-O notation for computational complexity, $\tilde{O}(\cdot)$ denote the soft-O notation for computational complexity, which suppresses logarithmic and polylogarithmic factors in input size and accuracy.

2 Preliminary Results and Approximation Scheme

In this section, we establish the theoretical foundation for our approximation framework, which generalizes the approach of [9]. We first recall the equivalence between the discrete Wasserstein barycenter problem and the multi-marginal optimal transport (MOT) problem, which implies that an optimal barycenter may require an exponentially large support set S^*. To address this intractability, we introduce a restricted MOT formulation in which the barycenter support is constrained to be a smaller candidate set S, and we show that this restricted problem admits a separation oracle with complexity polynomial in n, k, d, and $|S|$, making it tractable whenever $|S|$ is polynomial in the input size. We further show how to recover an explicit barycenter measure supported on S from the optimal solution of the restricted MOT problem, and we prove that if set S appropriately approximates the optimal support S^*, then the resulting restricted barycenter achieves an objective value within a provable multiplicative factor of the true optimum.

2.1 Optimal Support

A key ingredient of our results is the well-known equivalence between the Wasserstein barycenter problem and the multi-marginal optimal transport (MOT) problem. It has been shown (see Proposition 2.1 in [3] and Sect. 6 in [13]) that the Wasserstein barycenter problem (WBCenter) is equivalent to the following MOT formulation:

$$
\begin{aligned}
v^* = &\min_{\boldsymbol{\Pi} \in (\mathbb{R}^n_+)^{\otimes k}} \langle \boldsymbol{C}, \boldsymbol{\Pi} \rangle, \\
\text{s.t.} &\sum_{j_1 \in [|\Xi_1|]} \cdots \sum_{j_{i-1} \in [|\Xi_{i-1}|]} \sum_{j_{i+1} \in [|\Xi_{i+1}|]} \cdots \sum_{j_k \in [|\Xi_k|]} \Pi_{j_1,\ldots,j_k} = \hat{p}_{ij_i}, \forall i \in [k], j_i \in [|\Xi_i|],
\end{aligned} \tag{1}
$$

where $\boldsymbol{\Pi}$ is a k-way coupling tensor and the transportation cost is given by

$$
C_{\vec{j}} = \min_{\boldsymbol{w} \in \mathbb{R}^d} \sum_{i \in [k]} \lambda_i \|\hat{\boldsymbol{x}}_{i,j_i} - \boldsymbol{w}\|_2^2, \qquad \vec{j} = (j_1, \ldots, j_{i-1}, j_i, j_{i+1}, \ldots, j_k).
$$

In other words, each entry $\Pi_{j_1,\ldots,j_k}$ represents the joint mass assigned to the tuple of support points $(\hat{\boldsymbol{x}}_{1,j_1}, \ldots, \hat{\boldsymbol{x}}_{k,j_k})$, and $C_{\vec{j}}$ is the optimal cost of fitting a single barycenter location $\boldsymbol{w}$ to that tuple under the weighted squared Euclidean loss.

From the MOT formulation, we observe that the support of the optimal barycenter is finite and consists only of points that attain the minimum cost in the definition of $\boldsymbol{C}$.

Lemma 1. *The support S^* of the optimal barycenter of problem* (WBCenter) *is given by*

$$
S^* = \bigcup_{\boldsymbol{x}_i \in \Xi_i, i \in [k]} \left\{ \arg\min_{\boldsymbol{w} \in \mathbb{R}^d} \sum_{i \in [k]} \lambda_i \|\boldsymbol{x}_i - \boldsymbol{w}\|_2^2 \right\}.
$$

This characterization appears in the remark following Proposition 2.1 of [3] and in Sect. 6 of [13], and is consistent with known finiteness results for discrete barycenters of (WBCenter) (see, e.g., [4]). In particular, the optimal barycenter can be chosen to have finite support, and all its support points arise as minimizers of a weighted least-squares fit to some k-tuple of input support points.

By Lemma 1, we have $|S^*| = O(n^k)$, which is exponential in k. Using this notation, we may equivalently rewrite the transportation cost in the MOT formulation as

$$
C_{\vec{j}} = \min_{\boldsymbol{w} \in S^*} \sum_{i \in [k]} \lambda_i \|\hat{\boldsymbol{x}}_{i,j_i} - \boldsymbol{w}\|_2^2,
$$

for each index tuple $\vec{j} = (j_1, \ldots, j_k)$.

Lemma 1 also shows that, although an optimal barycenter exists and is finitely supported, its support S^* may be exponentially large. This motivates constructing a reduced candidate support within our approximation framework.

2.2 The Dual Representation of MOT Formulation and Candidate Support Reduction

Let $\{\gamma_{ij}\}_{i\in[k],j\in[|\Xi_i|]}$ be the dual variables associated with the marginal constraints in the MOT formulation (1). By LP strong duality, the optimal value v^* of the Wasserstein barycenter problem admits the dual representation:

$$v^* = \max_{\gamma} \left\{ \sum_{i\in[k]} \sum_{j\in[|\Xi_i|]} \hat{p}_{ij}\gamma_{ij} : C_{\vec{j}} - \sum_{i\in[k]} \gamma_{ij_i} \geq 0, \forall \vec{j} \in \otimes_{i\in[k]}[|\Xi_i|] \right\}. \tag{2}$$

This dual problem involves at most nk variables γ_{ij}, which are polynomial in input size. However, enforcing feasibility in the dual is still challenging, because each dual constraint depends on the transportation cost $C_{\vec{j}}$ and the optimal support set S^* may contain exponentially many candidates, i.e., $|S^*| = O(n^k)$. Thus, even though the dual has only $O(nk)$ variables, it is not directly tractable: (i) there are exponentially many dual constraints (one per k-tuple $\vec{j}$), and (ii) evaluating each left-hand side requires minimizing over S^*, which itself is exponentially large.

This motivates our approach. Rather than working with the full optimal support S^*, we seek a reduced candidate support $S \subseteq \mathbb{R}^d$ of manageable size such that replacing S^* by S perturbs the cost $C_{\vec{j}}$ by only a controlled multiplicative factor. This support reduction is the core mechanism that enables our polynomial-time approximation scheme.

For notational convenience, given a candidate support $S = \{\boldsymbol{w}_\ell\}_{\ell\in[|S|]} \subseteq \mathbb{R}^d$ for a barycenter $\mathbb{P}(S)$, we define its transportation cost as

$$C_{\vec{j}}(S) = \min_{\boldsymbol{w}\in S} \sum_{i\in[k]} \lambda_i \|\hat{\boldsymbol{x}}_{i,j_i} - \boldsymbol{w}\|_2^2,$$

for each multi-index $\vec{j} = (j_1, \ldots, j_k)$.

We then define the corresponding restricted dual objective value

$$v(S) = \max_{\gamma} \left\{ \sum_{i\in[k]} \sum_{j\in[|\Xi_i|]} \hat{p}_{ij}\gamma_{ij} : C_{\vec{j}}(S) - \sum_{i\in[k]} \gamma_{ij_i} \geq 0, \forall \vec{j} \in \otimes_{i\in[k]}[|\Xi_i|] \right\}. \tag{3}$$

In other words, $v(S)$ is the optimal dual objective value when we restrict all barycenter support to be S instead of the full optimal support.

When the candidate support size $|S|$ is polynomial in n and k, the restricted MOT dual (3) can be solved in polynomial time, since its separation oracle admits a closed-form evaluation.

Lemma 2. *Given a candidate dual vector* $\boldsymbol{\gamma}$ *for* (3)*, feasibility of* $\boldsymbol{\gamma}$ *can be checked in time* $O(kn|S|d)$*. Therefore, the restricted MOT dual* (3) *can be solved in time* $\tilde{O}(n^3k^3d|S|)$.

We also show that a feasible barycenter of a given support can be explicitly constructed from the MOT formulation (1).

Lemma 3. *Fix a candidate support $S \subseteq \mathbb{R}^d$. Let $\boldsymbol{\Pi}^*(S)$ be an optimal primal solution of the MOT formulation* (1) *in which the cost $C_{\vec{j}}$ is restricted to $C_{\vec{j}}(S)$. For each $\vec{j} \in \otimes_{i\in[k]}[|\Xi_i|]$ such that $\Pi^*_{\vec{j}}(S) > 0$, choose*

$$\boldsymbol{w}_{\vec{j}} \in \arg\min_{\boldsymbol{w}\in S} \sum_{i\in[k]} \lambda_i \|\hat{\boldsymbol{x}}_{i,j_i} - \boldsymbol{w}\|_2^2, \qquad m_{\boldsymbol{w}} = \sum_{\vec{j}\in\otimes_{i\in[k]}[|\Xi_i|]} \Pi^*_{\vec{j}}(S)\mathbb{I}_{\{\boldsymbol{w}_{\vec{j}}=\boldsymbol{w}\}}, \forall \boldsymbol{w} \in S,$$

where $\mathbb{I}_{(\cdot)}$ is the indicator function. Then the measure $\mathbb{P}(S) = \sum_{\boldsymbol{w}\in S} m_{\boldsymbol{w}} \delta_{\tilde{\boldsymbol{w}}=\boldsymbol{w}}$ is a feasible barycenter supported on S, where $\delta(\cdot)$ is the Dirac measure and the random variable $\tilde{\boldsymbol{w}}$ follows the distribution of $\mathbb{P}(S)$.

The MOT primal solution $\boldsymbol{\Pi}^*(S)$ tells how often each tuple of support points $(\hat{\boldsymbol{x}}_{1,j_1}, \ldots, \hat{\boldsymbol{x}}_{k,j_k})$ is jointly selected. For each such tuple, we attach the "best-fitting" barycenter location in S, i.e., the minimizer $\boldsymbol{w}_{\vec{j}}$. Aggregating the mass $\Pi^*_{\vec{j}}(S)$ onto these chosen points yields a discrete measure $\mathbb{P}(S)$ on S. This measure is therefore a feasible barycenter with support contained in S.

We conclude this section by stating an approximation guarantee that quantifies the effect of reducing the full optimal support S^* to a smaller candidate set $\tilde{S}$.

Theorem 1. *Let $\tilde{S} \subseteq \mathbb{R}^d$ be a (possibly random) candidate support set. Suppose there exists $\alpha \geq 0$ such that, for any choice of points $\boldsymbol{x}_i \in \Xi_i$ for all $i \in [k]$, we have*

$$\mathbb{E}_{\tilde{S}}\left[\min_{\boldsymbol{w}\in\tilde{S}} \sum_{i\in[k]} \lambda_i \|\boldsymbol{w} - \boldsymbol{x}_i\|_2^2\right] \leq (1+\alpha) \min_{\boldsymbol{w}\in\mathbb{R}^d} \sum_{i\in[k]} \lambda_i \|\boldsymbol{w} - \boldsymbol{x}_i\|_2^2. \tag{4}$$

Then the expected objective value of the restricted barycenter problem satisfies $\mathbb{E}_{\tilde{S}}[v(\tilde{S})] \leq (1+\alpha)v(S^*)$.

Remark 1. If the set $\tilde{S}$ is deterministic, then the same guarantee holds directly under condition (4) (i.e., the expectation over $\tilde{S}$ is not needed).

By Theorem 1, if every support point of the optimal barycenter has a "good enough" representative in the reduced candidate set, then solving the barycenter problem over that reduced set yields an objective value within a factor $(1+\alpha)$ of the optimal one. In the next section, we therefore construct a reduced support $\tilde{S}$ that satisfies (4), either deterministically or in expectation.

3 Approximation Algorithms for the Wasserstein Barycenter Problem with General Weights

By Lemma 1, the optimal barycenter of the Wasserstein barycenter problem admits a finite support $S^* \subseteq \mathbb{R}^d$ of the form

$$S^* := \left\{ \sum_{i\in[k]} \lambda_i \boldsymbol{x}_i : \boldsymbol{x}_i \in \Xi_i, \forall i \in [k] \right\}.$$

The cardinality of this set satisfies $|S^*| = \prod_{i\in[k]} |\Xi_i| = O(n^k)$, which is exponential in the number of measures k. In this section, we develop two approximation schemes for the case of general (i.e., not necessarily uniform) weights $\boldsymbol{\lambda} \in \mathbb{R}_+^k$: a randomized subset sampling algorithm and a deterministic subset enumeration algorithm. We show that both methods achieve the same approximation guarantee.

3.1 Sampling Algorithm with Repetition

We begin by drawing t indices $T_1, \ldots, T_t \in [k]$ independently according to the weight vector $\boldsymbol{\lambda}$; i.e.,

$$\text{Prob}\left[T_j = i\right] = \lambda_i, \forall i \in [k], j \in [t].$$

Let $\mathcal{T}$ denote the resulting multiset of selected indices. Based on $\mathcal{T}$, we construct the candidate support set

$$S_1^{\mathcal{T}} = \left\{ \frac{1}{t} \sum_{i\in[t]} \boldsymbol{x}_{T_i} : \boldsymbol{x}_{T_i} \in \Xi_{T_i}, \forall i \in [t] \right\}. \tag{5}$$

Note that each element of $S_1^{\mathcal{T}}$ is obtained by averaging t support points, where at each position $i \in [t]$ we select one atom from the support of the probability measure indexed by T_i. The cardinality of $S_1^{\mathcal{T}}$ satisfies $|S_1^{\mathcal{T}}| = O(n^t)$, which is polynomial in n for any fixed t. We then solve the restricted MOT dual (3) using $S_1^{\mathcal{T}}$ as the candidate support, and obtain an approximation barycenter $\mathbb{P}(S_1^{\mathcal{T}})$ based on Lemma 3. The complete procedure is summarized in [30].

The approximation guarantee is formally stated in the following theorem.

Theorem 2. *Let $t \in [k]$ be a fixed constant, and let $S_1^{\mathcal{T}}$ be the candidate support described in* (5). *Then*

$$\mathbb{E}_{\mathcal{T}}\left[v\left(S_1^{\mathcal{T}}\right)\right] \leq \left(1 + \frac{1}{t}\right) v^*.$$

Remark 2. To obtain a $(1+\alpha)$-approximation in expectation, it suffices to choose $t = \lceil 1/\alpha \rceil$, so that $\frac{1}{t} \leq \alpha$. Since $|S_1^{\mathcal{T}}| = O(n^t)$, Lemma 2 implies that solving the restricted problem over $S_1^{\mathcal{T}}$ has complexity $\tilde{O}(n^{3+\frac{1}{\alpha}} k^3 d)$, which is polynomial in $n^{1/\alpha}$, k, and d. The proofs of Theorems 2–5 correspond to a specific choice of parameter values that reflects the classic bias–variance tradeoff.

3.2 Deterministic Counterpart Based on Multi-subset Enumeration

A deterministic counterpart to the sampling algorithm can be obtained by enumerating all possible sampling outcomes of the random multiset $\mathcal{T}$ and aggregating all resulting averages into a single reduced candidate support. Specifically, we define

$$S_1^t := \left\{ \frac{1}{t} \sum_{i\in[t]} \boldsymbol{x}_{T_i} : \boldsymbol{x}_{T_i} \in \Xi_{T_i}, T_i \in [k], i \in [t] \right\}. \tag{6}$$

We then solve the restricted MOT dual (3) using S_1^t as the candidate support, and obtain an approximation barycenter $\mathbb{P}(S_1^t)$ based on Lemma 3. The full deterministic procedure is summarized in [30].

Since S_1^t contains the support points generated by all possible realizations of the random construction $S_1^{\mathcal{T}}$, we have

$$\min_{s \in S_1^t} \|s - c\|_2^2 \leq \mathbb{E}_{\mathcal{T}}\left[\min_{s \in S_1^{\mathcal{T}}} \|s - c\|_2^2\right] \leq \mathbb{E}_{\mathcal{T}}\left[\left\|\frac{\sum_{i \in \mathcal{T}} x_i}{t} - c\right\|_2^2\right],$$

for any reference point c. Using this observation and following the same argument as in the proof of Theorem 2, we obtain the following deterministic guarantee.

Theorem 3. *Let $t \in [k]$ be a fixed constant, and let S_1^t be the candidate support described in* (6). *Then*

$$v(S_1^t) \leq \left(1 + \frac{1}{t}\right) v^*,$$

and the support size satisfies $|S_1^t| = O((nk)^t)$.

Remark 3. When $t = 1$, the procedure enumerates all individual data points and thus reduces to the best-known 2-approximation method of [9]; our analysis recovers the same approximation guarantee for this special case. Since $|S_1^t| = O((nk)^t)$, achieving a $(1 + \alpha)$ approximation requires choosing $t = \lceil 1/\alpha \rceil$, which leads to an overall complexity of $\tilde{O}\left(n^{3+\frac{1}{\alpha}} k^{3+\frac{1}{\alpha}} d\right)$.

4 Approximation Algorithms for the Wasserstein Barycenter Problem with Equal Weights

In this section, we focus on the special case of (WBCenter) in which the input probability measures are equally weighted, i.e., $\lambda_i = \frac{1}{k}$ for each $i \in [k]$. In this setting, we show that the approximation ratio can be further improved by using a nonrepetitive sampling procedure, yielding a tighter guarantee than in the general-weight case.

4.1 Subset Sampling Algorithm Without Repetition

We uniformly sample a subset $\mathcal{T} \subseteq [k]$ of size t, without repetition. Based on this subset, we define the candidate support

$$S_2^{\mathcal{T}} := \left\{ \frac{1}{t} \sum_{i \in \mathcal{T}} x_i : x_i \in \Xi_i \right\}. \tag{7}$$

We then solve the restricted MOT dual (3) with $S_2^{\mathcal{T}}$ as the candidate support and recover an approximate barycenter $\mathbb{P}(S_2^{\mathcal{T}})$ using Lemma 3. A formal description of this procedure is given in [30].

The approximation guarantee is formally stated in the following theorem.

Theorem 4. *Let $t \in [k]$ be a fixed constant, and let $S_2^{\mathcal{T}}$ be the candidate support described in* (7). *Then*

$$\mathbb{E}_{\mathcal{T}}\left[v(S_2^{\mathcal{T}})\right] \leq \left(1 + \frac{k-t}{t(k-1)}\right) v^*.$$

Remark 4. To obtain a $(1+\alpha)$-approximation guarantee, it suffices to choose t such that $\frac{k-t}{t(k-1)} \leq \alpha$; equivalently, $t \geq \frac{k}{\alpha k - \alpha + 1}$. Since $|S_2^{\mathcal{T}}| = O(n^t)$, solving the restricted problem over $S_2^{\mathcal{T}}$ requires time $\tilde{O}\left(n^{3+\frac{k}{\alpha k-\alpha+1}} k^3 d\right)$ to achieve a $(1+\alpha)$-approximation.

4.2 Deterministic Counterpart Based on Subset Enumeration

Similar to Sect. 3.2, we obtain a deterministic counterpart to the sampling algorithm by enumerating all possible choices of the subset $\mathcal{T} \subseteq [k]$ of size t (without repetition), and aggregating all corresponding averaged points into a single reduced candidate support. Specifically, we define

$$S_2^t := \left\{ \frac{1}{t} \sum_{i \in \mathcal{T}} \boldsymbol{x}_i : \boldsymbol{x}_i \in \Xi_i, \mathcal{T} \subseteq [k], |\mathcal{T}| = t \right\}. \tag{8}$$

We then solve the restricted MOT dual (3) using S_2^t as the candidate support and recover an approximate barycenter $\mathbb{P}(S_2^t)$ via Lemma 3. The resulting deterministic procedure is summarized in [30].

Since S_2^t contains the support points generated by all possible realizations of the random construction $S_2^{\mathcal{T}}$, we have, for any reference point $\boldsymbol{c}$,

$$\min_{\boldsymbol{s} \in S_2^t} \|\boldsymbol{s} - \boldsymbol{c}\|_2^2 \leq \mathbb{E}_{\mathcal{T}}\left[\min_{\boldsymbol{s} \in S_2^{\mathcal{T}}} \|\boldsymbol{s} - \boldsymbol{c}\|_2^2\right] \leq \mathbb{E}_{\mathcal{T}}\left[\left\|\frac{\sum_{i \in \mathcal{T}} \boldsymbol{x}_i}{t} - \boldsymbol{c}\right\|_2^2\right].$$

Using this observation and following the same argument as in the proof of Theorem 4, we obtain the corresponding deterministic approximation guarantee.

Theorem 5. *Let $t \in [k]$ be a fixed constant, and let S_2^t be the candidate support described in* (8). *Then*

$$\min_{\mathcal{T}:|\mathcal{T}|=t} v(S_2^t) \leq \left(1 + \frac{k-t}{t(k-1)}\right) v^*,$$

and the support size satisfies $|S_2^t| = O((nk)^t)$

Remark 5. Note that

$$1 + \frac{k-t}{t(k-1)} \leq 1 + \frac{1}{t},$$

which implies that, for $t > 1$, the approximation ratio achieved by Sect. 4.2 (which applies to the equal-weight Wasserstein barycenter problem) is strictly better than that of Sect. 3.2 (which applies to the general-weight Wasserstein barycenter problem).

5 Numerical Experiments

In this section, we conduct numerical experiments to evaluate the computational efficiency and solution quality of the proposed algorithms. Our main goals are:

(i) To empirically validate the theoretical $(1+\alpha)$-approximation guarantee and the tunable accuracy-complexity trade-off;
(ii) To demonstrate that our algorithms scale to large instances that are computationally intractable for exact extensive LP formulations; and
(iii) To compare our results with the state-of-the-art 2-approximation algorithm [9], which is equivalent to Sect. 4.2 with $t = 1$.

All of our algorithms were implemented in Python using Gurobi 11.0.3 as the LP solver and executed on a MacBook Pro with an Apple M4 Pro Chip and 48 GB RAM. Due to the page limit, we report only the synthetic numerical study here; additional numerical results are provided in [30].

5.1 Synthetic Nested Ellipse Dataset

The synthetic dataset used in this subsection is a standard benchmark from the literature [2,16,20], for which the optimal Wasserstein barycenter is known [2]. It consists of ten probability measures $\{\mathbb{P}_i\}_{i\in[10]}$, each supported on a 60×60 grid and corresponding to a nested ellipse (see Fig. 1). Each measure has between 139 and 192 support points, all with equal mass. We emphasize that the exact barycenter may not, in general, be supported on the grid. Solving the full MOT formulation (1) over S^* would require on the order of $192^{10} \approx 10^{22}$ basic arithmetic operations, which is computationally intractable.

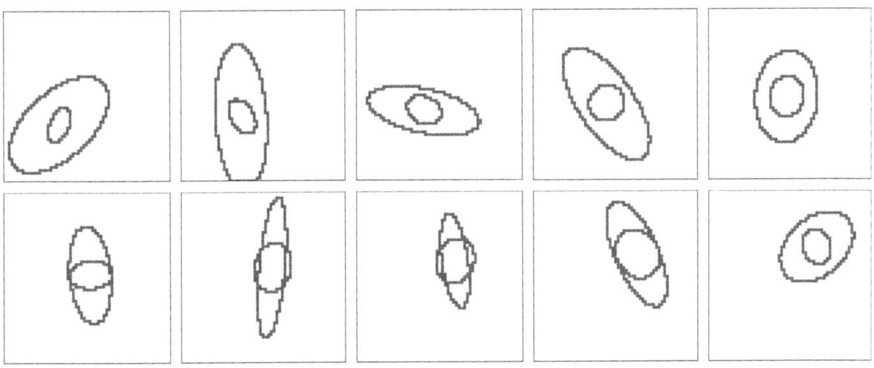

Fig. 1. Ten images from the nested ellipses dataset.

To ensure higher accuracy, we adopt the subset enumeration method (i.e., Sect. 4.2) rather than the subset sampling method (i.e., Sect. 4.1), even though the latter is more efficient and enjoys the same theoretical approximation ratio. The key reason is that sampling operates over the set of measures, so regardless of how many samples are drawn, the number of distinct candidate supports is finite. As illustrated in Fig. 2, even the best result among the ten possible samples for $t = 1$ is still substantially worse than the result obtained by deterministic enumeration. Consequently, in all subsequent experiments, we apply our deterministic subset enumeration algorithm with $t = 1$ and $t = 2$.

Additionally, as illustrated in Fig. 2, all methods produce solutions that are substantially better than their theoretical worst-case guarantees. We also observe that even small differences in objective values between $t = 1$ and $t = 2$ can correspond to visibly inferior barycenter quality. While the $t = 1$ setting offers faster computation, the $t = 2$ setting consistently yields barycenters of much higher visual fidelity.

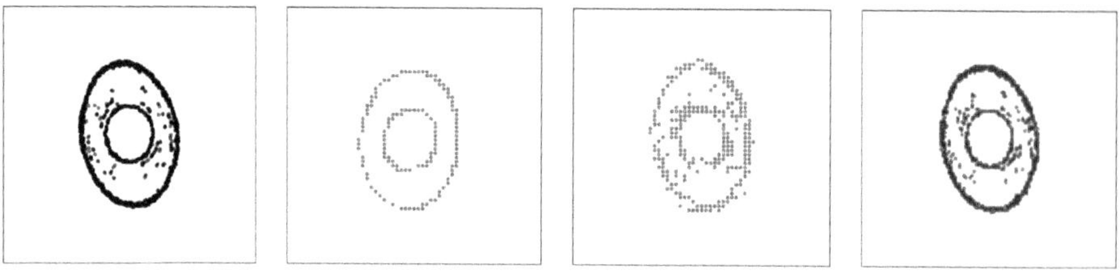

(a) Exact ellipse [2], objective=0.02666, time=8865.03s. (b) 10 runs of subset sampling method with $t = 1$, objective=0.02709, time=17.46s. (c) Subset enumeration method with $t = 1$. objective=0.02673, time=21.08s. (d) Subset enumeration method with $t = 2$. objective=0.02667, time=212.44s.

Fig. 2. Barycenters produced by different algorithms for nested ellipse.

6 Conclusion

In this paper, we present efficient approximation algorithms for computing discrete Wasserstein barycenters. Our approach generalizes the best-known 2-approximation algorithm and yields a polynomial-time approximation scheme (PTAS) that provides an explicit trade-off between solution quality and computational complexity. Our numerical results show that the proposed methods achieve near-optimal objective values on challenging datasets, including cases with a large number of measures and very large supports. In particular, our methods consistently outperform existing 2-approximation approaches in accuracy while maintaining reasonable running times. Several directions remain open.

Acknowledgments. This research has been supported in part by National Science Foundation grants 2246414 and the Office of Naval Research grant N00014-24-1-2066.

The second author would like to thank Ph.D. students Zedong Wang and Shuai Li for their discussions on the problem.

Disclosure of Interests. The authors have no competing interests to declare that are relevant to the content of this article.

References

1. Agueh, M., Carlier, G.: Barycenters in the Wasserstein space. SIAM J. Math. Anal. **43**(2), 904–924 (2011)
2. Altschuler, J.M., Boix-Adsera, E.: Wasserstein barycenters can be computed in polynomial time in fixed dimension. J. Mach. Learn. Res. **22**(44), 1–19 (2021). http://jmlr.org/papers/v22/20-588.html
3. Altschuler, J.M., Boix-Adsera, E.: Wasserstein barycenters are NP-hard to compute. SIAM J. Math. Data Sci. **4**(1), 179–203 (2022)
4. Anderes, E., Borgwardt, S., Miller, J.: Discrete Wasserstein barycenters: optimal transport for discrete data. Math. Methods Oper. Res. **84**(2), 389–409 (2016)
5. Benamou, J.D., Carlier, G., Cuturi, M., Nenna, L., Peyré, G.: Iterative Bregman projections for regularized transportation problems. SIAM J. Sci. Comput. **37**(2), A1111–A1138 (2015)
6. Bigot, J., Klein, T., et al.: Consistent estimation of a population barycenter in the Wasserstein space. ArXiv e-prints **49** (2012)
7. Bonneel, N., Peyré, G., Cuturi, M.: Wasserstein barycentric coordinates: histogram regression using optimal transport. ACM Trans. Graph. **35**(4), 71–1 (2016)
8. Bonneel, N., Rabin, J., Peyré, G., Pfister, H.: Sliced and radon Wasserstein barycenters of measures. J. Math. Imaging Vis. **51**(1), 22–45 (2015)
9. Borgwardt, S.: An LP-based, strongly-polynomial 2-approximation algorithm for sparse Wasserstein barycenters. Oper. Res. Int. Journal **22**(2), 1511–1551 (2022)
10. Borgwardt, S., Patterson, S.: Improved linear programs for discrete barycenters. Inf. J. Optim. **2**(1), 14–33 (2020)
11. Borgwardt, S., Patterson, S.: On the computational complexity of finding a sparse Wasserstein barycenter. J. Comb. Optim. **41**(3), 736–761 (2021)
12. Borgwardt, S., Patterson, S.: A column generation approach to the discrete barycenter problem. Discret. Optim. **43**, 100674 (2022)
13. Carlier, G., Ekeland, I.: Matching for teams. Econ. Theory **42**(2), 397–418 (2010). http://www.jstor.org/stable/25619994
14. Cheng, K., Aeron, S., Hughes, M.C., Miller, E.L.: Dynamical Wasserstein barycenters for time-series modeling. In: Advances in Neural Information Processing Systems, vol. 34, pp. 27991–28003 (2021)
15. Claici, S., Chien, E., Solomon, J.: Stochastic Wasserstein barycenters. In: International Conference on Machine Learning, pp. 999–1008. PMLR (2018)
16. Cuturi, M., Doucet, A.: Fast computation of Wasserstein barycenters. In: International Conference on Machine Learning, pp. 685–693. PMLR (2014)
17. Ge, D., Wang, H., Xiong, Z., Ye, Y.: Interior-point methods strike back: solving the Wasserstein barycenter problem. In: Advances in Neural Information Processing Systems, vol. 32 (2019)
18. Huang, M., Ma, S., Lai, L.: Projection robust Wasserstein barycenters. In: International Conference on Machine Learning, pp. 4456–4465. PMLR (2021)

19. Janati, H., Bazeille, T., Thirion, B., Cuturi, M., Gramfort, A.: Multi-subject MEG/EEG source imaging with sparse multi-task regression. Neuroimage **220**, 116847 (2020)
20. Janati, H., Cuturi, M., Gramfort, A.: Debiased Sinkhorn barycenters. In: International Conference on Machine Learning, pp. 4692–4701. PMLR (2020)
21. Kroshnin, A., Tupitsa, N., Dvinskikh, D., Dvurechensky, P., Gasnikov, A., Uribe, C.: On the complexity of approximating Wasserstein barycenters. In: International Conference on Machine Learning, pp. 3530–3540. PMLR (2019)
22. Lau, T.T.K., Liu, H.: Wasserstein distributionally robust optimization with Wasserstein barycenters (2022). arXiv preprint arXiv:2203.12136
23. Lin, T., Ho, N., Chen, X., Cuturi, M., Jordan, M.: Fixed-support Wasserstein barycenters: computational hardness and fast algorithm. In: Advances in Neural Information Processing Systems, vol. 33, pp. 5368–5380 (2020)
24. Lin, Z., Ruszczyński, A.: Federated calculation of the free-support transportation barycenter by single-loop dual decomposition (2025). arXiv preprint arXiv:2507.19627
25. Lindheim, J.V.: Simple approximative algorithms for free-support Wasserstein barycenters. Comput. Optim. Appl. **85**(1), 213–246 (2023)
26. Luise, G., Salzo, S., Pontil, M., Ciliberto, C.: Sinkhorn barycenters with free support via Frank-Wolfe algorithm. In: Advances in Neural Information Processing Systems, vol. 32 (2019)
27. Schmitz, M.A., et al.: Wasserstein dictionary learning: optimal transport-based unsupervised nonlinear dictionary learning. SIAM J. Imag. Sci. **11**(1), 643–678 (2018)
28. Simon, D., Aberdam, A.: Barycenters of natural images constrained Wasserstein barycenters for image morphing. In: Proceedings of the IEEE/CVF Conference on Computer Vision and Pattern Recognition, pp. 7910–7919 (2020)
29. Uribe, C.A., Dvinskikh, D., Dvurechensky, P., Gasnikov, A., Nedić, A.: Distributed computation of Wasserstein barycenters over networks. In: 2018 IEEE Conference on Decision and Control (CDC), pp. 6544–6549. IEEE (2018)
30. Wang, J., Xie, W.: Closing the gap: efficient algorithms for discrete Wasserstein barycenters (2025). arXiv preprint arXiv:2511.04607
31. Yan, Y., et al.: 2D Wasserstein loss for robust facial landmark detection. Pattern Recogn. **116**, 107945 (2021)
32. Yang, L., Li, J., Sun, D., Toh, K.C.: A fast globally linearly convergent algorithm for the computation of Wasserstein barycenters. J. Mach. Learn. Res. **22**(21), 1–37 (2021)
33. Ye, J., Wu, P., Wang, J.Z., Li, J.: Fast discrete distribution clustering using Wasserstein barycenter with sparse support. IEEE Trans. Signal Process. **65**(9), 2317–2332 (2017)

Author Index

S. S. Dey et al. (Eds.): IPCO 2026, LNCS 16588, pp. 521–522, 2026.
https://doi.org/10.1007/978-3-032-28691-8

www.ingramcontent.com/pod-product-compliance
Ingram Content Group UK Ltd.
Pitfield, Milton Keynes, MK11 3LW, UK
UKHW021642180926
475650UK00002B/83
9783032286901